CRC

11-94

For Reference

Not to be taken from this room

Practical Handbook *of* Marine Science

Second Edition

The CRC Marine Science Series provides publications that synthesize recent advances in Marine Science. Marine Science is at an exciting threshold where new developments are providing insights into how the biology of the ocean is integrated with its chemistry, geology, and physics.

CRC MARINE SCIENCE SERIES

SERIES EDITORS

Michael J. Kennish, Ph.D.
Peter L. Lutz, Ph.D.

PUBLISHED TITLES

Ecology of Estuaries: Anthropogenic Effects, Michael J. Kennish
Physical Oceanography of the Great Barrier Reef Processes, Eric Wolanski
Practical Handbook of Marine Science, 2nd Edition, Michael J. Kennish
The Physiology of Fishes, David H. Evans

FORTHCOMING TITLES

Benthic Microbial Ecology, Paul F. Kemp
Chemosynthetic Communities, James M. Brooks and Charles Fisher
Ecology of Marine Invertebrate Larvae, Larry McEdward
Effects of Coastal Pollution on Living Resources, Carl J. Sindermann
Major Marine Ecological Disturbances, Ernest H. Williams, Jr. and Lucy Bunkley-Williams
Marine Bivalves and Ecosystem Processes, Richard F. Dame
Morphodynamics of the Inner Continental Shelf, L. Donelson Wright
Seabed Instability, M. Shamim Rahman
Sediment Studies of River Mouths, Tidal Flats, and Coastal Lagoons, Doeke Eisma

---

Practical Handbook of Marine Science, 2nd Edition contains selected physical, chemical, and biological reference data on the ocean environment. Presented in written, tabular, and illustrative forms, this concise, easy-to-use handbook enables the reader to locate quickly the data relevant to his or her needs in the areas of marine biology, marine chemistry, marine geology, and physical oceanography. It addresses natural phenomena in estuaries, lagoons, continental shelves, continental slopes, continental rises, abyssal regions, mid-ocean ridges, along with neritic and pelagic waters of the ocean. It also examines in detail anthropogenic effects on estuarine and marine ecosystems from local, regional, and global perspectives.

Practical Handbook *of* Marine Science

Second Edition

Edited by

Michael J. Kennish, Ph.D.
Institute of Marine and Coastal Sciences
Cook College, Rutgers University
New Brunswick, New Jersey

CRC Press
Boca Raton Ann Arbor London Tokyo

BRIDGEWATER PUBLIC LIBRARY
15 SOUTH STREET
BRIDGEWATER, MA 02324

Library of Congress Cataloging-in-Publication Data

Practical handbook of marine science / edited by Michael J. Kennish. -
- 2nd ed.
p. cm.
Includes bibliographical references and index.
ISBN 0–8493–3712–7
1. Oceanography. 2. Marine biology. I. Kennish, Michael J.
GC11.2.P73 1994 551.46--dc20 93–3715
CIP

This book contains information obtained from authentic and highly regarded sources. Reprinted material is quoted with permission, and sources are indicated. A wide variety of references are listed. Reasonable efforts have been made to publish reliable data and information, but the author and the publisher cannot assume responsibility for the validity of all materials or for the consequences of their use.

Neither this book nor any part may be reproduced or transmitted in any form or by any means, electronic or mechanical, including photocopying, microfilming, and recording, or by any information storage or retrieval system, without prior permission in writing from the publisher.

CRC Press, Inc.'s consent does not extend to copying for general distribution, for promotion, for creating new works, or for resale. Specific permission must be obtained in writing from CRC Press for such copying.

Direct all inquiries to CRC Press, Inc., 2000 Corporate Blvd., N.W., Boca Raton, Florida 33431.

© 1994 by CRC Press, Inc.

No claim to original U.S. Government works
International Standard Book Number 0-8493-3712–7
Library of Congress Card Number 93–3715
Printed in the United States of America 1 2 3 4 5 6 7 8 9 0
Printed on acid-free paper

PURCHASED WITH STATE FUNDS

PREFACE

Practical Handbook of Marine Science, 2nd Edition provides the most up-to-date and comprehensive collection of ready-to-use reference material on the physical, chemical, and biological aspects of the ocean. As in the case of the 1st edition, the contents of this volume are arranged in a multisectional format, with each section consisting of expository, illustrative, and tabular reference material. However, the 2nd edition has been vastly improved by the addition of updated information on open ocean and estuarine circulation, elemental cycling, seafloor spreading centers, deep-sea hydrothermal vents, biological production, trophic relationships, and other subject areas. Moreover, a new section is presented on marine pollutants. This detailed treatment of anthropogenic impacts on marine and estuarine waters covers important topics such as organic loading, oil pollution, chemical contaminants, radioactivity, and dredging and dredged-spoil disposal.

The handbook is designed to serve the multidisciplinary research needs of contemporary marine biologists, marine chemists, marine geologists, and physical oceanographers. Perhaps the greatest value of this single-volume databook is that it offers the practicing marine scientist a wide spectrum of information covering all of the major disciplines as they relate to the ocean. It may be effectively used as a supplement to the extensive specialized literature already available in oceanography.

Although this publication has been prepared as a reference for marine scientists, it should also be of value to administrators and other professionals dealing in some way with the management of marine resources and the investigation of problems pertaining to the marine environment. Its primary objective, however, is to amass significant data that will have appeal and utility for practitioners and students of oceanography and related disciplines. The scope of this task, of course, has grown immensely commensurate with the rapidly expanding database in oceanography.

Because of the expanse of the marine science field, the editor is well aware of possible gaps and omissions in this volume. The significant limitations of space inherent in a single volume preclude equal coverage of all disciplines comprising this important field. Hence, constructive criticisms and suggestions for improving future editions of the handbook are urgently requested from the readership.

EDITOR

Michael J. Kennish, Ph.D., is a research marine scientist in the Institute of Marine and Coastal Sciences at Cook College and a member of the graduate faculty in the Geosciences Department of Rutgers University, New Brunswick, New Jersey.

He graduated in 1972 from Rutgers University, Camden, New Jersey, with a B.A. degree in geology and obtained his M.S. and Ph.D. degrees in the same discipline from Rutgers University, New Brunswick in 1974 and 1977, respectively.

Dr. Kennish's professional affiliations include the American Fisheries Society (Mid-Atlantic Chapter), American Geophysical Union, American Institute of Physics, New England Estuarine Research Society, Atlantic Estuarine Research Society, Southeastern Estuarine Research Society, Gulf Estuarine Research Society, Pacific Estuarine Research Society, National Shellfisheries Association, New Jersey Academy of Science, and Sigma Xi. He is also a member of the Advisory Board of the Fisheries and Aquaculture TEX Center of Rutgers University, overseeing the development of fisheries and shellfisheries in estuarine and marine waters of New Jersey.

Although maintaining research interests in broad areas of marine ecology and marine geology, Dr. Kennish has been most actively involved with investigations of anthropogenic effects on estuarine and coastal marine ecosystems and with studies of seafloor spreading centers. He is the author of *Ecology of Estuaries* (Volumes 1 and 2) and *Ecology of Estuaries: Anthropogenic Effects* published by CRC Press, the editor of *Practical Handbook of Marine Science* published by CRC Press, and the co-editor of *Ecology of Barnegat Bay, New Jersey,* published by Springer-Verlag. In addition to these five books, Dr. Kennish has published many articles in scientific journals and presented papers at numerous conferences. Currently, he is the marine science editor of the journal, *Bulletin of the New Jersey Academy of Science,* and series editor of Marine Science Books for CRC Press. His biogeographical profile appears in *Who's Who in Frontiers of Science and Technology* and *Who's Who Among Rising Young Americans.*

ACKNOWLEDGMENTS

I want to thank my colleagues at Rutgers University who have interacted with me during the preparation of this book. In the Institute of Marine and Coastal Sciences, I am especially grateful to R. A. Lutz, who has been an effective collaborator on many marine science projects, and K. W. Able, J. F. Grassle, J. N. Kraeuter, and N. P. Psuty. In the Department of Biological Sciences, I acknowledge the support of R. E. Loveland. In the Department of Geological Sciences, an expression of gratitude is extended to R. K. Olsson.

Appreciation is also expressed to the Editorial Department of CRC Press, especially P. Petralia, who coordinated the production of this book. I also thank C. Sweatman and C. Messing for assisting on editorial problems. All artwork was handled expeditiously by the Editing, Production, and Design Department.

Finally, I thank my wife, Jo-Ann, and sons, Shawn and Michael, for their patience and love during the preparation of this handbook.

INTRODUCTION

The 2nd edition of *Practical Handbook of Marine Science* consists of five major sections of expository, tabular, and illustrative material on the oceans. Four of these sections conform to the primary disciplines of oceanography — marine biology, marine chemistry, marine geology, and physical oceanography. The fifth section (marine pollution) is an outgrowth of heightened public concern regarding anthropogenic impacts on the marine environment. Unequivocally, this final section involves both scientific and societal issues.

Section 1 deals with physical oceanography, emphasizing open ocean, coastal, and estuarine circulation, as well as waves and wave-induced effects, tides and tidal currents, winds and air-sea interactions. In respect to the open ocean, information is provided on thermohaline circulation, water masses, and the physical properties of seawater. Tides and tidal currents supply much of the energy for turbulence and mixing in estuaries as recounted in this section. Important modifying forces are surface wind stress and meteorological forcing mechanisms which also play a critical role in the circulation of coastal ocean waters.

Section 2 details the chemistry of oceanic and estuarine waters. It focuses on the chemical composition of major, minor, trace, and nutrient elements in seawater. Because of the seemingly endless problems associated with heavy metals in estuarine and shallow coastal regimes, the physical and biogeochemical processes modulating their concentration and behavior are also investigated. Many studies of nutrient elements (nitrogen, phosphorus, and silicon) logically center on their importance as limiting factors for producers, consumers, and decomposers, and as electron donors for microbial decomposers. Nitrogen is the principal limiting element to primary production in estuarine and marine waters. Phosphorus may be a limiting nutrient to autotrophic growth during certain seasons of the year in some estuaries. Silicon in low concentrations can suppress metabolic activity of the cell, can limit phytoplankton production, and can reduce skeletal growth of diatoms, radiolarians, and siliceous sponges. The chemistry of seawater, therefore, is closely coupled to the production of organic matter in the sea. The composition and concentration of organic matter in estuarine and oceanic waters are likewise presented in Section 2.

Section 3 examines various aspects of marine geology. Included are data on the hypsometry of the ocean basins, major structural features of the seafloor (e.g., mid-ocean ridges, transform faults, seamounts), and mechanisms of plate tectonics. Crustal and lithospheric formation in the ocean occurs along extensive systems of mid-ocean ridges and rifts termed seafloor spreading centers as part of a magmatic/volcanic-hydrothermal-tectonic cycle. This is supported by a wealth of seismic, magnetic, gravity, and heat-flow data. The magmatic and volcanic processes operating at seafloor spreading centers are directly responsible for the formation of hydrothermal vents. Research on active and inactive hydrothermal vents has revealed the importance of these complex systems to the oceans of the world. For example, hydrothermal vent emissions account for an estimated 20% of the total ocean input of some chemical elements and, consequently, play a key role as modulator of global ocean chemistry. In addition, they may be directly responsible for the formation of stockworks and massive metal-rich sulfides that can have economic value as ore deposits. Heat discharged from hydrothermal vents also contributes significantly to the cooling of lithospheric plates. Calculations indicate that at least 40% of the earth's total heat loss occurs via hydrothermal circulation near seafloor spreading centers. Moreover, hydrothermal vents sustain lush and exotic communities of organisms not previously known to exist before 1977.

Section 4 covers marine biology. A brief account is given on the major faunal and floral groups found in estuarine and marine environments. Data are compiled on the distribution, density, and biomass of the biota in different habitats. Additional information chronicled in this section includes production estimates, assimilation efficiencies, and trophic relationships of marine organisms. The significance of the biota as habitat formers is clearly evident, especially in estuaries and shallow marine waters.

Section 5 addresses marine pollution. Nine main categories of pollution sources are described, notably organic loading, petroleum hydrocarbons, polynuclear aromatic hydrocarbons, chlorinated hydrocarbons, heavy metals, radioactivity, dredging and dredged-spoil disposal, electric generating stations, and air pollution. The toxic effects of many chemical contaminants released to marine environments (e.g., PAH, DDT, PCB, and heavy metals) are well known. These contaminants remain the target of many ongoing monitoring programs.

TABLE OF CONTENTS

Section 1
Physical Oceanography

Section 1

PHYSICAL OCEANOGRAPHY

OPEN OCEAN CIRCULATION

Wind stress, thermal forcing, and freshwater flux resulting from precipitation and evaporation drive ocean circulation. The ocean is a highly turbulent body of water whose circulation and mixing involve both advective and diffusive processes.[1] While the ocean is primarily an advective body, diffusion plays a role in establishing property fields.

Physical oceanography entails the study of the motions of ocean waters — currents, tides, and waves — and the forcing mechanisms responsible for these motions. Ocean circulation can be subdivided into two components: (1) surface circulation; and (2) thermohaline deep circulation. The radiation balance on earth, the rotation of the earth, and the relative motions of the earth, moon, and sun account for the major currents in the ocean which are manifested as large-scale horizontal water movements. The surface currents in the ocean are principally driven by the mean atmospheric circulation patterns.[2] The major current systems arise from global wind patterns. For example, the trade winds produce the northward-flowing western boundary currents, and the westerlies, the eastern boundary currents. It is these long-term winds (e.g., prevailing westerlies and trade winds), together with the Coriolis force, which generate a series of gyres or large circulating current systems in all ocean basins, centered at approximately 30°N and 30°S.[3] The gyres circulate clockwise in the North Atlantic and North Pacific Oceans and counterclockwise in the South Atlantic, South Pacific, and Indian Oceans. The rotation of the earth displaces the gyres toward the western boundary of the oceans, creating stronger currents along this perimeter and effectively separating coastal ocean waters from the open ocean. The displacement of ocean gyres toward the west resulting from the rotation of the earth causes a steeper slope of the sea surface than on the east. The steepness of the slopes controls current strength.[4] Thus, western boundary currents (Gulf Stream, Kiroshio, Brazil, and East Australian currents) are more intense, deeper, and narrower than eastern boundary currents (California, Humboldt, Canaries, and Benguela currents) which are characteristically slow, wide, shallow, and diffuse. As a consequence, the western boundary currents transport significant volumes of warm water northward, thereby transferring heat toward the poles. Surface currents flowing along the western coasts of continents toward the equator are deflected away from the coastline due to the Coriolis force. This upwelling process yields regions of exceptionally high biological productivity.[3,4]

A conspicuous feature of western boundary currents are meanders and rings. Commonly occurring in the western portions of ocean basins, meanders and rings cut off bodies of water which are enclosed by strong currents. Rings, observed on both sides of the Gulf Stream and along the North Atlantic Current in the Atlantic Ocean and the Kuroshio Current in the Pacific Ocean, measure about 100 to 300 km across; they are bounded by a nearly circular system of high-velocity currents (~90 cm/s or 80 km/d). The rings move with the waters around them at velocities of 5 to 10 km/d. Although some rings are rather quickly resorbed into the boundary currents, many travel with the currents and can be tracked over long distances and for significant spans of time. For instance, one cold-core ring was tracked for seven months as it traversed the Sargasso Sea from south of Cape Cod to near Cape Hatteras before it was resorbed into the Gulf Stream.[4]

Similar, albeit weaker, features along western boundary currents and elsewhere are eddies. Mesoscale eddies, time-dependent current and temperature patterns with horizontal dimensions of 100 to 500 km and periods of one to several months, are ubiquitous phenomena in the oceans of the world. However, eddy fields are much more dense in the North Atlantic and North Pacific than elsewhere.[5] Mesoscale ocean currents serve as a primary source of these eddies, being formed by instability mechanisms. Having rotary currents with velocities of approximately 0.1 m/s, mesoscale eddies travel at a few kilometers per day.[6]

Various aspects of ocean eddies are poorly understood. For example, the role that they play in the dynamics of the oceans remains largely unresolved. The coupling between the intensity of mesoscale eddies and energetic ocean currents at mid-latitudes has been of keen interest to physical oceanographers for about 2 decades.[5] The mesoscale data base for the oceans of the world indicates that mesoscale eddies are strongest near intense mesoscale currents.[7–11] The subject of ocean eddies continues to be a major research topic in physical oceanography.[12–14] Of particular value are eddy-resolving numerical circulation models, which are consistently becoming more realistic and useful to oceanographic studies.[15–19]

Other major current systems in the open ocean include the north and south equatorial currents. Positioned between them is the weakly flowing equatorial counter current. Also pronounced is the Antarctic circumpolar current (west wind drift), with a continuously eastward flow.

Wind stress on the sea surface, which owes its development to pressure gradients, is responsible for convergences and divergences of ocean waters. Where prevailing winds cause surface waters to converge along a line or front, sinking of dense water occurs. Large-scale oceanic convergences are identified by equatorial, subtropical, and polar convergence lines in the oceans of the world. However, small-scale convergences take place in the oceans as well. When wind stress leads to the horizontal flow of surface waters away from a common center or zone, divergence results. In this case, subsurface water rises to replace the diverging horizontal masses. The divergence of oceanic waters is often associated with the upwelling of more nutrient-rich, cold subsurface waters.

Langmuir circulation — helical vortices with axes parallel to the wind — arises in some converging areas of the ocean. Frequently marked by debris or foam floating on the sea surface, these corkscrewlike water motions appear to develop from instability in well-mixed surface waters. The large concentration of materials collecting at the ocean surface along these vortices usually attracts a wide variety of feeding organisms.[4,6]

As the wind blows over the sea surface, its energy passes down through the water column. However, because of the Coriolis force, each successively deeper layer of water is deflected toward the right of the wind direction in the northern hemisphere and toward the left in the southern hemisphere. The subsurface currents likewise decrease in velocity with increasing depth. Ultimately a depth is reached (the Ekman depth) where the water flows in a direction opposite to that of the surface current. When represented by vectors, the resultant change of current direction and velocity with depth follows a spiral pattern known as the Ekman spiral. Effects of the wind can be transmitted to depths of approximately 100 m to several hundred meters.

THERMOHALINE CIRCULATION

Whereas the planetary wind system drives the surface circulation of the ocean, the circulation in the deep sea depends on density differences largely determined by water temperature and salinity.[20] In profile, the ocean can be divided into three layers: (1) an upper, wind-mixed zone; (2) a thermocline; and (3) a uniform, deep-water region.[2] The upper, wind-mixed zone extends from the air-seawater interface to the depth of the thermocline beginning at about 100 to 300 m. The main thermocline is that portion of the water column having a well-developed vertical temperature gradient characterized by declining temperature with increasing depth appreciably greater than in overlying and underlying waters. Persisting for several hundred meters, this thermocline acts as a stably stratified layer limiting vertical water transfer, thereby precluding surface-induced mixing at greater depths. Below the main thermocline lies the bulk of ocean waters having uniform temperature and salinity. The dynamics of ocean circulation, incorporating numerical models, are treated comprehensively by Abarbanel and Young.[21]

Studies of thermohaline circulation in the deep ocean concentrate on identifiable water masses classified by their source region and depth of occurrence. The Antarctic bottom water (AABW), for example, originates in the Weddel Sea, flows down the Antarctic continental slope, and moves northward at great depths. Having temperature and salinity characteristics of 0.5°C and 34.7‰, respectively, this water mass can be traced into the Atlantic, Pacific, and Indian Oceans. Ocean bottom topography controls the deep-water movement through the ocean basins.[4] Another source region of deep oceanic waters is the Antarctic convergence, a convergence of surface waters in the Antarctic circumpolar current. This area represents the main supply of Antarctic intermediate water (AAIW) (4°C, 34.4‰) that also descends northward, where it overlies North Atlantic deep water (NADW) and AABW. The Norwegian and Greenland Seas are the major source regions of water below the thermocline in the northern hemisphere.[2] Water from these seas sinks and flows southward, mixing with overlying waters to form the NADW (2°C, 34.95‰). Originating at about 60°N latitude, the NADW spreads southward above the AABW to as far as 50°S latitude. The mixing of NADW, AAIW, and AABW produces the largest water mass in the deep sea, the Indian and Pacific Oceans common water (1.5°C, 34.7‰).

Oceanographers may identify and trace a deep oceanic water mass by its characteristic temperature and salinity acquired at the sea surface, together with oxygen measurements.[3] However, dissolved oxygen as a tracer gives only a general sense of the rates of water movements.[6] Radioactive tracers are now used most frequently to map deep-sea water movements and to determine current velocities. For example, the decay of the isotope carbon-14 (^{14}C) has proven to be valuable in this regard. Its application has revealed time scales of between 200 and 1000 years for thermohaline deep circulation of the oceans.[2] Another important radioactive tracer of deep oceanic water masses is tritium, a hydrogen isotope with a half-life of 12.3 years.

WAVES

KELVIN AND ROSSBY WAVES

Wind stress produces most surface waves in the ocean, although catastrophic events such as earthquakes and volcanic eruptions occasionally generate major disturbances of the sea surface (e.g., tsunamis). Other types of waves encountered in marine and estuarine waters include tidal and internal waves, as well as seiches (i.e., standing waves). In regard to ocean circulation, Kelvin waves and planetary or Rossby waves are of paramount importance. Kelvin waves, which can travel as surface (barotropic) waves or as baroclinic waves, represent perturbations of the sea surface or of the thermocline that propagate parallel to and close to the coast. As water piles up against a coastal boundary, an offshore horizontal pressure gradient force develops. The coast may act as a wave guide to constrain the way in which the water can move in response to the forces acting on it. The Coriolis force directed towards the coast and opposed by a horizontal pressure gradient force that is coupled to the slope of the sea surface is necessary condition for the propagation of Kelvin waves.[6]

As described by Rhines,[22] " . . . A Rossby wave is not a surface wave of the familiar sort, but a great undulation of the whole ocean mass that carries signals from one shore to another over weeks, months, and years." The need for the potential vorticity to be conserved accounts for their existence in the ocean.[6] They propagate zonally, along the equator, as well as along other lines of latitude.

SURFACE WAVES

Surface waves are classified by size, typically ranging from small capillary waves or ripples to large waves having periods of up to 5 s.[4] When wind velocity exceeds 4 km/h across the ocean, waves develop because of friction on the sea surface.[23] Bowden[24] defines wind stress on the sea surface as follows:

$$\tau_s = C_D \rho_a W^2 \tag{1}$$

where W is the wind speed measured at a standard height, ρ_a is the density of the air, and C_D is a drag coefficient dependent on the height at which the wind is measured, the roughness of the water surface, and the stability of the air in the first few meters above the surface.

Waves increase in size with increasing wind velocity, the length of time the wind blows in one direction, and the distance the wind blows over the water surface in a constant direction (i.e., fetch). Although the waveform advances laterally with little net movement of water, individual water particles oscillate in approximately circular paths, moving forward on the crest of the wave, then vertically, and finally backward under the trough. The diameter of the circular orbits decreases with increasing water depth, being nearly equal to the wave height at the surface and declining to near zero at a depth of one half the wavelength. Below this depth, there is little water movement with wave passage.

In deep water, the waveform is symmetrical and can be described by its length (the horizontal distance between adjacent crests), height (the vertical distance between a wave crest and adjacent trough), period (the time interval between successive waves), and velocity (wavelength/wave period). As shallow water is approached, water particles near the seafloor do not transcribe circular orbits, but move in a horizontal, back-and-forth motion along the bottom. Sediment is slowly transported in the direction of wave propagation as water particles move backward and forward along the seafloor.[25] When the water depth becomes less than one half the wavelength, the interference of the bottom with the motion of water particles affects wave characteristics, reducing the wave velocity and length and increasing the wave height. The seafloor strongly influences the motion of water particles at depths less than 1/20 of the wavelength, although at the surface, the orbits of water particles may be only slightly altered into the form of an ellipse.[6] Continued shoaling of water makes the wave unstable; the crest increases in steepness, can no longer support itself, and it breaks forward along the shore where rapid water movements release energy, transport sediments, and generate currents. Waves break when the wave steepness (the ratio of wave height to wavelength) exceeds 1/7[6] and the ratio of water depth to wave height is about 4/3.[25] Several types of breaking waves are recognized: spilling, plunging, collapsing, and surging. The type of breaker at a given locality depends on the types of waves approaching the shoreline and the complexity and slope of the seafloor.

The velocity of shallow water waves can be calculated by the following expression:

$$V \simeq \sqrt{gd} \qquad (2)$$

where g is the acceleration due to gravity and d is the depth to the seafloor. In comparison, the velocity of deep water waves may be determined by the equations:

$$V = \frac{gT}{2\pi} \qquad (3)$$

$$V = \sqrt{\frac{gL}{2\pi}} \qquad (4)$$

where T is the wave period, L is the wavelength, and g is the acceleration due to gravity.

Waves approaching a shoreline may be refracted or diffracted. As wavefronts enter shallow water at an oblique angle, they bend such that the waves strike nearly parallel to the shore. The portion of a wavefront initially advancing into shallow water "feels" bottom first, causing the circular orbits of water particles to become elliptical. This process of wave retardation from bottom effects slows down the segment of the wavefront closest to the shore, enabling the wavefront to change the direction of its approach to shore. Waves, although refracted as the shore is approached, usually break at a slight angle to the shore; this action produces longshore currents parallel to shore and an accompanying longshore drift (zigzag movement) of sediment. Water that accumulates on the shore from breaking waves returns rapidly offshore in rip currents that cut narrow, intermittent channels away from the shore. Longshore currents and longshore drift are more pronounced in coastal ocean waters than in estuaries, but have been documented in both environments. Rip currents are less obvious in estuaries, being important only in those systems having a wide mouth.

The refraction of waves on an irregular shoreline concentrates energy on headlands and disperses it in adjacent bays. Because water shoals near a headland, approaching waves refract and converge on the projecting landmass, eroding it more quickly than the neighboring shoreline. As waves enter a bay, they are refracted in shallow water along the perimeter, whereas in deep water they travel with greater velocity. The divergence of wave energy in bays reduces erosion and increases sedimentation. Through time, therefore, shoreline irregularities become smoother as a result of this nonuniform distribution of wave energy.

If waves impinge perpendicularly on an obstacle, for example a barrier beach, they may be reflected and their energy transferred to waves traveling in another direction.[6] Waves that pass the end of an obstacle, such as the terminus of a spit, may be diffracted into sheltered regions of an estuary. Subsequent to passage through the mouth of an estuary, waves generally decrease in velocity due to bottom effects and the opposing river flow.

Wave heights and wavelengths in estuaries are usually less than those in nearshore oceanic areas because of the reduced fetch of the wind. Wavelengths commonly range from 15 to 25 m, and wave periods are usually less than 5 s. In addition, waves tend to be more irregular in estuaries than in the ocean. For an ideal, triangular-shaped estuary in which the base opens to the ocean, the other two sides are long and bordered by land, and the apex forms the head, the fetch of the wind is very short from most directions. Only along the longitudinal axis is the fetch sufficiently great to allow the formation of large waves. Waves entering estuaries from the nearshore ocean will be diffracted as they travel through the mouth, dissipating much of their energy before the main body of the estuary is reached. Diffraction of ocean waves becomes accentuated in lagoon-type, bar-built estuaries having a narrow, shallow mouth. Although most estuaries are calm compared to the open ocean, wave action can be significant in systems with a wide, deep mouth where ocean waves meet with less interference during passage in an upestuary direction.

INTERNAL WAVES

Much less conspicuous than progressive surface waves, internal waves occur beneath the sea surface between water layers of different density. These gravity waves propagate along a pycnocline associated either with a halocline or thermocline, and are formed by multiple mechanisms, including traveling atmospheric pressure fields, interactions of currents with bottom topography, tidal oscillations, or second-order interactions between surface waves of different frequencies.[26–29] The shear of rapidly flowing surface layers also induces internal waves in stratified estuaries. Internal waves travel more slowly than surface waves, but can attain much greater heights. They mix waters below the surface and may be important in the movement of sediment.

SEICHES

Seiches are stationary waves in an enclosed or semienclosed basin which oscillate pendulum fashion around a point (node) on the water surface that does not move vertically. Maximum vertical movement takes place at points on the water surface termed "antinodes". Seiches develop in response to atmospheric, seismic, and tidal forces, and they have periods that are variable and strongly related to the basin geometry. The period of seiches, for example, range from a few minutes to up to a day. In some fjord-type estuaries (e.g., Oslo Fjord), tidal currents interact with a sill to form internal seiches.[30] These seiches usually have ranges and periods substantially greater than those on the surface.

Seiches sometimes act synergistically with progressive waves in estuaries to produce large water movements. In spite of the small magnitude of the tide-generating forces, resonance can generate large tidal movements.[31] Resonance is likely to take place when the natural period of oscillation of an estuary, which depends on its dimensions, is similar to the period of one of the tide-generating forces.

TSUNAMIS

Sudden displacements of the seafloor associated with earthquakes or volcanic eruptions are responsible for long-period (several minutes to several hours) gravity waves termed "tsunamis" or "seismic sea waves". In the open ocean, tsunamis may travel for thousands of kilometers at velocities of 700 to 900 km/h. In coastal waters, the velocities decrease, while the wave heights increase up to 30 to 70 m. Because of their great velocity and height, tsunamis are extremely destructive in coastal environments.

COASTAL AND ESTUARINE CIRCULATION

Two topics of coastal circulation receiving considerable attention during the 1970s, 1980s, and 1990s have been coastal upwelling and oceanic fronts.[4,6,32–34] Intense studies of coastal upwelling have been conducted in the eastern Pacific Ocean along the coasts of Oregon (U.S.) and Peru. Research on fronts expanded rapidly in the late 1970s. Part of this expansion can be ascribed to the utilization of satellite-borne infrared radiometry for sea surface monitoring, which prompted the discovery of new fronts.[35]

Circulation and mixing change abruptly at oceanic or estuarine fronts, which have been defined as boundaries between horizontally juxtaposed water masses of dissimilar properties.[36] Bowman and Esaias[37] and Mann[38] describe six types of fronts: (1) river plume fronts, (2) upwelling fronts, (3) shallow sea fronts, (4) shelf break fronts, (5) fronts at the edge of major western boundary currents, and (6) fronts of planetary scale, removed from major ocean boundaries. The first four types are coastal phenomena, and the last two types, open-ocean phenomena.

Fronts are pronounced in salt wedge estuaries. They mark areas of convergent flows having strong vertical motions and sharp gradients, approaching discontinuity, in physical parameters, especially velocity and density fields.[39] Along an estuarine front, isohalines rise steeply to the surface, and sharp horizontal gradients in salinity are apparent.[24] Surface water sinks at the frontal interface and, because of convergence, a line of foam or floating organic and detrital material generally collects at the surface.[40] Changes in color and turbidity of the water often delineate the site of the front as it advances downestuary over a lower layer of water.

Fronts are common in narrow estuaries having consistently high river discharge,[41–43] although episodic, heavy fluvial inflow may enhance their formation.[44] Tidal dynamics frequently control the development of an estuarine front,[45,46] with the bathymetry of the basin affecting their position.[47] An integral steady-state model formulated by Garvine[48] deals with the dynamics of small-scale oceanic fronts. Garvine[48] and Garvine and Monk[49] examine plumes and fronts formed by the Connecticut River flowing into Long Island Sound.

Water circulation in estuaries consists of tidal and nontidal components. The tidal component results from extrinsic forces, that is, tide-producing forces of the sun and moon.[31] These forces are responsible for the periodic vertical displacement of estuarine waters and the associated horizontal movement of tidal currents. The nontidal component originates from factors independent of astronomical forces, specifically water movements induced by the interaction of fresh and saline waters of different densities.[25]

The strength of tidal currents relative to river flow, in combination with the geometrical configuration of the estuary, largely determines circulation patterns. While tides and tidal currents provide the ultimate driving force for much of the turbulence and mixing in most stratified and well-mixed estuaries, other mechanisms (e.g., surface wind stress and meteorological forcing) can be important modifying factors, especially in shallow coastal bays.[50,51] Marked changes in estuarine circulation can be induced by nontidal forcing from the nearshore ocean (e.g., coastal Ekman convergence or divergence), storm surges, and variability in the neap-spring tidal cycle.

Fresh water (density = 1.0 gm/cm^3) is less dense than seawater (density = 1.025 gm/cm^3); consequently, these two fluids tend to form separate water masses in estuaries, with freshwater overlying seawater. Although water temperature influences the density value of both freshwater and seawater, it has a much smaller effect on density than the concentration of dissolved salts. In a density-stratified system lacking currents, the mixing between freshwater and seawater masses is ascribed to diffusive and advective processes. Diffusion in estuarine water is defined as a flux of salt, and advection as a flux of salt and a flux of water.[52] Vertical advection, the upward breaking of internal waves at the interface between the freshwater and seawater layers, is the primary mixing agent in the absence of currents, causing a diffuse boundary layer between the two water masses and a gradual increase in salinity in the freshwater layer downestuary.[52,53] While the upward flux of salt by diffusion does not produce the mass flux of water and salt as advection, it plays an important role in the mixing of waters along the vertical, lateral, and longitudinal axes of estuaries.

Currents greatly increase the mixing of estuarine waters. The interaction of tidal currents, wind stress, internal friction, and bottom friction can reduce or eliminate density stratification of the water column. The turbulence produced by both internal shear and bottom friction is responsible for mixing in estuaries. In well-mixed estuaries, turbulence generated by bottom friction predominates, whereas in highly stratified estuaries, turbulence produced by internal shear is paramount.[54] Both internal shear and bottom friction are important factors in the mixing of partially stratified estuaries.[55]

In summary, the magnitude of river discharge relative to tidal flow is the principal factor determining the type of water circulation in estuaries. The river discharge acting against tidal motion and interacting with wind stress, internal friction, and bottom friction controls the degree of turbulent mixing and consequently the vertical salinity and velocity structure.[56] The characteristics of this system vary as the volume of river discharge, range of the tide, and geomorphology of the estuarine basin change.[57] Water flow in the estuary is strongly three dimensional because of the generally irregular structure of the basin, spatial density currents, and the effect of the Coriolis force. These factors all contribute to the complexity of estuarine circulation patterns.

It is the temporal and spatial variability of forcing mechanisms that account for complicated estuarine responses. As far as temporal variability in forcing mechanisms is concerned, those at time scales between the tidal period and annual period are of most interest in terms of estuarywide response. In regard to spatial variability, those forcing mechanisms operating on the order of one to several kilometers may be most significant.[50]

Pritchard[58,59] classified estuaries into four types based on circulation: (1) type A, or salt wedge estuaries, which are highly stratified; (2) type B, or partially mixed estuaries, which are moderately stratified; (3) type C, or vertically homogeneous estuaries, with a lateral salinity gradient; and (4) type D, or sectionally homogeneous estuaries, with a longitudinal salinity gradient. Closely related to these circulation patterns are mixing and dispersion resulting from advective and diffuse processes. This classification scheme depends on the advection-diffusion equation for salt, the so-called salt-balance equation.[59,60]

Studies of spatial variability in circulation have concentrated on flow in the longitudinal sense, along the estuary, and its changes in the vertical plane, with increasing depth. Little or no consideration is often given to the variation of flow across the estuary in a lateral direction. In many cases, lateral differences in velocity or density fields produce transverse effects that can be large compared to vertical variations.[61] Secondary flows may arise when circulation is not evenly distributed across the estuary because of the lack of consistency in the cross-sectional form of the basin in a longitudinal direction, the effect of the Coriolis force, and differences in lateral density fields.[62]

In many stratified and well-mixed estuaries, longitudinal density-gradient circulation and mixing are common. The longitudinal surface slope, which acts in a downestuary direction, and the longitudinal, density-gradient force, which acts in an upestuary direction, drive this type of circulation.[63] According to Officer,[63] when river discharge is low, the surface slope force dominates in the upper part of the water column, with a net flow downestuary, and the density gradient force dominates in the lower part of the water column, with a net flow upestuary. However, particularly in larger estuaries, lateral effects, surface wind stress, meteorological forcing, and the basin geometry may obscure this circulation pattern, at least for certain periods of time. Thus, Smith and Takhar[64] conveyed the significance of wind stress interacting with tidal and river flows in governing the degree of turbulent mixing and controlling the vertical salinity and velocity structure. Elliott[65] and Elliott and Hendrix,[66] reporting on the Potomac estuary, documented short-term changes in circulation patterns attributable to the wind. Smith[67,68] recorded a local coupling between coastal winds and water levels at Corpus Christi Bay and Aransas Pass, Texas. Rattray and Hansen[69] presented theoretical models of wind effects on circulation in coastal plain estuaries. Clearly, these investigations reveal the complexities of the short- and long-term variations of estuarine circulation.

Quantitative treatment of estuarine circulation can be found in Bowden,[24] Dyer,[70] and Officer.[71] The three equations of motion plus the continuity equations for mass and salt represent the theoretical description of estuarine circulation and mixing. They are treated in detail by Officer.[63]

OBSERVATIONAL METHODS

In the shallow water environments of estuaries and continental shelves, substantial strides have been achieved in assessing current profiles and turbulence by deploying several field-measuring techniques. Electromagnetic flow meters, acoustic current meters, heated profiling thermistors and related instruments involving heat transfer, impellors, and laser-doppler current meters have proven to be very useful in coastal oceanographic research.[72] InterOcean S-4 current meters, high-technology sampling devices that record velocity, direction, depth, conductivity, salinity, and temperature, display powerful capabilities; yet, their small size makes them ideally suited for wide-range field applications.

A whole range of new devices for observing the ocean emerged with the advent of modern solid-state electronics, leading to several generations of internally recording current meters, neutrally buoyant floats, and profiling devices. More recently, satellite altimetry and other satellite observational tools may permit oceanographers to ascertain the sea in its global entirety and to great depths.[73] Thus, altimetric satellites are capable of observing Gulf Stream fluctuations, although not with the same spatial and temporal resolution of infrared instruments. Microwave sensors developed in the late 1970s have enabled the measurement of sea-surface fluctuations of several centimeters averaged over several kilometers. Mooring technology has also evolved over the past 3 decades into reliable multiyear settings of large instrument arrays.[74] Future advances in mooring and satellite technology promise to provide greater resolution of the circulation of the ocean and its influence on world climate.

GLOBAL STUDIES

The World Ocean Circulation Experiment (WOCE), conducted by scientists from oceanographic institutions and government agencies worldwide, represents the first unified research program of sufficient scope to mount a truly global investigation of the ocean. Satellite-borne sensors serve as vital components of WOCE to provide adequate global coverage. The overall goal of WOCE is to yield ocean models that will adequately predict decadal climate change. Commencing in 1990, the field program is scheduled to be completed by 1997. The early work on WOCE will be in the southern hemisphere, followed by efforts in the Indian, North Atlantic, and Pacific Oceans.[75]

Another major study in physical oceanography is the Tropical Ocean-Global Atmosphere (TOGA) program, which was initiated in 1985 to predict seasonal-to-interannual climate variability. Efforts in TOGA resulted in the successful prediction (using a coupled tropical Pacific ocean-atmosphere model) of El Nino/Southern Oscillation events in 1987 and 1992 1 year in advance.[76] The U.S. has successfully contributed to TOGA with the creation of the TOGA-TAO (Tropical Atmosphere-Ocean) array for measuring the internal thermal structure of the tropical Pacific Ocean.[77] TOGA, in addition to WOCE, should greatly advance both short- and long-term climate prediction and facilitate man's understanding of the atmosphere-ocean system.

SECTION PLAN

This section on physical oceanography deals with surface and deep-water circulation in the open ocean. In addition, it examines the identifiable water masses in the deep sea, focusing on temperature, salinity, and density distributions. Information is also given on surface currents, gyres, eddies, rings, and meanders in the open ocean. It likewise details water circulation in coastal and estuarine systems, examining various processes such as upwelling, advection-diffusion, and mixing.

REFERENCES

1. **Bowden, K. F.,** *Physical Oceanography of Coastal Waters,* Ellis Horwood, Chichester, England, 1983.
2. **Smith, D. G., Ed.,** *Cambridge Encyclopedia of Earth Sciences,* Cambridge University Press, London, 1981.
3. **Levinton, J. S.,** *Marine Ecology,* Prentice-Hall, Englewood Cliffs, NJ, 1982.
4. **Gross, M. G.,** *Oceanography: a View of the Earth,* 5th ed., Prentice-Hall, Englewood Cliffs, NJ, 1990.
5. **Schmitz, W. J., Jr.,** Mesoscale ocean currents, *Rev. Geophys.,* in press.
6. **Brown, J., Colling, A., Park, D., Phillips, J., Rothery, D., and Wright, J.,** *Ocean Circulation,* Pergamon Press, Oxford, 1989.
7. **Emery, W. J.,** On the geographical variability of the upper level mean and eddy fields in the North Atlantic and North Pacific, *J. Phys. Oceanogr.,* 13, 269, 1983.
8. **Krauss, W. and Kase, R. H.,** Mean circulation and eddy kinetic energy in the eastern North Atlantic, *J. Geophys. Res.,* 89, 3407, 1984.
9. **Schmitz, W. J., Jr., Holland, W. R., and Price, J. F.,** Mid-latitude mesoscale variability, *Res. Geophys. Space Phys.,* 21, 1109, 1983.
10. **Schmitz, W. J., Jr.,** Abyssal eddy kinetic energy in the North Atlantic, *J. Mar. Res.,* 42, 509, 1984.
11. **Kraus, W., Fahrbach, E., Aitsam, A., Elken, J., and Koske, P.,** The North Atlantic Current and its associated eddy field southeast of Flemish Cap, *Deep-Sea Res.,* 34, 1163, 1987.
12. **Holland, W. R., Harrison, D. E., and Semtner, A. J.,** Eddy-resolving numerical models of large-scale ocean circulation, in *Eddies in Marine Science,* Robinson, A. R., Ed., Springer-Verlag, New York, 1983, 329.
13. **Robinson, A. R., Ed.,** Overview and summary of eddy science, in *Eddies in Marine Science,* Springer-Verlag, New York, 1983, 3.
14. **Krauss, W. and Boning, C. W.,** Langrangian properties of eddy fields in the northern North Atlantic as deduced from satellite tracked buoys, *J. Mar. Res.,* 45, 259, 1987.
15. **Haidvogel, D. B. and Holland, W. R.,** The stability of ocean currents in eddy-resolving general circulation models, *J. Phys. Oceanogr.,* 8, 394, 1978.
16. **Schmitz, W. J., Jr. and Holland, W. R.,** Observed and modeled mesoscale variability near the Gulf Stream and Kuroshio Extension, *J. Geophys. Res.,* 91, 9624, 1986.
17. **Holland, W. R., Harrison, D. E., and Semtner, A. D., Jr.,** Eddy-resolving models of large-scale ocean circulation, in *Eddies in Marine Science,* Robinson, A. R., Ed., Springer-Verlag, Berlin, 1983, 379.
18. **Thompson, J. D. and Schmitz, W. J., Jr.,** A limited-area model of the Gulf Stream: design, initial experiments, and model-data intercomparison, *J. Phys. Oceanogr.,* 19, 791, 1989.
19. **Hurlburt, H. E., Wallcraft, A. J., Sirkes, Z., and Metzger, E. J.,** Modeling of the global and Pacific Oceans: on the path to eddy-resolving ocean prediction, *Oceanography,* in press.
20. **Price, J. F.,** Overflows, *Oceanus,* 36, 28, 1992.
21. **Abarbanel, I. and Young, W. R., Eds.,** *General Circulation of the Ocean,* Springer-Verlag, New York, 1987.
22. **Rhines, P. B.,** Physical oceanography, *Oceanus,* 36, 78, 1992.
23. **Lounsbury, J. F. and Ogden, L.,** *Earth Science,* 2nd ed., Harper & Row, New York, 1973.
24. **Bowden, K. F.,** Physical factors: salinity, temperature, circulation, and mixing processes, in *Chemistry and Biogeochemistry of Estuaries,* Olausson, E. and Cato, I., Eds., John Wiley & Sons, Chichester, England, 1980, 37.
25. **Perkins, E. J.,** *The Biology of Estuaries and Coastal Waters,* Academic Press, London, 1974.
26. **Thorpe, S. A.,** The excitation, dissipation, and interaction of internal waves in the deep ocean, *J. Geophys. Res.,* 80, 328, 1975.
27. **Wunsch, C.,** Deep ocean internal waves: what do we really know?, *J. Geophys. Res.,* 80, 339, 1975.
28. **Wright, L. D.,** Internal waves, in *The Encyclopedia of Beaches and Coastal Environments,* Schwartz, M. L., Ed., Hutchinson Ross, Stroudsburg, PA, 1982, 492.
29. **Haury, L. R., Wiebe, P. H., Orr, M. H., and Briscoe, M. G.,** Tidally generated high-frequency internal wave packets and their effects on plankton in Massachusetts Bay, *J. Mar. Res.,* 41, 65, 1983.
30. **Stigebrandt, A.,** Vertical diffusion driven by internal waves in a sill fjord, *J. Phys. Oceanogr.,* 6, 486, 1976.
31. **Glen, N. C.,** Tidal measurement, in *Estuarine Hydrography and Sedimentation,* Dyer, K. R., Ed., Cambridge University Press, Cambridge, 1979, 19.
32. **Brink, K. H., Halpern, D., and Smith, R. L.,** Circulation in the Peruvian upwelling system near 15°S, *J. Geophys. Res.,* 85, 4036, 1980.

33. **Shaffer, G.,** On the upwelling circulation over the wide shelf off Peru. 2. Vertical velocities, internal mixing, and heat balance, *J. Mar. Res.*, 44, 227, 1986.
34. **Le Fevre, J.,** Aspects of the biology of frontal systems, in *Advances in Marine Biology,* Vol. 23, Blaxter, J. H. S. and Southward, A. J., Eds., Academic Press, London, 1986, 163.
35. **Legeckis, R.,** A survey of worldwide sea surface temperature fronts detected by environmental satellites, *J. Geophys. Res.*, 83, 4501, 1978.
36. **Bowman, M. J.,** Introduction and historical perspective, in *Oceanic Fronts in Coastal Processes,* Bowman, M. J. and Esaias, W. E., Eds., Springer-Verlag, New York, 1978, 2.
37. **Bowman, M. J. and Esaias, W. E., Eds.,** *Oceanic Fronts in Coastal Processes,* Springer-Verlag, New York, 1978.
38. **Mann, K. H.,** *Ecology of Coastal Waters: a Systems Approach,* University of California Press, Berkeley, 1982.
39. **Huzzey, L. M.,** The dynamics of a bathymetrically arrested estuarine front, *Estuarine Coastal Shelf Sci.*, 15, 537, 1982.
40. **Garvine, R. W.,** Dynamics of small-scale oceanic fronts, *J. Phys. Oceanogr.*, 4, 557, 1974.
41. **Wright, L. D. and Coleman, J. M.,** Effluent expansion and interfacial mixing in the presence of a salt wedge, Mississippi River delta, *J. Geophys. Res.*, 76, 8649, 1971.
42. **Bowman, M. J.,** Spreading and mixing of the Hudson River effluent into the New York Bight, in *Hydrodynamics of Estuaries and Fjords,* Nihoul, J., Ed., Elsevier, Amsterdam, 1978, 373.
43. **Ingram, R. G.,** Characteristics of the Great Whale River plume, *J. Geophys. Res.*, 86, 2017, 1981.
44. **Wolanski, E. and Collis, P.,** Aspects of aquatic ecology of the Hawksbury River. I. Hydrodynamical processes, *Aust. J. Mar. Freshwater Res.*, 27, 565, 1976.
45. **Ingram, R. G.,** Characteristics of a tide-induced estuarine front, *J. Geophys. Res.*, 81, 1951, 1976.
46. **Nunes, R. A. and Simpson, J. H.,** Axial convergence in a well-mixed estuary, *Estuarine Coastal Shelf Sci.*, 20, 637, 1985.
47. **Klemas, V. and Polis, D. F.,** A study of density fronts and their effects on coastal pollutants, *Remote Sensing Environ.*, 6, 95, 1977.
48. **Garvine, R. W.,** An integral hydrodynamic model of upper ocean frontal dynamics. I. Development and analysis, *J. Phys. Oceanogr.*, 9, 1, 1979.
49. **Garvine, R. W. and Monk, J. D.,** Frontal structure of a river plume, *J. Geophys. Res.*, 79, 2251, 1974.
50. **National Academy of Sciences-National Research Council,** Variability of circulation and mixing in estuaries, in *Fundamental Research on Estuaries: The Importance of an Interdisciplinary Approach,* National Academy Press, Washington, D.C., 1983, 15.
51. **Kennish, M. J. and Lutz, R. A., Eds.,** *Ecology of Barnegat Bay, New Jersey,* Springer-Verlag, New York, 1984.
52. **Biggs, R. B. and Cronin, L. E.,** Special characteristics of estuaries, in *Estuaries and Nutrients,* Neilson, B. J. and Cronin, L. E., Eds., Humana Press, Clifton, NJ, 1981, 3.
53. **Biggs, R. B.,** Estuaries, in *The Encyclopedia of Beaches and Coastal Environments,* Schwartz, M. L., Ed., Hutchinson Ross, Stroudsburg, PA, 1982, 393.
54. **Abraham, G.,** On internally generated estuarine turbulence, Proc. 2nd I.A.H.R. Int. Symp. Stratified Flows, Trondheim, 344, 1980.
55. **Dyer, K. R.,** Mixing caused by lateral internal seiching within a partially mixed estuary, *Estuarine Coastal Shelf Sci.*, 15, 443, 1982.
56. **Smith, T. J. and Takhar, H. S.,** A mathematical model for partially mixed estuaries using the turbulence energy equation, *Estuarine Coastal Shelf Sci.*, 13, 27, 1981.
57. **Dyer, K. R.,** Localized mixing of low salinity patches in a partially mixed estuary (Southampton water, England), in *Estuarine Comparisons,* Kennedy, V. S., Ed., Academic Press, New York, 1982, 21.
58. **Pritchard, D. W.,** Observations of circulation in coastal plain estuaries, in *Estuaries,* Lauff, G. H., Ed., Publ. 83, American Association for the Advancement of Science, Washington, D.C., 1967, 37.
59. **Pritchard, D. W.,** Estuarine circulation patterns, *Proc. Am. Soc. Civ. Eng.*, 81, 1, 1955.
60. **Pritchard, D. W.,** A study of salt balance in a coastal plain estuary, *J. Mar. Res.*, 13, 133, 1954.
61. **Dyer, K. R.,** Lateral circulation effects in estuaries, in *Estuaries, Geophysics and the Environment,* National Academy of Science, Washington, D.C., 1977, 22.
62. **Dyer, K. R.,** Estuaries and estuarine sedimentation, in *Estuarine Hydrography and Sedimentation,* Dyer, K. R., Ed., Cambridge University Press, Cambridge, 1979, 1.
63. **Officer, C. B.,** Physics of estuarine circulation, in *Estuaries and Enclosed Seas,* Ketchum, B. H., Ed., Elsevier, Amsterdam, 1983, 15.
64. **Smith, T. J. and Takhar, H. S.,** A mathematical model for partially mixed estuaries using the turbulence energy equation, *Estuarine Coastal Shelf Sci.*, 13, 27, 1981.
65. **Elliott, A. J.,** Observations of the meteorologically-induced circulation in the Potomac estuary, *Estuarine Coastal Mar. Sci.*, 6, 285, 1978.
66. **Elliott, A. J. and Hendrix, T. E.,** Intensive Observations of the Circulation in the Potomac Estuary, Spec. Rept. 55, Chesapeake Bay Institute, The Johns Hopkins University, Baltimore, 1976.
67. **Smith, N. P.,** Tidal dynamics and low-frequency exchanges in the Aransas Pass, Texas, *Estuaries,* 2, 218, 1979.
68. **Smith, N. P.,** Meteorological and tidal exchanges between Corpus Christi Bay, Texas, and the northwestern Gulf of Mexico, *Estuarine Coastal Mar. Sci.*, 5, 511, 1977.
69. **Rattray, M. and Hansen, D. V.,** A similarity solution for circulation in an estuary, *J. Mar. Res.*, 20, 121, 1962.

70. **Dyer, K. R.,** *Estuaries: a Physical Introduction,* John Wiley & Sons, London, 1973.
71. **Officer, C. B.,** *Physical Oceanography of Estuaries and Associated Coastal Waters,* John Wiley & Sons, New York, 1976.
72. **Wright, L. D.,** Benthic boundary layers of estuarine and coastal environments, *Rev. Aquat. Sci.,* 1, 75, 1989.
73. **Wunsch, C.,** Observing ocean circulation from space, *Oceanus,* 36, 9, 1992.
74. **Hogg, N.,** The Gulf Stream and its recirculations, *Oceanus,* 36, 18, 1992.
75. **Needler, G. T.,** WOCE: The World Ocean Circulation Experiment, *Oceanus,* 36, 74, 1992.
76. **Lukas, R. and Webster, P. J.,** TOGA-COARE: Tropical Ocean-Global Atmosphere Program and Coupled Ocean-Atmosphere Response Experiment, *Oceanus,* 36, 62, 1992.
77. **Sarachik, E. S.,** Climate prediction and the ocean: modeling future conditions, *Oceanus,* 36, 66, 1992.

1.1 CONVERSION FACTORS, MEASURES, AND UNITS

Table 1.1—1
RECOMMENDED DECIMAL MULTIPLES AND SUBMULTIPLES

Multiples and submultiples	Prefixes	Symbols
10^{18}	exa	E
10^{15}	peca	P
10^{12}	tera	T
10^{9}	giga	G
10^{6}	mega	M
10^{3}	kilo	k
10^{2}	hecto	h
10	deca	da
10^{-1}	deci	d
10^{-2}	centi	c
10^{-3}	milli	m
10^{-4}	micro	μ (Greek mu)
10^{-9}	nano	n
10^{-12}	pico	p
10^{-15}	femto	f
10^{-18}	atto	a

From Beyer, W. H., Ed., *CRC Standard Mathematical Tables,* 28 ed, CRC Press, Boca Raton, FL, 1987, 1.

Table 1.1—2
CONVERSION FACTORS

Metric to English

To obtain	Multiply	By
Inches	Centimeters	0.3937007874
Feet	Meters	3.280839895
Yards	Meters	1.093613298
Miles	Kilometers	0.6213711922
Ounces	Grams	$3.527396195 \times 10^{-2}$
Pounds	Kilograms	2.204622622
Gallons (U.S. Liquid)	Liters	0.2641720524
Fluid ounces	Milliliters (cc)	$3.381402270 \times 10^{-2}$
Square inches	Square centimeters	0.1550003100
Square feet	Square meters	10.76391042
Square yards	Square meters	1.195990046
Cubic inches	Milliliters (cc)	$6.102374409 \times 10^{-2}$
Cubic feet	Cubic meters	35.31466672
Cubic yards	Cubic meters	1.307950619

Table 1.1—2 (continued)
CONVERSION FACTORS

English to Metric*

To obtain	Multiply	By
Microns	Mils	**25.4**
Centimeters	Inches	**2.54**
Meters	Feet	**0.3048**
Meters	Yards	**0.9144**
Kilometers	Miles	**1.609344**
Grams	Ounces	28.34952313
Kilograms	Pounds	**0.45359237**
Liters	Gallons (U.S. Liquid)	**3.785411784**
Milliliters (cc)	Fluid ounces	29.57352956
Square centimeters	Square inches	**6.4516**
Square meters	Square feet	**0.09290304**
Square meters	Square yards	**0.83612736**
Milliliters (cc)	Cubic inches	**16.387064**
Cubic meters	Cubic feet	$2.831684659 \times 10^{-2}$
Cubic meters	Cubic yards	0.764554858

Conversion Factors — General*

To obtain	Multiply	By
Atmospheres	Feet of water @ 4°C	2.950×10^{-2}
Atmospheres	Inches of mercury @ 0°C	3.342×10^{-2}
Atmospheres	Pounds per square inch	6.804×10^{-2}
BTU	Foot-pounds	1.285×10^{-3}
BTU	Joules	9.480×10^{-4}
Cubic feet	Cords	**128**
Degree (angle)	Radians	57.2958
Ergs	Foot-pounds	1.356×10^{7}
Feet	Miles	5280
Feet of water @ 4°C	Atmospheres	33.90
Foot-pounds	Horsepower-hours	1.98×10^{6}
Foot-pounds	Kilowatt-hours	2.655×10^{6}
Foot-pounds per min	Horsepower	3.3×10^{4}
Horsepower	Foot-pounds per sec	1.818×10^{-3}
Inches of mercury @ 0°C	Pounds per square inch	2.036
Joules	BTU	1054.8
Joules	Foot-pounds	1.35582
Kilowatts	BTU per min	1.758×10^{-2}
Kilowatts	Foot-pounds per min	2.26×10^{-5}
Kilowatts	Horsepower	0.745712
Knots	Miles per hour	0.86897624
Miles	Feet	1.894×10^{-4}
Nautical miles	Miles	0.86897624
Radians	Degrees	1.745×10^{-2}
Square feet	Acres	43560
Watts	BTU per min	17.5796

* Boldface numbers are exact; others are given to ten significant figures where so indicated by the multiplier factor.

Temperature Factors

$$°F = 9/5\,(°C) + 32$$

Fahrenheit temperature = 1.8 (temperature in kelvins) −459.67

$$°C = 5/9\,[(°F) - 32]$$

Fahrenheit temperature = 1.8 (Celsius temperature) + 32   Celsius temperature = temperature in kelvins −273.15

From Beyer, W. H., Ed., *CRC Standard Mathematical Tables,* 28 ed., CRC Press, Boca Raton, FL, 1987, 2.

Table 1.1—3
METRIC AND U.S. SYSTEM, MEASURES, UNITS AND CONVERSIONS

Lengths

Metric system		U.S. system
10^{-8} cm	1Å	$3.937 \cdot 10^{-9}$ in.
10^{-4} cm	1 μ	$3.937 \cdot 10^{-5}$ in.
	1 cm	0.3937 in.
	2.540 cm	1 in.
	0.3048	1 ft
	0.9144 m	1 yd
10^{2} cm	1 m	1.09361 yd
	1.8288 m	1 fath.
10^{5} cm	1 km	0.62137 mi
	1.60935 km	1 mi
	1.852 km	1 int. nautical mi

1 A.U. (astronomical unit) = $1.49598 \cdot 10^{8}$ km

Area

Metric system	U.S. system
1 mm^2	0.00155 $in.^3$ (sq. in.)
1 cm^2	0.155 $in.^2$
6.45163 cm^2	1 $in.^2$
0.0929 m^2	1 ft^2
0.83613 m^2	1 yd^2
1 m^2	10.7639 ft^2
1 km^2	0.3861 mi^2
2.58998 km^2	1 mi^2

Liquid measures

Metric system	U.S. system	
1 ml	0.0610 $in.^3$	
0.473 L	28.875 $in.^3$	1 pt
0.946 L	57.749 $in.^3$	1 qt
1 L	61.0 $in.^3$	1.0567 qt
3.7853 L	231 $in.^3$	1 gal

Volume

Metric system	U.S. system
1 mm^3	$0.6102 \cdot 10^{-4}$ $in.^3$ (cu. in.)
1 cm^3	0.06102 $in.^3$
16.3872 cm^3	1 in^3
0.02831 m^3	1 ft^3
0.76456 m^3	1 yd^3
1 m^3	1.30794 yd^3

Density

Metric system	U.S. system
1 g/cm^3	0.036127 $lb/in.^3$
27.68 g/cm^3	1 $lb/in.^3$
0.0160 g/cm^3	1 lb/ft^3

Mass

Metric system	U.S. system
1 g	0.035 oz av (ounce av)
28.349 g	1 oz av
453.59 g	1 lb av (lb av.)[a]
1 kg	2.20462 lb av
907.1848 kg	1 ton sh (short ton)
1 t	1.1023 ton sh
1016.047 kg	1 ton 1 (long ton)

[a] 1 lb av = 1 pound avoirdupois is the mass of 27.692 $in.^3$ of water weighed in air at 4°C, 760 mm pressure.

Table 1.1—3 (continued)
METRIC AND U.S. SYSTEM, MEASURES, UNITS AND CONVERSIONS

Energy

	erg	$Joule_{int}$	$k W_{int}h$	$kcal_l$	Liter-atmos.	BTU
erg	1	0.9997×10^{-7}	2.7769×10^{-14}	2.389×10^{-11}	9.8692×10^{-10}	9.4805×10^{-11}
$Joule_{int}$	1.0002×10^{7}	1	2.7778×10^{-7}	2.390×10^{-4}	9.8722×10^{-3}	9.480×10^{-4}
$kW_{int}h$	3.6011×10^{13}	3.6000×10^{6}	1	8.6041×10^{2}	3.5540×10^{4}	3.413×10^{3}
$kcal_{15}$	4.1853×10^{10}	4.186×10^{3}	1.1622×10^{-3}	1	4.1306×10^{1}	3.9685
Liter-atmos.	1.0133×10^{9}	1.0133×10^{2}	2.8137×10^{-5}	2.421×10^{-2}	1	9.607×10^{-2}
BTU	1.0548×10^{10}	1.0548×10^{3}	2.930×10^{-4}	2.5198×10^{-1}	1.0409×10^{1}	1

Pressure

	bar	Torr	atm.	at	lb/in.^{2}
1 bar (10^6 dynes/cm^2)	1	750	0.98692	1.0197	14.504
1 Torr	0.00133	1	0.00131	0.001359	0.01934
1 atm	1.0133	760	1	1.033	14.696
1 at (1 kg/cm^2)	0.98067	735.56	0.96784	1	14.223
1 lb/in.^{2}	0.06895	51.7144	0.068046	0.07031	1

Temperature

Absolute Centigrade or Kelvin (K)	x°K = T°C + 273.18
Degrees Centigrade (°C)	x°C = 5/9 (T°F − 32)
	x°C = 5/4 T°R
Degrees Fahrenheit (°F)	x°F = 9/4 T°R + 32
	x°F = 9/5 T°C + 32
Degrees Réaumur (°R)	x°R = 4/9 (T°F − 32)
	x°R = 4/5 T°C

Centigrade to Fahrenheit

°C	°F	°C	°F	°C	°F
−200	−328	60	140	200	392
−150	−238	70	158	250	482
−100	−148	80	176	300	572
−50	−58	90	194	400	752
0	+32	100	212	500	932
10	50	110	230	600	1112
20	68	120	248	700	1292
30	86	130	266	800	1472
40	104	140	284	900	1652
50	122	150	302	1000	1832

From Beyer, W. H., Ed., *CRC Standard Mathematical Tables*, 28th ed., CRC Press, Boca Raton, FL, 1987.

From Heydemann, A., *Handbook of Geochemistry, Vol. I*, Wedepohl, K. H., Ed., Springer Verlag, Berlin, 1969. With permission.

1.2 ASTRONOMICAL DATA

Table 1.2—1
ASTRONOMICAL CONSTANTS

Defining constants	
No. ephemeris s in 1 tropical yr (1900)	s = 31,556,925.9747
Gaussian gravitational constant, defining the A.U.	k = 0.01720209895
Primary constants	
Measure of the A.U. in meters	$A = 149{,}600 \times 10^6$
Velocity of light in m/s	$c = 299{,}792.5 \times 10^3$
Equatorial radius for Earth in meters	a_e = 6,378,160
Dynamical form-factor for Earth	J_2 = 0.0010827
Geocentric gravitational constant (units: $m^3\ s^{-2}$)	$GE = 398{,}603 \times 10^9$
Ratio of the masses of the Moon and Earth	μ = 1/81.30
Sidereal mean motion of Moon in radians/s (1900)	$n^* = 2.661699489 \times 10^{-6}$
General precession in longitude/tropical century (1900)	p = 5,025.″644
Obliquity of the ecliptic (1900)	ϵ = 23°27′08.″26
Constant of nutation (1900)	N = 9.″210
Derived constants	
Heliocentric gravitational constant (units:$m^3\ s^{-2}$)	$GS = 132{,}718 \times 10^{15}$
Ratio of masses of Sun and Earth	S/E = 322,958
Ratio of masses of Sun and Earth + Moon	$S/E\ (1 + \mu)$ = 328,912
Perturbed mean distance of Moon, in meters	$a = 384{,}400 \times 10^3$

From Wedepohl, K. H., *Handbook of Geochemistry,* Vol. 1, Springer-Verlag, Berlin, 1969. With permission.

REFERENCE

1. *Astronomer's Handbook,* Transactions of the International Astronomical Union, Vol. XIIC, Academic Press, London, 1966.

Table 1.2—2
SOLAR DIMENSIONS

Radius	6.960×10^{10} cm
Surface area	6.087×10^{22} cm^2
Volume	1.412×10^{33} cm^3
Mass	1.989×10^{33} g
Mean density	1.409 g cm^{-3}
Density at the center	98 g cm^{-3}
Gravitational acceleration at the solar surface	2.740×10^4 cm s^{-2}
Escape velocity at the surface	6.177×10^7 cm s^{-1}
Effective temperature	5,785°K
Temperature at the center	13.6×10^6 °K
Radiation	3.9×10^{33} erg s^{-1}
Specific surface emission	6.41×10^{10} erg cm^{-2} s^{-1}
Specific mean energy production	1.96 erg g^{-1} s^{-1}
Solar constant = extraterrestrial energy flux at the mean distance between Earth and Sun	1.39×10^6 erg cm^{-2} s^{-1} 2.00 cal cm^{-2} min^{-1}

From Wedepohl, K. H., *Handbook of Geochemistry,* Vol. I, Springer-Verlag, Berlin, 1969. With permission.

REFERENCE

1. **Waldmeier, M.,** The quiet sun, *Landolt-Bornstein,* New series, Group VI: Astronomy, astrophysics and space research, Vol. I, Springer-Verlag, Berlin, 1965.

Table 1.2—3
DIMENSIONS OF THE PLANETS AND THE MOON

Symbols:

- a = semimajor axis of the orbit.
- P = sidereal period = true period of the revolution of the planet around the Sun (with respect to the fixed star field).
- g_{Eq} = total acceleration, including centrifugal acceleration, at equator.
- v_e = velocity of escape at equator.
- A = Albedo = total reflectivity, wavelength λ_{eff} = 5,500 Å.
- T_{max} = max temp for the subsolar point of a slowly rotating planet or satellite (computed from the visual albedo).
- T_{av} = avg temp of a rapidly rotating sphere.
- Atm. constituents = main atmospheric constituents.

Name	a, 10^6 km	P a	Diameter, km	Mass,[a] 10^{26} g	Volume, 10^{10} km^3
Mercury	57.9	0.24085	4,840	3.333	5.958
Venus	108.2	0.61521	12,228	48.70	95.765
Earth	149.6	1.00004	12,742.06	59.76	108.332
Moon	0.384[b]	0.07480[c]	3,476	0.735	2.192
Mars	227.9	1.88089	6,770	6.443	16.250
Jupiter	778	11.86223	140,720	18,993	145,923.204
Saturn	1,427	29.4577	116,820	5,684	83,469.806
Uranus	2,870	84.0153	47,100	867.6	5,481.599
Neptune	4,496	164.7883	44,600	1,029	4,636.610
Pluto	5,881.9 to 5,946.5	247.7	6,000	55.3	10.833

[a] Mass without moons.
[b] Mean distance from Earth.
[c] True period of the revolution of the Moon around Earth (with respect to the fixed star field).

Table 1.2—3 (continued)
DIMENSIONS OF THE PLANETS AND THE MOON

Name	ρ g/cm³	g_{Eq}, cm/s²	v_e, km/s	A	T_{max}, °K	$T_{av.}$, °K	Atm. constituents
Mercury	5.62	380	4.29	0.056	625	—	((^{40}Ar))
Venus	5.09	869	10.3	0.76	324	229	CO_2, H_2O
Earth	5.517	978	11.2	0.39	349	246	N_2, O_2
Moon	3.35	162	2.37	0.067	387	274	—
Mars	3.97	372	5.03	0.16	306	216	N_2, CO_2, H_2O
Jupiter	1.30	2301	57.5	0.67	131	93	H_2, CH_4, NH_4
Saturn	0.68	906	33.1	0.69	95	68	H_2, CH_4, NH_3
Uranus	1.58	972	21.6	0.93[d]	67[d]	47[d]	He, H_2, CH_4
Neptune	2.22	1347	24.6	0.84[d]	53[d]	38[d]	He, H_2, CH_4
Pluto	—	—	—	0.14	60	43	Uncertain

[d] Since the albedos of Uranus and Neptune are very low in the red and infrared, an effective value A = 0.7 has been adopted for calculating the temperatures.

From Wedepohl, K. H., *Handbook of Geochemistry,* Vol. I, Springer-Verlag, Berlin, 1969. With permission.

REFERENCES

1. **Gondolatsch, F.,** Mechanical data of planets and satellites, *Landolt-Bornstein,* New series, Group VI, Astronomy, astrophysics and space research, Vol. I, Springer-Verlag, Berlin, 1965.
2. **Kuiper, G. P.,** Physics of planets and satellites, *Landolt-Bornstein,* New series, Group VI, Astronomy, astrophysics, and space research, Vol. I, Springer-Verlag, Berlin, 1965.

Table 1.2—4
CONVERSION FACTORS

Time

1 mean solar second (sec, s)
= 1.002738 sidereal seconds
1 mean solar minute (min, m)
= 60 sec (mean solar)
1 mean solar hour (hr, h)
= 3,600 sec (mean solar)
= 60 min (mean solar)
1 mean solar day (da., d)
= 86,400 sec (mean solar)
= 1,440 min (mean solar)
= 24 hr (mean solar)
= 24 hr, 3 min, 56,555 sec of mean sidereal time
1 tropical (mean solar, ordinary) year (yr)
= 31,5569 × 10⁶ sec (mean solar)
= 525,949 min (mean solar)
= 8,765.81 hr (mean solar)
= 365,2422 day (mean solar)
= 366.2422 sidereal days
1 sidereal second
= 0.997270 sec (mean solar)
1 sidereal day
= 86,164.1 sec (mean solar)
= 23 hr, 56 min, 4.091 sec (mean solar)

Reprinted from *Smithsonian Meteorological Tables,* 6th rev. ed., Robert J. List, Ed., (Washington, DC: Smithsonian Institution Press), by permission of the publisher. Copyright 1984 Smithsonian Institution.

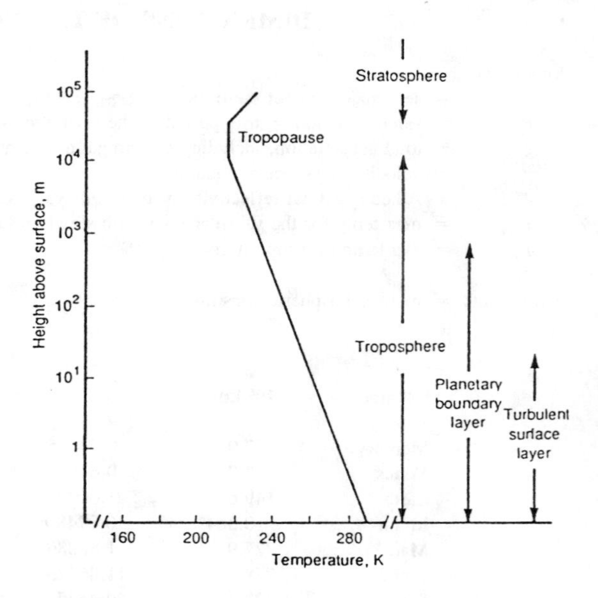

FIGURE 1.2—1. Stratification of the atmosphere. (From Richardson, M. L., Ed., *Chemistry, Agriculture, and the Environment,* The Royal Society of Chemistry, Cambridge, 1991. With permission.)

1.3 WIND

Table 1.3—1
BEAUFORT WIND SCALE

Beaufort Wind Scale

Table A

	Mean wind speeds at 10 m[a]				Limits of wind speed at 10 m[a]			
Force	Knots	m/sec	km/hr	mi/hr	Knots	m/sec	km/hr	mi/hr
0	0	0	0	0	<1	0–0.2	<1	<1
1	2	0.9	3	2	1–3	0.3–1.5	1–5	1–3
2	5	2.4	9	5	4–6	1.6–3.3	6–11	4–7
3	9	4.4	16	10	7–10	3.4–5.4	12–19	8–12
4	13	6.7	24	15	11–16	5.5–7.9	20–28	13–18
5	18	9.3	34	21	17–21	8.0–10.7	29–38	19–24
6	24	12.3	44	28	22–27	10.8–13.8	39–49	25–31
7	30	15.5	55	35	28–33	13.9–17.1	50–61	32–38
8	37	18.9	68	42	34–40	17.2–20.7	62–74	39–46
9	44	22.6	82	50	41–47	20.8–24.4	75–88	47–54
10	52	26.4	96	59	48–55	24.5–28.4	89–102	55–63
11	60	30.5	110	68	56–63	28.5–32.6	103–117	64–72
12	68	34.8	125	78	64–71	32.7–36.9	118–133	73–82
13	76	39.2	141	88	72–80	37.0–41.4	134–149	83–92
14	85	43.8	158	98	81–89	41.5–46.1	150–166	93–103
15	94	48.6	175	109	90–99	46.2–50.9	167–183	104–114
16	104	53.5	193	120	100–108	51.0–56.0	184–201	115–125
17	114	58.6	211	131	109–118	56.1–61.2	202–220	126–136

Table B

Force	Description of wind[b]	Specifications for use on land[b]
0	Calm	Calm, smoke rises vertically.
1	Light air	Direction of wind shown by smoke drift, but not by wind vanes.
2	Light breeze	Wind felt on face; leaves rustle; ordinary vane moved by wind.
3	Gentle breeze	Leaves and small twigs in constant motion; wind extends light flag.
4	Moderate breeze	Raises dust and loose paper; small branches are moved.
5	Fresh breeze	Small trees in leaf begin to sway; crested wavelets form on inland waters.
6	Strong breeze	Large branches in motion; whistling heard in telegraph wires; umbrellas used with difficulty.
7	Moderate gale	Whole trees in motion; inconvenience felt when walking against wind.
8	Fresh gale	Breaks twigs off trees; generally impedes progress.
9	Strong gale	Slight structural damage occurs (chimney pots and slate removed).

Table 1.3—1B (continued)
BEAUFORT WIND SCALE

10	Whole gale	Seldom experienced inland; trees uprooted; considerable structural damage occurs.
11	Storm	Very rarely experienced, accompanied by widespread damage.
12 or above	Hurricane	

[a] Resolution 9, International Meteorological Committee, Paris, 1946.
[b] Meteorological Office, *The Meteorological Observers Handbook,* London, 1939.

Reprinted from *Smithsonian Meteorological Tables,* 6th rev. ed., Robert J. List, Ed., (Washington, DC: Smithsonian Institution Press), by permission of the publisher. Copyright 1984 Smithsonian Institution.

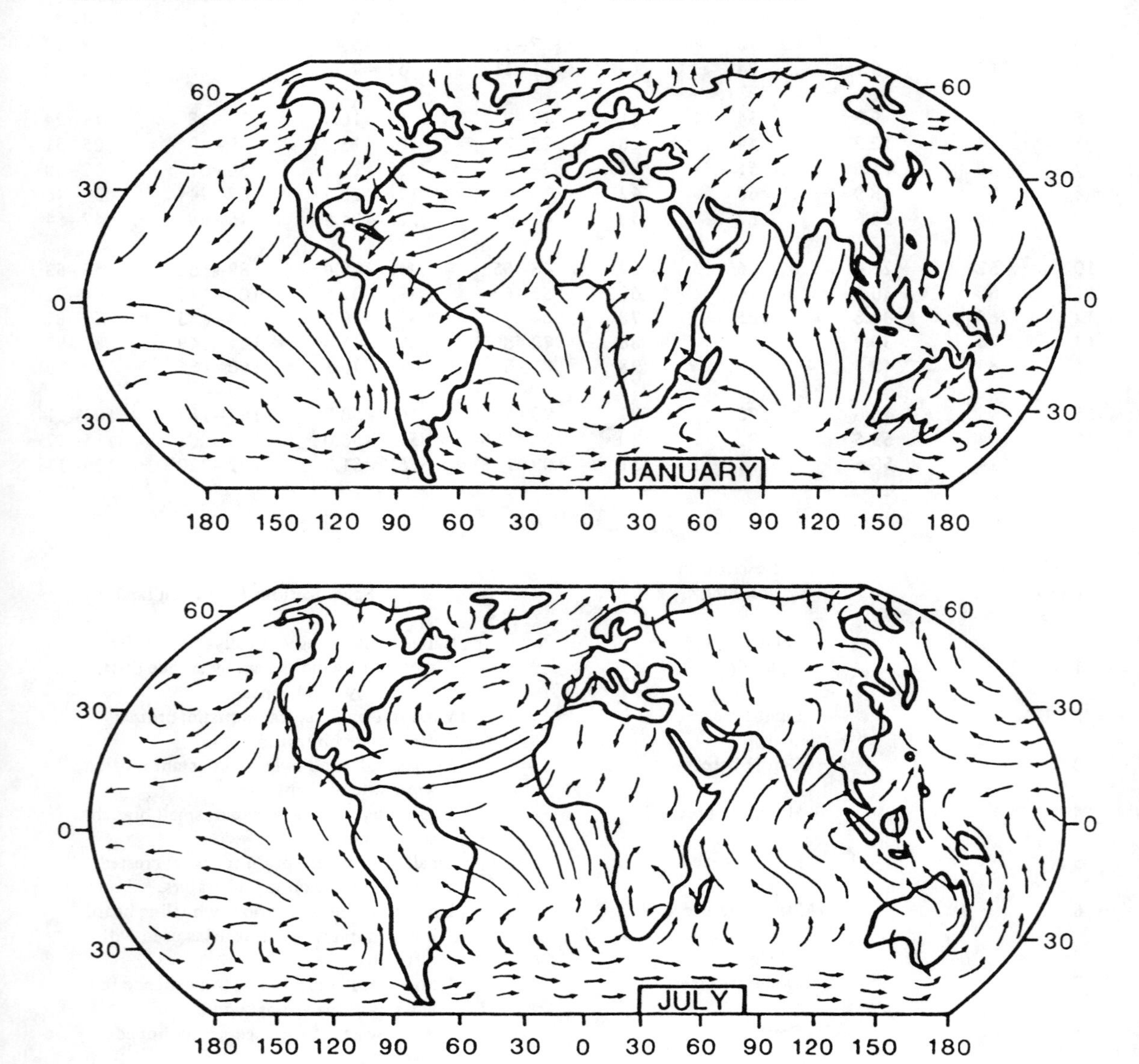

FIGURE 1.3—1. The global distribution of wind fields in January and July. (From Walsh, J. J., *On the Nature of Continental Shelves,* Academic Press, New York, 1989. With permission.)

1.4 CLIMATE

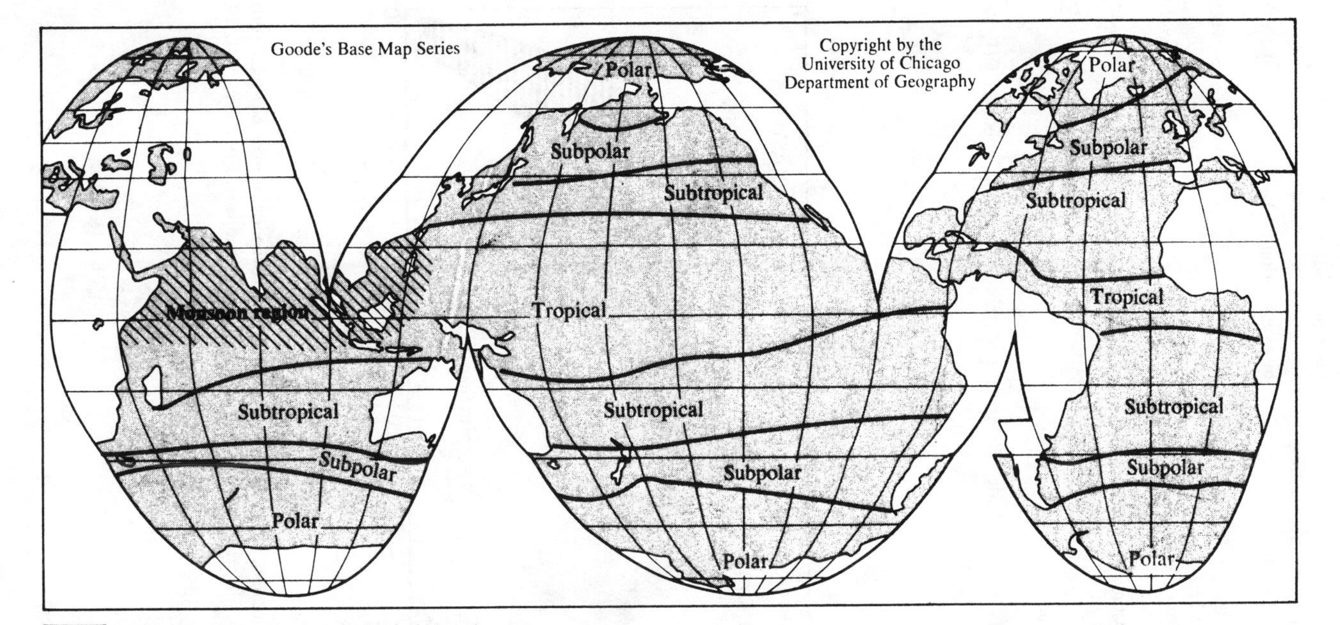

FIGURE 1.4—1. Climatic regions of the open ocean. (From Gross, M. G., *Oceanography,* 3rd ed., Charles E. Merrill Publishing Company, Columbus, OH, 1976. With permission.)

Table 1.4—1
OPEN-OCEAN CLIMATIC REGIONS IN THE SURFACE ZONE

Oceanic region	Surface heating and cooling cycle	Net precipitation (+) or evaporation (−)	Mixing processes	Seasonal temperature range (°C)
Polar	Seasonal melting & freezing of sea ice; net heat loss	+	Wind and convective mixing	Intermediate (5–9)
Subpolar (west wind drift)	Seasonal heating and cooling; net heat loss	+	Wind and convective mixing	Intermediate (2–8)
Subtropical	Seasonal heating and cooling; net heat gain	−	Wind and convective mixing	Large (6–18)
Tropical	Daily heating and cooling; net heat gain; no seasons	+	Wind	Small (<2)

From Gross, M. G., *Oceanography,* 3rd ed., Charles E. Merrill Publishing Company, Columbus, OH, 1976. With permission.

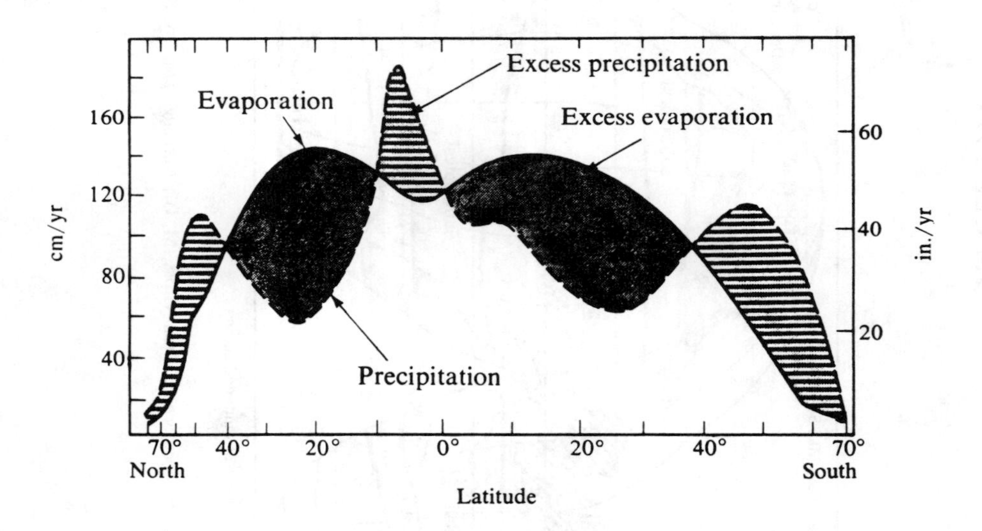

FIGURE 1.4—2. Latitudinal distribution of precipitation and evaporation over the ocean. (From Gross, M. G., *Oceanography,* 3rd ed., Charles E. Merrill Publishing Company, Columbus, OH, 1976. With permission.)

1.5 SEA LEVEL

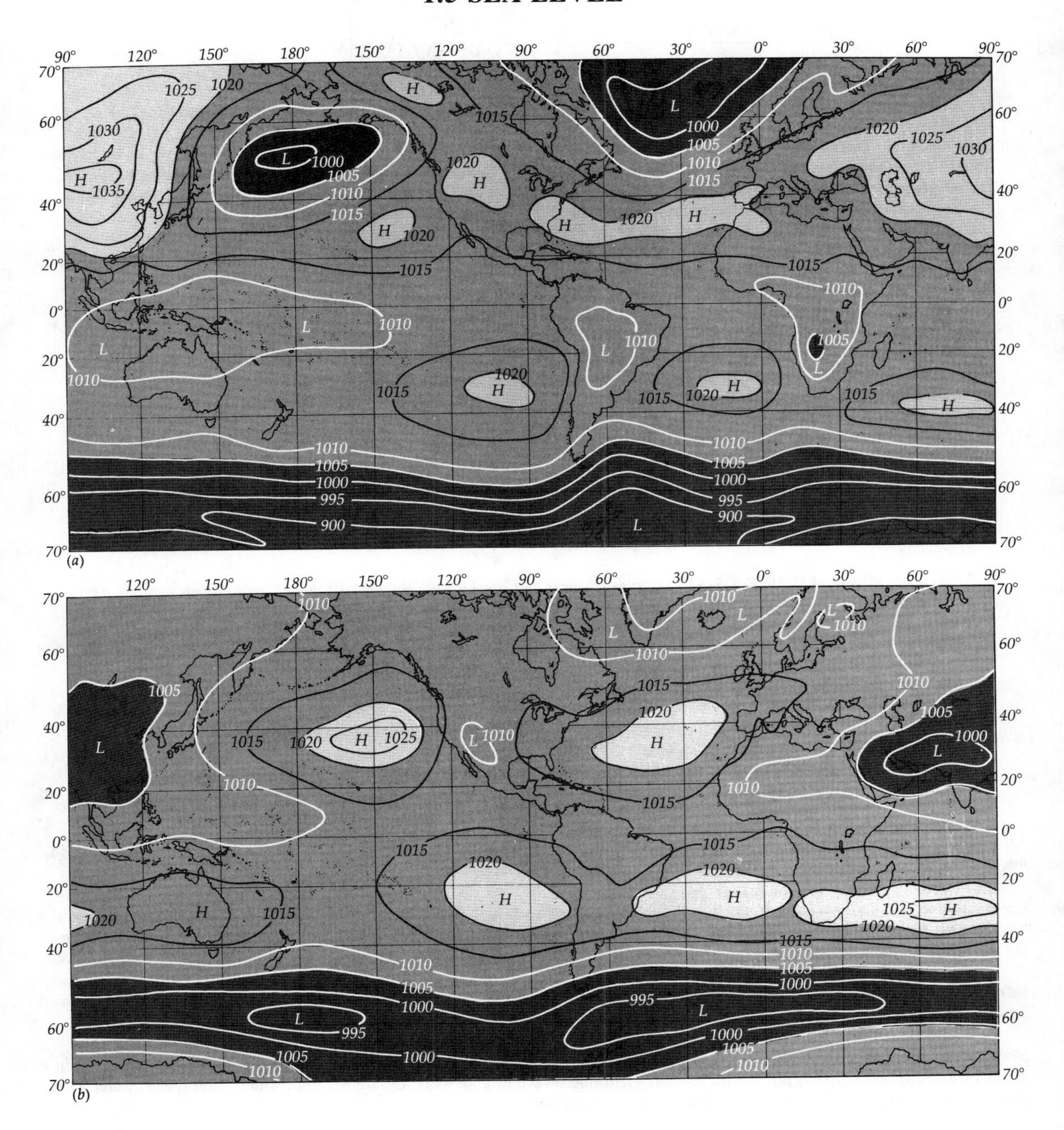

FIGURE 1.5—1. Mean sea level pressure, given in millibars during (a) January and (b) July. (From Ingmanson, D. E. and Wallace, W. J., *Oceanology: An Introduction*, 2nd ed., Wadsworth Publishing Company, Belmont, CA, 1979. With permission.)

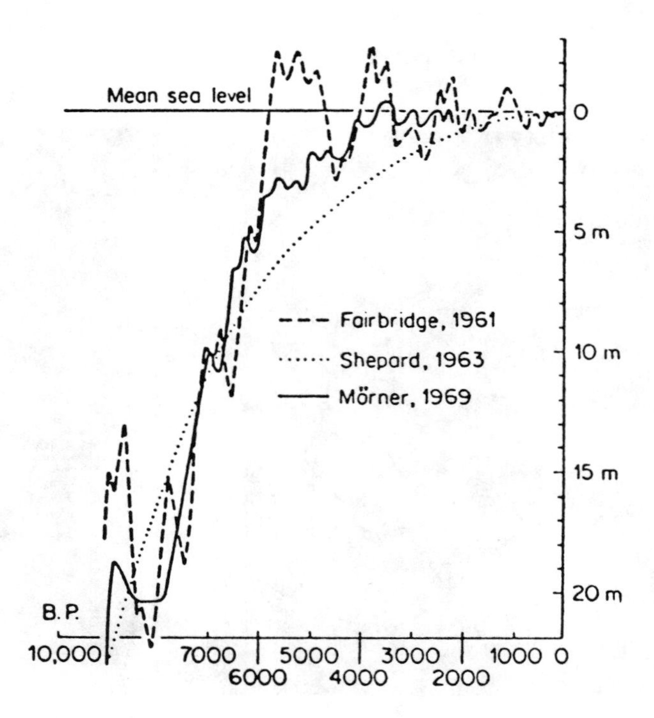

FIGURE 1.5—2. Curves of mean sea level during the past 10,000 years according to the studies of R. Fairbridge, F. P. Shepard, and N.-A. Mörner. (From Fairbridge, R. W., in *Chemistry and Biogeochemistry of Estuaries,* Olausson, E. and Cato, I., Eds., John Wiley & Sons, Chichester, England, 1980, 1. Reproduced by permission of copyright © owner, John Wiley & Sons, Ltd.)

Table 1.5—1
SATELLITE-BORNE REMOTE-SENSING TECHNIQUES USED TO MEASURE THE TOPOGRAPHY OF THE SEA SURFACE AND OTHER PHYSICAL PARAMETERS

Altimeter—A pencil-beam microwave radar that measures the distance between the spacecraft and the earth. Measurements yield the topography and roughness of the sea surface from which the surface current and average wave height can be estimated.

Color scanner—A radiometer that measure the intensity of radiation emitted from the sea in visible and near-infrared bands in a broad swath beneath the spacecraft. Measurements yield ocean color, from which chlorophyll concentration and the location of sediment-laden waters can be estimated.

Infrared radiometer—A radiometer that measures the intensity of radiation emitted from the sea in the infrared band in a broad swath beneath the spacecraft. Measurements yield estimates of sea-surface temperature.

Microwave radiometer—A radiometer that measure the intensity of radiation emitted from the sea in the microwave band in a broad swath beneath the spacecraft. Measurements yield microwave brightness temperatures, from which wind speed, water vapor, rain rate, sea-surface temperature, and ice cover can be estimated.

Scatterometer—A microwave radar that measures the roughness of the sea surface in a broad swath on either side of the spacecraft with a spatial resolution of 50 km. Measurements yield the amplitude of short surface waves that are approximately in equilibrium with the local wind and from which the surface wind velocity can be estimated.

Synthetic Aperture Radar—A microwave radar similar to the scatterometer except that it electronically synthesizes the equivalent of an antenna large enough to achieve a spatial resolution of 25 m. Measurements yield information on features (swell, internal waves, rain, current boundaries, etc.) that modulate the amplitude of the short surface waves; they also yield information on the position and character of sea ice from which, with successive views, the velocity of ice floes can be estimated.

From Walsh, J. J., *On the Nature of Continental Shelves,* Academic Press, New York, 1989. With permission.

1.6 SOLAR RADIATION

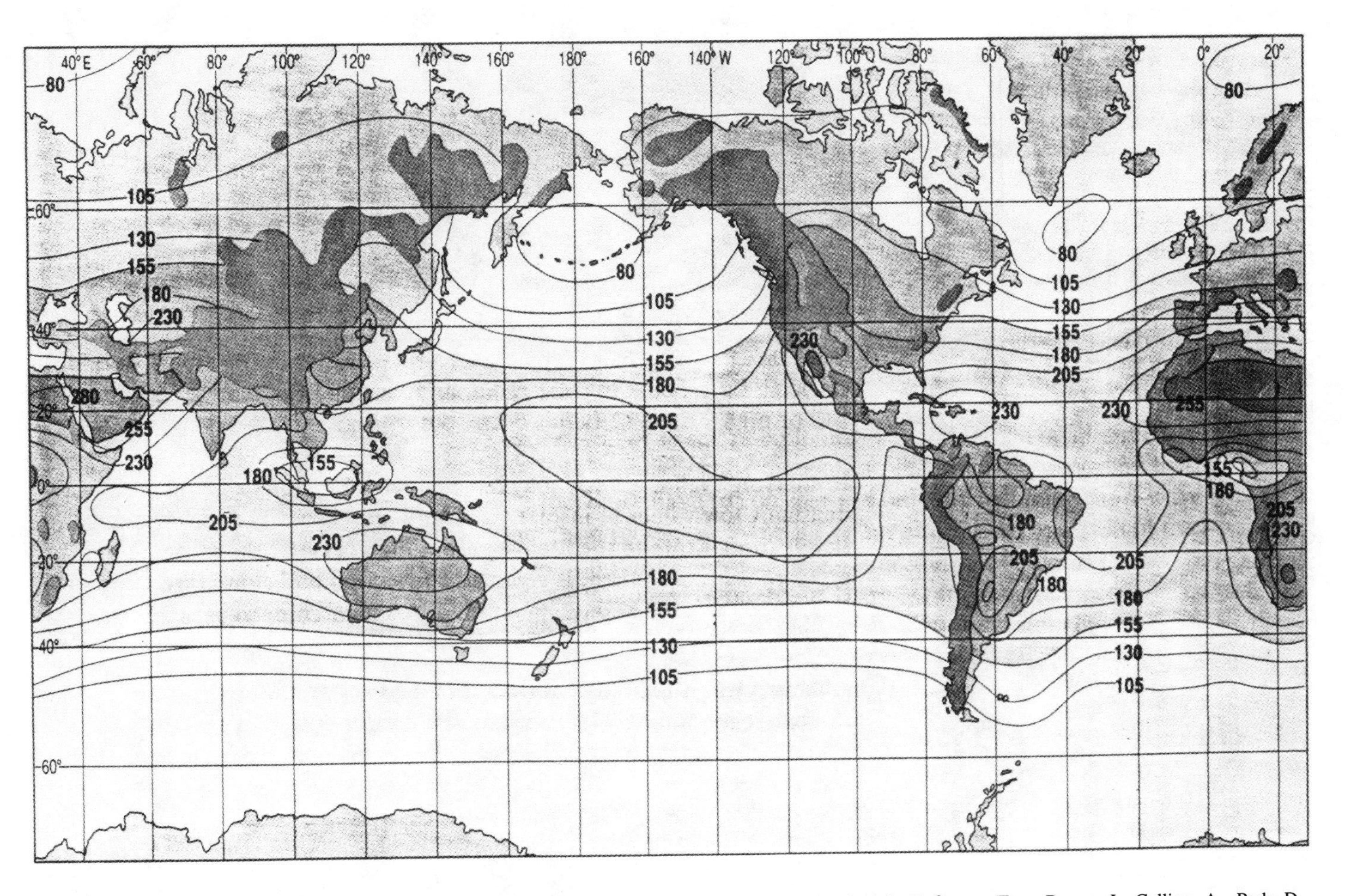

FIGURE 1.6—1. The amount of solar radiation received at the sea and land surface on Earth. Units given in J/m²/year. (From Brown, J., Colling, A., Park, D., Phillips, J., Rothery, D., and Wright, J., *Ocean Circulation*, 1989, 159, with kind permission from Pergamon Press Ltd., Headington Hill Hall, Oxford OX3 OBW, U.K.)

Table 1.6—1
ABSORPTION OF RADIATION BY SEAWATER

Utterback[1] has made observations of the *extinction coefficient* of typical oceanic waters, defined in the same manner as the absorption coefficient *k* of ozone. Sverdrup, Johnson, and Fleming[2] have summarized Utterback's observations as follows:

He has made numerous observations in the shallow waters near islands in the inner part of Juan de Fuca Strait and at four stations in the open oceanic waters off the coast of Washington, and these can be considered typical of coastal and oceanic waters, respectively. Table 1.6—1 contains the absorption coefficients of pure water at the wavelengths used by Utterback, the maximum average, and maximum coefficients observed in oceanic water, and the minimum, average, and maximum coefficients observed in coastal water. The minimum and maximum coefficients have all been computed from the four lowest and the four highest values in each group.

	Wavelength[a] – μ						
	0.46	0.48	0.515	0.53	0.565	0.60	0.66
Type water	cm^{-1}	cm^{-1}	cm^{-1}	cm^{-1}	cm^{-1}	cm^{-1}	cm^{-1}
Pure water	.00015	.00015	.00018	.00021	.00033	.00125	.00280
Ocean water							
Lowest	.00038	.00026	.00035	.00038	.00074	.00199	
Avg	.00086	.00076	.00078	.00084	.00108	.00272	
Highest	.00160	.00154	.00143	.00140	.00167	.00333	
Coastal water							
Lowest	.00224	.00230	.00192	.00169		.00375	.00477
Avg	.00362	.00334	.00276	.00269		.00437	.00623
Highest	.00510	.00454	.00398	.00348		.00489	.00760

[a] It should be understood that the wavelength actually stands for a spectral band of finite width.

Reprinted from *Smithsonian Meteorological Tables,* 6th rev. ed., Robert J. List, Ed., (Washington DC: Smithsonian Institution Press), by permission of the publisher. Copyright 1984 Smithsonian Institution.

REFERENCES

1. **Utterback, C. L.,** *Cons. Perm. Int. Explor. Mer.,* Rapp. et Proc.-Verb., 101 (4), 15, 1936.
2. **Sverdrup, H. U., Johnson, M. W., and Fleming, R. H.,** *The Oceans,* Prentice-Hall, Inc., New York, 1942, p. 84.

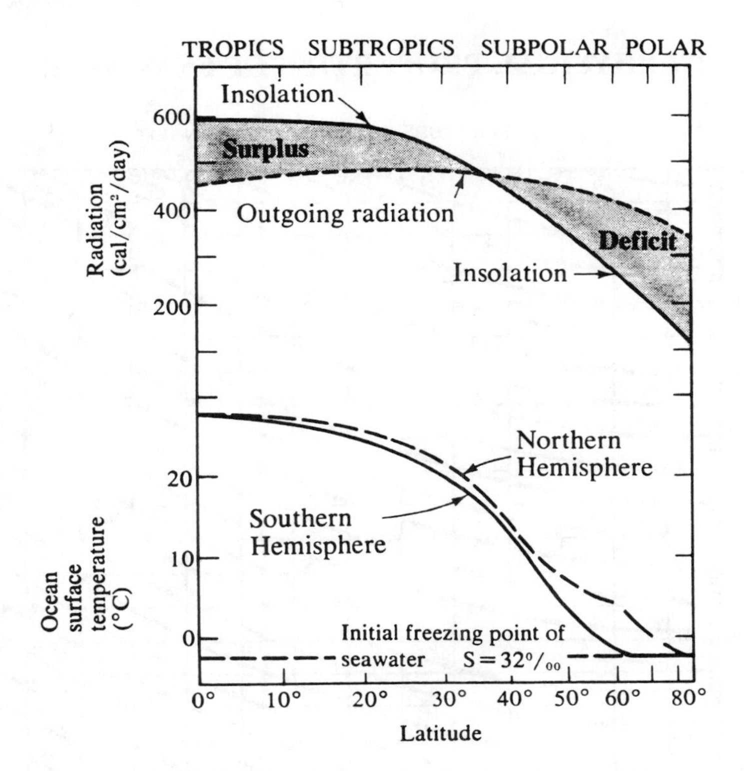

FIGURE 1.6—2. Radiation balance for the northern hemisphere and mean sea surface temperatures for the northern and southern hemispheres. (From Gross, M. G., *Oceanography,* 3rd ed., Charles E. Merrill Publishing, Columbus, OH, 1976. With permission.)

1.7 PHYSICAL PROPERTIES OF SEAWATER

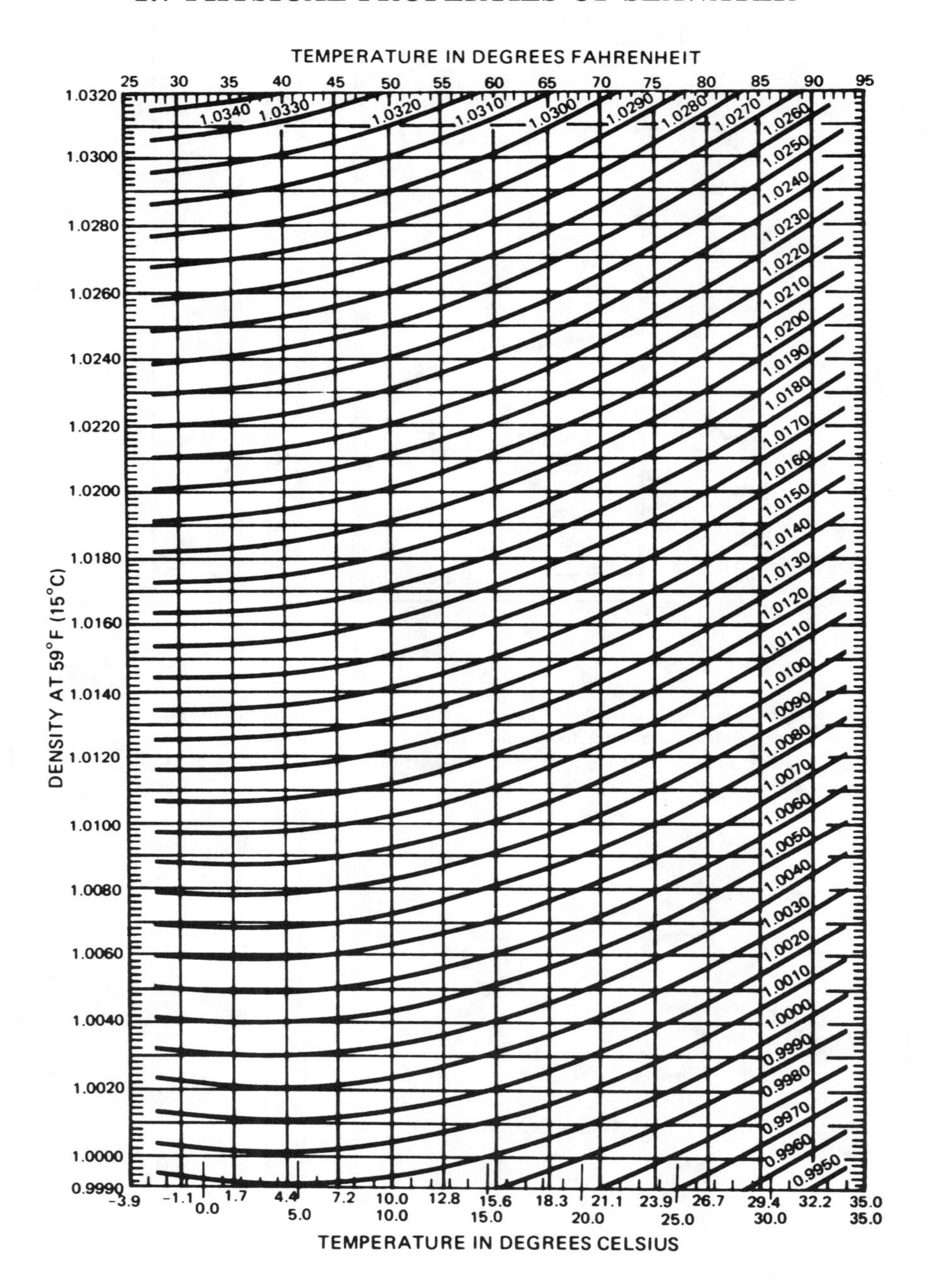

FIGURE 1.7—1. Seawater density at various temperatures. (From National Ocean Survey, National Oceanic and Atmospheric Administration, Rockville, MD. With permission.)

The purpose of this graph is to provide the density of seawater at any temperature apt to be encountered when the density at the standard temperature of 59°F(15°C) is known. To convert a density at 59°F(15°C) to density at another temperature, enter the graph horizontally from the left with the known density and downward from the top or upward from the bottom with the desired temperature; the position of the point of intersection with respect to the curves gives the density at the desired temperature. Interpolate between curves when necessary. For example, by this method, water having a density of 1.0162 at 59°F is found to have a density of 1.0124 at 85°F. The densities are referred to the density of fresh water at 4°C(39.2°F) as unity.

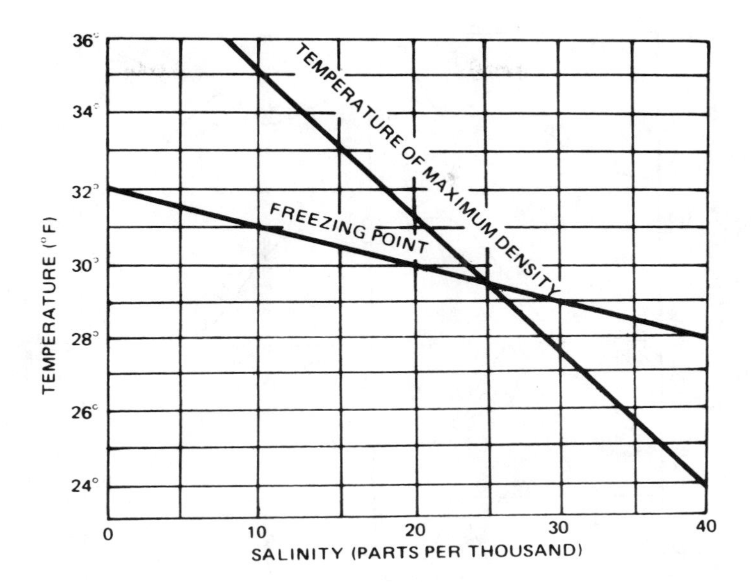

FIGURE 1.7—2. Relationship between temperature of maximum density and freezing point for water of varying salinity. (From *American Practical Navigator,* U.S. Naval Hydrographic Office, Washington, D.C., H.O. Publ. No. 9. With permission of NAVOCEANO.)

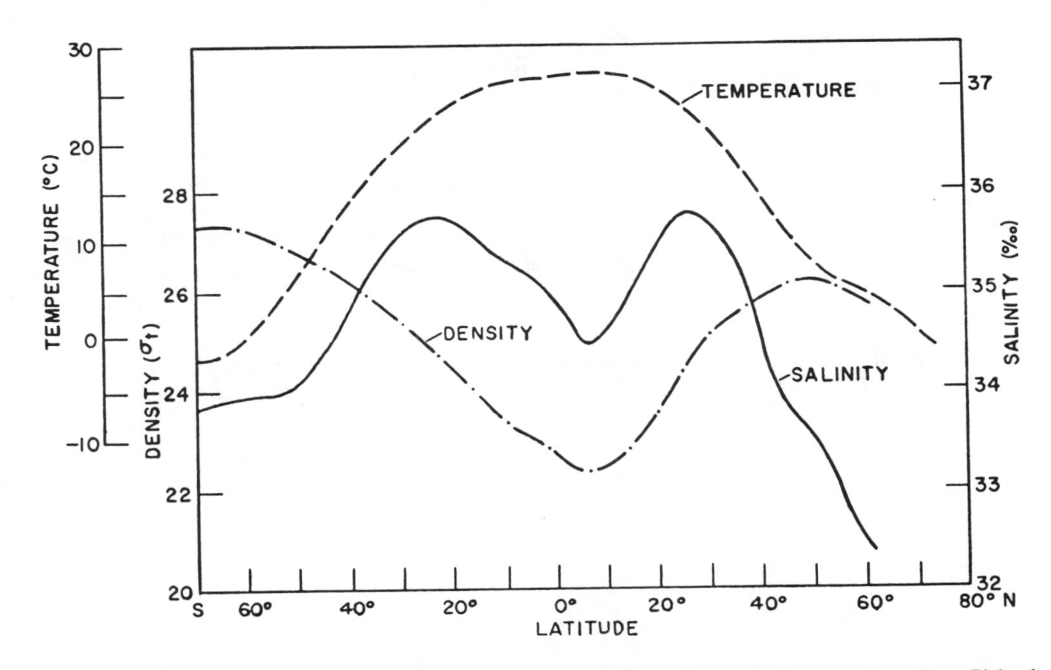

FIGURE 1.7—3. Average surface temperature, salinity, and density variation with latitude for all oceans. (From Pickard, G. L., *Descriptive Physical Oceanography,* 1964, with kind permission from Pergamon Press Ltd., Headington Hill Hall, Oxford OX3 OBW, U.K.)

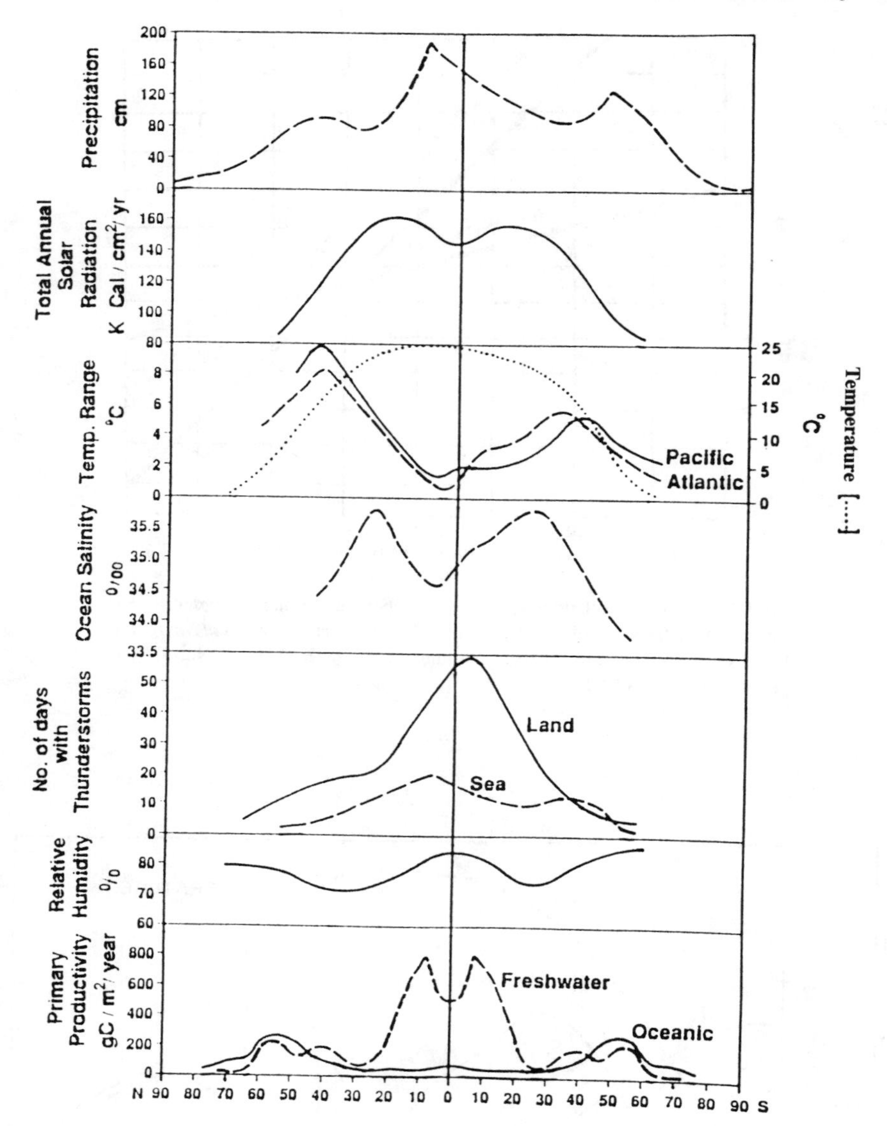

FIGURE 1.7—4. Latitudinal changes in various physical and biophysical parameters. (From Connell, D. W. and Hawker, D. W., Eds., Pollution in Tropical Aquatic Systems, CRC Press, Boca Raton, FL, 1992.)

Table 1.7—1
MEAN VERTICAL TEMPERATURE (°C) DISTRIBUTION IN THE THREE OCEANS BETWEEN 40°N AND 40°S

	Atlantic Ocean		Indian Ocean		Pacific Ocean		Mean	
Depth (m)	**°C**	**Δ°C/ 100 m**	**°C**	**Δ°C/ 100 m**	**°C**	**Δ°C/ 100 m**	**°C**	**Δ°C/ 100 m**
0	20.0		22.2		21.8		21.3	
		2.2		3.3		3.1		2.8
100	17.8		18.9		18.7		18.5	
		4.4[a]		4.7[a]		4.4[a]		4.5[a]
200	13.4		14.3		14.3		14.0	
		1.8		1.6		2.6		2.0
400	9.9		11.0		9.0		10.0	
		1.5		1.2		1.2		1.3
600	7.0		8.7		6.4		7.4	
		0.7		0.9		0.65		0.75
800	5.6		6.9		5.1		5.9	
		0.35		0.7		0.4		0.5
1000	4.9		5.5		4.3		4.9	
		0.20		0.4		0.4		0.35
1200	4.5		4.7		3.5		4.2	
		0.15		0.3		0.2		0.22
1600	3.9		3.4		2.6		3.3	
		0.12		0.15		0.1		0.12
2000	3.4		2.8		2.15		2.8	
		0.08		0.09		0.05		0.07
3000	2.6		1.9		1.7		2.1	
		0.08		0.03		0.03		0.05
4000	1.8		1.6		1.45		1.6	

[a] Maximum.

From Defant, A., *Physical Oceanography*, Vol. II, Pergamon Press, New York, 1961. With permission.

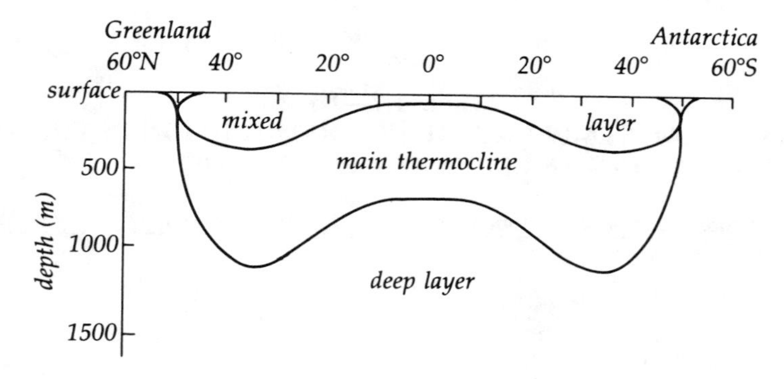

FIGURE 1.7—5. A three-layered ocean. (From Ingmanson, D. E. and Wallace, W. J., *Oceanology: an Introduction,* 2nd ed., Wadsworth Publishing Company, Belmont, CA, 1979. With permission.)

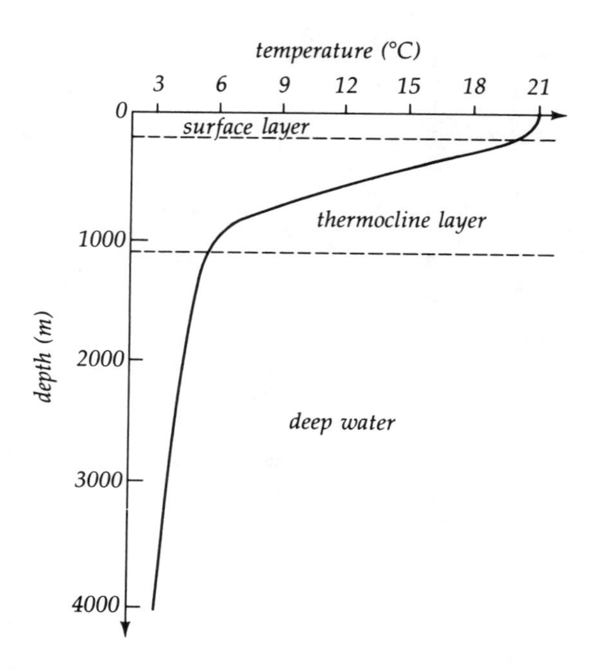

FIGURE 1.7—6. The vertical temperature of the oceans of the world. (From Ingmanson, D. E. and Wallace, W. J., *Oceanology: an Introduction,* 2nd ed., Wadsworth Publishing Company, Belmont, CA, 1979. With permission.)

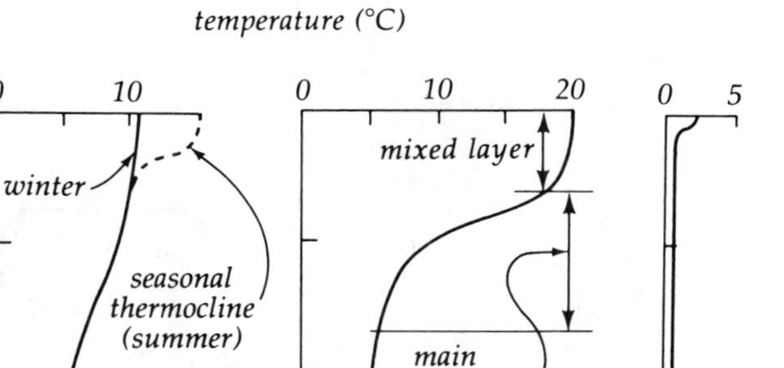

FIGURE 1.7—7. Average temperature profiles for the open ocean. (a) Mid-latitudes; (b) low latitudes; (c) high latitudes. (From Ingmanson, D. E. and Wallace, W. J., *Oceanology: an Introduction*, 2nd ed., Wadsworth Publishing Company, Belmont, CA, 1979. With permission.)

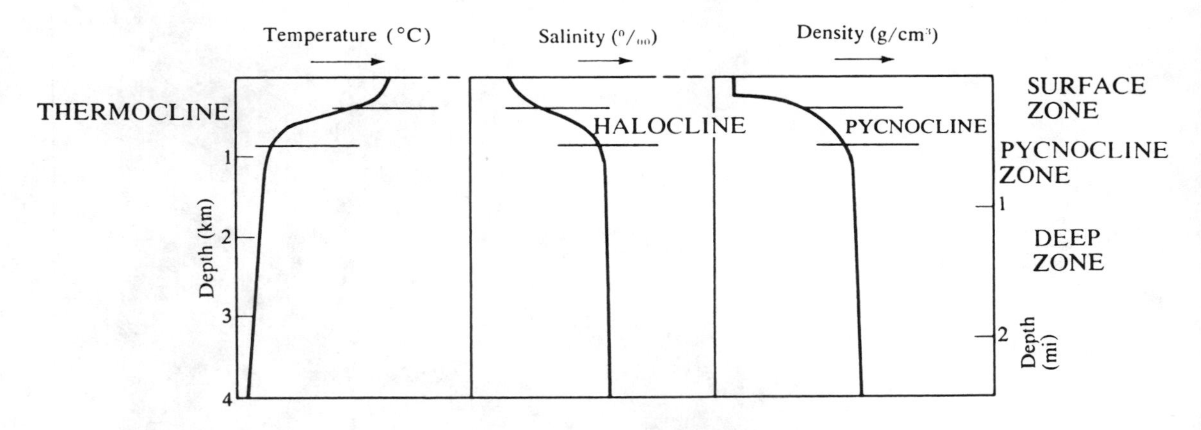

FIGURE 1.7—8. Temperature profile compared to salinity and density profiles for the open ocean. (From Gross, M. G., *Oceanography*, 3rd ed., Charles E. Merrill Publishing Company, Columbus, OH, 1976. With permission.)

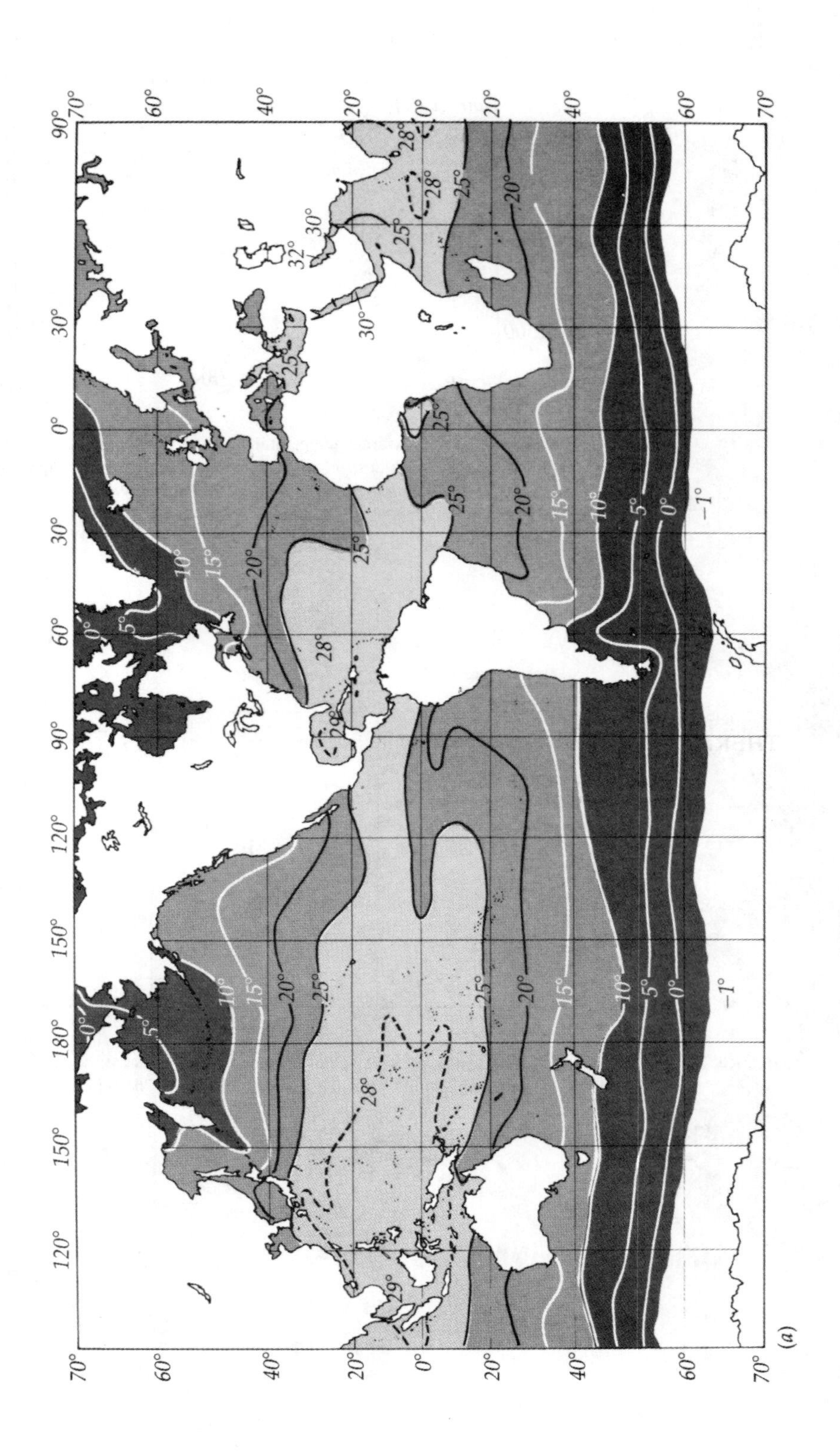

70°
60°
40°
20°
0°
20°
40°
60°
70°
120°
150°
180°
150°
120°
90°
60°
30°
0°
30°
90°
28°
29°
25°
20°
15°
10°
5°
0°
−1°
30°
32°

(a)

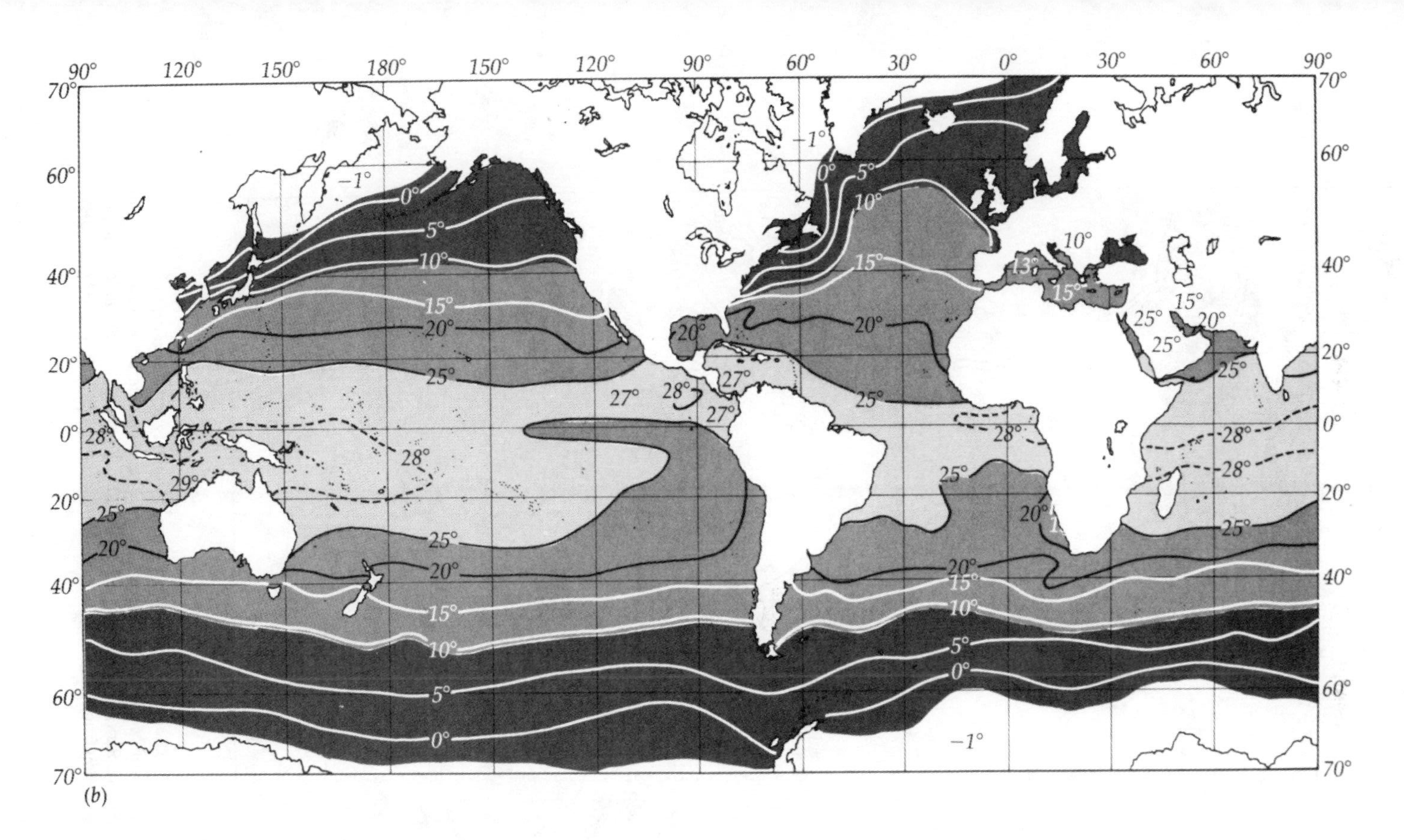

FIGURE 1.7—9. Surface temperature of the oceans (a) in August, (b) in February. (From Ingmanson, D. E. and Wallace, W. J., *Oceanology: an Introduction*, 2nd ed., Wadsworth Publishing Company, Belmont, CA, 1979. With permission.)

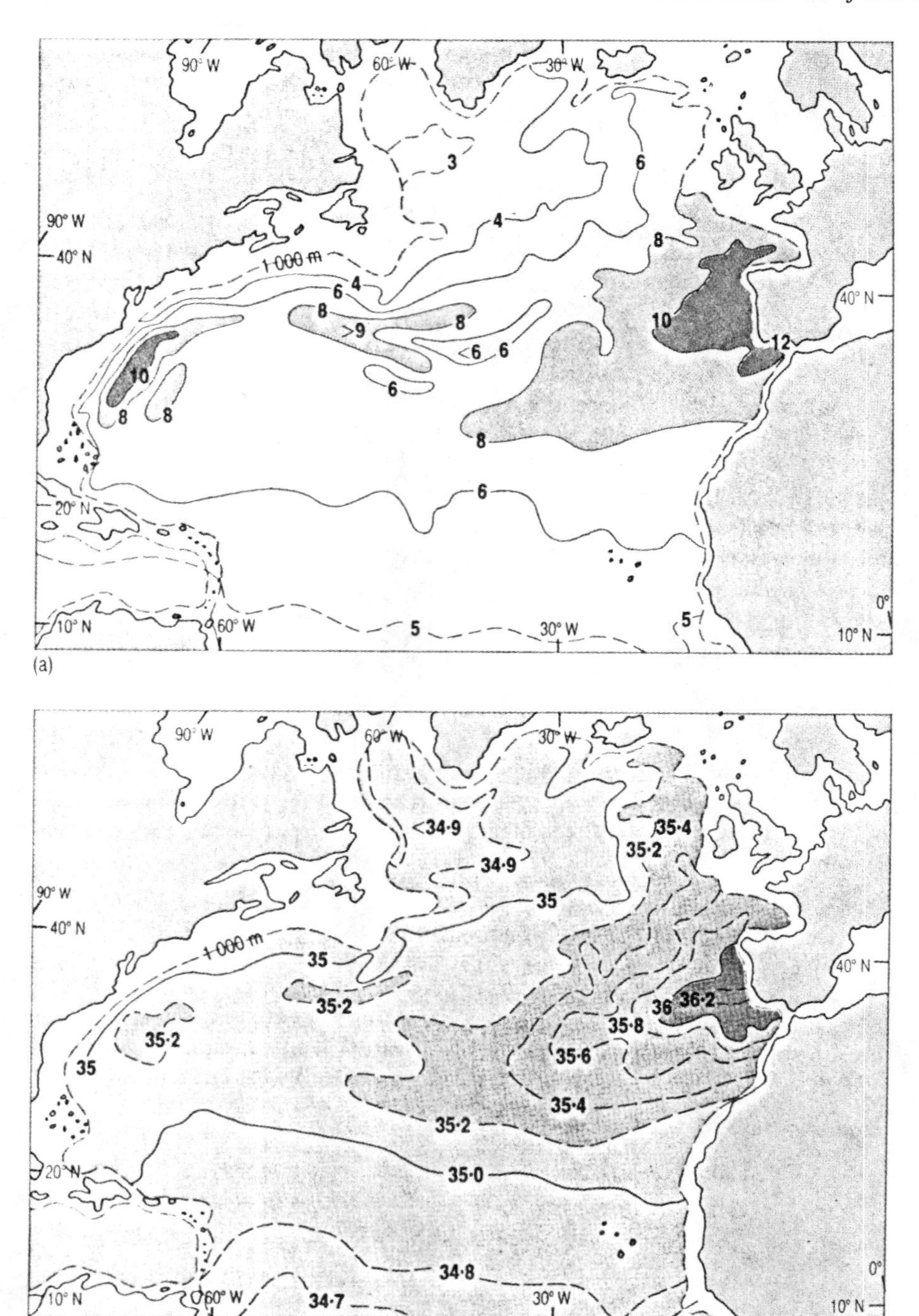

FIGURE 1.7—10. Distribution of (a) temperature (°C) and (b) salinity at 1000 m depth in the North Atlantic, showing the spread of Mediterranean water. The broken black line is the 1000-m isobath. (From Brown, J., Colling, A., Park, D., Phillips, J., Rothery, D., and Wright, J., *Ocean Circulation,* 1989, 173, with kind permission from Pergamon Press Ltd., Headington Hill Hall, Oxford OX3 OBW, U.K.)

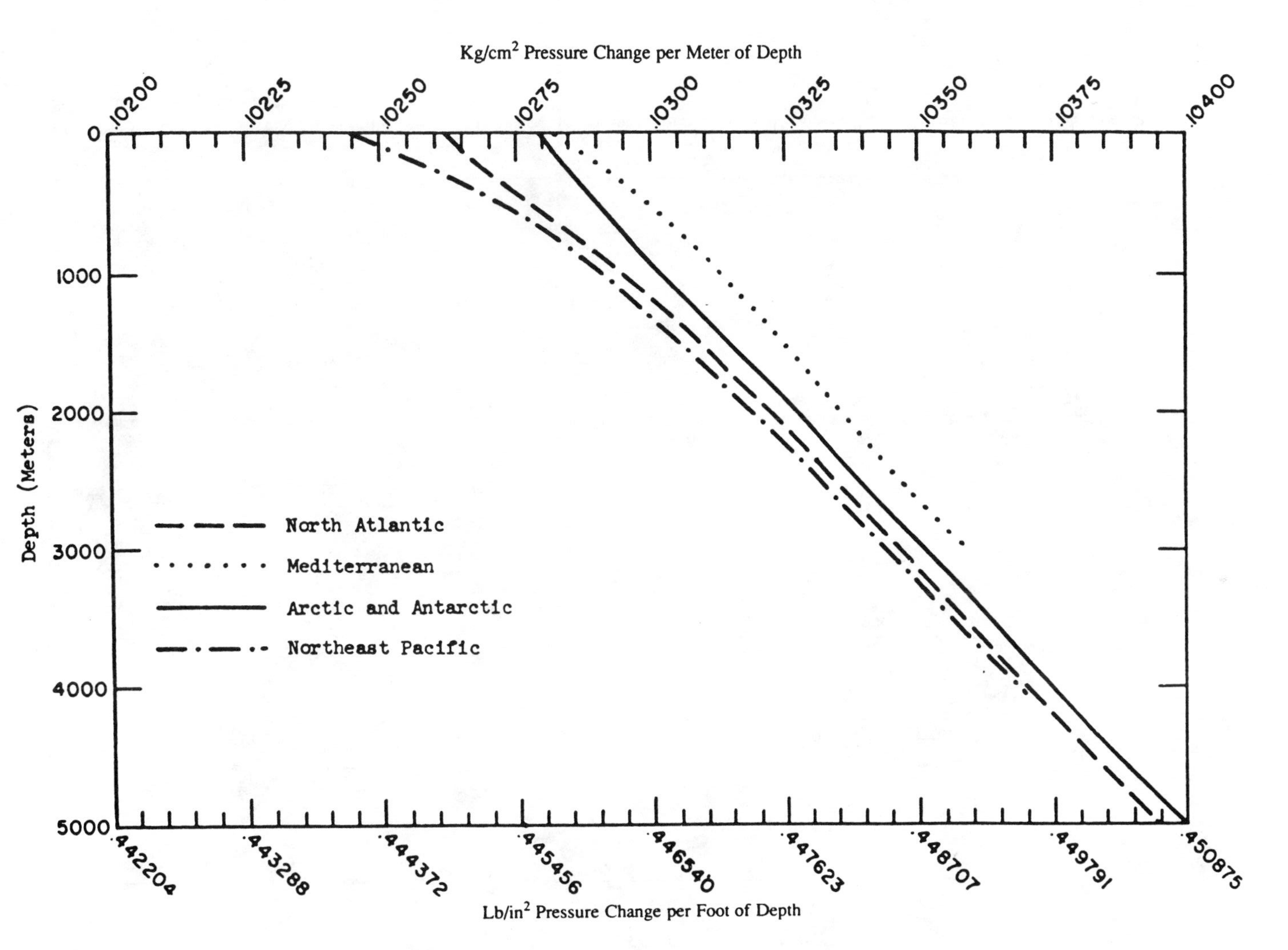

FIGURE 1.7—11. Pressure changes with depth. (From Bialek, E. L., Errors in the determination of depth, *Int. Hydrograph. Rev.*, 43, 1966; tables from *International Oceangraphic Tables*, Vol. I. ©UNESCO 1966. Reproduced by permission of UNESCO.)

1.8 OCEANIC CIRCULATION

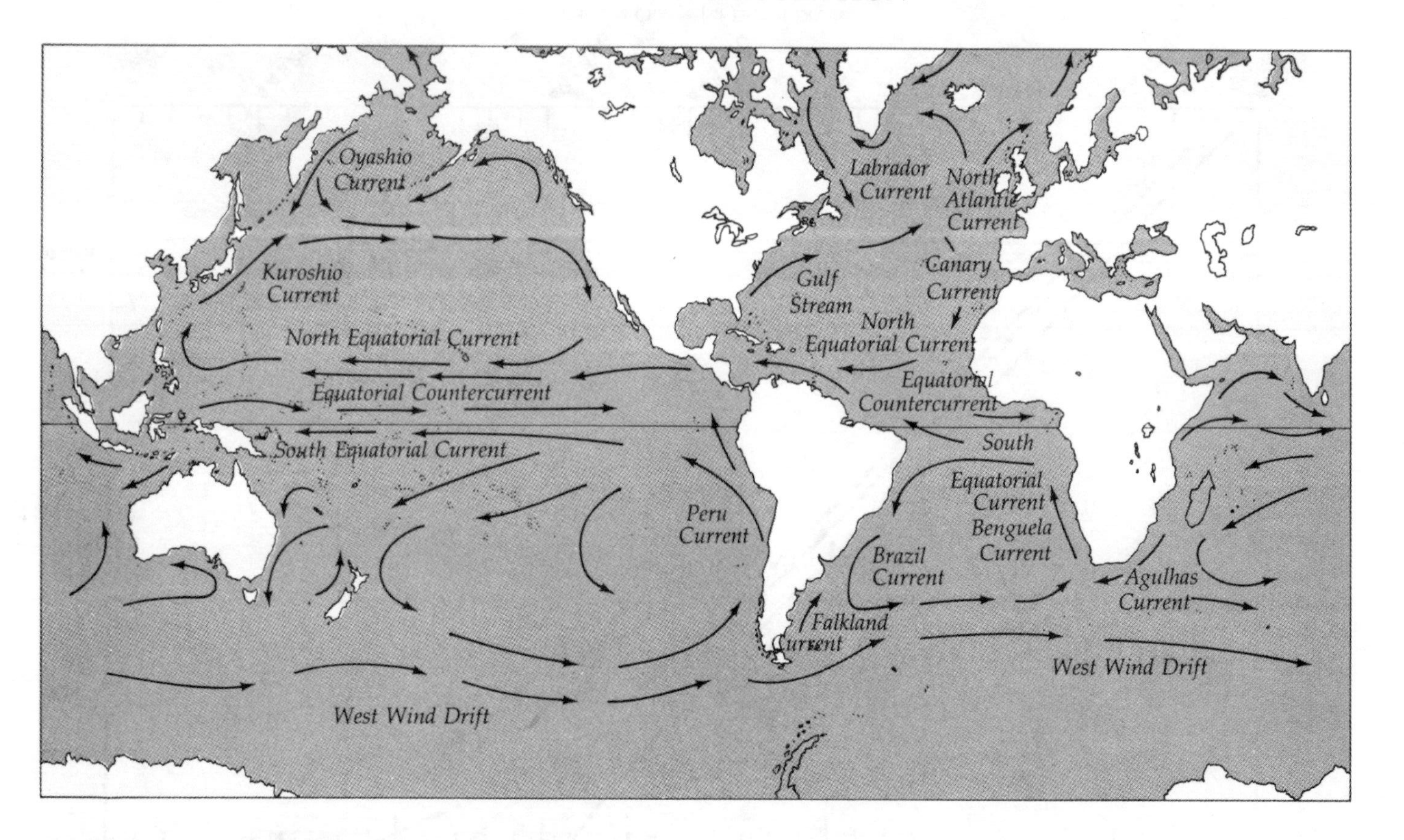

FIGURE 1.8—1. Surface circulation of the oceans of the world. (From Ingmanson, D. E. and Wallace, W. J., *Oceanology: an Introduction,* 2nd ed., Wadsworth Publishing Company, Belmont, CA, 1979. With permission.)

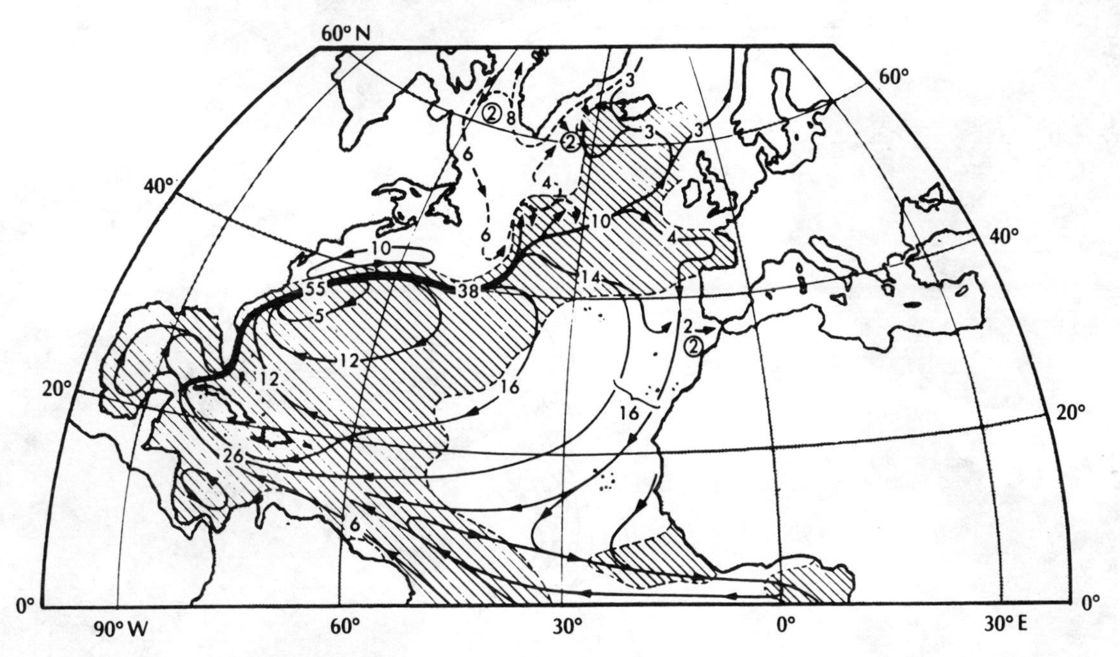

FIGURE 1.8—2. General circulation of the North Atlantic. Numbers give the transport volumes of seawater in millions of cubic meters per second, and the arrows show the flow direction. The intense dark line along the Atlantic seaboard of the U.S. is the Gulf Stream. It is the prime mover of warm water from the tropical to polar latitudes. (From Neshyba, S., *Oceanography: Perspectives on a Fluid Earth,* John Wiley & Sons, New York, 1987, 107. Reproduced by permission of copyright © owner, John Wiley & Sons, Inc.)

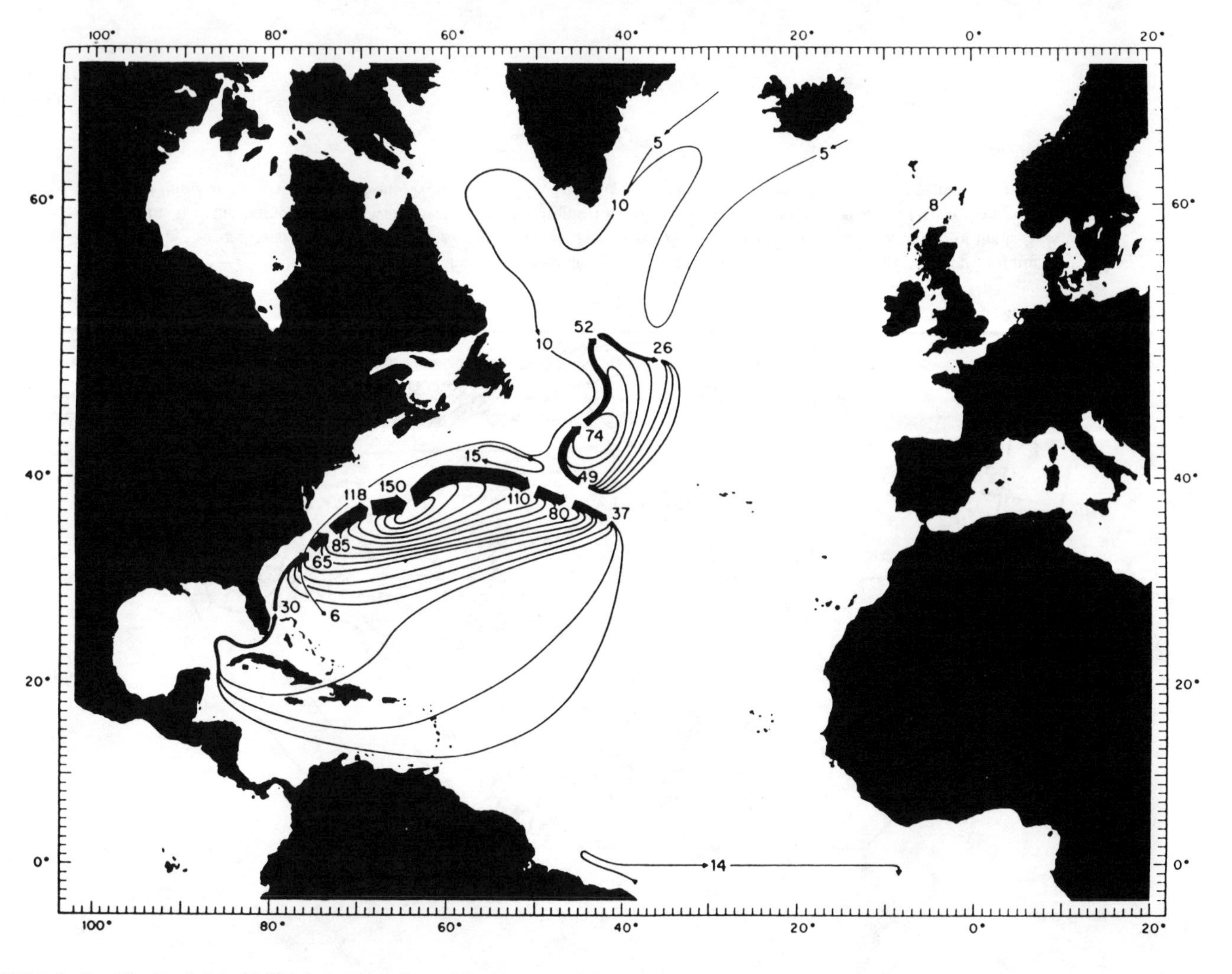

FIGURE 1.8—3. The North Atlantic circulation according to Worthington (1976): total transport, in Sverdrups (10^6 m^3/s). (From Schmitz, W. J., Jr. and McCartney, M. S., On the North Atlantic Circulation, *Rev. Geophys.*, 31, 29, 1993. With permission.)

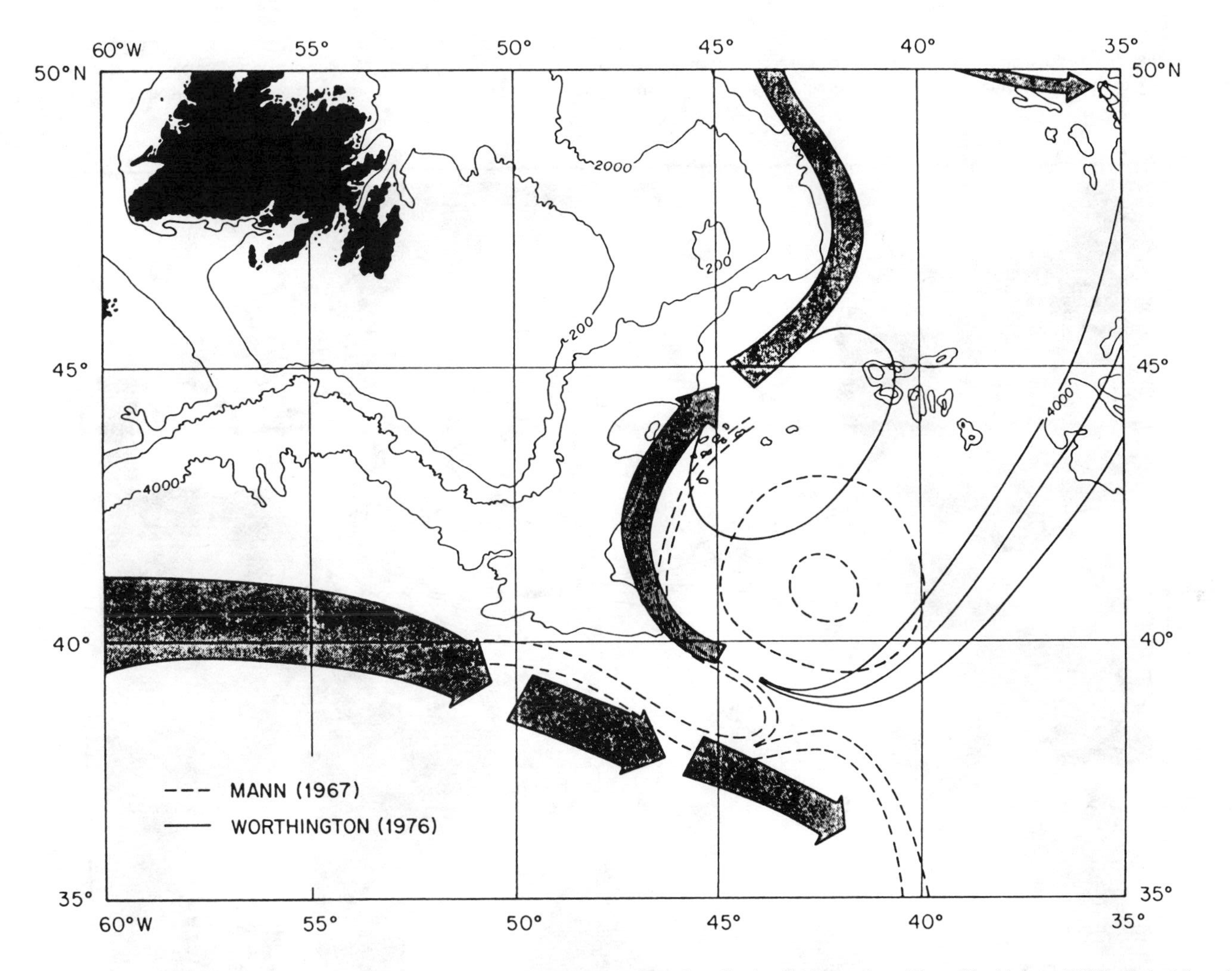

FIGURE 1.8—4. A schematic illustration of flow patterns near the Grand Banks of Newfoundland: solid lines adapted from Worthington (1976), dashed lines from Mann (1967). (From Schmitz, W. J., Jr. and McCartney, M. S., On the North Atlantic circulation, *Rev. Geophys.*, 31, 29, 1993. With permission.)

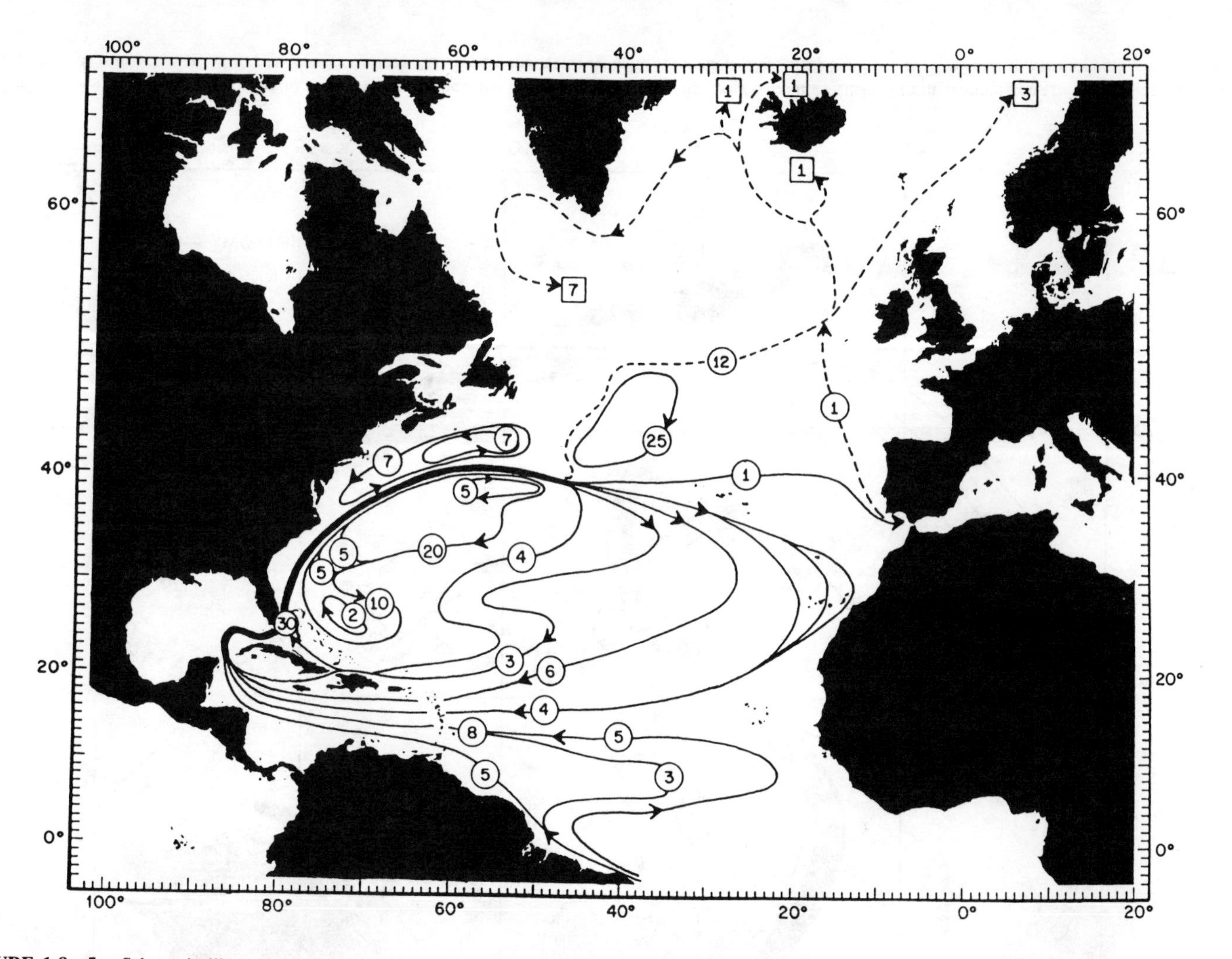

FIGURE 1.8—5. Schematic illustration of transport (values enclosed in circles are in Sverdrups) for the North Atlantic at temperatures above 7°C (nominal). Numbers in squares (in Sverdrups) denote sinking. (From Schmitz, W. J., Jr. and McCartney, M. S., On the North Atlantic circulation, *Rev. Geophys.*, 31, 29, 1993. With permission.)

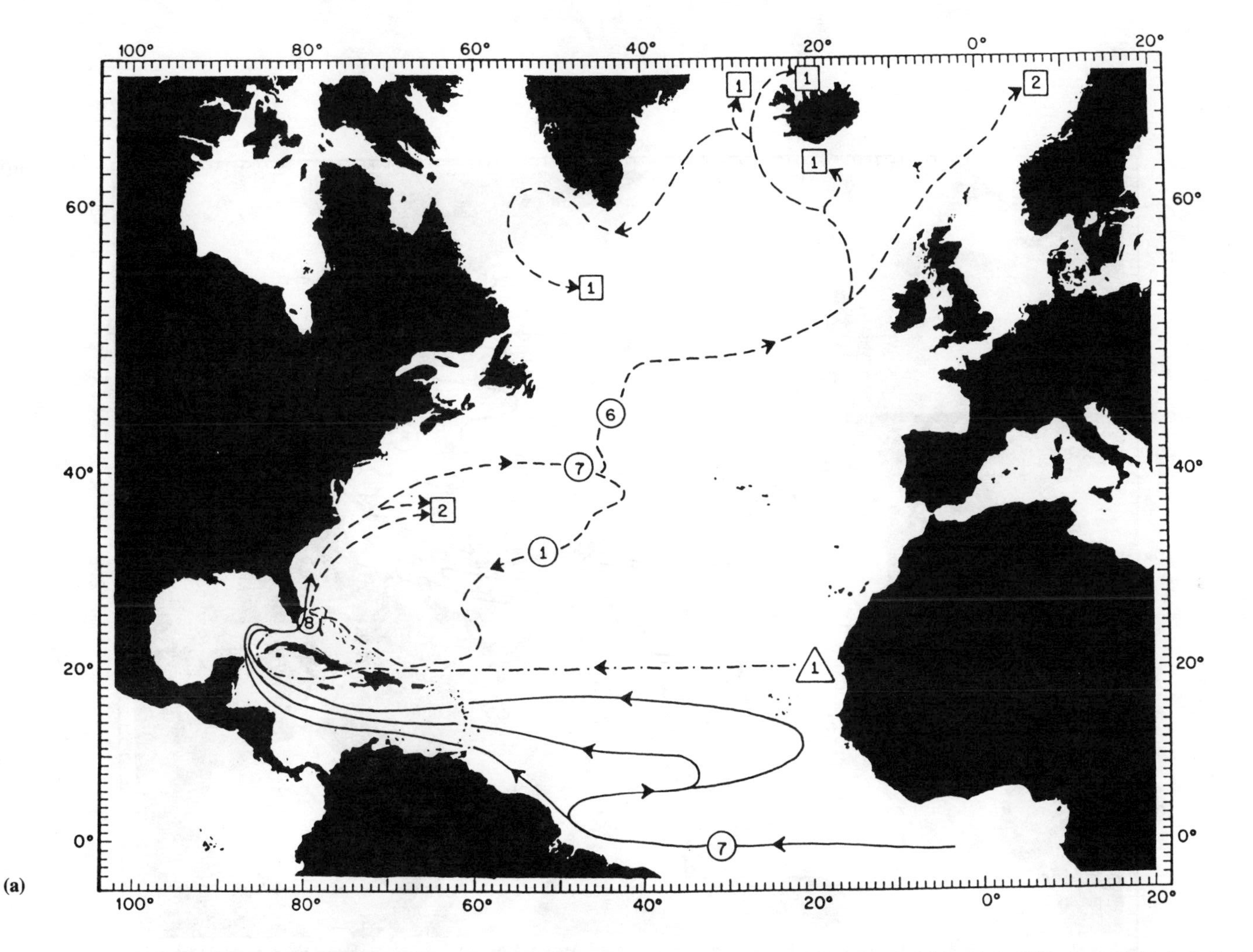

FIGURE 1.8—6. The composite circulation in Figure 1.8—5 is split into three layers: (a) near surface, (b) 12–24°C, (c) 7–12°C, (d) 7–12°C, including the circulation of Mediterranean water and subpolar mode water. Transports in circles are in Sverdrups. Squares represent sinking; triangles denote upwelling. (From Schmitz, W. J., Jr. and McCartney, M. S., On the North Atlantic circulation, *Rev. Geophys.*, 31, 29, 1993. With permission.)

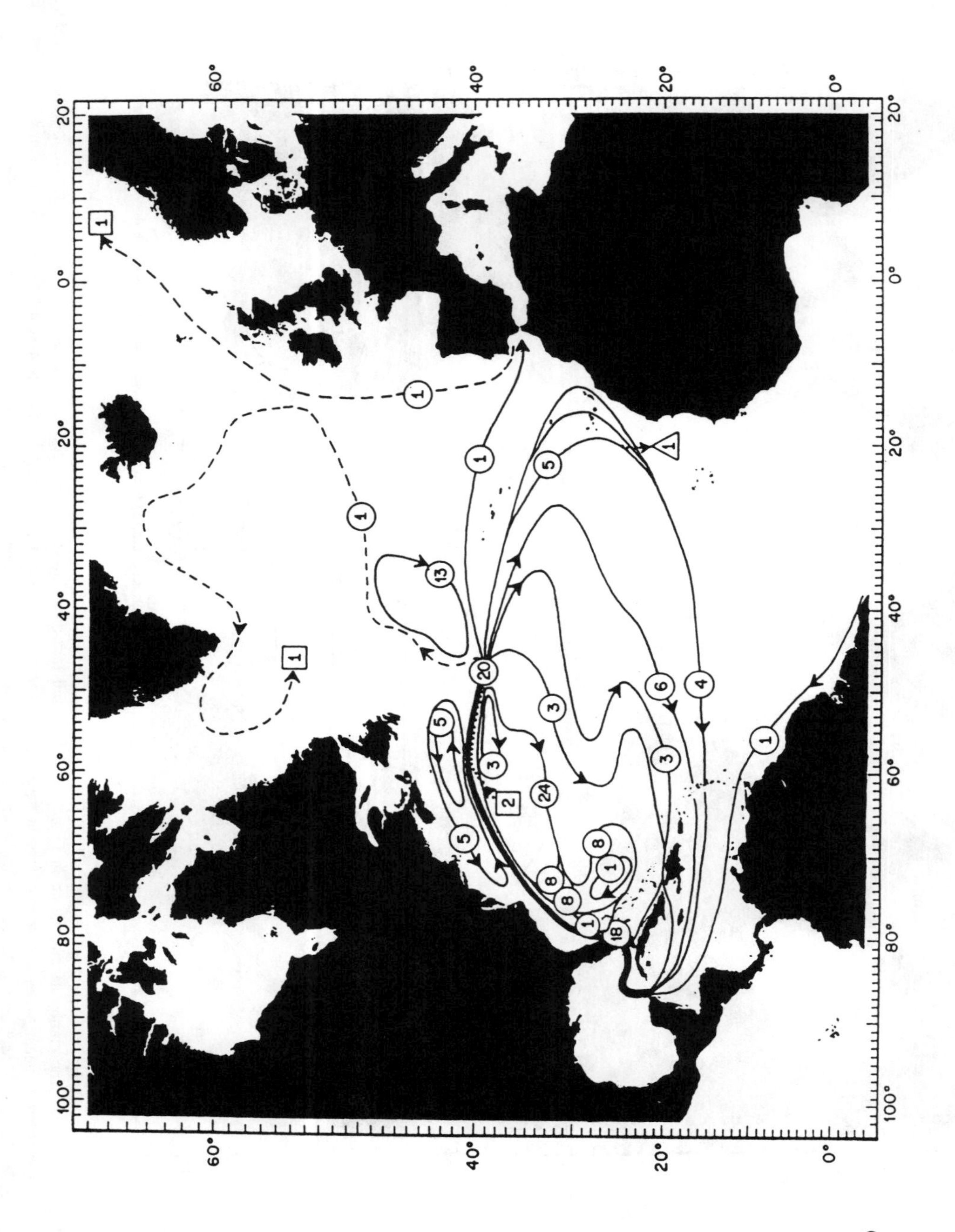

(b)

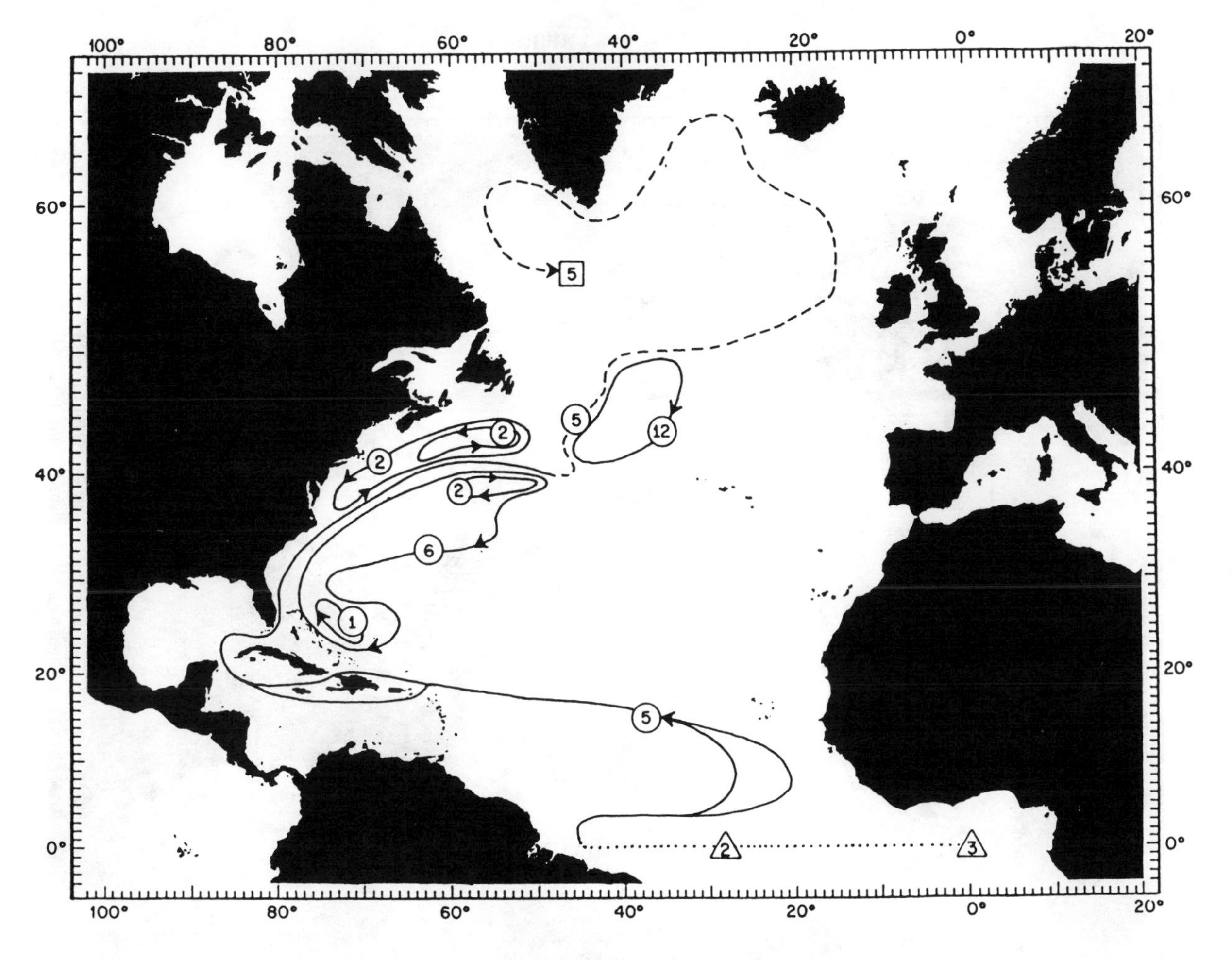

(c)

FIGURE 1.8—6. (continued)

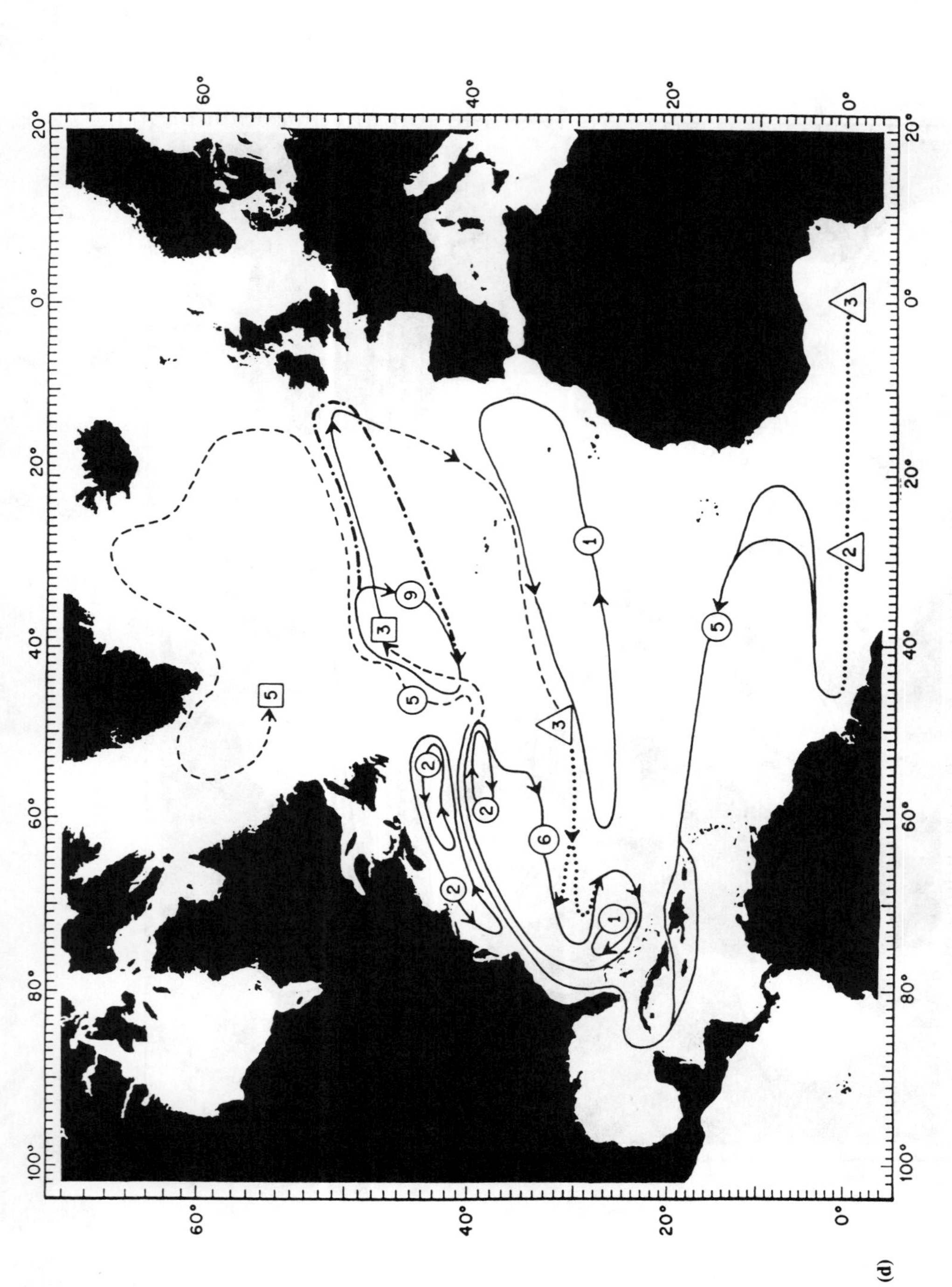

FIGURE 1.8—6 (continued).

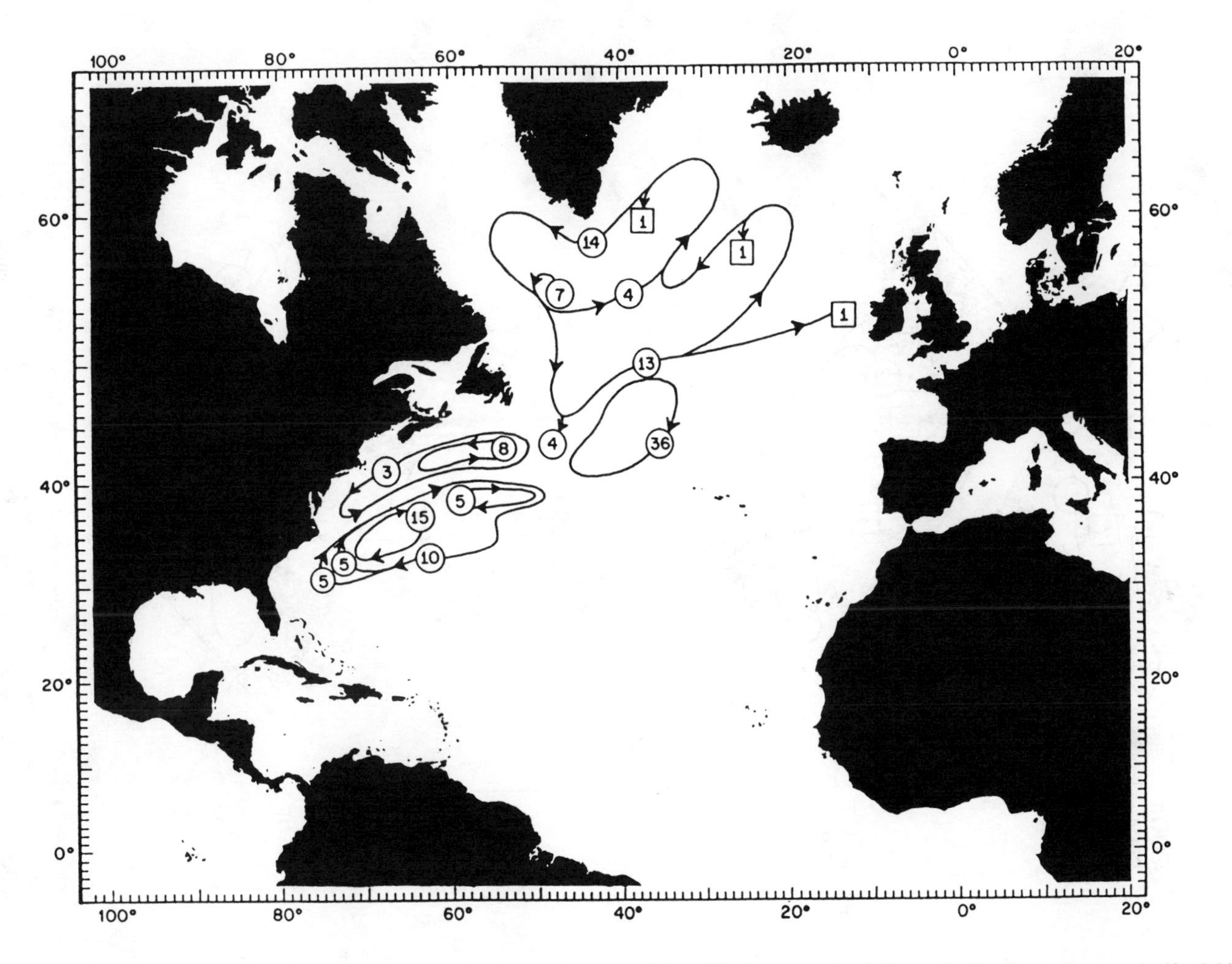

FIGURE 1.8—7. Selected circulation features in the North Atlantic at temperatures below 4°C. Transports in circles are in Sverdrups. Squares signify sinking. (From Schmitz, W. J., Jr. and McCartney, W. S., On the North Atlantic circulation, *Rev. Geophys.*, 31, 29, 1993. With permission.)

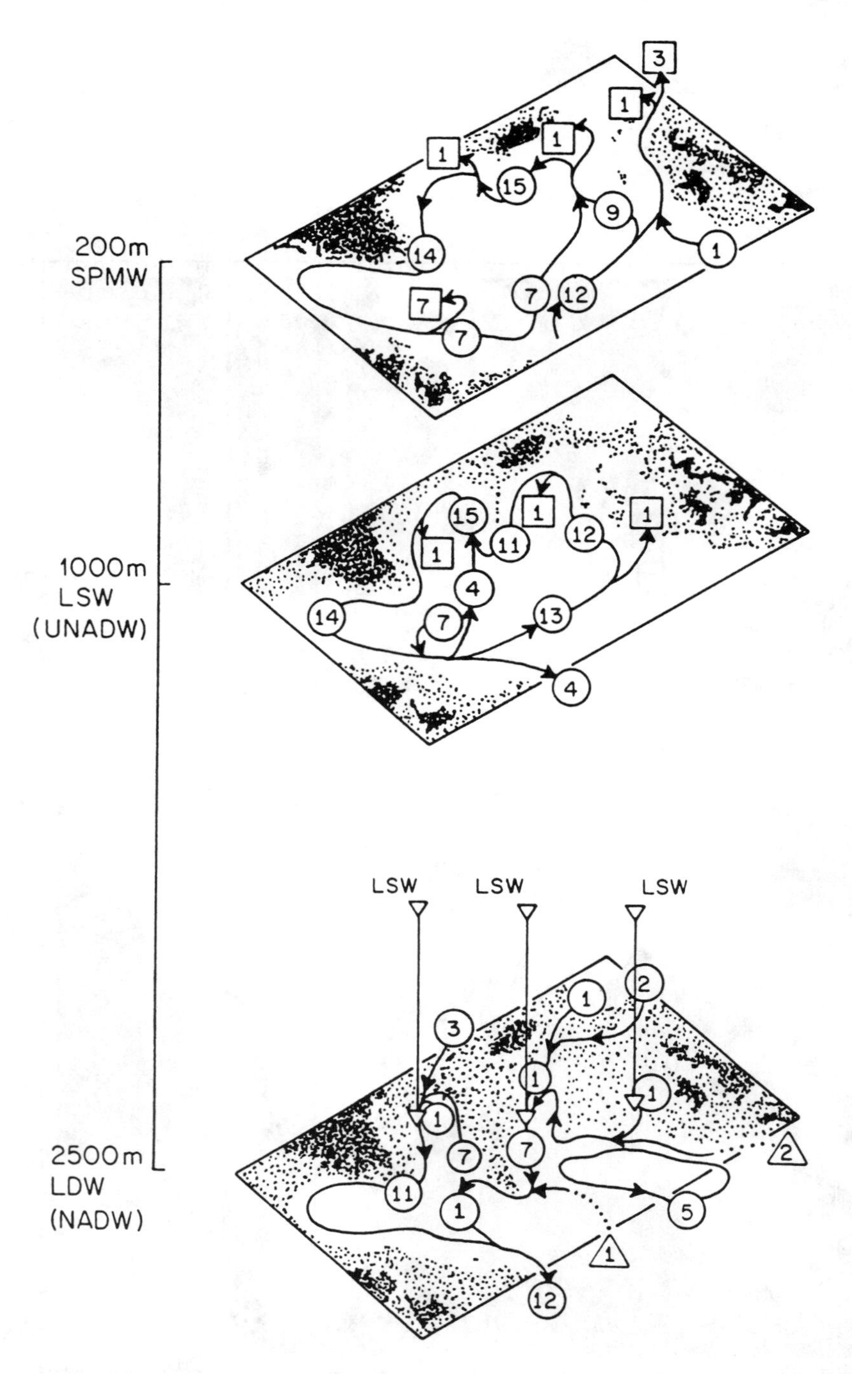

FIGURE 1.8—8. Circulation patterns for the northern North Atlantic. Transports in circles are in Sverdrups. Squares represent sinking; triangles denote upwelling. (From Schmitz, W. J., Jr. and McCartney, M. S., On the North Atlantic circulation, *Rev. Geophys.*, 31, 29, 1993. With permission.)

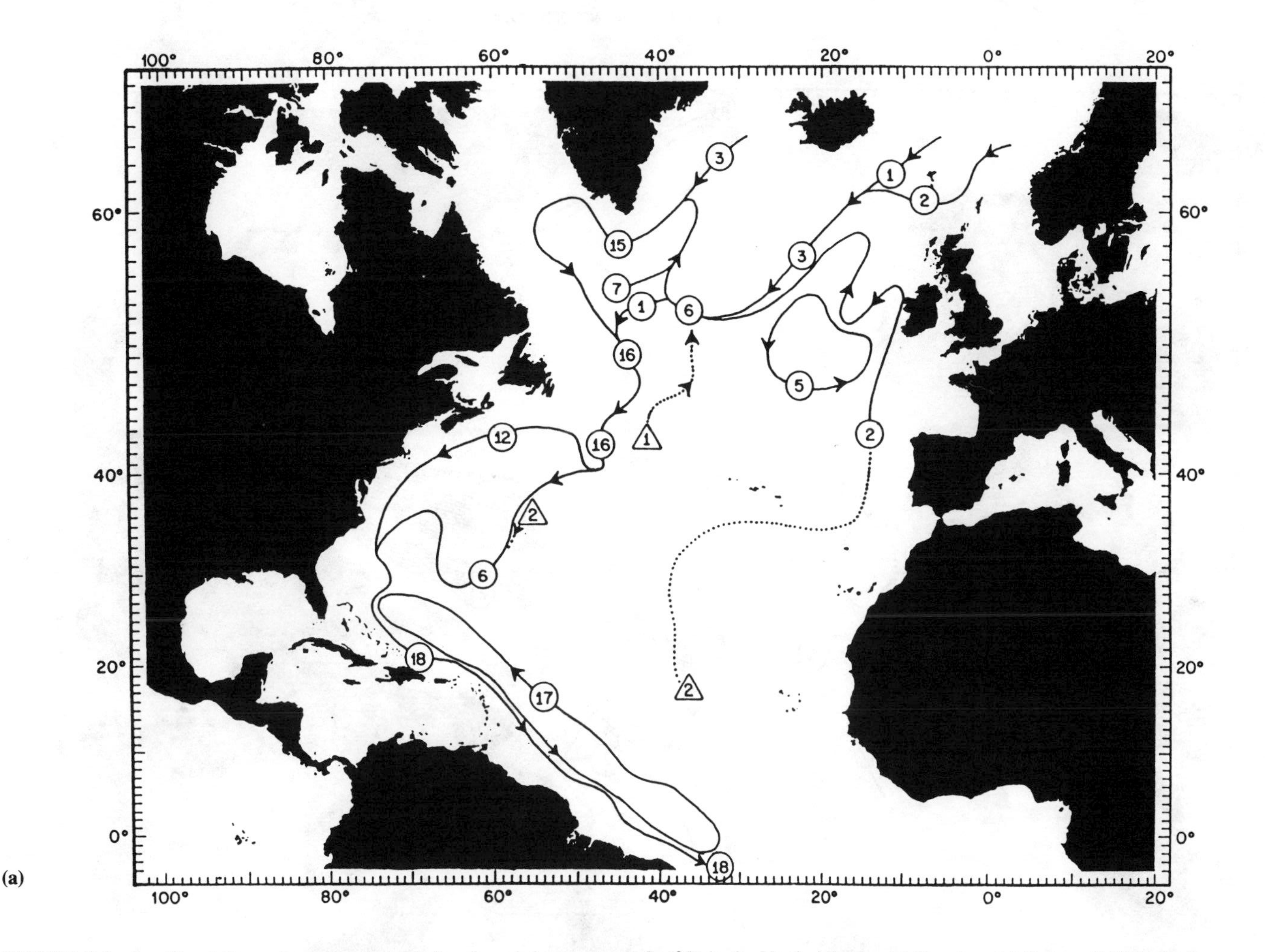

FIGURE 1.8—9. Circulation patterns for thermohaline forced deep water (1.8–4°C) in the North Atlantic. (a) Based on McCartney (1992). (b) Deep gyres in Figure 1.8—7 added as hatched areas. Transports in circles are in Sverdrups. Squares denote sinking; triangles represent upwelling. (From Schmitz, W. J., Jr. and McCartney, M. S., On the North Atlantic circulation, *Rev. Geophys.*, 31, 29, 1993. With permission.)

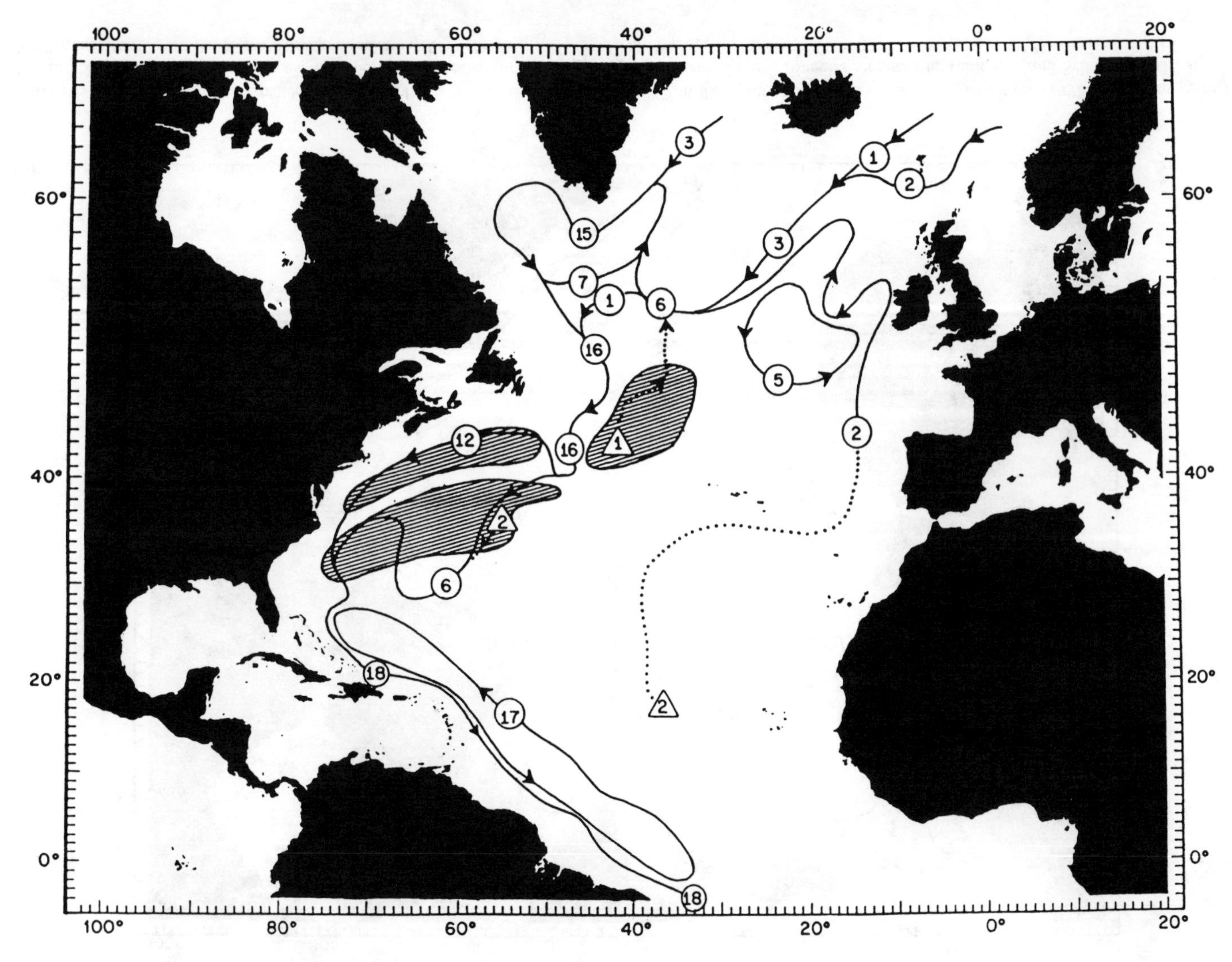

(b)

FIGURE 1.8—9 (continued).

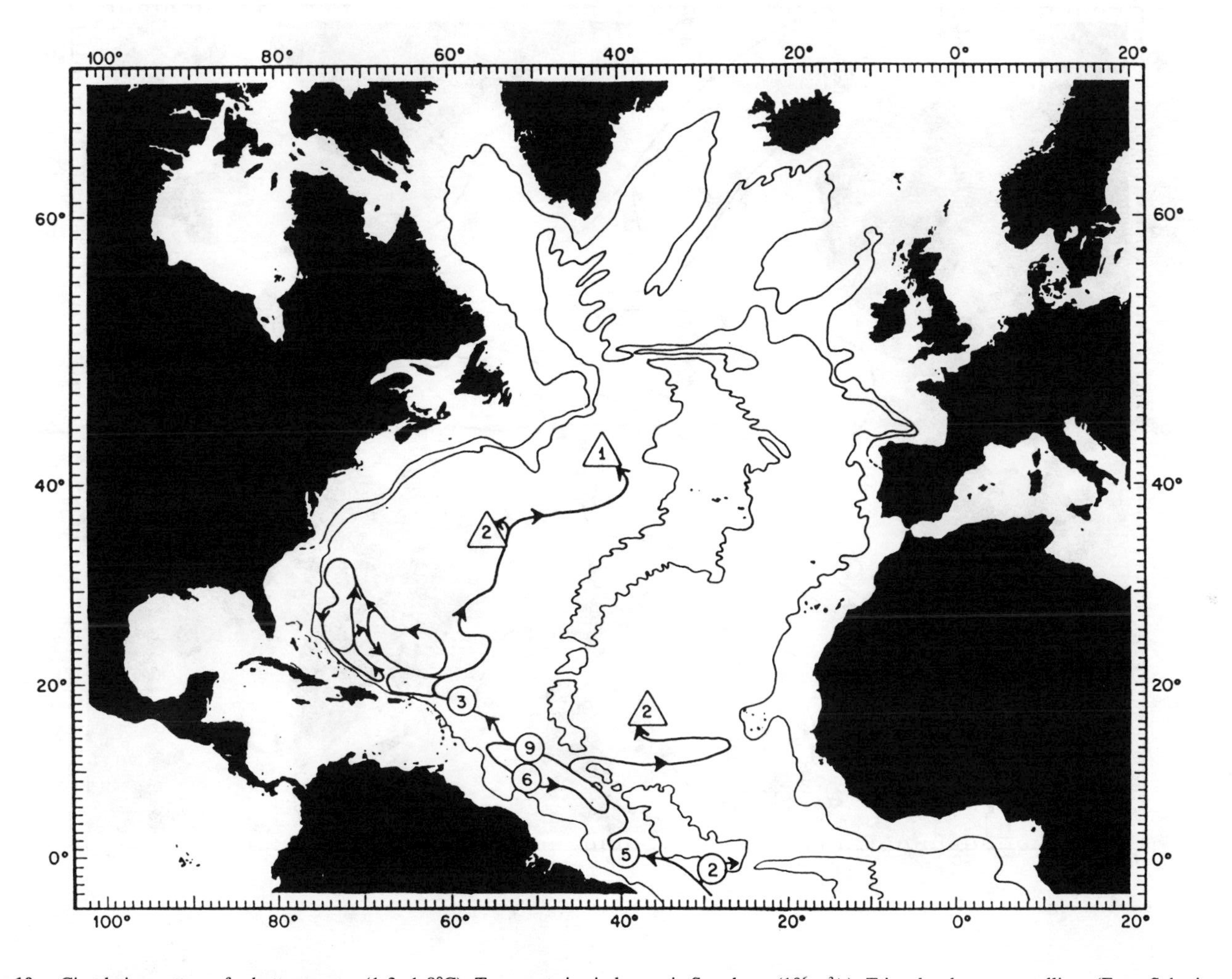

FIGURE 1.8—10. Circulation patterns for bottom water (1.3–1.8°C). Transports in circles are in Sverdrups (10^6 m^3/s). Triangles denote upwelling. (From Schmitz, W. J., Jr. and McCartney, M. S., On the North Atlantic circulation, *Rev. Geophys.*, 31, 29, 1993. With permission.)

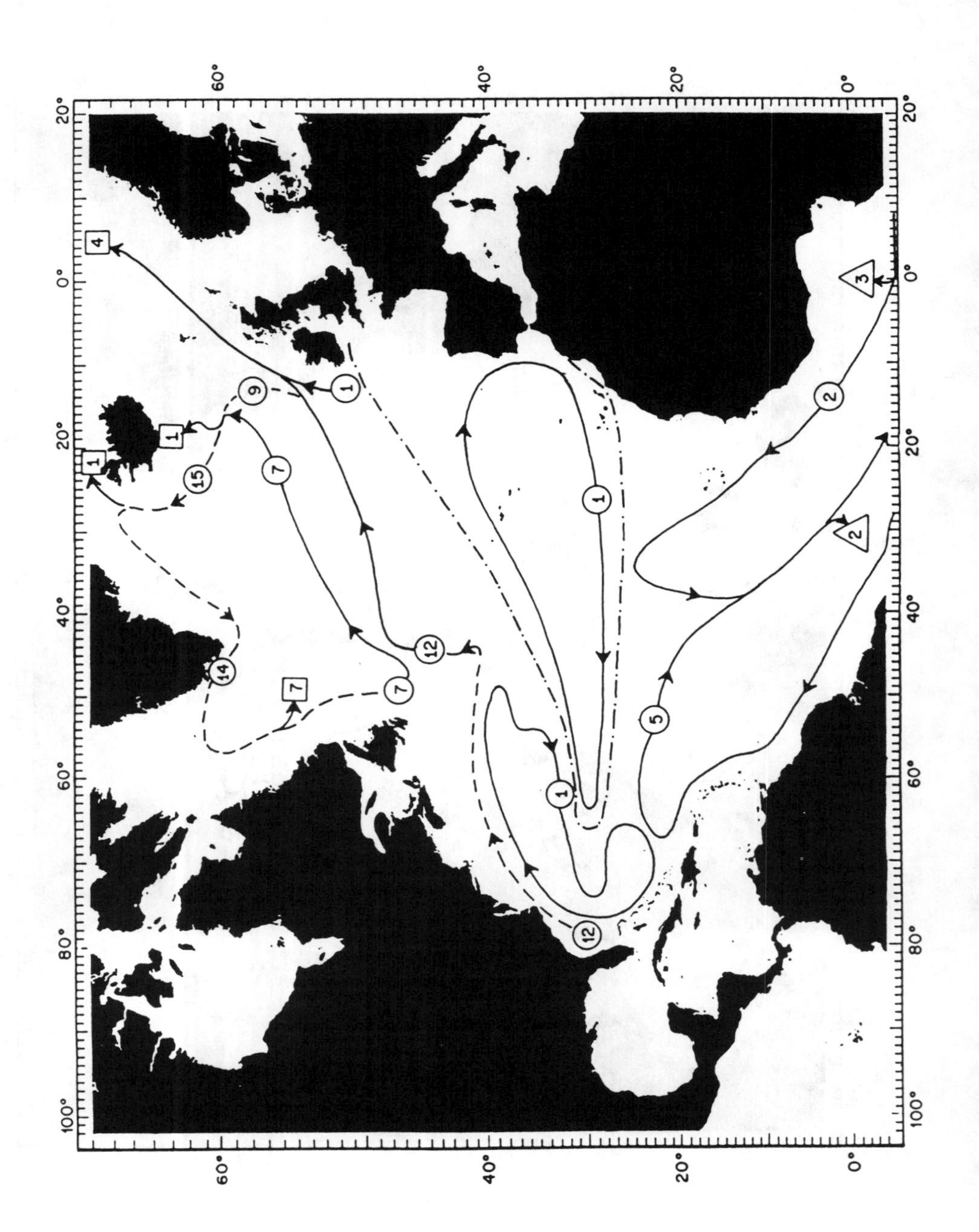

(a)

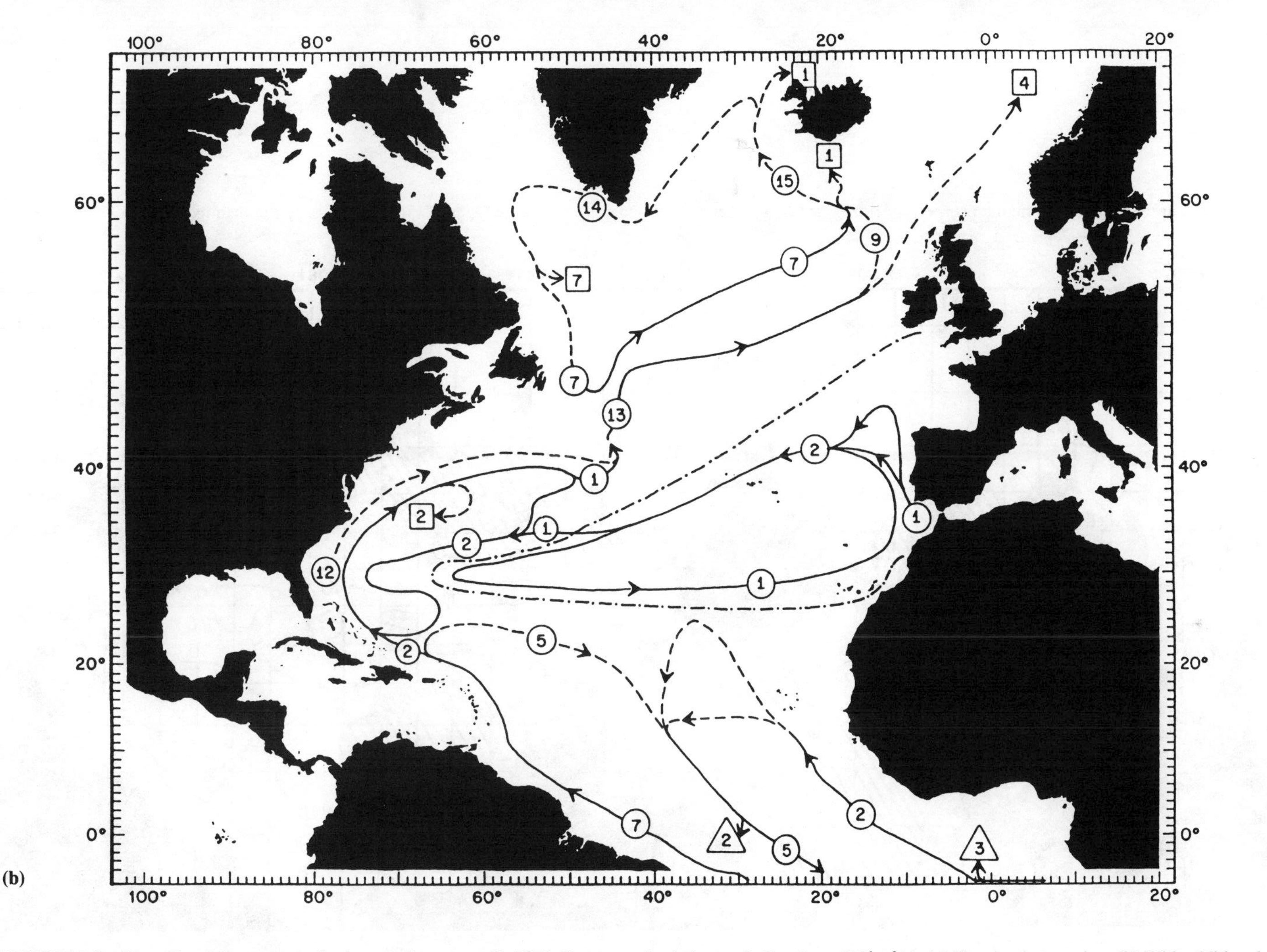

FIGURE 1.8—11. Circulation patterns for intermediate water (4–7°C). Transports in circles are in Sverdrups (10^6 m^3/s). (a) The simplest version. (b) With additional and modified features. Squares denote sinking; triangles signify upwelling. (From Schmitz, W. J., Jr. and McCartney, M. S., On the North Atlantic circulation, *Rev. Geophys.*, 31, 29, 1993. With permisson.)

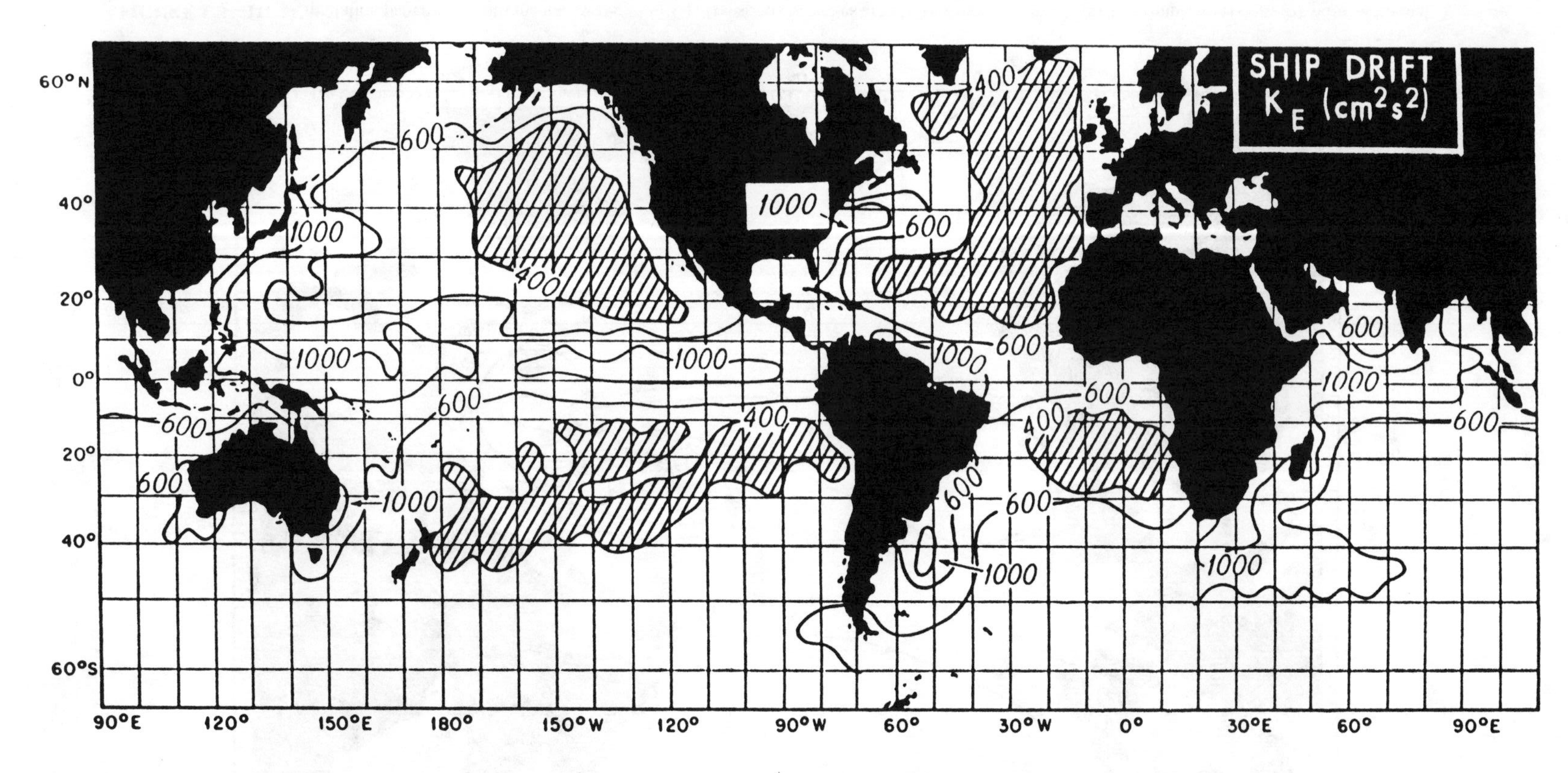

FIGURE 1.8—12. A smoothed global map of eddy kinetic energy (cm^2/s^2) at the sea surface. (From Schmitz, W. J., Jr., Mesoscale ocean currents, *Rev. Geophys.*, in press. With permission.)

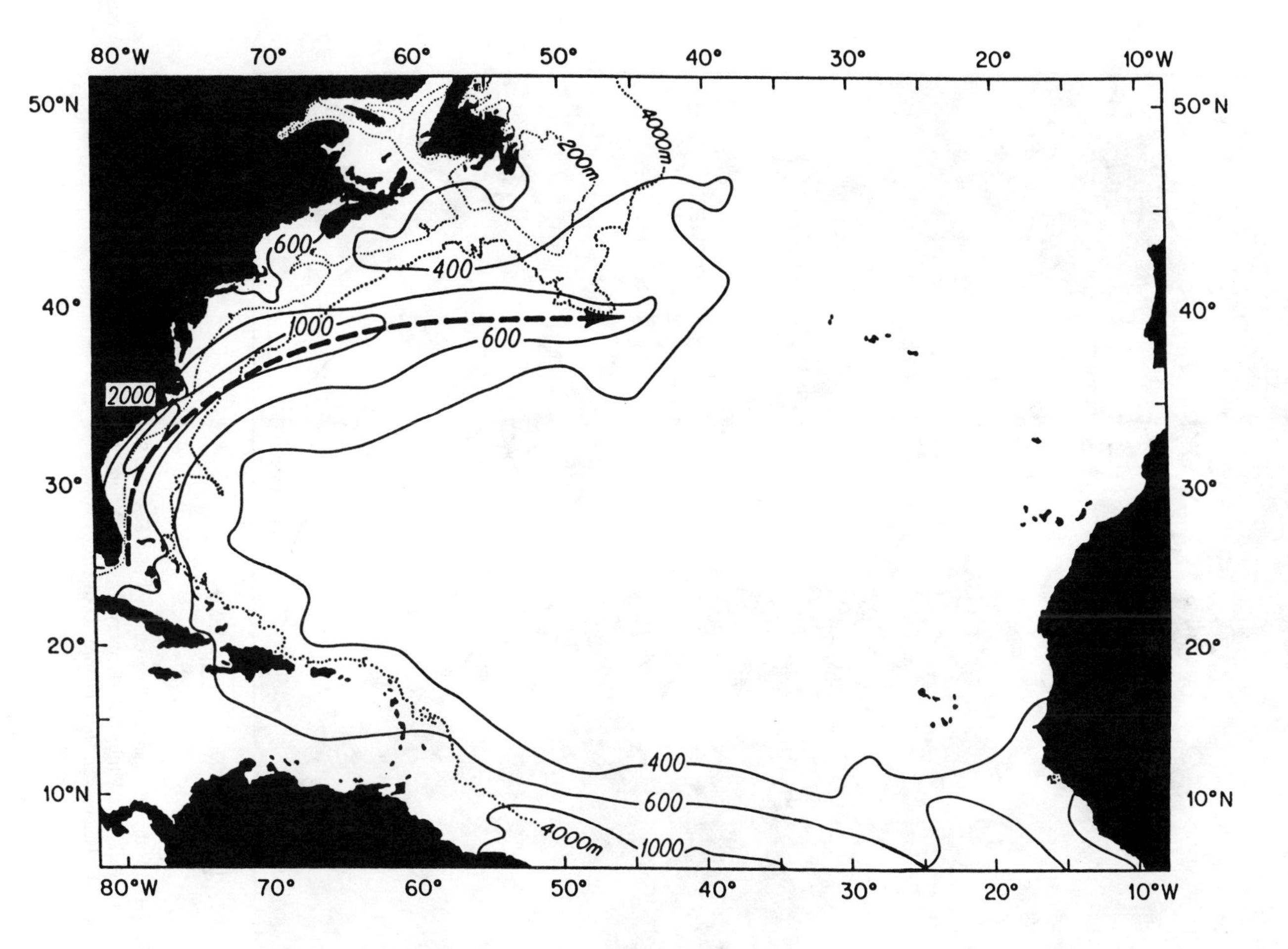

FIGURE 1.8—13. Contours of surface eddy kinetic energy (cm²/s²) for the North Atlantic, with the axis of the Gulf Stream superimposed as a heavy dashed line with arrowhead. (From Schmitz, W. J., Jr., Mesoscale ocean currents, *Rev. Geophys.*, in press. With permission.)

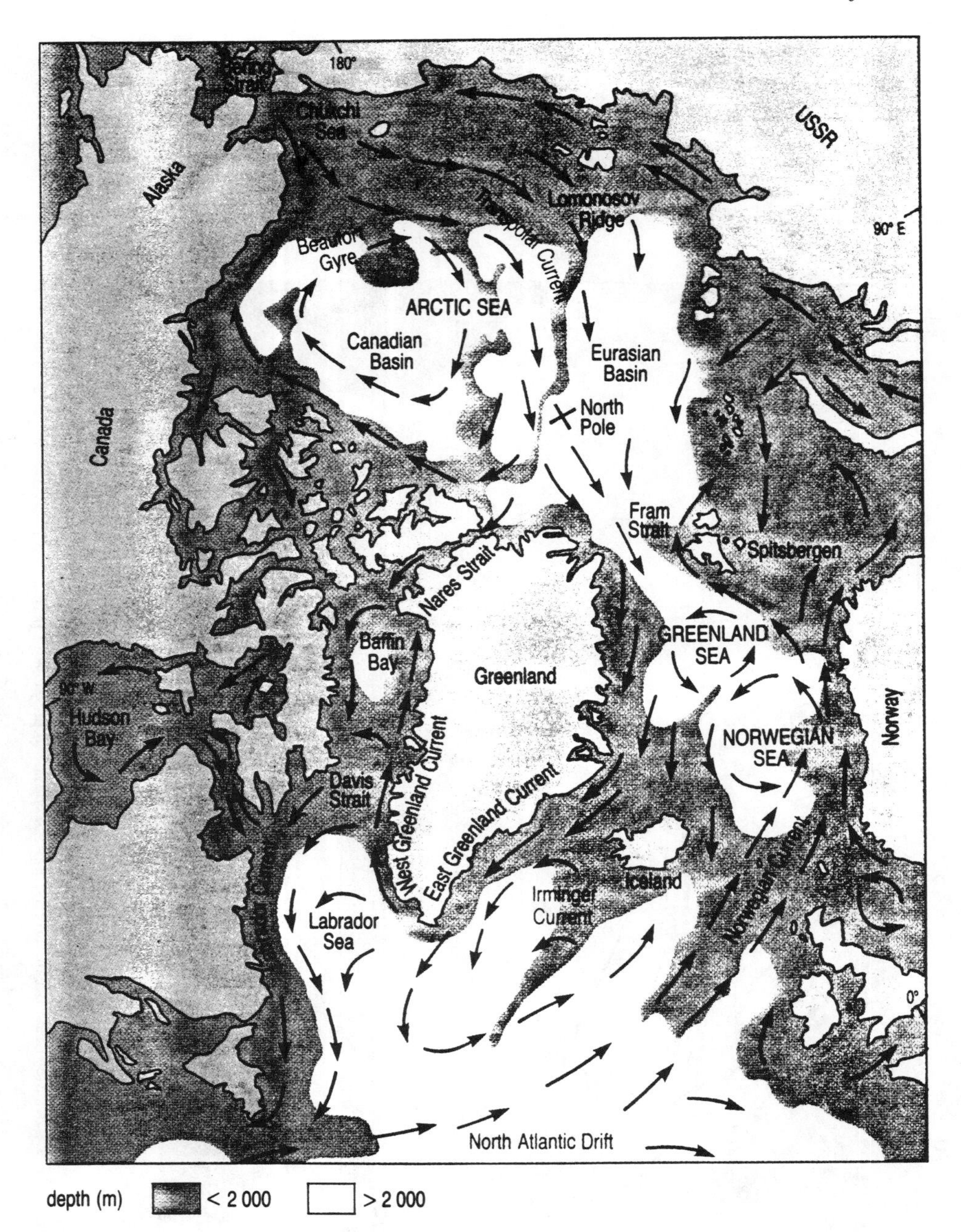

FIGURE 1.8—14. The bathymetry and surface currents of the Arctic Sea and the adjacent seas of the North Atlantic. (From Brown, J., Colling, A., Park, D., Phillips, J., Rothery, D., and Wright, J., *Ocean Circulation,* 1989, 149, with kind permission from Pergamon Press Ltd., Headington Hall, Oxford, OX3 OBW, U.K.)

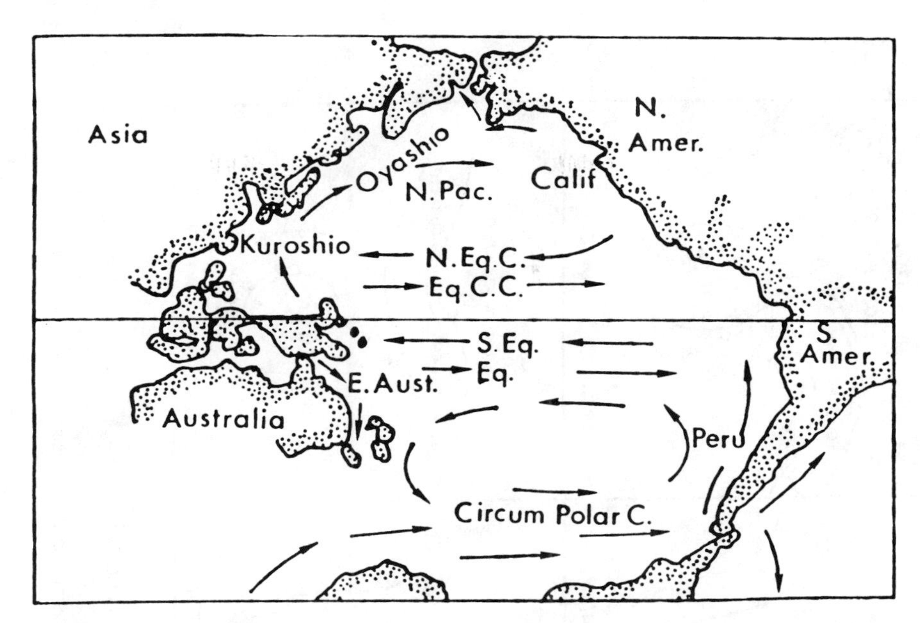

FIGURE 1.8—15. Surface currents of the Pacific Ocean. (From Millero, F. J. and Sohn, M. L., *Chemical Oceanography*, CRC Press, Boca Raton, FL, 1992.)

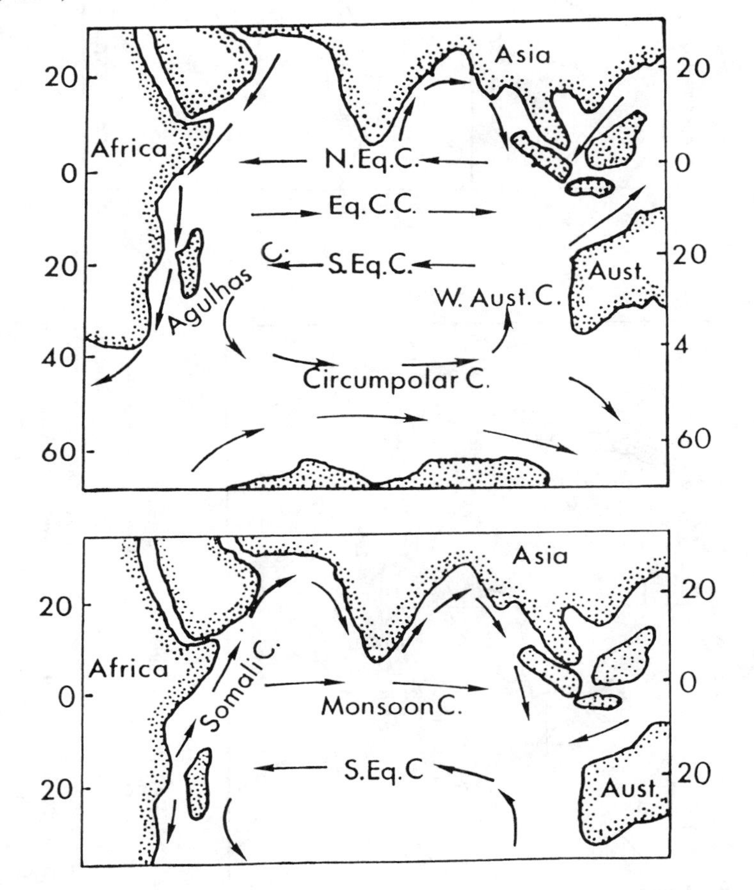

FIGURE 1.8—16. Surface currents of the Indian Ocean. (From Millero, F. J. and Sohn, M. L., *Chemical Oceanography*, CRC Press, Boca Raton, FL, 1992.)

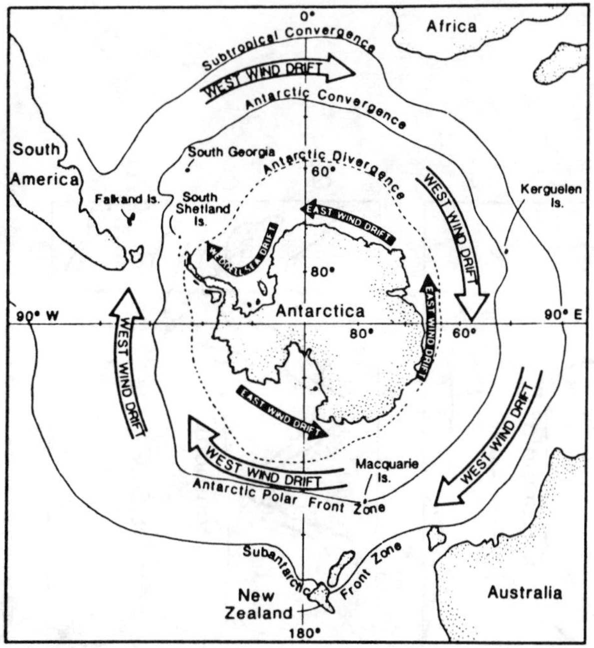

FIGURE 1.8—17. Circulation of the Southern Ocean which is bounded by the Antarctic continent and the seafloor south of the Subtropical Convergence (Subantarctic Front Zone). The predominant clockwise trajectory of the West Wind Drift (Antarctic Circumpolar Current) extends south of the Antarctic Convergence (Antarctic Polar Zone). South of the West Wind Drift is the counter-clockwise East Wind Drift and the Antarctic Divergence between them. (From Berkman, P. A., The Antarctic marine ecosystem and humankind, *Rev. Aquat. Sci.*, 6, 295, 1992.)

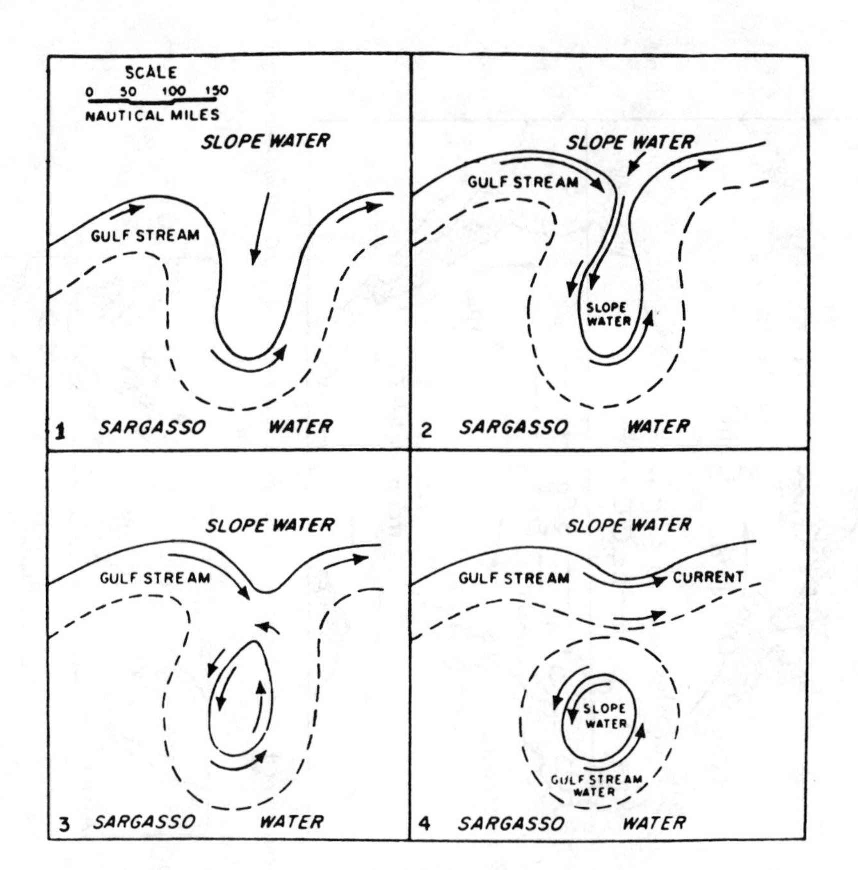

FIGURE 1.8—18. Diagram of Gulf Stream ring generation from meander formation to separation. (From Parker, C. E., Gulf Stream rings in the Sargasso Sea, *Deep-Sea Research*, 18, 981, 1971. With kind permission from Pergamon Press Ltd., Headington Hill Hall, Oxford OX3 OBW, U.K.)

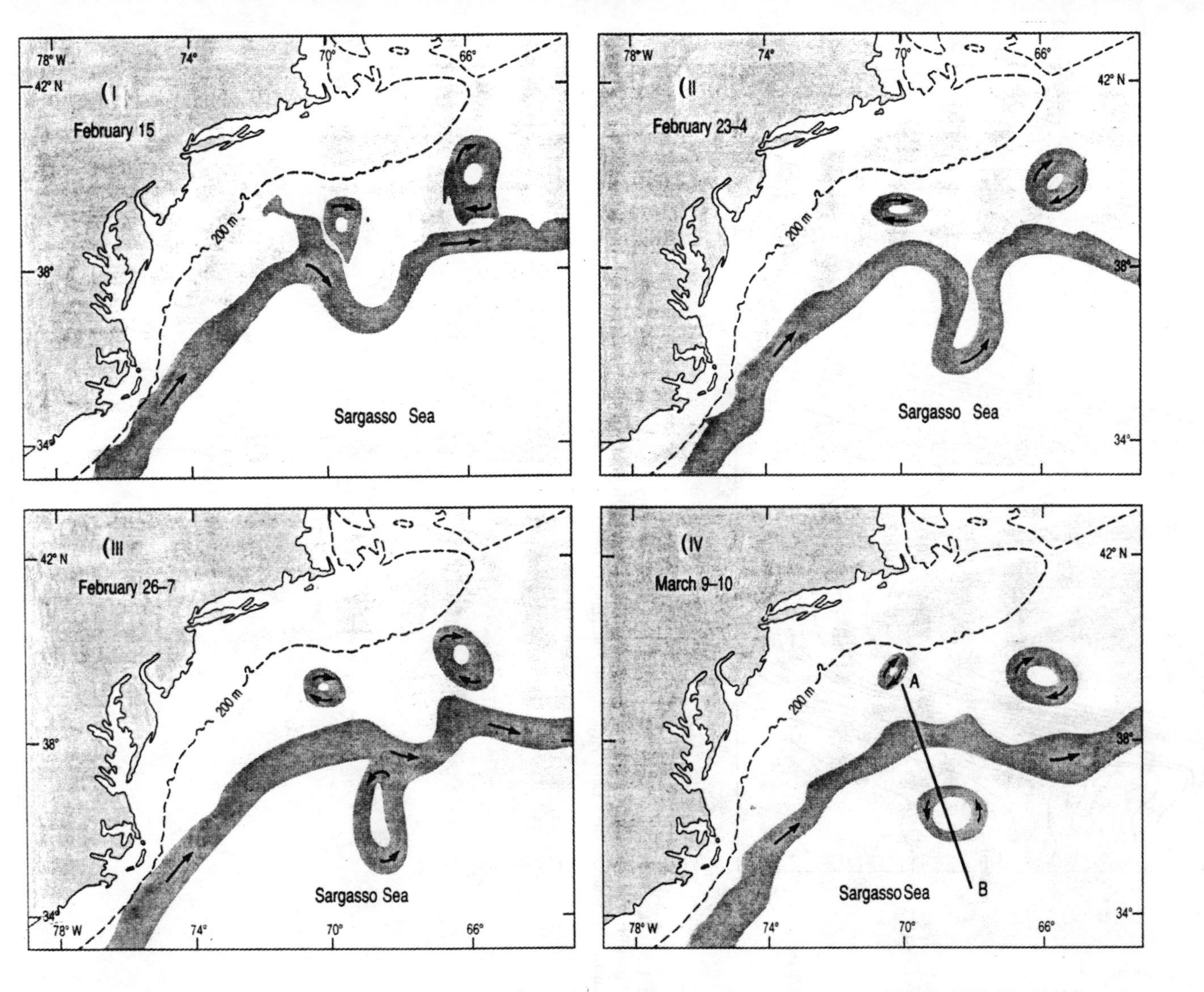

FIGURE 1.8—19. The evolution of Gulf Stream eddies as deduced from infrared satellite images made in February and March 1977. The Gulf Stream separates cool continental shelf water from warm Sargasso Sea water. (From Brown, J., Colling, A., Park, D., Phillips, J., Rothery, D., and Wright, J., *Ocean Circulation.* 1989, 110, with kind permission from Pergamon Press Ltd., Headington Hill Hall, Oxford OX3 OBW, U.K.)

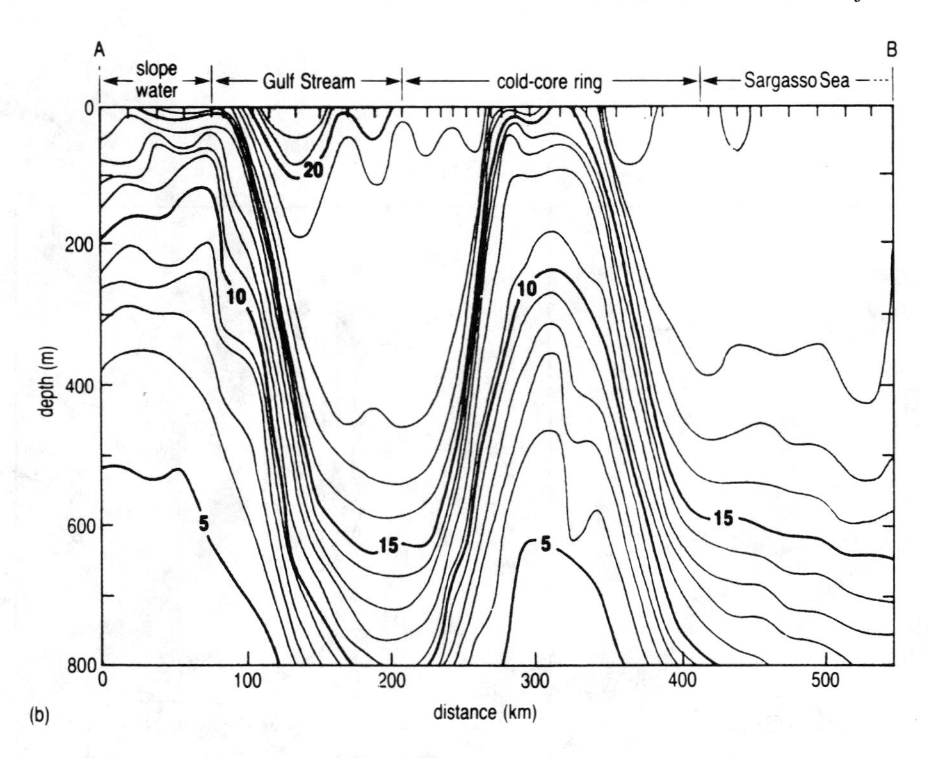

FIGURE 1.8—20. Temperature section along the line shown in Figure 1.8—19 (IV). Note that the eddies extend to significant depths. (From Brown, J., Colling, A., Park, D., Phillips, J., Rothery, D., and Wright, J., *Ocean Circulation,* 1989, 111, with kind permission from Pergamon Press Ltd., Headington Hill Hall, Oxford OX3 OBW, U.K.)

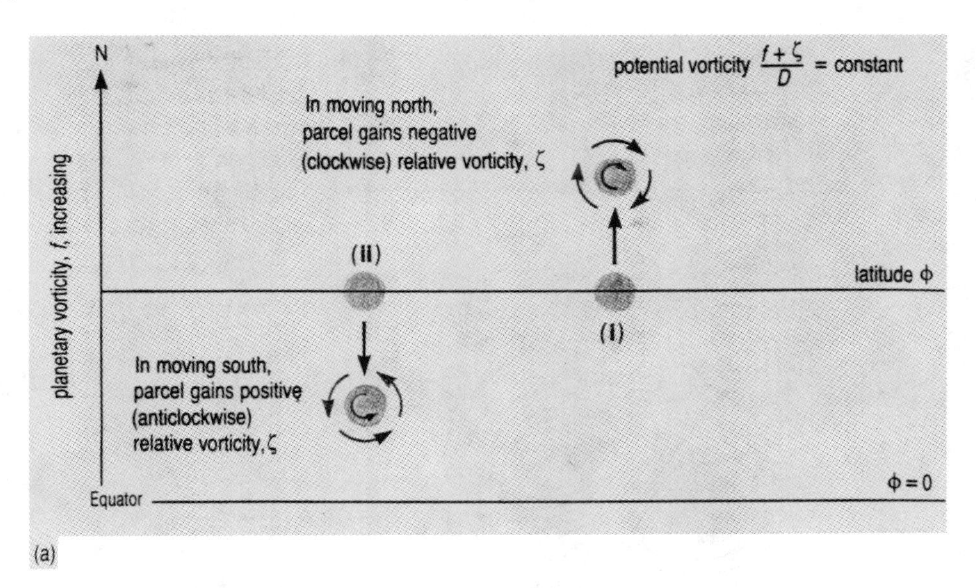

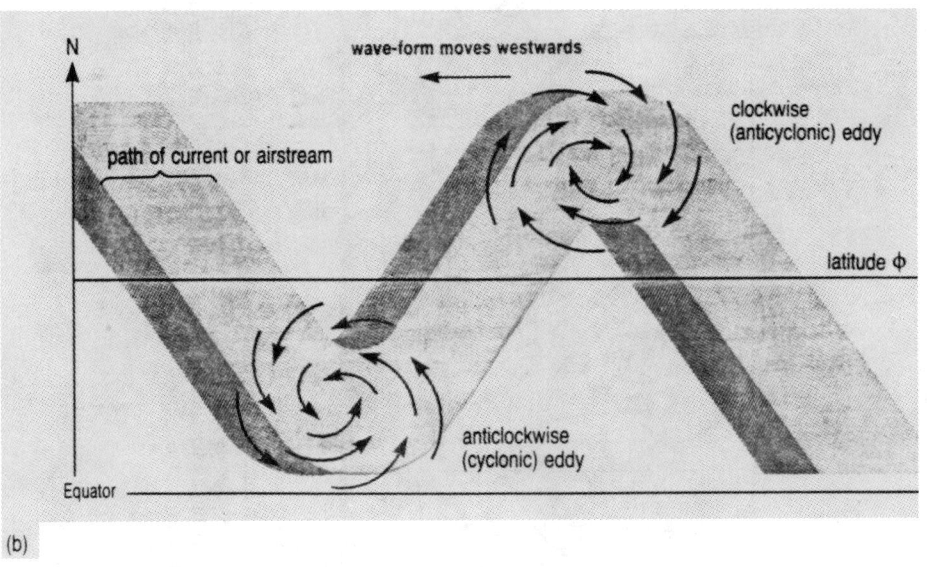

FIGURE 1.8—21. (a) Diagram showing how in a Rossby wave the need to conserve potential vorticity leads to a parcel of water oscillating about a line of latitude while alternately gaining and losing relative vorticity. (b) The path taken by a current or airstream affected by a Rossby wave. Note that the flow pattern is characterized by anticyclonic and cyclonic eddies, and that the waveform moves westwards relative to the current or airstream. (From Brown, J., Colling, A., Park, D., Phillips, J., Rothery, D., and Wright, J., *Ocean Circulation,* 1989, 144, with kind permission from Pergamon Press Ltd., Headington Hill Hall, Oxford OX3 OBW, U.K.)

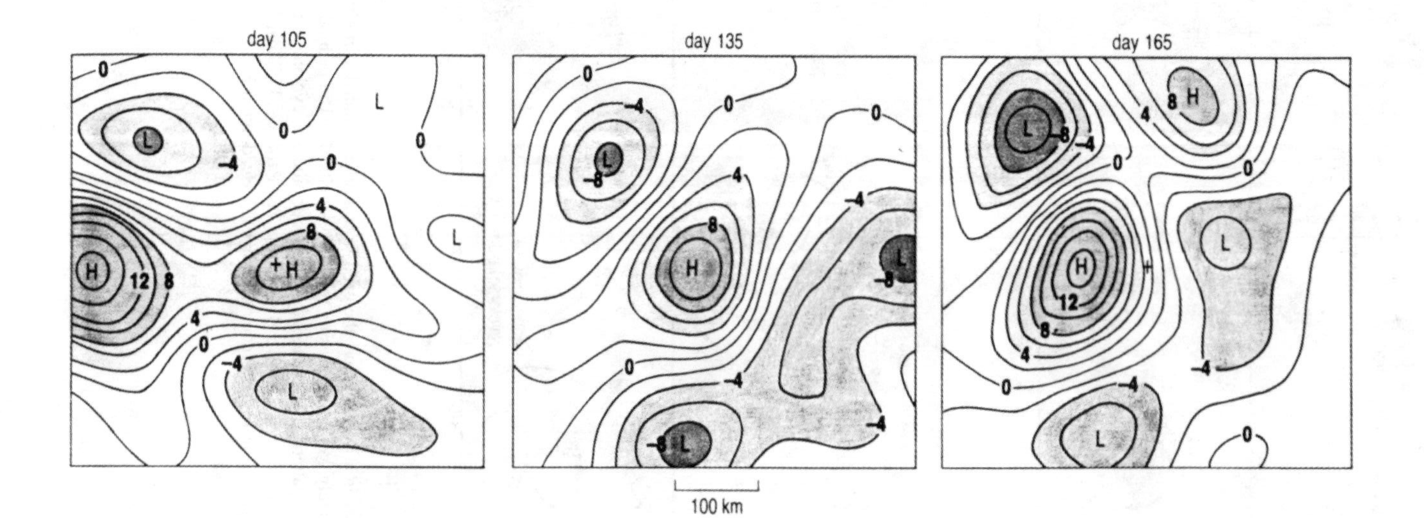

FIGURE 1.8—22. An example of mesoscale structure in the region of 69°40′W, 28°N (marked by the cross in the center of each map), as revealed by the dynamic topography at 150 m depth. The numbers on the contours are a measure of the total flow between the contour in question and the zero contour. The ''highs'' (H) correspond to warmer water and the ''lows'' (L) to cooler water. The three maps depict conditions at intervals of 30 d. (From Brown, J., Colling, A., Park, D., Phillips, J., Rothery, D., and Wright, J., *Ocean Circulation,* 1989, 67, with kind of permisson from Pergamon Press Ltd., Oxford OX3 OBW, U.K.)

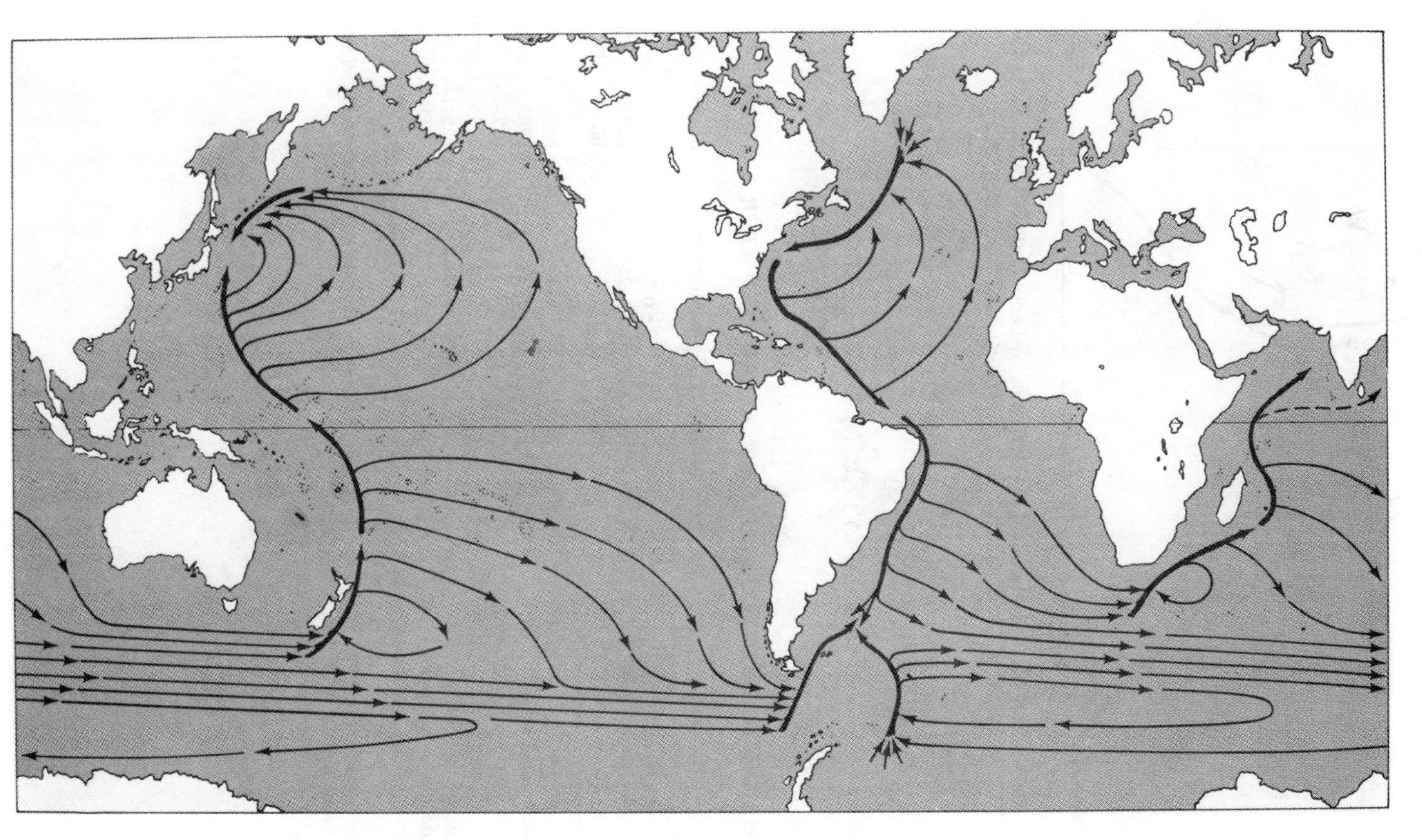

FIGURE 1.8—23. The deep ocean circulation below 2000 m as determined by Stommel (1958). The circulation is generated by sources in the North Atlantic and Weddell Sea with uniform upwelling in other areas. (From Ingmanson, D. E. and Wallace, W. J., *Oceanology: an Introduction,* 2nd ed., Wadsworth Publishing Company, Belmont, CA, 1979. With permission.)

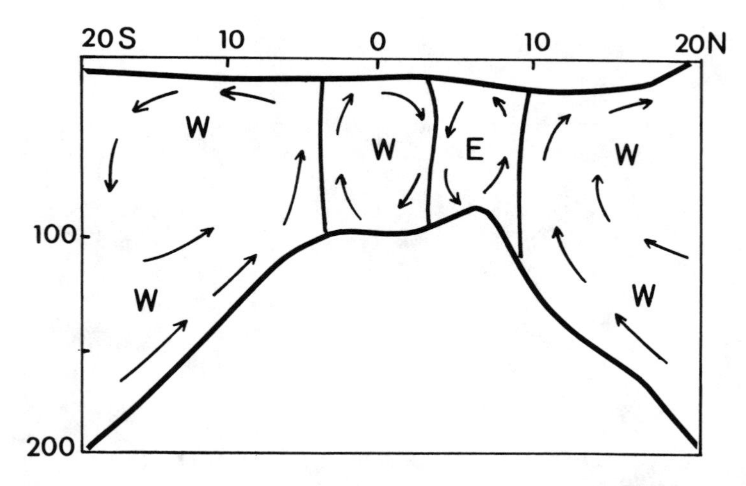

FIGURE 1.8—24. Vertical circulation in the equatorial Atlantic. (From Millero, F. J. and Sohn, M. L., *Chemical Oceanography,* CRC Press, Boca Raton, FL, 1992.)

1.9 WATER MASSES

Table 1.9—1 CHARACTERISTICS OF SEVERAL OCEANIC WATER MASSES

Water mass	Salinity (‰)	Temperature (°C)	Density (g/ml)
Antarctic bottom water	34.66	−0.4	1.02786
Antarctic circumpolar water	34.7	0–2	1.02775–1.02789
Antarctic intermediate water	34.2–34.4	3–7	1.02682–1.02743
Arctic intermediate water	34.8–34.9	3–4	1.02768–1.02783
Mediterranean water	36.5	8–17	1.02592–1.02690
North Atlantic central water	35.1–36.7	8–19	1.02630–1.02737
North Atlantic deep and bottom water	34.9	2.5–3.1	1.02781–1.02788
South Atlantic central water	34.5–36.0	6–18	1.02606–1.02719
North Pacific central water	34.2–34.9	10–18	1.02521–1.02634
North Pacific intermediate water	34.0–34.5	4–10	1.02619–1.02741
Red Sea water	35.5	9	1.02750

From Ingmanson, D. A. and Wallace, W. J., *Oceanology: An Introduction,* 2nd ed., Wadsworth Publishing Company, Belmont, CA, 1979. With permission.

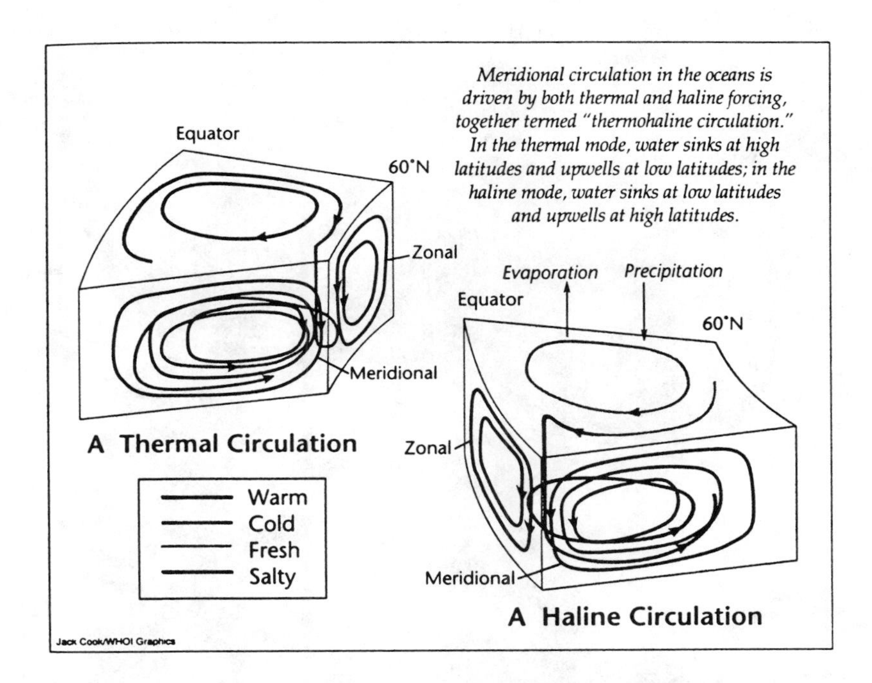

FIGURE 1.9—1. Meridional circulation in the oceans is driven by both thermal and haline forcing, together termed ''thermohaline circulation''. In the thermal mode, water sinks at high latitudes and upwells at low latitudes; in the haline mode, water sinks at low latitudes and upwells at high latitudes. (From Huang, R. X., Freshwater driving forces: a new look at an old theory, *Oceanus,* 36, 37, 1992. Courtesy of *Oceanus* Magazine/Woods Hole Oceanographic Institution.)

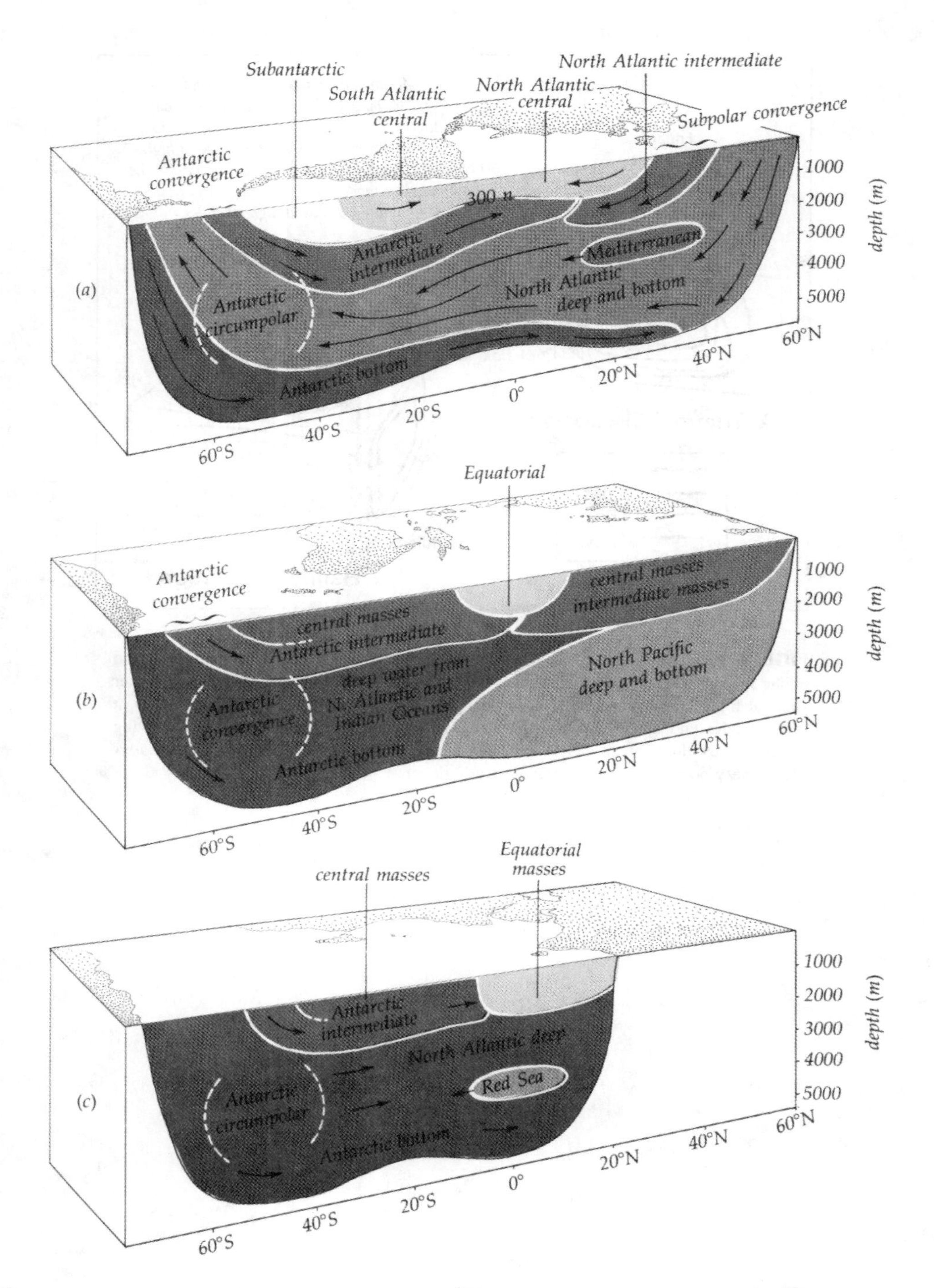

FIGURE 1.9—2. Subsurface oceanic circulation showing location of oceanic water masses. North-south cross sections of the (a) Atlantic, (b) Pacific, and (c) Indian oceans. (From Ingmanson, D. A. and Wallace, W. J., *Oceanology: an Introduction*, 2nd ed., Wadsworth Publishing Company, Belmont, CA, 1979. With permission.)

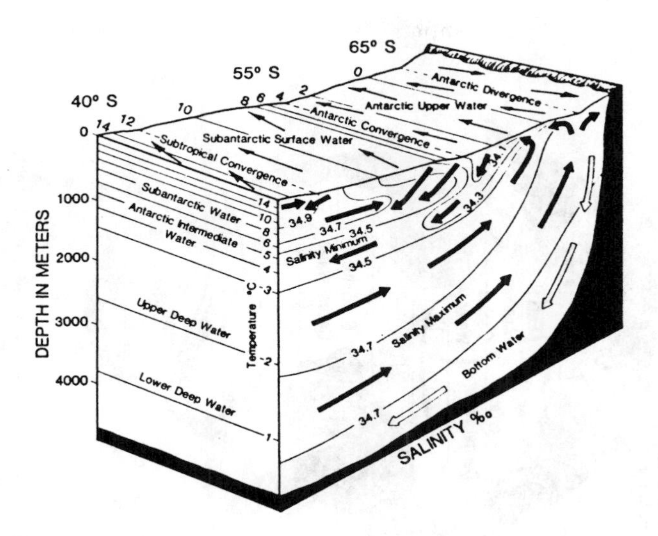

FIGURE 1.9—3. Vertical profile of the major water masses and thermohaline characteristics in the Southern Ocean including the flow of Antarctic bottom water (AABW), North Atlantic deep water (NADW), and Antarctic intermediate water (AAIW), which are predominant features of the oceans of the world. (From Berkman, P. A., The Antarctic marine ecosystem and humankind, *Rev. Aquat. Sci.*, 6, 295, 1992.)

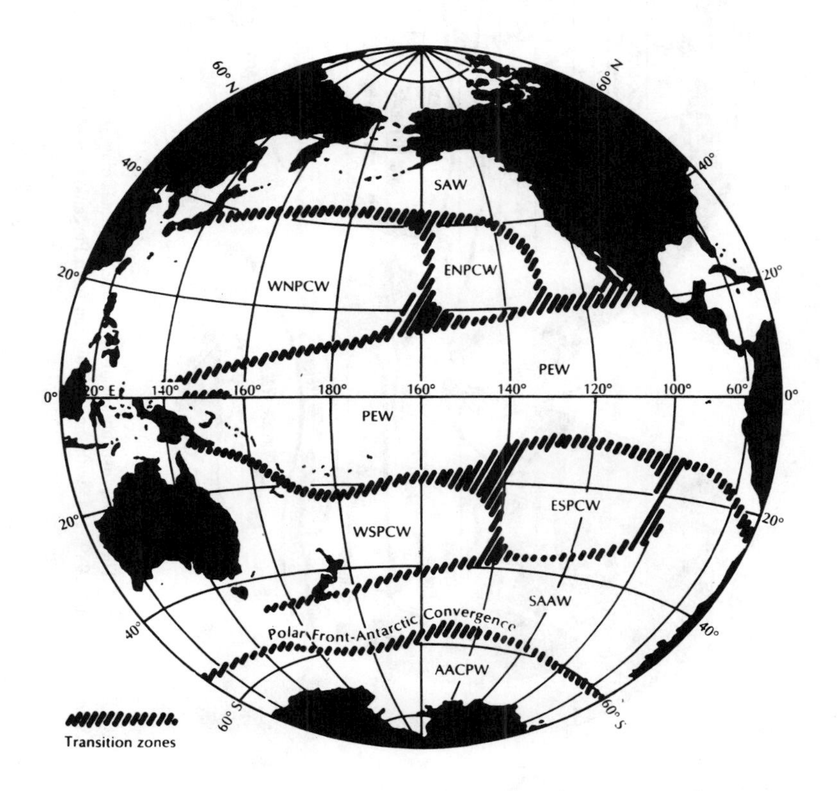

FIGURE 1.9—4. Classification of Pacific Ocean water masses. SAW, Subarctic water; WNPCW, Western North Pacific Central water; ENPCW, Eastern North Pacific Central water; PEW, Pacific Equatorial water; WSPCW, Western South Pacific Central water; ESPCW, Eastern South Pacific Central water; SAAW, Subantarctic water; AACPW, Antarctic Circumpolar water. (From Neshyba, S., *Oceanography: Perspectives on a Fluid Earth*, John Wiley & Sons, New York, 1987. With permission.)

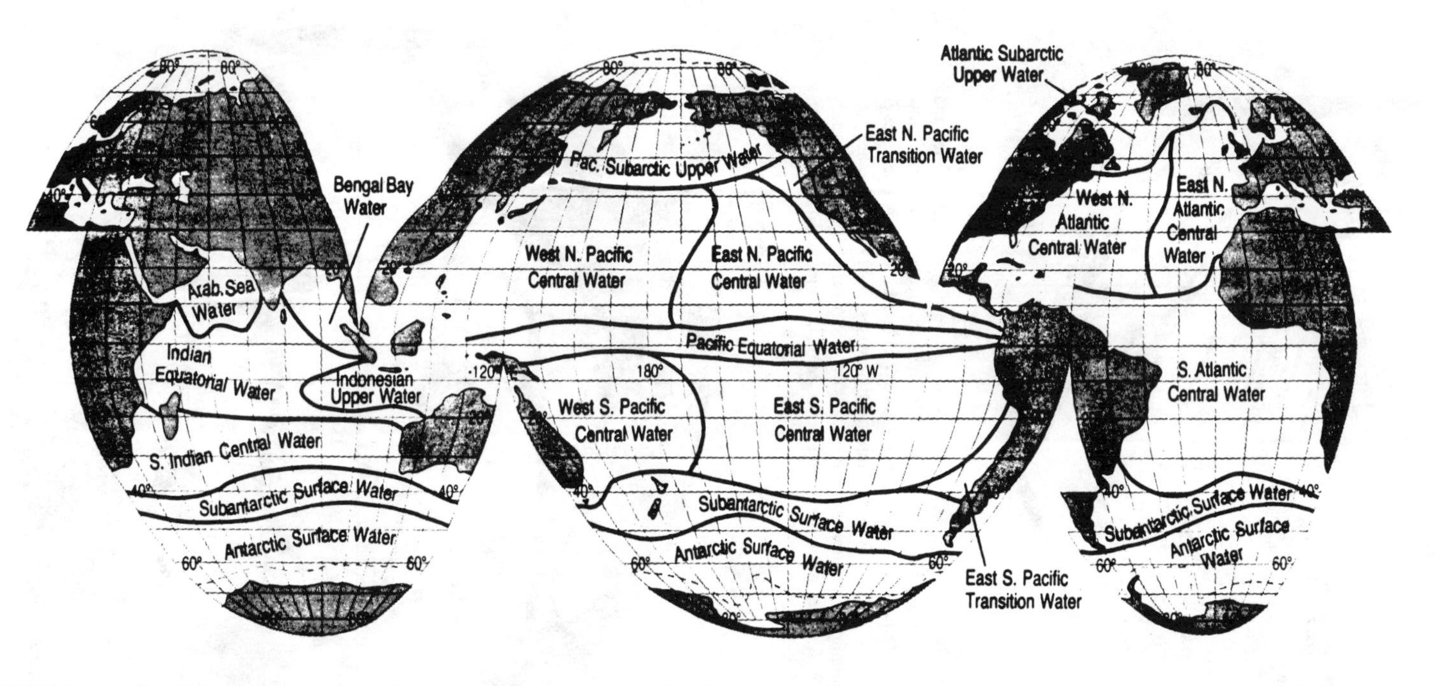

FIGURE 1.9—5. The global distribution of upper oceanic water masses. (From Brown, J., Colling, A., Park, D., Phillips, J., Rothery, D. M., and Wright, J., *Ocean Circulation*, 1989, 174, with kind permission from Pergamon Press Ltd., Headington Hill Hall, Oxford OX3 OBW, U.K.)

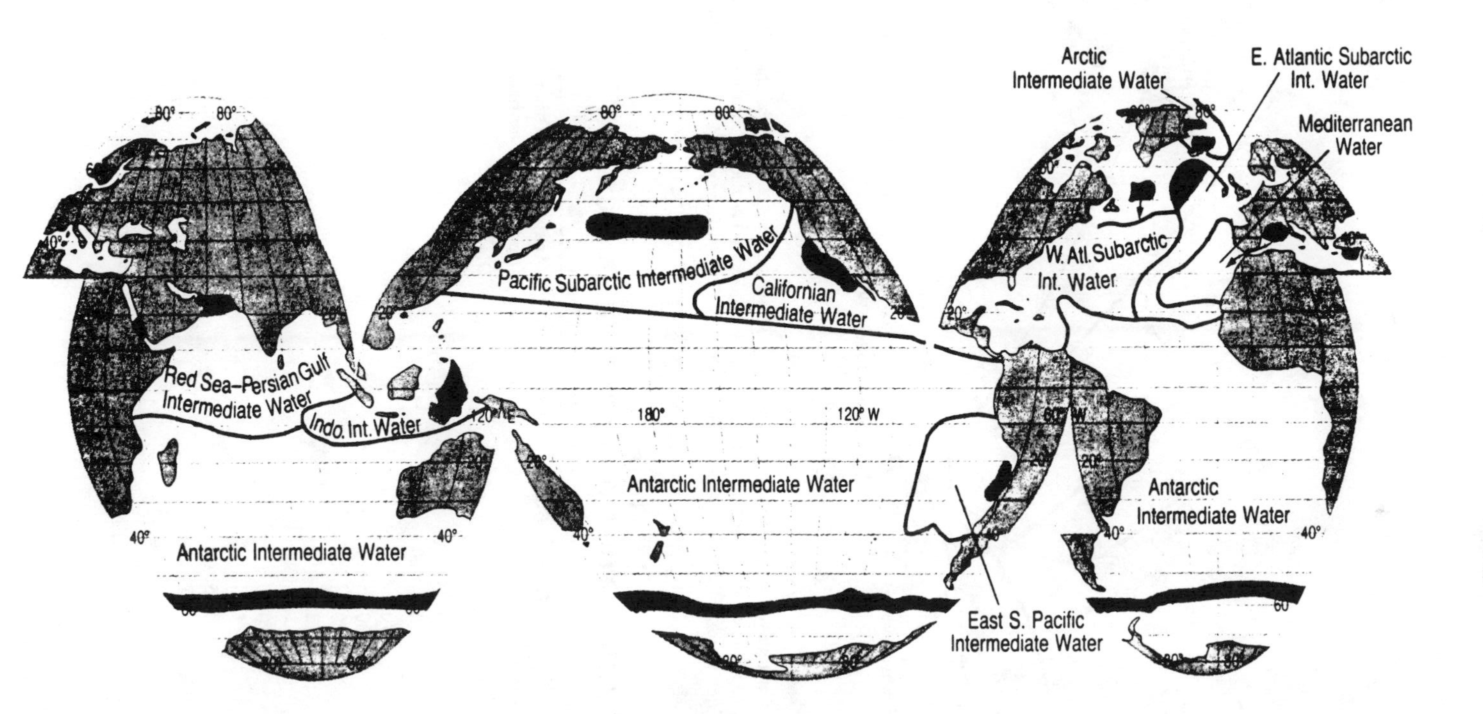

FIGURE 1.9—6. The global distribution of intermediate water masses (between about 550 and 1500 m depth). Note that Antarctic intermediate water is by far the most widespread intermediate water mass. (From Brown, J., Colling, A., Park, D., Phillips, J., Rothery, D. M., and Wright, J., *Ocean Circulation,* 1989, 175, with kind permission from Pergamon Press Ltd., Headington Hill Hall, Oxford OX3 OBW, U.K.)

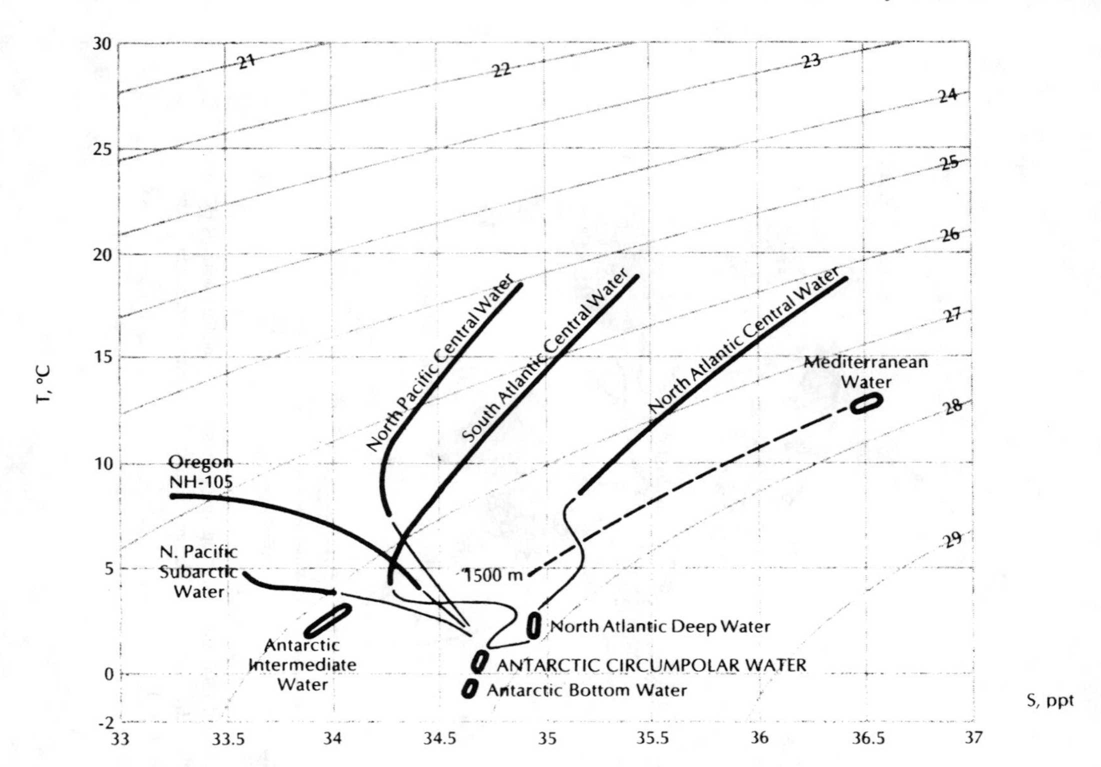

FIGURE 1.9—7. Comparison of water mass structures from different oceans. The broad lines represent the characteristics of the upper layer, from 100- to 1000-m depth. The thinner sections of the curves represent the deep-water density structure. (From Neshyba, S., *Oceanography: Perspectives on a Fluid Earth,* John Wiley & Sons, New York, 1987, 156. Reproduced by permission of copyright © owner, John Wiley & Sons, Inc.)

Table 1.9—2
WATER-TRANSPORT TRACERS IN OCEANOGRAPHY

		Origin			
Isotope[a]	**Half-life (years)**	**Cosmic rays**	**U + Th series**	**Weapons testing**	**Other anthro.**
^{14}C	5,730	√		√	
^{226}Ra	1,600		√		
^{32}Si	250	√			
^{39}Ar	270	√			
^{137}Cs	30.2			√	
^{90}Sr	28.6			√	
^{3}H	12.4	√		√	
^{3}He	—	√	√	√	
^{85}Kr	10.7			√	√
^{228}Ra	5.8		√		
^{7}Be	0.15	√			
^{222}Rn	0.01		√		
Freons	—				√
^{239}Pu	24,400			√	√
^{240}Pu	6,540			√	√
^{210}Pb	22.3		√		
^{228}Th	1.9		√		
^{210}Po	0.38		√		
^{234}Th	0.07		√		

[a] Freons and all entries above it are water tracers; all entries below freons are particulate tracers.

From Broecker, W. S., Geochemical tracers and ocean circulation, in *Evolution of Physical Oceanography,* Warren, B. A. and Wunsch, C., Eds., Massachusetts Institute of Technology, Cambridge, 1981, 434–461. With permission.

1.10 COASTAL CIRCULATION

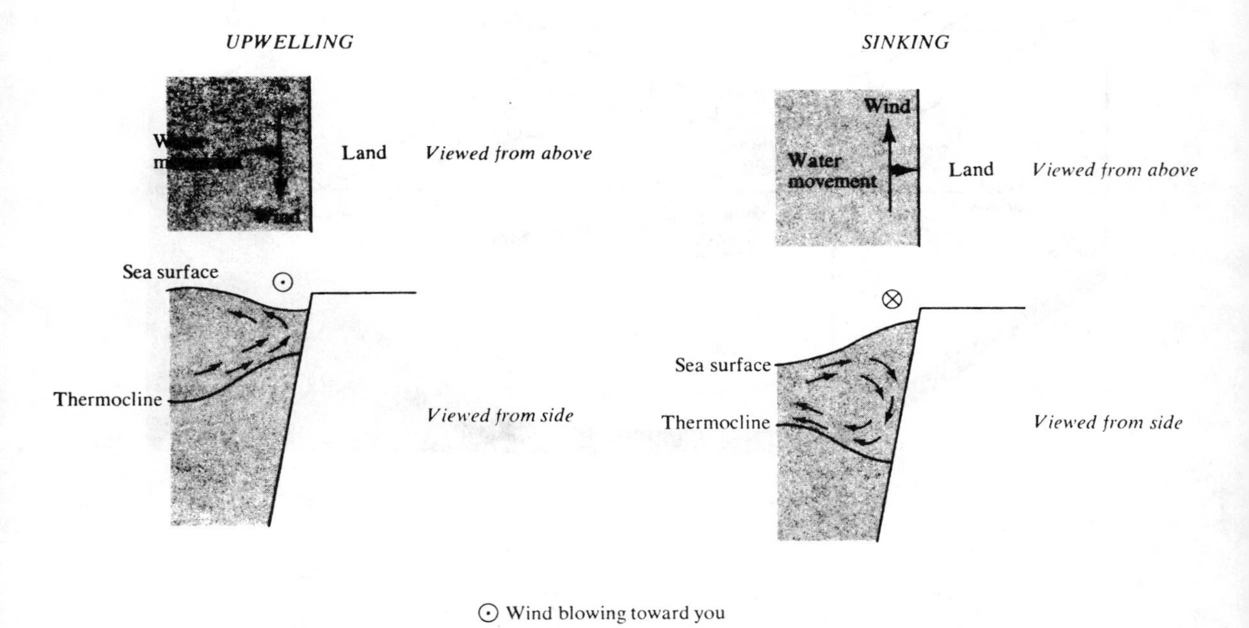

FIGURE 1.10—1. Wind-induced coastal upwelling and sinking in the northern hemisphere. The slopes of the sea surface and thermocline are greatly exaggerated. The arrows show the direction of water movement. (From Gross, M. G., *Oceanography,* 3rd ed., Charles E. Merrill Publishing Company, Columbus, OH, 1976. With permission.)

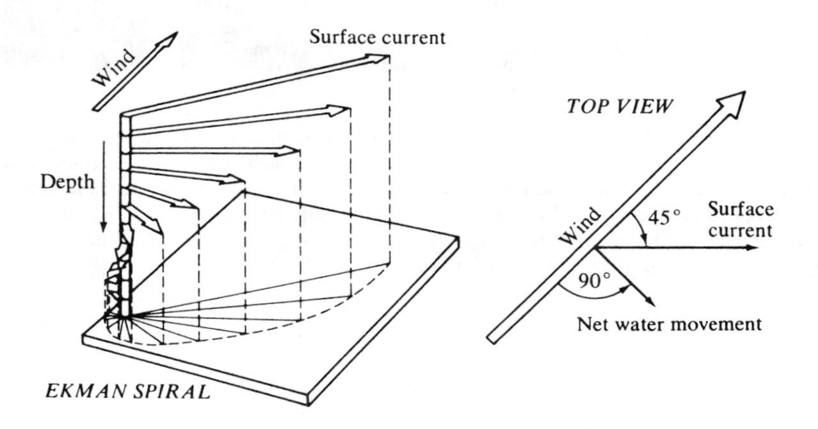

FIGURE 1.10—2. The Ekman spiral (northern hemisphere). (From Gross, M. G., *Oceanography,* 3rd ed., Charles E. Merrill Publishing Company, Columbus, OH, 1976. With permission.)

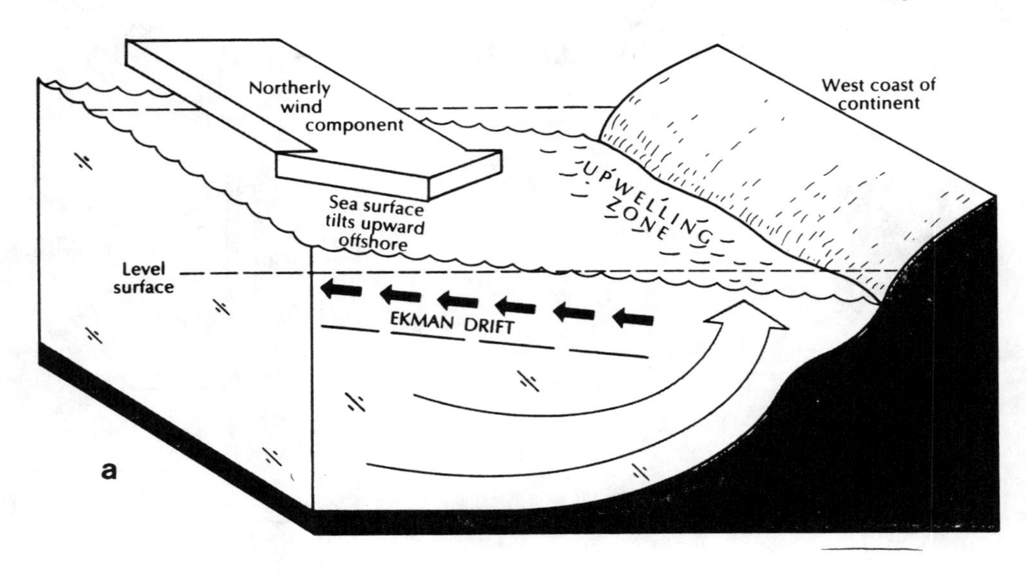

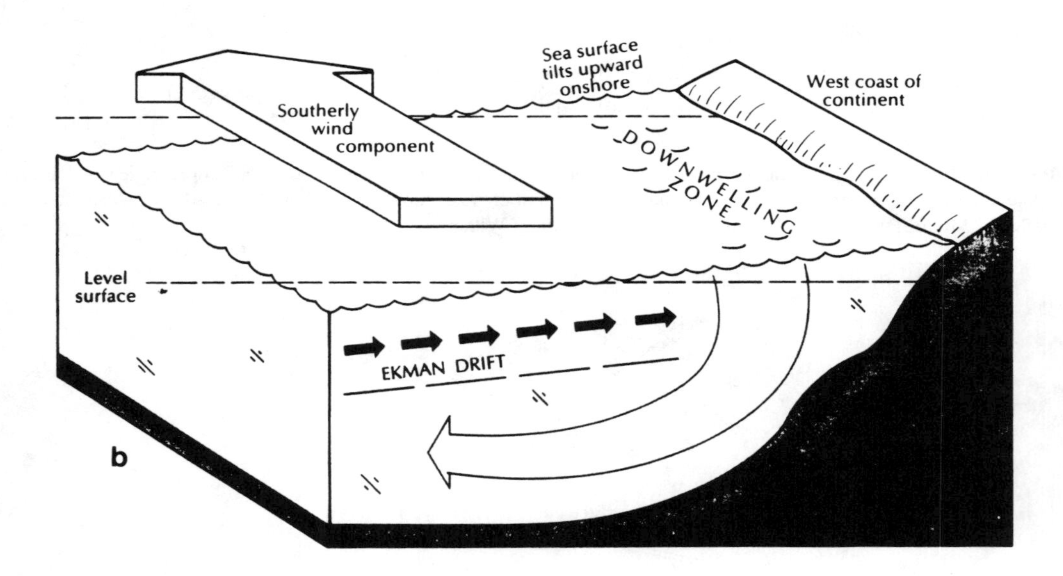

FIGURE 1.10—3. Upwelling and downwelling Ekman effects produced by surface winds along coastlines. (a) Ekman transport driven by a surface wind that has a northerly component is in the offshore direction. These are the necessary conditions for upwelling. Notice the upward tilt of sea level in the offshore direction. (b) Ekman transport is onshore, moving warm water against the coastline and creating a tilt in the sea surface near the shore. Strong winds will actually force enough water against the shore that the warm water sinks somewhat. (From Neshyba, S., *Oceanography: Perspectives on a Fluid Earth,* John Wiley & Sons, New York, 1987, 179. Reproduced by permission of copyright © owner, John Wiley & Sons, Inc.)

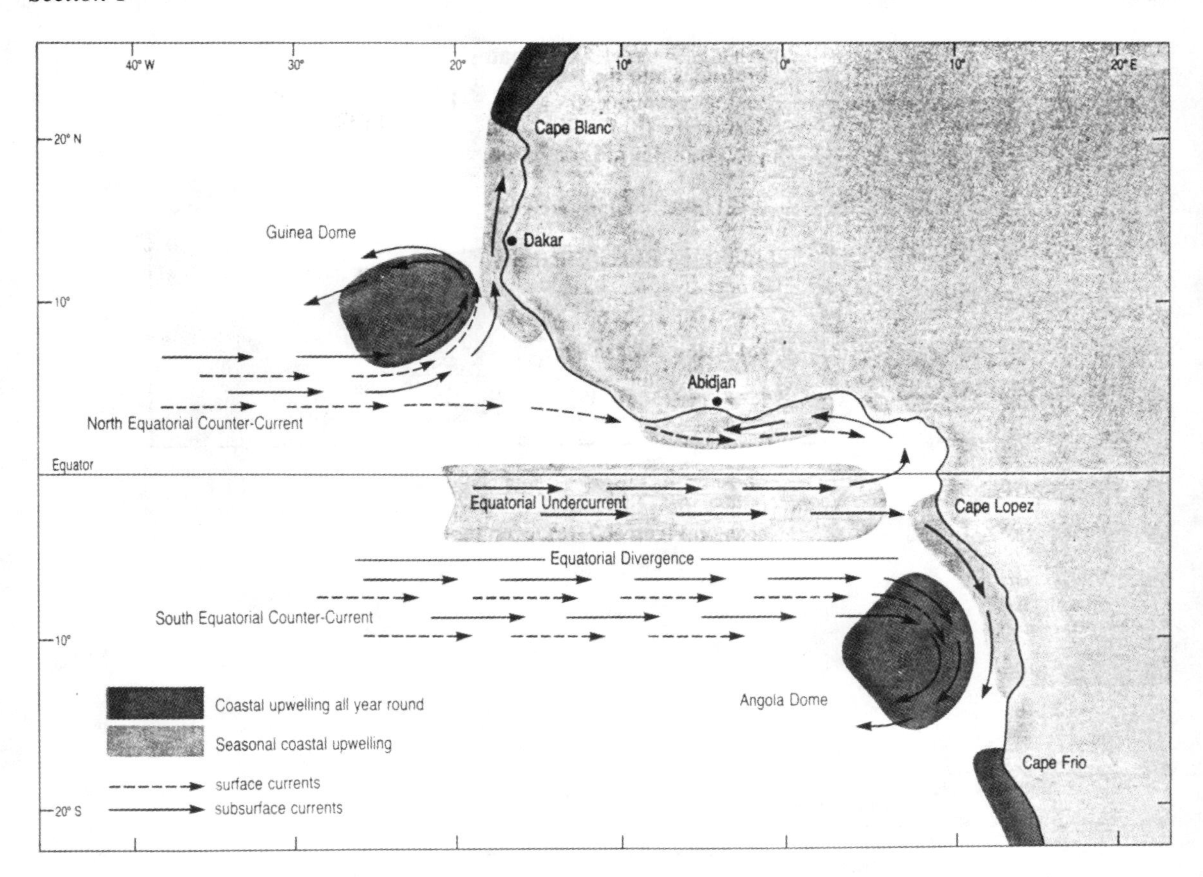

FIGURE 1.10—4. Positions of various upwelling areas in the tropical Atlantic Ocean in relation to the eastward flow in the equatorial current system. (From Brown, J., Colling, A., Park, D., Phillips, J., Rothery, D., and Wright, J., *Ocean Circulation,* 1989, 131, with kind permission from Pergamon Press Ltd., Headington Hill Hall, Oxford OX3 OBW, U.K.)

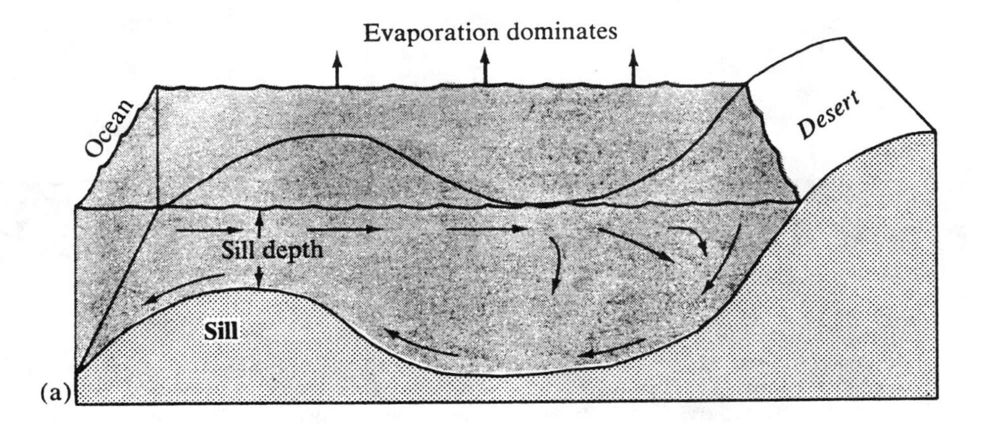

FIGURE 1.10—5. Water circulation in silled basins where (a) evaporation exceeds precipitation and runoff and (b) runoff and precipitation exceed evaporation. (From Gross, M. G., *Oceanography,* 3rd ed., Charles E. Merrill Publishing Company, Columbus, OH, 1976. With permission.)

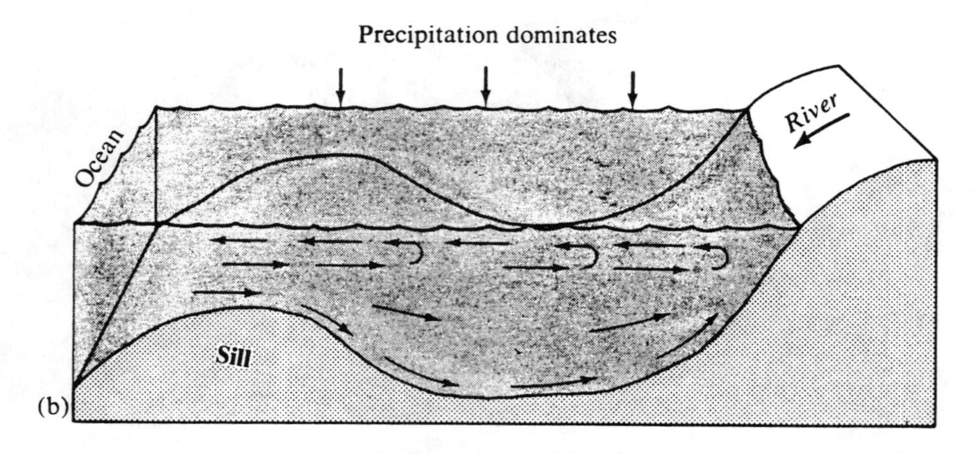

FIGURE 1.10—5 (continued).

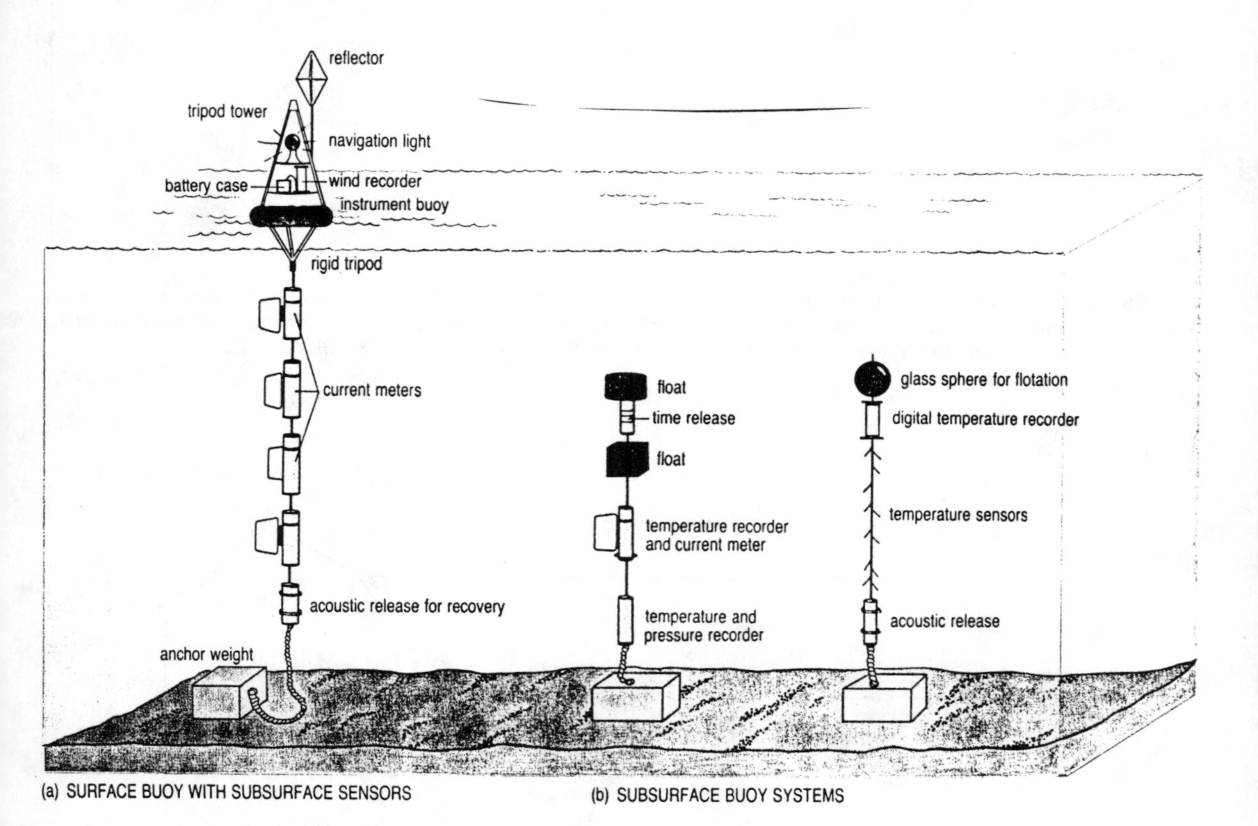

FIGURE 1.10—6. Moored current meters (not to scale) in shallow coastal waters showing how the mooring can be used to accommodate equipment for measuring other variables of interest, such as wind speed, water temperature, etc. (From Brown, J., Colling, A., Park, D., Phillips, J., Rothery, D., and Wright, J., *Ocean Circulation,* 1989, 104, with kind permission from Pergamon Press Ltd., Headington Hill Hall, Oxford OX3 OBW, U.K.)

SHELF CURRENTS

TIME SCALES OF VARIATION	WIND DRIVEN	WAVES	TIDES	DENSITY DRIVEN	DEEP OCEAN
Secs		Orbital Velocity			
Hours	Surges	Longshore and Rip Currents	Tidal Currents	Internal Tides	
Days Weeks	Upwelling Shelf Waves	Drift Velocity			
Months	Circulation		Spring-Neaps	Estuarine Circulation	Rings Eddies
Seasonal	Seasonal Patterns	Hurricanes Winter Storms	Low frequency variability	Seasonal Run-off.	Ocean Currents
Annual	Winter	Ice		Insolation	
Long Term	Extreme Storm	Conditions			Oceanic Circulation

FIGURE 1.10—7. Contributions to the velocity field on a continental shelf. (From Davies, A. M., Ed., *Modeling Marine Systems,* Vol. 1, CRC Press, Boca Raton, FL, 1990.)

1.11 TIDES AND TIDE-GENERATING FORCES

Table 1.11—1
MAIN CONSTITUENTS OF TIDE-GENERATING FORCES

Constituent	Symbol	Period (hr)	Relative amplitude	Description
Semidaily tides	M_2	12.4	100.00	Main lunar constituent
	S_2	12.0	46.6	Main solar constituent
	N_2	12.7	19.2	Lunar constituent due to changing distance between earth and moon
Daily tides	K_1	23.9	58.4	Solilunar constituent
	O_1	25.8	41.5	Main lunar constituent
	P_1	24.1	19.3	Main solar constituent
Long-period tides (spring and neap tides)	M_1	327.9	17.2	Main fortnightly constituent

From Defant, A., *Ebb and Flow: The Tides of Earth, Air, and Water*, University of Michigan Press, Ann Arbor, 1958. With permission.

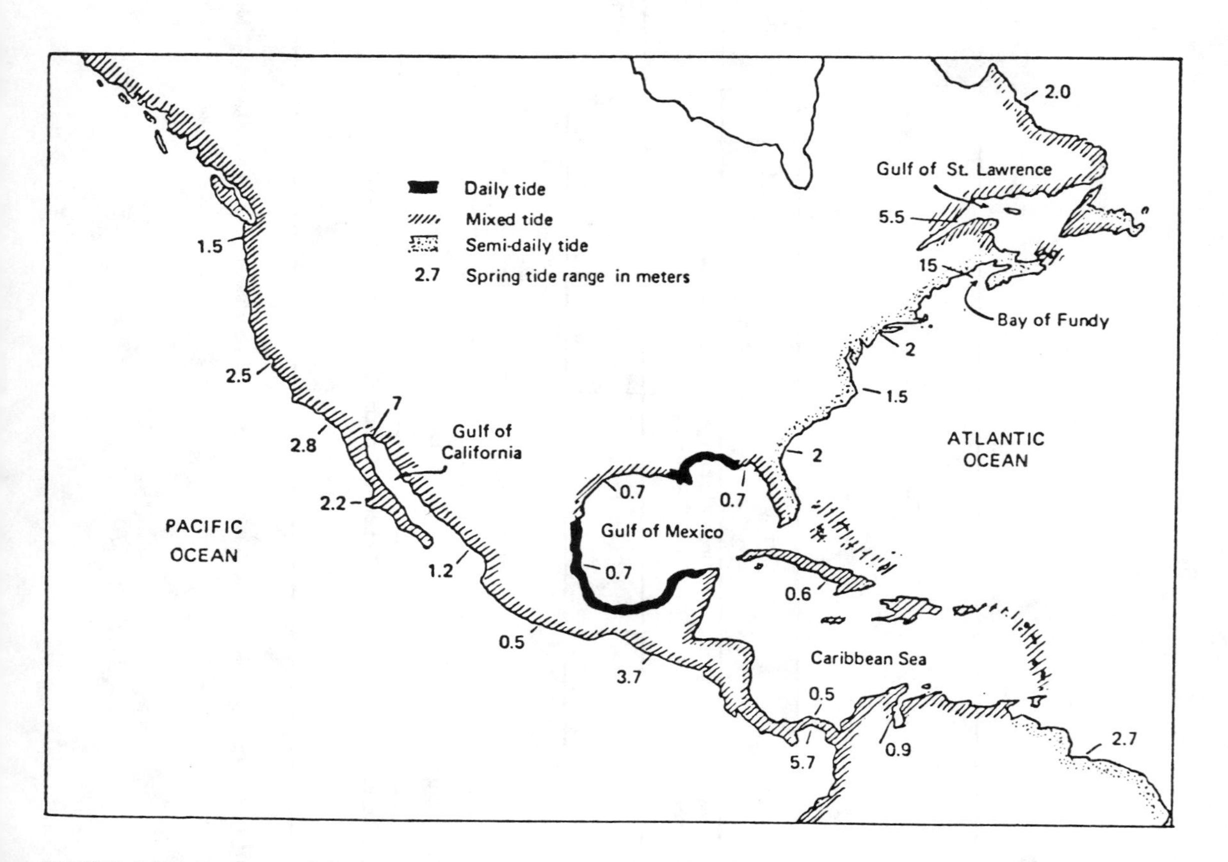

FIGURE 1.11—1. Types of tides and ranges of spring tides (in meters) on North American coastlines. (From U.S. Naval Oceanographic Office, Oceanographic Atlas of the North Atlantic Ocean, Section 1: Tides and Currents, H. O. Publ. 700, Washington, D.C., 1968. With permission of NAVOCEANO.)

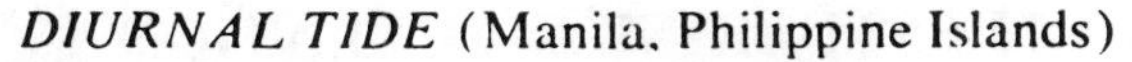

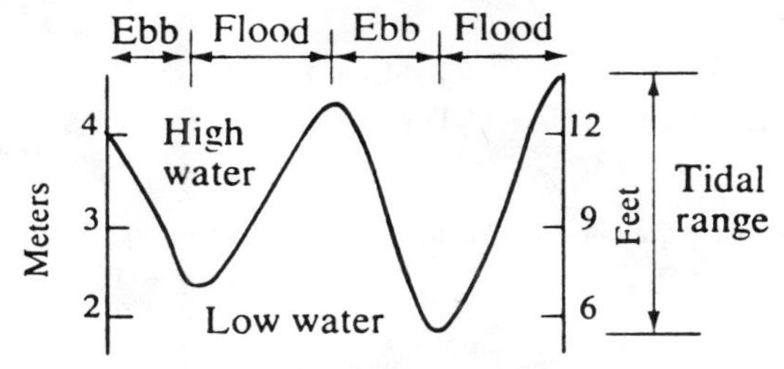

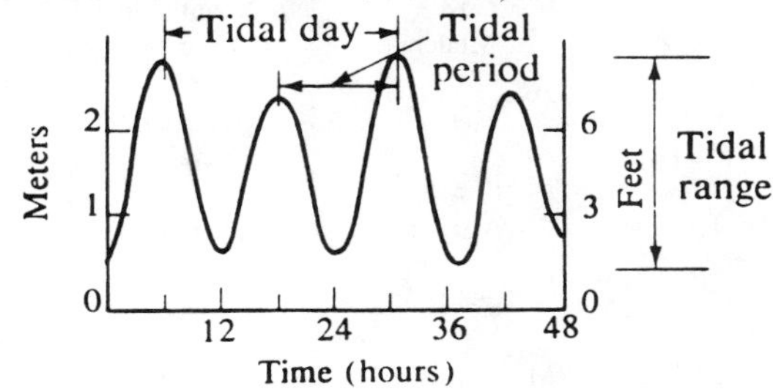

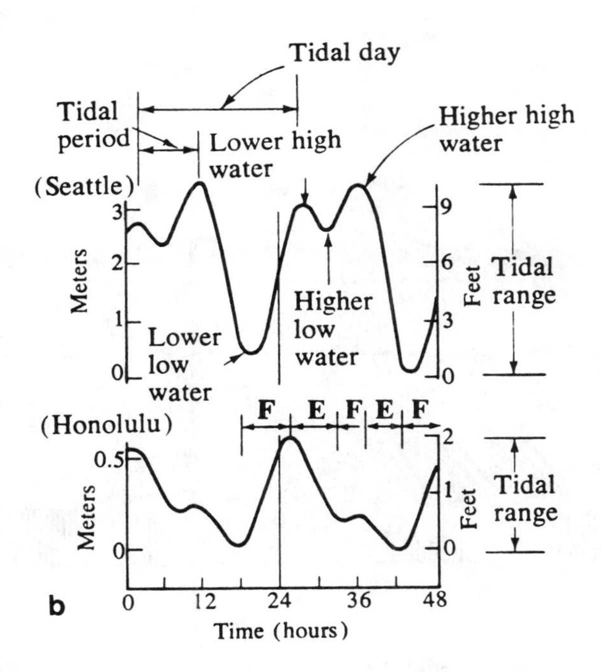

FIGURE 1.11—2. Diurnal, semidiurnal (a), and mixed type of tides (b). F, flood current; E, ebb current. (From Gross, M. G., *Oceanography,* 3rd ed., Charles E. Merrill Publishing Company, Columbus, OH, 1976. With permission.)

1.12 ESTUARINE CIRCULATION

Table 1.12—1
GENERAL ESTUARINE PHYSICAL CHARACTERISTICS

Estuarine type	Dominant mixing force	Width/depth ratio	Salinity gradient	Probable topographic categories	Example
A Highly stratified	River flow	Low	Longitudinal Vertical	Fluvial Deltaic	Columbia River Mississippi River
B Moderately stratified	River flow Tide Wind	Low Moderate	Longitudinal Vertical Lateral	Fjord Coastal plain Bar built	Puget Sound Chesapeake Bay Albemarle Sound
C Vertically homogeneous	Tide, wind	High	Longitudinal Lateral	Bar built (lagoon) Coastal plain	Baffin Bay Delaware Bay
D Vertically and laterally homogeneous	Tide, wind	Very high	Longitudinal	?	?

From Biggs, R. B., DeMoss, T. B., Carter, M. M., and Beasley, E. L., Susceptibility of U.S. estuaries to pollution, *Rev. Aquat. Sci.*, 1, 189, 1989. With permission.

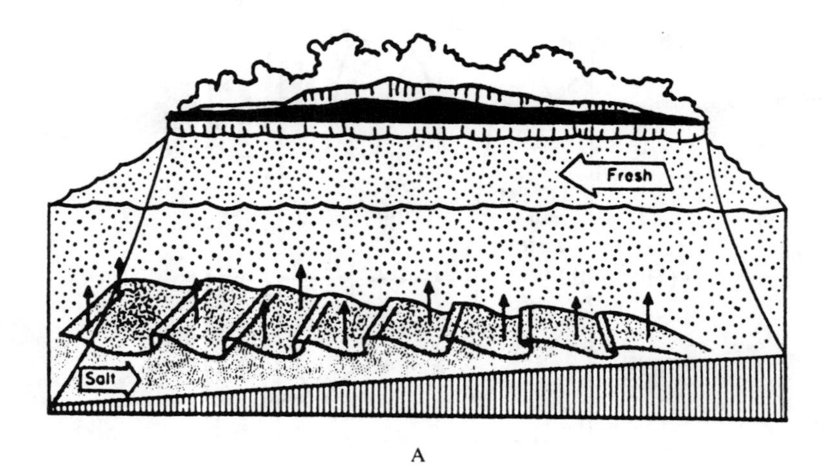

A

FIGURE 1.12—1. Classification of estuaries based on circulation. (A) Highly stratified, salt wedge estuary, type A; (B) partially mixed, moderately stratified estuary, type B; (C) completely mixed, vertically homogeneous estuary with lateral salinity gradient, type C; (D) completely mixed, sectionally homogeneous estuary with longitudinal salinity gradient, type D. (From Pritchard, D. W., *Proc. Am. Soc. Civil Eng.*, 81, 1, 1955. With permission.)

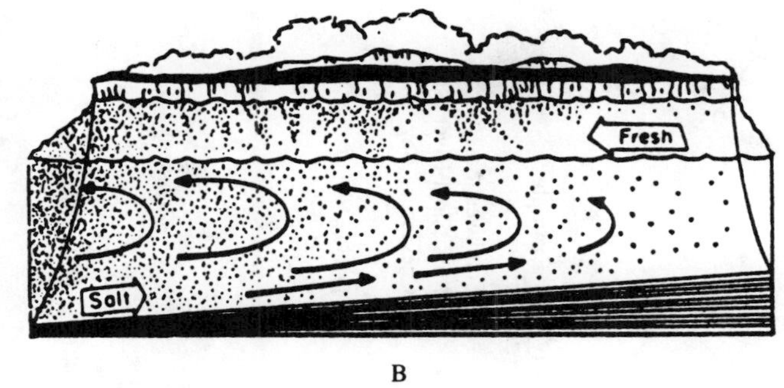

B

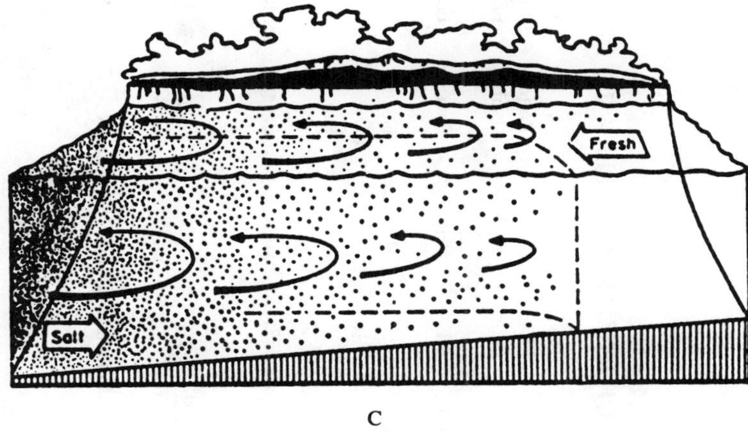

C

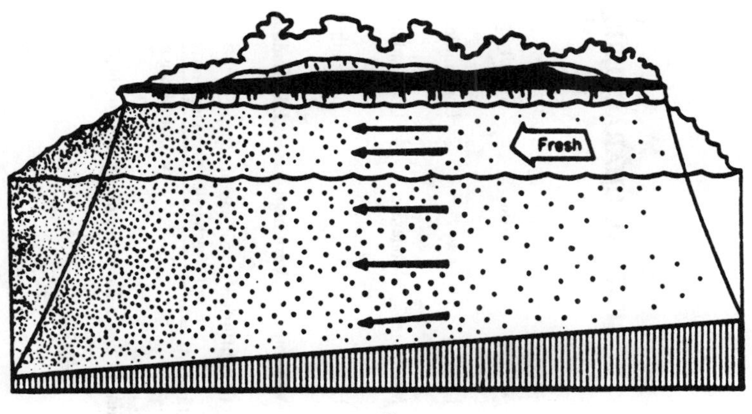

D

FIGURE 1.12—1 (continued).

Table 1.12—2
GENERAL CHARACTERISTICS OF ESTUARINE TYPES A TO D

Estuarine type[a]	Dominant mixing force	Mixing energy	Width/depth ratio	Salinity gradient	Mixing index[b]	Turbidity	Bottom stability	Biological productivity	Example
A	River flow	Low	Low	Longitudinal, vertical	$\geqslant 1$	Very high	Poor	Low	Southwest Pass, Mississippi River
B	River flow, tide	Moderate	Moderate	Longitudinal, vertical, lateral	$<\frac{1}{10}$	Moderate	Good	Very high	Chesapeake Bay
C	Tide, wind	High	High	Longitudinal, lateral	$<\frac{1}{20}$	High	Fair	High	Delaware Bay
D	Tide, wind	Very high	Very high	Longitudinal	?	High	Poor	Moderate	?

[a] Follows Pritchard's advection-diffusion classification scheme.[37]

[b] Follows Schubel's definition: $MI = \dfrac{\text{vol. freshwater discharge on 1/2 tidal period}^{48}}{\text{vol. tidal prism}}$

From Biggs, R. B. and Cronin, L. E., in *Estuaries and Nutrients*, Neilson, B. J. and Cronin, L. E., Eds., Humana Press, Clifton, N. J., 1981, 3. With permission.

Table 1.12—3
CHARACTERISTICS OF ESTUARINE CIRCULATION AND WATER QUALITY FOR DIFFERENT ENVIRONMENTS OF THE COASTAL BIOPHYSICAL REGIONS OF THE U.S.

Biophysical region	Smooth shoreline	Indented shoreline	Marshy shoreline	Unrestricted river entrance	Embayment, coastal drainage only	Embayment, continuous upland river flow
North Atlantic	Deep nearshore, oceanic water, longshore currents some suspended sand and clay	Deep nearshore, oceanic water, erratic tidal currents, eddies and tidal pools	Strong currents in many small channels through marsh, some turbidity, high oxygen	Highly stratified, some turbidity, high oxygen, temperatures warmer in summer, colder in winter than ocean	Little turbidity, water of oceanic character, strong tidal currents through inlets	Little turbidity, high oxygen, may be stratified, upper layer fresh, with temperatures warmer in summer, colder in winter than ocean
Middle Atlantic	Oceanic water, longshore currents, suspended mud, clay, silt	Generally shallow, suspended mud and sand, oceanic water	Moderate currents in well-defined channels, high dissolved organic material, little turbidity, high oxygen	Moderate stratification, suspended mud and silt, high oxygen, strong currents	Generally shallow, small tides, clear water with lowered salinity, high oxygen	Variable stratification, suspended mud and silt, high oxygen, small amounts of organic material
Chesapeake	Longshore tidal currents, highly variable salinities, small amounts of organic material	Moderate tidal currents, highly variable salinities, some turbidity	Poorly defined channels, small currents, dissolved organic material, moderate fluctuations of oxygen	Moderate stratification, suspended mud and silt, high oxygen, strong currents	Generally shallow, small tides, clear water with lowered salinity, high oxygen	Variable stratification, suspended mud and silt, high oxygen, small amounts of organic material
South Atlantic	Primarily tidal and and wave-induced currents, oceanic water with mud, clay, and silt	Moderate tidal currents, highly variable salinities, some turbidity	Small currents, high color, high dissolved organics, highly variable oxygen, high temperatures	Strong stratification, high suspended mud and clay, strong currents, dissolved organics, moderate oxygen	Some color, small currents, generally shallow, high dissolved organics, highly fluctuating oxygen	Slight and variable stratification, river water coolor than ocean, slight color, some oxygen fluctuation, moderate to high suspended sediment

Table 1.12—3 (continued)
CHARACTERISTICS OF ESTUARINE CIRCULATION AND WATER QUALITY FOR DIFFERENT ENVIRONMENTS OF THE COASTAL BIOPHYSICAL REGIONS OF THE U.S.

Biophysical region	Smooth shoreline	Indented shoreline	Marshy shoreline	Unrestricted river entrance	Embayment, coastal drainage only	Embayment, continuous upland river flow	Fjord
Caribbean	Clear oceanic water, gentle currents, warm temperature throughout the year	Clear oceanic water, gentle currents, eddies, warmer than ocean	High dissolved organics, high color, suspended mud, very small currents, hot	Slightly turbid, strong currents, river coolor than ocean water	Very small currents, generally shallow, quite warm, clear oceanic water	Slightly turbid, eddying currents, slight stratification, high oxygen	
Gulf of Mexico	Clear, generally warm oceanic water, longshore currents	Very small currents, oceanic water with slight turbidity, warmer than ocean	High dissolved organics, high color, very small currents, slightly to moderately turbid, high temperature	Slightly turbid, strong currents, river cooler than ocean water	Very small currents except in inlet, shallow, warm, slight turbidity from sand and silt, highly fluctuating oxygen	Slight and variable stratification, river water cooler than ocean, some oxygen fluctuation	
Southwest Pacific	Strong wave action, cool oceanic water, some silt and clay turbidity	Moderate suspended solids, erratic currents, high oxygen, cool	High suspended solids, erratic tidal currents, warmer than ocean and rivers	Strong stratification, offshore bar formation, cool, high oxygen	Some suspended silt, erratic currents, cool, high oxygen	Moderate to strong stratification, high suspended silt, strong currents, high oxygen, cool	
Northwest Pacific	Strong wave action, cold oceanic water, some silt and clay turbidity	Moderate suspended solids, erratic currents, high oxygen, cold	High suspended solids, erratic tidal currents, warmer than ocean and rivers	Strong stratification, offshore bar formation, cold, high oxygen	Some suspended silt, erratic currents, cold, high oxygen	Moderate to strong stratification, high suspended silt, strong currents, high oxygen, cold	
Alaska	Very cold oceanic water, usually ice, salinities slightly depressed	Very cold oceanic water, overlain by some freshwater, high oxygen	Very cold water, variable salinities, much fine silt, debris from freezing	Strong currents, high suspended solids, frequently glacial in origin, very cold	Very cold organic water, much ice, surface layer of freshwater, high oxygen	High turbidity with glacial debris, seasonal freeze-ups, strong currents during runoffs	Stagnant below sill depth, very little oxygen, high salinity, hydrogen sulfide
Pacific Islands	Clear, warm oceanic water, strong wave action	Clear oceanic water, gentle currents, eddies, warmer than ocean	High dissolved organics, color, high suspended mud, very small currents, hot	Slightly turbid, strong currents, river cooler than ocean water	Very small currents, generally shallow, quite warm, clear oceanic water	Slightly turbid, eddying currents, slight stratification, high oxygen	

From U.S. Department of Interior, The National Estuarine Pollution Study, Vol. 4, Federal Water Pollution Control Administration, Washington, D.C., 1969.

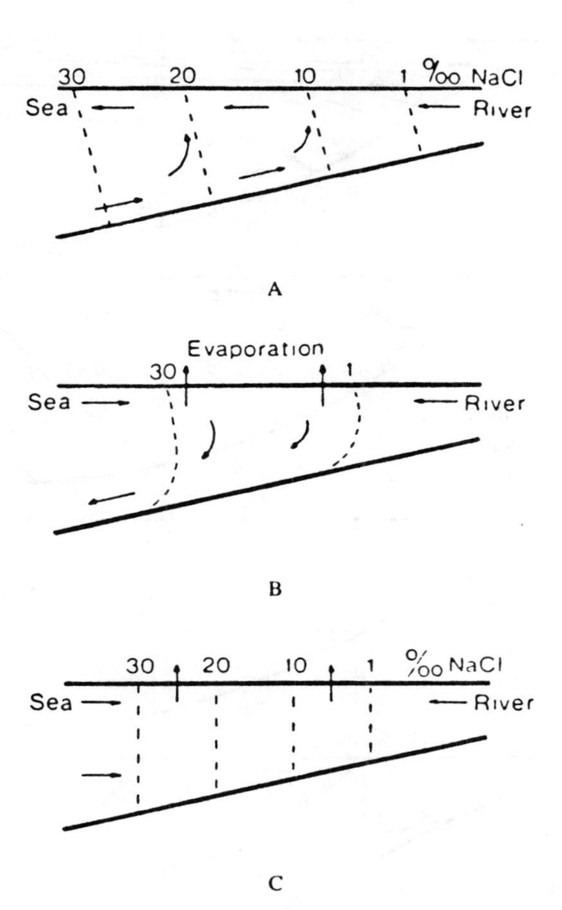

FIGURE 1.12—2. Classification of estuaries based on salinity distribution. (A) Positive estuary in which fresh water inflow and precipitation exceed evaporation. Arrows indicate circulation pattern with denser seawater entering along the bottom of the estuary and mixing upward with outgoing fresh water. (B) Negative estuary in which freshwater inflow and precipitation are less than evaporation. Both seawater and fresh water enter on the surface. Evaporation produces high-salinity water that sinks and flows out of the estuary as a bottom current. (C) Neutral estuary in which fresh water inflow and precipitation equal evaporation, generating a static salinity regime. (Reproduced with permission from McLusky, D. S., *The Estuarine Ecosystem,* Blackie Academic & Professional, an imprint of Chapman & Hall, 1981.

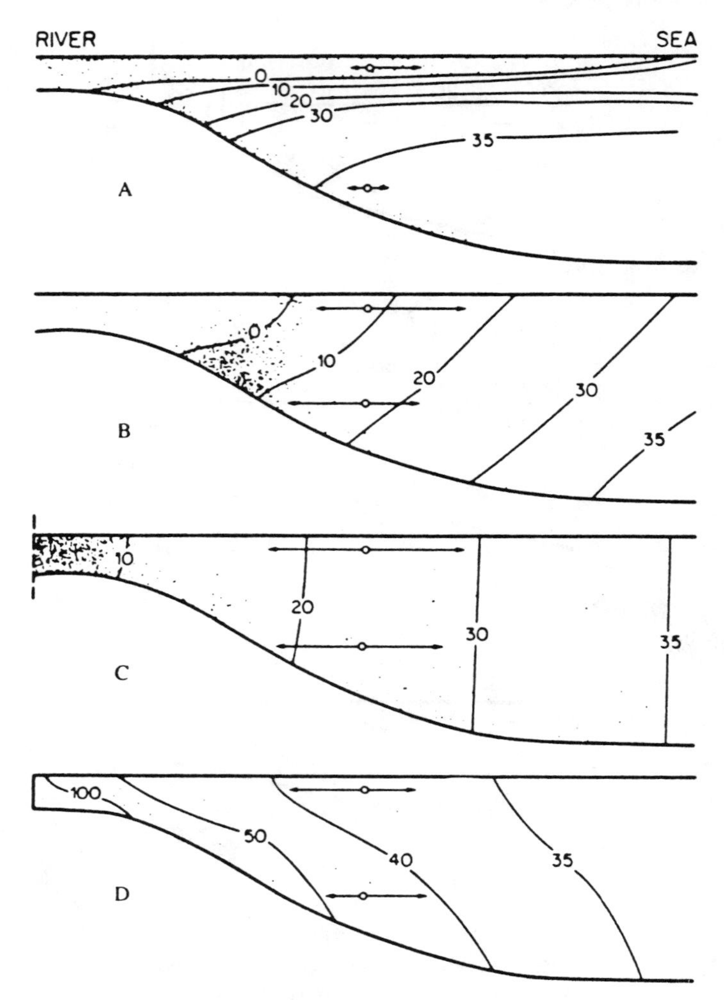

FIGURE 1.12—3. Schematic representation of types of estuaries. The dots indicate sediment concentration, and the arrows net water movements over ebb and flood. (A) Salt wedge estuary: most river sediment carried to the ocean in the upper layer; (B) partially mixed estuary: suspended sediment is concentrated in a turbidity maximum; (C) completely mixed estuary: suspended sediment concentrated near shore; and (D) negative estuary: net upestuary flow at the surface and downestuary flow near the bottom attributable to excessive evaporation and high salinities at the head. (From Postma, H., in *Chemistry and Biogeochemistry of Estuaries,* Olausson, E. and Cato, I., Eds., John Wiley & Sons, Chichester, 1980, 153. Reproduced by permission of copyright © owner, John Wiley & Sons, Ltd.)

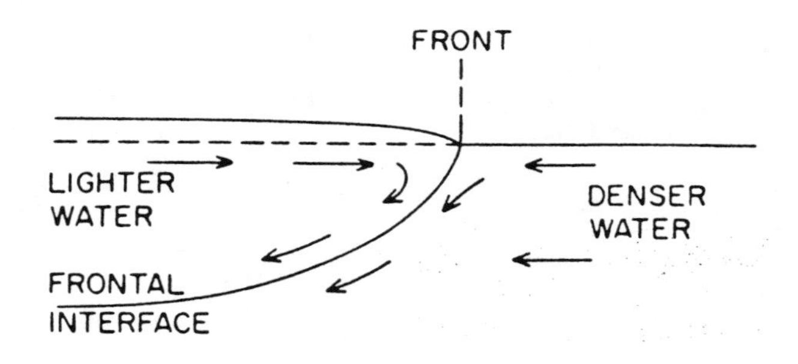

FIGURE 1.12—4. Schematic diagram of water movements at an estuarine front. (From Garvine, R. W., *J. Phys. Oceanogr.*, 4, 557, 1974. With permission.)

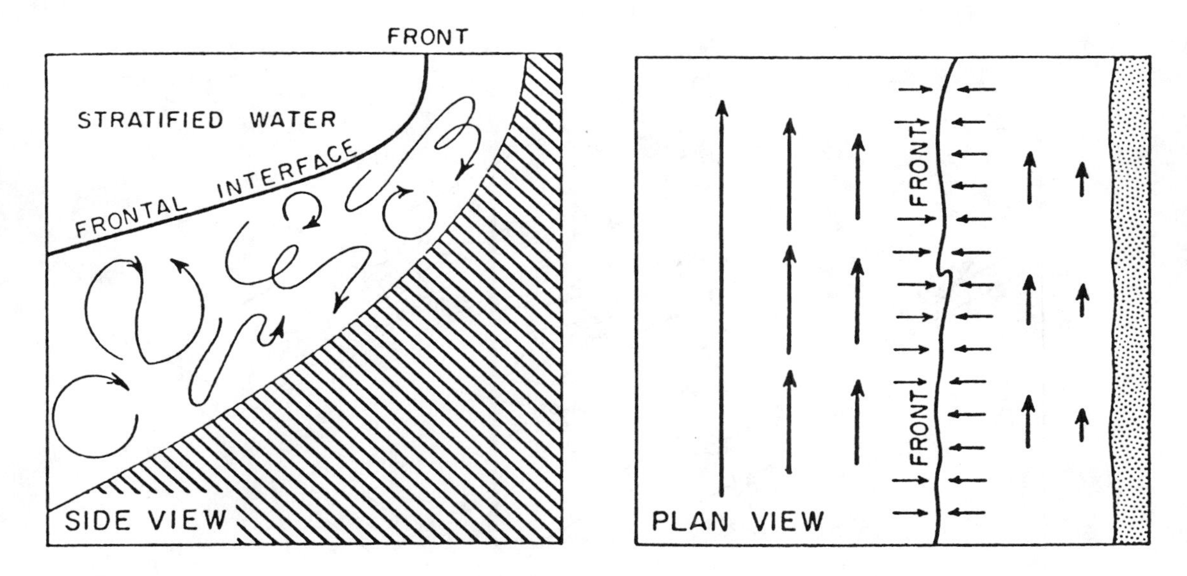

FIGURE 1.12—5. Schematic diagram of an estuarine frontal zone. The wiggly lines represent random turbulent motions and the straight lines, tidal currents. (From Kjerfve, B., Ed., *Hydrodynamics of Estuaries,* Vol. 1, Estuarine Physics, CRC Press, Boca Raton, FL, 1988.)

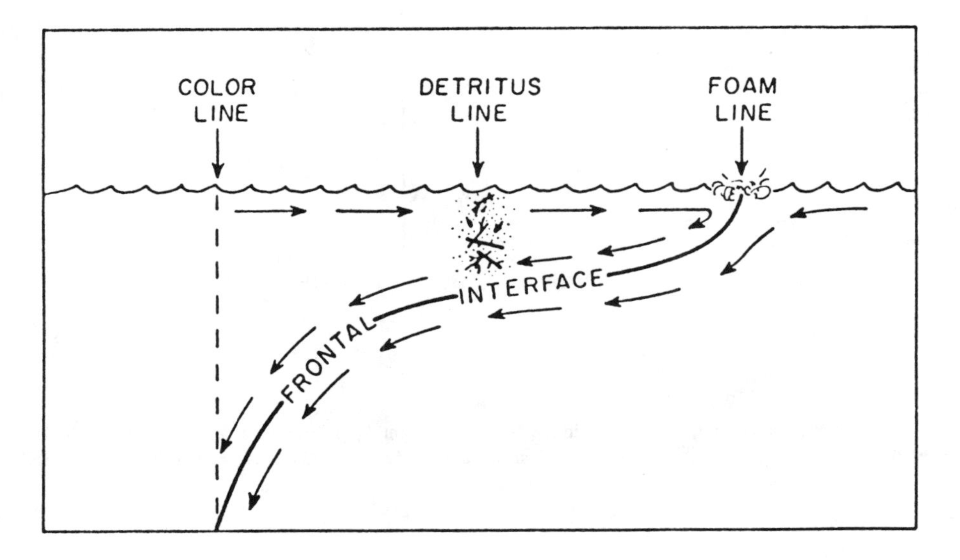

FIGURE 1.12—6. Schematic cross section of a shallow front. Three boundaries are often visible:

1. The color front, perceived to lie where the depth integrated upwelled light undergoes a distinct spectral shift in the region of rapidly descending isopycnals.
2. The detritus line, where large buoyant objects are trapped by oppositely directed currents at the surface and near the frontal interface.
3. The foam line, which is located at the surface convergence. Since the frontal slope may be $\sim 10^{-2}$, the three demarcations can be separated by several tens of meters.

(From Kjerfve, B., Ed., *Hydrodynamics of Estuaries,* Vol. 1, Estuarine Physics, CRC Press, Boca Raton, FL, 1988.)

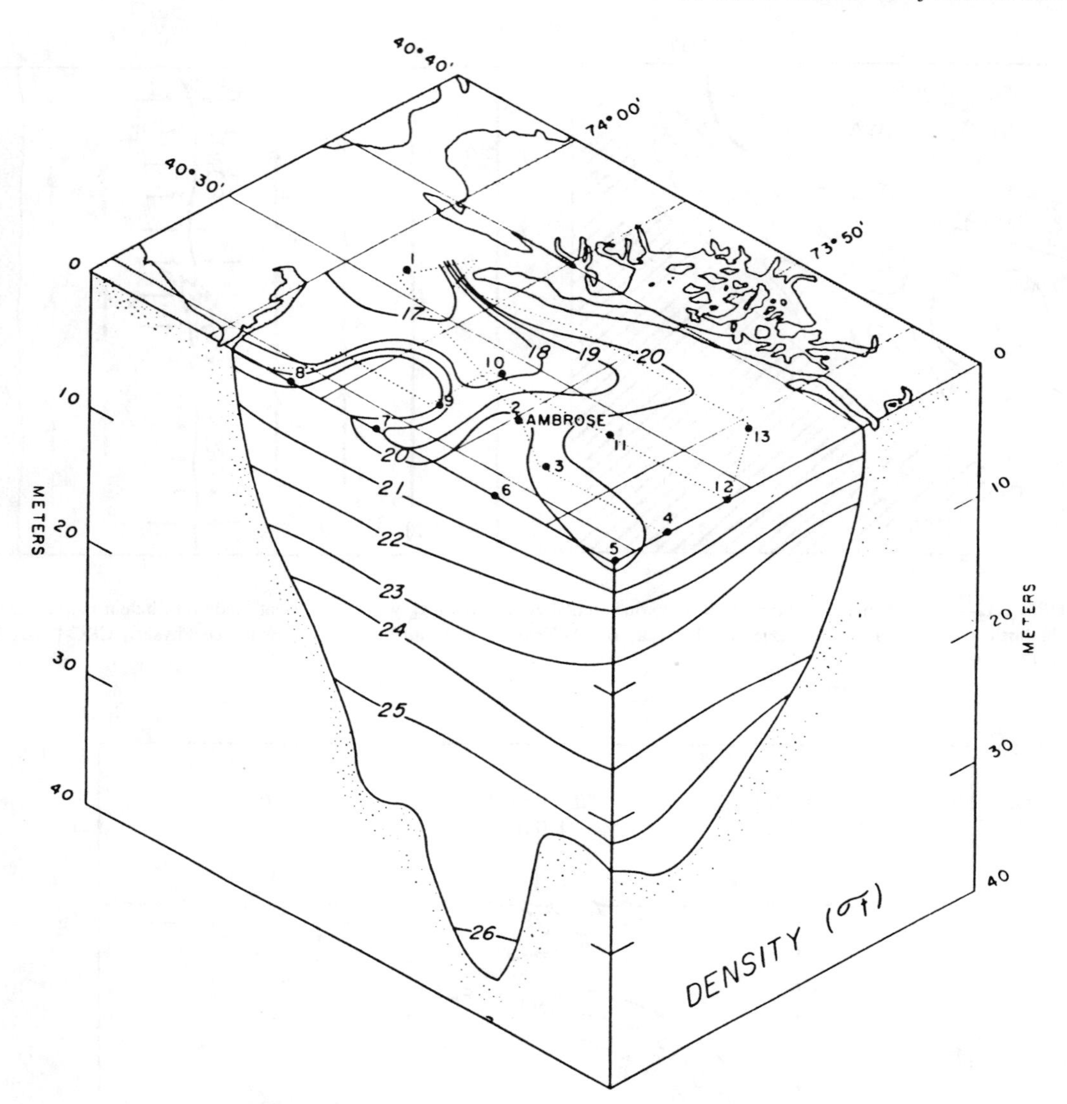

FIGURE 1.12—7. Isometric diagram of density (σ_t) in the New York Bight Apex, August 13, 1976, showing the Hudson Estuary plume with its associated fronts. (From Kjerfve, B., Ed., *Hydrodynamics of Estuaries,* Vol. 1, Estuarine Physics, CRC Press, Boca Raton, FL, 1988.)

Table 1.12—4
LARGE RIVERS OF THE WORLD RANKED BY LENGTH, MEAN ANNUAL DISCHARGE AT MOUTH, AND AVERAGE SUSPENDED LOAD

Length (km)

Nile	Africa	6650
Amazon	South America	6437
Mississippi/Missouri	North America	6020
Yangtze	Asia	5980
Yenisey	Asia	5540
Ob	Asia	5410
Hwang-Ho	Asia	4845
Zaire	Africa	4700
Amur	Asia	4444
Lena	Asia	4400

Mean Annual Discharge at Mouth (1000 m^3/s)

Amazon	South America	212.5
Zaire	Africa	39.7
Yangtze	Asia	21.8
Brahmaputra	Asia	19.8
Ganges	Asia	18.7
Yenisey	Asia	17.4
Mississippi/Missouri	North America	17.3
Orinoco	South America	17.0
Lena	Asia	15.5
Paraná	South America	14.9

Average Annual Suspended Load

	Total t × 106	Load/discharge
Ganges	1625.6	86.9
Brahmaputra	812.9	41.0
Yangtze	560.8	25.7
Indus	488.7	87.3
Amazon	406.4	1.9
Mississippi	349.5	20
Nile	123.8	44.2

From Kjerfve, B., Ed., *Hydrodynamics of Estuaries,* Vol. 1, Estuarine Physics, CRC Press, Boca Raton, FL, 1988.

Table 1.12—5
METHODS OF MEASURING RIVER FLOWS, INFLOWS, AND TRIBUTARY FLOWS

Method	Principle of operation used to define flow rate, Q	Measurement uncertainty	Flow range or criteria (m^3/s)	Min. depth or diameter (m)
Open-Channel Flows				
Velocity area	Q = AU. Velocity measured with Price or pygmy meters, pitot tubes, acoustic meter, electromagnetic meters, propeller meters, deflection vanes, or floats. Float and vane measurements and point measurements require correction to obtain the average velocity	2.2% using Price meters. 10–25% using floats	0.085–2.43 m/s using Price meters	0.18
Moving boat	Adaptation of the velocity-area method. Current meter moved across the stream at a fixed depth, and velocity is recorded for each subsection used to compute mean flow as $Q = C_{mb}\Sigma(A_i V_v \sin \alpha)$	Probably on the order of 3–7%	Larger flows accessible by boat	1.5–1.8
Slope-area	Flow related to estimates or measurements of slope, S, or head loss for pipes, channel geometry, and some form of the friction coefficient such as the Manning n (friction factor for pipes): $Q = (A/n)R^{2/3}S^{1/2}$	10–20% for pipe lengths of 61–305 m. 20–50% for less accurate estimates of R, A, S, and n	Used for high Q and preliminary estimates	Depth and width should be as uniform as possible over channel reaches
Tracers	Salt, dyes, and radioisotopes are used to measure flow from the dilution of a tracer: $Q = q_t (C_t - C_D)/(C_D - C_o)$ where C_t, C_D, and C_o = tracer concentration being injected and measured upstream and downstream of the injection or from the tracer travel time between two stations, Δt: $Q = (L/\Delta t)A$ where length between stations, L, and area, A, must be measured	2–3% in well defined small channels in treatment plants	Useful for shallow depth where current meters cannot be used. Not useful for large wide flows with limited lateral mixing	Limited by lateral mixing
Width contraction	Flow related to head loss through a bridge opening or channel constriction: $Q = C_c A_c[2g(\Delta h + \alpha(U_1^2/2g) + h_f)]^{1/2}$	Order of 10% or more	High Q or highly constricted with measurable $\Delta h \simeq 0.01$–0.08	Small to moderate where constriction is likely

	where α = velocity head coefficient, $\Delta h = h_1 - h_3$, and h_f = head loss through the constriction			
Flumes	Usually forces the flow into critical or supercritical flow in the throat by width contractions, free fall, or steepening of the flume bottom so that discharge can be empirically determined as a function of depth and flume characteristics. The approach channel must be straight and free of waves, eddies, and surging to provide a uniform velocity distribution across the throat. Has less head loss and solids deposition but is more expensive to construct than weirs. Most common device used in sewage plants	3–6% or better with calibration	Higher Q than similar size weir	—
Parshall	Scaled from standard dimensions of 1 in. to 50 ft (0.025–15 m) throat widths $Q = C_p W h_A^N$	3–6% (7% when submerged)	0.0001–93.9; see range for each size	0.03–0.09; see each size
Trapezoidal supercritical flow	Supercritical depth is maintained in three scaled sizes (throat widths = 1, 3, and 8 ft or 0.3, 0.91, and 2.44 m) $Q = (A_c)^{3/2}(T_c)^{1/2}g^{1/2}$ More sensitive to lower flow; less sensitive to submergence (<80%); less siltation; fits existing channels and ditches	~5% (Simlar Venturi flumes are error prone; 1–10% if head loss is small)	0.02–56 (See range for each size. Accurate at lower flows. Up to 736 m^3/s for 37 m wide flumes.)	0.03
Cutthroat	Flat bottom passes solids better than a Parshall flume and is better adapted for existing channels. Accurate at higher degrees of submergence. For free flow: $Q = KW^{1.025}h_A^{n'}$ (English units)	—	—	0.061
H, HS, and HL types	Simply constructed and installed agricultural runoff flumes that are reasonably accurate for a wide range of flows. Flumes can be easily attached to the end of a pipe. Approach flow must be subcritical and uniform. The flume slope must be $\leq 1\%$	1% for 30% submergence. 3% for 50% submergence	Max. flows HS:0.0023–0.023; H:0.010–0.89; and HL:0.586–3.31 m^3/s	0.061

Table 1.12—5 (continued)
METHODS OF MEASURING RIVER FLOWS, INFLOWS, AND TRIBUTARY FLOWS

Method	Principle of operation used to define flow rate, Q	Measurement uncertainty	Flow range or criteria (m^3/s)	Min. depth or diameter (m)
San Dimas	Developed for sediment-laden flows, having a 3% bottom slope to create supercritical flow. Not rated for submerged flow. $Q = 6.35W^{1.04}h^{1.5\text{-}n''}$ (English units)	Not sensitive or accurate at low Q.	0.0042–5.7	0.061
Weirs	Control of flow produces relationship between discharge and head. Inexpensive and accurate but requires at least 0.15 m (0.5 ft) head loss and must be maintained to clean weir plate and remove solids	Accurate under proper conditions	Most useful for smaller Q where 0.15 m (0.5 ft) head loss can be afforded	0.061 + P
Thin plate	Sharp crest exists when $h/L_c \geq 15$ and air is free to circulate under nappe (jet of water over weir). L_c = weir thickness	2–3% when calibrated and maintained. 5–10% when silted in	0.007–1.68. Best when 0.061 m ≤ h ≤ h/2b	h ≥ 0.061 m for nappe to spring free of weir
Rectangular	$Q = (2/3)C_1(2g)^{1/2}bh^{1.5}$ where b = horizontal length of crest; neglects correction of −0.001–0.004 m	3–4%	Best when Q = 0.0081 to 23 m^3/s	0.03 m or 15 L_c and P ≥ 0.1 m
V-Notch	More sensitive to low flows: $Q = (8/15)C_2(2g)^{1/2}[\tan(\theta/2]h_e^{5/2}$ Tailwater level lower than the vertex of the notch. Common angles are 22.5 45, 60, 90, and 120°	3–6%	Q = 0.0003–0.39 but usually limited to 0.028	0.049 ≤ h ≤ 0.061
Trapezoidal Cipolletti	$Q = C_3(2/3)(2g)^{1/2}[b + (4/5)h \tan(\theta/2)]h^{3/2}$	—	—	—
	Trapezoidal weir having same equation for discharge as rectangular weir and is designed with side slopes of 1 horizontal to 4 vertical lengths to avoid the need for contraction adjustments to C_1	Less accurate than rectangular or V-notch	Q = 0.0085 to 272	h = 0.061–0.61–0.5b
Compound weir	V-notch weir that changes side slope to a wider angle to handle higher flows at higher h. The Q vs. h relationship is ambiguous as the sides	Has not been fully investigated	—	—

	change slope. For a 1 ft (0.3 m) deep 90° cut into rectangular notches of widths 2, 4, and 6 ft. (0.61, 1.2, and 1.8 m): $Q = 3.9h_{vn}^{1.72} - \dot{1}.5 + 3.3bh_m^{1.5}$			
Other types	Sutro or proportional and approximate linear weirs have an approximately linear relationship between Q and h. Other types include approximate exponential and poebing weirs	Has not been fully investigated	—	—
Broad crested	$Q = C_B b(h + h_v)^{3/2}$; $h_t = h + h_v$ (Weir is broad enough to consider to the fluid above at hydrostatic pressure. Notches for low flow modify equation for Q. Typical streamwise cross sections are rectangular, triangular, trapezoidal, and rounded $C_B = C_b C_v K_n g^{1/2}$ where K_n is a constant depending on notch shape.)	More uncertainty for lower h. Difficult to predict without in place calibration	$0.08 \leq h_t/L \leq 0.50$ (No effect of submergence until $h_t/h > 0.65$ to 0.85 depending on weir profile.)	0.061
Rectangular notch	Flat crest: $Q = C_b C_v (2/3)2g/3)^{1/2} bh^{2/3}$ Trenton type with 1:1 slopes upstream and downstream of the flat crest: $Q = 3.5bh^{1.65}$ (English units) The Crump triangular weir with a 50% slope on the upstream face and 20% slope on the downstream face: $Q = 1.96bh_t^{3/2}$	—	$0.08 < h_t/L \leq 0.33$. $0.33. \leq h_t/L \leq 1.5$ to 1.8 is a short-crested weir that has not been fully investigated. $1.5 \leq h_t/L \leq 3$ is usually unstable. $h_t/L \geq 3$ similar to sharp crested weirs	0.061
Triangular notch	$Q = C_b\ C_v (16/25)(2g/5)^{1/2}[\tan(\theta/2)]h^{5/2}$ where $Q = 2.5bh^{1.65}$ for flat vee	—	Usually ≥ 28 (1000 cfs)	0.061
Truncated triangular notch	$Q = C_b\ C_v\ T\ (2/3)\ (2g/3)^{1/2}\ (h - h_b)^{3/2}$	—	$h_t \geq 1.25h_b$	—
Parabolic notch	$Q = C_b\ C_v\ (3fg/4)^{1/2} h_t^2$	—	—	—

Table 1.12—5 (continued)
METHODS OF MEASURING RIVER FLOWS, INFLOWS, AND TRIBUTARY FLOWS

Method	Principle of operation used to define flow rate, Q	Measurement uncertainty	Flow range or criteria (m^3/s)	Min. depth or diameter (m)
Trapezoidal notch	$Q = C_b (Th_c + mh_c^2) [(2g(h_t - h_c)]^{1/2}$ where T = top width of flow over weir	—	—	—
Columbus type control	Most frequently used gauging control in the US. Requires calibration for the full range of flows. Above h = 0.21 m: $Q = 8.5 (h - 0.2)^{3.3}$ (English units) Consists of an upward convex notch below h = 0.21 m and slopes 1:5 for 0.611 m on either side of the notch and 1:10 to the remaining distance to the banks	—	—	—
Submerged orifice	Flow is related to the head difference across the orifice: $Q = 0.61 (1 + 0.15r_o)A_d[2g(\Delta h)]^{1/2}$ for a contracted or suppressed rectangular orifice.	Properly operated meter gates: 2%, but as much as 18% noted	0.003–1.43	0.38
Acoustic meter	The difference in travel time, of an acoustic pu·e across the flow and back (t_{AB} and t_{BA}) over a path length, L_A, diagonal to the flow is related to average velocity along the path: $Q = AC_AL_A (1/t_{AB} - 1/t_{BA})/(2\cos\alpha)$ where α = angle of path with flow	1–7% for parallel flow. ≤14% in poorly developed flow	Used in large rivers. No practical upper limit for Q	Depends on W, density gradients, and allowable error
Electromagnetic coils	Experimental method involving coils buried in the bed or suspended in the flow. See pipe method $Q = C_eAE_m/WH$	—	Generally used in shallow flows	No practical limits
Rating navigation locks, dam crests, and gates	Q determined by calibration in model studies or by in place measurements. Generally locks behave like submerged orifices: $Q = C\Delta h$	5–30%	—	—

	and dams act like weirs: $Q = C_b bh_t^{3/2}$			
Superelevation in bends	The difference in water surface elevation across a bend is related to Q	—	Requires high vel. for measurable Δh	No constraint
	Partially Full Pipe, Culvert, and Sewer Line Flow			
Velocity-area	See pipe and open-channel methods: Q = UA. U related to point velocities in at least 4 ways: (1) $U = 0.9u_{max}$ (2) $U = u_{0.4}$ (3) $U = (u_{0.2} + u_{0.4} + u_{0.8})/3$ (4) U = average of 3 vertical profiles at the quarter points across the pipe plus 2 measurements at both walls 1/8 the distance across the flow	All methods are expected to be accurate to at least 10%	Should be applicable to all flows. Use method (1) or (2) if the flow is rapidly varying	Use method (1) if depth ≤5 cm (2 in.)
Tracers	See open-channel method	5%	Requires good mixing	—
Volumetric	Flow from a pipe or channel is diverted to a bucket, tank, sump, or pond of known dimensions and the increase in weight or volume is timed to measure the average flow rate. Orifice buckets may also be useful	1% or better depending on how well dimensions are known	Useful for smaller flows, but unique conditions may make application to any flow possible	No practical minimum for flows encountered
Slope-area	The Manning equation is applied to uniform pipe reaches. See channel method and culvert method. The effects of manholes (changing channel shape, slope, or direction) must be avoided or the increased friction losses taken into account. Design slopes should be verified	10–20% or 20–50% for less precise estimates of S and h	Requires a 200–1000 d uniform approach flow	—
Culvert equations	Q computed from continuity and energy equations	2–8%	Not readily available	0.15
Palmer-Bowlus flume	Constriction inserted into existing partially full pipes where relationship exists between Q and h: $Q = h^{3/2} [\frac{w^{(26 + h)}}{8(L_w + h)}]^3 g^{1/2}$ Does not have fully standardized design; more accurate but less resolution than Parshall flume. Fits pipe d = 0.10 to 1.1 m and larger. Portable. For full pipe flow: $Q = 8.335(\Delta h/d)^{0.512} d^{5/2}$ (English units)	3–5% depending on care in construction. 10% for low Q	0.0010 to 0.51 for d = 0.15 to 0.76 m. Submergence ≤85%. Requires uniform approach for 25 times diameter with slope ≤2% and 0.4 ≤ depth ≤ 0.9 d	$0.61\ h_{min} = 0.061$

Table 1.12—5 (continued)
METHODS OF MEASURING RIVER FLOWS, INFLOWS, AND TRIBUTARY FLOWS

Method	Principle of operation used to define flow rate, Q	Measurement uncertainty	Flow range or criteria (m^3/s)	Min. depth or diameter (m)
USGS sewer flow-meter	A U-shaped fiberglass constriction of standard dimensions is inserted in a pipe to form a Venturi flume where $Q = f(\Delta h, h)$. Full: $Q = 5.74d^{5/2} (\Delta h/d)^{0.52}$ Transition between channel and pipe flow: $Q = d^{5/2} [2.6 \pm (\lvert 0.590 - h_2/d\rvert /0.164)^{1/2}]$ Channel flow: (1) Supercritical: $Q = 5.58d^{5/2} (h_1/d)^{1.58}$ (2) Subcritical — culvert slope <0.020: (a) $h_1/d \geqslant 0.30$: $Q = 2.85d^{5/2} (h_1/d - 0.191)^{1.76}$ (b) $h_1/d < 0.30$: $Q = 1.15d^{5/2} (h_1/d - 0.177)^{1.38}$ (3)Subcritical — culvert slope $\geqslant 0.020$: $Q = 1.07ad^{5/2} (h_1/d)^{2.71}$ $a = 2.15 + (9.49)(10)^{11} (S - 0.008)^{6.76}$ (all equations in English units)	Depends on field calibration	Not available	0.61 and 1.52
Wenzel flume	Symmetrical or asymmetrical constrictions contoured to the side of the pipe result in unique Q vs. h for partially full and full flows. Open bottom does not trap solids. $Q = C_w\{2gA_2\Delta h/[1 - (A_2/A_1)^2]\}^{1/2}$ $C_w = \{1 + K_e + [A_2^2A_1^2/(A_1^2 - A_2^2)]$ $[(f_1L_1/4R_1A_1^2 + f_3L_3/4R_3A_3^2)\ 3/2]\{^{-1/2}$	At least 5%. 25% at low flow	As much as 30:1 variation in flows. Not valid for steep $S \geqslant 0.020$	0.20 (8 in.)
Trajectory methods	The extent of a jet leaving the end of a pipe, and pipe flow geometry are related to discharge	Generally not accurate enough for US EPA NPDES inspections	—	—

Purdue	For level pipe flowing full or partially full where the water surface elevation below the top of the pipe at the outlet (if $a/d < 0.8$), or at 6, 12, or 18 in (0.15, 0.30, or 0.46 m), the distance from the top of the pipe down to the water surface has been experimentally related to flow	Q will be underestimated if the pipe slopes downward	$Q = 0.00032$ to 0.10	$d = 0.051$ to 0.15
California pipe	For a level pipe of length 6d or more, flowing only partially full, discharging freely into air, and having a negligible approach, U: $Q = 8.69(1 - a/d)^{1.88}d^{2.48}$ (a and d measured in ft)	—	Confirmed for $d = 0.076$ to 0.25 m and $a/d \geqslant 0.5$, but probably useful for larger d	0.076
Vertical pipes	In gal/min, with d and height of jet, H′, in. and $H' > 1.4d$: $Q = 5.01d^{1.99}(H')^{0.53}$ For $H' < 0.37d$: $Q = 6.17d^{1.25}(H')^{1.35}$	—	$d = 0.051$ to 0.30 and $H' = 0.013$ to 1.52	0.051
Parabolic discharge nozzle	Attached to the end of a pipe with a free outfall. Q is related to h^2 by laboratory calibration. Nozzle length is $\simeq 4d$	5%, 1% at Q_{max}	0.0000–0.85	0.15, 0.20, 0.25, 0.3, 0.41, 0.51, 0.61, 0.76, 0.91
Kennison discharge nozzle	Attached to the end of a pipe with a free outfall. Q is related to h by laboratory calibration. Nozzle lengths is $\simeq 2d$	5%, 1% at Q_{max}	≤0.85	Same as parabolic nozzle
Acoustic meter	See pipe method			
Electromagnetic meter	See open-channel and pipe methods. Generally used to measure one or more point velocities for the velocity-area method	2% under optimine conditions	$U = -1.67$ to 6.1 m/s (−5.5–20 fps)	Insertable in most pipes
		Pressurized Pipe Flow		
Differential head meters	Flow constrictions produce pressure losses related to flow, Q	Better results when calibrated	Min. $\Delta h = 25.4$ mm (1 in) for water and 51 mm (2 in.) for sewage	—
Venturi throat	$Q = \dfrac{C_d A_d (2g\Delta h)^{1/2}}{(1 - r^4)^{1/2}}$	0.5–3% depending on calibration	Well tested for diameters up to 0.81 (32 in.)	0.051
Nozzle	Same as venturi throat	1–1.5%	—	0.051
Orifice	Same as venturi throat	0.5–4.4%	5:1 range for Q unless calibrated	0.038

Table 1.12—5 (continued)
METHODS OF MEASURING RIVER FLOWS, INFLOWS, AND TRIBUTARY FLOWS

Method	Principle of operation used to define flow rate, Q	Measurement uncertainty	Flow range or criteria (m^3/s)	Min. depth or diameter (m)
Centrifugal meter	Flow rate related to pressure difference between the inside and outside of a pipe bend: $Q = C_dA_d(2g\Delta h)^{1/2}$ For uncalibrated 90° bends with moderate or higher Reynolds no. $C_d = [r_b/(2d)]^{1/2}$	<10% if calibrated. Predicted C_d error ≃10%	May be installed in any pipe for which Δh in measurable (1 in H_2O)	0.038
Pipe friction meter	Flow rate related to friction loss in fully developed pipe flow: $Q = C_d(\Delta h)^{1/2}$ where $C_d = A_dd^{1/2}/fL$ and f is the friction factor	Unknown	Requires high flow rates to yield measurable Δh	Any pipe
Velocity area	$Q = A_dU$, A_d determined from inside diameter; U determined from point velocity measurements by pitot tubes, small propeller meters, electromagnetic probes, etc. Velocities best measured at 0.026, 0.082, 0.146, 0.226, 0.342, 0.6658, 0.774, 0.854, 0.918, and 0.974d and then averaged for round pipes. Relationships between U and a single point velocity can be derived. See channel method as well	0.5% for pitot tube meas. 5% for some propeller meters	$d \leqslant 1.5m$ and $U_{max} = 1.5$ to 6 m/s	Depends on pitot tube or velocimeter diameter. ≃0.10 is the limit for a 3.8 cm diameter pitot tube to make precise measurements
Tracers	Salt or fluorescent dyes are used to determine flow rate from the dilution of a known amount of tracer mass at a downstream location where the tracer is fully mixed or from the time of travel between two locations at which the tracer is well mixed. The latter method requires that the distance between station and flow cross-sectional area be measured. See channel method	3%	Requires good mixing	Should be applicable to any size pipe

Gibson	Pressure rise following valve closure is related to U: Q = (Ag/L)(area ABCA)	Believed to be very accurate	Requires at least 25d approach length	Unknown
Mechanical meters	—	—	0.001–0.28	Varies
Displacement	Flow displaces piston or disk. Oscillations are counted as a function time. Calibration required	1% new, more with age	Depends on manufacturer's specifications	Generally used for smaller pipes
Inferential	Flow rotates a turbine whose rate of rotation is related to flow rate by calibration. Requires an approach length of at least 20–30d to develop the velocity profile. Vanes control spiraling from bends	Needs constant checks: 2–5%	0.15–5.2 m/s (0.5 to 17 ft/sec) but inaccurate below 0.3–0.46 m/s (1–1.5 ft/s)	0.61–2.3 (2–7.5 ft)
Variable area	Rotometer consists of vertical tapered tube with a metal float that rises as flow increases	1% at Q_{max}	Small flows	Depends on manufacturer
Acoustic meter	See channel-flow method for time of travel: $Q = AC_AL_A(1/t_{AB} - 1/t_{BA})/2\cos\alpha$ Reflected doppler meters detect a frequency shift from a signal reflected back from a point where sound beams are crossed in the flow. The shift is related to the point velocity which must be related to the average velocity by calibration to calculate the flow rate. Time of flight acoustic meters determine a turbulent signature of the flow at a station and correlates that with a downstream turbulent signature to determine time of travel over a known distance through a known pipe volume: $Q = \pi d^2L/4\Delta t$ All types can be attached to outside of pipes	2–5% and higher at bends	Time of travel and doppler meters require approach lengths of at least 10–20d of straight pipe or must be calibrated to determine C_A. Time of flight meters require a limited approach of a few diameters	No practical limit
Electromagnetic meter	Induced voltage, E_m, caused by water flow perpendicular to magnetic field, proportional to velocity, U: $Q = AE_m/HdC_e$ Meter must be in contact with fluid. Regular checking of the electrodes is necessary to avoid fouling. Magnetic coils can be embedded in new pipe or in a ring insert, or a probe can be	Larger of 1–2% or 0.002 m/s	Depends on manufacturer. At least up to 16.7 (590 ft^3/sec)	At least 0.10–1.52 m diameter (4–60 in)

Table 1.12—5 (continued)
METHODS OF MEASURING RIVER FLOWS, INFLOWS, AND TRIBUTARY FLOWS

Method	Principle of operation used to define flow rate, Q	Measurement uncertainty	Flow range or criteria (m^3/s)	Min. depth or diameter (m)
	inserted to measure point velocities or profiles to be used in the velocity-area method			
Calibrated pumps, turbines, valves, and gates	Empirical relationships between Q and power, or valve or gate opening for appropriate heads are defined with laboratory models or by calibration in place	Depends on inplace calibration. Laboratory calibration ≃10% or more for accurate in place calibration	—	—

DEFINITION OF SYMBOLS USED IN THE TABLE

A = Cross-sectional area of channels and pipes. Subscript i refers to the area of individual subsections. Subscript c refers to area of a width constriction or area of a throat of a flume.
U = Cross-sectional average velocity.
C_{mb} = Coefficient relating point velocity measurements to the mean velocity in a cross section. For large rivers C_{mb} = 0.87 to 0.92 and 0.9 is typically used.
n = Manning's roughness coefficient describing friction loss in channels.
R = Hydraulic radius (A/wetted perimeter). Subscripts refer to section numbers such as 1 and 3 for the Wenzel flume.
C_c = Discharge coefficient for a width constriction.
g = Gravitational acceleration.
C_p = Discharge coefficient for Parshall flumes.
N = Exponent in the Parshall flume equation.
T_c = Top width of the water surface in the trapezoidal flume at the point where critical flow occurs.
K = Free flow discharge coefficient for cutthroat flume; the standard 1.5-, 3-, 4.5-, and 9-ft (0.46-, 0.91-, 1.37-, and 2.74-m)-long flumes varying from 6.1 to 3.5 (English units).
n′ = Free flow discharge exponent for a cutthroat flume; the standard 1.5-, 3-, 4.5-, and 9-ft (0.46-, 0.91-, 1.37-, and 2.74-m)-long flumes varying from 2.15 to 1.56 (English units).
n″ = Discharge exponent for the San Dimas flume = $0.179W^{0.32}$ (English units).
h = Water surface height of the approaching flow above the weir crest or bottom of the notch. For precise estimates 0.001 m is added to h to account for the effects of surface tension and viscosity. h is measured at least 3 to 4 times h_{max} upstream of the weir face for sharp-crested rectangular and V-notch weirs; 2 to 3 times h for Cipolletti weirs; and 2 to 3 times ht, for broad-crested weir. Generally the minimum distance upstream should be $4h_{max}$.[8]
P = Height of lip of weir above stream bottom.
C_1 = Discharge coefficient for rectangular thin-plate weir; f(h/P, b/B, E) for free falling, free discharge that springs free of the plate without clinging (thickness of the weir is ≤1/15 of the depth of flow over the weir).
C_s = $[1 - (h/D)^m]^{0.385}$ where m = 1.44 for a fully contracted weir and 1.50 for a suppressed weir.
C_2 = Discharge coefficient for a V-notch weir ≃0.58 for a fully contracted weir flow where $h/P \leq 0.4$, $h/B \leq 0.2$, $0.049 < h \leq 0.381$ m, $P \geq 0.46$ m, $B \geq 0.91$ m. Values cannot be predicted for partially contracted weir flow without calibration.

h_e = $h + K_h$. K_h varies from 0.0025 to 0.0085 depending on the angle of the notch of the weir.

C_3 = $0.63C_v$. C_v = velocity head correction factor.

C_b = Discharge coefficient for broad-crested weir. For flat-crested weirs, C_b depends on the rounding or slope of the upstream face, h_t/L, and $h_t/(h_t + P)$. For a horizontal upstream face with a sharp corner, the basic discharge coefficient is $C_b = 0.848$ if $0.08 \leq h/L \leq 0.33$ and $h/(h + P) \leq 0.35$. For $h/L > 0.33$ and $h/(h + P) > 0.35$. The value of C_b is corrected for $h/L \leq 1.5$ if $h/(h + P) \leq 0.35$. A final correction is made for the approach velocity head varying from 1.00 to 1.2 as a function of C_bA_b/A_a where A_b is the area of flow over the control structure and A_a is the area of flow in the approach section where h is measured. For rounding of the upstream corner: $C_b = [1 - 2x(L - r_t)/B][1 - x(L - r_t)/h]^{3/2}$ where x is a parameter that accounts for boundary-layer effects and r_r is the radius of rounding of the upstream corner. For field installation of well-finished concrete, $x \simeq 0.005$. Where clean water flows over precise cut blocks, $x \simeq 0.003$. C_b varies between 0.85 and 1.00 for triangular weirs having $h/L = 0.08$ to 0.7.

h_v = $\alpha U^2/2g$ for the approach flow.

C_v = Discharge coefficient correcting for the effect of an approach velocity for weirs and orifices.

r_o = Ratio of suppressed portion of the perimeter of a submerged orifice to the entire perimeter.

C_A = Discharge coefficient for acoustic gauging stations relating cross-sectional average velocity to average velocity along the acoustic path. C_A = f (water depth and tidal condition).

C_e = Discharge coefficient for electromagnetic meters determined in the laboratory for portable meters attached to the outside of a pipe. For open-channel coils, $C_e = [1 + (W\sigma_b/2h\sigma_w)]/\beta$ where σ_b and σ_w are conductivities of the bed and river water, respectively, and β is a correction factor for the end effects of the magnetic field and for incomplete coverage of the cross-sectional area by the field in case of limited coils.

H = Magnetic field intensity in Tesla.

E_m = Electromotive force in volts generated by the movement of a conduction fluid such as slightly contaminated water through a magnetic field.

A_d = Area of venturi throat, orifice, or flow nozzle opening or pipe diameter for bend meters.

d = Pipe diameter.

Δh = Pressure drop or hydraulic head difference measured across a venturi throat or flume, orifice, flow nozzle, bend meter, or along a straight length of pipe. Also the difference in water surface elevations across a submerged orifice or width contraction.

h = Water depth measurement in partially full pipes where subscripts 1 and 2 refer to locations upstream of and locations in the throat of a flume.

S = Slope of a pipe or channel.

C_w = Discharge coefficient for the Wenzel flume that accounts for energy loss and velocity head corrections. For full pipe flow, C_w is given above. K_e is an entrance loss coefficient of approximately 0.2. f is the Darcy-Weisbach friction factor. L_i is the length. Valid when backwater conditions do not exist. For partially full flow, C_w seems to be $\simeq 1$.

C_d = Discharge coefficient for venturi throat, orifice, and flow nozzle or bend meter. Function of Reynolds number, Re, for venturi throat when pressure taps are properly located in the straight pipe section of the throat and 0.5 to 0.25 pipe diameters upstream of the meter. $C_d = 0.96$ to 0.99 as a function of $Re \leq 10^6$ for flow nozzle when pressure taps are located one pipe diameter upstream of the beginning of the nozzle and before the end of the nozzle. For precise measurements, C_d is determined by calibration for bend meters and for straight pipe sections between pressure taps. The best results are obtained in prediction C_d when the flow meter is well maintained and correctly installed. Indirect flow measurements are used to check the meter raging and to determine if debris lodges in the meter. Proper installation places the meter downstream of at least 10 diameters of straightened flow and avoids solids deposition. Vanes are used to straighten flows if necessary.

Table 1.12—5 (continued)
METHODS OF MEASURING RIVER FLOWS, INFLOWS, AND TRIBUTARY FLOWS

Method	Principle of operation used to define flow rate, Q	Measurement uncertainty	Flow range or criteria (m^3/s)	Min. depth or diameter (m)

r = Ratio of throat diameter, d_2, to pipe diameter, d.
r_b = Radius of the center line of a pipe bend.
f = Friction factor for pipes.
L = Length of straight pipe between pressure tapes for a pipe friction meter.

From McCutcheon, S. C., *Water Quality Modeling, Vol. 1, Transport and Surface Exchange in Rivers,* CRC Press, Boca Raton, FL, 1989, 277.

Table 1.12—6
PROCEDURE FOR MEASURING AVERAGE VELOCITY AND DISCHARGE (VELOCITY- AREA METHOD)

Step	Procedure
1.	Select a cross section from a straight, uniform reach with parallel streamlines and a relatively uniform bottom that is at least 0.15 m (0.5 ft) deep, that has velocities of at least 0.15 m/s (0.5 ft/s), and where there is easy access from cableways, bridges, or by wading (otherwise measure from boats). If possible the section should be free of large eddies with upstream circulation near the banks, areas of slack water, or excessive turbulence caused by upstream bends, radical changes in cross-sectional shape, and irregular obstructions such as boulders, trees, vegetation, and other debris in the vicinity.
2.	Choose a time of measurement such that the discharge is steady or approximately steady during the period of measurement that usually ranges from 1 to 3 h, depending on the size of the river. If flow changes rapidly, short-cuts in the method will be necessary.
3.	Measure the cross-sectional area, A, by measuring depth with a sounding line or wading rod and width with hand lines or tapes. In large rivers, electronic depth sounders and triangulation with transits or laser distance measuring equipment are used.
4.	Divide the section into at least ten subsections based on the expected distribution of discharge over the section. Typically, 20 to 30 sections are required for precise measurements.
5.	Measure the vertical velocity profile in each subsection.
6.	Compute average velocity in each subsection, u_1, from the profile.
7.	Compute the subsection discharge, q: $$q_i = u_i\left[\frac{(b_i - b_{i-1})}{2} + \frac{(b_{i+1} - b_i)}{2}\right]d_i$$
8.	Compute the discharge from $Q = \Sigma q_i$.
9.	Compute the average velocity from Q and cross-sectional area, A, $U = Q/A$.

From McCutcheon, S. C., *Water Quality Modeling,* Vol. 1, Transport and Surface Exchange in Rivers, CRC Press, Boca Raton, FL, 1989.

1.13 BOTTOM CURRENTS AND THE BENTHIC BOUNDARY LAYER

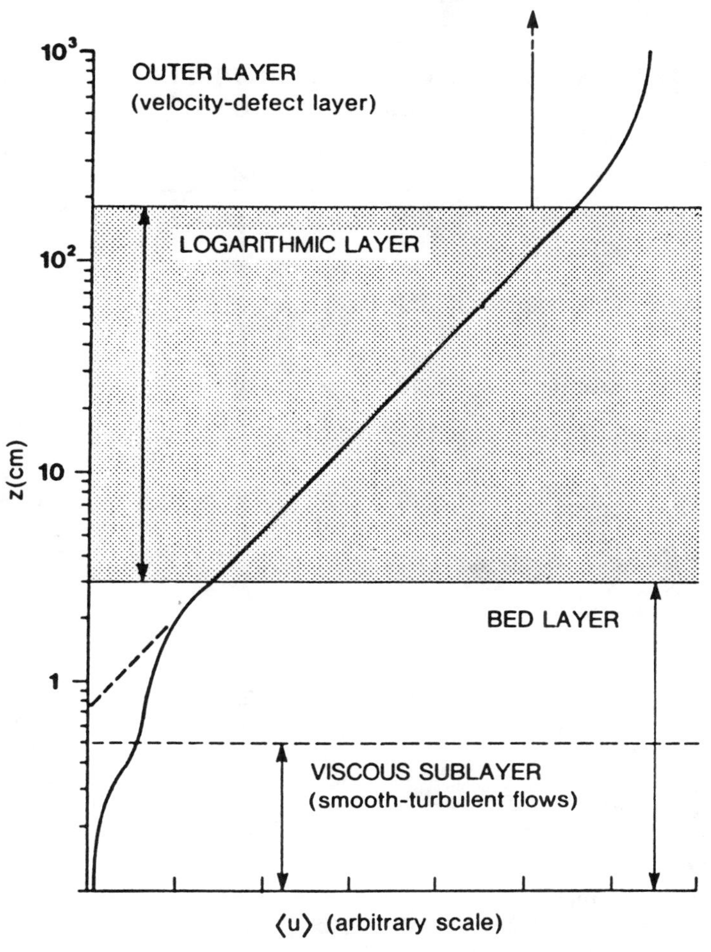

FIGURE 1.13—1. Regions of the lower benthic boundary layer influenced by bottom currents. (From Wright, L. D., Benthic boundary layers of estuarine and coastal environments, *Rev. Aquat. Sci.*, 1, 75, 1989.)

Table 1.13—1
SUMMARY OF THE RELATIVE EFFECTS OF UNSTEADY FLOW, WAVE-CURRENT INTERACTIONS, DISTRIBUTED BED ROUGHNESS, SEDIMENT TRANSPORT, AND STRATIFICATION ON BOUNDARY LAYER PROPERTIES (AS COMPARED TO STEADY, NEUTRALLY STRATIFIED FLOW OVER A FLAT FIXED BED)

	z_o	ξ_{xz} (z < 10 cm)	C_{100}	$\bar{\tau}$
Accelerating flow	Decreases	Decreases	Decreases	Decreases
Decelerating flow	Increases	Increases	Increases	Increases
Wave-current interaction	Increases	Increases	Increases	Increases
Distributed bed roughness	Increases	Increases	Increases	Local increases; space-averaged decreases
Sediment transport	Increases	Increases	Increases	Increases
Stratification	Increases	Decreases	Decreases	Decreases

From Wright, L. D., Benthic boundary layers of estuarine and coastal environments, *Rev. Aquat. Sci.*, 1, 75, 1989. With permission.

1.14 WAVES

Table 1.14—1
RELATIVE FREQUENCY OF WAVES OF DIFFERENT HEIGHTS IN DIFFERENT REGIONS

	Height of waves, ft					
Ocean region	0–3	3–4	4–7	7–12	12–20	≥20
	%	%	%	%	%	%
North Atlantic, between Newfoundland and England	20	20	20	15	10	15
Mid-equatorial Atlantic	20	30	25	15	5	5
South Atlantic, latitude of Argentina	10	20	20	20	15	10
North Pacific, latitude of Oregon and south of Alaskan Peninsula	25	20	20	15	10	10
East equatorial Pacific	25	35	25	10	5	5
West wind belt of South Pacific, latitude of southern Chile	5	20	20	20	15	15
North Indian Ocean, Northeast monsoon season	55	25	10	5	0	0
North Indian Ocean, Southwest monsoon season	15	15	25	20	15	10
Southern Indian Ocean, between Madagascar and northern Australia	35	25	20	15	5	5
West wind belt of southern Indian Ocean, on route between Cape of Good Hope and southern Australia	10	20	20	20	15	15

From Bigelow, H. B. and Edmondson, W. T., *Wind Waves at Sea, Breakers, and Surf,* U.S. Navy Hydrographic Office, Washington, D.C., Pub. 602, 1962.

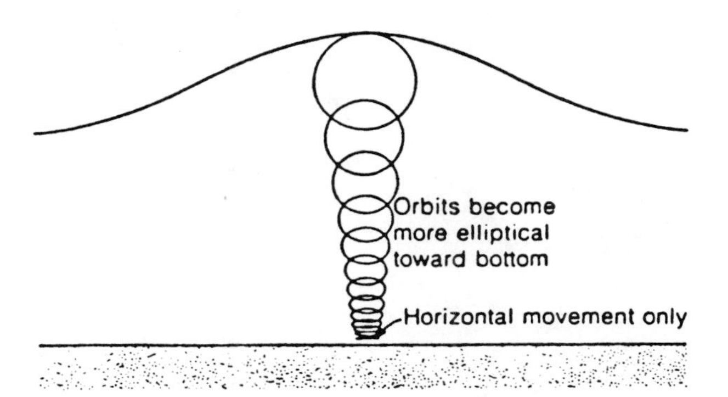

FIGURE 1.14—1. Motion of water particles in shallow-water waves. As the seafloor is approached, orbits of water particles change from circular, to elliptical, to horizontal (back and forth only). (From *Earth,* by Frank Press and Raymond Siever. Copyright ©1986 by W. H. Freeman and Company. Reprinted with permission.)

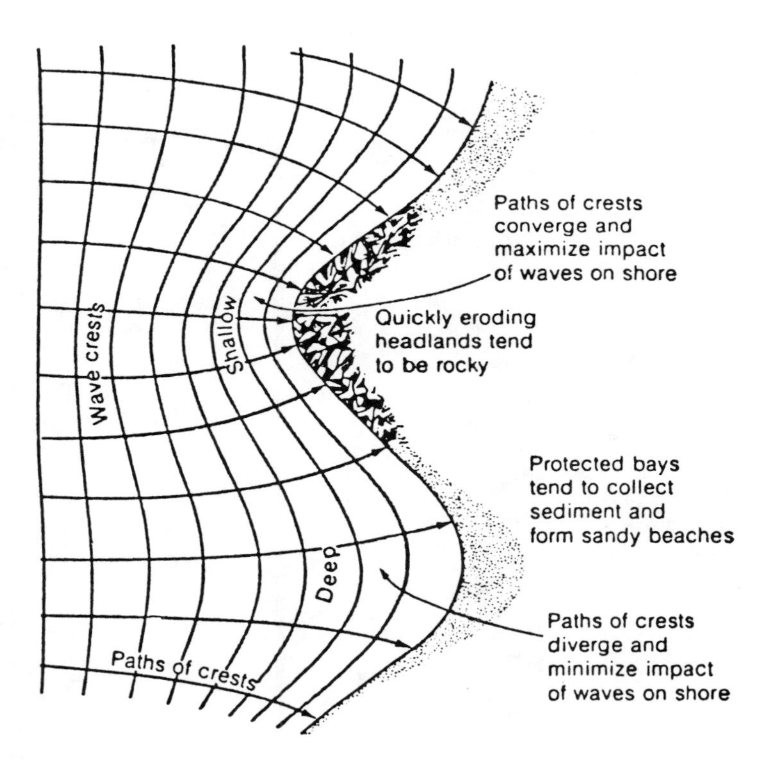

FIGURE 1.14—2. Effects of wave refraction on an irregular shoreline. Wave energy is concentrated on head-lands and dispersed in bays. Consequently, head-lands are sites of erosion, and bays, sites of deposition. Wave action works to smooth the shore-line. (From *Earth,* by Frank Press and Raymond Siever. Copyright ©1986 by W. H. Freeman and Company. Reprinted with permission.)

Section 2
Marine Chemistry

Section 2

MARINE CHEMISTRY

DISSOLVED CONSTITUENTS IN SEAWATER

SALINITY

Seawater consists of a solution of inorganic salts, atmospheric gases, traces of organic matter, and small amounts of particulate material.[1,2] The elements in seawater can be organized into four groups: major, minor, trace, and nutrient elements. Each group of elements is vital to the organisms inhabiting the oceanic environment (see Section 4).

Salinity is defined as the weight in grams of dissolved salts in 1 kg of seawater (after all bromine has been replaced by chlorine, all carbonate converted to oxide, and all organic matter destroyed). It is usually expressed in parts per thousand (‰), but may be reported in milligrams per liter (mg/l), milliequivalents per liter (meq/l), grams per kilogram (g/kg), or percent (%). The constancy of the proportions of the major dissolved constituents in seawater has enabled salinity to be determined by the concentration of the chloride ion which represents an index of salinity. By titrating a seawater sample with silver nitrate (Knudsen method), the amount of chloride, plus a chloride equivalent of bromide and iodide, can be obtained. Chlorinity (Cl) — the total weight of these elements in a 1-kg seawater sample — is related to salinity (S) according to the following equation:

$$S(‰) = 1.80655 \times Cl(‰) \quad (1)$$

Because seawater is an electrolyte and conducts an electric current, salinity can be measured by electrical conductivity. Chlorinity measurements are less accurate (0.02% for 32 to 38‰) than conductivity techniques (0.003‰) of salinity estimation.[3] Therefore, salinity is now more frequently measured by a salinometer, which records conductivity, rather than by volumetric titration of chlorinity. Measurements of density (with a hydrometer), the refractive index of seawater (with a refractometer), and freezing-point depression provide alternate methods of estimating salinity.

The salinity of seawater averages 35‰, but ranges from 33 to 38‰. Deviations from average salinity occur locally at ocean boundaries due to dilution from river runoff and at deep-sea hydrothermal vents along seafloor spreading centers due to the emission of mineral-laden water from the oceanic crust. On a regional scale, deviations from average salinity arise from differences in evaporation and precipitation. Evaporation and the formation of sea ice raise salinity in the ocean, whereas large rates of precipitation and river runoff reduce salinity.[3] High rates of precipitation associated with tropical rains depress salinity in equatorial seas, as does the melting of ice caps in polar seas. At high latitudes greater than 40°N and 40°S, precipitation in excess of evaporation also lowers salinity. Salinity maxima develop at 30°N and 30°S latitudes.[4]

Dissolved constituents in the ocean originate primarily from the weathering of rocks and leaching of soils on land. Solution, oxidation-reduction, the action of hydrogen ions, and the formation of complexes are weathering processes that control the supply of ions. Fluvial and glacial processes continually deliver most dissolved material to the ocean; of secondary importance are materials supplied from terrestrial and cosmic sources (e.g., aeolian, meteoritic, and volcanic particles) by atmospheric processes, as well as those originating from hydrothermal activity at midocean ridges. Equally important are biological, chemical, and sedimentological processes which remove dissolved constituents from seawater, tending to maintain equilibrium among the constituents. Included among the most important removal mechanisms are authigenesis, involving inorganic reactions in seawater, and biogenesis, involving organic reactions which use seawater constituents.[5] Biogenic processes in which marine organisms deposit skeletal material are important in the removal of chemical species, such as SiO_2, Ca^{2+}, and bicarbonate ions (HCO_3^-).

The length of time the major ions remain in the ocean, the residence time, ranges from approximately 10^5 years for bicarbonate, to greater than 10^8 years for sodium, and approaching infinity for chloride.[6] The residence time of many minor ions is significantly shorter. Because the residence time of most of the major ions is great relative to the mixing time of the ocean, the ratios of the concentrations of the major ions are effectively constant. Minor ions with residence times substantially shorter than the mixing time of the ocean have variable concentrations in seawater. Chemical constituents that have short residence times generally are more reactive than those that have long residence times. The major ions, therefore, behave conservatively in seawater; their concentrations are constant except at ocean boundaries where they are changed by input or output processes, such as dilution by river runoff, anthropogenic activity, or precipitation onto the ocean surface.[5] The minor constituents, however, are more reactive, being removed rather quickly. Hence, they often act nonconservatively.

CONSTITUENT CONCENTRATIONS

Nearly all natural elements exist in seawater, but the concentration of the four major constituents (sodium, magnesium, chloride, and sulfate) comprises more than 95% of the dissolved solids. The combined weight of the minor dissolved constituents of seawater amounts to less than 2 mg/kg (2 ppm).[7] According to Bruland,[8] who reviewed the concentration and speciation of elements in seawater, the five major cations Na^+, Mg^{2+}, K^+, Ca^{2+}, and Sr^{2+} and four major anions Cl^-, SO_4^{2-}, Br^-, and F^- have concentrations greater than 1 mg/l or 50 μmol/kg. Other elements occurring in concentrations greater than 50 μmol/kg are B (as H_3BO_3), C (as HCO_3^- and CO_3^{2-}), N (as dissolved N_2), O (as dissolved O_2), and Si (as H_4SiO_4). Minor elements with concentrations between 0.05 and 50 μmol/kg include the alkali metal cations Li^+ and Rb^+, the alkaline earth cation Ba^{2+}, the transition metal Mo (as MoO_4^{2-}), and the halogen I (as IO_3^- and I^-). The nutrient elements N (as NO_3^-) and P (as HPO_4^{2-}) also are found at concentrations between 0.05 and 50 μmol/kg. The average concentrations of 22 trace elements range from 0.05 to 50 μmol/kg, whereas 35 others exist at concentrations <50 pmol/kg.

In terms of total mass, 11 elements comprise more than 99.9% of the total dissolved material in seawater. These include in decreasing order of total mass: chlorine (2.5×10^{16} tonnes), sodium (1.4×10^{15} t), magnesium (1.7×10^{15} t), calcium (5.4×10^{14} t), potassium (5.0×10^{14} t), silicon (2.6×10^{12} t), zinc (6.4×10^{9} t), copper (2.6×10^{9} t), iron (2.6×10^{9} t), manganese (2.6×10^{8} t), and cobalt (6.6×10^{7} t).

In addition to the major and minor dissolved constituents in seawater, trace metals (e.g., antimony, arsenic, cadmium, lead, mercury, nickel, and silver), nutrient elements (e.g., nitrogen, phosphorus, and silicon), dissolved atmospheric gases (e.g., oxygen, nitrogen, and carbon dioxide), and organic matter (dissolved and particulate forms) are important components of seawater. Trace metals, although essential to the growth of many organisms at low concentrations, can be toxic when present at elevated levels.[9] Trace metals enter the ocean via riverine and atmospheric influx, anthropogenic input, and hydrothermal convection of seawater through new oceanic crust at seafloor spreading centers.[10–15] The major nutrients represent the chief limiting elements to phytoplankton production in the sea. Of all atmospheric gases dissolved in seawater, oxygen and carbon dioxide probably are most critical. Oxygen serves as an essential element in the metabolic processes of organisms, as an indicator of water quality of the marine environment, and as a tracer of the motion of deep water masses of the oceans (see Section 1).

DISSOLVED CONSTITUENTS IN ESTUARIES

SALINITY

The total concentration of dissolved solids in an estuary is far more variable than in the oceanic realm, changing temporally (daily, seasonally, and over a tidal cycle) and spatially (with increasing distance along vertical, transverse, and longitudinal planes). The mixing of river water and seawater, solutions significantly different with respect to physical-chemical properties and chemical composition, largely accounts for the variable salinity observed in estuaries.[16,17] With the exception of hypersaline systems, salinity of estuarine waters ranges from 0.5 to 35‰. The horizontal and vertical salinity gradients observed in estuaries are dependent, in part, on the relative balance of land runoff, precipitation, and evaporation.[18]

The mixing processes in estuaries are important, and strong gradients often occur in physical-chemical properties in these ecosystems.[19] Diverse chemical and biogenic reactions may take place independently and simultaneously; chemical constituents are added, removed, or altered in a particular system. Anthropogenic input of large volumes of waterwater discharges can mask contributions of chemical components from natural waters.[13] Additionally, local meteorological conditions, estuarine circulation patterns, and *in situ* biological activity generate temporal and spatial changes in water chemistry that can obfuscate the contributions of chemical elements from freshwater and marine sources.[16,17,19,20] Transport mechanisms and botton sediments, both of which influence the distribution, occurrence, and behavior of constituent forms, affect the budgets of chemical components. Topographic features of the estuarine basin may modify currents, restrict mixing, and enhance the development of unusual chemical environments (e.g., anoxic systems). Because of the great variability of physical, biological, and chemical characteristics of estuaries, no two estuaries are exactly alike in chemistry or biogeochemistry. Nevertheless, it is possible to identify general properties, processes, and trends that help describe the chemistry of this unique environment. Most notably, horizontal and vertical salinity gradients in estuaries result from interactive effects, including the relative volume of freshwater and seawater input, tidal mixing, currents, and topography of the estuarine basin.

MINOR AND TRACE ELEMENTS

Minor and trace elements, present in low, but variable dissolved concentrations in estuaries, are essential for the growth of many estuarine organisms.[21,22] Only a portion of all trace elements is transported in solution to estuaries; colloidal and particulate forms are also quantitatively significant sources. Dissolved and particulate matter may interact and alter the concentration of trace elements in solution.[23,24] Bottom sediments represent a reservoir for trace elements which may be released to overlying waters by dissolution, desorption, and autolytic biological processes. Changes in pH and redox potential in the sediment column can promote chemical reactions that mobilize trace elements, effecting their release to interstitial and overlying waters. Physical (e.g., tidal currents) and biological (e.g., bioturbation by organisms) factors facilitate exchange of elements between sediments and interstitial and overlying waters. Organisms are a pathway for transporting and accumulating trace elements.

NUTRIENT ELEMENTS

Nitrogen, phosphorus, and silicon are major nutrient elements in estuaries, occurring in both dissolved and particulate forms.[25–27] The concentrations of these elements in river water exceed those in seawater, since the weathering of rocks and leaching of soils on land and the addition of anthropogenic wastewater discharges deliver the nutrients directly to riverine systems. Land runoff is the principal source of nutrient elements in estuaries; the atmosphere and oceans supply secondary amounts.

Nitrogen is the major limiting nutrient to primary production in estuaries,[28] although phosphorus may be the limiting nutrient to plant growth during certain seasons in some systems. Silicon is needed for skeletal growth of diatoms (often influencing diatom abundance and productivity), radiolarians, certain sponges, and other siliceous organisms. Their concentrations are biologically modulated and therefore temporally and spatially variable. Bottom sediments act as a nutrient pool, frequently containing high levels of these elements, which may be regenerated by various physical and biological processes and reused by organisms on the estuarine seafloor and in overlying waters. Nutrient cycling is critical to the high biological productivity of estuaries.

The three primary dissolved inorganic forms of nitrogen in estuaries are ammonia (NH_3), nitrite (NO_2^-), and nitrate (NO_3^-), although dissolved organic and particulate forms also comprise useful and significant nitrogen sources. Dissolved silicon occurs mainly as silicic acid (H_4SiO_4). The particulate forms of silicon are mainly detrital quartz and clay minerals (alumino silicates). A rather complex cycling of nitrogen, phosphorus, and silicon takes place in the estuarine environment and, because the cycling processes are not completely understood, they remain the subject of continuing research.

DISSOLVED CONSTITUENT BEHAVIOR

During estuarine mixing, dissolved constituents behave either conservatively or nonconservatively.[29] The major dissolved constituents of seawater generally act conservatively — their concentrations vary only as a result of physical mixing and dilution. The concentration of a conservative constituent plotted against salinity or chlorinity (known as a mixing plot) will either increase or decrease linearly with increasing salinity or chlorinity, depending on whether its concentration is greater in river water or seawater. Bends in the mixing curves of conservative constituents are attributable to temporal variations in river or ocean constituent concentrations. The minor constituents often behave nonconservatively, being added or removed by biological or chemical processes. The mixing plot of a nonconservative constituent is curved, indicative of the reactive nature of the constituent, with the estuary serving as a constituent source or sink.

Speciation models have provided a useful mechanism for evaluating the fate of dissolved constituents in estuaries.[30–32] The chemical speciation of the major dissolved constituents is much better understood than that of minor and trace constituents. However, both field and laboratory mixing experiments are beginning to clarify some of the processes considered to be responsible for trace metal transformations in estuaries. Detailed information on the chemical speciation of minor constituents awaits further study on the effect of environmental factors (e.g., pH, Eh, and salinity) on speciation.

DISSOLVED GASES

Dissolved gases exist in bottom sediments, interstitial waters, and the water column of estuaries.[25] The solubilities of dissolved gases are a function of salinity, temperature, and partial pressure. As salinity and temperature increase, gaseous solubility decreases. At a given temperature, the solubility of a gas is directly proportional to the partial pressure it exerts (Henry's Law). Gaseous exchange between the atmosphere and estuary is increased by disturbance of the air-sea surface interface via strong winds and storms, whereas surface films exceeding 10^{-4} cm in thickness hinder gaseous exchange.

Dissolved oxygen and carbon dioxide differ from other dissolved gases because they are of fundamental importance to the life processes of estuarine organisms. Biological activity strongly influences the concentrations and distributions of these gases. Gaseous input from the atmosphere across the air-sea surface interface and photosynthetic activity yield most of the dissolved oxygen in estuaries; dissolved carbon dioxide is derived from river discharge, gaseous exchange with the atmosphere, inorganic chemical reactions, plant and animal respiration, and microbial decomposition. Both gases exhibit significant temporal and spatial variations due not only to changes in photosynthetic and respiratory rates, but also to fluxes in tidal and freshwater flows, irregular basin morphology, geographic location, anthropogenic enrichment of organic compounds, and other factors. Carbon dioxide, as part of the carbon dioxide-carbonate-bicarbonate chemical system, forms a buffer that regulates the environment against rapid shifts in acidity and alkalinity. By modulating pH, it therefore plays a key role in the chemical processes of the estuarine environment.

SECTION PLAN

This section details the chemical composition of seawater, concentrating on the major, minor, trace, and nutrient elements. Organic compounds occurring in seawater are also treated. Information is presented on the chemistry of both oceanic and estuarine waters.

REFERENCES

1. **Day, J. H.,** The chemistry and fertility of estuaries, in *Estuarine Ecology: with Particular Reference to Southern Africa,* Day, J. H., Ed., A. A. Balkema, Rotterdam, 1981, 57.
2. **Millero, F. J. and Sohn, M. L.,** *Chemical Oceanography,* CRC Press, Boca Raton, FL, 1992.
3. **Levinton, J. S.,** *Marine Ecology,* Prentice-Hall, Englewood Cliffs, NJ, 1982.
4. **Sverdrup, H. U., Johnson, M. W., and Fleming, R. H.,** *The Oceans, Their Physics, Chemistry, and General Biology,* Prentice-Hall, Englewood Cliffs, NJ, 1942.
5. **Smith, D. G., Ed.,** *The Cambridge Encyclopedia of Earth Sciences,* Cambridge University Press, Cambridge, 1981.
6. **Weyl, P. K.,** *Oceanography: An Introduction to the Marine Environment,* John Wiley & Sons, New York, 1970.
7. **Burton, J. D. and Liss, P. S., Eds.,** *Estuarine Chemistry,* Academic Press, London, 1978.
8. **Bruland, K. W.,** Trace elements in seawater, in *Chemical Oceanography,* Vol. 8, Riley, J. P. and Chester, R., Eds., Academic Press, London, 1983, 157.
9. **Spaargaren, D. H. and Ceccaldi, H. J.,** Some relations between the elementary chemical composition of marine organisms and that of seawater, *Oceanological Acta,* 7, 63, 1984.
10. **Wong, C. S., Boyle, E., Bruland, K. W., Burton, J. D., and Goldberg, E. D., Eds.,** *Trace Metals in Sea Water,* Plenum Press, New York, 1983.
11. **Mance, G.,** *Pollution Threat of Heavy Metals in Aquatic Environments,* Elsevier, New York, 1987.
12. **Cutter, G. A.,** Trace elements in estuarine and coastal waters, *Rev. Geophys. (Suppl.), Contrib. Oceanogr.,* 1991, 639.
13. **Kennish, M. J.,** *Ecology of Estuaries: Anthropogenic Effects,* CRC Press, Boca Raton, FL, 1992.
14. **Kennish, M. J.,** Geology of deep-sea hydrothermal vents and seafloor spreading centers, *Rev. Aquat. Sci.,* 6, 97, 1992.
15. **Kennish, M. J. and Lutz, R. A.,** Ocean crust formation, *Rev. Aquat. Sci.,* 6, 493, 1992.
16. **Kjerfve, B.,** *Hydrodynamics of Estuaries,* Vol. 1, CRC Press, Boca Raton, FL, 1988.
17. **Kjerfve, B.,** *Hydrodynamics of Estuaries,* Vol. 2, CRC Press, Boca Raton, FL, 1988.
18. **Reid, G. K. and Wood, R. D.,** *Ecology of Inland Waters and Estuaries,* 2nd ed., D. Van Nostrand, New York, 1976.
19. **Day, J. W., Jr., Hall, C. A. S., Kemp, W. M., and Yanez-Arancibia, A.,** *Estuarine Ecology,* John Wiley & Sons, New York, 1989.
20. **Neilson, B. J., Brubaker, J., and Kuo, A., Eds.,** *Estuarine Circulation,* Humana Press, Clifton, NJ, 1989.

21. **Rainbow, P. S.,** The biology of heavy metals in the sea, *Int. J. Environ. Stnd.,* 25, 195, 1985.
22. **Viarengo, A.,** Heavy metals in marine invertebrates: mechanisms of regulation and toxicity at the cellular level, *Rev. Aquat. Sci.,* 1, 295, 1989.
23. **Duinker, J. C. and Nolting, R. F.,** Mixing, removal, and mobilization of trace metals in the Rhine estuary, *Neth. J. Sea Res.,* 12, 205, 1978.
24. **Duinker, J. C.,** Suspended matter in estuaries: adsorption and desorption processes, in *Chemistry and Biogeochemistry of Estuaries,* Olausson, E. and Cato, I., Eds., John Wiley & Sons, Chichester, England, 1980, 121.
25. **Aston, S. R.,** Nutrients, dissolved gases, and general biogeochemistry in estuaries, in *Chemistry and Biogeochemistry of Estuaries,* Olausson, E. and Cato, I., Eds., John Wiley & Sons, Chichester, England, 1980, 233.
26. **Carpenter, E. J. and Capone, D. G., Eds.,** *Nitrogen in the Marine Environment,* Academic Press, New York, 1983.
27. **Neilson, B. J. and Cronin, L. E., Eds.,** *Estuaries and Nutrients,* Humana Press, Clifton, NJ, 1981.
28. **Hobbie, J. E.,** Nutrients in estuaries, *Oceanus,* 19, 41, 1976.
29. **Liss, P. S.,** Conservative and nonconservative behavior of dissolved constituents during estuarine mixing, in *Estuarine Chemistry,* Burton, J. D. and Liss, P. S., Eds., Academic Press, London, 1976, 93.
30. **Aston, S. R.,** Estuarine chemistry, in *Chemical Oceanography,* Vol. 7, Riley, J. P. and Skirrow, G., Eds., Academic Press, London, 1978, 361.
31. **Batley, G. E. and Gardner, D.,** A study of copper, lead, and cadmium speciation in some estuarine and coastal marine waters, *Estuarine Coastal Mar. Sci.,* 7, 59, 1978.
32. **Moore, R. M., Burton, J. D., Williams, P. J. le B., and Young, M. L.,** The behavior of dissolved organic material, iron, and manganese in estuarine mixing, *Geochim. Cosmochim. Acta,* 43, 919, 1979.

2.1. PERIODIC TABLE OF THE ELEMENTS

Group numbering: New notation (1–18); Previous IUPAC form; CAS version.

KEY TO CHART

Atomic Number → 50 +2 ← Oxidation States
Symbol → Sn +4
1983 Atomic Weight → 118.71
18 18 4 ← Electron Configuration

1 Group IA	2 IIA	3 IIIA IIIB	4 IVA IVB	5 VA VB	6 VIA VIB	7 VIIA VIIB	8	9 VIIIA VIII	10	11 IB	12 IIB	13 IIIB IIIA	14 IVB IVA	15 VB VA	16 VIB VIA	17 VIIB VIIA	18 VIIIA	Orbit
1 +1 −1 H 1.00794 1																	2 0 He 4.00260 2	K
3 +1 Li 6.941 2-1	4 +2 Be 9.01218 2-2											5 +3 B 10.81 2-3	6 +2 +4 −4 C 12.011 2-4	7 +1 +2 +3 +4 +5 −1 −2 −3 N 14.0067 2-5	8 −2 O 15.9994 2-6	9 −1 F 18.9984 2-7	10 0 Ne 20.179 2-8	K-L
11 +1 Na 22.9898 2-8-1	12 +2 Mg 24.305 2-8-2											13 +3 Al 26.9815 2-8-3	14 +2 +4 −4 Si 28.0855 2-8-4	15 +3 +5 −3 P 30.9738 2-8-5	16 +4 +6 −2 S 32.06 2-8-6	17 +1 +5 +7 −1 Cl 35.453 2-8-7	18 0 Ar 39.948 2-8-8	K-L-M
19 +1 K 39.0983 -8-8-1	20 +2 Ca 40.08 -8-8-2	21 +3 Sc 44.9559 -8-9-2	22 +2 +3 +4 Ti 47.88 -8-10-2	23 +2 +3 +4 +5 V 50.9415 -8-11-2	24 +2 +3 +6 Cr 51.996 -8-13-1	25 +2 +3 +4 +7 Mn 54.9380 -8-13-2	26 +2 +3 Fe 55.847 -8-14-2	27 +2 +3 Co 58.9332 -8-15-2	28 +2 +3 Ni 58.69 -8-16-2	29 +1 +2 Cu 63.546 -8-18-1	30 +2 Zn 65.39 -8-18-2	31 +3 Ga 69.72 -8-18-3	32 +2 +4 Ge 72.59 -8-18-4	33 +3 +5 −3 As 74.9216 -8-18-5	34 +4 +6 −2 Se 78.96 -8-18-6	35 +1 +5 −1 Br 79.904 -8-18-7	36 0 Kr 83.80 -8-18-8	-L-M-N
37 +1 Rb 85.4678 -18-8-1	38 +2 Sr 87.62 -18-8-2	39 +3 Y 88.9059 -18-9-2	40 +4 Zr 91.224 -18-10-2	41 +3 +5 Nb 92.9064 -18-12-1	42 +6 Mo 95.94 -18-13-1	43 +4 +6 +7 Tc (98) -18-13-2	44 +3 Ru 101.07 -18-15-1	45 +3 Rh 102.906 -18-16-1	46 +2 +4 Pd 106.42 -18-18-0	47 +1 Ag 107.868 -18-18-1	48 +2 Cd 112.41 -18-18-2	49 +3 In 114.82 -18-18-3	50 +2 +4 Sn 118.71 -18-18-4	51 +3 +5 −3 Sb 121.75 -18-18-5	52 +4 +6 −2 Te 127.60 -18-18-6	53 +1 +5 +7 −1 I 126.905 -18-18-7	54 0 Xe 131.29 -18-18-8	-M-N-O
55 +1 Cs 132.905 -18-8-1	56 +2 Ba 137.33 -18-8-2	57* +3 La 138.906 -18-9-2	72 +4 Hf 178.49 -32-10-2	73 +5 Ta 180.948 -32-11-2	74 +6 W 183.85 -32-12-2	75 +4 +6 +7 Re 186.207 -32-13-2	76 +3 +4 Os 190.2 -32-14-2	77 +3 +4 Ir 192.22 -32-15-2	78 +2 +4 Pt 195.08 -32-16-2	79 +1 +3 Au 196.967 -32-18-1	80 +1 +2 Hg 200.59 -32-18-2	81 +1 +3 Tl 204.383 -32-18-3	82 +2 +4 Pb 207.2 -32-18-4	83 +3 +5 Bi 208.980 -32-18-5	84 +2 +4 Po (209) -32-18-6	85 At (210) -32-18-7	86 0 Rn (222) -32-18-8	-N-O-P
87 +1 Fr (223) -18-8-1	88 +2 Ra 226.025 -18-8-2	89** +3 Ac 227.028 -18-9-2	104 +4 Unq (261) -32-10-2	105 Unp (262) -32-11-2	106 Unh (263) -32-12-2	107 Uns (262) -32-13-2												O P Q

															Orbit
*Lanthanides	58 +3 +4 Ce 140.12 -20-8-2	59 +3 Pr 140.908 -21-8-2	60 +3 Nd 144.24 -22-8-2	61 +3 Pm (145) -23-8-2	62 +2 +3 Sm 150.36 -24-8-2	63 +2 +3 Eu 151.96 -25-8-2	64 +3 Gd 157.25 -25-9-2	65 +3 Tb 158.925 -27-8-2	66 +3 Dy 162.50 -28-8-2	67 +3 Ho 164.930 -29-8-2	68 +3 Er 167.26 -30-8-2	69 +3 Tm 168.934 -31-8-2	70 +2 +3 Yb 173.04 -32-8-2	71 +3 Lu 174.967 -32-9-2	N O P
**Actinides	90 +4 Th 232.038 -18-10-2	91 +5 +4 Pa 231.036 -20-9-2	92 +3 +4 +5 +6 U 238.029 -21-9-2	93 +3 +4 +5 +6 Np 237.048 -22-9-2	94 +3 +4 +5 +6 Pu (244) -24-8-2	95 +3 +4 +5 +6 Am (243) -25-8-2	96 +3 Cm (247) -25-9-2	97 +3 +4 Bk (247) -27-8-2	98 +3 Cf (251) -28-8-2	99 +3 Es (252) -29-8-2	100 +3 Fm (257) -30-8-2	101 +2 +3 Md (258) -31-8-2	102 +2 +3 No (259) -32-8-2	103 +3 Lr (260) -32-9-2	O P Q

Note: Numbers in parentheses are mass numbers of most stable isotope of that element. From *Chemical and Engineering News,* 63(5), 27, 1985. This format numbers the group 1 to 18.

2.2. COMPOSITION OF AIR AND PRECIPITATION

Table 2.2—1
COMPOSITION OF CLEAN DRY AIR NEAR SEA LEVEL

Component	Content (% by vol)	Mol wt
Nitrogen	78.084	28.0134
Oxygen	20.9476	31.9988
Argon	0.934	39.948
Carbon dioxide	0.0314	44.00995
Neon	0.001818	20.183
Helium	0.000524	4.0026
Krypton	0.000114	83.80
Xenon	0.0000087	131.30
Hydrogen	0.00005	2.01594
Methane	0.0002	16.04303
Nitrous oxide	0.00005	44.0128
Ozone		47.9982
Summer	0–0.000007	
Winter	0–0.000002	
Sulfur dioxide	0–0.0001	64.0628
Nitrogen dioxide	0–0.000002	46.0055
Ammonia	0–trace	17.03061
Carbon monoxide	0–trace	28.01055
Iodine	0–0.000001	253.8088

From Wedepohl, K. H., Ed., *Handbook of Geochemistry,* Vol. I, Springer-Verlag, Berlin, 1969. With permission.

Table 2.2—2
COMPOSITION OF MINOR GASES IN THE ATMOSPHERE

Species	X_i actual	Reliability	Source	Sink
CH_4	1.6×10^{-6}	High	Biog.	PhotoChem.
CO	$0.5\text{-}2 \times 10^{-7}$	Fair	Photo., Anthr.	PhotoChem.
O_3	10^{-7}-10^{-8}	Fair	Photo.	PhotoChem.
$NO + NO_2$	10^{-8}-10^{-12}	Low	Lightn., Anthr., Photo.	PhotoChem.
HNO_3	10^{-9}-10^{-11}	Low	Photo.	Rainout
NH_3	10^{-9}-10^{-10}	Low	Biog.	Photo., rainout
N_2O	3×10^{-7}	High	Biog.	Photo.
H_2	5×10^{-7}	High	Biog., Photo	Photo.
OH	10^{-15}-10^{-12}	Very low	Photo.	Photo.
HO_2	10^{-11}-10^{-13}	Very low	Photo.	Photo.
H_2O_2	10^{-10}-10^{-18}	Very low	Photo.	Rainout
H_2CO	10^{-10}-10^{-9}	Low	Photo.	Photo.
SO_2	10^{-11}-10^{-10}	Fair	Anthr., Photo., Volcanic	Photo.
CS_2	10^{-11}-10^{-10}	Low	Anthr., Biol.,	Photo.
OCS	10^{-10}	Fair	Anthr., Biol., Photo.	Photo.
CH_3CCl_3	$0.7^{-2} \times 10^{-10}$	Fair	Anthr.	Photo.

From Millero, F. J. and Sohn, M. L., *Chemical Oceanography,* CRC Press, Boca Raton, FL, 1992.

Table 2.2—3
CONCENTRATIONS OF ELEMENTS IN PRECIPITATION OVER MARINE REGIONS (UNITS, μg/l)

	North Sea	Bay of Bengal	Bermuda	N. Pacific (Enewetak)	S. Pacific (Samoa)
Na	19,000	14,000	3,400	1,100	2,500
Mg	—	1,600	490	170	270
K	—	1,400	160	39	88
Ca	—	1,700	310	50	75
Cl	32,000	25,000	6,800	2,000	4,700
Br	—	—	—	7.1	12
Al	105	—	—	2.1	16
Fe	84	30	4.8	1.0	0.42
Mn	<12	—	0.27	0.012	0.020
Sc	0.010	0.016	—	0.00023	—
Th	—	0.12	—	0.00091	—
Co	0.17	0.95	—	—	—
I	—	—	—	1.2	0.021
V	3.7	—	—	0.018	—
Zn	35	100	1.15	0.052	1.6
Cd	—	—	0.06	0.0021	—
Cu	15	—	0.66	0.013	0.021
Pb	—	—	0.77	0.035	0.014
Ag	—	—	—	0.0056	—
Se	—	—	—	0.021	0.026

From Furness, R. W. and Rainbow, P. S., Eds., *Heavy Metals in the Marine Environment,* CRC Press, Boca Raton, FL, 1990, 36.

Table 2.2—4
ATMOSPHERIC DEPOSITION FLUXES TO THE SEA SURFACE OVER A NUMBER OF MARINE REGIONS (μg/cm²/year)

Element	North Atlantic[a]	North Pacific[b,d]	South Pacific[c,d]
Al	5.0	1.8	0.13
Fe	3.2	0.82	0.047
Mn	0.07	0.0113	0.0036
Ni	0.02	—	—
Co	0.0027	0.00016	0.000025
Cr	0.014	—	—
V	0.017	0.0069	—
Cu	0.025	0.0071	0.0044
Pb	0.31	0.0074	0.0014
Zn	0.13	0.065	0.0058
Cd	0.002	0.00029	—

[a] Data for the tropical North Atlantic. Deposition fluxes for all elements calculated on the basis of a global total deposition (wet and dry) rate of 1 cm/s.

[b] Data for Enewetak, North Pacific.

[c] Data for Samoa, South Pacific.

[d] Wet deposition fluxes were calculated using two methods: (1) those based on the concentration of elements in rain and the total rainfall at the two islands, and (2) those based on scavenging ratios. In order to avoid any bias which may be introduced by unrepresentative total rainfall over the islands compared to the open-ocean regions, wet deposition rates for both Enewetak and Samoa have been estimated from scavenging ratios only; both wet and dry deposition rates have been adjusted to take account of sea-surface recycling.

From Furness, R. W. and Rainbow, P. S., Eds., *Heavy Metals in the Marine Environment,* CRC Press, Boca Raton, FL, 1990, 45.

Table 2.2—5
NET FLUVIAL AND ATMOSPHERIC FLUXES TO THE GLOBAL SEA SURFACE (μg/cm²/year)

Element	Net global fluvial dissolved flux		Net global total atmospheric input		Estimated average seawater solubility from aerosols (%)	Net global dissolved atmospheric flux[d]
	Present work[a]	Collier and Edmond[b]	Present work[c]	Collier and Edmond[b]		
Al	0.27	0.13—0.67	1.85	0.27—5.4	5	0.088
Fe	0.085	<0.73	1.0	1.12—3.35	7.5	0.075
Mn	0.085	0.093	0.021	0.016—0.044	35	0.0074
Ni	0.0082	0.0035	<0.02	0.003	40	<0.008
Co	0.0021	—	0.00068	—	22.5	0.00015
Cr	0.0052	—	<0.014	—	10	<0.0014
V	0.010	—	<0.010	—	25	<0.0028
Cu	0.018	0.019	0.010	0.00064—0.0095	30	0.0033
Pb	0.001	—	0.074	—	30	0.022
Zn	0.0079	<0.0065	0.055	0.022—0.013	45	0.025
Cd	0.00088	0.00022	<0.00096	0.0023	80	<0.00077

[a] Calculated from net fluvial flux data assuming a global ocean area of 352.6×10^6 km²; at a global river discharge of 37,400 km³/year, which is included in the net flux estimates, this would yield a layer 10.6 cm spread evenly over the whole ocean surface.

[b] Collier, R. and Edmond, J., The trace element geochemistry of marine biogenic particulate matter, *Prog. Oceanog.*, 13, 113, 1984.

[c] Calculated from the tropical N. Atlantic (column 1), N. Pacific (column 2), and S. Pacific (column 3) atmospheric deposition fluxes by scaling them to global fluxes weighted on an areal basis for each oceanic region.

[d] Net total atmospheric flux adjusted for the average seawater solubility of the individual elements.

From Furness, R. W. and Rainbow, P. S., Eds., *Heavy Metals in the Marine Environment,* CRC Press, Boca Raton, FL, 1990, 46.

Table 2.2—6
ATMOSPHERIC DISSOLVED VS. FLUVIAL DISSOLVED INPUTS TO REGIONS OF THE WORLD OCEAN

Element	Oceanic Region					
	North Atlantic		North Pacific		South Pacific	
	Fluvial dissolved flux[a]	Atmospheric dissolved flux	Fluvial dissolved flux	Atmospheric dissolved flux	Fluvial dissolved flux	Atmospheric dissolved flux
Al	0.56	0.25	0.24	0.09	0.12	0.0066
Fe	0.18	0.24	0.076	0.062	0.038	0.0035
Mn	0.18	0.025	0.076	0.004	0.038	0.0013
Ni	0.017	0.008	0.0074	—	0.0037	—
Co	0.0044	0.00061	0.0019	0.00004	0.0009	0.000006
Cr	0.011	0.0014	0.0047	—	0.0023	—
V	0.022	0.0043	0.0094	0.0017	0.0047	—
Cu	0.037	0.0075	0.016	0.0021	0.0079	0.0013
Pb	0.0022	0.093	0.0009	0.0022	0.0005	0.0004
Zn	0.016	0.059	0.0071	0.029	0.0035	0.0026
Cd	0.0018	0.0016	0.0008	0.00023	0.0004	—

[a] No attempt has been made to adjust the North Atlantic fluvial flux for European and North American anthropogenic inputs.

From Furness, R. W. and Rainbow, P. S., Eds., *Heavy Metals in the Marine Environment,* CRC Press, Boca Raton, FL, 1990, 46.

Table 2.2—7
RELATIVE FLUVIAL AND ATMOSPHERIC FLUXES TO SOME COASTAL OCEANIC REGIONS

A. Ratio of Total Atmospheric Flux to Fluvial Flux

Element	South Atlantic Bight	New York Bight	North Sea	Western Mediterranean
Fe	5.8	6.4	1.7	—
Mn	0.6	—	0.8	—
Cu	1.9	—	1.9	—
Ni	1.7	—	1.3	—
Pb	9.5	20	6.8	6.2
Zn	2.3	3.1	1.9	0.8
Cd	2.7	3.1	1.1	—
As	2.1	1.0	1.7	—
Hg	22	—	2.1	0.8

B. Estimated Fluvial and "Soluble" Atmospheric Fluxes to the South Atlantic Bight (Units, g × 10^6)

Element	Fluvial flux	Total atmospheric flux	Soluble atmospheric flux	Ratio: soluble atmospheric flux to fluvial flux
Fe	950	5500	413	0.43
Mn	91	57	20	0.29
Cu	110	210	63	0.57
Ni	220	370	148	0.85
Pb	65	620	186	4.8
Zn	310	710	320	1.6
Cd	3	8	6.4	2.3

From Furness, R. W. and Rainbow, P. S., Eds., *Heavy Metals in the Marine Environment,* CRC Press, Boca Raton, FL, 1990, 42.

Table 2.2—8
THE SOLUBILITY OF ATMOSPHERIC TRACE METALS IN SEAWATER

Atmospheric Population	Al	Fe	Mn	Ni	Co	Cr	V	Ag	Zn	Cu	Pb	Cd
Anthropogenic-rich[a]												
Mean concentration (ng/m^3 of air)	900	610	11	11	<0.4	<1.6	13	<0.05	25	16	560	0.25
Mean % soluble in seawater	0.56	1.1	47	47	25	12.5	31	80	68	28	39	84
Mean EF_{crust}	1.0	0.99	1.0	14	1.5	1.5	9.0	—	33	26	4148	16
Dust-rich[b]												
Mean concentration (ng/m^3 of air)	3380	2100	46	2.8	<0.5	4.0	<11	<0.03	18	20	150	0.20
Mean % soluble in seawater	0.09	0.19	34	28	20	10	18	33	24	14.5	13	80
Mean EF_{crust}	1.0	0.91	1.2	0.92	0.49	0.99	2.0	—	6.2	8.4	296	25

[a] Southern California.
[b] Baja, California.

From Furness, R. W. and Rainbow, P. S., Eds., *Heavy Metals in the Marine Environment*, CRC Press, Boca Raton, FL, 1990, 41.

2.3 COMPOSITION OF SEAWATER

Table 2.3—1
MAJOR COMPONENTS OF SEAWATER

1. Solids (material that does not pass through a 0.45 μm filter)
   a. Particulate organic material (plant detritus)
   b. Particulate inorganic material (minerals)
2. Gases
   a. Conservative (N_2, Ar, Xe)
   b. Nonconservative (O_2 and CO_2)
3. Colloids (passes through 0.45 μm filter, but is not dissolved)
   a. Organic
   b. Inorganic
4. Dissolved Solutes
   a. Inorganic solutes
      1. Major ($>$ 1 ppm)
      2. Minor ($<$ 1 ppm)
   b. Organic solutes

From Millero, F. J. and Sohn, M. L., *Chemical Oceanography,* CRC Press, Boca Raton, FL, 1992, 59.

Table 2.3—2
ELEMENTS IN SEAWATER

Atomic Number	Element	Species	Behavior	Predicted mean water concentration	Residence time years: Based on sediment accumulation	Residence time years: Based on river input
1	Hydrogen	H_2	Biogenic or hydrothermal origin	108 g/kg		
2	Helium		Nonnutrient gas	1.9 nmol/kg		
3	Lithium		Conservative	178 μg/kg	2.0×10^7	
4	Beryllium		Nutrientlike, but increases with depth	0.2 ng/kg	150	
5	*Boron	Inorganic boron	Conservative	4.4 mg/kg		
6	†Carbon	ΣCO_2	Nutrient	2200 μmol/kg		
7	†Nitrogen	N_2	Nonnutrient gas	590 μmol/kg		
		Nitrate	Nutrient	30 μmol/kg		
8	Oxygen	Dissolved O_2	Biological dependence	857 g/kg		
9	Fluorine		Conservative	1.3 mg/kg		
10	Neon		Nonnutrient gas	8 nmol/kg		
11	*Sodium		Conservative	10.781 g/kg	2.6×10^8	2.1×10^8
12	*Magnesium		Conservative	1.28 g/kg	4.5×10^7	2.2×10^7
13	Aluminum			1 μg/kg	100	
14	†Silicon	Silicate	Nutrient	110 μmol/kg	8×10^3	3.5×10^4
15	†Phosphorus	Phosphate	Nutrient	2 μmol/kg		
16	*Sulfur	Sulfate	Conservative	2.712 g/kg		
17	*Chlorine	Chloride	Conservative	19.353 g/kg		
18	Argon		Nonnutrient gas	15.6 μmol/kg		
19	*Potassium		Conservative	399 mg/kg	1.1×10^7	1×10^7
20	*Calcium		Correlates with carbonate alkalinity	415 mg/kg	8.0×10^6	1×10^6
21	Scandium			<1 ng/kg	5.6×10^3	
22	Titanium			<1 ng/kg	160	
23	Vanadium		Conservative	<1 μg/kg	1.0×10^4	
24	Chromium		Nutrient-correlated; silicate and phosphate or nitrate	330 ng/kg	350	
25	Manganese		Surface maximum; at depth, correlated with the labile nutrients and negatively correlated with dissolved oxygen	10 ng/kg	1400	
26	Iron		Correlated with the nutrients; negatively correlated with dissolved oxygen	40 ng/kg	140	
27	Cobalt		Similar to manganese	2ng/kg	1.8×10^4	
28	Nickel		Nutrient-correlated; phosphate and silicate	480 ng/kg	1.8×10^4	
29	Copper		Resembles nutrients with sedimentary release; scavenging at intermediate depths	120 ng/kg	5.0×10^4	

30	Zinc		Nutrient-correlated; silicate	390 ng/kg	1.8×10^5
31	Gallium			10 – 20 ng/kg	1.4×10^3
32	Germanium		Nutrient-correlated; silicate	5 ng/kg	7×10^3
33	Arsenic	As(V)	Nutrient-correlated; phosphate	2 μg/kg	
34	Selenium	Se(IV)	Nutrient-correlated; silicate and phosphate	170 ng/kg	
35	*Bromine	Bromide	Conservative	67 mg/kg	
36	Krypton		Nonnutrient gas	3.7 nmol/kg	
37	Rubidium		Conservative	124 μg/kg	2.7×10^5
38	Strontium		Nutrient-correlated; phosphate	7.8 mg/kg	1.9×10^7
39	Yttrium		First approximation: conservative	13 ng/kg	7.5×10^3
40	Zirconium			<1 μg/kg	
41	Niobium			'1 ng/kg	300
42	Molybdenum		Conservative	11 μg/kg	5.0×10^5
44	Ruthenium			0.5 ng/kg	
45	Rhodium				
46	Palladium				
47	Silver			3 ng/kg	2.1×10^6
48	Cadmium		Nutrient-correlated; phosphate	70 ng/kg	5.0×10^5
49	Indium			0.2 ng/kg	
50	Tin		Nonconservative; anthropogenic	0.5 ng/kg	5.0×10^5
51	Antimony		Conservative	0.2 μg/kg	3.5×10^5
52	Tellurium				
53	Iodine	Iodate	Nutrient-correlated; nitrate and phosphate	59 μg/kg	
54	Xenon		Nonnutrient gas	0.5 nmol/kg	
55	Cesium		Conservative	0.3 ng/kg	4×10^4
56	Barium		Nutrient-correlated; silicate, alkalinity	11.7 μg/kg	8.4×10^4
57–71	Lanthanum and the Lanthanides		Nutrient- or depth- correlated		1.1×10^4
72	Hafnium			<8 ng/kg	
73	Tantalum			<2.5 ng/kg	
74	Tungsten			<1 ng/kg	1.0×10^3
75	Rhenium			4 ng/kg	
76	Osmium				
77	Iridium				
78	Platinum				
79	Gold			11 ng/kg	5.6×10^5
80	Mercury		Nutrient-correlated; silicate	6 ng/kg	4.2×10^4
81	Thallium		Conservative	12 ng/kg	
82	Lead		Nonconservative; anthropogenic	1 ng/kg	2×10^3
83	Bismuth			10 ng/kg	4.5×10^4
84	Polonium				
85	(Astatine)				

Table 2.3—2 (continued)
ELEMENTS IN SEAWATER

Atomic Number	Element	Species	Behavior	Predicted mean water concentration	Residence time years: Based on sediment accumulation	Residence time years: Based on river input
86	Radon					
87	(Francium)					
88	Radium					
89	Actinium					
90	Thorium			<0.7 ng/kg	350	
91	Protactinium					
92	Uranium		Conservative	3.2 μg/kg	5×10^5	

* These eight elements are called conservative; they exist in uniform relative concentrations throughout all oceans.

† These are the commonly labeled plant *nutrient* elements.

From Neshyba, S., *Oceanography: Perspectives on a Fluid Earth,* John Wiley & Sons, New York, 1987, 118. Reproduce by permission of copyright © owner, John Wiley & Sons, Inc.

Table 2.3—3
SPECIATION OF ELEMENTS IN SEAWATER

Element	Probable main species in oxygenated seawater
Li	Li^+
Be	$BeOH^+$, $Be(OH)_2^0$
B	H_3BO_3
C	HCO_3^-, CO_3^{2-}
N	NO_3^- (also as N_2)
O	O_2 (also as H_2O)
F	F, MgF^+
Na	Na^+
Mg	Mg^{2+}
Al	$Al(OH)_4^-$, $Al(OH)_3^0$
Si	H_4SiO_4
P	HPO_4^{2-}, $NaHPO_4^-$, $MgHPO_4^0$
S	SO_4^{2-}, $NaSO_4^-$, $MgSO_4^0$
Cl	Cl^-
K	K^+
Ca	Ca^{2+}
Sc	$Sc(OH)_3^0$
Ti	$Ti(OH)_4^0$
V	HVO_4^{2-}, $H_2VO_4^-$, $NaHVO_4^-$
Cr	CrO_4^{2-}, $NaCrO_4^-$
Mn	Mn^{2+}, $MnCl^+$
Fe	$Fe(OH)_3^0$
Co	Co^{2+}, $CoCO_3^0$, $CoCl^+$
Ni	Ni^{2+}, $NiCO_3^0$, $NiCl^+$
Cu	$CuCO_3^0$, $CuOH^+$, Cu^{2+}
Zn	Zn^{2+}, $ZnOH^+$, $ZnCO_3^0$, $ZnCl^+$
Ga	$Ga(OH)_4^-$
Ge	H_4GeO_4, $H_3GeO_4^-$
As	$HAsO_4^{2-}$
Se	SeO_4^{2-}, SeO_3^{2-}, $HSeO_3^-$
Br	Br^-
Rb	Rb^+
Sr	Sr^{2+}
Y	YCO_3^+, YOH^{2+}, Y^{3+}
Zr	$Zr(OH)_4^0$, $Zr(OH)_5^-$
Nb	$Nb(OH)_6^-$, $Nb(OH)_5^0$
Mo	MoO_4^{2-}
(Tc)	TcO_4^-
Ru	?
Rh	?
Pd	?
Ag	$AgCl_2^-$
Cd	$CdCl_2^0$
In	$In(OH)_3^0$
Sn	$SnO(OH)_3^-$
Sb	$Sb(OH)_6^-$
Te	TeO_3^{2-}, $HTeO_3^-$
I	IO_3^-
Cs	Cs^+
Ba	Ba^{2+}
La	La^{3+}, $LaCO_3^+$, $LaCl^{2+}$
Ce	$CeCO_3^+$, Ce^{3+}, $CeCl^{2+}$
Pr	$PrCO_3^+$, Pr^{3+}, $PrSO_4^+$
Nd	$NdCO_3^+$, Nd^{3+}, $NdSO_4^+$
Sm	$SmCO_3^+$, Sm^{3+}, $SmSO_4^+$
Eu	$EuCO_3^+$, Eu^{3+}, $EuOH^{2+}$
Gd	$GdCO_3^+$, Gd^{3+}
Tb	$TbCO_3^+$, Tb^{3+}, $TbOH^{2+}$
Dy	$DyCO_3^+$, Dy^{3+}, $DyOH^{2+}$
Ho	$HoCO_3^+$, Ho^{3+}, $HoOH^{2+}$
Er	$ErCO_3^+$, $ErOH^{2+}$, Er^{3+}
Tm	$TmCO_3^+$, $TmOH^{2+}$, Tm^{3+}
Yb	$YbCO_3^+$, $YbOH^{2+}$
Lu	$LuCO_3^+$, $LuOH^{2+}$
Hf	$Hf(OH)_4^0$, $Hf(OH)_5^-$
Ta	$Ta(OH)_5^0$
W	WO_4^{2-}
Re	ReO_4^-
Os	?
Ir	?
Pt	?
Au	$AuCl_2^-$
Hg	$HgCl_4^{2-}$
Tl	Tl^+, $TlCl^0$; or $Tl(OH)_3^0$
Pb	$PbCO_3^0$, $Pb(CO_3)_2^{2-}$, $PbCl^+$
Bi	BiO^+, $Bi(OH)_2^+$

From Bruland, K. W., in *Chemical Oceanography,* Vol. 8, Riley, J. P. and Chester, R., Eds., Academic Press, London, 1983, 157.
Copyright: 1983 Academic Press Inc. (London) Ltd.

Table 2.3—4
CONCENTRATION OF THE MAJOR IONS IN SEAWATER (g/kg SEAWATER) NORMALIZED TO 35‰ SALINITY

Ion	Average value	Range	Ref.
Chloride	19.353		
Sodium	10.76	10.72–10.80	2
	10.79	10.76–10.80	9
Magnesium	1.297	1.292–1.301	2
	1.292	1.296–1.287	9
Sulfate	2.712	2.701–2.724	5
Calcium	0.4119	0.4098–0.4134	2
	0.4123	0.4088–0.4165	9
Potassium	0.399	0.393–0.405	2, 9
Bicarbonate[a]	0.145	0.137–0.153	4, 7, 6
Bromide	0.0673	0.0666–0.0680	5
Boron	0.0046	0.0043–0.0051	1
Strontium	0.0078	0.0074–0.0079	2
	0.0081	0.0078–0.0085	9
Fluoride	0.0013	0.0012–0.0017	3, 8

[a] The values reported for bicarbonate are actually titration alkalinities.

Compiled by D. R. Kester, Graduate School of Oceanography, University of Rhode Island.

REFERENCES

1. **Culkin, F.,** The major constituents of sea water, in *Chemical Oceanography,* Vol. 1, Riley, J. P. and Skirrow, G., Eds., Academic Press, London, 1965, 121–161.
2. **Culkin, F. and Cox, R. A.,** Sodium, potassium, magnesium, calcium, and strontium in seawater, *Deep-Sea Res. Oceanogr. Abstr.,* 13, 789, 1966.
3. **Greenhalgh, R. and Riley, J. P.,** Occurrence of abnormally high fluoride concentrations at depth in the oceans, *Nature,* 197, 371, 1963.
4. **Koczy, F. F.,** The specific alkalinity, *Deep-Sea Res. Oceanogr. Abstr.,* 3, 279, 1956.
5. **Morris, A. W. and Riley, J. P.,** The bromide/chlorinity and sulphate/chlorinity ratio in sea water, *Deep-Sea Res. Oceanogr. Abstr.,* 3, 699, 1966.
6. **Park, Kilho,** Deep-sea pH, *Science,* 154, 1540, 1966.
7. **Postma, H.,** The exchange of oxygen and carbon dioxide between the ocean and the atmosphere, *Neth. J. Sea Res.,* 2, 258, 1964.
8. **Riley, J. P.,** The occurrence of anomalously high fluoride concentrations in the North Atlantic, *Deep-Sea Res. Oceanogr. Abstr.,* 12, 219, 1965.
9. **Riley, J. P. and Tongudai, M.,** The major cation/chlorinity ratios in sea water, *Chem. Geol.,* 2, 263, 1967.

Table 2.3—5
MOLALITY OF THE MAJOR CONSTITUENTS IN SEAWATER AT VARIOUS SALINITIES

Constituent	Salinity 30.0°/₀₀	34.8°/₀₀	40.0°/₀₀
Cl^-	0.48243	0.56241	0.64997
Na^+	0.41417	0.48284	0.55801
Mg^{2+}	0.04666	0.05440	0.06287
SO_4^{2-}	0.02495	0.02909	0.03362
Ca^{2+}	0.00909	0.01059	0.01224
K^+	0.00902	0.01052	0.01215
HCO_3^-	0.00211	0.00245	0.00284
Br^-	0.00074	0.00087	0.00100
$B(OH)_3$	0.00038	0.00044	0.00051
Sr^{2+}	0.00008	0.00009	0.00011

Note: These molalities are based on the concentrations in Table 2.3—4. The averages of the two sets of analyses were used for Mg, Ca, K, and Sr. The Na^+ was calculated to preserve electroneutrality; the resulting value is within 0.1% of the analytical values of Na^+ reported in Table 2.3—4.

Compiled by D. R. Kester, Graduate School of Oceanography, University of Rhode Island.

Table 2.3—6
COMPOSITION OF 1 kg OF NATURAL SEAWATER AS A FUNCTION OF CHLORINITY

Species	g_i/Cl	n_i/Cl	e_i/Cl	$n_iZ_i^2/Cl$
Na^+	0.55653	0.0242077	0.0242077	0.0242077
Mg^{2+}	0.06626	0.0027262	0.0054524	0.0109048
Ca^{2+}	0.02127	0.0005307	0.0010614	0.0021228
K^+	0.02060	0.0005268	0.0005368	0.0005268
Sr^{2+}	0.00041	0.0000047	0.0000094	0.0000187
Cl^-	0.99891	0.0281756	0.0281756	0.0281756
SO_4^{2-}	0.14000	0.0014575	0.0029149	0.0058298
HCO_3^-	0.00586	0.0000960	0.0000960	0.0000960
Br^-	0.00347	0.0000434	0.0000434	0.0000434
CO_3^{2-}	0.00060	0.0000100	0.0000200	0.0000400
$B(OH)_4^-$	0.00034	0.0000043	0.0000043	0.0000043
F^-	0.00006	0.0000035	0.0000035	0.0000035
1/2 =	0.0288932	0.0312577	0.0359867	
$B(OH)_3$	0.00105	0.0000170	0.0000170	
	1.815367	0.0289102	0.0312577	

Note: pH = 8.1 and t = 25°C.

Table 2.3—7
MAJOR CONSTITUENT CONCENTRATION-TO-CHLORINITY RATIOS FOR VARIOUS OCEANS AND SEAS

Ocean or Sea	Na °/oo *Cl*	Mg °/oo *Cl*	K °/oo *Cl*	Ca °/oo *Cl*	Sr °/oo *Cl*	SO_4 °/oo *Cl*	Br °/oo *Cl*
N. Atlantic	–	–	0.02026	–	–	–	0.00337 – 0.00341
Atlantic	0.5544 – 0.5567	0.0667	0.01953 – 0.0263	0.02122 – 0.02126	0.000420	0.1393	0.00325 – 0.0038
N. Pacific	0.5553	0.06632 – 0.06695	0.02096	0.02154	–	0.1396 – 0.1397	0.00348
W. Pacific	0.5497 – 0.5561	0.06627 – 0.0676	0.02125	0.02058 – 0.02128	0.000413 – 0.000420	0.1399	0.0033
Indian	–	–	–	0.02099	0.000445	0.1399	0.0038
Mediterranean	0.5310 – 0.5528	0.06785	0.02008	–	–	0.1396	0.0034 – 0.0038
Baltic	0.5536	0.06693	–	0.02156		0.1414	0.00316 – 0.00344
Black	0.55184	–	0.0210	–	–	–	–
Irish	0.5573	–	–	–	–	0.1397	0.0033
Puget Sound	0.5495 – 0.5562	–	0.0191	–	–	–	–
Siberian	0.5484	–	0.0211	–	–	–	–
Antarctic	–	–	–	0.02120	0.000467	–	0.00347
Tokyo Bay	–	0.0676	–	0.02130	–	0.1394	–
Barents	–	0.06742	–	0.02085	–	–	–
Arctic	–	–	–	–	0.000424	–	–
Red	–	–	–	–	–	0.1395	0.0043
Japan	–	–	–	–	–	–	0.00327 – 0.00347
Bering	–	–	–	–	–	–	0.00341
Adriatic	–	–	–	–	–	–	0.00341

From Culkin, F. and Cox, R. A., *Deep-Sea Research*, 13, 789, 1966. With kind permission of Pergamon Press, Ltd., Headington Hill Hall, Oxford OX3 OBW, U.K.

Table 2.3—8
MAJOR CHEMICAL SPECIES IN SEAWATER

Constituent	Percentage of the constituent present in each species at 25°C, 19.375‰ chlorinity, 1 atm, and pH 8.0
Chloride	—
Sodium	Na^+ (97.7%); $NaSO_4^-$ (2.2%); $NaHCO_3^\circ$ (0.03%)
Magnesium	Mg^{2+} (89%); $MgSO_4^\circ$ (10%); $MgHCO_3^+$ (0.6%); $MgCO_3^\circ$ (0.1%)
Sulfate	SO_4^{2-} (39%); $NaSO_4^-$ (37%); $MgSO_4^\circ$ (19%); $CaSO_4^\circ$ (4%)
Calcium	Ca^{2+} (88%); $CaSO_4^\circ$ (11%); $CaHCO_3^+$ (0.6%); $CaCO_3^\circ$ (0.1%)
Potassium	K^+ (98.8%); KSO_4^- (1.2%)
Bicarbonate	HCO_3^- (64%) $MgHCO_3^+$ (16%); $NaHCO_3^\circ$ (8%); $CaHCO_3^+$ (3%); CO_3^{2-} (0.8%); $MgCO_3^\circ$ (6%); $NaCO_3^-$ (1%); $CaCO_3^\circ$ (0.5%)
Bromide	—
Boron	$B(OH)_3$ (84%); $B(OH)_4^-$ (16%)
Strontium	—
Fluoride	F^- (50-80%); MgF^+ (20-50%)

From Pytkowicz, R. M. and Kester, D. R., The physical chemistry of seawater, in *Oceanogr. Mar. Biol. Ann. Rev.*, Barnes, H., Ed., 9, 11, 1971. With permission of George Allen and Unwin, Ltd., London.

Table 2.3—9
TOTAL ACTIVITY COEFFICIENTS OF IONS IN SEAWATER

Na^+	K^+	Mg^{2+}	Ca^{2+}	Cl^-	SO_4^{2-}	HCO_3^-	CO_3^{2-}
0.75	0.63	0.32	0.25	0.64	0.065	0.47	0.02
0.70	0.62	0.25	0.23	0.63	0.07	0.51	0.02
0.68	0.63	0.23	0.21	0.66	0.11	0.59	0.03
0.65	0.62	0.22	0.20	0.69	0.12	—	—
0.69	0.62	0.26	0.23	0.63	0.066	0.55	0.016
0.68	0.64	0.23	0.21	0.68	0.11	0.55	0.02

Note: All values, except those on the last line, are estimated using the approaches outlined in Burton.[1] The calculated values are generally for an ionic strength of 0.7, but those of Leyendekkers[2] for HCO_3^- and CO_3^{2-} are for an ionic strength of 0.5.

From Burton, J. D., in *Estuarine Chemistry,* Copyright: 1976 Academic Press Inc. (London).

REFERENCES

1. **Burton, J. D.,** Basic properties and processes in estuarine chemistry, in *Estuarine Chemistry,* Burton, J. D. and Liss, P. S., Eds., Academic Press, London, 1976, 1.
2. **Leyendekkers, J. V.,** The chemical potentials of seawater components, *Mar. Chem.*, 1, 75, 1973.

Table 2.3—10
MINOR CONSTITUENTS OF SEAWATER EXCLUDING THE DISSOLVED GASES*

Element	Concentration μg/l Average	Range	References on the distribution in the oceans
Lithium	185 (2, 18, 29, 52, 62)	180–195 (2, 18, 29, 52, 62)	(2, 18, 29, 52, 62)
Beryllium	5.7 x 10^{-4} (48)		
Nitrogen	280 (82)	0–560 (82)	(82)
Aluminum	2 (65, 70, 10, 31)	0–7 (65, 70)	(65)
Silicon	2000 (5)	0–4900 (5)	(5)
Phosphorus	30 (4)	0–90 (4)	(4)
Scandium	0.04 (31)		
	<0.004 (68)	0.1–18 x 10^{-4} (37)	
	9.6 x 10^{-4} (37)		
Titanium	1 (34)		
Vanadium	2.5 (11, 12)	2.0–3.0 (11, 12)	(11)
Chromium	0.3 (30)	0.23–0.43 (30)	
	0.05 (41)	0.04–0.07 (41)	
Manganese	1.5 (64)	0.2–8.6 (64)	(64, 78)
	0.9 (78)	0.7–1.3 (78)	
		3.0–4.4 (28)	
Iron	6.6 (78)	0.1–62 (78)	(3, 7, 22, 78)
	2.6 (70)	8–13 (22)	
	0.2 (7)	0–7 (28, 70)	
		0.03–2.56 (7)	
Cobalt	0.27 (68)	0.035–4.1 (68)	(68, 69, 78)
	0.032 (78)	<0.005–0.092 (78)	
Nickel	5.4 (68)	0.43–43 (22, 68)	(68, 69, 73)
	1.7 (73)	0.8–2.4 (73)	
		0.13–0.37 (28)	
Copper	2 (28, 80)	0.2–4 (28, 73, 78, 80)	(1, 73, 78, 80)
	1.2 (78)	0.5–27 (1, 9, 22, 38)	
	0.7 (73)		
Zinc	12.3 (78)	3.9–48.4 (78)	(64, 73, 78, 80)
	6.5 (64, 80)	2–18 (64, 80)	
	2 (73)	1–8 (73)	
		29–50 (9)	
Gallium	0.03 (23)	0.023–0.037 (23)	
Germanium	0.05 (12, 27)	0.05–0.06 (12)	
Arsenic	4 (39)	3–6 (39)	
	0.46 (41)	2–35 (61)	
Selenium	0.2 (15, 68)	0.34–0.50 (15)	(68)
		0.052–0.12 (68)	
Rubidium	120 (8, 52, 63, 71)	112–134 (8, 52, 63, 71)	(8, 29, 52, 71)
		86–119 (29)	
Yttrium	0.03 (31)	0.0112–0.0163 (37)	
	0.0133 (37)		
Zirconium	2.6 x 10^{-2} (88)		
Niobium	0.01 (13)	0.01–0.02 (13)	
Molybdenum	10 (41)	0.24–12.2 (9, 85)	
	1 (9)		
Technetium			

Table 2.3—10 (continued)
MINOR CONSTITUENTS OF SEAWATER EXCLUDING THE DISSOLVED GASES*

Element	Concentration μg/l Average	Range	References on the distribution in the oceans
Ruthenium	0.0007 (88)		
Rhodium			
Palladium			
Silver	0.29 (68)	0.055–1.5 (68)	(68, 69)
	0.04 (31)		
Cadmium	0.113 (53)	0.02–0.25 (53)	
Indium	<20 (31)		
Tin	0.8 (31)		
Antimony	0.33 (68)	0.18–1.1 (68)	(68)
Tellurium			
Iodine	63 (6)	48–80 (6)	(6)
	44 (41)		
Cesium	0.4 (8, 63)	0.27–0.33 (8)	(8)
		0.48–0.58 (63)	
Barium	20 (8, 17, 19, 81)	5–93 (8, 17, 19, 81)	(19, 81)
Lanthanum	3×10^{-3} (33, 37)	$1–6 \times 10^{-3}$ (37)	(35–37)
Cerium	14×10^{-3} (14)	$4–850 \times 10^{-3}$ (14)	(35–37)
	1×10^{-3} 37)	$0.6–2.8 \times 10^{-3}$ (37)	
Praseodymium	6.4×10^{-4} (33, 37)	$4.1–15.8 \times 10^{-4}$ (37)	(35–37)
Neodymium	23×10^{-4} (33)	$13–65 \times 10^{-4}$ (37)	(35–37)
	28×10^{-4} (37)		
Promethium			
Samarium	4.2×10^{-4} (33)	$2.6–10 \times 10^{-4}$ (37)	(35–37)
	4.5×10^{-4} (37)		
Europium	1.14×10^{-4}	$0.9–7.9 \times 10^{-4}$ (37)	(35–37)
	1.3×10^{-4} (37)		
Gadolinium	6.0×10^{-4} (33)	$5.2–11.5 \times 10^{-4}$ (37)	(35–37)
	7.0×10^{-4} (37)		
Terbium	1.4×10^{-4} (37)	$0.6–3.6 \times 10^{-4}$ (37)	(35–37)
Dysprosium	7.3×10^{-4} (33)	$5.2–14.0 \times 10^{-4}$ (37)	(35–37)
	9.1×10^{-4} (37)		
Holmium	2.2×10^{-4} (33, 37)	$1.2–7.2 \times 10^{-4}$ (33, 37)	(33, 35–37)
Erbium	6.1×10^{-4} (33)	$6.6–12.4 \times 10^{-4}$ (37)	(35–37)
	8.7×10^{-4} (37)		
Thulium	1.3×10^{-4} (33)	$0.9–3.7 \times 10^{-4}$ (37)	(35–37)
	1.7×10^{-4} (37)		
Ytterbium	5.2×10^{-4} (33)	$4.8–28 \times 10^{-4}$ (33, 37)	(33, 35–37)
	8.2×10^{-4} (37)		
Lutetium	2.0×10^{-4} (33)	$1.2–7.5 \times 10^{-4}$ (33, 37)	(33, 35–37)
	1.5×10^{-4} (37)		
Hafnium	80×10^{-4} (68)		
Tantalum	25×10^{-4} (68)		
Tungsten	0.1 (41)		
Rhenium	8.4×10^{-3} (66)		
Osmium			
Indium	1×10^{-4} (88)		
Platinum			

Table 2.3—10 (continued)
MINOR CONSTITUENTS OF SEAWATER EXCLUDING THE DISSOLVED GASES*

Element	Concentration μg/l — Average	Concentration μg/l — Range	References on the distribution in the oceans
Gold	0.068 (86)	0.004–0.027 (68)	
Mercury	0.03 (31)		
Thallium	<0.01 (31)		
Lead	0.05 (19)	0.02–0.4 (19, 76, 77)	(19, 76, 77)
Bismuth	0.02 (56)	0.015–0.033 (56)	
Polonium			
Astatine			
Francium			
Radium	8×10^{-8} (55)	$4–15 \times 10^{-8}$ (45, 49, 75)	(45, 49, 75)
Actinium			
Thorium	0.05 (31)	$2–40 \times 10^{-4}$ (51, 72)	
	0.02 (55)		
	6×10^{-4} (51, 72)		
	$<7 \times 10^{-5}$ (42)		
Protactinium	2×10^{-6} (31)		
	5×10^{-8} (55)		
Uranium	3 (50, 55, 79)	2–4.7 (50, 55, 79)	(50, 55, 79)

* The numbers in parentheses refer to the citations listed after the table. The concentrations represent the dissolved and particulate forms of the elements.

(Based on compilations of Pytkowicz, R. M. and Kester, D. R., The physical chemistry of seawater, in *Oceanogr. Mar. Biol. Ann. Rev.*, Barnes, H., Ed., 9, 11, 1971. With permission of George Allen and Unwin, Ltd., London.)

REFERENCES

1. **Alexander, J. E. and Corcoran, E. F.,** *Limnol. Oceanogr.*, 12, 236, 1967.
2. **Angino, E. E. and Billings, G. K.,** *Geochim. Cosmochim. Acta*, 30, 153, 1966.
3. **Armstrong, F. A. J.,** *J. Mar. Biol. Assoc. U.K.*, 36, 509, 1957.
4. **Armstrong, F. A. J.,** in *Chemical Oceanography*, Vol. 1, Riley, J. P. and Skirrow, G., Eds., Academic Press, London, 1965, 323-364.
5. **Armstrong, F. A. J.,** in *Chemical Oceanography*, Vol. 1, Riley, J. P. and Skirrow, G., Eds., Academic Press, London, 1965, 409-432.
6. **Barkley, R. A. and Thompson, T. G.,** *Deep Sea Res.*, 7, 24, 1960.
7. **Betzer, P. and Pilson, M. E. Q.,** *J. Mar. Res.*, 28, 251, 1970.
8. **Bolter, E., Turekian, K. K., and Schutz, D. F.,** *Geochim. Cosmochim. Acta*, 28, 1459, 1964.
9. **Brooks, R. R.,** *Geochim. Cosmochim. Acta*, 29, 1369, 1965.
10. **Burton, J. D.,** *Nature*, 212, 976, 1966.
11. **Burton, J. D. and Krishnamurty, K.,** *Rep. Challenger Soc.*, 3, 24, 1967.
12. **Burton, J. D. and Riley, J. P.,** *Nature*, 181, 179, 1958.
13. **Carlisle, D. B. and Hummerstone, L. G.,** *Nature*, 181, 1002, 1958.
14. **Carpenter, J. H. and Grant, V. E.,** *J. Mar. Res.*, 25, 228, 1967.
15. **Chau, Y. K. and Riley, J. P.,** *Anal. Chim. Acta*, 33, 36, 1965.
16. **Chester, R.,** *Nature*, 206, 884, 1965.
17. **Chow, T. J. and Goldberg, E. D.,** *Geochim. Cosmochim. Acta*, 20, 192, 1960.

18. **Chow, T. J. and Goldberg, E. D.,** *J. Mar. Res.*, 20, 163, 1962.
19. **Chow, T. J. and Patterson, C. C.,** *Earth and Planet. Sci. Lett.*, 1, 397, 1966.
20. **Chow, T. J. and Tatsumoto, M.,** in *Recent Researches in the Fields of Hydrosphere, Atmosphere, and Nuclear Geochemistry,* Miyake, Y. and Koyama, T., Eds., Maruzen Co., Tokyo, 1964, 179-183.
21. **Chuecas, L. and Riley, J. P.,** *Anal. Chim. Acta,* 35, 240, 1966.
22. **Corcoran, E. F. and Alexander, J. E.,** *Bull. Mar. Sci. Gulf Caribbean,* 14, 594, 1964.
23. **Culkin, F. and Riley, J. P.,** *Nature,* 181, 180, 1958.
24. **Curl, H., Cutshall, N., and Osterberg, C.,** *Nature,* 205, 275, 1965.
25. **Cutshall, N., Johnson, V., and Osterberg, C.,** *Science,* 152, 202, 1966.
26. **Duursma, E. K. and Sevenhuysen, W.,** *Neth. J. Sea Res.*, 3, 95, 1966.
27. **El Wardani, S. A.,** *Geochim. Cosmochim. Acta,* 15, 237, 1958.
28. **Fabricand, B. P., Sawyer, R. R., Ungar, S. G., and Adler, S.,** *Geochim. Cosmochim. Acta,* 26, 1023, 1962.
29. **Fabricand, B. P., Imbimbo, E. S., Brey, M. E., and Weston, J. A.,** *J. Geophys. Res.*, 71, 3917, 1966.
30. **Fukai, R.,** *Nature,* 213, 901, 1967.
31. **Goldberg, E. D.,** in *Chemical Oceanography,* Vol. 1, Riley, J. P. and Skirrow, G., Eds., Academic Press, London, 1965, 163-196.
32. **Goldberg, E. D. and Arrhenius, G. S.,** *Geochim. Cosmochim. Acta,* 13, 153, 1958.
33. **Goldberg, E. D., Koide, M., Schmitt, R. A., and Smith, R. H.,** *J. Geophys. Res.*, 68, 4209, 1963.
34. **Griel, J. V. and Robinson, R. J.,** *J. Mar. Res.*, 11, 173, 1952.
35. **Høgdahl, O.,** Semi Annual Progress Report No. 5, NATO Scientific Affairs Div., Brussels, 1967.
36. **Høgdahl, O.,** Semi Annual Progress Report No. 6, NATO Scientific Affairs Div., Brussels, 1968.
37. **Høgdahl, O., Melsom, S., and Bowen, V. T.,** Trace inorganics in water, in *Advances in Chemistry Series,* No. 73, American Chemical Society, Washington, D.C., 1968, 308-325.
38. **Hood, D. W.,** in *Oceanogr. Mar. Biol. Annu. Rev.*, Vol. 1, Barnes, H., Ed., George Allen and Unwin, Ltd., London, 1963, 129-155.
39. **Ishibashi, M.,** *Rec. Oceanogr. Works Jap.*, 1, 88, 1953.
40. **Johnson, V., Cutshall, N., and Osterberg, C.,** *Water Resour. Res.*, 3, 99, 1967.
41. **Kappanna, A. N., Gadre, G. T., Bhavnagary, H. M., and Joshi, J. M.,** *Curr. Sci. (India),* 31, 273, 1962.
42. **Kaufman, A.,** *Geochim. Cosmochim. Acta,* 33, 717, 1969.
43. **Kester, D. R. and Pytkowicz, R. M.,** *Limnol. Oceanogr.*, 12, 243, 1967.
44. **Kharkar, D. P., Turekian, K. K., and Bertine, K. K.,** *Geochim. Cosmochim. Acta,* 32, 285, 1968.
45. **Koczy, F. F.,** *Proc. Second U.N. Internat. Conf. Peaceful Uses Atomic Energy,* 18, 351, 1958.
46. **Krauskopf, K. B.,** *Geochim. Cosmochim. Acta,* 9, 1, 1956.
47. **Menzel, D. W. and Ryther, J. H.,** *Deep Sea Res.*, 7, 276, 1961.
48. **Merrill, J. R., Lyden, E. F. X., Honda, M., and Arnold, J.,** *Geochim. Cosmochim. Acta,* 18, 108, 1960.
49. **Miyake, Y. and Sugimura, Y.,** in *Studies on Oceanography,* Yoshida, K., Ed., Univ. of Washington Press, Seattle, 1964, 274.
50. **Miyake, Y., Sugimura, Y., and Uchida, T.,** *J. Geophys. Res.*, 71, 3083, 1966.
51. **Moore, W. S. and Sackett, W. M.,** *J. Geophys. Res.*, 69, 5401, 1964.
52. **Morozov, N. P.,** *Oceanology,* 8, 169, 1968.
53. **Mullin, J. B. and Riley, J. P.,** *J. Mar. Res.*, 15, 103, 1956.
54. **Peshchevitskiy, B. I., Anoshin, G. N., and Yereburg, A. M.,** *Dokl. Earth Sci. Sect.*, 162, 205, 1965.
55. **Picciotto, E. E.,** in *Oceanography,* Sears, M., Ed., Amer. Assoc. Adv. Sci., Washington, D.C., 1961, 367.
56. **Portmann, J. E. and Riley, J. P.,** *Anal. Chim. Acta,* 34, 201, 1966.
57. **Putnam, G. L.,** *J. Chem. Educ.*, 30, 576, 1953.
58. **Pytkowicz, R. M.,** *J. Oceanogr. Soc. Jap.*, 24, 21, 1968.
59. **Pytkowicz, R. M. and Kester, D. R.,** *Deep Sea Res.*, 13, 373, 1966.
60. **Pytkowicz, R. M. and Kester, D. R.,** *Limnol. Oceanogr.*, 12, 714, 1967.

61. **Richards, F. A.,** in *Physics and Chemistry of the Earth,* Vol. 2, Ahrens, L. H., Press, F., Rankama, K., and Runcorn, S. K., Eds., Pergamon Press, New York, 1957, 77-128.
62. **Riley, J. P. and Tongudai, M.,** *Deep Sea Res.,* 11, 563, 1964.
63. **Riley, J. P. and Tongudai, M.,** *Chem. Geol.,* 1, 291, 1966.
64. **Rona, E., Hood, D. W., Muse, L., and Buglio, B.,** *Limnol. Oceanogr.,* 7, 201, 1962.
65. **Sackett, W. and Arrhenius, G.,** *Geochim. Cosmochim. Acta,* 26, 955, 1962.
66. **Scadden, E. M.,** *Geochim. Cosmochim. Acta,* 33, 633, 1969.
67. **Schink, D. R.,** *Geochim. Cosmochim. Acta,* 31, 987, 1967.
68. **Schutz, D. F. and Turekian, K. K.,** *Geochim. Cosmochim. Acta,* 29, 259, 1965.
69. **Schutz, D. F. and Turekian, K. K.,** *J. Geophys. Res.,* 70, 5519, 1965.
70. **Simmons, L. H., Monaghan, P. H., and Taggart, M. S.,** *Anal. Chem.,* 25, 989, 1953.
71. **Smith, R. C., Pillai, K. C., Chow, T. J., and Folson, T. R.,** *Limnol. Oceanogr.,* 10, 226, 1965.
72. **Somayajulu, B. L. K. and Goldberg, E. D.,** *Earth Planet. Sci. Lett.,* 1, 102, 1966.
73. **Spencer, D. W. and Brewer, P. G.,** *Geochim. Cosmochim. Acta,* 33, 325, 1969.
74. **Sugawara, K. and Terada, K.,** *Nature,* 182, 250, 1958.
75. **Szabo, B. J.,** *Geochim. Cosmochim. Acta,* 31, 1321, 1967.
76. **Tatsumoto, M. and Patterson, C. C.,** *Nature,* 199, 350, 1963.
77. **Tatsumoto, M. and Patterson, C. C.,** in *Earth Sciences and Meteoritics,* Geiss, J. and Goldberg, E. D., Compilers, North Holland Publ. Co., Amsterdam, 1963, 74-89.
78. **Topping, G.,** *J. Mar. Res.,* 27, 318, 1969.
79. **Torii, T. and Murata, S.,** in *Recent Researches in the Fields of Hydrosphere, Atmosphere, and Nuclear Geochemistry,* Miyake, Y. and Koyama, T., Eds., Maruzen Co., Tokyo, 1964.
80. **Torii, T. and Murata, S.,** *J. Oceanogr. Soc. Jap.,* 22, 56, 1966.
81. **Turekian, K. K. and Johnson, D. G.,** *Geochim. Cosmochim. Acta,* 30, 1153, 1966.
82. **Vaccaro, R. F.,** in *Chemical Oceanography,* Vol. 1, Riley, J. P. and Skirrow, G., Eds., Academic Press, London, 1965, 365-408.
83. **Veeh, H. H.,** *Earth and Planet. Sci. Lett.,* 3, 145, 1967.
84. **Wangersky, P. J. and Gordon, D. C., Jr.,** *Limnol. Oceanogr.,* 10, 544, 1965.
85. **Weiss, H. V. and Lai, M. G.,** *Talanta,* 8, 72, 1961.
86. **Weiss, H. V. and Lai, M. G.,** *Anal. Chim. Acta,* 28, 242, 1963.
87. **Williams, P. M.,** *Limnol. Oceanogr.,* 14, 156, 1969.
88. **Riley, J. P. and Chester, R.,** *Introduction to Marine Chemistry,* Academic Press, London, 1971.

Table 2.3—11
RESIDENCE TIME OF SOME MAJOR AND MINOR CONSTITUENTS OF SEAWATER

Element	Residence time (years)	Element	Residence time (years)	Element	Residence time (years)	Element	Residence time (years)
		Ga	1.4×10^3	K	1.1×10^7	La	1.1×10^4
		Ge	7.0×10^3	Ca	8.0×10^6	Ce	6.1×10^3
Li	2.0×10^7			Sc	5.6×10^3	W	1.0×10^3
Be	1.5×10^2			Ti	1.6×10^2	Au	5.6×10^5
		Rb	2.7×10^5	V	1.0×10^4	Hg	4.2×10^4
		Sr	1.9×10^7	Cr	3.5×10^2		
		Y	7.5×10^3	Mn	1.4×10^3	Pb	2.0×10^3
		Nb	3.0×10^2	Fe	1.4×10^2	Bi	4.5×10^5
		Mo	5.0×10^5	Co	1.8×10^4		
		Ag	2.1×10^6	Ni	1.8×10^4		
Na	2.6×10^8	Cd	5.0×10^5	Cu	5.0×10^4	Th	3.5×10^2
Mg	4.5×10^7			Zn	1.8×10^5		
Al	1.0×10^2	Sn	5.0×10^5			U	5.0×10^5
Si	8.0×10^3	Sb	3.5×10^5				
		Cs	4.0×10^4				
		Ba	8.4×10^4				

From Goldberg, E. D., in *The Sea: Ideas and Observations on the Progress in the Study of the Sea,* Vol. 1, Hill, M. N., Ed., Wiley-Interscience, New York, 1963, 3. Copyright © by John Wiley & Sons, Inc. Reprinted by permission of John Wiley & Sons, Inc.

Table 2.3—12
CONDUCTIVE SALINITY, DENSITY, AND MAJOR CHEMICAL COMPOSITION OF THE ORCA BASIN BRINE AND THE RED SEA BRINE COMPARED TO AVERAGE SEAWATER

	Orca Basin brine	Red Sea brine	Average seawater
Salinity (g/kg)	258.1	256.4	35.4
Density[a] (g/ml)	1.185	1.199	1.025
Na^+ (g/kg)	91.5	92.9	10.8
K^+ (g/kg)	0.63	2.16	0.40
Ca^{2+} (g/kg)	1.09	4.71	0.41
Mg^{2+} (g/kg)	1.05	0.81	1.29
Cl^- (g/kg)	149.5	155.3	19.4
SO_4^{2-} (g/kg)	3.66	0.75	2.72
NO_3^- (μM)	0	0.8	23[b]
PO_4^{3-} (μM)	81.5		2.5[b]
Si (μM)	235	235	25[b]
O_2 (ml/l)	0	0	4—5

[a] Measured at 22.5°C.
[b] Measured in water from 1900 m at the Orca Basin site.

From Sheu, D. D., The anoxic Orca Basin (Gulf of Mexico): geochemistry of brines and sediments, *Rev. Aquat. Sci.,* 2, 491, 1990.

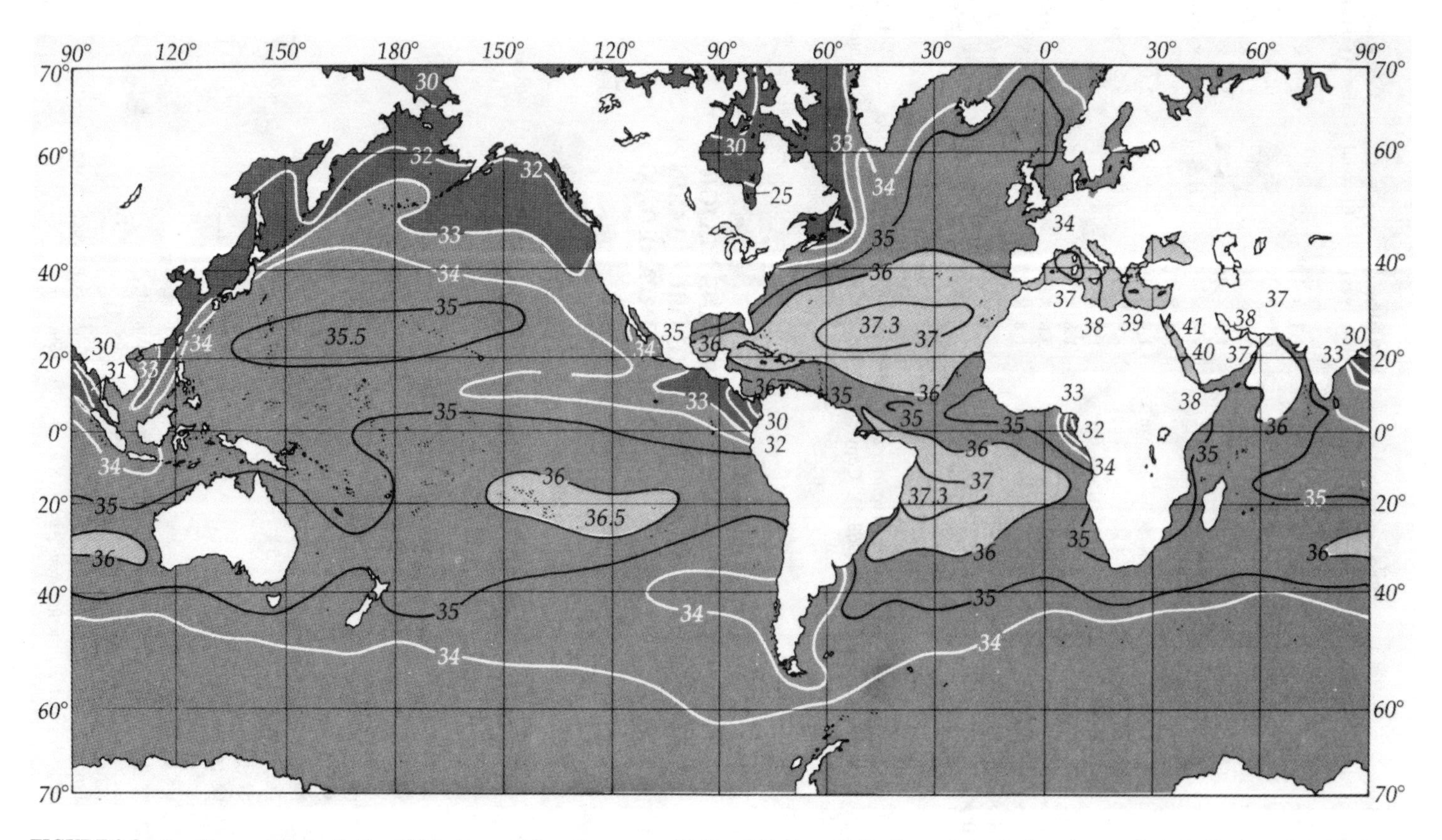

FIGURE 2.3—1. Ocean surface salinities (‰) in August. (From Ingmanson, D. E. and Wallace, W. J., *Oceanology: An Introduction*, 2nd ed., Wadsworth Publishing Company, Belmont, CA, 1979, 80. With permission.)

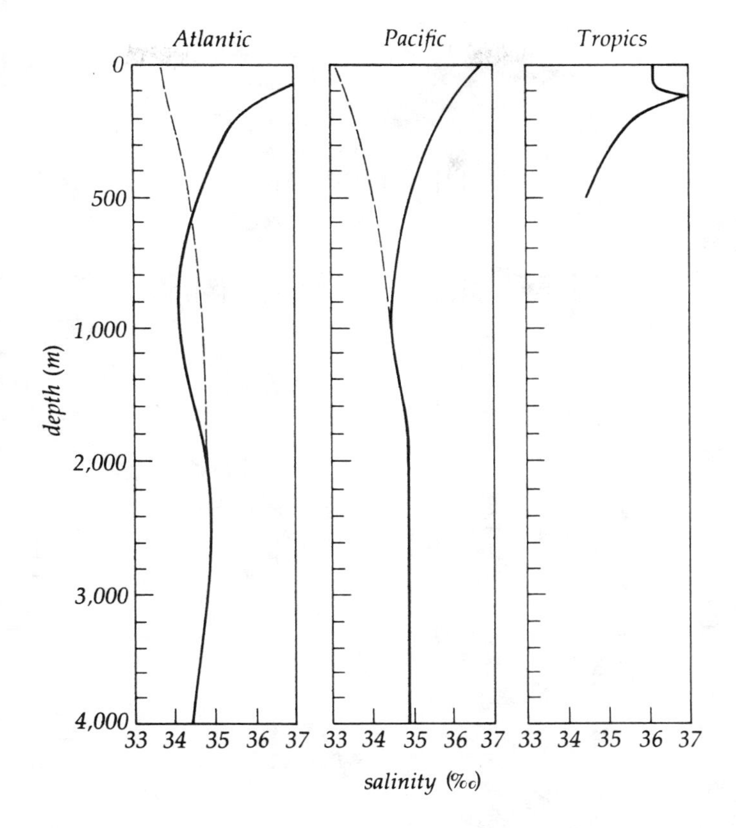

FIGURE 2.3—2. Typical salinity profiles for the open ocean. Solid lines indicate low and middle latitudes; dashed lines indicate high latitudes. (From Ingmanson, D. E. and Wallace, W. J., *Oceanology: An Introduction,* 2nd ed., Wadsworth Publishing Company, Belmont, CA, 1979, 83. With permission.)

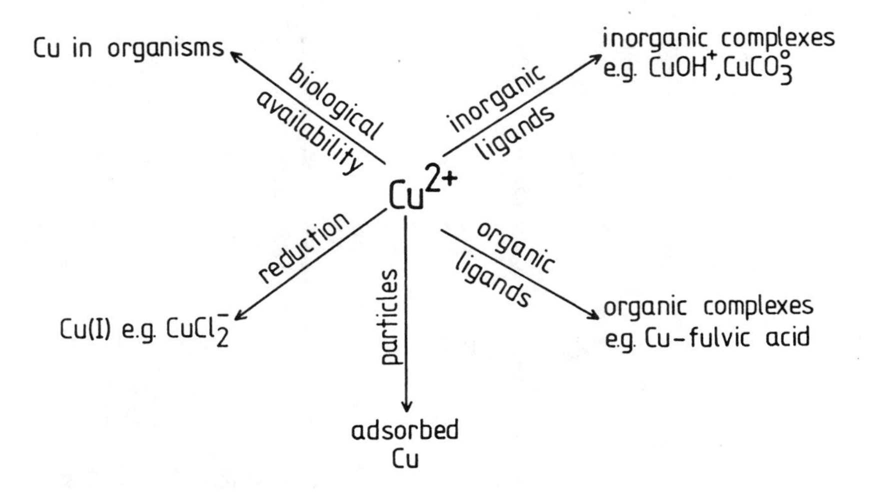

FIGURE 2.3—3. Major processes involving trace metals in water, taking copper in seawater as an example. (From Harrison, R. M., Ed., *Pollution: Causes, Effects, and Control,* 2nd ed., Royal Society of Chemistry, Cambridge, 1990, 22. With permission.)

Table 2.3—13
FLUXES OF TRACE METALS TO THE SEA SURFACE (units ng/cm^2 y)

	New York Bight	North Sea	Western Med.	South Atlantic Bight	Bermuda	North Atlantic northeast trades	Tropical North Atlantic	Tropical North Pacific; total net deposition	South Pacific; total net deposition	North Atlantic; westerlies	North Pacific; westerlies	South Pacific; westerlies
Al	6,000	30,000	5,000	2,900	3,900	97,000	5,000	1,200	132—1,800	—	—	—
Sc	—	5	1	—	0.6	—	1.1	0.18	0.06	—	—	—
V	—	480	—	—	5	—	17	7.8	—	—	—	—
Cr	—	210	49	—	9	111	14	—	—	—	—	—
Mn	—	920	—	60	45	570	70	9.0	3.6	—	—	—
Fe	5,700	25,500	5,100	5,900	3,000	48,000	3,200	560	47—337	—	—	—
Co	—	39	3.5	—	1.2	12	2.7	—	0.25	—	—	—
Ni	—	260	—	390	3	67	20	—	—	—	—	—
Cu	—	1,300	96	220	30	48	25	8.9	4.4—7.9	—	—	—
Zn	1,400	8,950	1,080	750	75	152	130	67	2.4—5.8	—	—	—
As	—	280	54	45	3	—	—	—	—	—	—	—
Se	—	22	48	—	3	—	14	4.2	0.8	—	—	—
Ag	—	—	3	—	—	—	0.9	—	—	—	—	—
Cd	30	43	13	9	4.5	—	5	0.35	—	—	—	—
Sb	—	58	48	—	1.0	—	3.5	—	—	—	—	—
Au	—	—	0.05	—	—	—	0.1	—	—	—	—	—
Hg	—	—	5	24	—	—	2.1	—	—	—	—	—
Pb	3.900	2,650	1,050	660	100	32	310	7.0	1.4—2.8	170	50	3
Th	—	4	1.2	—	—	—	0.9	0.61	0.036	—	—	—

From Furness, R. W. and Rainbow, P. S., Eds., *Heavy Metals in the Marine Environment,* CRC Press, Boca Raton, FL, 1990, 37.

Table 2.3—14
THE SOLUBILITY OF ATMOSPHERIC TRACE METALS IN SEAWATER

Atmospheric Population	Al	Fe	Mn	Ni	Co	Cr	V	Ag	Zn	Cu	Pb	Cd
Anthropogenic-rich[a]												
Mean concentration (ng/m^3 of air)	900	610	11	11	<0.4	<1.6	13	<0.05	25	16	560	0.25
Mean % soluble in seawater	0.56	1.1	47	47	25	12.5	31	80	68	28	39	84
Mean EF_{crust}	1.0	0.99	1.0	14	1.5	1.5	9.0	—	33	26	4148	16
Dust-rich[b]												
Mean concentration (ng/m^3 of air)	3380	2100	46	2.8	<0.5	4.0	<11	<0.03	18	20	150	0.20
Mean % soluble in seawater	0.09	0.19	34	28	20	10	18	33	24	14.5	13	80
Mean EF_{crust}	1.0	0.91	1.2	0.92	0.49	0.99	2.0	—	6.2	8.4	296	25

[a] Southern California.
[b] Baja, California.

From Furness, R. W. and Rainbow, P. S., Eds., *Heavy Metals in the Marine Environment,* CRC Press, Boca Raton, FL, 1990, 41.

Table 2.3—15
A COMPARISON OF HYDROTHERMAL AND RIVER FLUXES OF ELEMENTS INTO THE OCEANS (mol/year)

Elements	21° North	GSC	River water
Li	$1.2 \rightarrow 1.9 \times 10^{11}$	$9.5 \rightarrow 16 \times 10^{10}$	1.4×10^{10}
Na	$-8.6 \rightarrow 1.9 \times 10^{12}$	+, −	6.9×10^{12}
K	$1.9 \rightarrow 2.3 \times 10^{12}$	1.3×10^{12}	1.9×10^{12}
Rb	$3.7 \rightarrow 4.6 \times 10^{9}$	$1.7 \rightarrow 2.8 \times 10^{9}$	5.0×10^{6}
Be	$1.4 \rightarrow 5.3 \times 10^{6}$	$1.6 \rightarrow 5.3 \times 10^{6}$	3.3×10^{7}
Mg	-7.5×10^{12}	-7.7×10^{12}	5.3×10^{12}
Ca	$2.4 \rightarrow 15 \times 10^{11}$	$2.1 \rightarrow 4.3 \times 10^{12}$	1.2×10^{13}
Sr	$-3.1 \rightarrow +1.4 \times 10^{9}$	0	2.2×10^{10}
Ba	$1.1 \rightarrow 2.3 \times 10^{9}$	$2.5 \rightarrow 6.1 \times 10^{9}$	1.0×0^{10}
F	$-1.0 \rightarrow 2.3 \times 10^{9}$	$2.5 \rightarrow 6.1 \times 10^{9}$	1.0×10^{10}
Cl	$0 \rightarrow 1.2 \times 10^{13}$	$-31 \rightarrow +7.8 \times 10^{12}$	6.9×10^{12}
SiO_2	$2.2 \rightarrow 2.8 \times 10^{12}$	3.1×10^{12}	6.4×10^{12}
Al	$5.7 \rightarrow 7.4 \times 10^{8}$	na	6.0×10^{10}
SO_4	-4.0×10^{12}	-3.8×10^{12}	3.7×10^{12}
H_2S	$9.4 \rightarrow 12 \times 10^{11}$	+	
ΣS	$-2.8 \rightarrow 3.1 \times 10^{11}$	−	
Mn	$1.0 \rightarrow 1.4 \times 10^{11}$	$5.1 \rightarrow 16 \times 10^{10}$	4.9×10^{9}
Fe	$1.1 \rightarrow 3.5 \times 10^{11}$	+	2.3×10^{10}
Co	$3.1 \rightarrow 32 \times 10^{6}$	na	1.1×10^{8}
Cu	$0 \rightarrow 6.3 \times 10^{9}$	−	5.0×10^{9}
Zn	$5.7 \rightarrow 15 \times 10^{9}$	na	1.4×10^{10}
Ag	$0 \rightarrow 5.4 \times 10^{6}$	na	8.8×10^{7}
Cd	$2.3 \rightarrow 26 \times 10^{6}$	−	
Pb	$2.6 \rightarrow 5.1 \times 10^{7}$	na	1.5×10^{8}
As	$0 \rightarrow 6.5 \times 10^{7}$	na	7.2×10^{8}
Se	$0 \rightarrow 1.0 \times 10^{7}$	na	7.9×10^{7}

Note: GSC, Galapagos Spreading Center; +, gain; −, loss; na, not analyzed.

From Millero, F. J. and Sohn, M. L., *Chemical Oceanography,* CRC Press, Boca Raton, FL, 1992, 470.

2.4 ORGANIC MATTER

Table 2.4—1
CONCENTRATIONS OF ORGANIC CARBON IN NATURAL WATERS

Conc. (mg/ℓ) of organic carbon	River	Estuary	Coastal sea	Open sea: Surface	Open sea: Deep	Sewage
Dissolved	10—20 (50)	1—5 (20)	1—5 (20)	1—1.5	0.5—0.8	100
Particulate	5—10	0.5—5	0.1—1.0	0.01—1.0	0.003—0.01	200
Total	15—30 (60)	1—10 (25)	1—6 (21)	1—2.5	0.5—0.8	300

Note: Numbers in parentheses represent extreme values.

With permission from Head, P. C., in *Estuarine Chemistry*, Burton, J. D. and Liss, P. S., Eds., Academic Press, London, 1976, 53. Copyright: 1976 Academic Press, Inc. (London) Ltd.

Table 2.4—2
ORGANIC COMPOUNDS IDENTIFIED IN SEAWATER

Substances	Quantities	Locality	Method	Ref.
Rhamnoside, dehydroascorbic acid	Up to 0.1 g/l present	Inshore waters, Gulf of Mexico	Activated charcoal absorption, ethanol elution	1
Carbohydrates, arabinose equivalents	0.0–20 mg/l	Estuary, Gulf of Mexico	*N*-Ethyl carbazole	2
Carbohydrates, sucrose equivalents	0.14–0.45 mg/l	Pacific Coast, U.S.A.	Anthrone and *N*-ethyl carbazole	3
Carbohydrates, arabinose equivalents	0.0–2.6 mg/l (max. of 12 mg/l at surface, 29°N, 80°, 31′W)	South Atlantic (30°N–25°N)	*N*-Ethyl carbazole	4
	0.0–3.0 mg/l (23% = 0.0; 50% = 0.2–1 mg/l)	Continental Shelf, Gulf of Mexico (50 mg/l in red tides of *G. breve*)	*N*-Ethyl carbazole	5
Citric acid	0.025–0.145 mg/l	Littoral Atlantic French Coast		6
Malic acid	0.028–0.277 mg/l	Northeast Pacific, surface and inshore	Chloroform or ether extraction at pH 3; partition chromatography on silica gel column	7
Acetic and formic acids[a]	<0.1 mg/l			
Fatty acids (up to 20 carbons)	0.4–0.5 mg/l (weight of methyl esters)	Gulf of Mexico	Ethyl acetate extraction at pH 2; Gas-liquid chromatography	8
Amino acids, hydrolyzed proteins	Traces to 13 mg/m^3[b]	Gulf of Mexico, Yucatan, Strait Reef (British Honduras), Caribbean	Coprecipitation of organic material with $FeCl_3$ + NaOH; acid hydrolysis; paper and ion exchange chromatography	9

Vitamin B_{12}	Present			10
Plant hormones	Present	North Sea	Chloroform extraction at pH 5; ether extract of residue, measured biologically	

[a] Acetic, formic, lactic, and glycolic (up to 1.4 mg/l) acids are liberated from breakdown of larger organic molecules during the long extraction procedure (4–5 weeks).

[b] 18 Amino acids were found in hydrolysates. The amounts and kind of amino acids vary widely in samples.

From Provasoli, L., Organic regulation of phytoplankton fertility, *The Sea,* Vol. 2, Hill, M. N., Ed., Interscience, New York, 1963, 1974. With permission of John Wiley & Sons, copyright©, owner.

REFERENCES

1. **Wangersky, P. J.,** *Science,* 115, 685, 1960.
2. **Collier, A., Ray, S. M., Magritsky, A. W., and Bell, J. O.,** *U.S. Dept. Int. Fish and Wildlife Service, Fish Bull.,* 84, 167, 1953.
3. **Lewis, G. J. and Rakestraw, N. W.,** *J. Mar. Res.,* 14, 253, 1955.
4. **Anderson, W. W. and Gehringer, J. W.,** *Spec. Sci. Rep.* (Fisheries), 265, 1, 1953; 303, 1, 1958.
5. **Collier, A.,** *Limnol. Oceanogr.,* 3, 33, 1958.
6. **Creac'h, P.,** *C.R. Acad. Sci. Paris,* 240, 2551, 1955.
7. **Koyama, T. and Thompson, T. G.,** *Preprints Intern. Oceanogr. Cong. A.A.A.S.,* p. 925, 1959.
8. **Slowey, J. F., Jeffrey, L. M., and Hood, D. W.,** *Geochim. Cosmochim. Acta,* 26, 607, 1962.
9. **Tatsumoto, M., Williams, W. T., Prescott, J. M., and Hood, D. W.,** *J. Mar. Res.,* 19, 89, 1961.
10. **Bentley, J. A.,** *Preprints Intern. Oceanogr. Cong. A.A.A.S.,* p. 910, 1959.

Table 2.4—3
THE DISSOLVED ORGANIC CONSTITUENTS OF SEAWATER

Specific Dissolved Organic Compounds Identified in Seawater

Name of compound and chemical formula	Concentration				Author(s)	Locality
		I. Carbohydrates				
Pentoses $C_5H_{10}O_5$		0–8	mg/l		Collier et al. (1950, 1956)	Gulf of Mexico
Pentoses $C_5H_{10}O_5$		0.5	μg/l		Degens et al. (1964)	Pacific off California
Hexoses		14–36	μg/l		Degens et al. (1964)	Pacific off California
Rhamnosides $C_6H_{12}O_5$ Rhamnosides		0.1–0.4	mg/l		Lewis and Rakestraw (1955)	Pacific Ocean coast U.S.A.
Dehydroascorbic acid (ring: O bridging) COCOCOCHCH(OH)CH_2OH		0.1	mg/l		Wangersky (1952)	Gulf of Mexico inshore water
		II. Proteins and Their Derivatives				
Peptides C:N ratio = 13.8:1					Jeffrey and Hood (1958)	Gulf of Mexico
Polypeptides and polycondensates of:	(a) μg/l	(b) μg/l	(c) μg/l	(d) μg/l		
Glutamic acid COOH$(CH_2)_2$CH(NH_2)COOH		8–13	8–13	0.1–1.8	(a) Park et al. (1962) (by ion-exchange)	Gulf of Mexico
Lysine $NH_2(CH_2)_4$CH(NH_2)COOH	<1	?	trace–3	0.1–0.9	(b) Tatsumoto et al. (1961) (by paper chromatography)	Gulf of Mexico
Glycine NH_2CH_2COOH		–	trace–3	1.2–3.7	(c) Tatsumoto et al. (1961) (by ion-exchange)	Gulf of Mexico
Aspartic acid COOHCH_2CH(NH_2)COOH		3–8	trace–3	0.1–1.0	(d) Degens et al. (1964)	Pacific off California

Serine $CH_2OHCH(NH_2)COOH$		?	trace–3	1.8–5.6		
Alanine $CH_3CH(NH_2)COOH$		3–8	trace–3	0.7–3.1	(a) Park et al. (1962) (by ion-exchange)	Gulf of Mexico
Leucine $(CH_3)_2CHCH_2CH(NH_2)COOH$	0.5–1	8–13	trace–3	0.9–3.8	(b) Tatsumoto et al. (1961) (by paper chromatography)	Gulf of Mexico
Valine $(CH_3)_2CHCH(NH_2)COOH$		trace–3	trace–3	0.1–1.7	(c) Tatsumoto et al. (1961) (by ion-exchange)	Gulf of Mexico
Cystine $[SCH_2CH(NH_2)COOH]_2$		trace–3	–	0.0–3.8	(d) Degens et al. (1964)	Pacific off California
Isoleucine $CH_3CH_2CH(CH_3)CH(NH_2)COOH$		8–13	trace–3	–		
Leucine $(CH_3)_2CHCH_2CH(NH_2)COOH$		–	–	0.9–3.8		
Ornithine $NH_2(CH_2)_3CH(NH_2)COOH$		–	trace–3	0.2–2.4		
Methionine sulphoxide $CH_3S(:O)CH_2CH_2CH(NH_2)COOH$				–		
Threonine $CH_3CHOHCH(NH_2)COOH$		–	3–8	0.3–1.3		
Tyrosine $HOC_6H_4CH_2CH(NH_2)COOH$		–	trace–3	tr.–0.5		
Phenylalanine $C_6H_5CH_2CH(NH_2)COOH$	<0.5	–	–	0.1–0.9		
Histidine $C_3H_3N_2CH_2CH(NH_2)COOH$		?	trace–3	tr.–2.4		
Arginine $NH_2C(:NH)NH(CH_2)_3CH(NH_2)COOH$		?	trace–3	0.1–0.6		
Proline C_4H_8NCOOH		?	–	0.3–1.4		
Methionine $CH_3SCH_2CH_2CH(NH_2)COOH$		–	trace–3	tr.–0.4		
Tryptophan $C_8H_6NCH_2CH(NH_2)COOH$		–	trace–3	–		
Glucosamine $C_6H_{13}NO_5$		–	trace–3			
Free amino acids		(e)	(f) μg/l			
Cystine $[SCH_2CH(NH_2)COOH]_2$		det	–		(e) Palmork (1963a)	Norwegian coastal water

Table 2.4—3 (continued) THE DISSOLVED ORGANIC CONSTITUENTS OF SEAWATER

Name of compound and chemical formula	Concentration		Author(s)	Locality
Lysine $NH_2(CH_2)_4CH(NH_2)COOH$	det.	0.2–3.1	(f) Degens et al. (1964)	Pacific off California
Histidine $C_3H_3N_2CH_2CH(NH_2)COOH$	det.	0.5–1.7		
Arginine $NH_2C(:NH)NH(CH_2)_3CH(NH_2)COOH$	det.	0.0		
Serine $CH_2OHCH(NH_2)COOH$	det.	2.3–28.4		
Aspartic acid $COOHCH_2CH(NH_2)COOH$	det.	tr.–9.6		
Glycine NH_2CH_2COOH	det.	tr.–37.6		
Hydroxyproline $C_4H_7N(OH)COOH$	det.	tr.–2.8		
Glutamic acid $COOH(CH_2)_2CH(NH_2)COOH$	det.	1.4–6.8		
Threonine $CH_3CHOHCH(NH_2)COOH$	det.	2.8–11.8		
α-Alanine $CH_3CH(NH_2)COOH$	det.			
Proline C_4H_8NCOOH	det.	0.0		
Tyrosine $HOC_6H_4CH_2CH(NH_2)COOH$	det.	tr.–5.0		
Tryptophan $C_8H_6NCH_2CH(NH_2)COOH$	det.	–		
Methionine $CH_3SCH_2CH_2CH(NH_2)COOH$	det.	–		
Valine $(CH_3)_2CHCH(NH_2)COOH$	det.	0.3–2.7		
Phenylalanine $C_6H_5CH_2(NH_2)COOH$	det.	tr.–2.4		
Isoleucine $CH_3CH_2CH(CH_3)CH(NH_2)COOH$	det.	–		
Leucine $(CH_3)_2CHCH_2CH(NH_2)COOH$	det.	0.5–5.5		

Free compounds			
Uracil	det.	Belser (1959, 1963)	Pacific coast near La Jolla
NHCONHCOCH:CH (ring)			
Isoleucine	det.		
$CH_3CH_2CH(CH_3)CH(NH_2)COOH$			
Methionine	det.		
$CH_3SCH_2CH_2CH(NH_2)COOH$			
Histidine	det.		
$C_3H_3N_2CH_2CH(NH_2)COOH$			
Adenine	det.		
$C_5H_3N_4NH_2$			
Peptone	det.		
Threonine	det.		
$CH_3CHOHCH(NH_2)COOH$			
Tryptophan	det.		
$C_8H_6NCH_2CH(NH_2)COOH$			
Glycine	det.		
NH_2CH_2COOH			
Purine	det.		
$C_5H_4N_4$			
Urea	det.	Degens et al. (1964)	Pacific off California
CH_4ON_2			

III. Aliphatic Carboxylic and Hydroxycarboxylic Acids

	mg/l (0–200 m)	mg/l (200–600 m)	mg/l (>600 m)		
Lauric acid	0.01–0.32	0.01–0.28	0–0.28	Slowey et al. (1962)	Coastal waters of Gulf of Mexico
$CH_3(CH_2)_{10}COOH$					
Myristic acid	0.01–0.10	0.01–0.05	0–0.07		
$CH_3(CH_2)_{12}COOH$					
Myristoleic acid	traces–0.02	0.01–0.03	0–0.05		
$CH_3(CH_2)_3CH{:}CH(CH_2)_7COOH$					
Palmitic acid	0.01–0.17	0.03–0.42	0–0.38		
$CH_3(CH_2)_{14}COOH$					
Palmitoleic acid	0.02–0.16	0.02–0.16	0–0.21		
$CH_3(CH_2)_5CH{:}CH(CH_2)_7COOH$					
Stearic acid	0.04–0.09	0.02–0.13	0–0.10		
$CH_3(CH_2)_{16}COOH$					
Oleic acid	0.01	0.02	0		
$CH_3(CH_2)_7CH{:}CH(CH_2)_7COOH$					
Linoleic acid	0.01	0.01	0		
$CH_3(CH_2)_4CH{:}CHCH_2CH{:}CH(CH_2)_7COOH$					

Table 2.4—3 (continued)
THE DISSOLVED ORGANIC CONSTITUENTS OF SEAWATER

Name of compound and chemical formula	Concentration	Author(s)	Locality
	mg/l (1000–2500 m)		
Fatty acids with:			
12 C-atoms	0.0003–0.02	Williams (1961)	Pacific Ocean coastal water
14 C-atoms	0.0004–0.043		
16 C-atoms	0.0027–0.0209		
16 C-atoms + 1 double bond	0.0003–0.003		
18 C-atoms	0.0037–0.0222		
18 C-atoms + 1 double bond	0.0083		
18 C-atoms + 2 double bonds	0.0000–0.0029		
20 C-atoms	traces–0.0081		
22 C-atoms	traces–0.0014		
	mg/l		
Acetic acid CH_3COOH	<1.0	Koyama and Thompson (1959)	Pacific Ocean
Lactic acid $CH_3CH(OH)COOH$			
Glycolic acid $HOCH_2COOH$			
Malic acid $HOOCCH(OH)CH_2COOH$	0.28	Creac'h (1955)	Atlantic coastal water
Citric acid $HOOCCH_2C(OH)(COOH)CH_2COOH$	0.14		
Carotenoids and brownish-waxy or fatty matter	2.5	Johnston (1955)	North Sea
		Wilson and Armstrong (1955)	English Channel
IV. Biologically Active Compounds (see also Provasoli, 1963)			
Organic Fe compound(s)	3.4–1.6 mμg/l	Harvey (1925)	Deep seawater
Vitamin B_{12} (Cobalamin) $C_{63}H_{88}O_{14}N_{14}PCo$		Vishniak and Riley (1961)	Long Island Sound

Compound	Concentration	Units	Reference	Location
Vitamin B_{12}	0.2	mμg/l (summer)	Cowey (1956)	Oceanic surface water
	2.0	mμg/l (winter)		
Vitamin B_{12}	0.2–5.0	mμg/l	Daisley and Fisher (1958)	
Vitamin B_{12}	0–2.6	mμg/l	Kashiwada et al. (1957)	North Pacific Ocean
Vitamin B_{12}	0–0.03	mμg/l	Menzel and Spaeth (1962)	Sargasso Sea 0–05 m.
Thiamine (Vitamin B_1) $C_{12}H_{17}ON_4SCl_2$	0–20	mμg/l	Cowey (1956)	Surface water, possibly from land drainage
Plant hormones (auxins)	3.41	mμg/l	Bentley (1960)	North Sea near Scotland
V. Humic Acids				
"Gelbstoffe" (Yellow substances) Melanoidin-like			Kalle (1949, 1962) Jerlov (1955) Armstrong and Boalch (1961a,b)	Coastal waters
VI. Phenolic Compounds				
p-Hydroxybenzoic acid HOC_6H_5COOH		1–3 μg/l	Degens et al. (1964)	Pacific off California
Vanillic acid $CH_3(HO)C_6H_3COOH$		1–3 μg/l		
Syringic acid $(CH_3O)_2(HO)C_6H_2COOH$		1–3 μg/l		
VII. Hydrocarbons				
Pristane: (2, 6, 10, 14-tetramethylpentadecane)		trace	Blumer et al. (1963)	Cape Cod Bay

From Duursma, E. K., in *Chemical Oceanography,* Vol. 1, Riley, J. P. and Skirrow, G., Eds., Academic Press, London, 1965, 450. With permission.

Note: tr. = trace; — = not detected; ? = possibly present; det. = detected.

REFERENCES

1. **Armstrong, F. A. J. and Boalch, G. T.,** *Nature (Lond.),* 192, 858, 1961a.
2. **Armstrong, F. A. J. and Boalch, G. T.,** *J. Mar. Biol. Assoc. U.K.,* 41, 591, 1961b.
3. **Belser, W. L.,** *Proc. Natl. Acad. Sci., Wash.,* 45, 1533, 1959.
4. **Belser, W. L.,** in *The Sea,* Hill, M. N., Ed., Vol. II, Wiley-Interscience, New York, 1963, 220-231.
5. **Bentley, Joyce A.,** *J. Mar. Biol. Assoc. U.K.,* 39, 433, 1960.
6. **Blumer, M., Mullin, M. M., and Thomas, D. W.,** *Science,* 140, 974, 1963.
7. **Collier, A.,** *Spec. Sci. Rep. U.S. Fish Wildl.,* 178, 7, 1956.

Table 2.4—3
REFERENCES (continued)

8. **Collier, A., Ray, S. M., and Magnitzky, A. W.,** *Science,* 111, 151, 1950.
9. **Cowey, C. B.,** *J. Mar. Biol. Assoc. U.K.,* 35, 609, 1956.
10. **Creac'h, P.,** *C. R. Acad. Sci., (Paris),* 240, 2551, 1955.
11. **Daisley, K. W. and Fisher, L. R.,** *J. Mar. Biol. Assoc. U.K.,* 37, 683, 1958.
12. **Degens, E. T., Reuter, J. H., and Shaw, K. N. F.,** *Geochim. Cosmochim. Acta,* 28, 45, 1964.
13. **Harvey, H. W.,** *J. Mar. Biol. Assoc. U.K.,* 13, 953, 1925.
14. **Jeffrey, L. M. and Hood, D. W.,** *J. Mar. Res.,* 17, 247, 1958.
15. **Jerlov, N. G.,** *Göteb. Vetensk Samh. Handl.,* F.6. B.6. (14), 1955.
16. **Johnston, R.,** *J. Mar. Biol. Assoc. U.K.,* 34, 185, 1955.
17. **Kalle, K.,** *Dtsch. Hydrogr. Z.,* 2, 117, 1949.
18. **Kalle, K.,** *Kiel. Meeresforsch.,* 18, 128, 1962.
19. **Kashiwada, K., Kakimoto, D., Morita, T., Kanazawa, A., and Kawagoe, K.,** *Bull. Jap. Soc. Sci. Fish.,* 22, 637, 1957.
20. **Koyama, T. and Thompson, T. G.,** *Preprints International Oceanographic Congress, 1959,* American Association for Advancement of Science, Washington, D.C., 1959, 925.
21. **Lewis, G. J. and Rakestraw, N. W.,** *J. Mar. Res.,* 14, 253, 1955.
22. **Menzel, D. W. and Spaeth, J. P.,** *Limnol. Oceanogr.,* 7, 151, 1962.
23. **Palmork, K. H.,** *Acta Chem. Scand.,* 17, 1456, 1963a.
24. **Park, K., Williams, W. T., Prescott, J. M., and Hood, D. W.,** *Science,* 138, 531, 1962.
25. **Provasoli, L.,** in *The Sea,* Hill, M. N., Ed., Vol. II, Wiley-Interscience, New York, 1963, 165-219.
26. **Slowey, J. F., Jeffrey, L. M., and Hood, D. W.,** *Geochim. Cosmochim. Acta,* 26, 607, 1962.
27. **Tatsumoto, M., Williams, W. T., Prescott, J. M., and Hood, D. W.,** *J. Mar. Res.,* 19, 89, 1961.
28. **Vishniac, H. S. and Riley, G. A.,** *Limnol. Oceanogr.,* 6, 36, 1961.
29. **Wangersky, P. J.,** *Science,* 115, 685, 1952.
30. **Williams, P. M.,** *Nature (Lond.),* 189, 219, 1961.
31. **Wilson, D. P. and Armstrong, F. A. J.,** *J. Mar. Biol. Assoc. U.K.,* 31, 335, 1952.

Table 2.4—4
CONCENTRATION RANGES (μg/l) OF THE MAJOR IDENTIFIED GROUPS OF DISSOLVED ORGANIC SUBSTANCES IN NATURAL WATERS

	Rain	Groundwater	River	Lake	Sea
Volatile fatty acids	10	40	100	100	40
Nonvolatile fatty acids	5—17	5—50	50—500	50—200	5—50
Amino Acids	—	20—350	50—1000	30—6000	20—250
Carbohydrates	—	65—125	100—2000	100—3000	100—1000
Aldehydes	—	—	~0.1	—	.01—0.1

From Wotton, R. S., Ed., *The Biology of Particles in Aquatic Systems,* CRC Press, Boca Raton, FL, 1990, 119.

Table 2.4—5
SEASONAL MAXIMA AND MINIMA OF DISSOLVED ORGANIC MATTER IN ESTUARIES AND COASTAL MARINE WATERS

Location	Measurement	Maximum (Time, conc.)	Minimum (Time, conc.)	Remarks
Estuary, Netherlands	DOC	Summer 4-5 mg l^{-1}	Fall 2 mg l^{-1}	Near mouth
Estuary, Netherlands	Amino acids	March 0.8 mg C l^{-1}	Sep. 0.1 mg C l^{-1}	Near mouth
Estuary, Netherlands	DOC	No pattern 13 mg l^{-1}	No pattern 8 mg l^{-1}	
Estuary, Netherlands	DOC	June & July 2 mg l^{-1}	Mar. & Nov. 0.3-0.5 mg l^{-1}	Autochthonous
Estuary, Netherlands	Amino acids	July 0.6-0.8 mg C l^{-1}	Feb.-Mar. 0.1 mg C l^{-1}	Highest near mouth
Nile estuary	DOM (O_2 demand)	August 9.05 mg O l^{-1}	Dec. 2.37 mg O l^{-1}	Temp.-dependent changes
Maine estuary	DOC	Summer 5-7 mg l^{-1}	Winter 1-2 mg l^{-1}	Up to 12.6 mg l^{-1}
Main estuary	DFAA	No pattern 300 nM l^{-1}	No pattern 100 nM l^{-1}	Some summer low values
English Channel	DOC	Mar., Jul.-Aug. & Oct. 1.8-2.2 mg l^{-1}	Mar.-Apr. & Jun. 0.8-1.2 mg l^{-1}	Lagged Chl *a*
English Channel	DON	Late summer 7 μmol N l^{-1}	Winter undetectable	Summarized Station E1
English Channel	DFAA	Winter 3-4 μmol N l^{-1}	Summer 1 μmol N l^{-1}	
English Channel	DOC	June & Sep. .88 & .96 mg l^{-1}	March .56 mg l^{-1}	
English Channel	DON	Apr. & Aug. 5-6 μg-atN l^{-1}	Oct. 2.5 μg-atN l^{-1}	
Southern North Sea	DOC	Spring 1.7 mg l^{-1}	Winter 0.4 mg l^{-1}	Summer-up autumn-down
Dutch Wadden Sea	DOC	May-June 4-5 mg l^{-1}	Winter 1-2 mg l^{-1}	Tidal inlet
Maine coast	DFAA	May 88 μg C l^{-1}	Dec. 10 μg C l^{-1}	Jul. & Oct. peaks
Louisiana coast	DOC	Oct. 3 mg l^{-1}	Jan. 1.5 mg l^{-1}	<Full year
Gulf of Naples	DOC	No pattern 31.4 mg l^{-1}	No pattern 0.4 mg l^{-1}	*Posidonia* sea grass bed
Strait of Georgia, B.C.	DOC	Summer 3 mg l^{-1}	Nov. 1 mg l^{-1}	
Coastal pond, Cape Cod	Carbohydrate	April 2.9 mg l^{-1}	Feb. 1.3 mg l^{-1}	Inverse to chlorophyll
Menai Strait, England	DOC	Autumn 3-4 mg l^{-1}	Winter 1 mg l^{-1}	Spring-fall increase
Irish Sea	DFAA	May & Sep. 31 & 25 μg l^{-1}	Feb. & Aug. 5-10 μg l^{-1}	Late bloom maximum
Irish Sea	Total AA	Jan. & July 120 & 111 μg l^{-1}	Feb. 2 μg l^{-1}	Little pattern

Note: DOC, dissolved organic carbon; DOM, dissolved organic material; DFAA, dissolved free amino acid; DON, dissolved organic nitrogen; AA, amino acid.

From Wotton, R. S., Ed., *The Biology of Particles in Aquatic Systems,* CRC Press, Boca Raton, FL, 1990, 95.

Table 2.4—6
SEASONAL MAXIMA AND MINIMA OF PARTICULATE ORGANIC MATTER IN ESTUARINE AND COASTAL MARINE WATERS

Location	Measurement	Maximum (Time conc.)	Minimum (Time conc.)	Remarks
Estuary, Netherlands	Amino acids	May 1.5-1.6 mg C l^{-1}	Feb. near zero	Lower estuary
Estuary, Netherlands	Amino acids	July 4.2 mg C l^{-1}	Feb. .1-.2 mg C l^{-1}	Upper estuary
Estuary, Netherlands	POC	Summer 3 mg l^{-1}	Sep.-Feb. 1 mg l^{-1}	Lower estuary
Estuary, Netherlands	POC	No pattern 8-9 mg l^{-1}	No pattern near zero	Upper estuary
Estuary, Netherlands	Amino acids	May 1.7 mg C l^{-1}	Oct. <0.1 mg C l^{-1}	Lower estuary
Estuary, Netherlands	POC	Aug.-Oct. 10-20 mg l^{-1}	Nov.-July 2-8 mg l^{-1}	
Arabian Sea estuary	POC	June & Aug. 5.24 mg l^{-1}	Oct.-Jan. 0.28 mg l^{-1}	Max. during monsoons
Estuary, Brazil	TSM	Summer 78.2 mg l^{-1}	Winter 6.9 mg l^{-1}	TSM about 60% organic
Estuary, Brazil	POC	Jan. & Nov. 2.1 mg l^{-1}	July & Oct. .6 & .3 mg l^{-1}	Near mouth
Estuary, Brazil	POC	Feb. 3.2 mg l^{-1}	July 0.5 mg l^{-1}	Upper estuary
Dutch Wadden Sea	POC	May 4-6 mg l^{-1}	Winter 0.5-1.0 mg l^{-1}	Tidal inlet
English Channel	POC	Apr. & July 386 & 320 μg l^{-1}	Sep.-May 120-144 μg l^{-1}	
English Channel	PON	Apr. & July 50 & 37 μg N l^{-1}	Sep.-Mar. 8-17 μg N l^{-1}	
Galveston Bay, TX	Protein	May-Jun. & Aug. 1.7 mg l^{-1}	Jan. 0.2 mg l^{-1}	0.6 mg l^{-1} mid Aug.
Galveston Bay, TX	Carbohydrate	Sep. 530 μg l^{-1}	Dec. & Aug. near zero	Weak seasonal pattern
Galveston Bay, TX	Lipid	Sep. 0.8 mg l^{-1}	Feb. 0.1 mg l^{-1}	Similar to CHO pattern
Gulf of Naples	POM dry wt.	No pattern 30 mg l^{-1}	No pattern 3.5 mg l^{-1}	*Posidonia* beds
Coastal Arabian Sea	POC	Nov. (30m) 2.51 mg l^{-1}	May (surface) 0.52 mg l^{-1}	
Puget Sound	POC	July 400 μg l^{-1}	Sep.-Mar. 50-100 μg l^{-1}	Surface
N. Dawes Inlet, AK	POC	July 1.22 mg l^{-1}	Nov. 0.24 mg l^{-1}	
Funka Bay, Japan	POC	July 11.5 g C m^{-2}	Early Oct. 3.5 g C m^{-2}	Integrated upper 50 m
Funka Bay, Japan	Protein	Feb. 3 g C m^{-2}	Early Oct. 1.5 g C m^{-2}	Integrated upper 50 m
Tripoli Harbor	POC	August 1.3 mg l^{-1}	Nov.-Apr. .9-1.1 mg l^{-1}	Surface
Strait of Georgia, B.C.	POC	Summer 400 μg l^{-1}	Nov. 170 μg l^{-1}	Seasonal means
Coastal S. California	POC	Feb. & Mar. 351 & 235 μg l^{-1}	Sep. 191 μg l^{-1}	Coastal stations

Note: POC, particulate organic carbon; TSM, total suspended material; PON, particulate organic nitrogen.

From Wotton, R. S., Ed., *The Biology of Particles in Aquatic Systems,* CRC Press, Boca Raton, FL, 1990, 97.

Table 2.4—7
SEASONAL MAXIMA AND MINIMA OF DISSOLVED ORGANIC MATTER IN SALT-MARSH WATERS

Location	Measurement	Maximum (Time, conc.)	Minimum (Time, conc.)	Remarks
Texas marsh	DOC	July 37 mg l^{-1}	Winter 11-12 mg l^{-1}	
Louisiana bays	DOC	Early Jan. 11.5 mg l^{-1}	Lt. Jan.-Nov. 5 mg l^{-1}	Minor changes after Jan.
Estuary, SC	DOC	Feb. 10.7 mg l^{-1}	July 3.2 mg l^{-1}	Flow-related variability
Estuary, SC	DON	Jan. & Aug. 32-38 μg-atN l^{-1}	Sep., Mar. & May 18-19 μg-atN l^{-1}	
Estuary, SC	DOP	July 0.7 μg-atP l^{-1}	Nov.-Feb. & Sep. undetect.	
Marsh, SC	DOC	Feb.-May 18.6 mg l^{-1}	Dec. 3.1 mg l^{-1}	Runoff-related maximum
Marsh, VA	DON	Summer 1 mg l^{-1}	Winter 0.1 mg l^{-1}	
Tidal creek, SC	DOC	Fall 3.9 mg l^{-1}	Summer 2.8 mg l^{-1}	Seasonal means
Marsh, VA	DOP	July 0.25 mg l^{-1}	Jan. <0.05 mg l^{-1}	
Duplin R., GA	DOC	Summer 9.6 & 9.2 mg l^{-1}	Winters 4.9 & 3.6 mg l^{-1}	Surface & bottom
Flax Pond, NY	DOC	July 2.5 mg l^{-1}	Jan.-Mar. 1.5 mg l^{-1}	Extreme values

Note: DOC, dissolved organic carbon; DON, dissolved organic nitrogen; DOP, dissolved organic particles.

From Wotton, R. S., Ed., *The Biology of Particles in Aquatic Systems,* CRC Press, Boca Raton, FL, 1990, 92.

Table 2.4—8
SEASONAL MAXIMA AND MINIMA OF PARTICULATE ORGANIC MATTER IN SALT-MARSH WATERS

Location	Measurement	Maximum (Time, conc.)	Minimum (Time, conc.)	Remarks
Texas marsh	POC	July 5-7 mg l^{-1}	Winter 3-4 mg l^{-1}	Usual values, not extremes
Louisiana bays	POC	Feb.-Mar. 5-6 mg l^{-1}	Rest of year negligible	
Marsh, SC	POC	Summer 4.6 mg l^{-1}	Dec. 0.7 mg l^{-1}	
Marsh, SC	POC	Summer 2-3 mg l^{-1}	Winter ca. 1 mg l^{-1}	
Marsh, SC	Partic. N	July 17 μg-atN l^{-1}	Jan. 2 μg-atN l^{-1}	
Marsh, SC	Partic. P	Sep. & Aug. 1.2-1.3 μg-atP l^{-1}	Jan. 0.1 μg-atP l^{-1}	
Duplin R., GA	POC	Summer 5.9 & 4.3 mg l^{-1}	Winter 1.8 & 1.5 mg l^{-1}	Surface & bottom
Flax Pond, NY	POC	Mar. & May 2.0 & 2.1 mg l^{-1}	Jan. & Dec. 0.2 mg l^{-1}	
Dill Creek, SC	POM	Aug. 58.5 mg l^{-1}	Sep. & Dec. 0.4 & 0.5 mg l^{-1}	
Marsh, MA	POC	Mar. & Sep. 1.0 & 1.6 mg l^{-1}	Apr.-Aug. & Nov. .03-0.6 mg l^{-1}	Monthly averages

From Wotton, R. S., Ed., *The Biology of Particles in Aquatic Systems,* CRC Press, Boca Raton, FL, 1990, 93.

Table 2.4—9
PARTICULATE ORGANIC CARBON BUDGET FOR THE GREVELINGEN ESTUARY, THE NETHERLANDS

Source	C/m²/year (g)
Export from salt marshes	0.3—7
Seagrass production	5—30
Benthic microalgae production	25—57
Phytoplankton production	130
Import from land runoff	2
Import from North Sea	155—225
Total	317—451

From Wolff, W. J., in *Ecology of Marine Benthos*, Coull, B. C., Ed., University of South Carolina Press, Columbia, 1977, 267. With permission.

Table 2.4—10
PARTICULATE ORGANIC CARBON BUDGET FOR BARATARIA BAY, LOUISIANA

Source	C/m²/year (g)	Sink	C/m²/year (g)
Export from salt marshes	297	Consumption in estuary	432
Benthic algae production	244	Export to Gulf of Mexico	318
Phytoplankton production	209		
Total	750		750

From Wolff, W. J., in *Ecology of Marine Benthos*, Coull, B. C., Ed., University of South Carolina Press, Columbia, 1977, 267. With permission.

Table 2.4—11
ORGANIC CARBON BUDGET FOR THE DOLLARD ESTUARY

Import/production		Export/utilization	
Particulate C from North Sea and River Ems	37.1	Dissolved C to North Sea	?
From potato flour mill	33.0	Utilization in water	7.2
From salt marshes	0.5	Utilization in sediment	18.2
Phytoplankton production	0.7	Buried in sediment	9.9
Benthic algae production	9.3	Bird feeding	0.26
Total	80.6		35.56

Note: Units are $\times 10^6$ kg C/year for the entire area of nearly 100 km².

From Van Es, F. B., *Helgol. Wiss. Meeresunters.*, 30, 283, 1977. With permission.

Table 2.4—12
SEASONAL MAXIMA AND MINIMA OF DISSOLVED ORGANIC MATTER IN RIVERS

Location	Measurement	Maximum (Time, conc.)	Minimum (Time, conc.)	Remarks
Ganges R., Bangladesh	DOC	July 9.3 mg l^{-1}	June 1.3 mg l^{-1}	Max near crest
Ganges R., Bangladesh	Carbohydrate	June 1120 μg l^{-1}	Oct. 141 μg l^{-1}	
Ganges R., Bangladesh	Amino acids	July 616 μg l^{-1}	March 150 μg l^{-1}	
Brahmaputra, R.	DOC	July 6.5 mg l^{-1}	Rest of year 1.3-2.6 mg l^{-1}	Rising water maximum
Brahmaputra R.	Amino acids	Aug. 262 μg l^{-1}	March 79 μg l^{-1}	
Brahmaputra R.	Carbohydrate	July 985 μg l^{-1}	Oct. 155 μg l^{-1}	
Indus R.	DOC	Aug.-Sep. 22 mg l^{-1}	Low-flow period 1.2 mg l^{-1} min	Maximum range
Amazon R.	DOC	May-June 6.5 mg l^{-1}	Feb.-Mar. 4.2 mg l^{-1}	
Orinoco R., Venezuela	DOC	May & Dec. 5 & 4 mg l^{-1}	Jun.-Nov. 2-3 mg l^{-1}	
Gambia R., West Africa	DOC	September 3.7 mg l^{-1}	Dec. 1.3 mg l^{-1}	
Columbia River	TOC	Spring-Summer 3.2 mg l^{-1}	Late fall 1.8 mg l^{-1}	ca. 89% Dissolved
Tigris R., Iraq	DOM	April 13.7-21.5 mg O l^{-1}	Oct. 0.3-1.6 mg O l^{-1}	As O_2 demand
Caroni R., Venezuela	DOC	Aug. & Jan. 8 mg l^{-1}	Nov. 4 mg l^{-1}	Humic rich
Ems River, Netherlands	DOC	Feb. 12 mg l^{-1}	Aug.-Sep. 4-5 mg l^{-1}	Increase in fall-winter
Guatemalan Rivers	DOM	June-July 4-36 mg l^{-1}	June or Oct. 3-5 mg l^{-1}	Peak discharge Jul.-Aug.
Alaskan Rivers	DOC	Variable 4-6 mg l^{-1}	Aug.-Sep. 1-2 mg l^{-1}	
Shetucket R., CT	DOC	May & Sep. 6.2-10 mg l^{-1}	Jan.-Apr. 2-4 mg l^{-1}	Max. 26.4 mg l^{-1} in runoff
Ogeechee R., GA	DOC	Jan.-May 12-15 mg l^{-1}	Aug.-Dec. 6-8 mg l^{-1}	To 17 mg l^{-1} July storm
Westerwoldse, Aa, Neth.	DOC	Dec. 48 mg l^{-1}	July 13 mg l^{-1}	Winter pollution
Black Creek, GA	DOC	Jan.-May 31-38 mg l^{-1}	June-Dec. 14-28 mg l^{-1}	To 42 mg l^{-1} in storm
N. Carolina stream	DOC	July & Oct. 1.1 & 1.3 mg l^{-1}	April 0.4 mg l^{-1}	Undisturbed watershed
N. Carolina stream	DOC	July & Oct. 0.5 & 0.6 mg l^{-1}	Apr.-May 0.3 mg l^{-1}	Clear-cut watershed
Little Miami R., OH	DOC	No pattern 12.5 mg l^{-1}	No pattern 2.5 mg l^{-1}	Pollution related
Loire R., France	DOC	Jan.-Apr. & Dec. 5-6 mg l^{-1}	May 2.5 mg l^{-1}	
White Clay, Cr., PA	DOC	Summer-fall 2-2.5 mg l^{-1}	Winter 1-1.5 mg l^{-1}	Higher DOC downstream
White Clay Cr., PA	DOC	Autumn 9-12 mg l^{-1}	Mid-winter 2-4 mg l^{-1}	To 18 mg l^{-1} fall peak

Table 2.4—12 (continued)
SEASONAL MAXIMA AND MINIMA OF DISSOLVED ORGANIC MATTER IN RIVERS

Location	Measurement	Maximum (Time, conc.)	Minimum (Time, conc.)	Remarks
Hubbard Br., NH	DOC	No pattern 2 mg l^{-1}	No pattern <0.1 mg l^{-1}	Max. & min. in fall
Moorland Stream, U.K.	DOM	August up to 30 mg l^{-1}	Feb. 0-3 mg l^{-1}	Peak flow in Aug.

Note: DOC, dissolved organic carbon; DOM, dissolved organic material; TOC, total organic carbon.

From Wotton, R. S., Ed., *The Biology of Particles in Aquatic Systems,* CRC Press, Boca Raton, FL, 1990, 85.

Table 2.4—13
SEASONAL MAXIMA AND MINIMA OF PARTICULATE ORGANIC MATTER (POC) IN RIVERS

Location	Measurement	Maximum (Time, conc.)	Minimum (Time, conc.)	Remarks
Ganges R., Bangladesh	Amino acids	July 2395 $\mu g\ l^{-1}$	March 24 $\mu g\ l^{-1}$	
Ganges R., Bangladesh	Carbohydrate	July 1672 $\mu g\ l^{-1}$	March 46 $\mu g\ l^{-1}$	
Indus R., Pakistan	POC	Aug. (1981-82) up to 16 mg l^{-1}	Feb. or Apr. 0.3 mg l^{-1}	Jun. & Sep. peaks (1983)
Indus R., Pakistan	Amino acids	Aug. or Sep. 659-2009 $\mu g\ l^{-1}$	Oct. or Apr. 127-277 $\mu g\ l^{-1}$	Irregular sampling
Indus R., Pakistan	Carbohydrate	Aug. or Sep. 412-1105 $\mu g\ l^{-1}$	Feb./Apr./May 58-122 $\mu g\ l^{-1}$	Irregular sampling
Yangtze R., China	POC	July 17 mg l^{-1}	Jan.-Feb. 3 mg l^{-1}	Averages
Amazon R., near mouth	POC	Feb.-Mar. 8.2 mg l^{-1}	May-June 1-2 mg l^{-1}	
Amazon R., up stream	POC	Feb.-Mar. 15-20 mg l^{-1}	May-June 3.7 mg l^{-1}	
Guatemalan Rivers	POC	June 7 mg l^{-1}	Nov.-Apr. undet.	Tributaries to 15 mg l^{-1}
Orinoco R., Venezuela	POC	No pattern 2.5 mg l^{-1}	No pattern 1 mg l^{-1}	
Gambia R., West Africa	POC	Early Sep. 2 mg l^{-1}	Oct.-Nov. 0.3 mg l^{-1}	No relation to discharge
Columbia River	POC	Summer 860 $\mu g\ l^{-1}$	Winter 2 $\mu g\ l^{-1}$	Highly variable
St. Lawrence River	POC (means)	May & Nov. .67 & .69 $\mu g\ l^{-1}$	Feb. .24 mg l^{-1}	May-terrigenous, Nov.-*in situ*
Erriff R., Ireland	POC	No pattern 4 mg l^{-1}	No pattern 0.25 mg l^{-1}	Runoff-related peaks
Bundorragha R., Ireland	POC	No pattern 0.9 mg l^{-1}	No pattern 0.3 mg l^{-1}	Steeper basin than Erriff
Loire R., France	POC	May-Aug. 6 mg l^{-1}	Nov. & Mar. 2-3 mg l^{-1}	Increased at low water
Chalk Stream, England	POM	Fall-winter 10-12 mg l^{-1}	Summer 1-2 mg l^{-1}	Dry weight

From Wotton, R. S., Ed., *The Biology of Particles in Aquatic Systems,* CRC Press, Boca Raton, FL, 1990, 88.

Table 2.4—14
PREDOMINANT ORGANIC COMPOUNDS IN SOME CANADIAN RIVERS

Sample location	DOC (mg/l)	Rel. color (col. units)	Humic acid (mg/l)	Fulvic acid (mg/l)	Tannins and lignins (mg/l)	Phenols (mg/l)	Carbohydrate (mg/l)	Amino acids (μg/l)	Total pesticides (μg/l)
Thompson River @ Savona (B.C.)	6	17	N.D.	2.1	0.4	N.D.	0.03	172	0.006
Saskatchewan River above Carrot River	13	10	1.7	8.0	1.3	N.D.	0.1	high	0.006
Yukon River @ Dawson (Yukon)	5		N.D.	2.7	0.6	N.D.	0.2	676	0.002
Qu'Appelle River @ Welby (Sask.)	12	20	0.8	12.0	1.4	N.D.	0.07	low	0.004
Moose River @ Mouth (Ont.)	18		7.1	11.3	4.0		N.D.	291	0.017
Richelieu River @ St.Helaire (Que.)	8	10	N.D.	7.7	0.3	N.D.	0.09	189	0.006
St. Lawrence River @ Levis (Que.)		10	0.3	6.8	0.7	0.003	0.08	low	0.004
Annapolis River @ Wilmot (NS)	7		0.6	5.2	1.2	N.D.		253	N.D.
St. John River @ Woodstock (NB)	14	30	N.D.	12.6	4.0	N.D.	0.1	203	0.001
Exploits River @ Millertown (Newf.)	7		0.6	5.5	1.4	N.D.	0.11	100	0.004

From Afghan, B. K. and Chau, A. S. Y., *Analysis of Trace Organics in the Aquatic Environment,* CRC Press, Boca Raton, FL, 1989, 316.

2.5 DECOMPOSITION OF ORGANIC MATTER

Table 2.5—1
IDEALIZED REPRESENTATIONS OF THE MAJOR DECOMPOSITION PATHWAYS OF ORGANIC MATTER

1. Aerobic respiration
   $(CH_2O)_x(NH_3)_y(H_3PO_4)_z + (x + 2y)O_2 \rightarrow xCO_2 + (x + y)H_2O + yHNO_3 + zH_3PO_4$

2. Nitrate reduction
   $5(CH_2O)_x(NH_3)_y(H_3PO_4)_z + 4xNO_3^- \rightarrow xCO_2 + 3xH_2O + xHCO_3^- + 2xN_2 + 5yNH_3 + 5zH_3PO_4$

3. Manganese reduction
   $(CH_2O)_x(NH_3)_y(H_3PO_4)_z + 2xMnO_2 + 3xCO_2 + xH_2O \rightarrow 4xHCO_3^- + 2xMn^{2+} + yNH_3 + zH_3PO_4$

4. Iron reduction
   $(CH_2O)_x(NH_3)_y(H_3PO_4)_z + 4xFe(OH)_3 + 7xCO_2 \rightarrow 8xHCO_3^- + 3xH_2O + 4xFe^{2+} + yNH_3 + zH_3PO_4$

5. Sulfate reduction
   $2(CH_2O)_x(NH_3)_y(H_3PO_4)_z + xSO_4^{2-} \rightarrow 2xHCO_3^- + xH_2S + yNH_3 + 2zH_3PO_4$

6. Methane production
   $2(CH_2O)_x(NH_3)_y(H_3PO_4)_z \rightarrow xCO_2 + xCH_4 + 2yNH_3 + 2zH_3PO_4$

7. Fermentation (generalized)
   $12(CH_2O)_x(NH_3)_y(H_3PO_4)_z \rightarrow xCH_3CH_2COOH + xCH_3COOH + 2xCH_3CH_2OH + 3xCO_2 + xH_2 + 12yNH_3 + 12zH_3PO_4$

From Aller, R. C., in *Animal-Sediment Relations*, McCall, P. L. and Tevesz, M. J. S., Eds., Plenum Press, New York, 1982, 53. With permission.

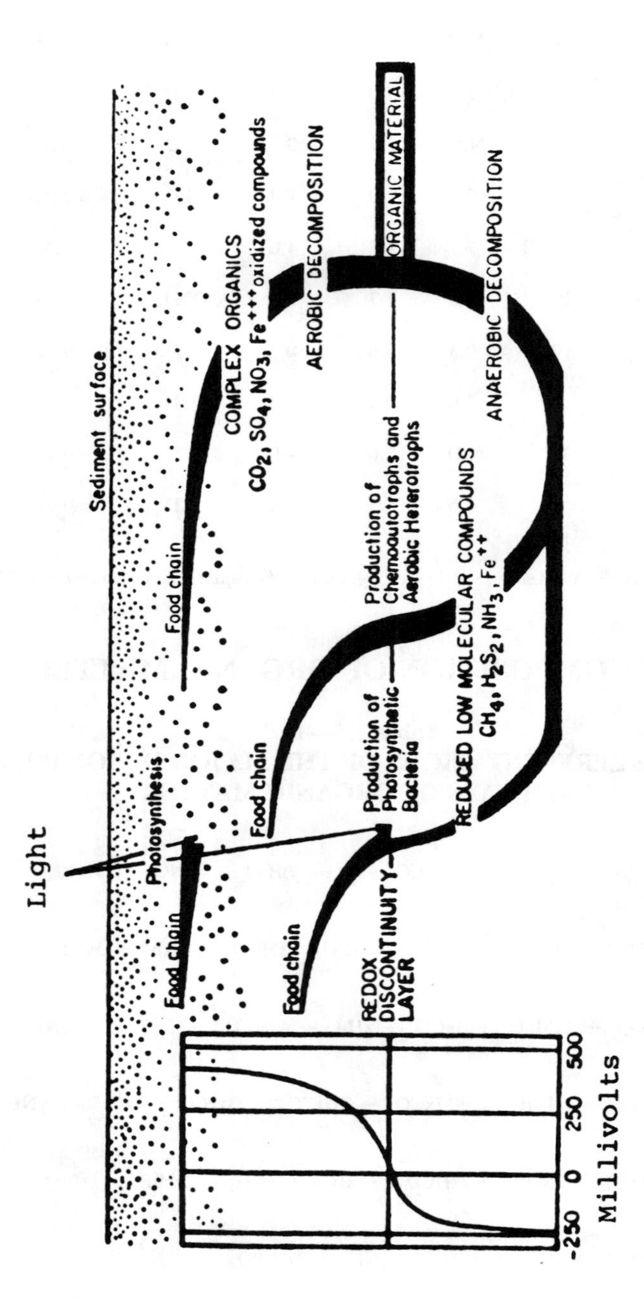

FIGURE 2.5—1. Aerobic and anaerobic decomposition processes in estuarine sediments. (From Fenchel, T. and Riedl, R. J., *Mar. Biol.*, 7, 255, 1970. With permission.)

Table 2.5—2
CONTRIBUTION OF VARIOUS AEROBIC AND ANAEROBIC PROCESSES TO CARBON MINERALIZATION IN A SPECTRUM OF MARINE AND FRESHWATER SEDIMENTS

Site	Rate of carbon oxidation (mmol C m^{-2} d^{-1})	Respiratory mode (1% of C oxidation)		
		O_2 Resp.	SO_4^{2-} Resp.	CH_4 Prod.
Marine				
Limfjorden	36	47	53	—
Sippewissett saltmarsh	458	10	90	(0)
Sapelo I. saltmarsh	200	20	70	(10)
Sippewissett saltmarsh	180	50	50	(0)
Sulfate-Depleted Marine				
Cape Lookout Bight	100	(0)	68	32
Lacustrine				
Blelham Tarn	—	42	2	25
Wintergreen Lake — A	3.1[a]	—	13[b]	87[b]
Wintergreen Lake — B	14.4[a]	—	13[b]	87[b]
Wintergreen	108[c]	—	30	71
Lake Vechten	23	—		70
Lawrence Lake	3[a]	—	30—81[c]	19—70[c]
Experimental Lakes Area	9.5—12	—	16—20[c]	72—82[c]

[a] Assuming 2 cm active depth in sediment
[b] Percentage of total anoxic mineralization only
[c] 0 to 15 cm.

From Wotton, R. S., Ed., *The Biology of Particles in Aquatic Systems,* CRC Press, Boca Raton, FL, 1990, 133.

Table 2.5—3
SOME REPRESENTATIVE REACTIONS ILLUSTRATING PATHWAYS OF MICROBIAL METABOLISM AND THEIR ENERGY YIELDS[a]

		Energy yield (kcal)
Aerobic respiration	$C_6H_{12}O_6$ (glucose) + $6O_2$ = $4CO_2$ + $4H_2O$	686
Fermentation	$C_6H_{12}O_6$ = $2CH_3CHOCOOH$ (lactic acid)	58
	$C_6H_{12}O_6$ = $2CH_2CH_2OH$ (ethanol) + $2CO_2$	57
Nitrate reduction and denitrification	$C_6H_{12}O_6$ + 24/6 NO_3^- + 24/5 H^+ = $6CO_2$ + 12/5 N_2 + 42/5 H_2O	649
Sulfate reduction	$CH_3CHOHCOO^-$ (lactate) + $^1/_2$ SO_4 + $^3/_2$ H^+ = CH_3COO^- (acetate) + CO_2 + H_2O + $^1/_2$ HS^-	8.9
	CH_3COO^- + $SO_4^=$ = $2CO_2$ + $2H_2O$ + HS^-	9.7
Methanogenesis	H_2 + $^1/_4$ CO_2 = $^1/_4$ CH_4 + $^1/_2$ H_2O	8.3
	CH_3COO + $4H_2$ = $2CH_4$ + $2H_2O$	39
	CH_3COO = CH_4 + CO_2[b]	6.6
Methane oxidation	CH_4 + $SO_4^=$ + $2H^+$ = CO_2 + $2H_2O$ + HS^-	3.1
	CH_4 + $2O_2$ = CO_2 + $2H_2O$	193.5
Sulfide oxidation	HS^- + $2O_2$ = $SO_4^=$ + H^+	190.4
	HS^- + 8/5 NO_3^- + 3/5 H^+ = $SO_4^=$ + 4/5 N_2 + 4/5 H_2O	177.9

[a] Energy yields vary depending on the conditions, so different measurements may be found in different references. The values reported here are representative.
[b] This reaction is sometimes considered fermentation.

From Valiela, I., *Marine Ecological Processes,* Springer-Verlag, New York, 1984. With permission.

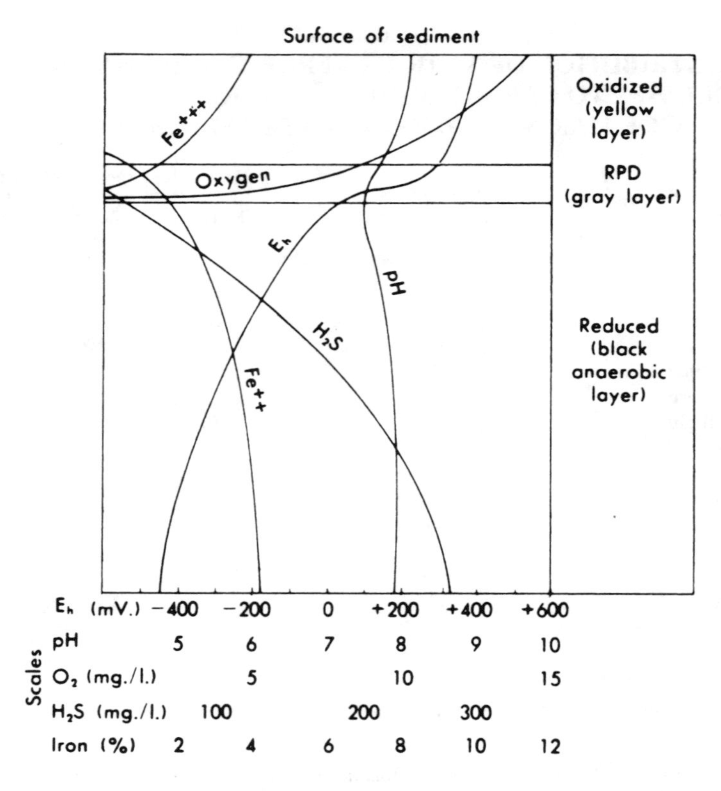

FIGURE 2.5—2. Profile of marine sediment reflecting color zonation and the change of chemical properties with increasing depth below the sediment-water interface. (From McConnaughey, B. H. and Zottoli, R., *Introduction to Marine Biology,* 4th ed., C. V. Mosby, St. Louis, 1983; adapted from Fenchel, T., *Ophelia,* 6, 1, 1969. With permission.)

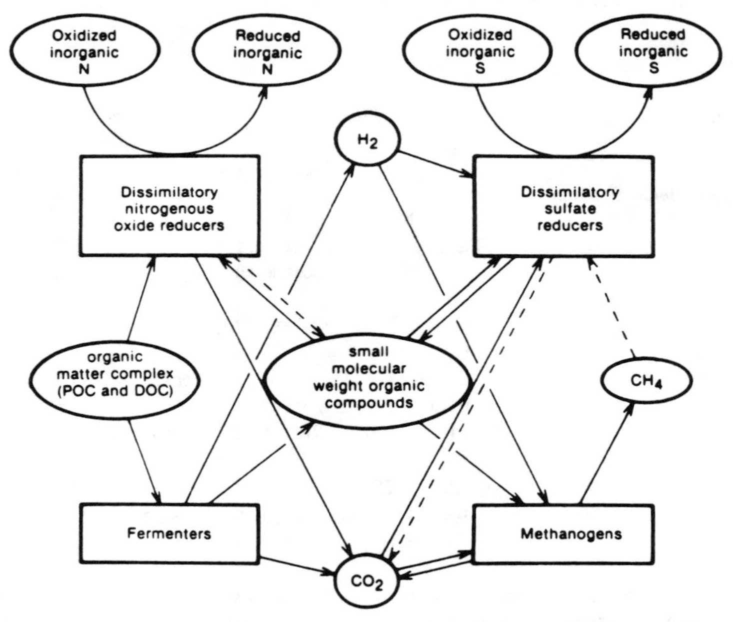

FIGURE 2.5—3. Conceptual model of anaerobic microbial processes in salt-marsh sediments. Solid lines are confirmed fluxes. Dashed lines are possible fluxes. (From Wiebe, W. J., Christian, R. R., Hansen, J. A., King, G., Sherr, B., and Skyring, G., in *The Ecology of a Salt Marsh,* Pomeroy, L. R. and Wiegert, R. G., Eds., Ecological Studies, Vol. 38, Springer-Verlag, New York, 1981, 137. With permission.)

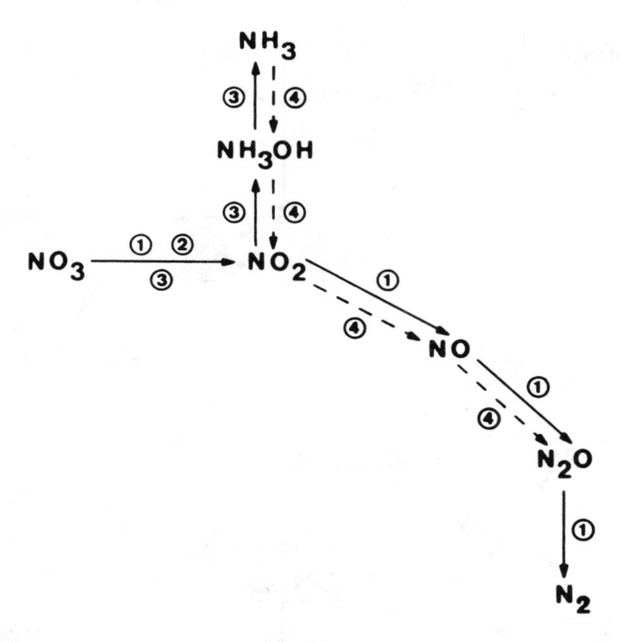

FIGURE 2.5—4. Pathways of dissimilatory nitrogenous oxide reduction. (1) Denitrification; (2) dissimilatory reduction (terminates at NO_2); (3) dissimilatory ammonia production; (4) "nitrification" N_2O pathway: ammonia to nitrous oxide. (From Wiebe, W. J., Christian, R. R., Hansen, J. A., King, G., Sherr, B., and Skyring, G., *The Ecology of a Salt Marsh,* Pomeroy, L. R. and Wiegert, R. G., Eds., Springer-Verlag, New York, 1981, 137. With permission.)

2.6 CARBON CYCLE

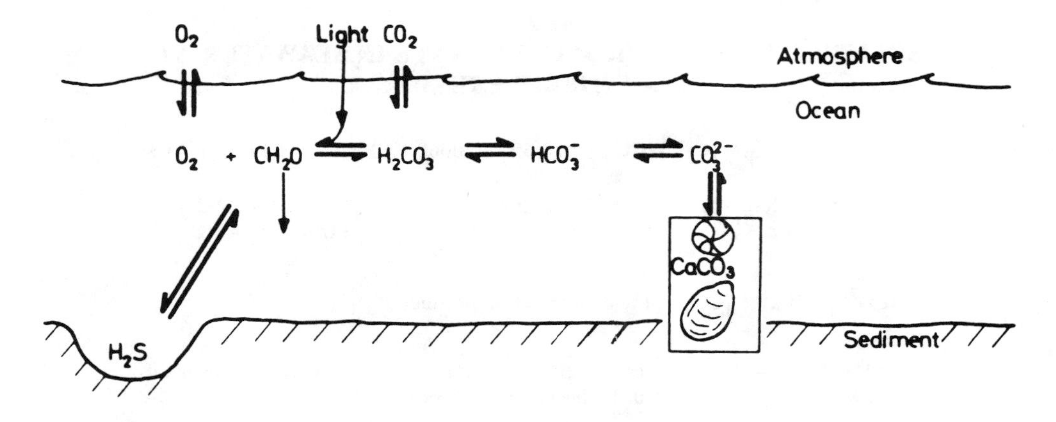

FIGURE 2.6—1. The CO_2-$CaCO_3$ cycle. (From Olausson, E., in *Chemistry and Biogeochemistry of Estuaries,* Olausson, E. and Cato, I., Eds., John Wiley & Sons, Chichester, 1980, 297. Reproduced by permission of copyright © owner, John Wiley & Sons, Ltd.)

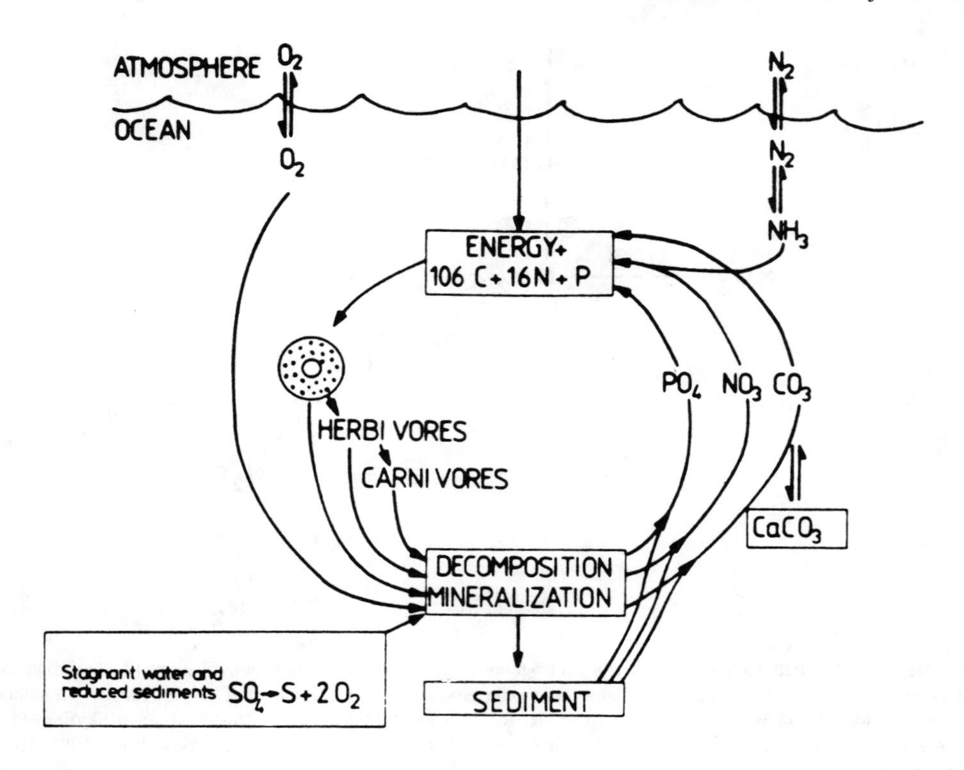

FIGURE 2.6—2. The carbon cycle in the estuarine environment. (From Olausson, E., in *Chemistry and Biogeochemistry of Estuaries,* Olausson, E. and Cato, I., Eds., John Wiley & Sons, Chichester, 1980, 297. Reproduced by permission of copyright© owner, John Wiley & Sons, Ltd.)

Table 2.6—1
SOLUBILITY OF CALCIUM CARBONATE IN SEAWATER AT ATMOSPHERIC PRESSURE

Calcite	$K'_{sp} = [Ca^{2+}][CO_3^{2-}] = (0.69 - 0.0063\ t°C) \times 10^{-6} \times S\ (°/_{oo})/34.3$
Aragonite	$K'_{sp} = [Ca^{2+}][CO_3^{2-}] = 0.90 \times 10^{-6}$ at 19°/oo Cl and t = 25°C $\Delta K'_{sp}/\Delta\ t°C = 0.0078 \times 10^{-6}/°C$ at 19°/oo Cl from 0 to 40°C

Note: $K'_{sp} = [Ca^{2+}]\ [CO_3^{2-}]$, the concentration product.

From MacIntyre, W. G., Fisheries Res. Bd. of Canada, Oceanographic and Limnological Series, Manuscript Rep. Ser. 200, Dartmouth, Nova Scotia, 1965.

Table 2.6—2
EFFECT OF PRESSURE ON THE SOLUBILITY OF CALCIUM CARBONATE IN SEAWATER*

Mineral	t°C	$(K'_{sp})_{500}/(K'_{sp})_1$	$(K'_{sp})_{1000}/K'_{sp})_1$
Aragonite	2	2.11 ± 0.06	4.23 ± 0.27
	22	1.80 ± 0.01	3.16 ± 0.02
Calcite	2	2.18	4.79
	22	1.88	3.56

Note: $K'_{sp} = [Ca^{2+}]\ [CO_3^{2-}]$

*The subscripts 1, 500, and 1000 refer to the pressure (atm).

Reprinted from Hawley, J. and Pytkowicz, R. M., *Geochim. Cosmochim. Acta,* 33, 1557, 1969. With kind permission from Pergamon Press, Ltd., Headington Hill Hall, Oxford OX3 OBW, U.K.

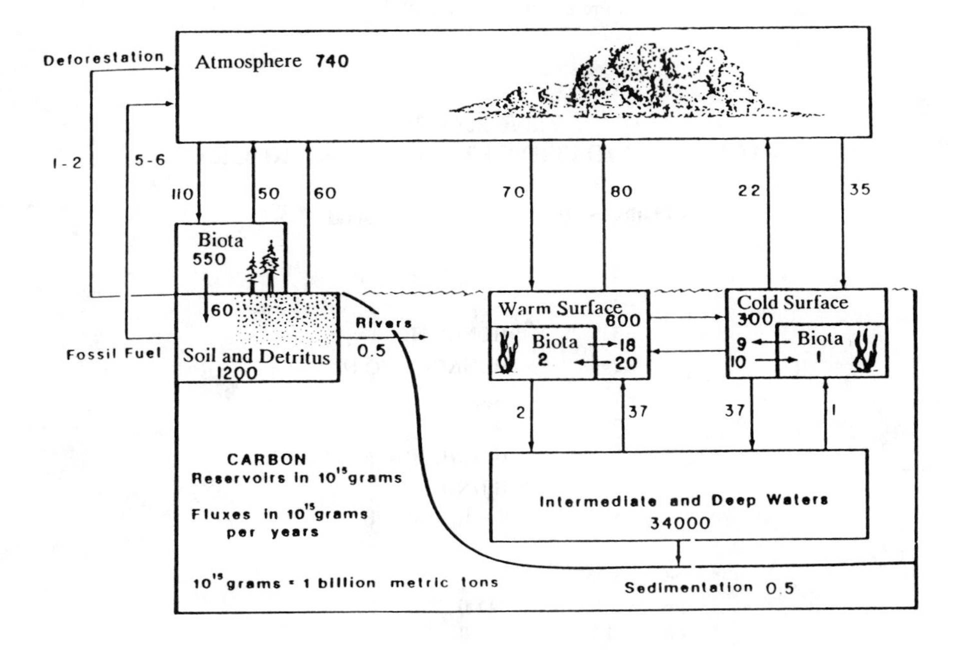

FIGURE 2.6—3. The reservoirs of carbon dioxide on the earth. (From Millero, F. J. and Sohn, M. L., *Chemical Oceanography,* CRC Press, Boca Raton, FL, 1992, 272.)

2.7 NUTRIENT ELEMENTS

Table 2.7—1
TRANSFORMATION OF NITROGEN FORMS IN THE NITROGEN CYCLE OF ESTUARIES

Process	Transformation
A. Nitrogen fixation (oxygen sensitive)	$N_2 \rightarrow NH_3$
B. Dissimilatory reduction (oxygen sensitive)	
1. Respiratory reduction (nitrate respiration)	$NO_3^- \rightarrow NO_2^-$
2. Denitrification	NO_3^-, NO_2^-, $N_2O \rightarrow$ gaseous products (N_2O, N_2, NH_3)
C. Assimilatory reduction (ammonia sensitive)	NO_3^-, etc. $\rightarrow NH_3$
D. Nitrification	$NH_3 \rightarrow NO_3^-$
E. Ammonification	"R-NH_2" $\rightarrow NH_3$

From Webb, K. L., in *Estuaries and Nutrients*, Neilson, B. J. and Cronin, L. E., Eds., Humana Press, Clifton, N.J., 1981, 25. With permission.

Table 2.7—2
VARIOUS OXIDATION STATES OF NITROGEN

Oxidation State	Compound
+5	NO_3^-, N_2O_5
+4	NO_2
+3	$HONO$[a], NO_2^-, N_2O_3
+2	$HONNOH$[b], $HO_2N_2^-$, $N_2O_2^{2-}$
+1	N_2O
0	N_2
−1	H_2NOH, HN_3, N_3^-, NH_2OH
−2	H_2NNH_2
−3	RNH_4, NH_3[c], NH_4^+ [c]

[a] $pK = 3.35$.
[b] $pK_1 = 7.05$, $pK_2 = 11.0$.
[c] $pK_B = 4.75$, $pK_A = 9.48$.

From Millero, F. J. and Sohn, M. L., *Chemical Oceanography*, CRC Press, Boca Raton, FL, 1992, 332.

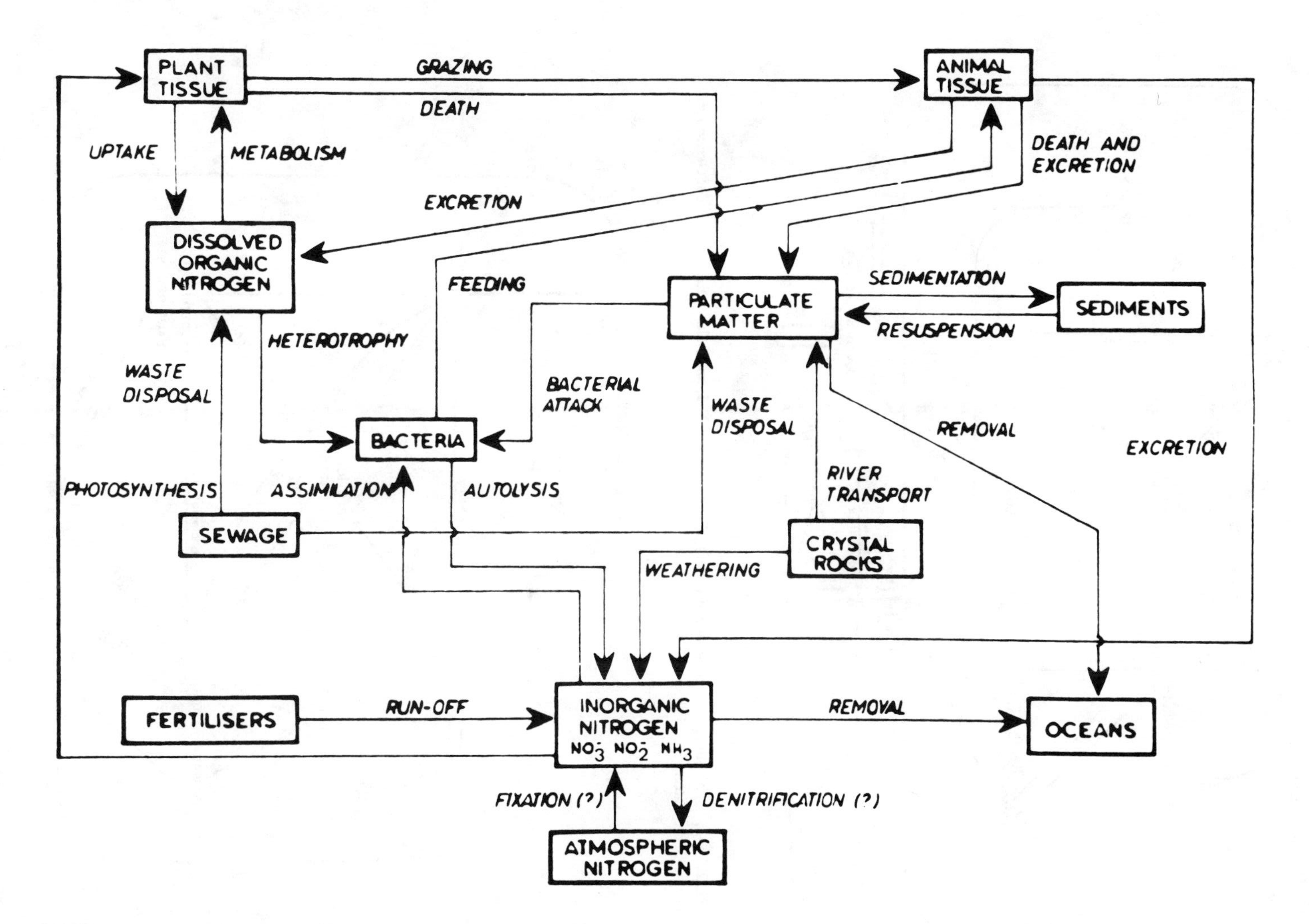

FIGURE 2.7—1. The nitrogen cycle in the estuarine environment. (From Aston, S. R., in *Chemistry and Biogeochemistry of Estuaries,* Olausson, E. and Cato, I., Eds., John Wiley & Sons, Chichester, 1980, 233. Reproduced by permission of copyright© owner, John Wiley & Sons, Ltd.)

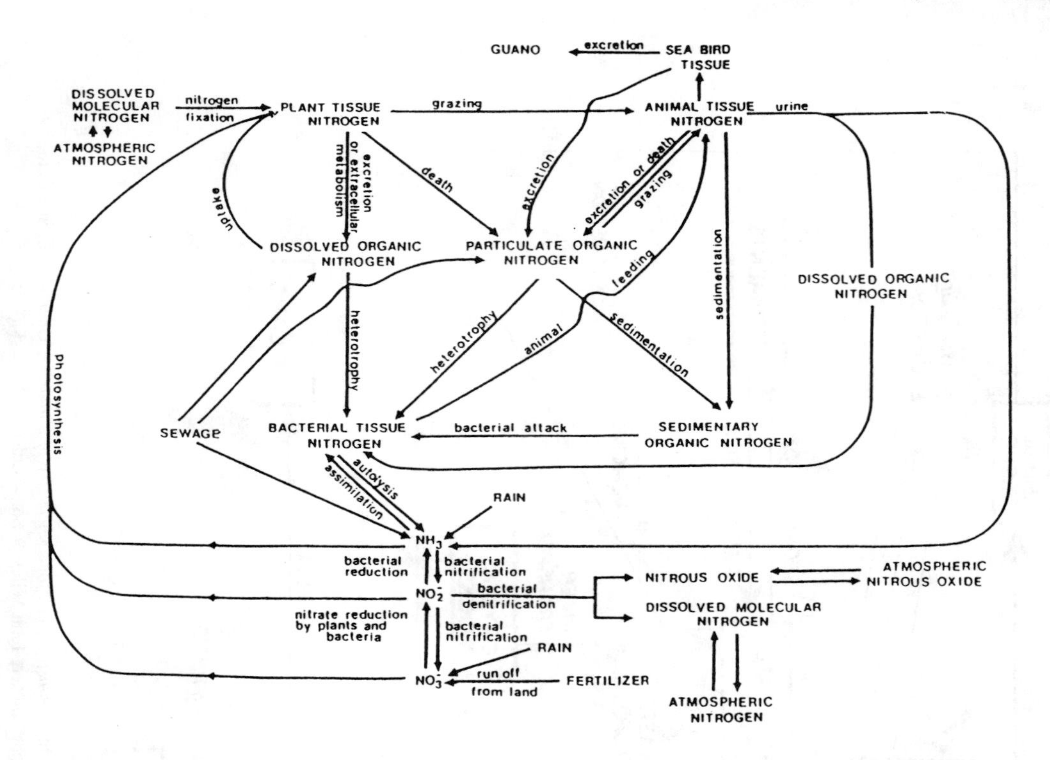

FIGURE 2.7—2. The nitrogen cycle in ocean waters. (From Millero, F. J. and Sohn, M. L., *Chemical Oceanography,* CRC Press, Boca Raton, FL, 1992, 336.)

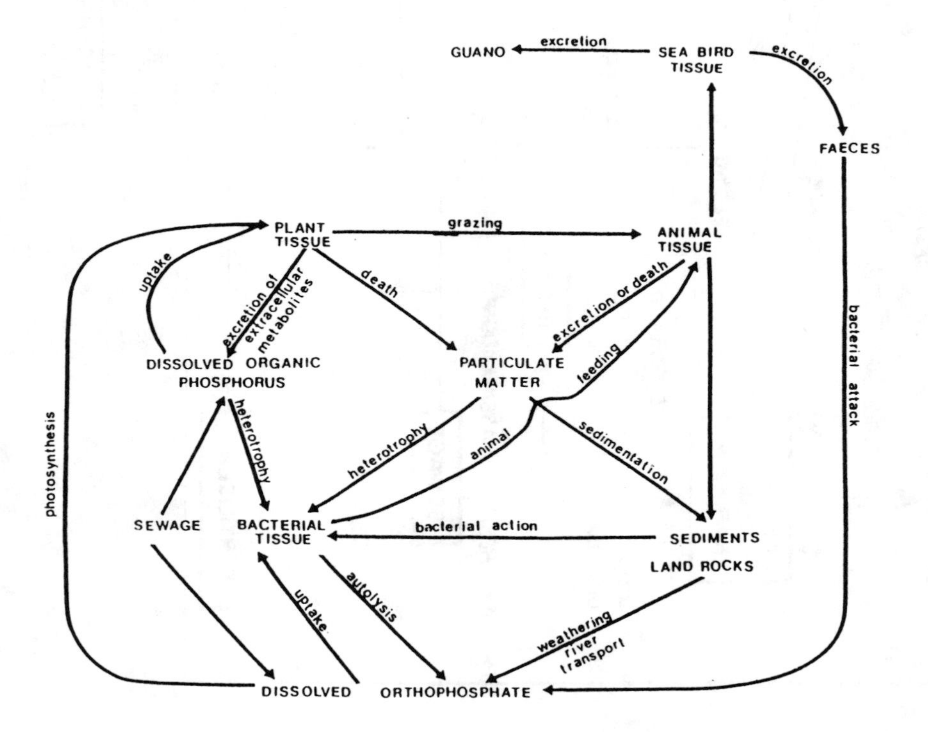

FIGURE 2.7—3. The phosphate cycle in ocean waters. (From Millero, F. J. and Sohn, M. L., *Chemical Oceanography,* CRC Press, Boca Raton, FL, 1992, 331.)

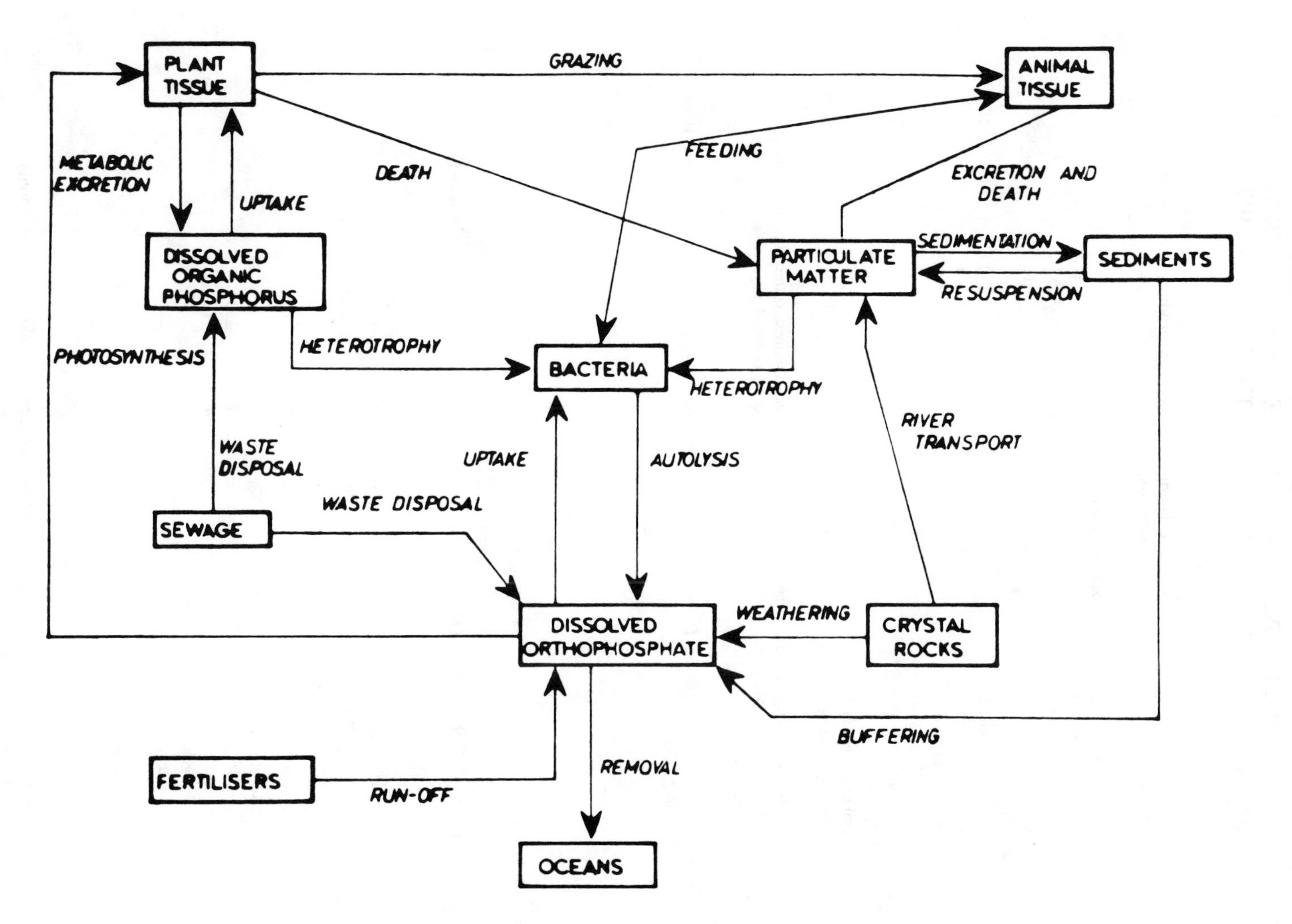

FIGURE 2.7—4. The phosphorus cycle in the estuarine environment. (From Aston, S. R., in *Chemistry and Biogeochemistry of Estuaries,* Olausson, E. and Cato, I., Eds., John Wiley & Sons, Chichester, 1980, 233. Reproduced by permission of copyright © owner, John Wiley & Sons, Ltd.)

Table 2.7—3
NET ANNUAL FLUXES OF PARTICULATE AND DISSOLVED CARBON, NITROGEN, AND PHOSPHORUS FROM THE CORAL CREEK SYSTEM (NORTHEASTERN AUSTRALIA) VIA TIDAL TRANSPORT

Component	Net annual exchange (g C, N or P m^{-2} $year^{-1}$)	Porportion of forest primary production requirements (%)
Particulate matter (mainly intact mangrove plant detritus)		
Particulate organic C	−327	−35.9
Particulate organic N	−3.7	−13.4
Particulate organic P	−0.25	−12.2
Dissolved materials		
Dissolved organic C	7.3	0.8
Dissolved organic N	1.3	4.7
Dissolved organic P	0.37	17.9
NH_4	0.15	0.6
NO_3+NO_2	−0.03	−0.1
PO_4	0.13	6.3
Total dissolved N	1.45	5.4
Total dissolved P	0.50	24.2

Note: The exchange rates are given in terms of forest area contained within the coral creek basin and in terms of net forest primary production requirements. A negative sign denotes net export.

From Connell, D. W. and Hawker, D. W., Eds., *Pollution in Tropical Aquatic Systems,* CRC Press, Boca Raton, FL, 1992, 135.

Table 2.7—4
CONCENTRATION OF NUTRIENTS IN SELECTED ESTUARIES

Estuary-sample	Nutrient (μmol)			
	NO_3^-	NH_4^+	$Si(OH)_4$	PO_4^{3-}
Scheldt (Belgium)				
0‰	0	600	230	40
30‰	30	40	10	2
Potomac-Chesapeake (U.S.)				
0‰	110	200	—	32
10‰	1	1	—	0.2
Hudson (U.S.)				
0‰	40	30	100	5
30‰	5	5	40	1.5

From Biggs, R. B. and Cronin, L. E., in *Estuaries and Nutrients,* Neilson, B. J. and Cronin, L. E., Eds., Humana Press, Clifton, N.J., 1981, 3. With permission.

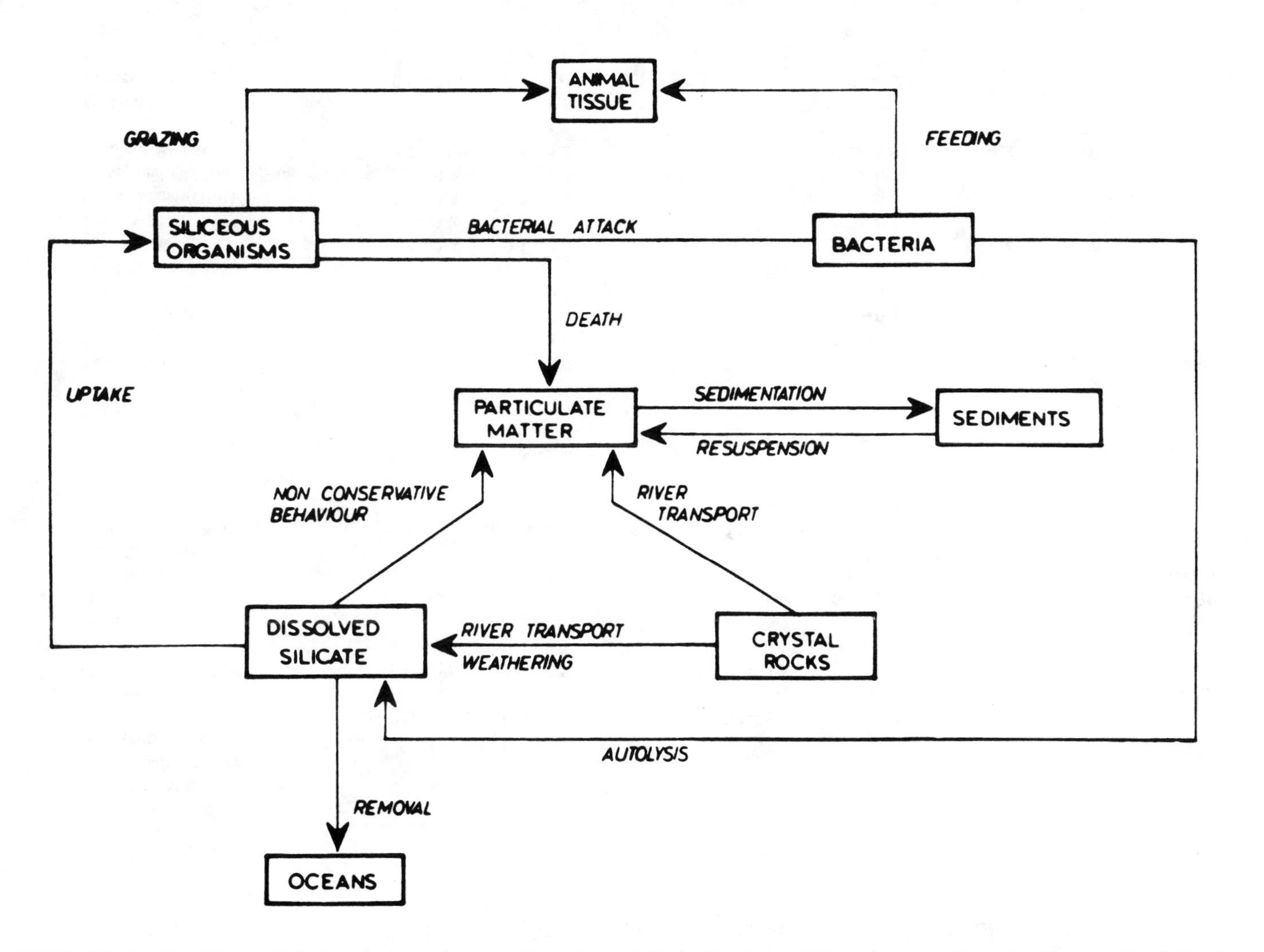

FIGURE 2.7—5. The silicon cycle in the estuarine environment. (From Aston, S. R., in *Chemistry and Biogeochemistry of Estuaries,* Olausson, E. and Cato, I., Eds., John Wiley & Sons, Chichester, 1980, 233. Reproduced by permission of copyright © owner, John Wiley & Sons, Ltd.)

Table 2.7—5
DISSOLVED AND PARTICULATE NUTRIENT CONCENTRATIONS IN TROPICAL AQUATIC ECOSYSTEMS

	NH_4	NO_2—NO_3	DON	PON (μmol/l)	PO_4	DOP	POP	Si
Oceanic (near surface)								
Gulf of Mexico	0.0—0.7	0.0—0.1		0.1—0.6	0.0—0.5			0—3
Subtropical North Pacific		0.0—0.1		0.2—0.3	0.1—0.1			2—3
Equatorial Pacific			6.2—13.8					
Equatorial Atlantic	0.0—0.1	0.0—2.0		−0.3			0.0—0.1	
South Pacific (Moorea)		−0.1			−0.4			−2
Coastal (near surface)								
Campeche Bank	0.0—2.7	0.0—0.3		0.1—5.1	0.0—0.4		0.0—0.1	0—3
Gulf of Papua/Torres Strait	0.0—3.2	0.0—3.1	0.9—13.7		0.3—0.6	0.0—0.2		1—30
Central GBR (18-20S)	0.0—0.5	0.0—0.5	2.4—14.8	1.0—3.8	0.0—0.3	0.0—0.8	0.0—0.2	0—2
Barbados	0.5—2.7	0.4—5.1			0.1—0.2			
Upwelling								
Peru	0.0—3+	0.0—20+		3.0—14+	0.0—2.5+			0—25
Arabian Sea		0.0—20+			0.0—2.0+			0—16
Estuary (near surface)								
Cochin Backwater (India)		0.0—20.3		0.0—150.0				
Missionary Bay (Australia)	0.1—0.4	2.0—7.0		0.1—0.4	0.2—0.6			
Coral reefs, Oceanic and atolls								
Canton Atoll	0.1—1.3	0.0—2.4			0.0—0.5			2—3
Enewetak Atoll	0.2—0.3	0.1—0.3	1.7—2.3		−0.2	−0.2		
Tonga Lagoon	0.1—0.7	0.1—1.0	1—23.0	3.0—10.0	0.1—0.9			17—91
Gilbert Islands	0.3—0.5	0.0—2.6	3.8—5.6		0.0—0.4			
Takapoto Atoll	−0.1	−0.2			−0.1			−0
Moorea		−0.1			−0.5			−2
Coral reefs, Shelf and fringing								
Kaneohe Bay	0.4—2.4	0.1—2.6	3.4—7.5		0.2—1.0	−0.4		
Jamaica	0.1—3.8				0.0—0.7			
Lizard Island (GBR)	0.1—0.2	0.2—1.0	3—5.0		0.2—0.4			1—2
Davies and Old Reef lagoons (GBR)	−0.2	−0.4			−0.2			−1
Abrolhos Islands	0.1—11.0	0.8—5.2			0.2—2.9	0.0—4.9		1—7
Rivers								
Amazon		−8.5			−0.2			−128
Ganges	0.7—31.4	35.7—770.0			1.9—652.0			21—2869
Maroni (S. America)		5.0—9.0			−0.2			−191
Niger	−1.0	−7.6						−250

Bermejo (S. America)	0.8—35.7	5.5—52.1			0.8—3.3	0.1—0.8	1.5—230.0	
Orinoco					−0.2			−91
Zaire	−0.4	6.7—7.0			0.1—0.8			165—342
Streams								
Savannah Rivers (Uganda)	<0.4—3.9	0.0—61.0			1.1—30.3			87—603
Wet forest streams (Uganda)	<0.4—9.7	0.0—171.0			0.1—8.2			118—825
Lesser Antilles	<2.8—62.8	<8.1—179.0			0.1—1.5			42—533
Lakes								
Lake Waigani (PNG)	−4.4	−21.6	−723.0		−42.6	−22.0		
Lake Victoria (Uganda)	−28.6	−0.4			−0.3			
Lake Nabugado (Kenya)	−<7				−21.0			
Lake Kyoga (Kenya)	−14.0	14.0			−16.8			−326
Lake Chilwa (Malawi)		2180.0—5114.0			0.0—229.0			
Lagartijo Reservoir		−4.3			−0.3			−478
Lakes Robertson/Mcllwaine (Zimbabwe)	0.4—8.8	1.3—28.2			0.0—11.8			
Lake Calado (Brazil)	0.0—1.4	0.0—3.2			0.1—0.3			
Lake La Plata (Puerto Rico)	0.0—3.5	0.0—53.0	−46.0		1.5—6.8			
Swamps and Wetlands								
Kawaga Swamp (Uganda)	−8.6	−0.3			−0.9			
North Swamp (Kenya)	−3.2	−4.6	−63.0	−871.0		−1.9	−30.6	
Tasek Bara (Malaysia)	0.0—54.7	0.7—20.7	4—109.0		0.0—3.4	1.1—25.2		−52
Papyrus swamps (Uganda)	−<0.4	0.0—30.3			0.2—1.3			34—558

Note: 0.0 = nondetectable.

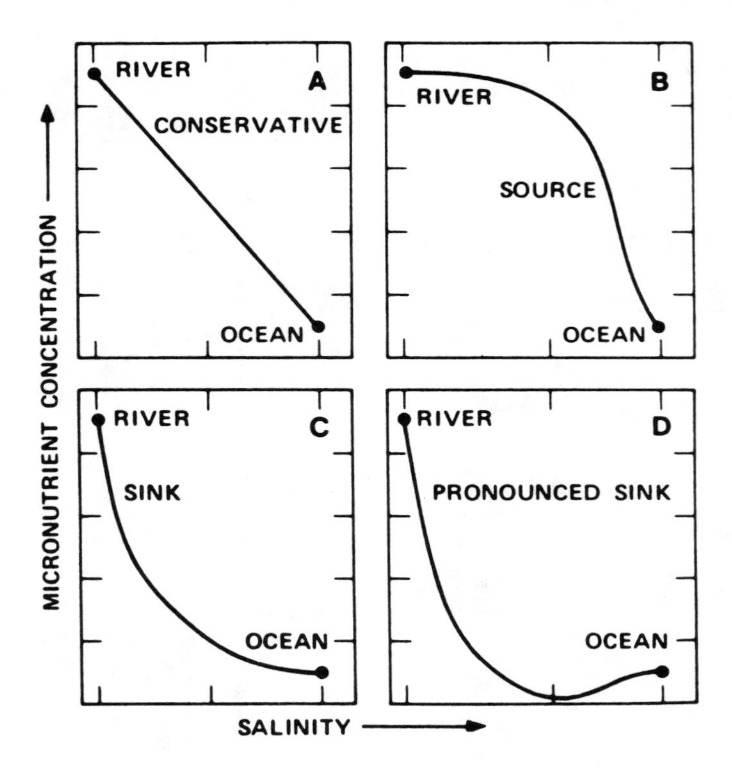

FIGURE 2.7—6. Idealized micronutrient-salinity relations showing concentration and mixing of nutrient-rich river water with nutrient-poor seawater. (A) Expected concentration-salinity distribution of a substance behaving in a conservative manner (for example, chloride) in an estuary; (B) expected concentration-salinity distribution of a substance for which the estuary is a source (for example, particulate carbon); (C) expected concentration-salinity distribution of a substance for which the estuary is a sink (for example, phosphorus); and (D) expected concentration-salinity distribution of a substance for which the estuary is a pronounced sink; that is, where the concentration of the substance in the estuary is lower than in the river and the ocean (for example, Si). (From Biggs, R. B. and Cronin, L. E., in *Estuaries and Nutrients,* Neilson, B. J. and Cronin, L. E., Eds., Humana Press, Clifton, NJ, 1981, 3. With permission.)

2.8 COMPOSITION OF STREAMS

Table 2.8—1
AVERAGE COMPOSITION OF STREAMS

	ppm
HCO_3^-	58.4
$SO_4^=$	11.2
Cl^-	7.8
NO_3^-	1.0
Ca^{++}	15.0
Mg^{++}	4.1
Na^+	6.3
K^+	2.3
(Fe)	(0.67)
SiO_2	13.1
Total	120

From Livingstone, D., Chemical composition of rivers and lakes, U.S. Geol. Surv. Profess. paper 440-G, 1963.

Table 2.8—2
THE TRACE ELEMENT COMPOSITION OF STREAMS

	ppb = μg/l	Approximate estimate, ppb	Region
Lithium	3.3	3	North America
Boron	13	10	USSR
Fluorine	88	100	USSR
	150		Japan
Aluminum	360	400	Japan
Phosphorus	19	20	Columbia R.
Scandium	0.004	0.004	Columbia R.
Titanium	2.7	3	Maine (US) lakes and streams
Vanadium	0.9	0.9	Japan
Chromium	1.4	1	US streams, Rhone, Amazon
	0.3		Maine (US) lakes and streams
Manganese	12	7	USSR,
	4.0		Maine (US) lakes and streams,
	4.8		Columbia R.
Cobalt	0.19	0.2	US streams, Rhone, Amazon
Nickel	0.3	0.3	Maine (US) lakes and streams
Copper	0.9	7	Japan,
	10		USSR,
	12		Maine (US) lakes and streams
	4.4		Columbia R.
Zinc	5.0	20	Japan,
	45		USSR,
	16		Columbia R.
Gallium	0.089	0.09	Saale and Elbe (Germany)
Germanium			
Arsenic	1.7	2	Japan,
	1.6		Columbia R.
Selenium	0.20	0.2	US streams, Rhone, Amazon
Bromine	19	20	USSR
Rubidium	1.1	1	US streams, Rhone, Amazon
Strontium	46	50	Eastern US
Yttrium		0.7	
Zirconium			
Niobium			
Molybdenum	0.6	1	Japan
	1.8		US streams, Amazon
Ruthenium			
Rhodium			
Palladium			
Silver	0.39	0.3	US streams, Rhone, Amazon
Cadmium			
Indium			

Table 2.8—2 (continued)
THE TRACE ELEMENT COMPOSITION OF STREAMS

	ppb = μg/l	Approximate estimate, ppb	Region
Tin			
Antimony	1.1	1	US streams, Rhone, Amazon
Tellerium			
Iodine	7.1	7	USSR
Cesium	0.020	0.02	US streams, Rhone, Amazon
Barium	11	10	Eastern US
Lanthanum	0.2	0.2	Sweden
		0.19	Columbia R.
Cerium		0.06	
Praseodymium		0.03	
Neodymium		0.2	
Samarium		0.03	
Europium		0.007	
Gadolinium		0.04	
Terbium		0.008	
Dysprosium		0.05	
Holmium		0.01	
Erbium		0.05	
Thulium		0.009	
Ytterbium		0.05	
Lutetium		0.008	
Hafnium			
Tantalum			
Tungsten	0.03	0.03	Sweden
Rhenium			
Osmium			
Iridium			
Platinum			

From Turekian, K. K., *Handbook of Geochemistry,* Vol. I, Wedepohl, K. H., Ed., Springer-Verlag, Berlin, 1969. With permission.

2.9 ELEMENTAL COMPOSITION OF ORGANISMS

Table 2.9—1
ELEMENTAL RATIOS (BY ATOMS) IN AQUATIC ORGANISMS OR BIOMASS[a]

Organism	C	N	P
Marine phytoplankton	106	16	1
Marine bacteria	47	7	1
	45	9	1
Tropical zooplankton	144	29	1
Aloricate ciliates	242—56	16	1
Oceanic seston (<200 μm)	122	15	1
(<3 μm)	181	20	1
(Total, 0—250 m)	120	16	
(Total, surface)	132	15	1
Atoll lagoon seston	490—55	64—9	1
Sedimenting detritus	410	29	1
Dissolved organics	300—400	19	1
Turf algae (prokaryote)	432	36	1
Macroalgae (eukaryotic)	696	36	1
Seagrasses (leaves)	458	21	1
(Roots and rhizomes)	596	12	1
Mangrove leaves (live)	1133	29	1
Mangrove litterfall	4567	24	1
Mangrove creek DOM	700—250	35—18	1
Nonmarine emergent macrophytes			
Cyperus papyrus (live)		45	1
(papyrus detritus)	4775	577	1
Typha domingensis (live)	715	24	1
T. domingensis (dead)	1107	4	1
Lepironia articulata	1920	16	
Freshwater nonemergent macrophytes			
Utricularia flexuosa	480	16	
Rainforest leaves (live)	516	20	1
	516—401	38—26	1
Rainforest litterfall	678	25	1
Floating macrophytes	480	27	1
Salvinia molesta (in sewage lagoon)		14—10	1
Azolla pinnata	329	40	1
Paspalum repens	880—416	16	
Eichhornia crassipes	267—166	16	
Amazon floodplain lake seston	253	27	1
Macrophyte epiphytes	230—85	25—12	1

[a] C:N and N:P ratios were calculated relative to N = 16 and P = 1 where data for full C:N:P ratios could not be obtained.

From Connell, D. W. and Hawker, D. W., Eds., *Pollution in Tropical Aquatic Systems*, CRC Press, Boca Raton, FL, 1992, 32.

Section 3
Marine Geology

Section 3

MARINE GEOLOGY

The theory of plate tectonics, the concept that the surface of the earth is comprised of a mosaic of eight large and several small rigid, dynamic lithospheric plates which move on a partially molten, plastic asthenosphere, has revolutionized the field of marine geology since the late 1960s. According to this theory, which forms the conceptual framework of much earth science literature today, individual lithospheric plates behave as rigid bodies, but at their boundaries the plates interact in three ways: (1) by spreading or diverging, as at mid-ocean ridges; (2) by converging, as at deep-sea trenches; and (3) by sliding past each other, as at transform faults or fracture zones. Forces operating at the plate boundaries include tension (at areas of divergence), compression (at zones of convergence), and shearing (at transform faults). The relative motions of the plates account for tectonic activity at their margins, concentrating most earthquakes and volcanoes on the surface of the earth at these perimeters.

The largest and most volcanically active chain of mountains on earth occurs along a globally encircling mid-ocean ridge (MOR) and rift system that extends through all the major ocean basins as seafloor spreading centers. It is along the 75,000-km global length of MORs where new oceanic crust forms through an interplay of magmatic construction, hydrothermal convection, and tectonic extension.[1–4] These spreading centers are sites of active basaltic volcanism, shallow-focus earthquakes, and high rates of heat flow. The rate of formation of new seafloor varies at fast-, intermediate-, and slow-spreading centers. At fast-spreading centers, seafloor genesis generally exceeds 9 cm/year. The East Pacific Rise at 13°N latitude (11 to 12 cm/year) and 20°S latitude (16 to 18 cm/year) provides examples.[5] Intermediate rates of seafloor spreading range from 5 to 9 cm/year;[6] for example, the Galapagos Rift yields values of 6 to 7 cm/year.[7] Low rates of seafloor spreading (≤2 cm/year) have been discerned on the Mid-Atlantic Ridge near 26°N latitude.[8] Central rift valleys of 50 to 200 m depth develop at slow and intermediate rates of seafloor spreading, but not at fast-spreading rates.[6]

The MOR axis marks the site where seafloor spreading widens ocean basins by the accretion of basaltic rock in a narrow lithospheric plate boundary zone. The most widespread view holds that the melts from which these basaltic rocks form are produced at depths of 20 to 80 km below the seafloor at temperatures of 1150 to 1400°C by adiabatic decompression and partial melting of peridotite mantle.[9] According to the pressure release melting hypothesis, the divergence of lithospheric plates at the MOR reduces downward pressure on the asthenosphere below, resulting in the upwelling of peridotite, which begins to melt as it rises. The partial melt fractions segregate from the residual solid mantle and migrate upward to fill shallow crustal magma chambers. As inferred from theoretical models, the rising asthenosphere attains maximum velocity in the partial melting zone. Approximately 10 to 20% of the upwelling mantle rock undergoes pressure release melting associated with ascending mantle convection. Geophysical models of upwelling-induced melting predict significant melting within a region about 100 km wide beneath a ridge axis.[10] The buoyant forces associated with the melt that collects in the magma chamber, in conjunction with the surrounding hot rock reservoir, cause shoaling of the ridge crest. The axial magma chamber expands along the strike of the MOR as repeated injections of melt drive magma away from the upwelling center and foster local eruptions.[2]

Most models of ocean crust formation require the occurrence, at least intermittently, of a molten reservoir, or shallow crustal chamber, beneath the MOR axis.[11,12] An impressive confluence of bathymetric, geophysical, and geochemical data collected during the past 2 decades has supported the basic tenets of these models. In general, the magma chamber has been envisioned as the focal point in the oceanic crust where igneous differentiation occurs, and fractionation processes construct a vertically heterogeneous crustal column.[13–20] As such, it directly controls the igneous structure and composition of the oceanic crust.[14]

The vertical structure of the ocean crust consists of three principal layers: (1) layer 1, the sedimentary layer; (2) layer 2, the basement; and (3) layer 3, the oceanic layer.[21] At the top of the oceanic crustal sequence is a layer of predominantly pelagic sediments in different stages of consolidation which increases in thickness away from the neovolcanic zone along the ridge crest. Off-axis accumulations not only consist of biogenous sediment, but also mixtures of terrigenous, hydrogenous, and hydrothermal components.[22] The thickness of the layer varies with the age of the crust, the source of the sediments, and the biological productivity of overlying waters,[23,24] typically ranging from a thin veneer beyond the narrow confines of the neovolcanic zone to perhaps 0.5 to 1 km or more in areas far removed from the MOR axis. Underlying the sedimentary layer is the oceanic basement composed of interspersed basalt pillows and sheet flows grading downward into a complex of vertical dikes, all of which may be heavily fractured. The extrusive basalts and diabase dikes constitute the chilled roof zone or lid to the subaxial magma chamber.[25] The basement layer averages about 1.6 km in thickness[22] and typically exhibits significant alteration from circulating hydrothermal fluids. The oceanic layer, with a gabbroic composition that may be altered considerably by metamorphism, lies below the basement and averages about 4 to 5 km in thickness. It crystallizes from basaltic melts in the magma chamber. The Mohorovicic discontinuity, or Moho, is a seismic boundary marking the base of the oceanic layer.

Magma chamber processes govern the stratigraphy of the oceanic crust. Mantle-derived melt accumulates and undergoes magmatic differentiation in the magma chamber prior to its emplacement in the crust and eruption onto the seafloor. Gabbros form in the chamber through slow cooling and crystal fractionation of the melt. Lateral and vertical injection of dikes from the molten reservoir into existing crust generates sheeted diabase dikes. Volcanic eruptions spill pillow and sheet flows across the seafloor. Lithospheric extension of the newly formed crust creates faults and fissures which, together with thermal contraction cracks, serve as conduits for the deep penetration of seawater that cools the dike complex and underlying magma chamber. Convective circulation returns the fluids to the seafloor through hydrothermal vents. Seawater-basalt interactions culminate in metamorphic alteration of rocks in the lower crust and rapid chemical change in the upper crust. Renewed rifting and divergence of lithospheric plates at MORs commence a new cycle of mantle upwelling, peridotite melting, diapir formation, and melt intrusion and extrusion. Older crust derived from the previous cycle of mantle upwelling accelerates away from the MOR axis, ultimately to be covered by a layer of deep-sea sediments.

There is a lack of consensus regarding the size, shape, longevity, and along-strike variability of ridge crest magma chambers.[6,11,12,20,26] Until rather recently, models of crustal magma chambers pictured the molten reservoirs as relatively large features (~10 to 20 km wide) in which melt accumulates and undergoes differentiation prior to being erupted onto the seafloor or emplaced in the crust. The "infinite onion" model gained widespread appeal in the 1970s because it provided a conceptually attractive explanation for the stratigraphy of oceanic crust (i.e., upper unit of pillow basalts and dikes, middle unit of isotopic gabbro, and lower unit of cumulate gabbro and ultramafics) derived from seismic studies of the MOR and investigations of ophiolite complexes.[27,28] However, recent geophysical studies[11,16,18,19,29,30] and petrological investigations[20,31,32] do not favor large, well-mixed magma chambers. Instead, in some areas the magma chambers appear to be only about 2 to 4 km wide and less than 1 km thick,[1] whereas in other regions they seem to be even smaller, essentially thin (tens to hundreds of meters high), narrow (<1 to 2 km wide) melt lenses overlying a zone of crystal mush in the mid-crust. Seismic tomography of fast-spreading centers (e.g., East Pacific Rise) presents an image of a small, steady-state melt storage region surrounded by a wider reservoir of very hot rock.[16,19,30] At slow-spreading centers (e.g., Mid-Atlantic Ridge), in contrast, magma supply rates are much less and no evidence exists for the presence of a steady-state axial magma chamber in the crust or upper mantle.[33–35] Here, melt replenishment and storage probably take place episodically.

Buoyant diapirism of partially molten asthenosphere has been proposed as a driving force of mantle upwelling and magma chamber genesis at seafloor spreading centers.[36] The greater density and viscosity of the overlying lithosphere can create gravitational instability in the partial melting zone of the underlying asthenosphere, similar to the classical Rayleigh-Taylor instability in fluid mechanics, inducing the development of regularly spaced melt diapirs beneath the MOR. The diapirs then rise viscously toward the ridge crest, feeding discrete magma chambers centered beneath individual ridge segments. Magmatic accretion is highly focused at these upwelling centers along the ridge axis.[37]

Uniformly spaced Rayleigh-Taylor-type gravitational instabilities in the less viscous and less dense partial melting zone of the upper mantle have been linked to the development of stable spreading cells approximately 50 km long on the Mid-Atlantic Ridge and 80 km long on the East Pacific Rise.[38–40] In this model of lithospheric formation, mantle buoyancy rather than passive, plate-controlled spreading, is the driving mechanism for first-order ridge segmentation.[2] Hence, the fundamental unit of crustal accretion of the global ridge system, owing to buoyant diapirism, is the ridge segment. The spacing, migration, and longevity of ridge segments along axis control the architecture of oceanic crust generated at MOR systems.[20] Two-dimensional geophysical profiling, high-resolution mapping (e.g., SEABEAM), and petrological observations reveal a hierarchy of ridge-segmentation patterns at a scale finer than that predicted by considerations of gravitational instability, which can subdivide the global ridge system into discrete accretionary units separated by various types of ridge axis discontinuities (e.g., transform faults, overlapping spreading centers, oblique shear zones, intervolcano gaps, and deviations from axial linearity).[20,41,42] Langmuir et al.[31] advocated that axial ridge discontinuities, regardless of their scale, represent boundaries between segments with separate magma sources and, consequently, are basic divisions of the magmatic, crustal-generation process. In effect, these discontinuities partition diapir-induced segments into units that behave independently of one another. Thus, they are deemed to be much more than just surface crustal features along diverging spreading boundaries between lithospheric plates. Langmuir et al.,[31] Macdonald et al.,[2,42] and Macdonald and Fox[1] provide detailed overviews of ridge segmentation.

In summary, all new oceanic crust bears the imprint of a complex cycle of magmatic/volcanic, hydrothermal, and tectonic processes within the spreading center plate boundary zone which varies significantly in space and time. While magmatic/volcanic processes are directly responsible for the emplacement and accretion of hot rock at divergent plate boundaries, tectonic processes transport the ocean crust away from the site of its genesis and, together with thermal contractive forces, produce an elaborate network of fractures and fissures that enable seawater to circulate through the crust, thereby cooling it. Convective hydrothermal circulation at mid-ocean ridges culminates in the release of heated fluids or hot springs on the seafloor through diffuse, white smoker, and black smoker vents. Hydrothermal convection not only plays an important role in heat transfer at oceanic ridge crests, but also in mass transfer, manifested in the metal-rich sulfide deposits concentrated in stockworks and mounds along upwelling zones, and on the oceanic crust at and adjacent to discharge zones. In addition to cooling oceanic crust and generating potential ore deposits, hydrothermal vents support lush and exotic animal communities and appear to be a chief factor modulating global ocean chemistry.

The lithosphere, while being generated at mid-ocean ridges, is resorbed along subduction zones. At boundaries of convergence, lithospheric plates collide, with one plate overriding another. The overriden plate sinks into the mantle and remelts. Deep-sea trenches mark areas of subducted oceanic crust beneath less dense continental lithosphere. By consumption of lithosphere at convergence boundaries, the earth maintains constancy in size. Characteristic features of convergence boundaries are deep-sea trenches, volcanic island arcs, shallow- and deep-focus earthquakes, adjacent mountain ranges of folded and faulted rocks, and basaltic and andesitic volcanism.[43]

The topography of the oceans can now be explained in light of the mechanisms of plate tectonics. As explained above, the 75,000-km global length of MORs encircling the earth arises from the injection of mantle material at diverging plate boundaries. These volcanic ridges comprise a mountain system comparable in physical dimensions to those on the continents. As the lithosphere cools and subsides on either side of the mid-ocean ridge, the elevations of the submarine volcanic mountains decline, and the topography becomes less rugged. These abyssal hills typically are thinly veneered with marine sediment. Further removed from the MOR system toward the continents, broad abyssal plains are found. These level segments of the seafloor form by the slow deposition of sand, silt, and clay (turbidites) transported via turbidity currents off of the outer continental shelves and continental slopes, ultimately burying hilly terrane.

Also protruding from the seafloor are isolated volcanic mountains (seamounts) rising 1 km or more above their surroundings. Occasionally, the seamounts merge into a chain of aseismic ridges. Flat-topped seamounts, referred to as guyots, develop as the volcanic peaks are eroded during emergence. Thick turbidite deposits underlie the continental rise, a gently sloping topographic feature that extends from the abyssal plain to the continental slope. The original basalt topography, therefore, is buried under a thick apron of sedimentary layers as the lithosphere ages and gradually moves away from the MOR. Proceeding toward a mainland, the relatively steeply inclined continental slope grades into the continental shelf, a gently sloping submerged edge of a continent. Continental shelves characteristically exhibit uneven topography consisting of small hills, basinlike depressions, troughs, and steep-walled submarine canyons cut by turbidity currents. Immediately seaward of submarine canyons lie thick sediments constituting submarine fans. Conspicuous features at the site of a subducting plate are deep-sea trenches, the deepest parts of the oceans reaching 10 km or more below sea level. These long, narrow, and deep troughs caused by the downward thrusting of a lithospheric plate are bordered by volcanic island arcs or a continental margin magmatic belt. A volcanic island arc (e.g., Japan) occurs when the overriding lithospheric plate is oceanic. A continental-margin magmatic belt (e.g., the Andes mountains) exists when the overriding plate is a continent.[43]

In this section, hypsometric data are provided on the ocean basins. Information is given on the location, size, and depths of deep-sea trenches, submarine canyons, and other oceanic areas. Heat flow measurements of deep-sea sediments are also presented.

In regard to ocean crustal accretion, details are given on subaxial magma chambers and their role in crustal genesis. The seismic, gravitational, petrological, and thermal constraints on the crustal structure of MORs are likewise examined. In addition, the ridge segmentation patterns observed along seafloor spreading centers are delineated.

REFERENCES

1. **Macdonald, K. C. and Fox, P. J.,** The mid-ocean ridge, *Sci. Am.,* 262, 72, 1990.
2. **Macdonald, K. C., Scheirer, D. S., and Carbotte, S. M.,** Mid-ocean ridges: discontinuities, segments, and giant cracks, *Science,* 253, 986, 1991.
3. **Kennish, M. J. and Lutz, R. A.,** Geology of deep-sea hydrothermal vents and seafloor spreading centers, *Rev. Aquat. Sci.,* 6, 97, 1992.
4. **Kennish, M. J. and Lutz, R. A.,** Ocean crust formation, *Rev. Aquat. Sci.,* 6, 247, 1992.
5. **Francheteau, J. and Ballard, R. D.,** The East Pacific Rise near 21°N, 13°N, and 20°S: inferences for along-strike variability of axial processes of the mid-ocean ridge, *Earth Planet. Sci. Lett.,* 64, 93, 1983.
6. **Macdonald, K. C.,** Mid-ocean ridges: fine scale tectonic, volcanic, and hydrothermal processes within the plate boundary zone, *Annu. Rev. Earth Planet. Sci.,* 10, 155, 1982.
7. **Grassle, J. F.,** The ecology of deep-sea hydrothermal vent communities, in *Advances in Marine Biology,* Vol. 23, Blaxter, J. H. S. and Southward, A. J., Eds., Academic Press, London, 1986, 301.
8. **Rona, P. A., Klinkhammer, G., Nelsen, T. A., Trefry, J. H., and Elderfield, H.,** Black smokers, massive sulphides, and vent biota at the Mid-Atlantic Ridge, *Nature,* 321, 33, 1986.
9. **Lin, J.,** The segmented Mid-Atlantic Ridge, *Oceanus,* 34, 11, 1991.
10. **Morgan, J. P.,** Melt migration beneath mid-ocean spreading centers, *Geophys. Res. Lett.,* 14, 1238, 1987.
11. **Detrick, R. S., Buhl, P., Vera, E., Mutter, J., Orcutt, J., Madsen, J., and Brocher, T.,** Multi-channel seismic imaging of a crustal magma chamber along the East Pacific Rise, *Nature,* 326, 35, 1987.
12. **Sinton, J. M. and Detrick, R. S.,** Mid-ocean ridge magma chambers, *J. Geophys. Res.,* 97, 197, 1992.
13. **Morgan, J. P. and Chen, Y. J.,** The genesis of oceanic crust: magma injection, hydrothermal circulation, and crustal flow, *J. Geophys. Res.,* in press.
14. **Detrick, R. S. and Langmuir, C. H.,** The geometry and dynamics of magma chambers, in *Mid-Ocean Ridge — a Dynamic Global System,* National Academy Press, Washington, D.C., 1988, 123.
15. **Nicolas, A., Reuber, I., and Benn, K.,** A new magma chamber model based on structural studies in the Oman Ophiolite, *Tectonophysics,* 151, 87, 1988.
16. **Burnett, M. S., Caress, D. W., and Orcutt, J. A.,** Tomographic image of the magma chamber at 12°50′N on the East Pacific Rise, *Nature,* 339, 206, 1989.
17. **Marsh, B. D.,** Magma chambers, *Annu. Rev. Earth Planet. Sci.,* 17, 439, 1989.
18. **Kent, G. M., Harding, A. J., and Orcutt, J. A.,** Evidence for a smaller magma chamber beneath the East Pacific Rise at 9°30′N, *Nature,* 344, 650, 1990.
19. **Toomey, D. R., Purdy, G. M., Solomon, S., and Wilcox, W.,** The three dimensional seismic velocity structure of the East Pacific Rise near latitude 9°30′N, *Nature,* 347, 639, 1990.
20. **Sinton, J. M., Smaglik, S. M., Mahoney, J. J., and Macdonald, K. C.,** Magmatic processes at superfast spreading oceanic ridges: glass compositional variations along the East Pacific Rise, 13°-23°S, *J. Geophys. Res.,* 96, 6133, 1991.
21. **Nicolas, A.,** *Structures of Ophiolites and Dynamics of Oceanic Lithosphere,* Kluwer Academic Publishers, Dordrecht, The Netherlands, 1989.
22. **Rona, P. A.,** Oceanic ridge crest processes, *Rev. Geophys.,* 25, 1089, 1987.
23. **Morse, J. W. and Mackenzie, F. T.,** *Geochemistry of Sedimentary Carbonates,* Elsevier, Amsterdam, 1991.

24. **Kennish, M. J. and Lutz, R. A.,** Calcium carbonate in surface sediments of the deep ocean floor, *Rev. Aquat. Sci.*, 6, 183, 1992.
25. **Rosencrantz, E.,** Formation of uppermost oceanic crust, *Tectonics,* 1, 471, 1982.
26. **Lewis, B. T. R.,** The process of formation of ocean crust, *Science,* 220, 151, 1983.
27. **Cann, J. R.,** A model for oceanic crustal structure developed, *Geophys. Trans. R. Soc. London,* 268, 495, 1974.
28. **Cann, J. R.,** Onions and leaks: magma at mid-ocean ridges, *Oceanus,* 34, 36, 1991.
29. **Macdonald, K. C.,** Anatomy of the magma reservoir, *Nature,* 339, 178, 1989.
30. **Toomey, D. R.,** Tomographic imaging of spreading centers, *Oceanus,* 34, 92, 1991.
31. **Langmuir, C. H., Bender, J. F., and Batiza, R.,** Petrologic and tectonic segmentation of the East Pacific Rise 5°30′-14°30′N, *Nature,* 322, 422, 1986.
32. **Bloomer, S. H., Natland, J. H., and Fisher, R. L.,** Mineral relationships in gabbroic rocks from fracture zones of Indian Ocean ridges: evidence for extensive fractionation, parental diversity, and boundary-layer recrystallization, in *Magmatism in the Ocean Basins,* Saunders, A. D. and Norry, M. J., Eds., Spec. Publ. 42, Geological Society of London, London, 1989, 107.
33. **Purdy, G. M. and Detrick, R. S.,** Crustal structure of the Mid-Atlantic Ridge at 23°N from seismic refraction studies, *J. Geophys. Res.*, 91, 3739, 1986.
34. **Detrick, R. S., Mutter, J. C., Buhl, P., and Kim, I. I.,** No evidence from multichannel reflection data for a crustal magma chamber in the MARK area of the Mid-Atlantic Ridge, *Nature,* 347, 61, 1990.
35. **Cashman, K. V.,** Magmatic processes in ridge environments, *Ridge Events,* 2, 20, 1991.
36. **Whitehead, J. A., Jr., Dick, H. J. B., and Schouten, H.,** A mechanism for magmatic accretion under spreading centers, *Nature,* 312, 146, 1984.
37. **Lin, J., Purdy, G. M., Schouten, H., Sempere, J.-C., and Zervas, C.,** Evidence from gravity data for focused magmatic accretion along the Mid-Atlantic Ridge, *Nature,* 344, 627, 1990.
38. **Schouten, H. and Klitgord, K. D.,** The memory of the accreting plate boundary and the continuity of fracture zones, *Earth Planet. Sci. Lett.*, 59, 255, 1982.
39. **Crane, K.,** The spacing of rift axis highs: dependence upon diapiric processes in the underlying asthenosphere?, *Earth Planet. Sci. Lett.*, 72, 405, 1985.
40. **Schouten, H., Klitgord, K. D., and Whitehead, J. A.,** Segmentation of mid-ocean ridges, *Nature,* 317, 225, 1985.
41. **Fox, P. J., Macdonald, K. C., and Batiza, R.,** Tectonic cycles and ridge crest segmentation, in *Mid-Ocean Ridge: a Dynamic Global System,* National Academy Press, Washington, D.C., 1988, 115.
42. **Macdonald, K. C., Fox, P. J., Perram, L. J., Eisen, M. F., Haymon, R. M., Miller, S. P., Carbotte, S. M., Cormier, M.-H., and Shor, A. N.,** A new view of the mid-ocean ridge from the behaviour of ridge-axis discontinuities, *Nature,* 335, 217, 1988.
43. **Press, F. and Siever, R.,** *Earth,* 4th ed., W. H. Freeman, New York, 1986.

3.1 GENERAL TABLES

Table 3.1—1
DIMENSIONS OF THE EARTH

	I.E.R.[a]	I.A.U.[b]
Equatorial radius a_e	6,378.388 km	6,378.160 km
Polar radius a_p	6,356.912 km	6,356.775 km
Flattening factor f	1/297	1/298.25

Radius of sphere of equal volume a_0	6,371 km
Area of surface	5.101×10^{8} km^2
Volume	1.083×10^{12} km^3
Mass	5.976×10^{27} g
Mean density	5.517 g cm^{-3}
Gravitational constant G	6.670×10^{-8} dynes cm^{-2} g^{-2}
Normal acceleration of gravity at equator g_e (based on Potsdam standard)	978.0436 cm sec^{-2}
Mean solar day d	86,400 sec = 24^{hr}
Sidereal day S	86,164.09 sec = 23^{hr} 56^{m} 4.09^{sec}
Velocity of rotation at equator	465.12 m sec^{-1}
Mean moment of inertia C_0	8.02×10^{44} g cm^2

[a] I.E.R. International Ellipsoid of Reference, 1924.
[b] I.A.U. International Astronomical Union, 1966.

(From Wedepohl, K.H., *Handbook of Geochemistry*, Vol. I, Springer-Verlag, Berlin, 1969. With permission.)

REFERENCES

1. *Astronomer's Handbook*, Transactions of the International Astronomical Union, Vol. XIIC, Academic Press, London, 1966.
2. **Gondolatsch, F.**, Mechanical data of planets and satellites, *Landolt-Bornstein*, New series, Group VI: Astronomy, astrophysics and space research, Vol. I, Springer-Verlag, Berlin, 1965.
3. **MacDonald, G. J. F.**, in Geodetic data, *Handbook of Physical Constants*, Clark, S. P., Jr., Ed., Geological Society of America, Memoir 97, 1966.

Table 3.1—2
INTERIOR, MASSES, AND DIMENSIONS OF THE PRINCIPAL SUBDIVISIONS OF THE EARTH

	Mass (10^{25} g)	Mean density (g/cm^3)	Surface area (10^6 km^2)	Radius or thickness (km)	Volume (10^9 km^3)	Mean moment of inertia (spherical symmetry) (10^{42} g cm^2)
Core	192	11.0	151	3471	1175	90
Mantle	403	4.5			898	705
Below 1000 km	240	5.1	362	1900	474	333
Above 1000 km	163	3.9	505	970–990	424	372
Crust	2.5	2.8	510		8.9	(7)
Continental	2.0	2.75	242[a]	30	7.3	
Oceanic	0.5	2.9	268[a]	6	1.6	
Oceans and marginal seas	0.14	1.03	361[a]	3.8	1.4[a]	(<1)
Whole earth	597.6	5.52	510	6371	1083	802
Atmosphere	0.00051	0.0013[b]	—	8[c]	—	—

[a] See Reference 2.
[b] Surface value.
[c] Scale height of the ''homogeneous'' atmosphere.

From Wedepohl, K. H., *Handbook of Geochemistry,* Vol. I, Springer-Verlag, Berlin, 1969. With permission.

REFERENCES

1. **MacDonald, G. J. F.,** in Geodetic data, *Handbook of Physical Constants,* Clark, S. P., Jr., Ed., Geological Society of America, Memoir 97, 1966.
2. **Poldervaart, A.,** Chemistry of the earth crust, *Geol. Soc. Am. Spec. Pap.,* No. 62, p. 119, 1955.
2. **Schmucker, U.,** in *Handbook of Geochemistry,* Vol. I, Wedepohl, K. H., Ed., Springer-Verlag, Berlin, 1969, chap. 6.

Table 3.1—3
THE SURFACE AREAS OF THE EARTH

	10^6 km^2		10^6 km^2
Continental shield region	105	Land about	
Region of young folded belts	42	29 2% of total	149
Volcanic islands in deep oceanic and suboceanic region	2		
Shelves and continental slopes region	93	Ocean about 70.8% of total	361
Deep oceanic region	268	Total surface	510

From Poldervaart, A., Chemistry of the earth crust, *Geol. Soc. Am. Spec. Pap.,* 62, 119, 1955. With permission.

Table 3.1—4
GEOLOGICAL TIME SCALE

ERA	DURATION (millions of years)	PERIOD	EPOCH	DURATION (millions of years)	MILLIONS OF YEARS AGO	EVOLUTION OF LIFE AS RECORDED BY FOSSILS
CENOZOIC	65	Quaternary; Neogene	Holocene (Recent)	.01	.01	First humans
		Tertiary; Neogene	Pleistocene	2.5	2.5	
		Tertiary; Neogene	Pliocene	4.5	7	
		Tertiary; Neogene	Miocene	19	26	Dominance of mammals and flowering plants
		Tertiary; Paleogene	Oligocene	12	38	
		Tertiary; Paleogene	Eocene	16	54	
		Tertiary; Paleogene	Paleocene	11	65	
MESOZOIC	160	Cretaceous		70	135	Last dinosaurs First flowering plants
		Jurassic		55	190	First birds Dominance of reptiles and conifers
		Triassic		35	225	First dinosaurs First mammals Wide extinctions
PALEOZOIC	345	Permian		55	280	Mammal-like reptiles
		Carboniferous: Pennsylvanian		45	325	First reptiles
		Carboniferous: Mississippian		20	345	Dominance of amphibians and ferns First seed plants
		Devonian		55	400	First amphibians Air-breathing fishes
		Silurian		30	430	First jawed fishes First vascular plants First land-dwelling invertebrates
		Ordovician		70	500	Jawless fishes
		Cambrian		70	570	First vertebrates Invertebrates widely established Appearance of numerous invertebrate fossils
PRECAMBRIAN				4 billion years		Fossils rare Algae

From Barnhart, R. K., *Dictionary of Science,* Houghton Mifflin, Boston, 1986.

3.2 TOPOGRAPHIC DATA

Table 3.2—1
MASS AND DISTRIBUTION OF THE HYDROSPHERE

	Mass (10^{15} tons)	Relative abundance (%)
Seawater	1410	86.5
Lakes, rivers	0.5	0.03
Continental ice	22	1.3
Water vapor in atmosphere	0.013	0.001
Water in sediments and sedimentary rocks	200	12.2
Totals	1632	100

From Poldervaart, A., *Chemistry of the Earth's Crust,* Geological Society of America Special Paper 62, 1955, 121.

Table 3.2—2
AREA, VOLUME, MEAN, AND MAXIMUM DEPTHS OF THE OCEANS AND THEIR ADJACENT SEAS

Sea	Area[a] (10^6 km²)	Volume[a] (10^6 km³)	Depth: Mean[a] (m)	Depth: Maximum[b] (m)
Oceans without adjacent seas				
Pacific Ocean	166.24	696.19	4,188	11,022[a]
Atlantic Ocean	84.11	322.98	3,844	9,219[b]
Indian Ocean	73.43	284.34	3,872	7,455[c]
Total	323.78	1,303.51	4,026	—
Mediterranean seas				
Arctic[d]	12.26	13.70	1,117	5,449
Austral-Asiatic[e]	9.08	11.37	1,252	7,440
American	4.36	9.43	2,164	7,680
European[f]	3.02	4.38	1,450	5,092
Total	28.72	38.88	1,354	—
Intracontinental Mediterranean seas				
Hudson Bay	1.23	0.16	128	218
Red Sea	0.45	0.24	538	2,604
Baltic Sea	0.39	0.02	55	459
Persian Gulf	0.24	0.01	25	170
Total	2.31	0.43	184	—
Marginal seas				
Bering Sea	2.26	3.37	1,491	4,096
Sea of Okhotsk	1.39	1.35	971	3,372
East China Sea	1.20	0.33	275	2,719
Sea of Japan	1.01	1.69	1,673	4,225
Gulf of California	0.15	0.11	733	3,127
North Sea	0.58	0.05	93	725[g]
Gulf of St. Lawrence	0.24	0.03	125	549
Irish Sea	0.10	0.01	60	272
Remaining seas	0.30	0.15	470	—
Total	7.23	7.09	979	—
Oceans, including adjacent seas				
Pacific Ocean	181.34	714.41	3,940	11,022[a]
Atlantic Ocean	106.57	350.91	3,293	9,219[b]
Indian Ocean	74.12	284.61	3,840	7,455[c]
World ocean	362.03	1,349.93	3,729	11,022[a]

[a] Vitiaz Depth in the Mariana Trench.
[b] Milwaukee Depth in the Puerto Rico Trench.
[c] Planet Depth in the Sunda Trench.
[d] Consisting of Artic Ocean, Barents Sea, Canadian Archipelago, Baffin Bay, and Hudson Bay.
[e] Including Andaman Sea.
[f] Including Black Sea.
[g] In the Skagerrak area.

From Millaro, F. J. and Sohn, M. L., *Chemical Oceanography,* 1992, 6.

Table 3.2—3
DEPTH ZONES IN THE OCEANS*

Table A

Ocean	Depth interval in kilometers												Total area
	0–0.2	0.2–1	1–2	2–3	3–4	4–5	5–6	6–7	7–8	8–9	9–10	10–11	(ocean)
Pacific Ocean	2.712	4.294	5.403	11.397	36.233	58.162	44.691	2.896	0.313	0.105	0.032	0.002	166.241
Asiatic Mediterranean	4.715	0.841	0.948	1.104	0.608	0.707	0.149	0.007	0.005	0	0	0	9.082
Bering Sea	1.050	0.135	0.172	0.234	0.670	0	0	0	0	0	0	0	2.261
Sea of Okhotsk	0.368	0.549	0.311	0.047	0.115	0	0	0	0	0	0	0	1.392
Yellow and East China seas	0.977	0.137	0.072	0.015	0.001	0	0	0	0	0	0	0	1.202
Sea of Japan	0.238	0.154	0.199	0.204	0.218	0	0	0	0	0	0	0	1.013
Gulf of California	0.071	0.032	0.040	0.010	0	0	0	0	0	0	0	0	0.153
Atlantic Ocean	6.080	4.474	3.718	7.436	16.729	28.090	19.324	0.639	0.058	0.010	0	0	86.557
American Mediterranean	1.021	0.465	0.589	0.667	0.906	0.586	0.112	0.008	0.002	0	0	0	4.357
Mediterranean	0.513	0.564	0.437	0.766	0.224	0.006	0	0	0	0	0	0	2.510
Black Sea	0.177	0.064	0.117	0.149	0	0	0	0	0	0	0	0	0.508
Baltic Sea	0.381	0.001	0	0	0	0	0	0	0	0	0	0	0.382
Indian Ocean	2.622	1.971	2.628	7.364	18.547	26.906	12.476	0.911	0.001	0	0	0	73.427
Red Sea	0.188	0.195	0.068	0.003	0	0	0	0	0	0	0	0	0.453
Persian Gulf	0.238	0	0	0	0	0	0	0	0	0	0	0	0.238
Arctic Ocean	3.858	1.569	0.968	1.249	1.573	0.269	0	0	0	0	0	0	9.485
Arctic Mediterranean	1.913	0.567	0.174	0.118	0	0	0	0	0	0	0	0	2.772
Total each depth	27.123	16.012	15.844	30.762	75.824	114.725	76.753	4.461	0.380	0.115	0.032	0.002	362.033

Table B

Ocean	Depth interval in kilometers												Percent of world ocean in each ocean
	0–0.2	0.2–1	1–2	2–3	3–4	4–5	5–6	6–7	7–8	8–9	9–10	10–11	
Pacific Ocean	1.631	2.583	3.250	6.856	21.796	34.987	26.884	1.742	0.188	0.063	0.019	0.001	45.919
Asiatic Mediterranean	51.913	9.255	10.433	12.151	6.698	7.780	1.636	0.076	0.058	0	0	0	2.509
Bering Sea	46.443	5.975	7.623	10.330	29.629	0	0	0	0	0	0	0	0.625
Sea of Okhotsk	26.475	39.479	22.383	3.403	8.260	0	0	0	0	0	0	0	0.384
Yellow and East China seas	81.305	11.427	5.974	1.239	0.055	0	0	0	0	0	0	0	0.332
Sea of Japan	23.498	15.176	19.646	20.096	21.551	0.033	0	0	0	0	0	0	0.280
Gulf of California	46.705	20.848	25.891	6.556	0	0	0	0	0	0	0	0	0.042
Atlantic Ocean	7.025	5.169	4.295	8.590	19.327	32.452	22.326	0.738	0.067	0.012	0	0	23.909
American Mediterranean	23.443	10.674	13.518	15.313	20.796	13.440	2.572	0.193	0.051	0	0	0	1.203
Mediterranean	20.436	22.475	17.413	30.515	8.940	0.221	0	0	0	0	0	0	0.693
Black Sea	34.965	12.587	23.077	29.371	0	0	0	0	0	0	0	0	0.140
Baltic Sea	99.832	0.168	0	0	0	0	0	0	0	0	0	0	0.105
Indian Ocean	3.570	2.685	3.580	10.029	25.259	36.643	16.991	1.241	0.001	0	0	0	20.282
Red Sea	41.454	43.058	14.920	0.568	0	0	0	0	0	0	0	0	0.125
Persian Gulf	100.000	0	0	0	0	0	0	0	0	0	0	0	0.066
Arctic Ocean	40.673	16.539	10.209	13.167	16.580	2.834	0	0	0	0	0	0	2.620
Arctic Mediterranean	69.013	20.454	6.274	4.260	0	0	0	0	0	0	0	0	0.766
Percent of world ocean in each depth interval	7.492	4.423	4.376	8.497	20.944	31.689	21.201	1.232	0.105	0.032	0.009	0.001	

*Area in percent of each ocean.

From Menard, H. W. and Smith, S. M., Hypsometry of ocean basin provinces, *J. Geophys. Res.*, 71, 4305, 1966. With permission of American Geophysical Union.

HYPSOMETRY OF OCEAN BASIN PROVINCES

Physiographic provinces — These 'provinces' are regions or groups of features that have distinctive topography and usually characteristic structures and relations to other provinces. Province boundaries are based on detailed physiographic diagrams where available (Heezen and Tharp, 1961, 1964; Heezen et al., 1959; Menard, 1964) supplemented by more generalized physiographic and bathymetric charts. Provinces do not overlap nor are they superimposed in this study. Thus the area of a volcano rising from an ocean basin is included only in province VOLCANO and excluded from province OCEAN BASIN. The provinces identified in this study, the capitalized province names used in the text, the abbreviations used in data processing, and the corresponding numbers appearing in illustrations are

1. Continental SHELF AND SLOPE (CONS), the whole region from the shoreline to the base of the steep continental slope. Shelf and slope are grouped because they are merely the top and front of the margins of continental blocks.
2. CONTINENTAL RISE and partially filled sedimentary basins (CNRI). Gently sloping or almost flat, they appear to have characteristic features resulting from the accumulation of a thick fill of sediment eroded from an adjacent continent and overlying an otherwise relatively normal oceanic crust. In this respect, the Gulf of Mexico and the western basin of the Mediterranean differ from the continental rise off the eastern U.S. only because they are relatively enclosed.
3. OCEAN BASIN (OCBN), the remainder after removing all other provinces. Abyssal plains and abyssal hills and archipelagic aprons are common features of low relief.
4. Oceanic RISE AND RIDGE (RISE), commonly called "mid-ocean ridges" despite the fact that they continue across ocean margins. They form one worldwide system with many branches. Boundaries are taken in most places as outer limit of essentially continuous slopes from crest.
5. RIDGE NOT KNOWN TO BE VOLCANIC (RIDG), relatively long and narrow and with steep sides. Most have unknown structure and some or most may be volcanic.
6. Individual VOLCANO (VOLC), with a boundary defined as the base of steep side slopes.
7. Island ARC AND TRENCH (TNCH), includes whole system of low swells and swales subparallel to trenches. Continental equivalents or extensions of island arcs, such as Japan, are excluded.
8. Composite VOLCANIC RIDGE (VRCM), formed by overlapping volcanoes and with a boundary at the base of steep side slopes.
9. POORLY DEFINED ELEVATION (BLOB), with nondescript side slopes and length no more than about twice width. Crustal structure unknown; may be thin continental type.

Tabulation of data and measuring procedure — Data were tabulated by 10° squares of latitude and longitude. Squares containing more than one ocean were split, and each ocean was treated separately. Within a square, the areas between the depth intervals 0–200 m, 200–1,000 m, and between 1-km contours down to 11 km were compiled for each physiographic province.

The polar planimeters (Keuffel and Esser models 4236 and 4242) used for measuring areas were read to the nearest unit on the vernier scale, and measurements were tabulated directly for card punching. These values were converted to square kilometers during computer processing by a scale factor derived from a measurement of the total number of units in the square. The area of a square was calculated assuming a spherical earth with a radius of 6,371.22 km.

Depth distribution in different oceans as a function of provinces — The depth distribution in provinces in different ocean basins, as seen in Figure 3.2—12, closely resembles the composite distribution in the world ocean (Figure 3.2—8). The sum of the depth distribution of all provinces in an ocean basin is double peaked, but for the individual provinces it is single peaked and relatively symmetrical. However, the depth distributions are sufficiently different to warrant some discussion. The mean depths of all provinces in the three major ocean basins, including marginal seas, range from 3575 m for the Atlantic Ocean to 3940 m for the Pacific (Table 3.2—7). The range of mean depths of the OCEAN BASIN province in each of these ocean basins is similar. The smallest mean OCEAN BASIN depth of 4530 m in the Indian Ocean may be the result of epirogenic movement of the oceanic crust, but it is also partially attributable to sedimentation. The eastern and southwestern parts of the Indian Ocean are deeper than 5000 m and are thus below the mean depth of the world ocean. The northwestern and southeastern parts, however, are exceptionally shallow. Seismic stations and topography show that the northwestern region has been shoaled by deposition of turbidities spreading from the mouths of the great Indian and east African rivers (Menard, 1961; Heezen and Tharp, 1964).

The mean depths of RISE and RIDGE have a limited range: from 3945 m in the Indian Ocean to 4008 m in the Atlantic Ocean (Table 3.2—7). This uniformity seems remarkable considering the widespread and diverse evidence that oceanic rises and ridges are tectonically among the more unstable features of the surface of the earth. It is all the more remarkable because the local, relief, or elevation above the adjacent OCEAN BASIN differ substantially in different oceans. The relief of RISE AND RIDGE in an ocean basin can be estimated by subtracting the mean depth from that of OCEAN BASIN or by determining the deepening required to give a best fit of individual hypsometric curves for each province. Comparing the means gives the relief of RISE AND RIDGE ABOVE OCEAN BASIN as 585 m in the Indian Ocean, 662 m in the Atlantic, and 928 m in the Pacific. The reliefs from matching curves are 800, 900, and 1200 m, respectively. The greater relief obtained by the curve-matching method results from ignoring the shallow tails of the depth distribution. Thus the range in relief is about six times as great as the range in mean depths of RISE AND RIDGE, which may be explained if the seafloor is not only elevated by epirogeny, but is also depressed. It seems reasonable to assume that the depth intervals in OCEAN BASIN with the largest areas (4 to 5 and 5 to 6 km) are those underlain by normal crust and mantle. A uniform oceanic process in the mantle acting on a uniform oceanic crust at a uniform depth may produce oceanic rises and ridges of uniform depth. Many current hypotheses for the origin of rises and ridges suggest just such an elevation. However, it is at least implicit that the mantle under the ocean basins cannot become denser and thus epirogenically depress the crust. Moreover, it is assumed by advocates of convection that if the crust is dragged down dynamically it forms a long narrow oceanic trench. The symmetrical distribution curves for OCEAN BASIN indicate that a considerable area is below the most common depth interval. Very extensive regions deeper than 6000 m exist in the northwestern Pacific and eastern Indian oceans, and there may be places where the normal oceanic crust is epirogenically depressed by more than a kilometer below the 4753-m mean depth of the OCEAN BASIN for the world ocean. Formation of broad depressions would alter the depth distribution in OCEAN BASIN and thereby vary the relief of RISE AND RIDGE in different oceans. If these broad epirogenic depressions exist, they may have a significant effect on the possible range of sea level changes relative to continents. This will be considered under ''Discussion''.

Depth distribution in island arc and trench provinces — These provinces have been defined to include not only trenches, but also the subparallel low swells and the island arcs which rise above some of them. The justification for this definition is that these features probably are caused by the same process; one question that can be answered by this type of study is whether the process elevates or depresses the seafloor. The volcanoes, some capped with limestone, which form most islands in this province, have a rather minor volume and have hardly any effect on the hypsometry.

The median depth for island ARC AND TRENCH is somewhat less than 4 km, which is less than the median depth for all ocean basins and considerably less than for the OCEAN BASIN province. The average depth would be much shallower if it were possible in some simple way to include the elevations above normal continents of the mountain ranges parallel to the Peru-Chile, Central America, Japan, and Java trenches. This would require some elaborate assumptions, but it is clear that the process which forms trenches and related features generally elevates the crust.

Volume of the ocean — Murray (1888) calculated the volume of the ocean at 323,722,150 cubic miles, which equals about 1.325×10^9 km^3. Kossinna (1921) obtained 1.370×10^9 km^3, and we obtain 1.350×10^9 km^3. It appears unlikely that this value is in error by more than a few percent. Our method of calculation is essentially the same as that of Murray and Kossinna. The midpoint value of a depth interval is multiplied by the area of that interval, and the volumes of the intervals are summed.

DISCUSSION

Seafloor epirogeny and sea level changes — Seafloor epirogeny is only one of a multitude of causes of sea level change of which the wax and wane of glaciers is probably the most intense. Epirogeny is especially important because it may have occurred at any time in the history of the earth in contrast to relatively brief periods of glaciation. That eustatic changes in sea level have occurred during geological time is suggested by widespread epicontinental seas alternating with apparently high continents.

The hypothesis that oceanic rises are ephemeral (Menard, 1958) provides a basis for quantitative estimates of epirogenic effects on sea level. If the approximate volume of existing rises and ridges is compared with the area of the oceans, it appears that uplift of the existing rises has elevated the sea level 300 m. Likewise, subsidence of the ancient Darwin rise has lowered it by 100 m (Menard, 1964).

The present study suggests that the seafloor may be depressed epirogenically in places where this movement does not merely restore the equilibrium disturbed by a previous uplift. The argument derives from the fact that the mean depth in the OCEAN BASIN province is about 4700 m. Considering that the crust has about the same thickness everywhere in the province, variations from this depth generally are caused by differences in density in the upper mantle. (We assume that where the mean depth of the crust is ''normal'' it is underlain by a ''normal'' mantle.) Thus the deeper regions, which are roughly 70 million km^2 in area, have been depressed by a density increase in the mantle. If large areas of the seafloor can be depressed as well as elevated, the resulting changes in sea level would be highly complex.

Only the most general conclusions can be drawn from this analysis, but they may be significant. First, a plausible mechanism is available to explain the eustatic changes in sea level observed in the geological record. At present, the mechanism places no constraints on the sign of a change, but appears to limit the amount to a few hundred meters. Second, in large regions the upper mantle may possibly become denser than normal. Substantial evidence exists that it is less dense than normal under rises and ridges (Le Pichon et al., 1965). If it can also be more dense than normal in large regions, these facts can provide very useful clues regarding the composition of the upper mantle and processes acting below the crust. The implications of possible densification of normal mantle can be avoided by defining the ''normal'' depth as the deepest that is at all widespread. If this definition is accepted as reasonable (it does not appear so to us), small decreases in density of the upper mantle occur under most of the world ocean. The volume of ocean basin elevated above normal is consequently large, and the possible range of sea level changes is thus at least 1 km.

Seafloor spreading and continental drift — Several aspects of our data appear to have some bearing on modern hypotheses of global tectonics. The relationships are not definitive, however, and at this time we prefer merely to indicate some of the equations which have arisen.

1. The proportion of RISE AND RIDGE to OCEAN BASIN in a basin could range from zero to infinity, but it is 0.84 for the Pacific, 0.82 for the Atlantic, and 0.61 for the Indian Ocean. The sample is very small, and consequently the similarity of the proportions may be coincidental. However, it suggests that the area of RISE AND RIDGE is proportional to the whole area of an ocean basin. This in turn suggests that the size of the basin is related to the existence of rises and ridges.

2. The proportion of SHELF AND SLOPE to OCEAN BASIN plus RISE AND RIDGE is relatively constant for large ocean basins and quite different from the proportion for small ocean basins. This relationship may require modification of at least many of the details of the hypothesis that the Atlantic Ocean basin was formed when an ancient continent split. When the supposed splitting began, the whole basin was SHELF AND SLOPE. Consequently, the proportion of SHELF AND SLOPE has since decreased. In the Pacific basin, on the other hand, the proportion of SHELF AND SLOPE to OCEAN BASIN plus RISE AND RIDGE was smaller than now and has since increased. If the Atlantic split apart at a constant rate and is still splitting as the Pacific contracts, the present equality of the proportions of SHELF AND SLOPE in the two ocean basins requires a striking coincidence. No coincidence is necessary if the splitting occurred relatively rapidly until it reached some dynamic equilibrium state, perhaps when the proportion of RISE AND RIDGE to OCEAN BASIN in each ocean basin reached about 0.8 to 0.9.

Acknowledgments — Some of these data were compiled by Isabel Taylor, Surendra Mathur, and Sarah Buffington. We wish especially to thank Mrs. Taylor for her careful rechecking of the measurements. We are indebted to Dr. G. B. Udintsev for providing Russian bathymetric charts from the first press runs.

This research was supported by Office of Naval Research Long Range Research Contract 2216(12) and National Science Foundation Grant NSF gp-4235.

From Menard, H. W. and Smith, S. M., Hypsometry of ocean basin provinces, *J. Geophys. Res.*, 71, 4305, 1966. With permission of the American Geophysical Union.

REFERENCES

Heezen, B. C. and Tharp, M., Physiographic diagram of the South Atlantic, Geological Society of America, New York, 1961.

Heezen, B. C. and Tharp, M., Physiographic diagram of the Indian Ocean, Geological Society of America, New York, 1964.

Heezen B. C., Tharp, M., and Ewing, M., The floors of the ocean. 1. The North Atlantic, Geological Society of America, New York, Special paper 65, 1959.

Kossinna, E., Die Tiefen des Weltmeeres, *Inst. Meereskunde, Veroff., Georg.-naturwiss.*, 9, 70, 1921.

Le Pichon, X., Houtz, R. E., Drake, C. L., and Nafe, J. E., Crustal structure of the mid-ocean ridges, 1, Seismic refraction measurements, *J. Geophys. Res.*, 70(2), 319, 1965.

Menard, H. W., Development of median elevations in ocean basins, *Bull. Geol. Soc. Am.*, 69(9), 1179, 1958.

Menard, H. W., *Marine Geology of the Pacific,* McGraw-Hill, New York, 1964.

Murray, J., On the height of the land and the depth of the ocean, *Scot. Geogr. Mag.*, 4, S. 1, 1888.

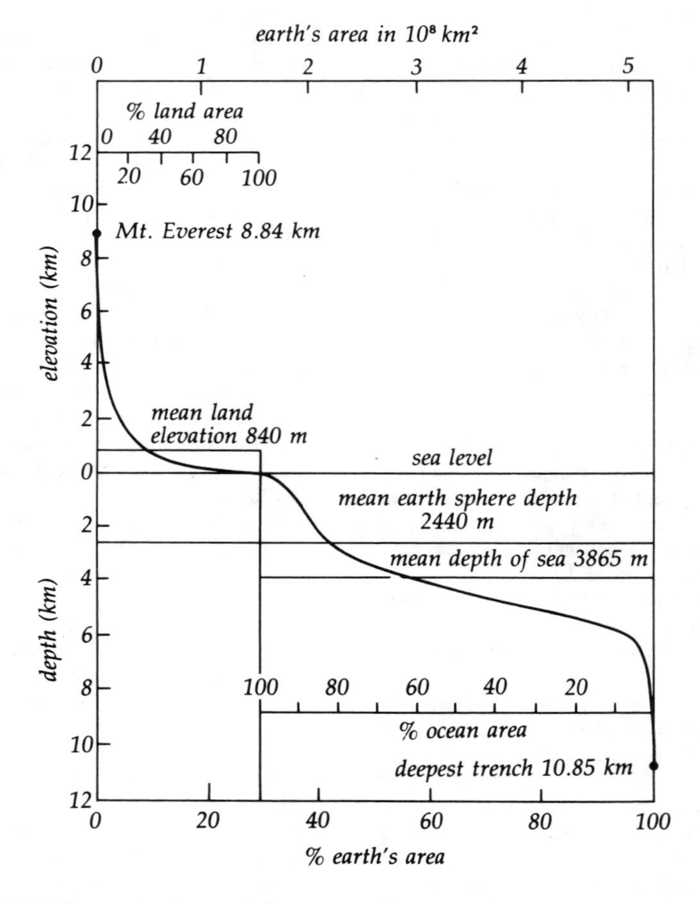

FIGURE 3.2—1. Hypsographic curve, showing abundance distribution of elevations and depths. (From Ingmanson, D. A. and Wallace, W. J., *Oceanology: an Introduction,* 2nd ed., Wadsworth Publishing Company, Belmont, CA, 1979. With permission.)

Table 3.2—4
PROVINCE AREAS IN EACH OCEAN AND TOTAL AREAS OF PROVINCES AND OCEANS (10^6 km^2)

Oceans and adjacent seas	RISE	OCBN	VOLC	CONS	TNCH	CNRI	VRCM	RIDG	BLOB	Total area of each ocean
Pacific Ocean	65.109	77.951	2.127	11.299	4.757	2.690	1.589	0.494	0.227	166.241
Asiatic Mediterranean	0	0	0.003	7.824	0.023	1.233	0	0	0	9.082
Bering Sea	0	0	0	1.286	0.281	0.694	0	0	0	2.261
Sea of Okhotsk	0	0	0	1.254	0.023	0.115	0	0	0	1.392
Yellow and East China seas	0	0	0	1.119	0.082	0	0	0	0	1.202
Sea of Japan	0	0	0.005	0.798	0	0.210	0	0	0	1.013
Gulf of California	0.042	0	0	0.111	0	0	0	0	0	0.153
Atlantic Ocean	30.519	35.728	0.882	12.658	0.447	5.381	0	0.412	0.530	86.557
American Mediterranean	0	1.346	0.060	1.889	0.201	0.861	0	0	0	4.357
Mediterranean	0	0	0	1.465	0	1.046	0	0	0	2.510
Black Sea	0	0	0	0.263	0	0.245	0	0	0	0.508
Baltic Sea	0	0	0	0.382	0	0	0	0	0	0.382
Indian Ocean	22.426	36.426	0.358	6.097	0.256	4.212	0.407	2.567	0.679	73.427
Red Sea	0	0.070	0	0	0.383	0	0	0	0	0.453
Persian Gulf	0	0	0	0.238	0	0	0	0	0	0.238
Arctic Ocean	0.513	0	0	5.874	0	2.267	0.302	0	0.528	9.485
Arctic Mediterranean	0	0	0	2.483	0	0.289	0	0	0	2.772
Total area each province	118.607	151.522	3.435	55.421	6.070	19.242	2.298	3.473	1.965	362.033

Note: CONS, Continental shelf and slope; CNRI, continental rise and partially filled sedimentary basins; OCBN, ocean basin; RISE, oceanic rise and ridge; RIDG, ridge not known to be volcanic; VOLC, individual volcano; TNCH, island arc and trench; VRCM, composite volcanic ridge; BLOB, poorly defined elevation.

From Menard, H. W. and Smith, S. M., Hypsometry of ocean basin provinces, *J. Geophys. Res.*, 71, 4305, 1966. With permission of American Geophysical Union.

Table 3.2—5
PERCENT OF PROVINCES IN OCEANS AND ADJACENT SEAS

Oceans and adjacent seas	RISE	OCBN	VOLC VRCM RIDG BLOB	CONS	TNCH	CNRI	Percent of world ocean in each ocean group
Pacific and adjacent seas	35.9	43.0	2.5	13.1	2.9	2.7	50.1
Atlantic and adjacent seas	32.3	39.3	2.0	17.7	0.7	8.0	26.0
Indian and adjacent seas	30.2	49.2	5.4	9.1	0.3	5.7	20.5
Arctic and adjacent seas	4.2	0	6.8	68.2	0	20.8	3.4
Percent of world ocean in each province	32.7	41.8	3.1	15.3	1.7	5.3	

Note: CONS, Continental shelf and slope; CNRI, continental rise and partially filled sedimentary basins; OCBN, ocean basin; RISE, oceanic rise and ridge; RIDG, ridge not known to be volcanic; VOLC, individual volcano; TNCH, island arc and trench; VRCM, composite volcanic ridge; BLOB, poorly defined elevation.

From Menard, H. W. and Smith, S. M., Hypsometry of ocean basin provinces, *J. Geophys. Res.*, 71, 4305, 1966. With permission of American Geophysical Union.

FIGURE 3.2—2. Pacific Ocean — physiographic provinces. Text contains key to province numbers. Individual volcanoes (VOLC) in black. (From Menard, H. W. and Smith, S. M., Hypsometry of ocean basin provinces, *J. Geophys. Res.*, 71, 4305, 1966. With permission of American Geophysical Union.)

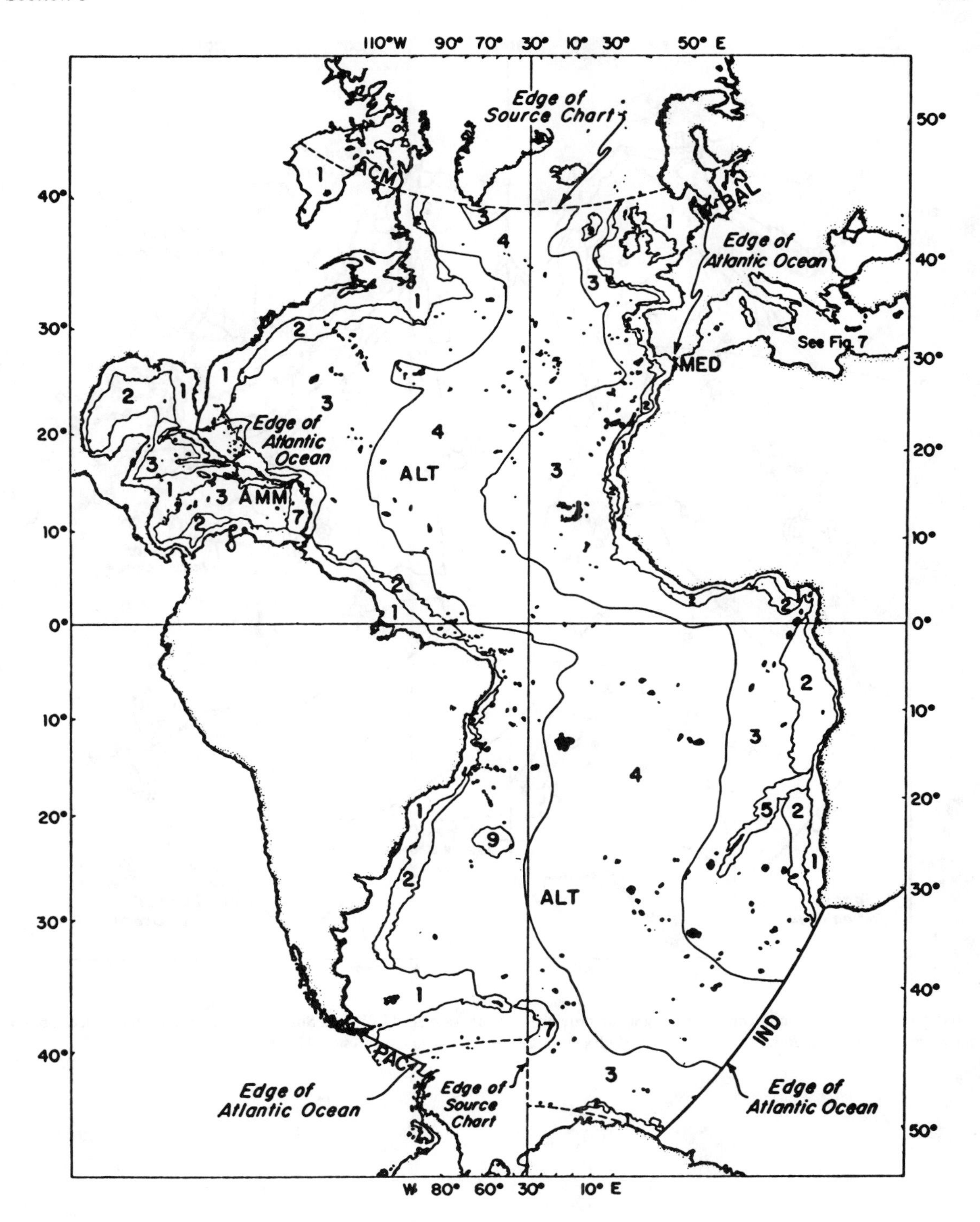

FIGURE 3.2—3. Atlantic Ocean — physiographic provinces. (From Menard, H. W. and Smith, S. M., Hypsometry of ocean basin provinces, *J. Geophys. Res.*, 71, 4305, 1966. With permission of American Geophysical Union.)

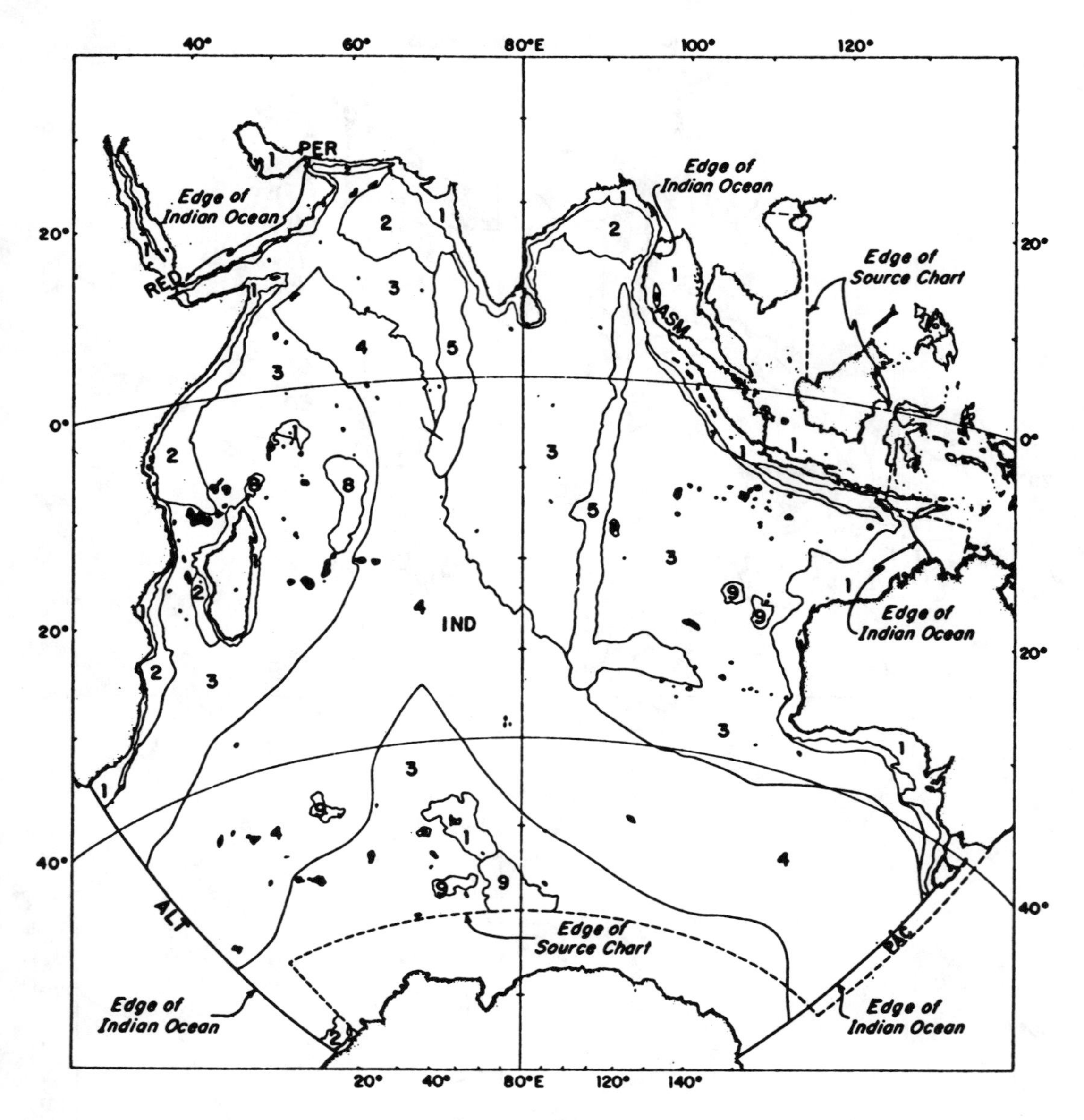

FIGURE 3.2—4. Indian Ocean — physiographic provinces. (From Menard, H. W. and Smith, S. M., Hypsometry of ocean basin provinces, *J. Geophys. Res.*, 71, 4305, 1966. With permission of American Geophysical Union.)

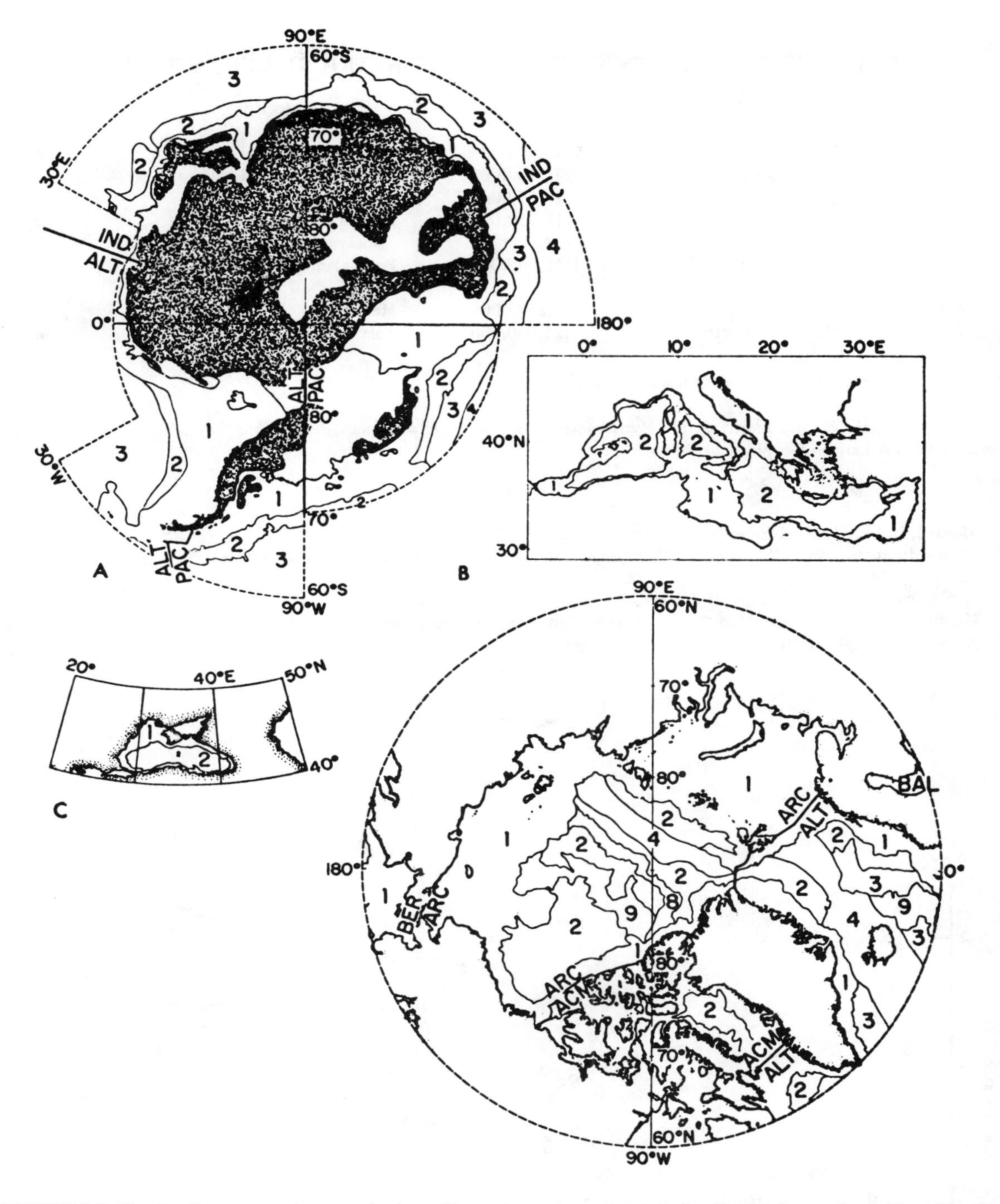

FIGURE 3.2—5. Smaller oceans and seas — physiographic provinces. Antarctic sub-ice in white is below sea level. (From Menard, H. W. and Smith, S. M., Hypsometry of ocean basin provinces, *J. Geophys. Res.*, 71, 4305, 1966. With permission of American Geophysical Union.)

Table 3.2—6
BATHYMETRIC CHARTS USED FOR HYPSOMETRIC CALCULATIONS

Source No.	Title	Scale	Projection	Reference
1	Pacific Ocean	1:7,270,000*	Lambert azimuthal equal-area	A
2	Indian Ocean	1:7,510,000*	Lambert azimuthal equal-area	B
3	Antarctica	1:9,667,000	Polar azimuthal equal-area	C
4	Atlantic Ocean	1:10,150,000*	Lateral projection with oval isoclines	B
5	Tectonic Chart of the Arctic	1:10,000,000	Polar azimuthal equal-area	D
6	Mediterranean Sea	1:2,259,000	Mercator	E
7	Northern Hemisphere	1:25,000,000	Polar azimuthal equal-area	B

* Scale of photographic enlargement used for measuring.

From Menard, H. W. and Smith, S. M., Hypsometry of ocean basin provinces, *J. Geophys. Res.*, 71, 4305, 1966. With permission of American Geophysical Union.

REFERENCES

A. **Menard, H. W.,** *Marine Geology of the Pacific,* McGraw-Hill, New York, 1964.
B. Main Administration in Geodesy and Cartography of the Government Geological Committee, USSR.
C. American Geographical Society, New York.
D. Geological Institute, Academy of Science, Moscow.
E. Unpublished chart of the Mediterranean, modified from contours compiled by R. Nason from various sources. U.S. Navy Hydrographic Office chart 4300 used as base.

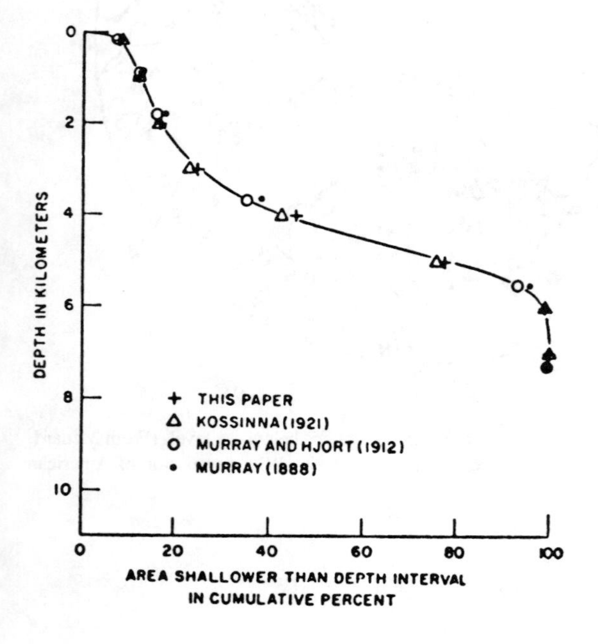

REFERENCES

1. **Menard, H. W. and Smith, S. M.,** Hypsometry of ocean basin provinces, *J. Geophys. Res.*, 71, 4305, 1966.
2. **Kossinna, E.,** Die Tiefen des Weltmeeres, *Inst. Meereskunde, Veroff., Geogr.-naturwiss.*, 9, 70, 1921.
3. **Murray, John and Hjort, J.,** *The Depths of the Ocean,* Macmillan and Co., London, 1912.
4. **Murray, John,** On the height of the land and the depth of the ocean, *Scot. Geogr. Mag.*, 4, S. 1, 1888.

FIGURE 3.2—6. Hypsometry of all ocean basins according to various studies (see References). (From Menard, H. W. and Smith, S. M., Hypsometry of ocean basin provinces, *J. Geophys. Res.*, 71, 4305, 1966. With permission of American Geophysical Union.)

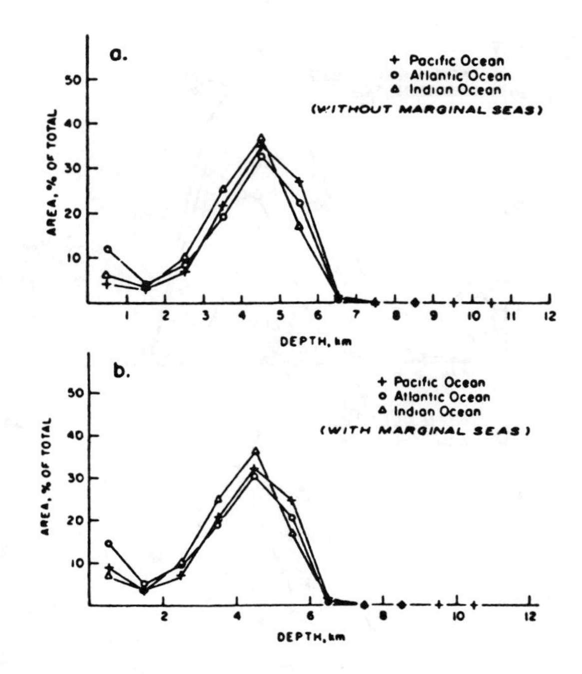

FIGURE 3.2—7. Hypsometry of individual major ocean basins. (From Menard, H. W. and Smith, S. M., Hypsometry of ocean basin provinces, *J. Geophys. Res.*, 71, 4305, 1966. With permission of American Geophysical Union.)

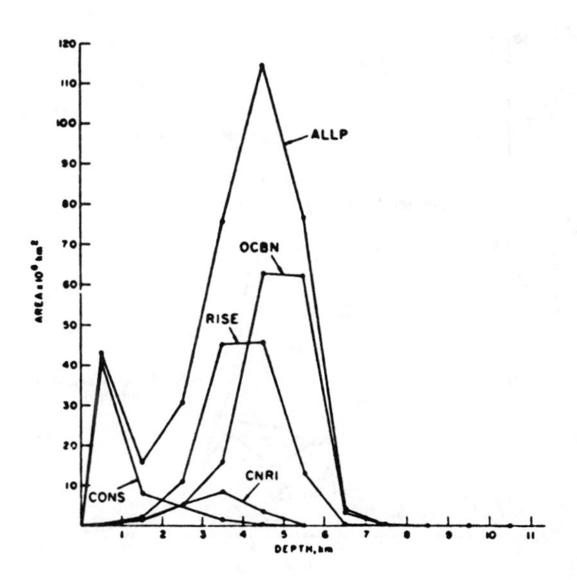

FIGURE 3.2—8. Hypsometry of all ocean basins. This diagram is for all provinces combined (ALLP) and for individual major provinces: CONS, continental shelf and slope; RISE, oceanic rise and ridge; OCBN, ocean basin. (From Menard, H. W. and Smith, S. M., Hypsometry of ocean basin provinces, *J. Geophys. Res.*, 71, 4305, 1966. With permission of American Geophysical Union.)

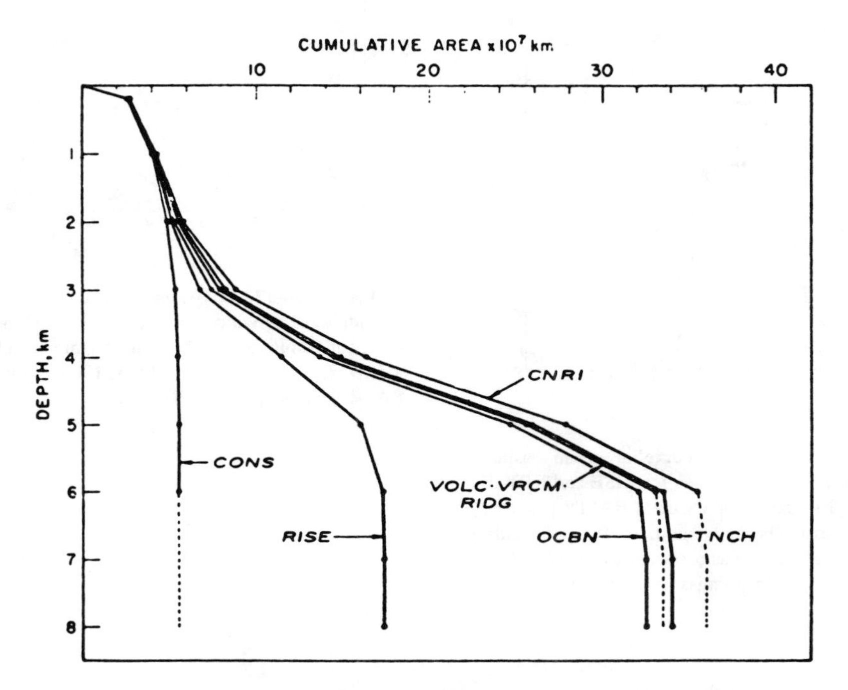

FIGURE 3.2—9. Hypsometry of ocean basins plotted cumulatively by provinces. CONS, continental shelf and slopes; RISE, oceanic rise and ridge; CNRI, continental rise; VOLC, individual volcano; VRCM, composite volcanic ridge; RIDG, ridge not known to be volcanic, OCBN, ocean basin; TNCH, island arc and trench. (From Menhard, H. W. and Smith, S. M., Hypsometry of ocean basin provinces, *J. Geophys. Res.*, 71, 4305, 1966. With permission of American Geophysical Union.)

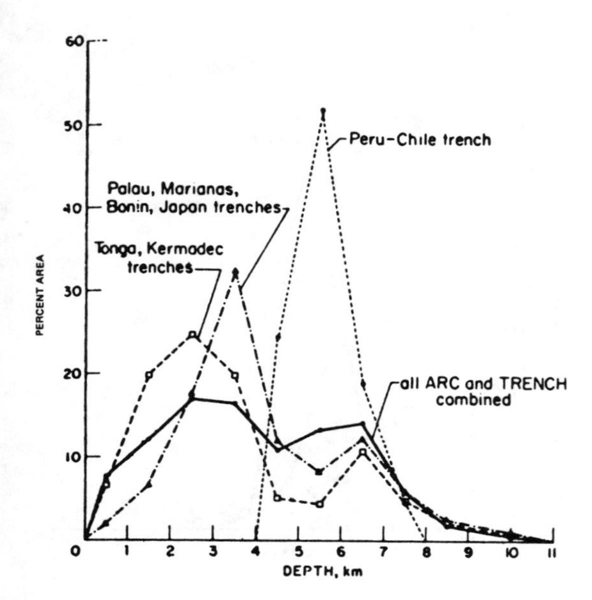

FIGURE 3.2—10. Hypsometry of all arc and trench provinces and of some groups of arcs and trenches. (From Menard, H. W. and Smith, S. M., Hypsometry of ocean basin provinces, *J. Geophys. Res.*, 71, 4305, 1966. With permission of American Geophysical Union.)

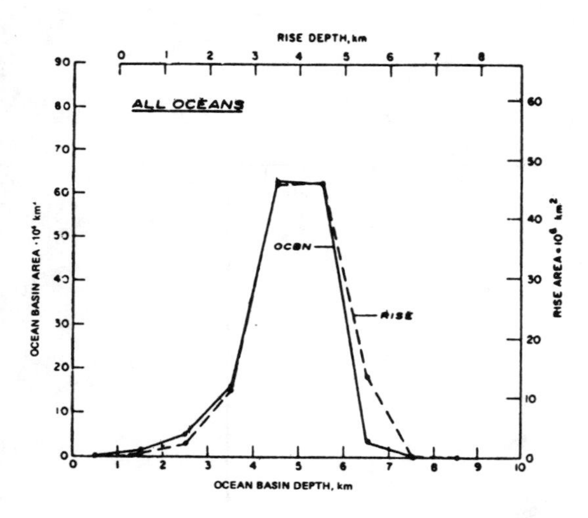

FIGURE 3.2—11. Hypsometric curve of ocean basins. Hypsometric curve of all ocean basins for RISE and RIDGE province normalized to curve for OCEAN BASIN province to show close similarity. (From Menard, H. W. and Smith, S. M., Hypsometry of ocean basin provinces, *J. Geophys. Res.*, 71, 4305, 1966. With permission of American Geophysical Union.)

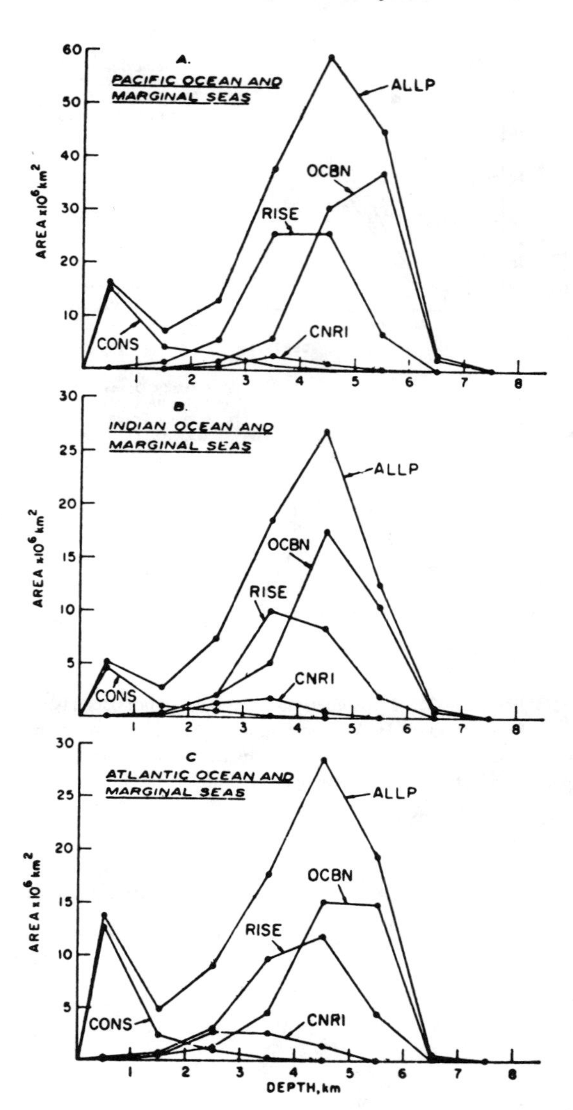

FIGURE 3.2—12. Hypsometry of all provinces (ALLP) and individual provinces in major basins. (From Menard, H. W. and Smith, S. M., Hypsometry of ocean basin provinces, *J. Geophys. Res.*, 71, 4305, 1966. With permission of American Geophysical Union.)

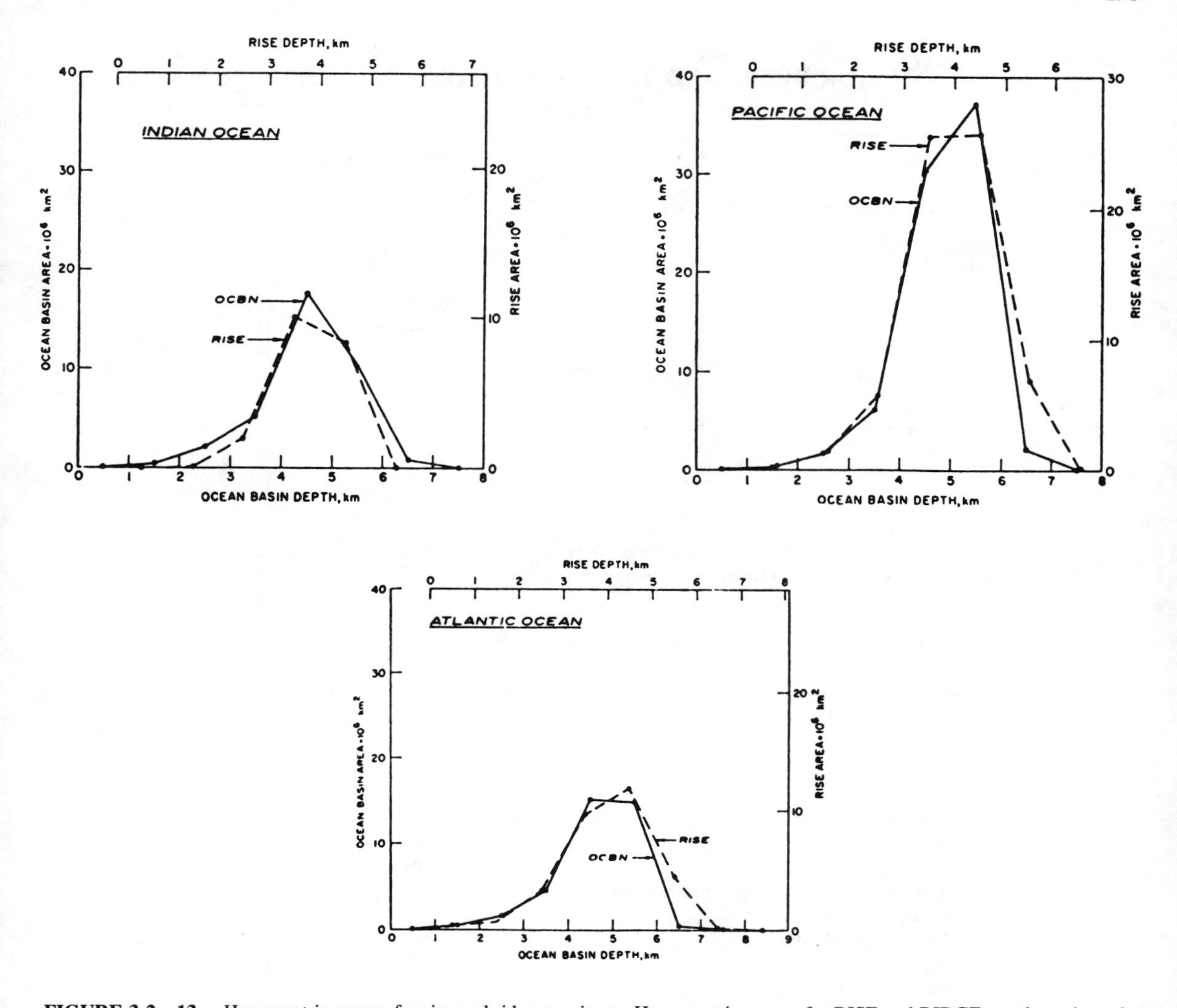

FIGURE 3.2—13. Hypsometric curves for rise and ridge provinces. Hypsometric curves for RISE and RIDGE provinces in major ocean basins normalized to curves for OCEAN BASIN provinces. (From Menard, H. W. and Smith, S. M., Hypsometry of ocean basin provinces, *J. Geophys. Res.*, 71, 4305, 1966. With permission of the American Geophysical Union.)

Table 3.2—7
CHARACTERISTICS OF OCEANIC RISES

	Mean depth, m			Relief, m	
	All provinces	OCBN (1)	RISE (2)	Mean (1) − Mean (2)	Shift of distribution curves*
World ocean	3729	4753	3970	783	1000
Pacific Ocean and marginal seas	3940	4896	3968	928	1100
Atlantic Ocean and marginal seas	3575	4670	4008	662	900
Indian Ocean and marginal seas	3840	4530	3945	585	800

* See Figure 3.2—11.

From Menard, H. W. and Smith, S. M., Hypsometry of ocean basin provinces, *J. Geophys. Res.*, 71, 4305, 1966. With permission of American Geophysical Union.

Table 3.2—8
MAJOR DEEPS AND THEIR LOCATION, SIZE, AND DEPTHS

Trench	Max depths, m	Ref.
Marianas Trench (specifically Challenger Deep)	11,034 ± 50	5
	10,915 ± 20[a]	
	10,915 ±	8
	10,863 ± 35	1
	10,850 ± 20[a]	
Tonga	10,882 ± 50	5
	10,800 ± 100	4
Kuril-Kamchatka	10,542 ± 100	13
	9,750 ± 100[b]	2
Philippine (vicinity of Cape Johnson Deep)	10,497 ± 100	6
	10,265 ± 45	16
	10,030 ± 10[a]	
Kermadec	10,047 ±	14
Idzu-Bonin		
(includes "Ramapo Deep" of the Japan Trench)	9.810	13
(vicinity of Ramapo Depth)	9,695	11
Puerto Rico	9,200 ± 20	9
New Hebrides (North)	9,165 ± 20[a]	
North Solomons (Bougainville)	9,103 ±	14
	8,940 ± 20[a]	
Yap (West Caroline)	8,527 ±	7
New Britain	8,320 ±	14
	8,245 ± 20[a]	
South Solomons	8,310 ± 20[a]	
South Sandwich	8,264	10
Peru-Chile	8,055 ± 10	3
Palau	8,054 ±	14
	8,050 ± 10[a]	
Aleutian (uncorrected, taken with nominal sounding velocity of 1,500 m/sec)	7,679	
Nansei Shoto (Ryuku)	7,507	
Java	7,450	15
New Hebrides (South)	7,070 ± 20[a]	
Middle America	6,662 ± 10	

[a] These soundings were taken during Proa Expedition, April–June, 1962, aboard R. V. *Spencer F. Baird.* A Precision Depth Recorder was employed, and the ship's track crossed over (within the limits of celestial navigation) points from which maximum depths had been reported.

[b] This is the maximum sounding obtained in the vicinity of the Vitiaz Depth (Udintsev, 1959) by French and Japanese vessels in connection with dives of the bathyscaph *Archimède,* July, 1962.

REFERENCES

1. **Carruthers, J. N. and Lawford, A. L.,** The deepest oceanic sounding, *Nature,* 169, 601, 1952.
2. **Delauze,** personal communication, 1962.
3. **Fisher, R. L.,** in *Preliminary Report on Expedition Downwind,* I.G.Y. General Report Ser., 2, I.G.Y. World Data Center A, Washington, D.C., 1958.
4. **Fisher, R. L. and Revelle, R.,** A deep sounding from the southern hemisphere, *Nature,* 174, 469, 1954.
5. **Hanson, P. P., Zenkevich, N. L., Sergeev, U. V., and Udintsev, G. B.,** Maximum depths of the Pacific Ocean, *Priroda (Mosk),* 6, 84, 1959 (in Russian).

Table 3.2—8 (continued)
MAJOR DEEPS AND THEIR LOCATION, SIZE, AND DEPTHS

6. **Hess, H. H. and Buell, M. W.,** The greatest depth in the oceans, *Trans. Am. Geophys. Un.*, 31, 401, 1950.
7. **Kanaev, V. F.,** New data on the bottom relief of the western part of the Pacific Ocean, *Oceanological Researches*, 2, 33, 1960 (in Russian).
8. **Lyman, J.,** personal communication, 1960.
9. **Lyman, J.,** The deepest sounding in the North Atlantic, *Proc. R. Soc. London,* A222, 334, 1954.
10. **Maurer, H. and Stocks, T.,** *Die Echolötungen des Meteor. Wiss. Ergebn. Deut. Atlant. Exped. 'Meteor', 1925–27,* 2, 1, 1933.
11. **Nasu, N., Iijima, A., and Kagami, H.,** Geological results in the Japanese Deep Sea Expedition in 1959, *Oceanog. Mag.*, 11, 201, 1960.
12. **Udintsev, G. B.,** Discovery of a deep-sea trough in the western part of the Pacific Ocean, *Priroda (Mosk),* 7, 85, 1958 (in Russian).
13. **Udintsev, G. B.,** Relief of abyssal trenches in the Pacific Ocean (abstr.), *Intern. Oceanog. Cong. Preprints,* Am. Assoc. Adv. Sci., Washington, D.C., 1959.
14. **Udintsev, G. B.,** Bottom relief of the western part of the Pacific Ocean, *Oceanological Researches,* 2, 5, 1960 (in Russian).
15. **van Riel, P. M.,** The bottom configuration in relation to the flow of bottom water, *The 'Snellius' Exped.*, E. J. Brill, Leiden Neth., 2(2), chap. 2, 1933.
16. **Wiseman, J. D. H. and Ovey, C. D.,** Proposed names of features on the deep-sea floor, *Deep-Sea Res.*, 2, 93, 1955.

From Hill, M. N., Ed., *The Sea,* Vol. III, Wiley-Interscience, New York, 1963. With permission. In part after the compilation of Wiseman and Ovey,[16] 1955.

Table 3.2—9
SUBMARINE CANYONS: LOCATION, SIZE, DEPTH

Symbol Key for Table

1. Length of canyon measured along axis (nautical miles)
2. Depth at canyon head (feet)
3. Depth at canyon terminus (feet)
4. Character of coast inside canyon head
   - A. Heads in estuary
   - B. Heads off embayment
   - C. Heads off straight beach or barrier
   - D. Heads off relatively straight cliff
   - E. Uncertain
5. Relation of canyon head to points of land
   - A. On upcurrent side of point
   - B. Relatively near upcurrent side of point
   - C. No relation to point
6. Relation of canyon head to river valleys
   - A. Probable connection
   - B. No connection
   - C. Uncertain
7. Source of sediments to canyon head
   - A. Receives good supply
   - B. Supply restricted now, greater during lowered sea level stages
   - C. Little known supply of sediment because of depth
8. Gradient of axis in meters per kilometer
9. Nature of longitudinal profile
   - A. General concave upward
   - B. Generally convex upward
   - C. Relatively even slope
   - D. Local steplike steepening along axis
10. Maximum height of walls in feet
11. Channel curvature
   - A. Straight
   - B. Slightly curving
   - C. Twisting or winding
   - D. Meandering
   - E. One meandering bend
   - F. Right-angled bends
12. Abundance of tributaries
   - A. As common as typical land valleys
   - B. Less common than typical land valleys
   - C. Confined to canyon head
   - D. No known tributaries
13. Character of transverse profile
   - A. Predominantly V-shaped
   - B. V-shaped inner canyon, trough-shaped outer canyon
   - C. Predominantly trough-shaped
   - D. Uncertain
14. Nature of canyon wall material
   - A. Crystalline rock dredged
   - B. Rock dredged, but all sedimentary
   - C. Mud only dredged on wall
   - D. Unknown
15. Nature of core sediment from axis
   - A. Includes sand layers
   - B. Includes sand and gravel layers
   - C. Mud cores only
   - D. Unknown

Canyon name and location	1 Canyon length	2 Depth at canyon head	3 Depth at canyon terminus	4 Coast character	5 Relation to points	6 Relation to river valleys	7 Sediment sources at head	8 Gradient in m/km
California								
Coronado	8.0	240	5,580	C	C	C	B	58
La Jolla	7.3	50	1,800	C	B	A	A	40
Scripps (tributary)	1.45	60	900	D	C	A	A	97
Redondo	8.0	30	1,920	C	B	B	A	39
Dume	3.0	120	1,860	C	A	A	A	97
Mugu	8.0	40	2,400	C	B	C	A	49
Sur Partington	49.0	300	10,200	D	C	C	A	34
Carmel (tributary)	15.0	30	6,600	A	A	A	A	73
Monterey	60.0	50?	9,600?	C	C	B	A	26.5
Delgada	55.0	90	8,400	D	C	B	A	25
Mattole	16.0	60	5,720	B	B	A	A	59
Eel	27.0	250	8,500	C	C	A	B	51
Total or Average (12)	21.5	110	5,290	1A, 1B, 7C, 3D	2A, 4B, 6C	6A, 3B, 3C	10A, 2B	54
Oregon-Washington								
Columbia	37.0?	360	6,130?	B	B	A	C	26
Willapa	60.0	500+	7,000	B	C	A	C	24
Gray	30.0	500	6,440	B	C	A	C	33
Quinault	25.0	500±	5,750	C	C	A	C	35
Juan de Fuca	31.0+	800±	4,520+	B	B	C	C	20
Total or Average (5)	36.6	532	5,968	4B, 1C	2B, 3C	4A, 1C	5C	27.6
Bering Sea								
Umnak	160.0	900	10,850	B	C	C	C	10.4
Bering	220.0	600±	11,160	B	C	C	C	8
Pribilof	86.0	500±	10,700	C?	C	C	C	20
Total or Average (3)	155.3	667	10,903	2B, 1C	3C	3C	3C	12.8
U.S. East Coast								
Corsair	14+	360	5,400	E	C	C	B	23
Lydonia	16+	370	4,400+	E	C	C	B	42
Gilbert	20+	480	7,680+	E	C	C	C	60
Oceanographer	17+	600+	7,230+	E	C	C	C	65

Table 3.2—9 (continued) SUBMARINE CANYONS: LOCATION, SIZE, DEPTH

	1	2	3	4	5	6	7	8
Canyon name and location	Canyon length	Depth at canyon head	Depth at canyon terminus	Coast character	Relation to points	Relation to river valleys	Sediment sources at head	Gradient in m/km
Welker	27+	400	6,450+	E	C	C	B	38
Hydrographer	27+	450+	6,600+	E	C	C	B	37
Hudson	50	300	7,000	B	C	A	B	25
Wilmington	23+	320	6,940+	E	C	C	B	48
Baltimore	28+	400	6,110+	B?	C	C	B	34
Washington	28+	360	6,740+	E	C	C	B	38
Norfolk	38	320	8,300	E	C	C	B	35
Total or Average (11)	26.2	395	6,623	2B, 9E	11C	1A, 10C	9B, 2C	40
Hawaiian-Molokai								
Halawai	6.0+	300±	3,540+	B	C	A	B	90
Naiwa	7.5	380	4,880	D	C	A	B	100
Waikolu	9.0	<600	6,540	B	C	A	B	110
Pelekunu	10.0	<320	6,320	B	C	A	A	100
Hawaiian-Kauai								
Hanakapiai	6.0+	280	7,480	D	C	A	B	200
Hanakoa	3.7	600±	4,820	D	C	C	C	190
Hanopu	3.6	300	5,100	D	C	C	B	220
Total or Average (7)	6.5	397	5,526	3B, 4D	7C	5A, 2C	1A, 5B, 1C	144
Western Europe								
Shamrock	30+	1,200±	14,400	B?	C	C	C	28
Black Mud	30+	900±	12,200	E	C	C	B	57
Audierne	27	600±	10,500	B	C	C	C	60
Cap Ferret	50+	800	11,647	C	C	C	C	31
Cap Breton	135 or 70	400±	13,100	D	C	C	B	58
Aviles	65	60±	8,000	C	B	A (old river)	A	20 or 16
Llanes	38	450	13,300	D	B	C	B	45
Nazare	93	200 ±	14,764	C	A	A	A	36
Lisbon	21	400	6,450	B	B	A?	B	48

Setubal	33+	350	6,880	C	C	A	B	33
Total or Average (10)	52.2 or 45.7	536	11,224	3B, 4C, 2D	1A, 3B, 5C	4A, 5C	2A, 4B, 3C	41.6 or 41.2
Mediterranean mainland								
Grand Rhone	15+	600±	5,550	C	C	A	C	55
Marseille	20+	600±	6,840±	B	C	C	C	52
Canon de la Cassidaigne	19+	360	6,630+	B	B	A	B	55
Toulon	12+	260	6,600	A borderline	C	C	B	110
Stoechades	17+	300	4,380+	A	C	C	A	40
St. Tropez	25+	60?	5,750	A borderline	C	A	A	38
Cannes	17+	100?	6,600?	C or A	C	B	A	65
Var	15	160	6,550	D delta	A	A	A	71
Nice	12	150	5,840	D	C	B	A	79
Cap d'Ail	14	320	6,870	B	C	C	B	78
Nervia	16	330	6,280	C	C	A	B	62
Taggia	12	300±	7,500	C	C	A	B	100
Mele	31	200	6,150	C curving	C	B	B	32
Noli	14	120	4,990	D curving	C	A?	A	58
Polcevera	49	300	8,830	C	C	A?	B	29
Genoa	20	260	6,260	B	C	A	B	50
Total or Average (16)	17.4	276.5	6,351	3A, 4B, 5C, 3D	1A, 1B, 14C	9A, 3B, 4C	7A, 7B, 2C	60.9
Mediterranean Islands								
Crete	4	<300	3,300	B?	C	A	A	200
West Corsica								
St. Florent	25	150±	7,850	A	A	A	A	51
Calvi	13	200±	7,800	B	C	A	B	97
Porto	20	150±	8,200	A	C	A	A	67
Sagone	29	150±	6,200	A	C	A	A	35
Ajaccio	34	150±	8,200	A	C	A	A	39
Valinco	35	150±	8,000	A	C	A	A	37
Total or Average (7)	22.8	178	7,078	5A, 2B	1A, 6C	7A	6A, 1B	75
Baja, California								
San Pablo	20+	<400	8,400+	D	A	B	A	67
Cardonal	16+	<450	7,500+	C	B	B	B	73
Vigia	10	?	7,200	C + D	C	A or B	A	115
San Lucas –	19	30	6,900?	A-bay	C	A	A	70
Santa Maria	(24)		8,000					56
San Jose	32	50	7,200?	C	C	A?	A	41

Table 3.2–9 (continued)
SUBMARINE CANYONS: LOCATION, SIZE, DEPTH

	1	2	3	4	5	6	7	8
Canyon name and location	Canyon length	Depth at canyon head	Depth at canyon terminus	Coast character	Relation to points	Relation to river valleys	Sediment sources at head	Gradient in m/km
Vinorama – Salado	9	200	6,300	C	A	A	B	113
Los Frailes	9.5	10	5,200	A-bay	A (S. wind)	A	A	91
Saltito	6	1,200	5,100	B?	C	B	C	108
Palmas – Pescadero	13 9.3	100?	5,300	C	A (S. wind)	C	A	65 or 91
Total or Average (9)	15.2	305	6,710	2A, 1B, 4½C, 1½D	4A, 1B, 4C	4½A, 3½B, 1C	6A, 2B, 1C	81
East Honshu								
Ninomiya	4.8	400	2,600	C	C	C	B	77
Sagami	5.0	310	3,300	C	C	C	B	100
Enoshima	6.7	450	3,250	E	C	C	B	70
Hayama	13	300	4,600 (Hayama and Miura combined)	E	C	C	B	54
Miura	15	173		E	C	C	B	48
Misaki	14.5	330	4,600	E	C	C	B	49
Jogashima	10	330		E	C	C	B	71
Tokyo	30	300	4,900	A	C	C	B	26
Mera	20	190	5,500	B (bay)	B	C	B	44
Kamogawa	25	200	9,100	B	C	A	B	59
Total or Average (10)	29.25	298	4,731	1A, 2B, 2C, 5E	1B, 9C	1A, 9C	10B	60
Miscellaneous								
Great Bahama	125	4,800	14,060	B	C	C	C	13
Congo	120	80	7,000	A	C	A	A	96
Ceylon Trincomalee	20+	30+	9,500+	A	A?	A	A	79
Manila	31+	300	7,800+	B	C	A	B	40
Bacarra NW Luzon	15	300	6,000+	C + D	C	A	B	63
San Antonio, Chile	20+	<150	2,700+	C?	C	A?	A	32
Total or Average (6)	55	943	7,843	2A, 2B, 1½C, ½D	1A, 5C	5A, 1C	3A, 2B, 1C	54

	9	10	11	12	13	14	15
Canyon name and location	**Nature of long profile**	**Max wall heights to nearest 1000 ft**	**Channel curvature**	**Abundance of tributaries**	**Transverse profile character**	**Name of canyon wall material**	**Sediment found in axial cores**
California							
Coronado	A	1,000	C	B	A	B	A
La Jolla	A	1,000	C	A	A	B	B
Scripps (tributary)	A	<1,000	B	C	A	B	B
Redondo	A	1,000	C	B	C	B	A
Dume	A	1,000	C	B	A	A	D
Mugu	A	<1,000	D	A	A	B	A
Sur Partington	A	2,000	C	A	A	B	D
Carmel (tributary)	A	2,000	C	A	A	A	A
Monterey	A + D	6,000	C or D	A	B	A	B
Delgada	A	2,000	D	B	A	D	D
Mattole	A	3,000	C	B	A	B	D
Eel	B	4,000	E	B	A	D	D
Total or Average (12)	10A, 1B, 1D	2,083	½B, 8½C, 2½D, 1E	5A, 6B 1C	10A, 1B, 1C	3A, 7B, 2D	4A, 3B, 5D
Orgeon-Washington							
Columbia	A	2,000	C	B	D	D	D
Willapa	A	2,000	C	A	D	B	A
Gray	A	1,000	C	A	D	D	D
Quinault	A	3,000	C	A	D	D	D
Juan de Fuca	A	2,000	C	D	D	D	D
Total or Average (5)	5A	2,400	5C	3A, 1B, 1D	5D	4D, 1B	4D, 1A
Bering Sea							
Umnak	A	4,000	C	A	D	D	D
Bering	A + D	6,000	C	A	A	B?	D
Pribilof	A	7,000	C	A	D	D	D
Total or Average (3)	2½A ½D	5,667	3C	3A	1A, 2D	1B, 2D	3D
U.S. East Coast							
Corsair	A	2,000	B	B	A	B	D
Lydonia	A	3,000	B	B	A	B	C
Gilbert	A + D	3,000	B	A	A	B	D
Oceanographer	A	2,000	B	B	A	D	D

Table 3.2—9 (continued)
SUBMARINE CANYONS: LOCATION, SIZE, DEPTH

	9	10	11	12	13	14	15
Canyon name and location	Nature of long profile	Max wall heights to nearest 1000 ft	Channel curvature	Abundance of tributaries	Transverse profile character	Nature of canyon wall material	Sediment found in axial cores
Welker	A + D	4,000	B	B	A	B	A
Hydrographer	A + D	3,000	B	B	A	C	A
Hudson	A + D	4,000	B	A	A	B	A
Wilmington	A	3,000	C	A	A	C	C
Baltimore	A + D	3,000	B	B	A	C	C
Washington	A + D	2,000	B	A	A	C	C
Norfolk	A + D	3,000	B	B	A	B	A
Total or Average (11)	7½A, 3½D	2,900	10B, 1C	4A, 7B	11A	6B, 4C, 1D	4A, 4C, 3D
Hawaiian-Molokai							
Halawai	C	1,000	C	D	A	D	D
Naiwa	C	1,000	B	D	A	A	A
Waikolu	A	2,000	B	B	A	A	B
Pelekunu	A	1,000	C	B	A	D	D
Hawaiian-Kauai							
Hanakapiai	B + D	1,000	B	B?	A	A	D
Hanakoa	B	1,000	B	B?	A	A	A
Hanopu	B + D	2,000	B	B?	A	B	A
Total or Average (7)	2A, 2B 2C, 1D	1,286	5B, 2C	5B, 2D	7A	4A, 1B, 2D	3A, 1B, 3D
Western Europe							
Shamrock	C + D	3,000	B	B?	D	D	D
Black Mud	D	3,000	B	B	B	B	A
Audierne	D	4,000	B	D	D	D	D
Cap Ferret	A	3,000	C	A	D	D	D
Cap Breton	A	6,000	B	B	D	D	D
Aviles	C + D	5,000 or 6,000	B	A?	B	B	D
Llanes	C	5,000	C	A	D	D	D
Nazare	D	5,000	F	B	D	A?	D
Lisbon	B + D	4,000	B	D	D	B	D

Setubal	C?	2,000	C?	B	D	B?	D
Total or Average (10)	2A, ½B, 3C, 3½D	4,000	5B, 3C, 1F	3A, 4B, 2D	1B, 8D	1A, 3B, 5D	9D
Mediterranean mainland							
Grand Rhone	C?	2,000	B	C	A	D	D
Marseille	A + D	2,000	C	A	A	D	D
Canon de la Cassidaigne	D	3,000	C	B	A	D	D
Toulon	D?	4,000	B	A	A	D	D
Stoechades	A	4,000	B	A	A	B?	D
St. Topez	C?	3,000	C	A	A	D	D
Cannes	A	3,000	C	A	A	D	D
Var	A	3,000	C?	A	B	D	B
Nice	A + D	2,000	C?	A	B	B?	B
Cap d'Ail	D	1,000	B	B	B	D	D
Nervia	D	2,000	B	B	B	D	B
Taggia	A	2,000	C	B	A	D	D
Mele	A + D	1,000	C	B	B	D	D
Noli	A	2.000	C	B	B	D	D
Polcevera	A + D	3,000	B	B	B	D	D
Genoa	D	2,000	B	B	C?	D	D
Total or Average (16)	7A, 2C, 7D	2,400	7B, 9C	7A, 8B, 1C	8A, 7B, 1C	2B, 15D	3B, 13D
Mediterranean Islands							
Crete	D	1,000	C	B	A	D	B
West Corsica							
St. Florent	**A**	**3,000**	**C**	**A**	**A**	**D**	**A**
Calvi	**A + D**	**3,000**	**C**	**Bor A**	**A**	**D**	**D**
Porto	**A**	**4,000**	**C**	**A**	**A**	**A?**	**D**
Sagone	**A**	**3,000**	**C**	**A**	**A**	**D**	**D**
Ajaccio	**A**	**4,000**	**B**	**A**	**A**	**D**	**D**
Valinco	**A + D**	**4,000**	**B**	**A**	**A**	**D**	**D**
Total or Average (7)	**6A, 1D**	**3,100**	**2B, 5C**	**5½A, 1½B**	**7A**	**1A, 6D**	**1A, 1B, 5D**
Baja California							
San Pablo	**B + D**	**3,000**	**C**	**B**	**A**	**B**	**D?**
Cardonal	**A + D**	**3,000**	**C**	**A**	**A**	**B**	**D**
Vigia	**A**	**3,000**	**B**	**C**	**B**	**A**	**B**
San Lucas – Santa Maria	**A + D**	**3,000**	**C**	**A**	**A**	**A**	**B**
San Jose	**A + D**	**3,000**	**C**	**A**	**A**	**A (sed.)**	**B**

Table 3.2—9 (continued)
SUBMARINE CANYONS: LOCATION, SIZE, DEPTH

Canyon name and location	9 Nature of long profile	10 Max wall heights to nearest 1000 ft	11 Channel curvature	12 Abundance of tributaries	13 Transverse profile character	14 Nature of canyon wall material	15 Sediment found in axial cores
Vinorama – Salado	A	1,000	C	A	A	A	B
Los Frailes	A + D	2,000	C	C	A	A	B
Saltito	A	1,000	C?	A	A	A	D
Palmas – Pescadero	A	2,000	C	A	A	A	A
Total or Average (9)	6A, ½B, 2½D	2,333	1B, 8C	6A, 1B, 2C	8A, 1B	7A, 2B	1A, 5B, 3D
East Honshu							
Ninomiya	A	<1,000	B	C	A	B	D
Sagami	A	<1,000	C	A	B	B	D
Enoshima	A	<1,000	B	C	A	B	B
Hayama	A	2,000	C	B	A	A	D
Miura	A	2,000	C	B	A	B	B
Misaki	A	2,000	C	A	A	B	A
Jogashima	D	1,000	C	A	A	A	D
Tokyo	A + D	3,000	C	A	B	A	B
Mera	C + D	2,000	C	A	A	B	B
Kamogawa	A + D	5,000	B	B	B	B	A
Total or Average (10)	7A, ½C, 2½D	2,000	3B, 7C	5A, 3B, 2C	7A, 3B	3A, 7B	2A, 4B, 4D
Miscellaneous							
Great Bahama	A	14,000	C	A	A	C	B
Congo	A	4,000	B	C	A?	D	A sand
Ceylon Trincomalee	A	4,000	F	A	A	A?	D
Manila	A + D	6,000	C	A	D	D	D
Bacarra NW Luzon	C	3,000	C	A	B?	D	D
San Antonio, Chile	A	3,000	B + C	B	D	D	D
Total or Average (6)	4½A, 1C, ½D	5,666	1½B, 3½C, 1F	4A, 1B, 1C	3A, 1B, 2D	1A, 1C, 4D	1A, 1B, 4D

From Shepard, F. P. and Dill, R. F., *Submarine Canyons and Other Sea Valleys*, Rand McNally & Co., Chicago, 1966. With permission.

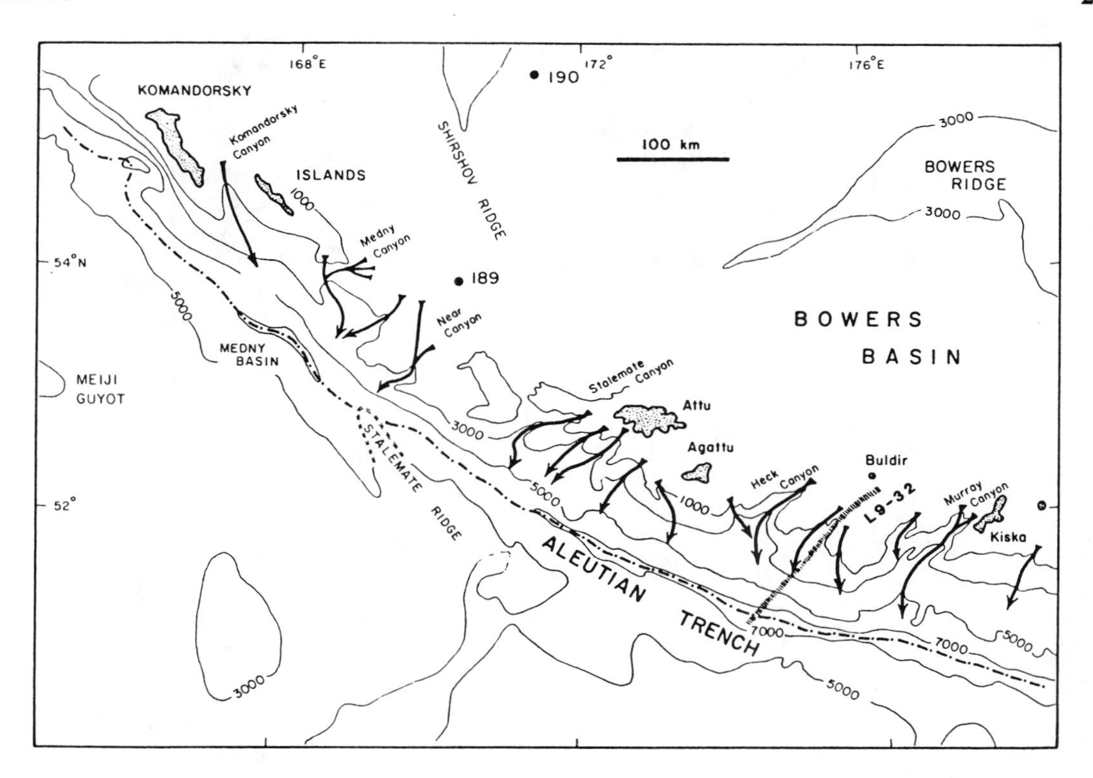

FIGURE 3.2—14. Simplified bathymetric map (in meters) of the Western segment of the Aleutian Trench and forearc. Submarine canyons are highlighted by arrows; as in the Central segment, none of these features can be traced all the way from the arc platform to the trench floor. Note the collision zone between Stalemate Ridge and the subduction front, sites drilled during DSDP Leg 19, and the trackline for USGS seismic-reflection profile L9–32. (From Underwood, M. B., Submarine canyons, unconfined turbidity currents, and sedimentary bypassing of forearc regions, *Rev. Aquat. Sci.*, 4, 149, 1991.)

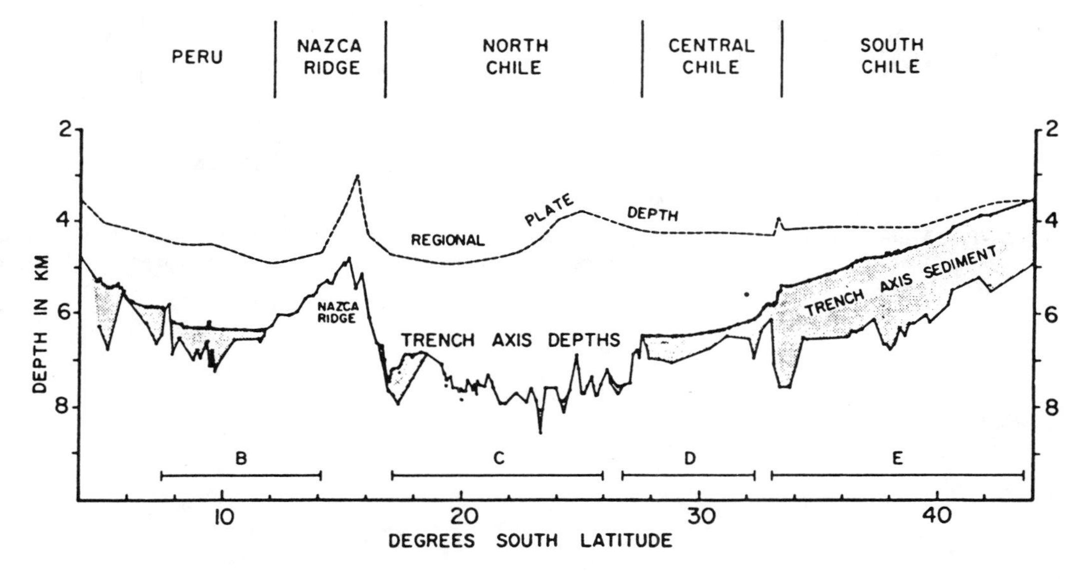

FIGURE 3.2—15. Axial profile of the Peru-Chile Trench, eastern Pacific Ocean. (From Schweller, W. J., Kulm, L. D., and Prince, R. A., Tectonics, structure, and sedimentary framework of the Peru-Chile Trench, *Geol. Soc. Am. Mem.*, 154, 323, 1981. With permission.)

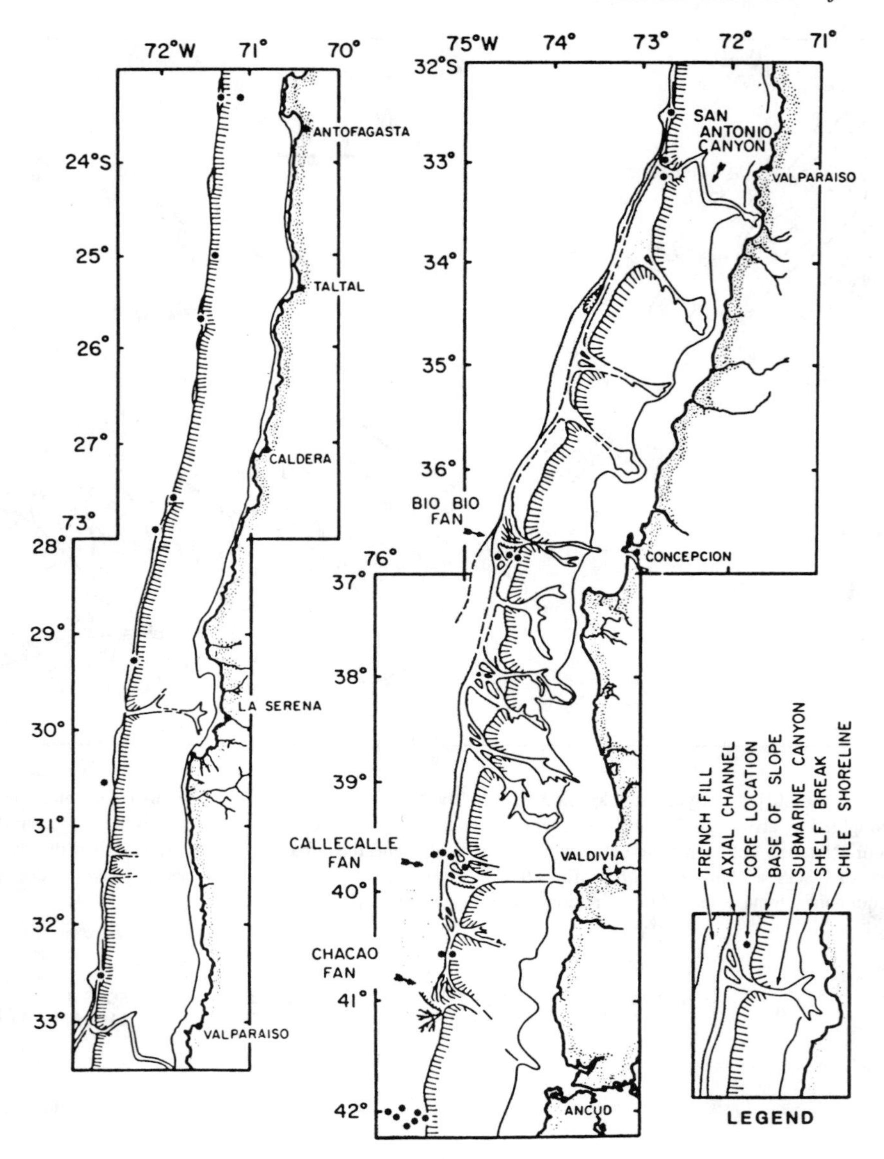

FIGURE 3.2—16. Map showing major submarine canyons, trench fans, and axial channel of the Chile Trench. (From Thornburg, T. M. and Kulm, L. D., Sedimentation in the Chile Trench; depositional morphologies, lithofacies, and stratigraphy, *Geol. Soc. Am. Bull.*, 98, 33, 1987. With permission.)

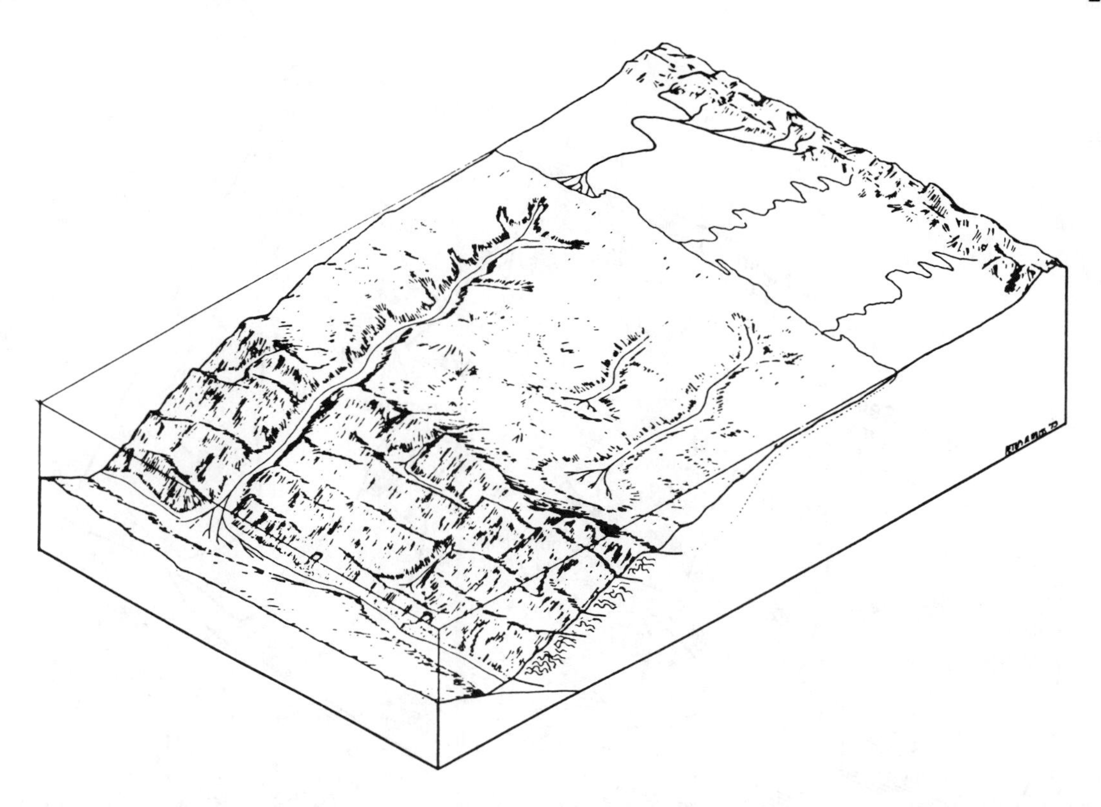

FIGURE 3.2—17. Diagram showing large submarine canyons as agents of sedimentary bypassing in an accretionary forearc terrane. (From Underwood, M. B. and Karig, D. E., Role of submarine canyons in trench and trench-slope sedimentation, *Geology,* 8, 432, 1980. With permission.)

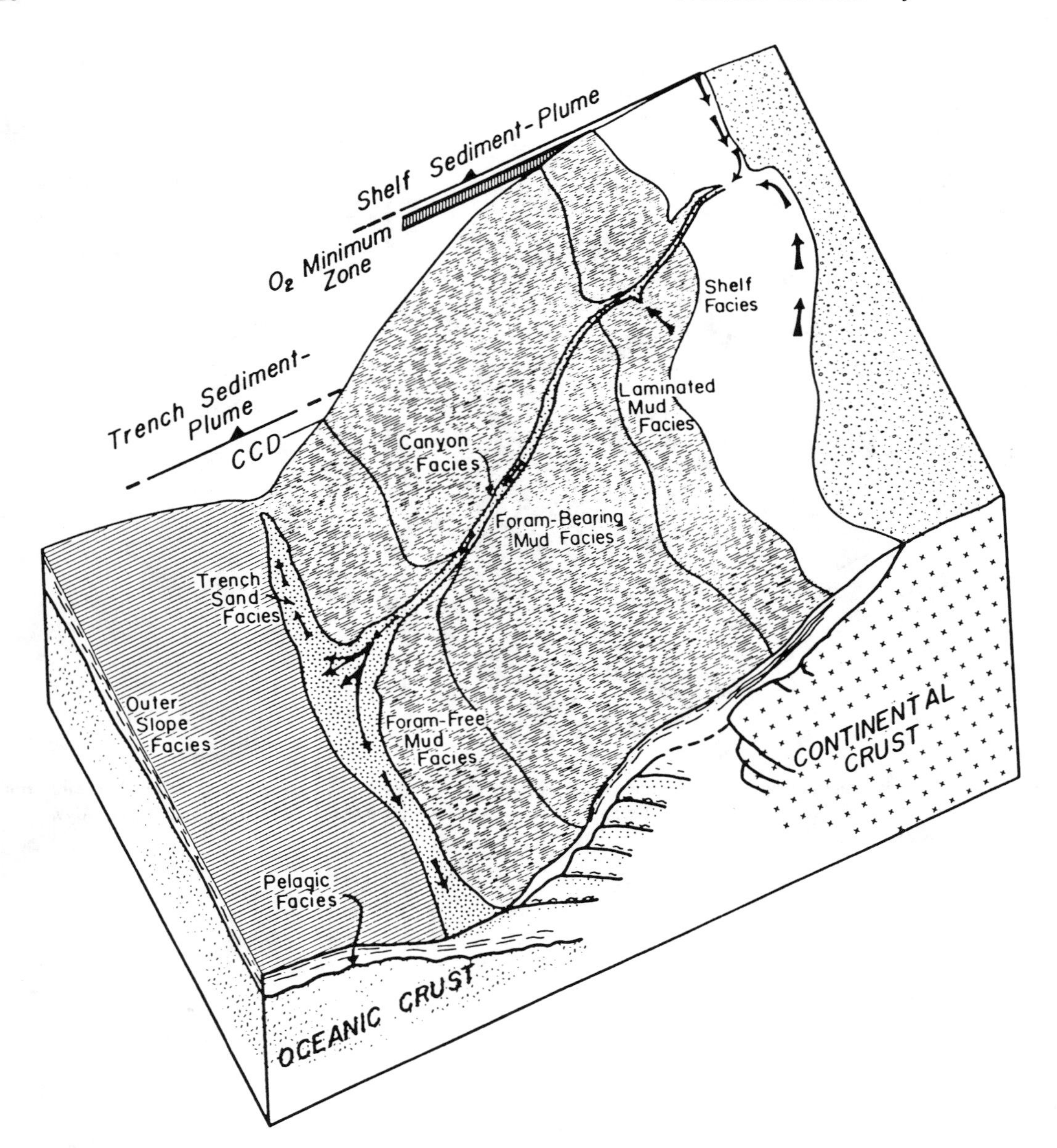

FIGURE 3.2—18. Sedimentary facies model for the Middle America Trench showing the influence of Ometepec Canyon on forearc bypassing, the formation of a trench fan, and the development of a plume of suspended sediment above the trench floor. (From Underwood, M. B., Submarine canyons, unconfined turbidity currents, and sedimentary bypassing of forearc regions, *Rev. Aquat. Sci.*, 4, 149, 1991.)

3.3 PLATE TECTONICS

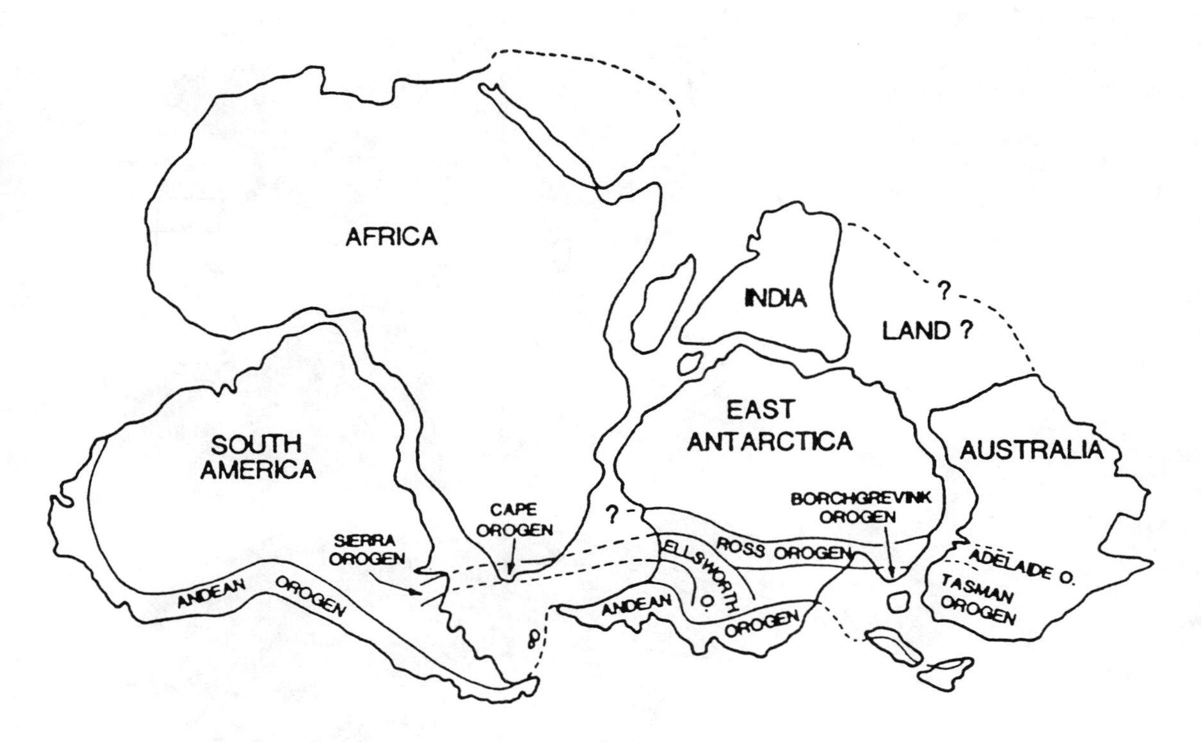

FIGURE 3.3—1. Illustration of the tectonic arrangement between the continents, land masses, and orogens (including the late Mesozoic-early Cenozoic Andean Orogen) that composed the supercontinent of Gondwana before it fragmented during the Mesozoic and Cenozoic. (From Berkman, P. A., The Antarctic marine ecosystem and humankind, *Rev. Aquat. Sci.*, 6, 295, 1992.)

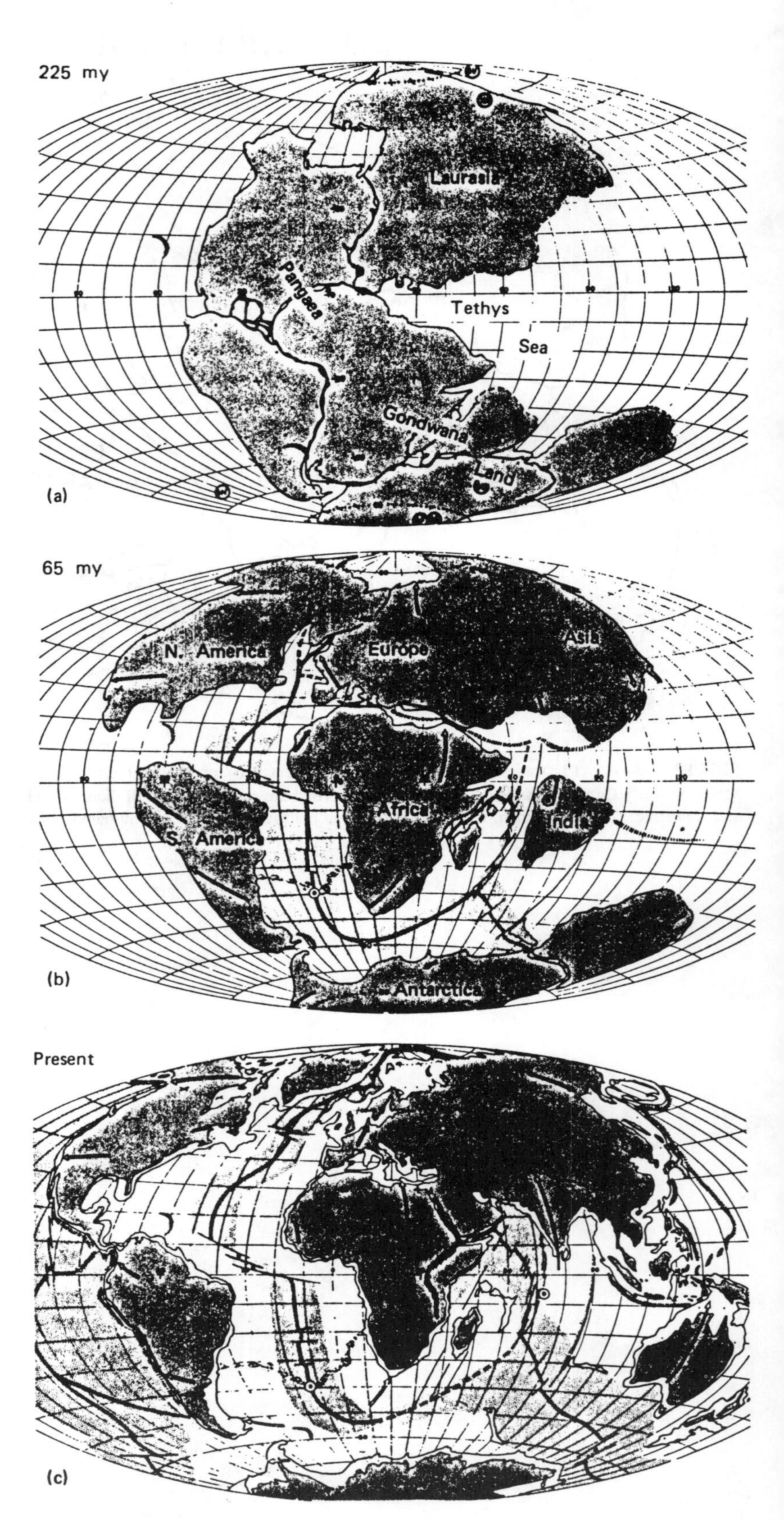

FIGURE 3.3—2. Movement of lithospheric plates subsequent to the breakup of Pangaea 225 million years ago (a), the opening of the Atlantic Ocean (b), and the formation of the Indian Ocean (c). (From Dietz, R. S. and Holden, J. C., Reconstruction of Pangaea: breakup and dispersion of continents, Permian to present, *J. Geophys. Res.*, 75, 4939, 1970. With permission.)

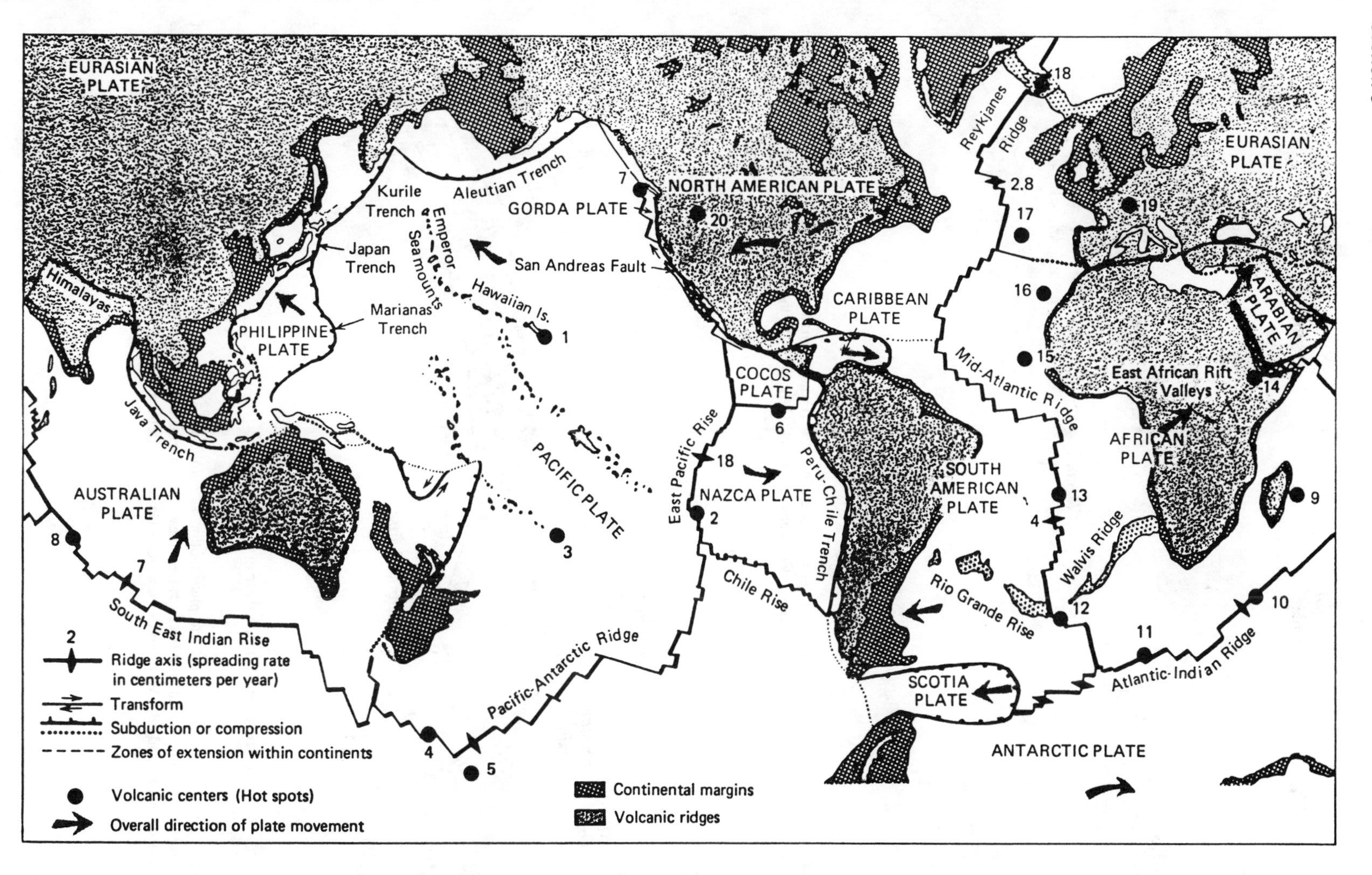

FIGURE 3.3—3. Major lithospheric plates and their boundaries, showing principal hot spots. Legend: 1. Hawaii, 2. Easter Island, 3. Macdonald Seamount, 4. Bellany Island, 5. Mt. Erebus, 6. Galapagos Islands, 7. Cobb Seamount, 8. Amsterdam Island, 9. Reunion Island, 10. Prince Edward Island, 11. Bouvet Island, 12. Tristan da Cunha, 13. St. Helena, 14. Afar, 15. Cape Verde Islands, 16. Canary Islands, 17. Azores, 18. Iceland, 19. Eifel, 20. Yellowstone. (From Brown, J., Colling, A., Park, D., Phillips, J., Rothery, D., and Wright, J., in *Ocean Circulation,* 1989. With kind permission of Pergagnon Press Ltd., Headington Hill-Hall, Oxford, OX3 OBW, UK.)

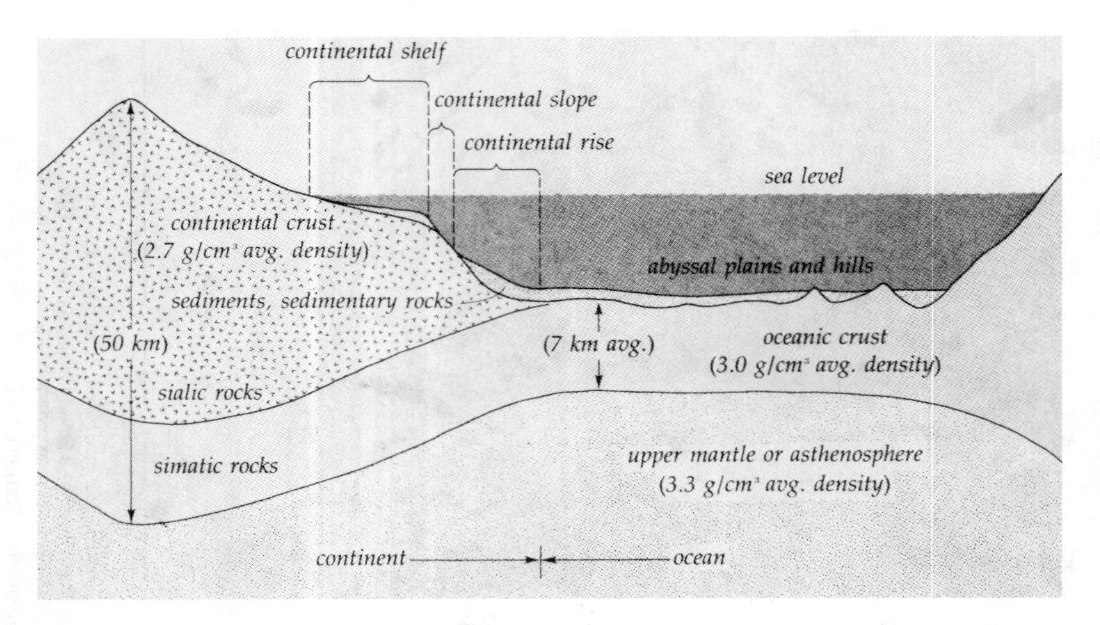

FIGURE 3.3—4. Differences between continental crust and oceanic crust. (From Ingmanson, D. A. and Wallace, W. J., *Oceanology: an Introduction,* 2nd ed., Wadsworth Publishing Company, Belmont, CA, 1979. With permission.)

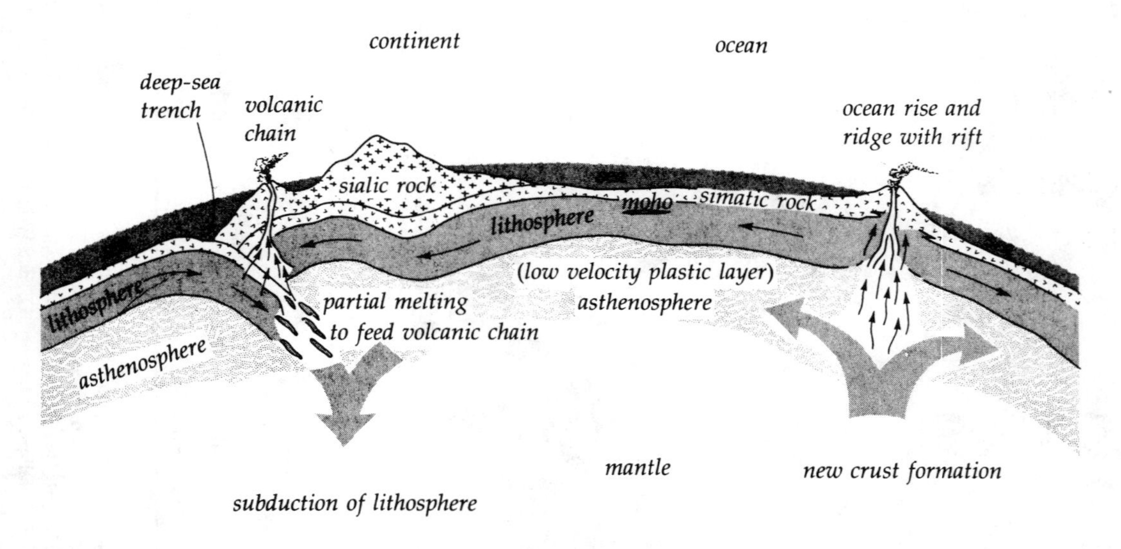

FIGURE 3.3—5. The dynamics of lithospheric plate motion as manifested by zones of convergence at trenches (subduction of lithosphere) and zones of divergence at mid-ocean ridges (new crust formation). (From Ingmanson, D. A. and Wallace, W. J., *Oceanology: an Introduction,* 2nd ed., Wadsworth Publishing Company, Belmont, CA, 1979. With permission.)

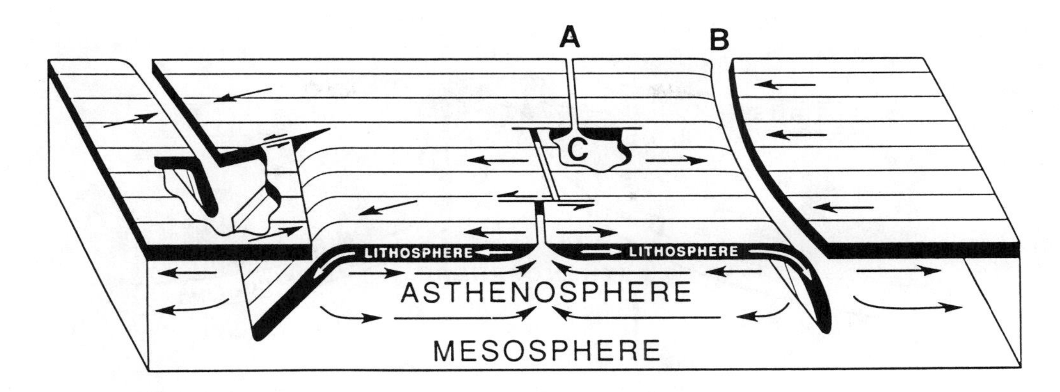

FIGURE 3.3—6. According to the theory of plate tectonics, the lithosphere of the earth is broken into a mosaic of a dozen or so dynamic rigid plates which move with respect to each other on a partially molten asthenosphere. The plates originate at mid-ocean ridges (A), subduct into the underlying asthenosphere at trenches (B), and slide by each other at transform faults (C). (From Isacks, B. L., Oliver, J., and Sykes, L. R., Seismology and the new global tectonics, *J. Geophys. Res.*, 73, 5855, 1968. With permission.)

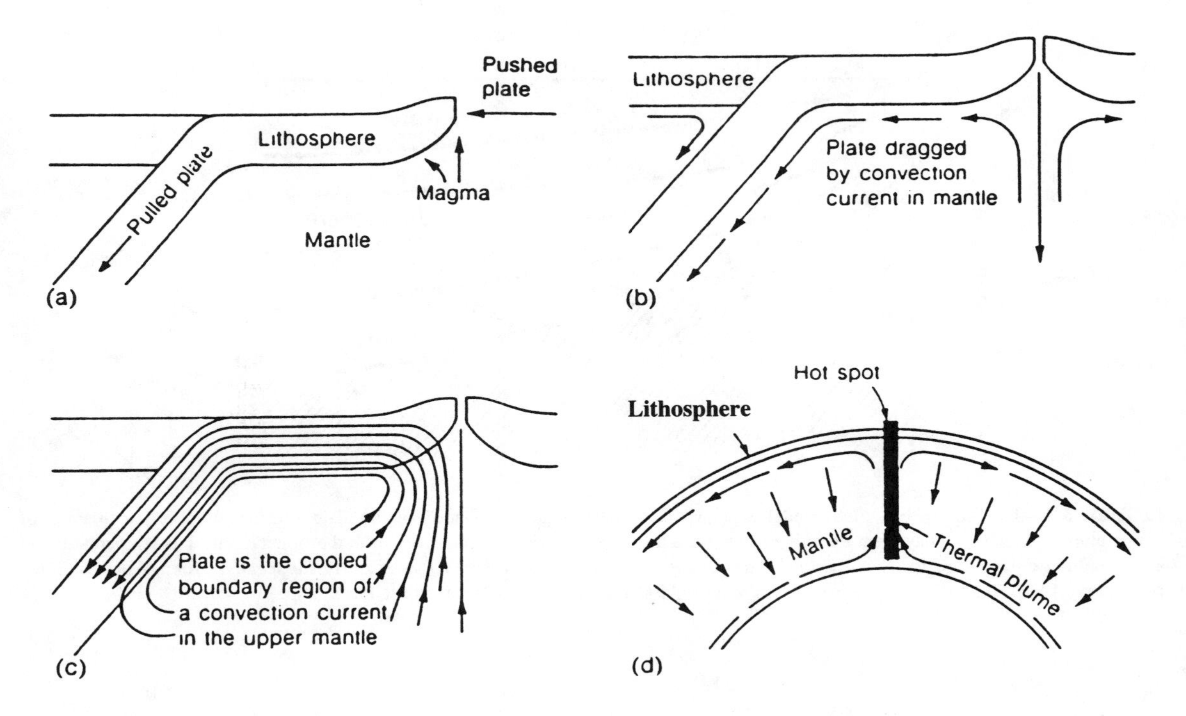

FIGURE 3.3—7. Schematic illustration of possible driving mechanism of plate tectonics. (a) Plate is pushed by the weight of the ridges at centers of spreading. (b) Plate is pulled by heavy downgoing slab of subduction zone. (c) Plate is dragged by convection current in mantle. (d) Plate is cooled, brittle, boundary region of a convection current involving the hot, low-viscosity upper mantle. (From Press, F. and Siever, R., *Earth,* 4th ed., W. H. Freeman, New York, 1986. With permission.)

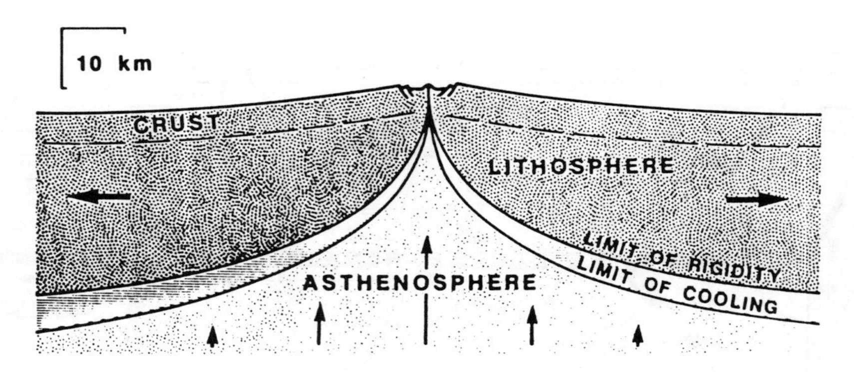

FIGURE 3.3—8. Schematic cross section through a seafloor spreading center. A distinction is made between the mechanical and thermal boundary layers produced by the progressive cooling and thickening of the lithosphere with age. (From Wright, J. A. and Louden, K. E., Eds., *CRC Handbook of Seafloor Heat Flow,* CRC Press, Boca Raton, FL, 1989.)

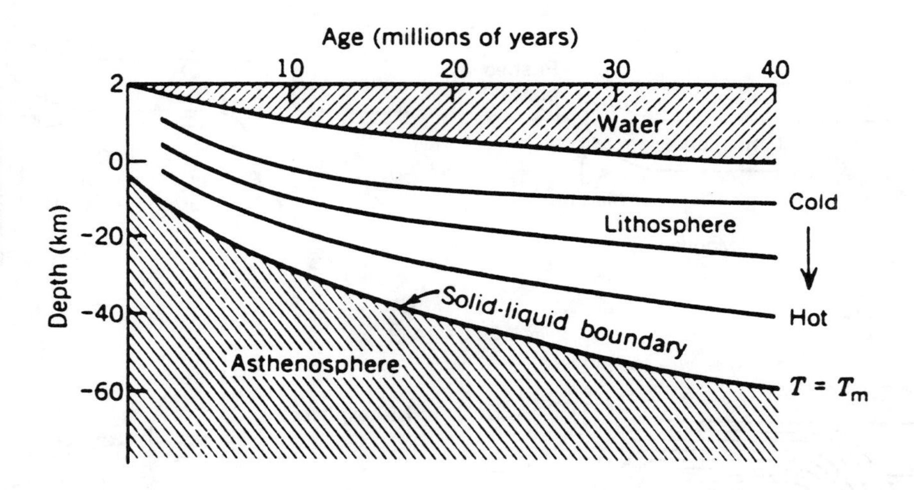

FIGURE 3.3—9. Lithospheric plate model explains form and change in elevation of the plate with age as purely a function of temperature and cooling away from the ridge crest. Note the increased water depth away from the ridge at left. T is temperature and T_m is melting temperature for rock. (From Anderson, R. N., *Marine Geology: A Planet Earth Perspective,* John Wiley & Sons, New York, 1986. Reproduced by permission of copyright © owner, John Wiley & Sons, Inc.)

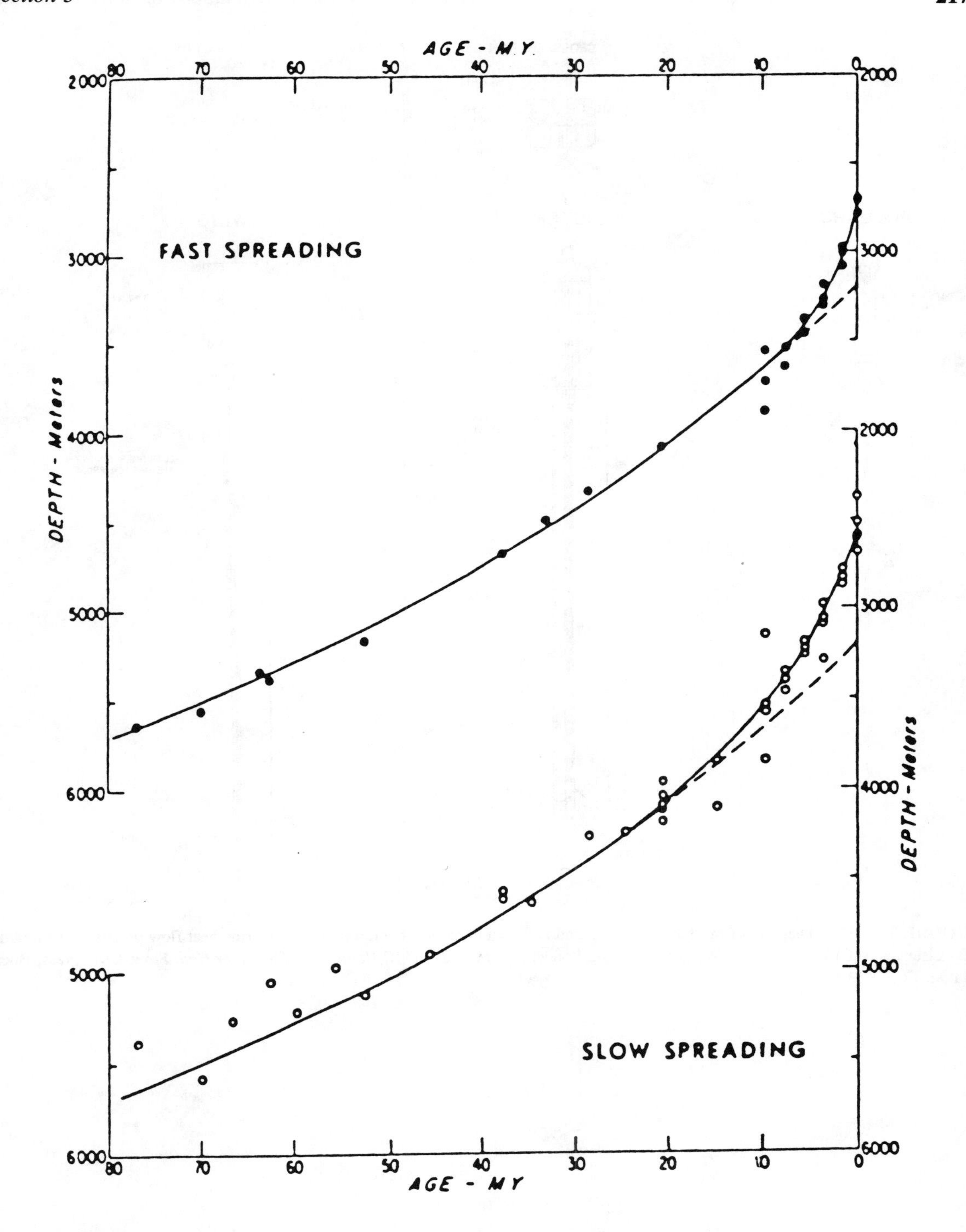

FIGURE 3.3—10. Both fast- and slow-spreading rate ridges subside exponentially away from the volcanic centers as the rock slowly cools. Eighty million years after formation, the lithosphere is still cooling and sinking because of thermal contraction. (From Anderson, R. N., *Marine Geology: A Planet Earth Perspective*, John Wiley & Sons, New York, 1986, 73. Reproduced by permission of copyright © owner, John Wiley & Sons, Inc.)

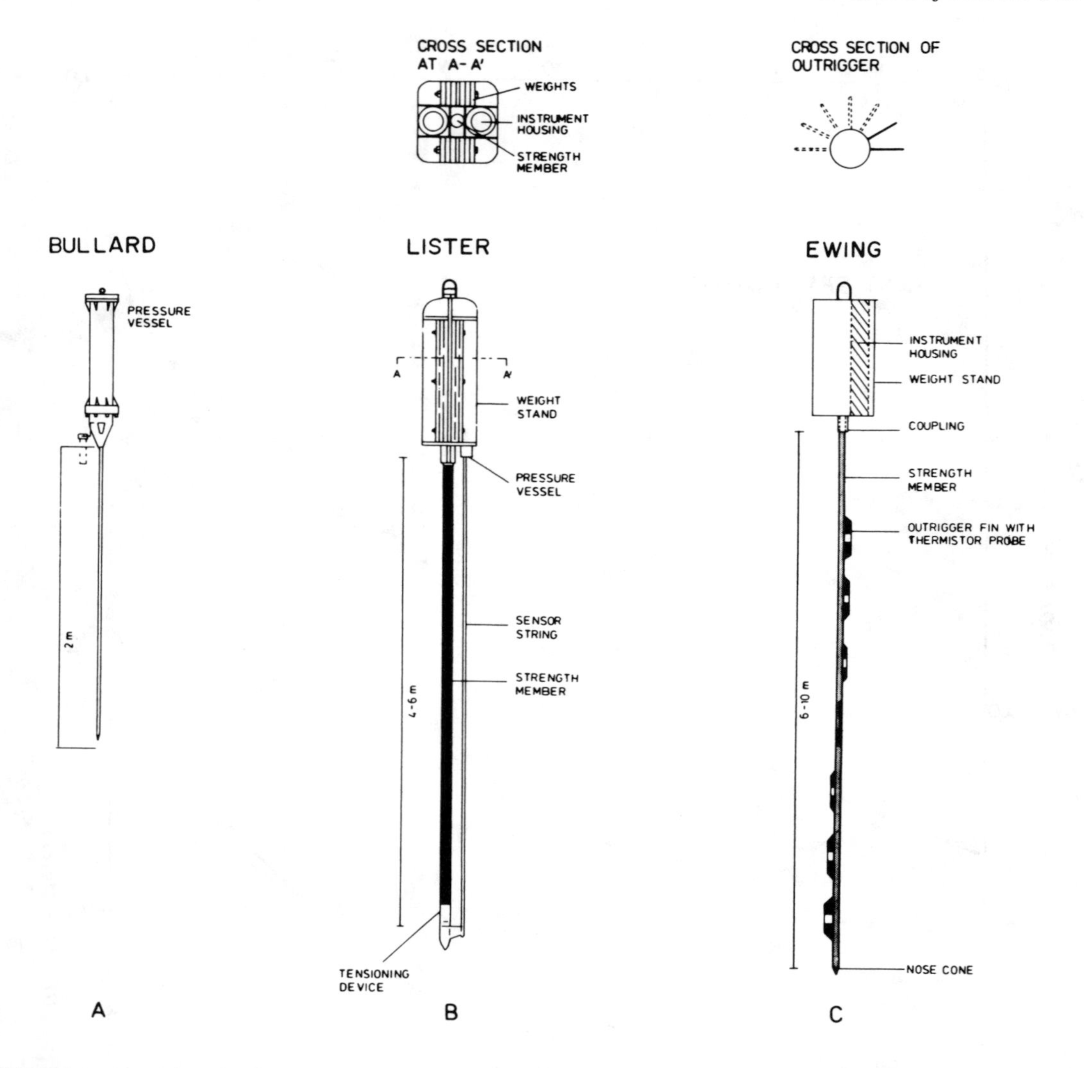

FIGURE 3.3—11. Diagram of mechanical configuration for the three most commonly used marine heat flow probes: (A) Bullard, (B) Lister, and (C) Ewing. (From Wright, J. A. and Louden, K. E., Eds., *CRC Handbook of Seafloor Heat Flow,* CRC Press, Boca Raton, FL, 1989.)

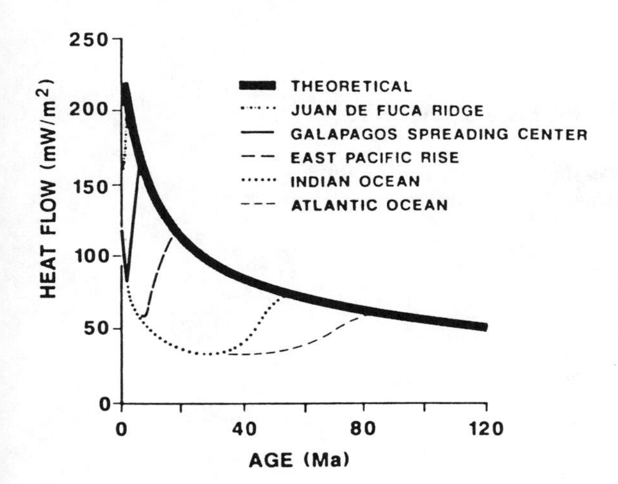

FIGURE 3.3—12. Schematic representation of the average heat flow as a function of age observed on the flanks of a variety of seafloor spreading ridges. Average heat flow through young seafloor is lower than expected due to the effects of hydrothermal circulation. Values approach the level predicted by simple cooling theory when sedimentation hydrologically seals the crust. The time when this occurs depends primarily on the local sedimentation rate, and in part on the roughness of the seafloor and the permeability of the sediments. (From Wright, J. A. and Louden, K. E., Eds., *CRC Handbook of Seafloor Heat Flow*, CRC Press, Boca Raton, FL, 1989.)

Table 3.3—1
MEAN HEAT FLOW AS A FUNCTION OF WATER DEPTH BY OCEAN BASIN

Water depth	Pacific			Atlantic			Indian			Marginal basins		
(m)	q[a]	σ[b]	N[c]	q	σ	N	q	σ	N	q	σ	N
1000–1500	45	79	3	135	117	97	116	53	4	81	30	70
1500–2000	191	116	101	88	85	63	98	79	14	170	268	81
2000–2500	233	201	212	98	73	47	104	74	28	107	85	84
2500–3000	198	156	452	104	72	106	80	66	40	80	39	94
3000–3500	109	81	306	76	73	141	51	32	77	78	39	151
3500–4000	82	59	282	67	41	223	54	44	84	75	39	120
4000–4500	58	39	357	56	35	188	57	36	134	55	51	64
4500–5000	57	35	168	53	24	219	66	41	117	65	33	65
5000–5500	58	22	135	61	35	190	61	35	113	62	37	37
5500–6000	52	16	62	57	57	215	51	27	41	64	33	33

[a] q = mean heat flow in mW/m². [b] σ = standard deviation. [c] N = number of values.

Table 3.3—2
HEAT FLOW AND AGE ESTIMATES IN MARGINAL BASINS

Basin		Heat flow (mW/m²)	Age (Ma)
Aleutian Basin (AL)		55.2 ± 0.4	117—132
Balearic Basin (BL)		92 ± 10	20—25
Caroline Basin (CA)		85 ± 31	28—36
Celebes Sea (CE)		56 ± 22	65—72
Coral Sea Basin (CO)		72 ± 11	56—64
Parece Vela Basin (PV)		88 ± 21	20—30
Shikoku Basin (SH)		82 ± 29	14—24
South China Sea (SC)	(1)	88 ± 6	27.5—33
	(2)	107 ± 4	19—23
Sulu Sea (SU)		89 ± 7	41—47
Tyrrhenian Sea (TY)	(1)	134 ± 8	7—12
	(2)	151 ± 10	5—8
West Philippine Basin (WP)		68 ± 22	39—50

Table 3.3—3
CONDUCTIVITY OF DEEP-SEA SEDIMENTS

Sediment type	Water content, % wet wt	Density, gm/cm³	Conductivity, 10^{-3} cal/cm sec °C	Ref.
Red clay	52	1.43	1.93	3
	54	1.39	1.93	
	56.5	1.38	1.93	
	50	1.47	2.17	
	50	1.47	2.20	
	42.5	1.58	2.37	
	43.5	1.57	2.43	
	52.5	1.41	1.91	
	52	1.40	1.96	
	69.5	1.20	1.68	
	61.8	1.27	1.73	
Mud	55	1.32	1.91	
	52.5	1.36	1.90	
	56.5	1.31	1.88	
	51.5	1.37	1.94	
	46	1.47	2.06	
Globigerina ooze and glacial clay	41.3	1.58	2.31	1
	39.8	1.62	2.40	
	44.7	1.52	2.24	
	43.8	1.56	2.23	
	40.5	1.55	2.52	
	37.5	1.61	2.60	
	31.5	1.83	2.72	
	50.0	1.44	2.04	
	47.0	1.50	2.19	
	43.1	1.55	2.27	
	20.2	2.14	3.24	
	38.2	1.59	2.54	
	43.7	1.46	2.27	
	40.3	1.56	2.44	
	32.2	1.72	2.68	
Globigerina ooze	37.8	1.54	2.33	2
	43.8	1.47	2.07	
	43.4	1.47	2.22	
	36.9	1.55	2.55	
	38.5	1.54	2.52	
Dark mud	46.8	1.43	2.08	
	45.7	1.45	2.17	
	44.4	1.47	2.24	
	44.6	1.47	2.24	
	42.6	1.49	2.24	
	38.0	1.57	2.39	
	37.8	1.57	2.30	
	38.9	1.56	2.44	

Note: These measurements show a close correspondence between conductivity and water content and little dependence on type of sediment.[4]

From Clark, S. P., Jr., *Handbook of Physical Constants,* Geological Society of America, Memoir 97, 1966. With permission.

REFERENCES

1. **Bullard, E.,** *Proc. R. Soc. London,* A222, 408, 1954.
2. **Bullard, E.,** unpublished.
3. **Butler, D. W.,** unpublished.
4. **Ratcliffe, E. H.,** *J. Geophys. Res.*, 65, 1535, 1960.

Table 3.3—4
COMPARISON OF CALCULATED HEAT-FLUX VALUES FROM THE JUAN DE FUCA RIDGE

Vent field	Calculated heat flux (MW)				
	1	2	3	4	5
SSR	2,096	30,240	1.6	814	580 ± 351
Axial	7,337	81,890			
NSR	2,266	28,220			
Surveyor smt	172	3,020			
Endeavour	1,133	12,600			1700 ± 1100
Total	13,004	155,970			

Note: SSR = Southern symmetrical ridge, NSR = northern symmetrical ridge.
1. Model 1 line source.
2. Model 2 flow model at a velocity of 1 cm/s.
3. Detailed survey, point-source model.
4. Detailed survey, line-source model.
5. 3D survey flow model type.

From Tivey, M. A. and Johnson, H. P., High resolution geophysical studies of oceanic hydrothermal systems, *Rev. Aquat. Sci.*, 1, 473, 1989. With permission.

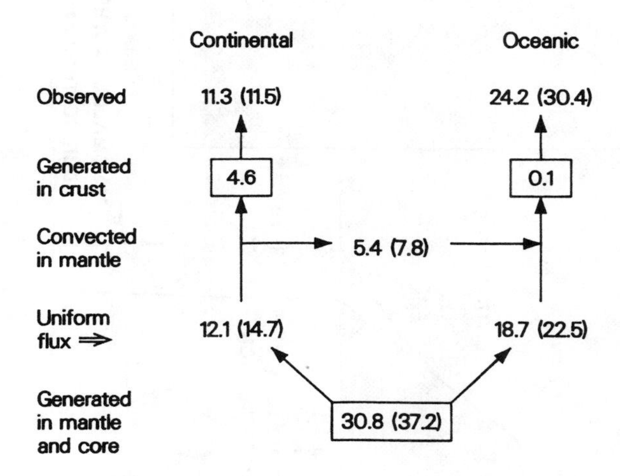

FIGURE 3.3—13. Heat flow bookkeeping. Note world average heat flows upward from continental and oceanic crust (in W/m^2) are rather similar, although there is much more heat generation by radioactivity in continental crust, which is much thicker. (From Harper, J. F., Forces driving plate tectonics: the use of simple dynamical models, *Rev. Aquat. Sci.*, 1, 322, 1989.)

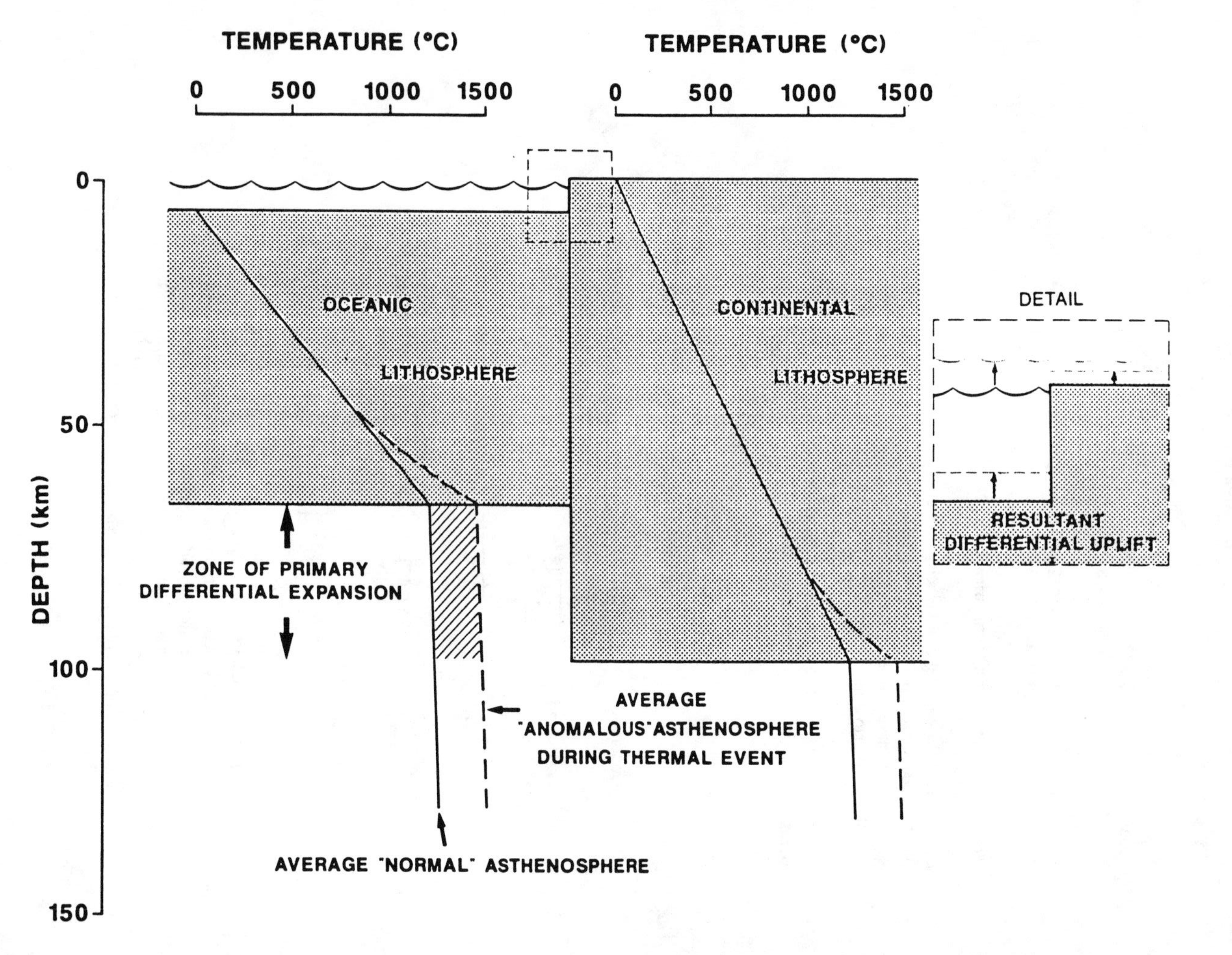

FIGURE 3.3—14. Schematic representation of typical continental and oceanic geotherms before and during a thermal event during which the convective heat supply in the asthenosphere, and thus the average temperature, is increased. A resultant decrease in continental freeboard will occur simultaneously with the increase in the average temperature in the asthenosphere. (From Wright, J. A. and Louden, K. E., Eds., *CRC Handbook of Seafloor Heat Flow,* CRC Press, Boca Raton, FL, 1989.)

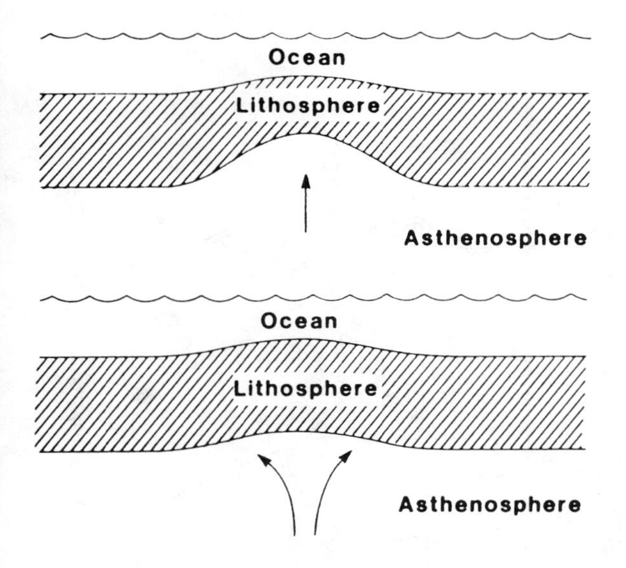

FIGURE 3.3—15. Two alternative models for the origin of midplate swells: (top) lithospheric-thinning model in which swells form as a result of a broad-scale reheating and thinning of the lithosphere as it passes over a mantle hot spot; (bottom) mantle-convection model in which swells are dynamically supported by temperature variations occurring within the upper thermal boundary layer of convection. (From Wright, J. A. and Louden, K. E., Eds., *CRC Handbook of Seafloor Heat Flow,* CRC Press, Boca Raton, FL, 1989.)

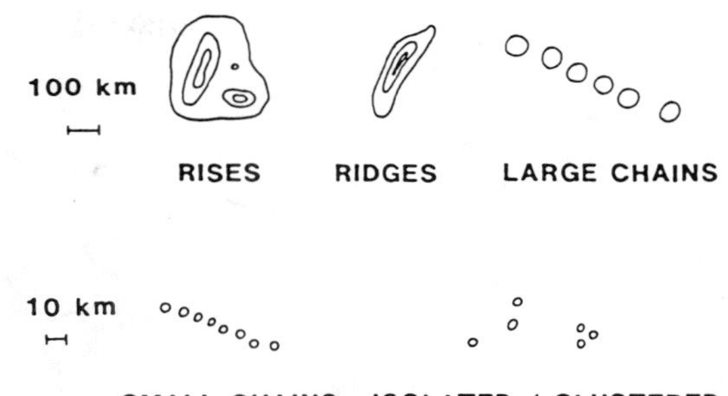

FIGURE 3.3—16. Schematic drawing depicting the types of volcanic landforms present in the eastern Pacific. (From Batiza, R., Seamounts and seamount chains of the eastern Pacific, in *The Geology of North America,* Vol. N, The Eastern Pacific Ocean and Hawaii, Winterer, E., Hussong, D. M., and Decker, R. W., Eds., Geological Society of America, Boulder, CO, 1989, 289. With permission.)

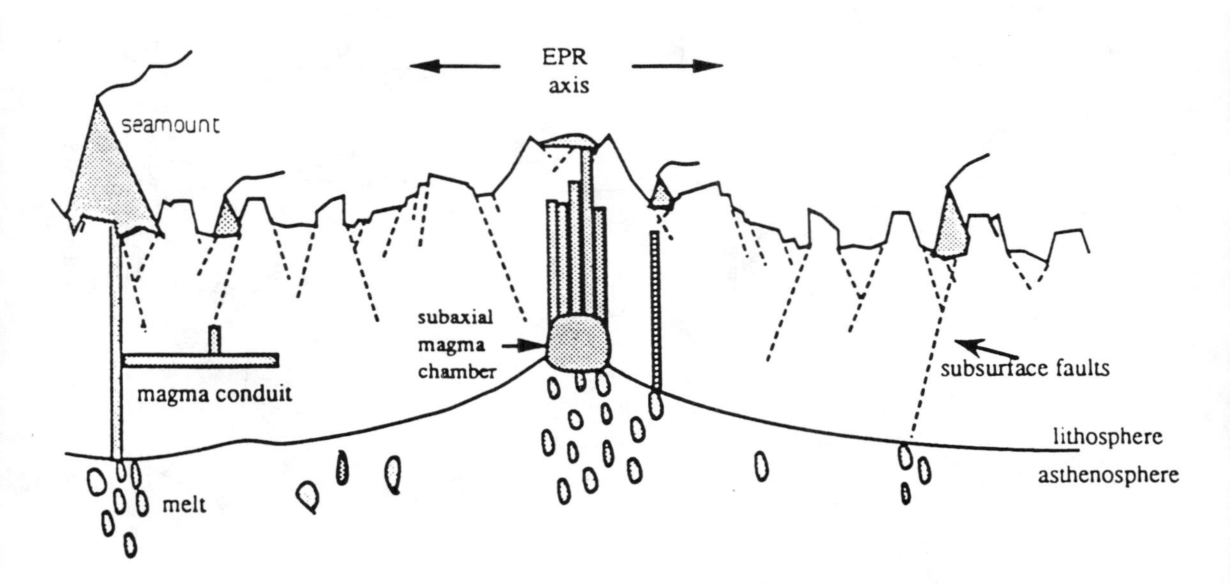

FIGURE 3.3—17. Formation of seamounts in proximity to the fast-spreading East Pacific Rise axis. (From Smith, D. K., Seamount abundances, size distributions, and their geographic variations, *Rev. Aquat. Sci.,* 5, 197, 1991.)

3.4 DEEP-SEA HYDROTHERMAL VENTS

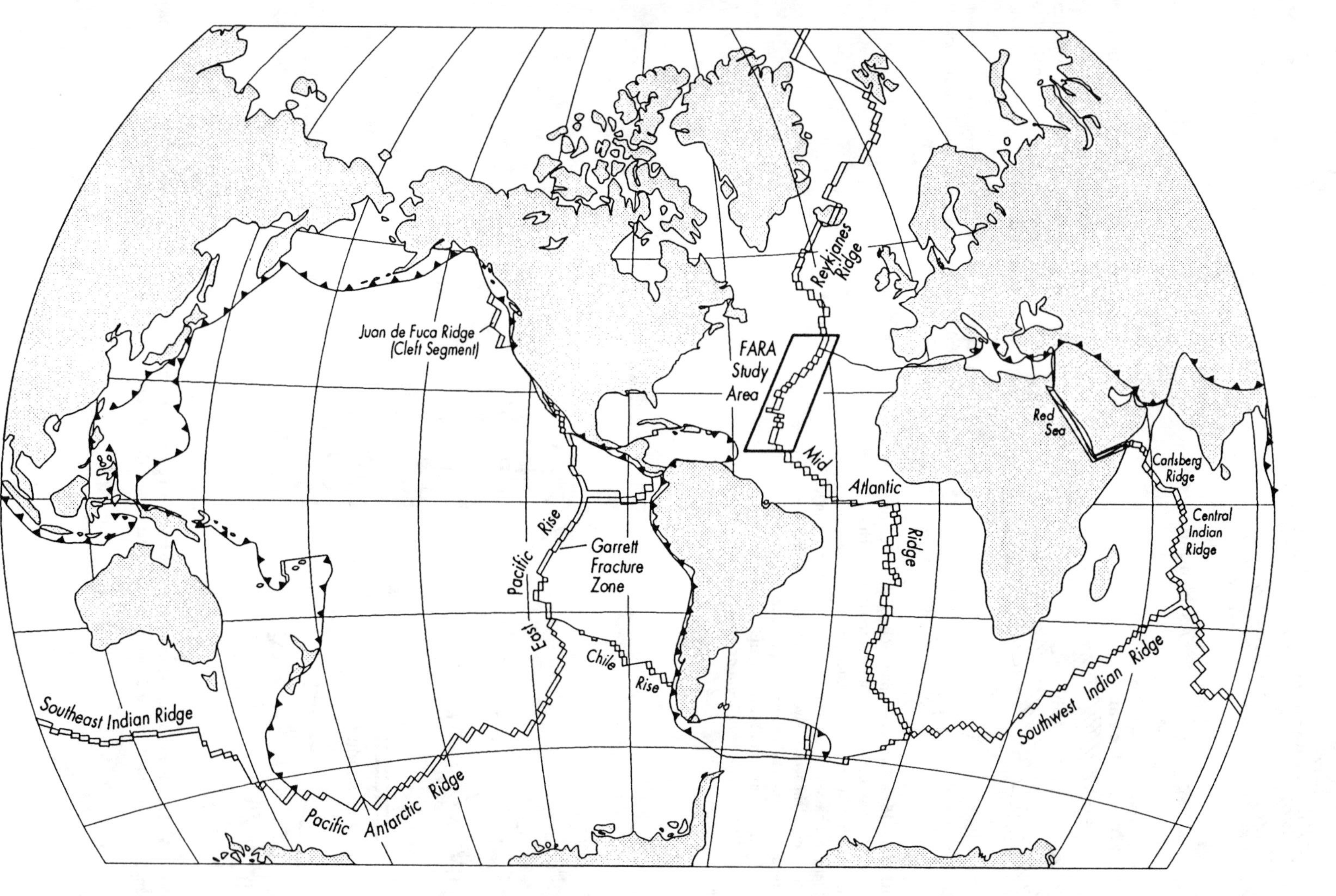

FIGURE 3.4—1. The global mid-ocean ridge system. (From Ridge Science Plan: 1993–1997, Woods Hole Oceanographic Institution, Woods Hole, MA, 1992.)

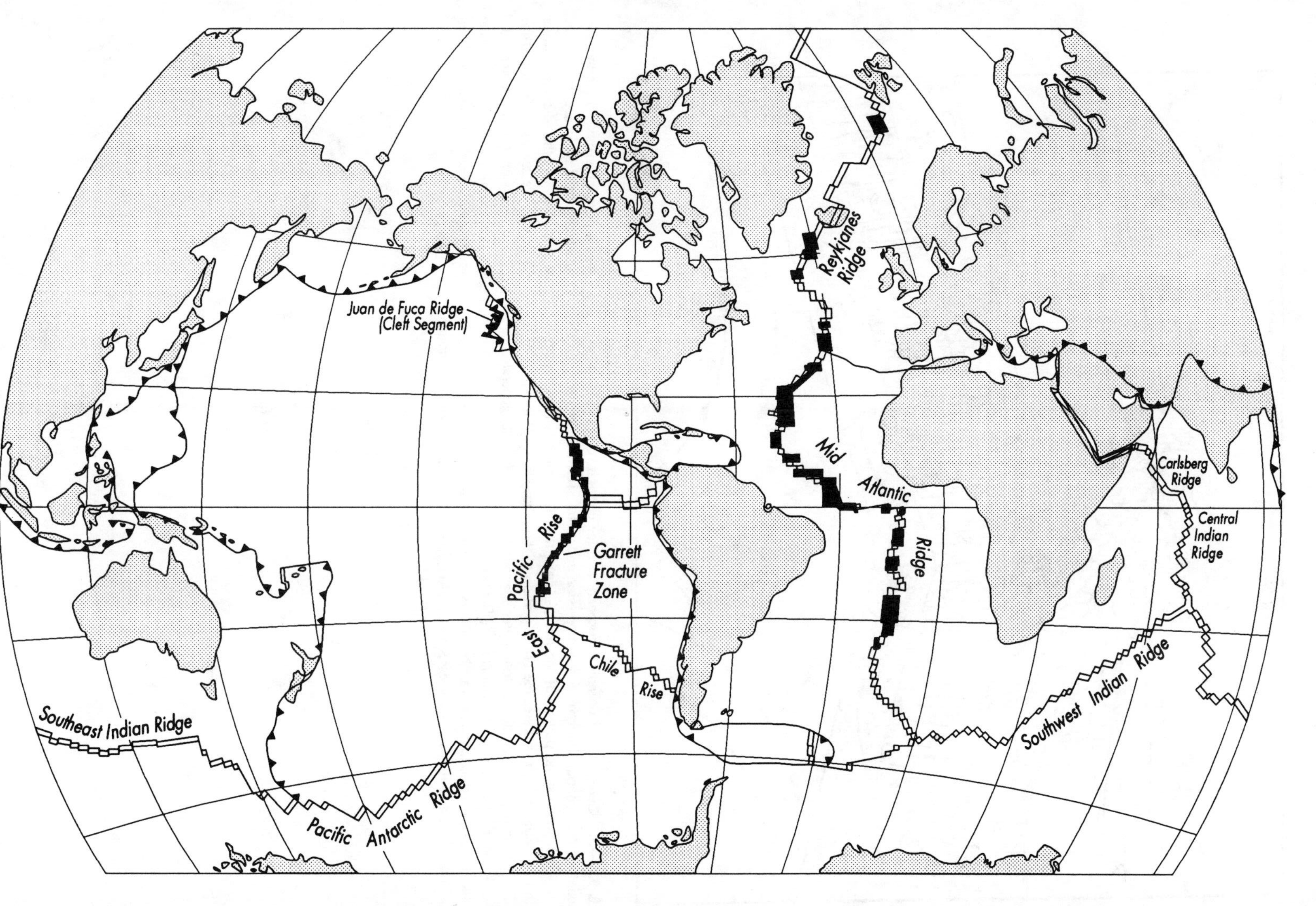

FIGURE 3.4—2. Distribution of multibeam bathymetric data currently available along the global mid-ocean ridge system. (From Ridge Science Plan: 1993–1997, Woods Hole Oceanographic Institution, Woods Hole, MA, 1992.)

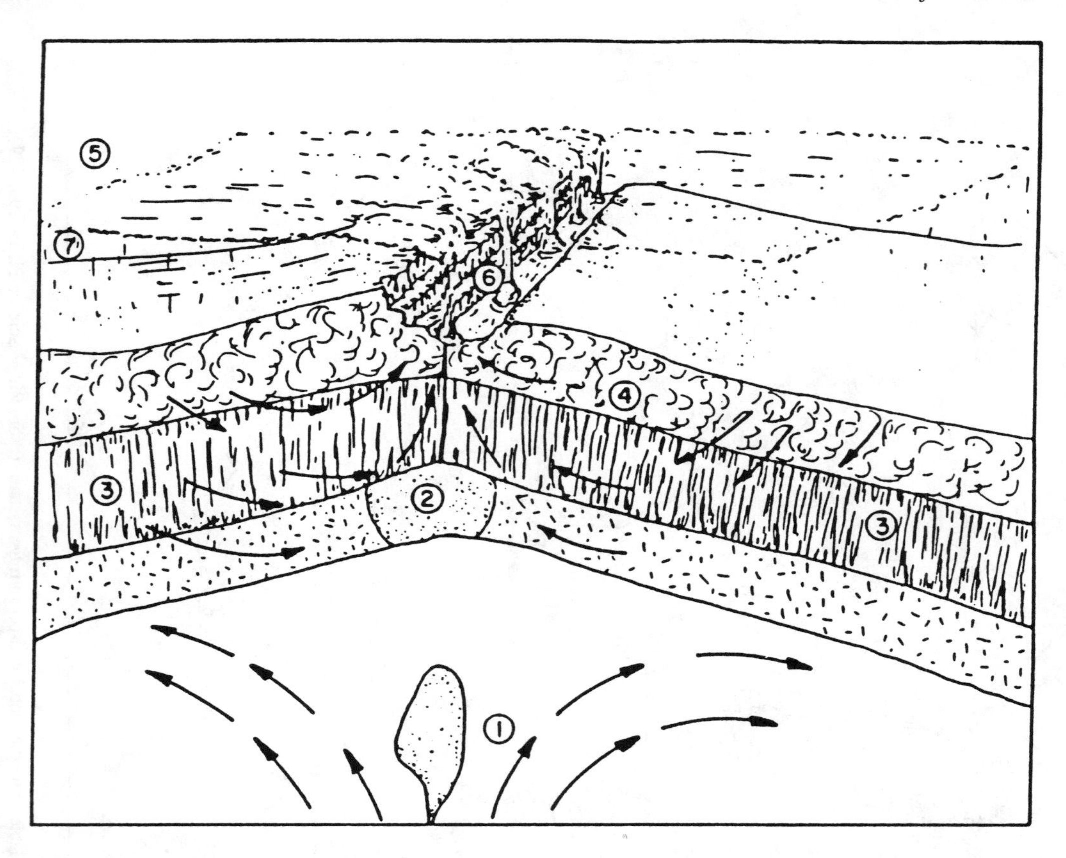

FIGURE 3.4—3. Generation of oceanic lithosphere at ridge crests involves at least seven sub-systems or components. These include: (1) sub-axial mantle material undergoing decompression and partial melting; (2) a magma chamber in which partial melts accumulate and undergo modification; (3) the portion of the oceanic crust and lithosphere which can sustain brittle fracturing; (4) the interstitial water derived from seawater which saturates the brittle portion of the lithosphere; (5) the mass of seawater overlying the seafloor; (6) organisms which inhabit near bottom and sub-seafloor water masses; and (7) sediments generated in the overlying seawater. (From Delaney, J., Spiess, F., Solomon, S., Hessler, R., Karsten, J., Baross, J., Norton, D., McDuff, R., Sayles, F., and Whitehead, J., Scientific rationale for establishing long-term ocean bottom observatory/laboratory systems, in *Mid-Ocean Ridge — a Dynamic Global System,* National Research Council, National Academy Press, Washington, D.C., 1988, 240. With permission.)

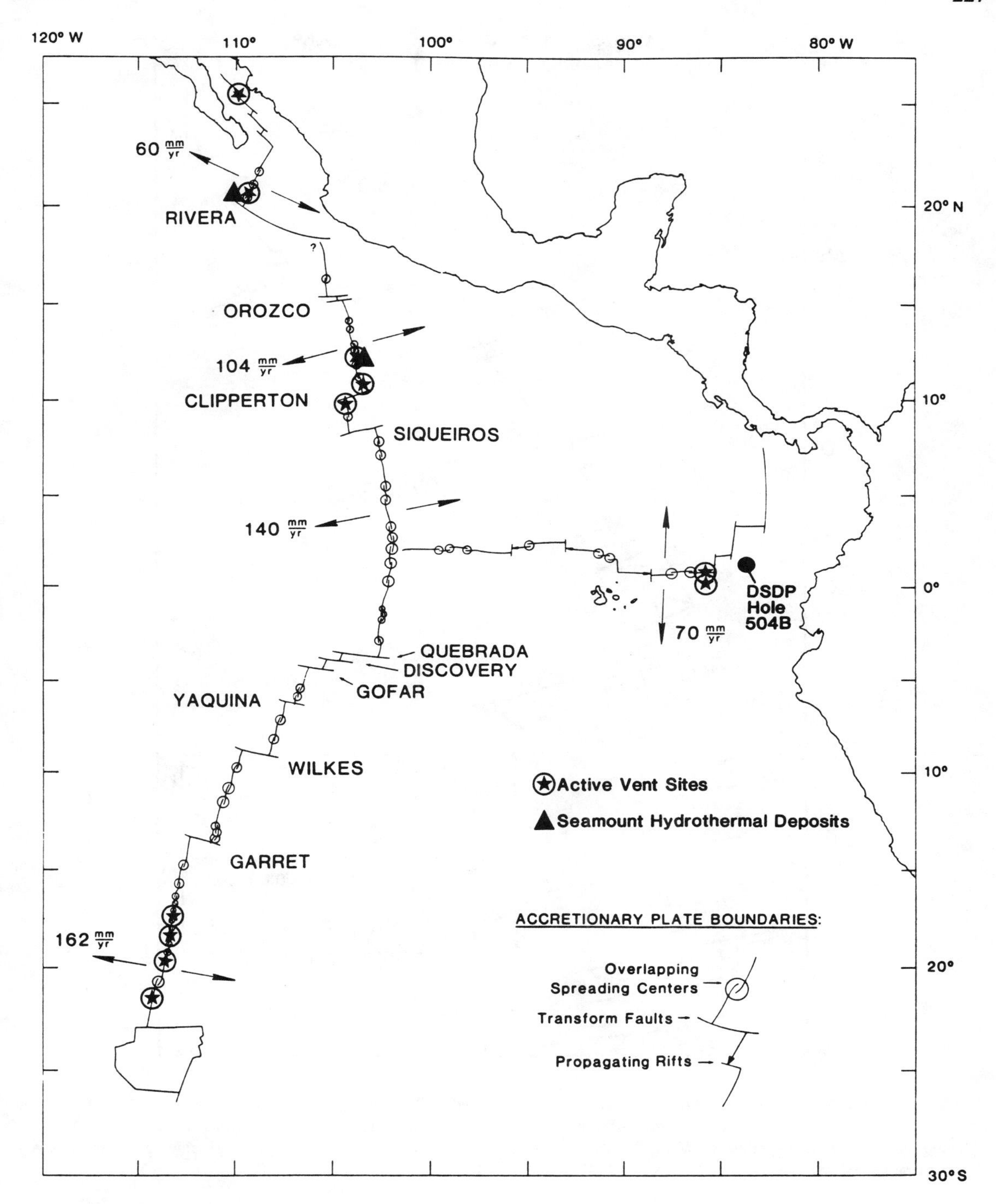

FIGURE 3.4—4. Distribution of known sites of active hydrothermal venting on the East Pacific Rise and Galapagos Rift. Solid stars mark sites of active venting; solid triangles mark locations of off-axis seamounts associated with hydrothermal mineral deposits. Solid circle shows location of DSDP Hole 504B. (From Haymon, R. M., Hydrothermal processes and products on the Galapagos Rift and East Pacific Rise, in *The Geology of North America,* Vol. N, The Eastern Pacific Ocean and Hawaii, Winterer, E., Hussong, D. M., and Decker, R. W., Eds., Geological Society of America, Boulder, CO, 1989, 173. With permission.)

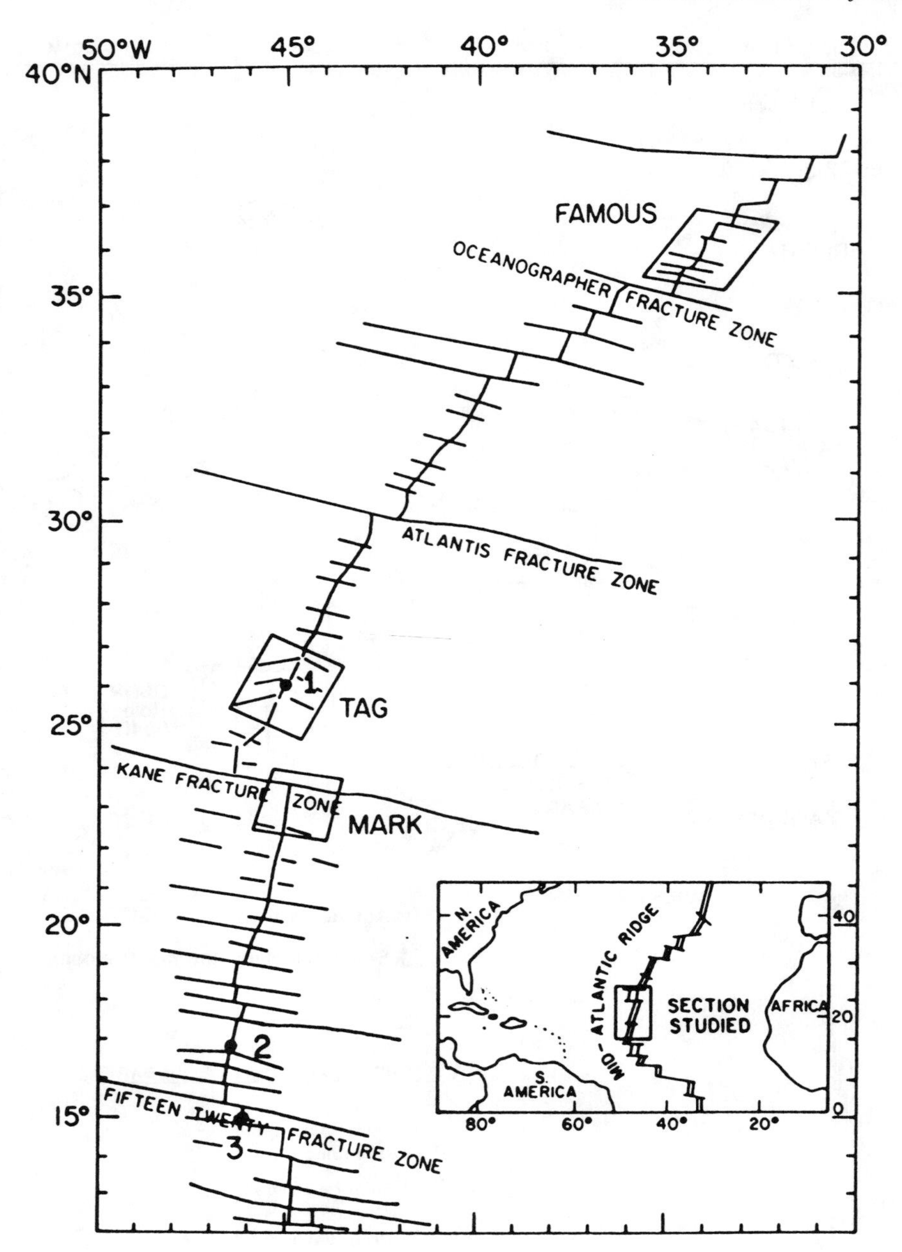

FIGURE 3.4—5. Map of the Mid-Atlantic Ridge between 12°N and 40°N showing major fracture zones, small-offset transform faults, and hydrothermal vent locations (e.g., TAG) along the slow-spreading ridge system. (From Rona, P. A., The Central North Atlantic Ocean Basin and Continental Margins: Geology, Geophysics, Geochemistry, and Resources, Including the Trans-Atlantic Geotraverse [TAG], NOAA Atlas 3, NOAA/ERL, Washington, D.C., 1980.)

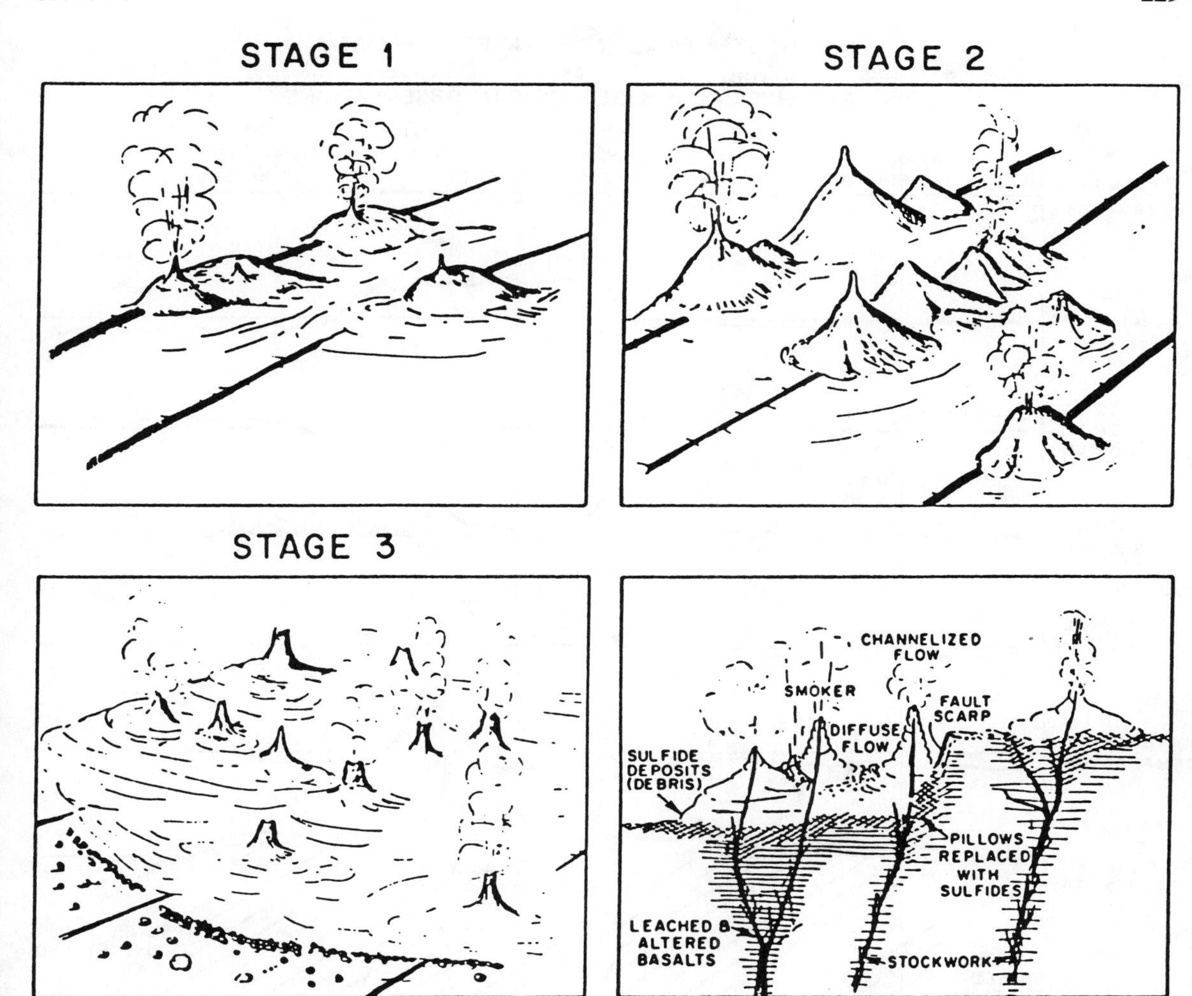

FIGURE 3.4—6. Sequence of development of a hydrothermal vent field at an unsedimented spreading center. (1) Initial stage showing small discrete vents over existing fissures. (2) Established stage illustrating coalescing hydrothermal vent deposits along fissures. (3) Mature vent field depicting large equidimensional deposits. (4) Profile of an active hydrothermal vent system. (From Tivey, M. A. and Johnson, H. P., High resolution geophysical studies of oceanic hydrothermal systems, *Rev. Aquat. Sci.*, 1, 473, 1989.)

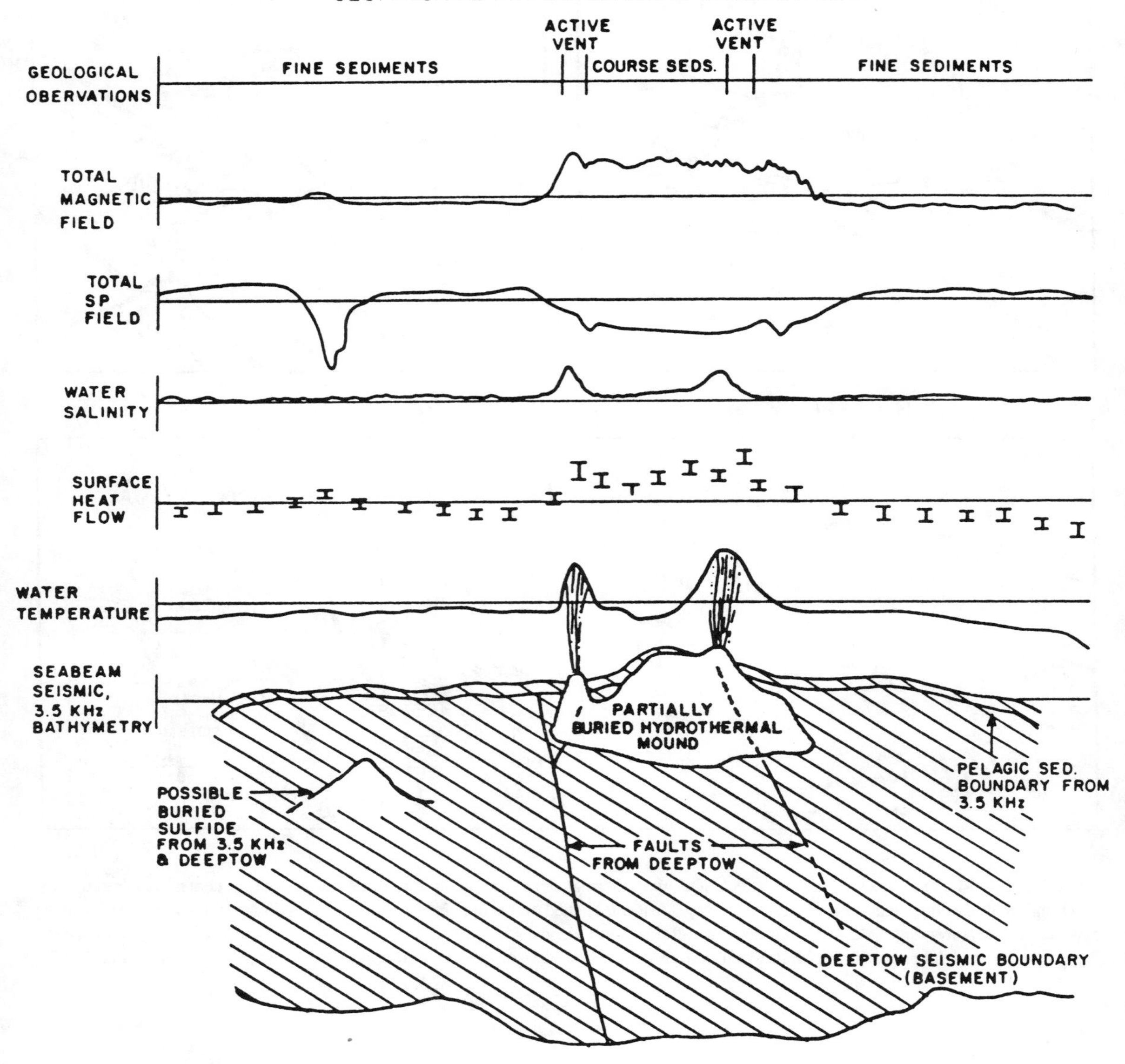

FIGURE 3.4—7. Profile of a sedimented, spreading-center mound showing possible geophysical anomalies in response to buried sulfide deposits. (From Tivey, M. A. and Johnson, H. P., High resolution geophysical studies of oceanic hydrothermal systems, *Rev. Aquat. Sci.*, 1, 473, 1989.)

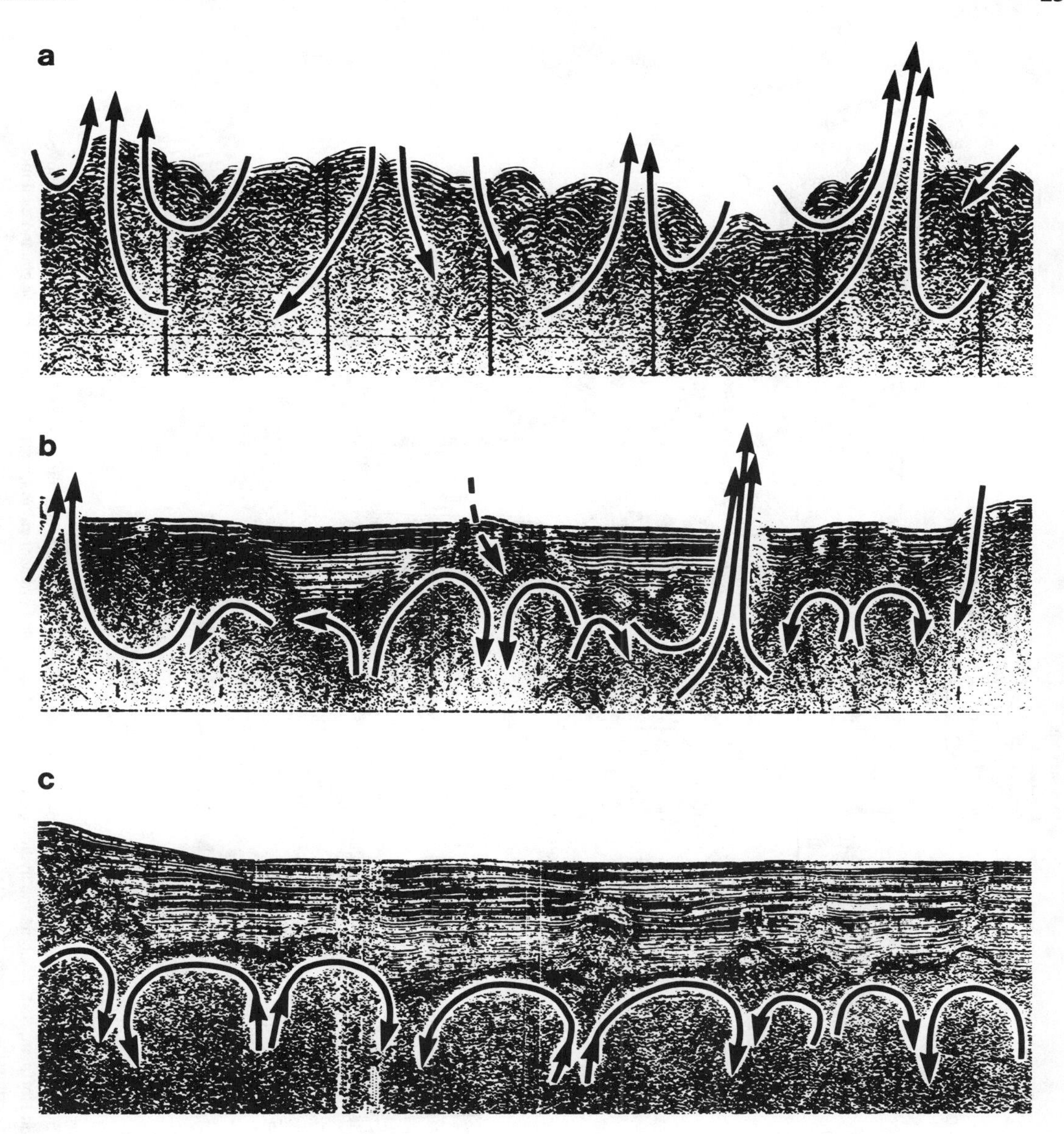

FIGURE 3.4—8. Schematic representation of hydrothermal circulation in young oceanic crust in areas having hydrologically (a) sparse, (b) incomplete, and (c) extensive sediment cover. (From Wright, J. A. and Louden, K. E., Eds., *CRC Handbook of Seafloor Heat Flow,* CRC Press, Boca Raton, FL, 1989.)

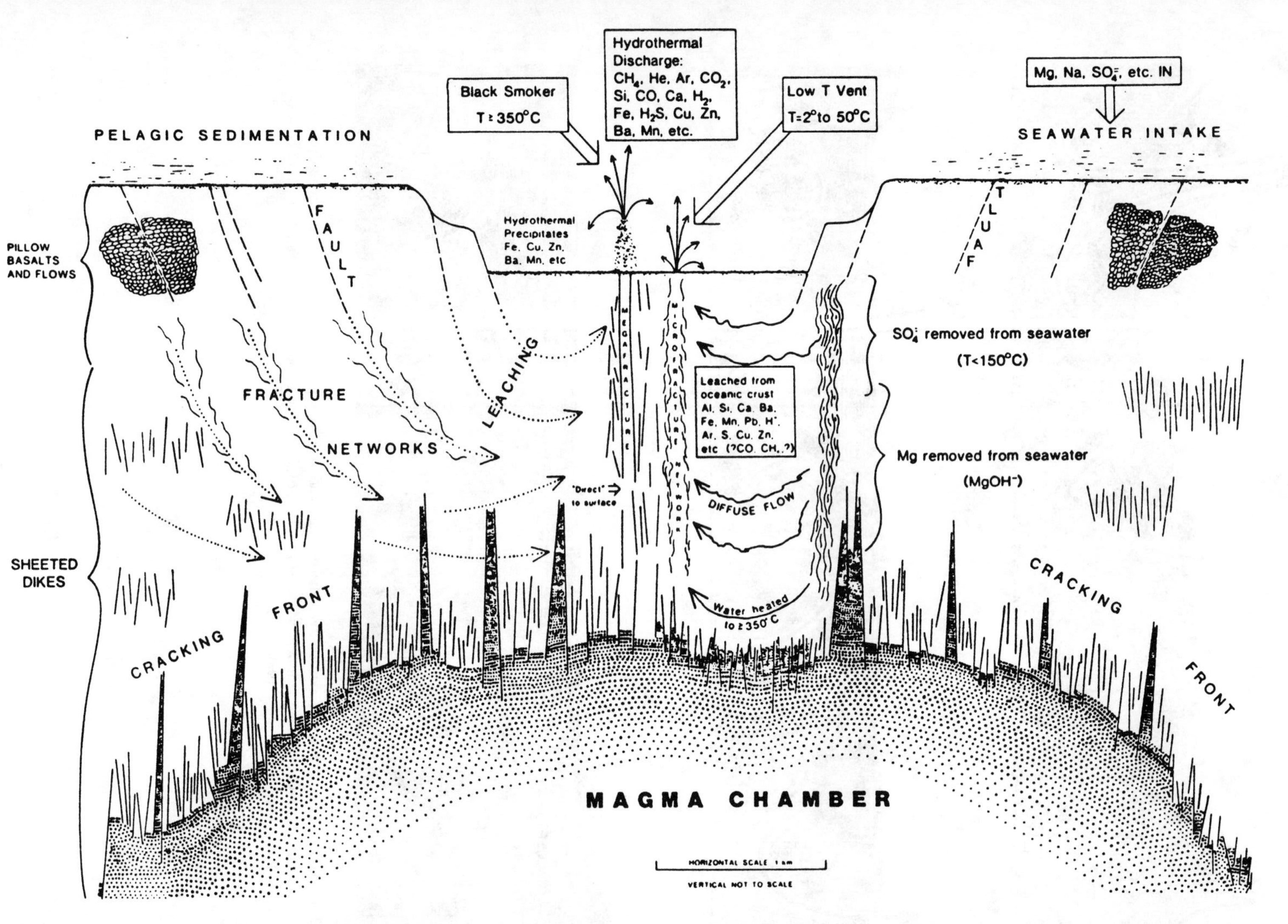

FIGURE 3.4—9. Profile of the central 6 km of a seafloor spreading center depicting hydrothermal systems, faults and fractures, and water/rock interactions above a magma chamber. (From Baross, J. A. and Hoffman, S. E., Submarine hydrothermal vents and associated gradient environments as sites for the origin and evolution of life, *Origins Life,* 15, 327, 1985. With permission.)

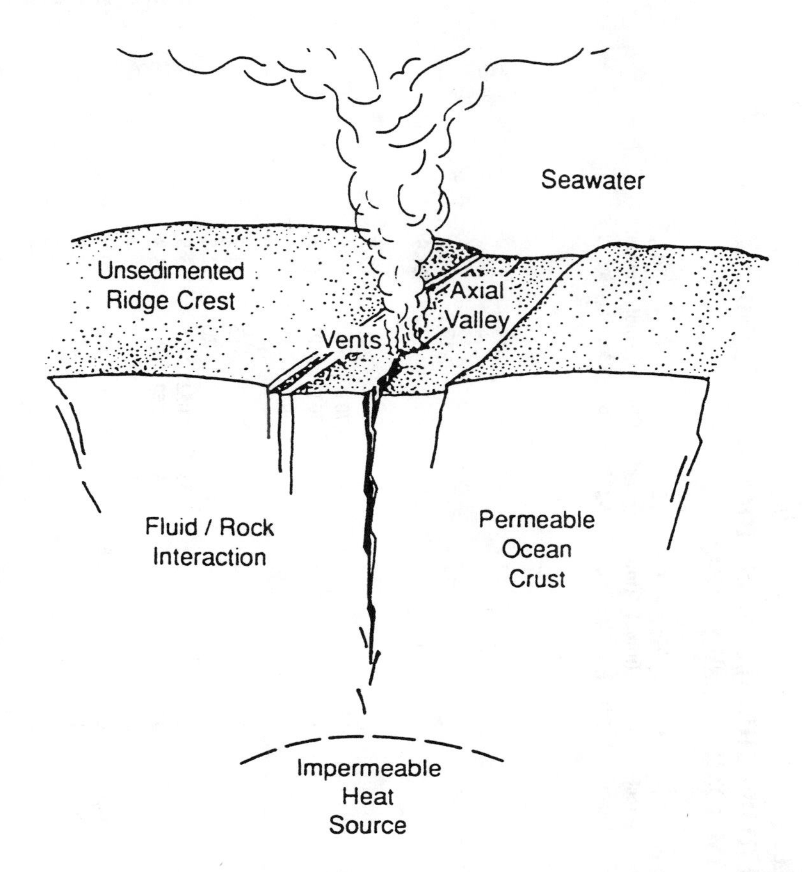

FIGURE 3.4—10. A schematic cross section of a seafloor hydrothermal system illustrating an impermeable heat source (magma chamber or hot rocks) overlain by permeable ocean crust at an unsedimented ridge crest. Fluid circulates within the crust, driven by temperature differences. During this circulation, seawater is modified by fluid/rock interaction to hot, metal-rich fluid that is buoyant, and vents on the seafloor. (From Tivey, M. K., Hydrothermal vent systems, *Oceanus,* 34, 68, 1991. Courtesy of *Oceanus* Magazine/ Woods Hole Oceanographic Institution.)

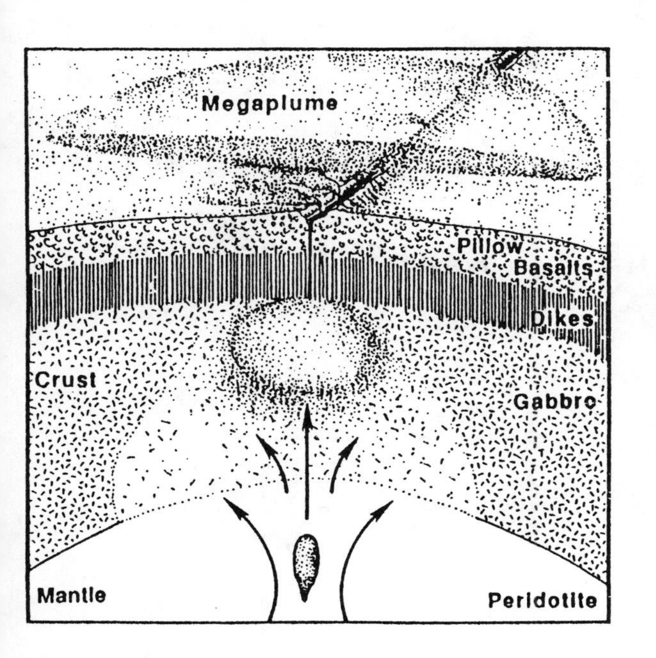

FIGURE 3.4—11. Model of oceanic crustal accretion at a seafloor spreading center showing mantle-derived basaltic melt either reestablishing or replenishing a mid-crustal magma chamber that solidifies to produce gabbro, injects melt into the upper crust to form diabase dikes, and erupts basaltic lava flows onto the seafloor. A megaplume generated by hydrothermal venting (discovered along the Juan de Fuca Ridge in 1986) rises above the ridge axis. (From Blackman, D. and Stroh, T., RIDGE: cooperative studies of mid-ocean ridges, *Oceanus,* 34, 21, 1991. Courtesy of *Oceanus* Magazine/Woods Hole Oceanographic Institution.)

Table 3.4—1
SUMMARY OF CHEMICAL DATA FOR SEAFLOOR HYDROTHERMAL SOLUTIONS. ALKALI AND ALKALINE EARTH METALS, AMMONIUM, AND SILICA[a]

Vent	Temp. (°C)	Li (μmol kg^{-1})	Na (mmol kg^{-1})	K (mmol kg^{-1})	Rb (μmol kg^{-1})	Cs (nmol kg^{-1})	NH_4 (mmol kg^{-1})	Be (nmol kg^{-1})	Mg (mmol kg^{-1})	Ca (mmol kg^{-1})	Sr (μmol kg^{-1})	$^{87}Sr/^{86}Sr$	Ba (μmol kg^{-1})	SiO_2 (mmol kg^{-1})
Galapagos spreading center								11–37						
CB	<13	1142	487	18.7	20.3				0	40.2	87		>42.6	21.9
GE	<13	1142	451	18.8	21.2				0	34.3	87		>17.2	21.9
DL	<13	1142	313	18.8	17.3				0	34.3	87		>17.2	21.9
OB	<13	689	259	18.8	13.4				0	24.6	87		>17.2	21.9
21°N EPR														
NGS	273	1033	510	25.8	31.0		<0.01	37	0	20.8	97	0.7030	>16	19.5
OBS	350	891	432	23.2	28.0	202	<0.01	15	0	15.6	81	0.7031	>8	17.6
SW	355	899	439	23.2	27.0		<0.01	10	0	16.6	83	0.7033	>10	17.3
HG	351	1322	443	23.9	33.0		<0.01	13	0	11.7	65	0.7030	>11	15.6
Guaymas Basin														
1	291	1054	489	48.5	85.0		15.6	12	0	29.0	202		>12	12.9
2	291	954	478	46.3	77.0		15.3	18	0	28.7	184		>20	12.5
3	285	720	513	37.1	57.0		10.3	42	0	41.5	253	0.7052	>15	13.5
4	315	873	485	40.1	66.0		12.9	29	0	34.0	226	0.7052	>54	13.8
5	287	933	488	43.1	74.0		14.5	29	0	30.9	211		>13	12.4
6	264	896	475	45.1	74.0		14.5	60	0	26.6	172	0.7059	>16	10.8
7	300	1076	490	49.2	86.0		15.2	17	0	29.5	212		>24	12.8
9	100	630	480	32.5	57.0		10.7	91	0	30.2	160			9.3
Southern Juan de Fuca												0.7034		
Plume	224	1718	796	51.6	37.0			95	0	96.4	312			23.3
Vent 1	285	1108	661	37.3	28.0			150	0	84.7	230			22.8
Vent 3		1808	784	45.6	32.0			150	0	77.3	267			22.7

11–13°N EPR														
N & S–13°N	317	688	560	29.6	14.1				0	55.0	175	0.7041		22.0
1–13°N		614	587	29.8	18.0				0	44.6	171			21.9
2–13°N	354	592	551	27.5	19.0				0	53.7	182			19.4
3–13°N	(380)	591	596	28.8	20.0				0	54.8	168			17.9
4–11°N	347	884	472	32.0	24.0				0	22.5	80			18.8
5–11°N		623	577	32.9	25.0				0	35.2	135			20.6
6–11°N		484	290	18.7	15.0				0	10.6	38			14.3
Mid-Atlantic Ridge														
TAG	290.321	411	584	17.0	10.0	100			0	26.0	99	0.7029		22.0
MARK-1	350	843	510	23.6	10.5	177		38.5	0	9.9	50	0.7028		18.2
MARK-2	335	849	509	23.9	10.8	181		38.0	0	10.5	51	0.7028		18.3
Axial Volcano														
HE, HI, MR	136–323	512	415	22.0					0	37.3				15.1
Inferno	149–328	637	500	27.5					0	46.8				15.1
VM, Crack	5–299	204	159	7.6					0	10.2				13.5
Seawater	2	26	464	9.8	1.3	2.0	<0.01	0.0	52.7	10.2	87	0.7090	0.14	0.16

[a] Data references listed at end of Table 3.4—3.

Reproduced with permission from *Annu. Rev. of Earth Planetary Science,* Vol. 18, ©1990 by Annual Reviews Inc.

Table 3.4—2
SUMMARY OF CHEMICAL DATA FOR SEAFLOOR HYDROTHERMAL SOLUTIONS. pH, CARBON, AND SULFUR SYSTEMS, HALOGENS, BORON, ALUMINUM, AND WATER ISOTOPES[a]

Vent	pH	Alk_T (meq kg^{-1})	Total CO_2 (mmol kg^{-1})	SO_4 (mmol kg^{-1})	H_2S (mmol kg^{-1})	$\delta^{14}S$	As (nmol kg^{-1})	Se (nmol kg^{-1})	Cl (mmol kg^{-1})	Br (μmol kg^{-1})	B (μmol kg^{-1})	$\delta^{11}B$	Al (μmol kg^{-1})	$\delta^{18}O$	δD
Galapagos spreading center			9.3–11.3							832–835					
CB	—	<0		0	+				595						
GE	—	<0		0	+				543						
DL	—	<0		0	+				395						
OB	—	<0		0	+				322						
21°N EPR			5.72											1.6–2.0	2.5
NGS	3.8	−0.19		0.00	6.6	3.4	30	<0.6	579	929	507	32.7	4.0		
OBS	3.4	−0.40		0.50	7.3	1.3–1.5	247	72	489	802	505	32.2	5.2		
SW	3.6	−0.30		0.60	7.5	2.7–5.5	214	70	496	877	500	31.5	4.7		
HG	3.3	−0.50		0.40	8.4	2.3–3.2	452	61	496	855	548	30.0	4.5		
Guaymas Basin															
1	5.9	10.60		−0.15	5.8		283	82	601	1054–1117			0.9		
2	5.9	9.60		−0.09	4.0		732	87	589		1630	17.4	1.2		
3	5.9	6.50		−0.34	5.2		1071	38	637				6.7		
4	5.9	8.10		0.06	4.8		1074	103	599	1063	1570	23.2	3.7		
5	5.9	9.70		−0.07	4.1		516		599				3.0		
6	5.9	7.30		−0.32	3.8		669	49	582				3.9		
7	5.9	10.50		−0.06	6.0		711	92	606	1054–1117	1730	19.6	1.0		
9	5.9	2.80		−4.20	4.6		577		581				7.9		
Southern Juan de Fuca			3.92–4.46												−2.5–+0.5
Plume	3.2			−0.50	3.5	4.2–7.3		<1	1087	1832	496	34.2		0.65	
Vent 1	3.2			−1.30	(3.0)	4.0–6.4		<1	896	1580				0.60	
Vent 3	3.2			−1.70	(4.4)	7.2–7.4		<1	951	1422				0.80	
11–13°N EPR															
N & S—13°N	3.2		10.8–16.7	0.00					740	1163				0.39–0.69	0.62–1.49
1–13°N	3.2	−0.64			2.9				718	1131			13.3		
2–13°N	3.1	−0.74			8.2	4.7			712	1158	467	34.9	19.8		
3–13°N	3.3	−0.40			4.5	2.3–3.5			760	1242			20.0		
4–11°N	3.1	−1.02			8.0	4.6			563	940			12.9		
5–11°N	3.7	−0.28			4.4	4.7–4.9			686	1105	451	36.8	13.5		
6–11°N	3.1	−0.88			12.2	4.1–5.2			338	533	493	31.5	13.6		

Mid-Atlantic Ridge															
TAG									659						
MARK-1	3.9	−0.06			5.9	4.9			559	847	518	26.8	5.3	2.37	
MARK-2	3.7	−0.24			5.9	5.0			559	847	530	26.5	5.0	2.37	
Axial Volcano											476	34.7			
HE, HI, MR	3.5	−0.52			8.1				515	760	565			0.8–1.1	
Inferno	3.5	−0.45			7.0				625	950	565			0.7–1.1	
VM, Crack	4.4	0.58	150–170		(19.5)				188	240	503			0.7–0.9	
Seawater	7.8	2.3	2.3	27.9	0.0		27	2.5	541	840	416	39.5	0.020	0.0	0.0

[a] Data references listed at end of Table 3.4—3.

Reproduced with permission from *Annu. Rev. of Earth Planetary Science,* Vol. 18, © 1990 by Annual Reviews Inc.

Table 3.4—3
SUMMARY OF CHEMICAL DATA FOR SEAFLOOR HYDROTHERMAL SOLUTIONS. TRACE METALS[a]

Vent	Mn (μmol kg⁻¹)	Fe (μmol kg⁻¹)	Co (nmol kg⁻¹)	Cu (μmol kg⁻¹)	Zn (μmol kg⁻¹)	Ag (nmol kg⁻¹)	Cd (nmol kg⁻¹)	Pb (nmol kg⁻¹)
Galapagos spreading center								
CB	1,140	+		0			0	
GE	390	+		0			0	
DL	480	+		0			0	
OB	360	+		0			0	
21°N EPR								
NGS	1,002	871	22	<0.02	40	<1	17	183
OBS	960	1,664	213	35.00	106	38	155	308
SW	699	750	66	9.70	89	26	144	194
HG	878	2,429	227	44.00	104	37	180	359
Guaymas Basin								
1	139	56	<5	<0.02	4.2	230	<10	265
2	222	49	<5	<0.02	1.8	<1	<10	304
3	236	180	<5	<0.02	40.0	24	46	652
4	139	77	<5	1.10	19.0	2	27	230
5	128	33	<5	0.10	2.2	<1	<10	<20
6	148	17	<5	<0.02	0.1	<1	<10	<20
7	139	37	<5	<0.02	2.2	<1	<10	<20
9	132	83	<5	<0.02	21.0	<1	<10	<20
Southern Juan de Fuca								
Plume 1	3,585	18,739		<2	900			900
Vent 1	2,611	10,349		<2	<600			
Vent 3	4,480	17,770		<2				

11–13°N EPR								
N & S–13°N	1,000	1,450						
1–13°N	1,689	3,980			102.0		55	135
2–13°N	2,932	10,370			2.0		70	27
3–13°N	2,035	10,760			2.0		65	14
4–11°N	766	6,470			105.0		30	50
5–11°N	742	1,640			73.0		43	270
6–11°N	925	2,640			44.0		1	9
Mid-Atlantic Ridge								
TAG	1,000	1,640						
MARK-1	491	2,180		17.0	50.0			
MARK-2	493	1,832		10.0	47.0			
Axial Volcano								
HE, HI, MR	1,081	1,006		12.0	113.0			302
Inferno	1,081	1,006		12.0	113.0			302
VM, Crack	162	9		0.70	2.3			101
Seawater	<0.001	<0.001	0.03	0.007	0.01	0.02	1.0	0.0100

[a] Data from following references:

Galapagos: Edmond et al., 1979a, b, Welhan, 1981.

21°N EPR: Craig et al., 1980. Welhan, 1981. Von Damm et al., 1985a. Spivack & Edmond, 1987. Woodruff & Shanks, 1988. Campbell & Edmond, 1989.

Guaymas: Von Damm et al., 1985b. Spivack et al., 1987. Campbell & Edmond, 1989.

Juan de Fuca: Von Damm & Bischoff, 1987. Shanks & Seyfried, 1987. Hinkley & Tatsumoto, 1987. Evans et al., 1988. Campbell & Edmond, 1989.

11–13°N EPR: Michard et al., 1984. Merlivat et al., 1987. Bowers et al., 1988. Bluth & Ohmoto, 1988. Campbell & Edmond, 1989.

Mid-Atlantic Ridge: Campbell et al., 1988b. Campbell & Edmond, 1989.

Axial: Butterfield et al., 1988. Massoth et al., 1989.

Reproduced with permission from *Annu. Rev. of Earth Planetary Science,* Vol. 18, © 1990 by Annual Reviews Inc.

Table 3.4—4
MEASURED COMPOSITIONS OF END MEMBER HYDROTHERMAL VENT FLUIDS*

Sample		EPR, 21°N	EPR, 13°N	Galapagos Rift	Seawater
Alkalis					
Li	u	891–1,033	820	689–1,142	26
Na	m	432–510	450	+, −	464
K	m	23.2–25.8	24	18.7–18.8	9.79
Rb	u	27–33	26	13.4–21.2	1.3
Alkaline earths					
Be	n	10–37	NA	11–37	0.02
Mg	m	0	0	0	52.7
Ca	m	11.7–20.8	16	24.6–40.2	10.2
Sr	u	65–97	90	87	87
Ba	u	7	NA	17.2–42.6	0.14
$^{87}Sr/^{86}Sr$		0.703019–0.703345	0.7041	NA	0.7091
pH		3.3–3.8	3.8	NA	7.8
Alk	meq	(−0.5)–(−0.19)	NA	0	2.3
Cl	m	489–579	550	+, −	541
SiO_2	m	15.6–19.5	22	21.9	0.16
Al	u	4.0–5.2	NA	NA	0.005
SO_4	m	0	0	0	28.6
$\delta^{18}O$	‰	+1.6	NA	NA	0
$\delta^{34}S$	‰	+2.3–+5.0	NA	NA	+18.86
Trace metals					
Mn	u	699–1,002	610	360–1,140	<0.001
Fe	u	750–2,429	1,800	+	<0.001
Fe/Mn		0.9–2.9	2.9		
Co	u	22–227	NA	NA	0.03
Cu	u	0.02–44	NA	0	0.007
Zn	u	40–106	NA	NA	0.01
Ag	u	1–38	NA	NA	0.02
Cd	u	17–180	NA	0	1
Pb	u	183–359	NA	NA	0.01
Gases					
$\Sigma\ CO_2$	m	5.7–6.4	NA	NA	2.33
CH_4	m	0.04–0.08	NA	NA	0
H_2S	m	6.5–8.7	NA	NA	0
H_2(aq)	m	0.36–1.69	NA	NA	0
He	u	3.75	NA	NA	0.002

* Units: m = millimoles/kilogram; u = micromoles/kilogram; n = nanomoles/kilogram; meq = milliequivalents/kilogram; Galapagos data is /liter; + = gain, − = loss.

Notes: EPR, 21°N data and strontium isotope data from von Damm and others (1985). EPR, 13°N data from Michard and others (1984). Galapagos data from Edmond and others (1979). Trace element data for seawater from Quinby-Hunt and Turekian (1983). Oxygen isotope data for EPR, 21°N from Craig and others (1980). Sulfur isotope data for EPR, 21°N from Kerridge and others (1983) and Bluth and Ohmoto (1986). ΣCO_2, H_2, and CH_4 data from Welhan and Craig (1979), Welhan and Craig (1983), Welhan (1980), and Craig and others (1980). H_2S data from Edmond and others (1982).

From Haymon, R. M., Hydrothermal processes and products on the Galapagos Rift and East Pacific Rise, in *The Geology of North America,* Vol. N, The Eastern Pacific Ocean and Hawaii, Winterer, E., Hussong, D. M., and Decker, R. W., Eds., Geological Society of America, Boulder, CO, 1989, 173. With permission.

Table 3.4—5
MINERAL DISTRIBUTION IN EPR HYDROTHERMAL DEPOSITS*

	Axis deposits								
		Chimneys			Particulates		Sediments		
	Mounds	Black smokers	White smokers	Inactive chimneys	Black smokers	White smokers	Proximal	Distal	Off-axis deposits
Sulfides/Sulfosalts									
(Most Abundant)									
Sphalerite (Zn(Fe)S)	A	C	R-A	A	A		A		R-T
Wurtzite (Zn(Fe)S)	A	R-C	R-A	A			C		
Pyrite (Fe_2S)	A	C	A	A	R-C	C-A	A		A
Chalcopyrite ($CuFeS_2$)	C	C-A	R	R-C	T		C-A		C-A
(Less Abundant)									
Iss-isocubanite (variable-$CuFe_2S_3$)	R-C	C-A	R	R-C	R				T
Marcasite (Fe_2S)	C	C	C-A	C-A			R-T		C-A
Melnicovite (FeS_{2-x})	C	C	C	C-A					C-A
Pyrrhotite ($Fe_{1-x}S$)		R-C			A		T-A		R-T
Bornite-chalcocite (Cu_5FeS_4-Cu_2S)		R-C							R-T
Covellite (CuS)		R							R
Digenite (Cu_9S_5)		R					T		R-T
Idaite ($Cu_{5.5}FeS_{6.5}$)		T							
Galena (PbS)			T	T					
Jordanite ($Pb_9As_4S_{15}$)			T	T					
Tennantite $(Cu, Ag)_{10}(Fe,Zn,Cu)_2As_4S_{23}$				T					
Valleriite $(2(Cu,Fe)_2S_23(Mg,Al)(OH)_2)$		T							
Sulfates									
Anhydrite ($CaSO_4$)		A	A						
Gypsum ($CaSO_4 \cdot 2H_2O$)		R	R					T-R	
Caminite ($MgSO_4 \cdot xMg(OH)_2 \cdot (1-2x)H_2O$)		R							
Barite ($BaSO_4$)	R-C	T	R-C	R-C		C			T-A
Jarosite-natrojarosite $(K,Na)Fe_3(SO_4)_2(OH)_6$	R			R					R
Chalcanthite ($CuSO_4 \cdot 5H_2O$)									R
Carbonate									
Magnesite ($MgCO_3$)		T							

Table 3.4—5 (continued)
MINERAL DISTRIBUTION IN EPR HYDROTHERMAL DEPOSITS*

	Axis deposits								
		Chimneys			Particulates		Sediments		
	Mounds	Black smokers	White smokers	Inactive chimneys	Black smokers	White smokers	Proximal	Distal	Off-axis deposits
Elements									
Sulfur (S)	R		R	C-A			R-T	T	T
Oxides/Oxyhydroxides									
Goethite (FeO(OH))	C			R-C					C
Lepidocrocite (FeO(OH))	R			T-R					
Hematite (Fe_2O_3)		T-R							
Magnetite (Fe_3O_4)		T-R							
"Amorphous" Fe compounds	C-A	R-C	R-C	C-A			T-A	C-A	R-A
"Amorphous" Mn compounds								C-A	T-A
Psilomelane ($(Ba, H_2O)_2Mn_5O_{10}$)									R-T
Silicates									
Amorphous silica ($SiO_2 \cdot nH_2O$)	C	R	C	C		A		A	R-A
Quartz (SiO_2)									A-T
Talc ($Mg_3Si_4O_{10}(OH)_2$)	R	R-C							
Nontronite ($(Fe,Al,Mg)_2(Si_{3.66}Al_{0.34})O_{10}(OH)_2$)	R			T-R				A	T-A
Illite-smectite									R-T
Aluminosilicate gel	R	R	R-C	R-C					
Hydroxychlorides									
Atacamite ($Cu_2Cl(OH)_3$)									T-C

* Key: A = abundant > C = common > R = rare > T = trace.

From Haymon, R. M., Hydrothermal processes and products on the Galapagos Rift and East Pacific Rise, in *The Geology of North America*, Vol. N, The Eastern Pacific Ocean and Hawaii, Winterer, E., Hussong, D. M., and Decker, R. W., Eds., Geological Society of America, Boulder, CO, 1989, 173. With permission.

Table 3.4—6
CHEMICAL COMPOSITION OF MASSIVE SULFIDE SAMPLES FROM THE EAST PACIFIC RISE, 21°N AND GALAPAGOS RIFT SPREADING CENTERS

	East Pacific Rise, 21°N				Galapagos Rift	
	Sphalerite/Wurtzite-rich			Silica-rich	Sphalerite/Wurtzite-rich	Pyrite-rich
				(wt. %)		
Fe	14.7	26.2	16.7	4.0	15.6	44.1
Zn	34.9	20.3	41.8	1.7	46.9	0.14
Cu	0.23	1.3	0.89	0.13	0.35	4.98
Pb	0.61	0.07	0.29	0.06	0.30	<0.07
S	31.3	39.7	34.9	4.3	36.8	52.2
SO_3	<0.01	7.6	<0.01	49.4	<0.03	<0.03
SiO_2	19.0	<0.5	4.3	2.8	1.5	<0.1
Al_2O_3	0.3	0.11	0.77	0.04	0.15	<0.06
MgO	<0.03	0.07	0.02	0.03	<0.05	<0.05
CaO	<0.01	5.42	<0.01	35.1	<0.03	<0.03
Sum	101.4	100.77	99.67	97.53	101.60	101.42
				(ppm)		
Ag	241	34	202	9.6	290	<10
As	483	770	215	13.2	411	125
Au	NA	0.17	<0.2	0.025	0.13	0.05
B	<7*	<7*	<7*	<7*	40*	<7*
Ba	6030	65	850	225	19*	16*
Bi	<0.2*	2*	0.2*	2*	<0.2*	<10*
Cd	120	890	790	60	490	<32*
Co	<2.0	2.5	6	37	24	482
Cr	8	16	<30	<6	<8*	55
Cs	6.7	<5.0	6.6	<2	<9	<3
Ga	3.3*	18*	21*	1.5*	<20*	15*
Ge	96*	<1.5*	100*	<1.5*	270	<1*
Hg	2*	<1*	<1*	2*	<1*	NA
Mn	570*	91*	500*	52*	720*	140*
Mo	16	78	13	2	3	170*
Ni	2*	5*	2*	<1.5*	NA	3.1*
Pd	0.001	0.001	0.001	0.001	<0.002	<0.002
Pt	0.002	0.002	<0.001	0.003	<0.005	<0.005
Rh	0.002	0.0010	0.0022	0.0007	0.003	<0.001
Sb	45.0	13	52.9	3.2	34	1.8
Sc	<0.4	0.2	0.25	<0.3	<1	<0.3
Se	7*	172	10*	5	29	100
Sr	220	9	19	3965	<10*	<1*
Te	<1*	2*	<1*	2*	<1*	NA
Tl	40*	20*	<1*	2*	10*	<5*
U	6.0	1.3	3.1	<2.0	10	1.0
Y	4*	<1.5*	3*	2*	<2*	<2*
W	1.0	<2.0	<3.0	<1.0	<10*	<10*
Zr	9*	<3*	43*	14*	28*	<3*

Note: NA = not analyzed.

* Semiquantitative optical emission spectroscopy. The following elements are below their respective detection limits (ppm) for all samples (by semiquantitative emission spectroscopy): Na (500), K (700), Ti (30), P (500), Be (1), Ce (40), La (10), Pr (20), Nd (10), Sm (20), Eu (8), Gd (20), Tb (100), Dy (10), Ho (8), Er (10), Yb (10), Li (20), Nb (3), Sn (1), Ta (400), Th (20).

From Haymon, R. M., Hydrothermal processes and products on the Galapagos Rift and East Pacific Rise, in *The Geology of North America,* Vol. N, The Eastern Pacific Ocean and Hawaii, Winterer, E. Hussong, D. M., and Decker, R. W., Eds., Geological Society of America, Boulder, CO, 1989, 173. With permission.

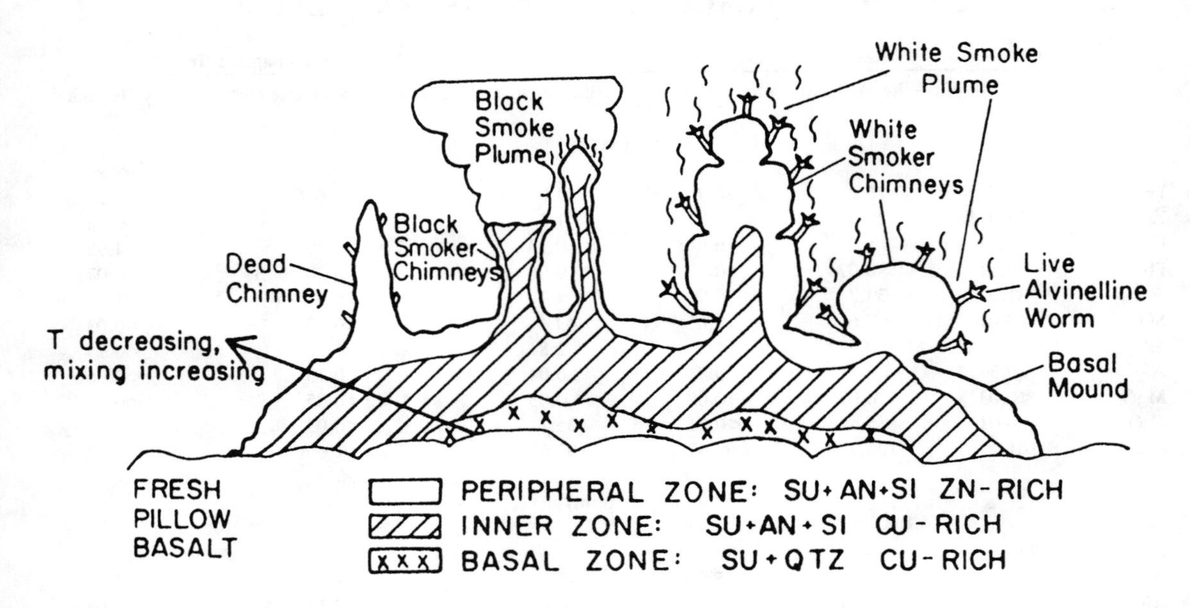

FIGURE 3.4—12. Composite sketch illustrating features of hydrothermal vent deposits on the East Pacific Rise. Mounds of hydrothermal precipitates, accumulated on top of basalt, are surmounted by chimney edifices. Simplified mound zonation includes an outer peripheral zone of anhydrite (AN) + amorphous silica (SI) + Zn-rich sulfide (SU); dominantly, ZnS + FeS_2 is replaced in the interior by an inner zone (cross hatched) of Cu-rich sulfide ($CuFeS_2$ + FeS_2) + minor anhydrite and amorphous silica. The inner zone may be replaced by a basal zone (cross pattern of sulfide ($CuFeS_2$ + FeS_2) + quartz (QTZ). These zones can migrate contingent upon thermochemical conditions. (From Haymon, R. M., Hydrothermal processes and products on the Galapagos Rift and East Pacific Rise, in *The Geology of North America,* Vol. N, The Eastern Pacific Ocean and Hawaii, Winterer, E., Hussong, D. M., and Decker, R. W., Eds., Geological Society of America, Boulder, CO, 1989, 173. With permission.)

STAGE I: SULFATE-DOMINATED STAGE

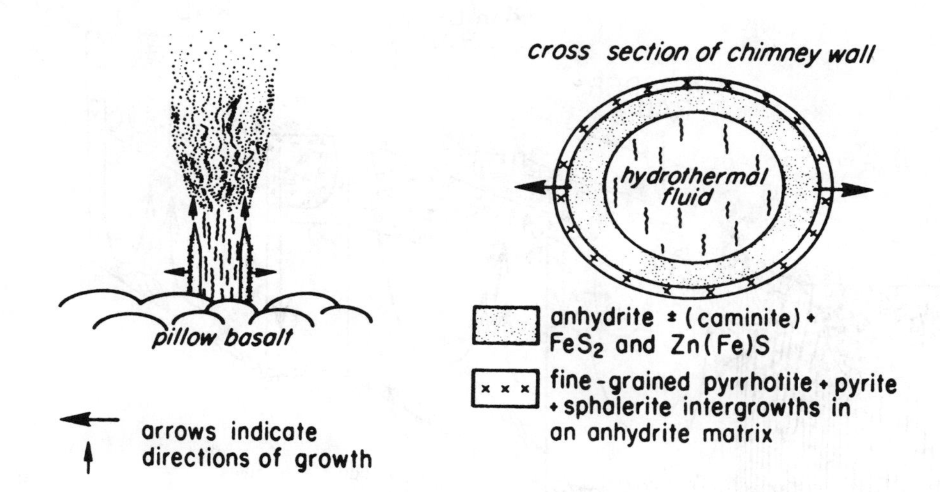

STAGE II: SULFIDE REPLACEMENT STAGE

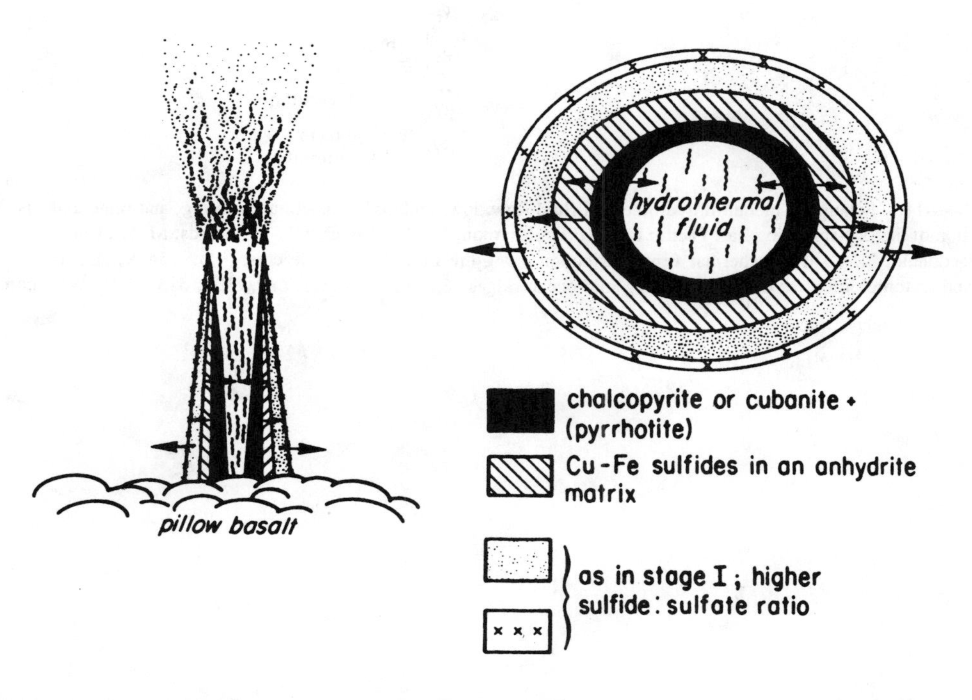

FIGURE 3.4—13. Formation of black smoker (hydrothermal) vents. The first stage of growth involves anhydrite (calcium sulfate) precipitation from seawater to build chimney walls. The second stage entails the sulfide replacement of anhydrite precipitated earlier in the walls. Mineral zonation in the black smoker is evident in the cross section of the chimney wall. (From Haymon, R. M., Hydrothermal processes and products on the Galapagos Rift and East Pacific Rise, in *The Geology of North America,* Vol. N, The Eastern Pacific Ocean and Hawaii, Winterer, E., Hussong, D. M., and Decker, R. W., Eds., Geological Society of America, Boulder, CO, 1989, 173. With permission.)

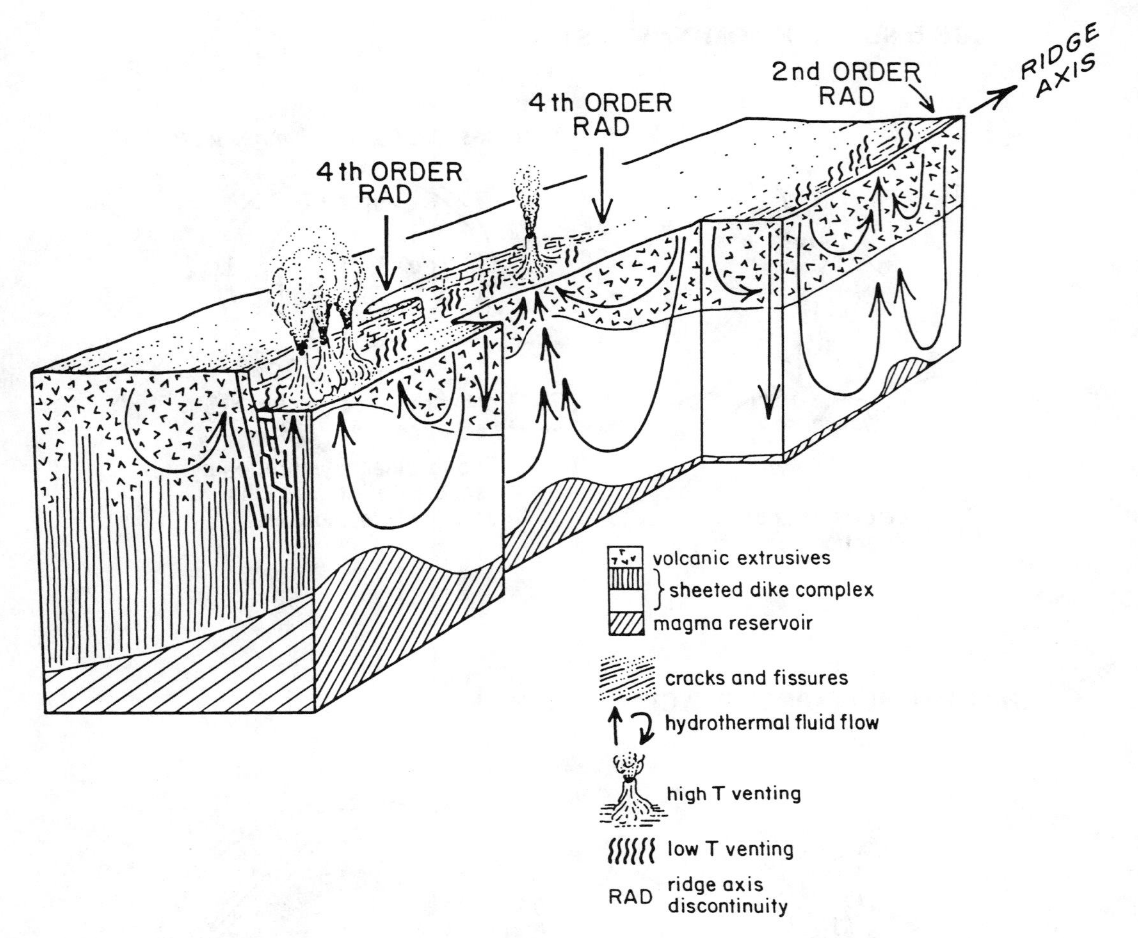

FIGURE 3.4—14. Schematic model of hydrothermal fluid flow, low- and high-temperature venting, and ridge axis discontinuities along a portion of the fast-spreading East Pacific Rise. (From Haymon, R. M., Fornari, D. J., Edwards, M. H., Carbotte, S., Wright, D., and Macdonald, K. C., Hydrothermal vent distribution along the East Pacific Rise crest (9°09′–54′N) and its relationship to magmatic and tectonic processes on fast-spreading mid-ocean ridges, *Earth Planet. Sci. Lett.*, 104, 513, 1991. With permission.)

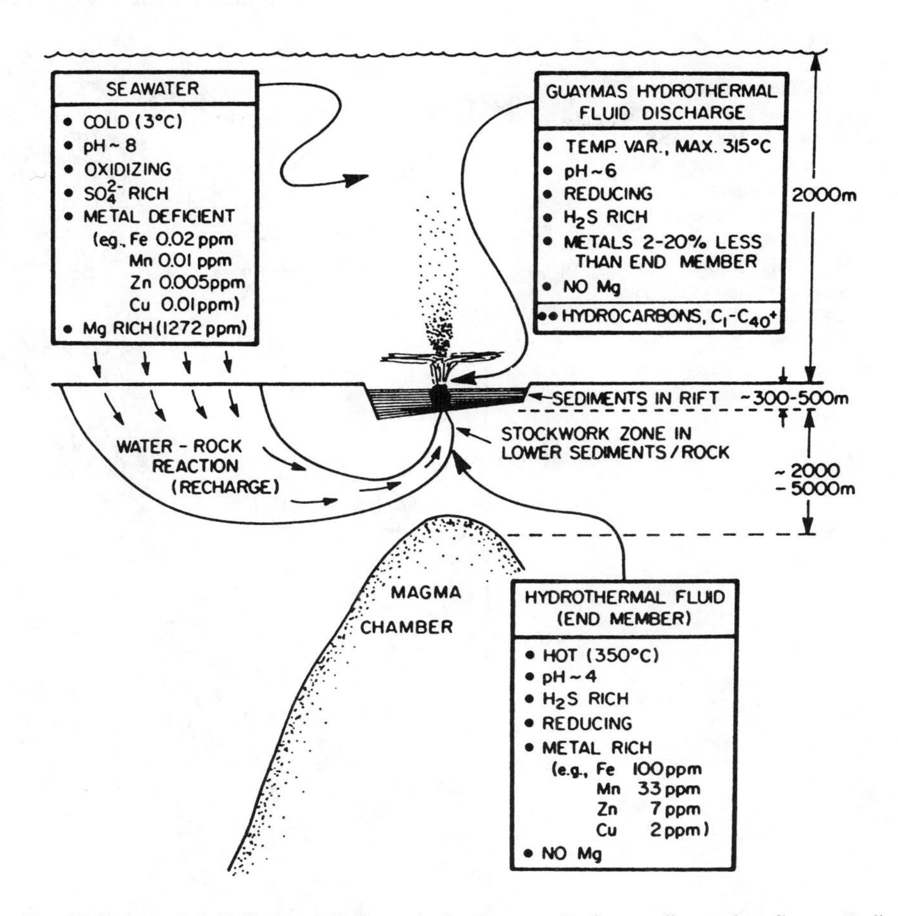

FIGURE 3.4—15. Hydrothermal fluid flow and discharge in the Guaymas Basin, a sedimented seafloor spreading center in the Gulf of California. Hydrothermal fluid compositions and hydrothermal petroleum generation and migration also depicted. (From Simoneit, B. R. T., Hydrothermal petroleum: genesis, migration, and deposition in Guaymas Basin, Gulf of California, *Can. J. Earth Sci.*, 22, 1919, 1985. With permission.)

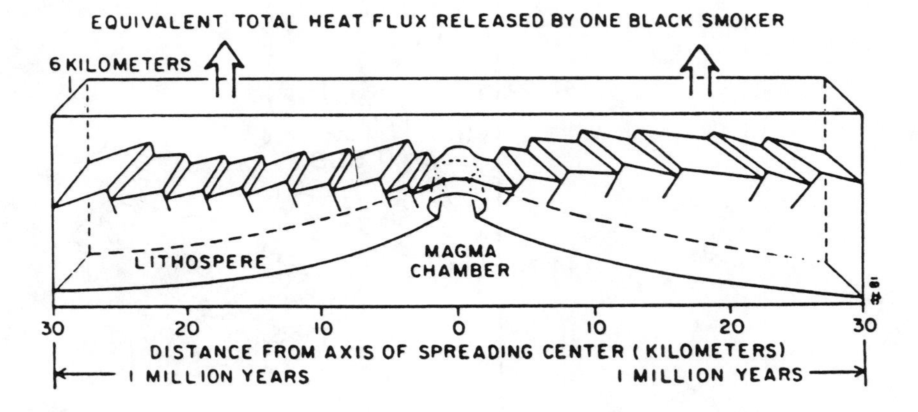

FIGURE 3.4—16. Heat fluxes from high-temperature axial hydrothermal vents make a significant contribution to the total budget of the earth. The heat output of a single "black smoker" vent is equivalent to the total theoretical conductive heat flux for a spreading segment 6 km along axis out to 1 million years age to either side (±30 km at 6 cm/yr). (From Macdonald, K. C., Mid-ocean ridges: fine scale tectonic, volcanic, and hydrothermal processes within the plate boundary zone, *Annu. Rev. Earth Planet. Sci.*, 10, 155, 1982. With permission of Annual Reviews Inc., © 1982.)

FIGURE 3.4—17. Seabeam bathymetry of Juan de Fuca Ridge, showing the location of hydrothermal fields mapped in 1982. (SSS = southern symmetrical segment; AXIAL = Axial seamount; NSR = northern symmetrical ridge; COBB = Cobb overlapper or Surveyor Seamount; END = Endeavour Ridge). (From Tivey, M. A. and Johnson, H. P., High resolution geophysical studies of oceanic hydrothermal systems, *Rev. Aquat. Sci.*, 1, 473, 1989.)

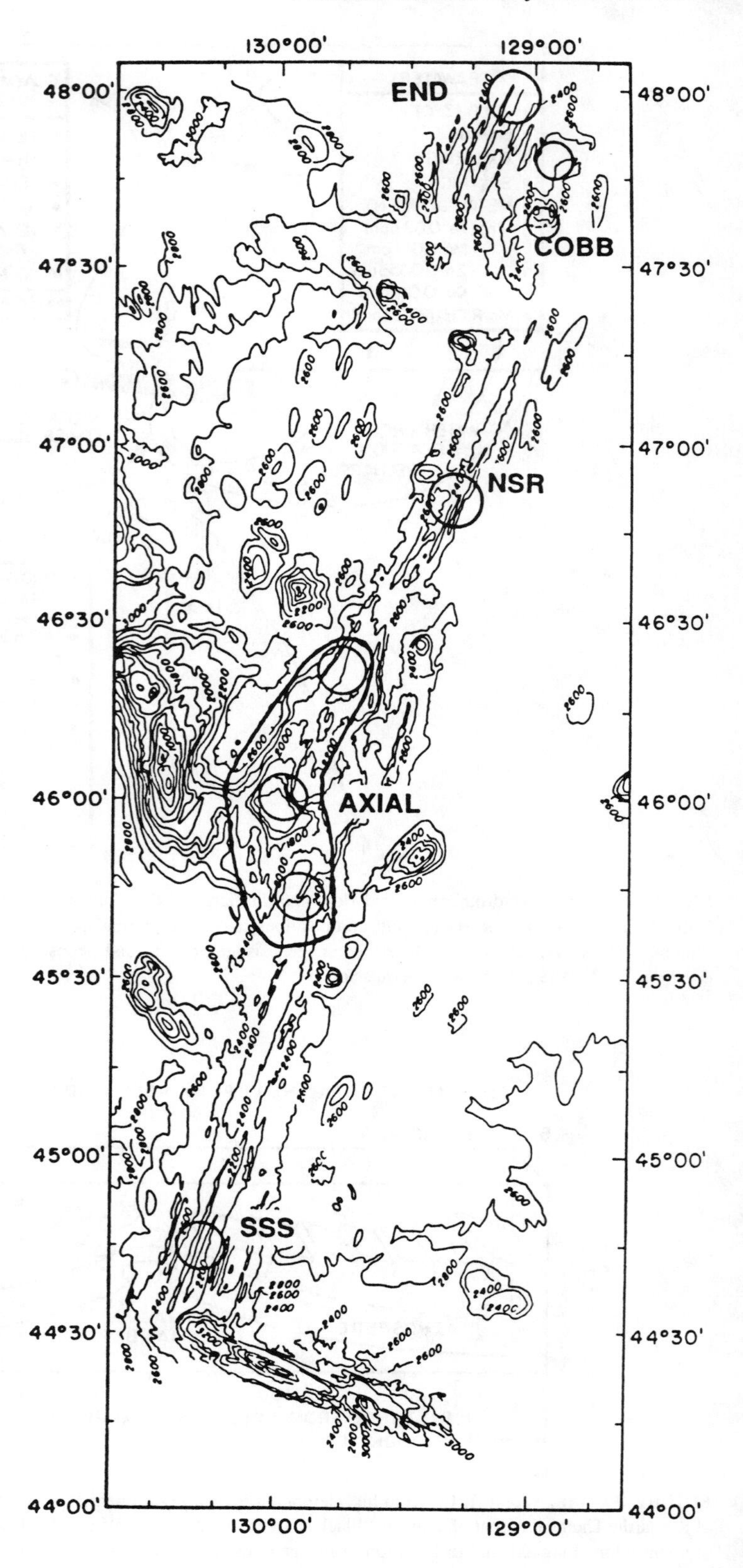

3.5 MID-OCEAN RIDGES: STRUCTURE AND DEVELOPMENT

Table 3.5—1
COMPARISON OF GEOPHYSICAL CHARACTERISTICS OF FAST, INTERMEDIATE, AND SLOW SPREADING CENTERS

	Fast, Intermediate	Slow
1. Axial low-velocity zone	Yes	No
2. Seismic reflection off magma chamber roof	Yes	No measurement
3. Axial low Q zone	Yes	No
4. Microearthquake focal depths	Max. 3 km along axis	Max. 10 km along axis
5. Harmonic tremors	Observed beneath axial zone	Not observed yet
6. Axial gravity anomaly	Magma chamber interpretation	No magma chamber interpretation, mechanically strong lithosphere in axial zone
7. Magnetic anomaly quality	Clear, two dimensional	Unclear, variable, very limited two-dimensionality
8. Magnetic polarity transition widths	Narrow, sharp	Highly variable
9. Inferred eruption rates	~50–500 yrs	~1,000–10,000 yrs
10. High-temp. (300–350°C) hydrothermal activity	Commonly observed on seafloor	Rare, not observed yet (may occur at depth within crust)
11. E-M sounding, resistivity	Pillow layer conductivity 10^3 greater than lower crust, vigorous shallow convection inferred	No deep measurements
12. Thermal models	Most models support a steady-state magma chamber	Most models suggest a transient magma chamber
13. Depth of axial zone	Smoothly varying	Highly variable, rough
14. Axial neovolcanic zone	High degree of continuity along strike (en echelon for intermediate rates)	Highly discontinuous
15. Topographic amplitude in axial zone	Small, 50–100 m	Large, 100–2,000 m
16. Transform fault spacing offset >30 km	90 km (most are OSCs) 500 km	50 km 170 km
17. Maintenance of transform fault pattern	Unstable for small offsets overlapping spreading centers common	Stable
18. Propagating rifts	Common	Rare
19. Petrologic data	Steady-state magma chamber inferred, in agreement with most geophysical data	Steady-state magma chamber inferred, in conflict with most geophysical data

From Macdonald, K. C., A geophysical comparison between fast- and slow-spreading centers: constraints on magma chamber formation and hydrothermal activity, in *Hydrothermal Processes at Seafloor Spreading Centers,* Rona, P. A., Bostrom, K., Laubier, L., and Smith, K. L., Jr., Eds., NATO Conference Series, Plenum Press, New York, 1983, 53. With permission.

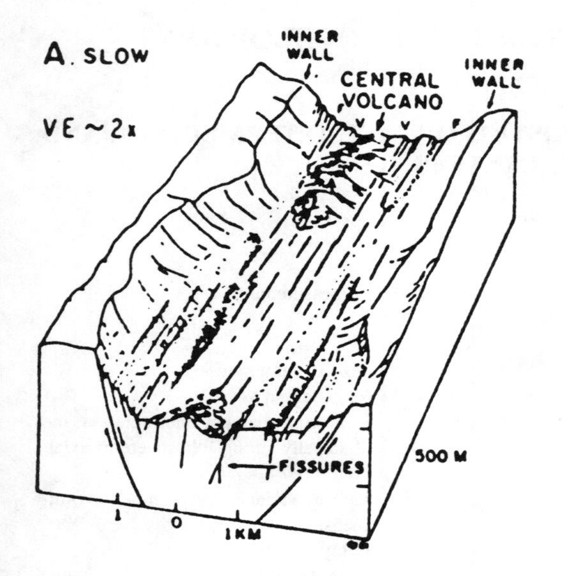

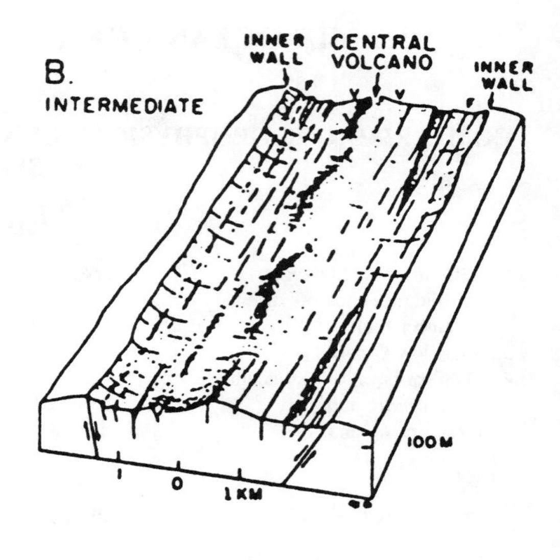

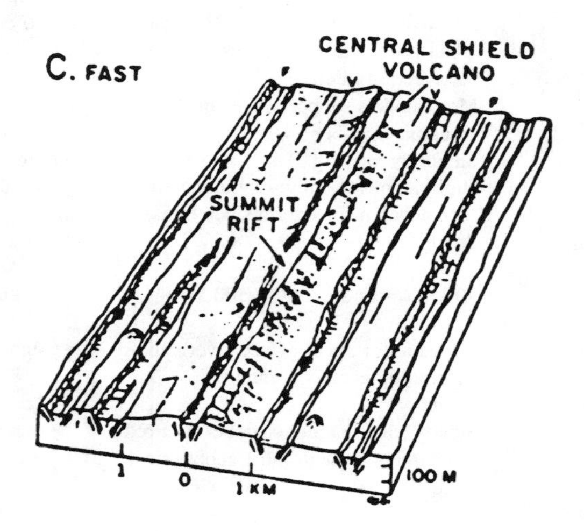

FIGURE 3.5—1. Morphology of the neovolcanic zone at different seafloor spreading centers. The central volcano is highly discontinuous at slow-spreading centers, moderately discontinuous with *en echelon* offsets at intermediate-spreading centers, and essentially continuous at fast-spreading centers. Pillow lavas characterize the central volcanic region at slow- and intermediate-spreading centers. The form of the central volcano at fast-spreading centers resembles a Hawaiian shield volcano with a summit rift. Crustal fissuring occurs most commonly adjacent to the neovolcanic zone, but may be found in the neovolcanic zone as well. (From Macdonald, K. C., Mid-ocean ridges: fine scale tectonic, volcanic, and hydrothermal processes within the plate boundary zone, *Annu. Rev. Earth Planet. Sci.*, 10, 155, 1982. With permission of Annual Reviews Inc., © 1982.)

Table 3.5—2
CHARACTERISTICS OF MID-OCEAN RIDGE SEGMENTATION

	Order 1	Order 2	Order 3	Order 4
		Segments		
Segment length (km)	600 ± 300[a] (400 ± 200)[b]	140 ± 90 (50 ± 30)	50 ± 30 (15 ± 10?)	14 ± 8 (7 ± 5?)
Segment longevity (years)	$>5 \times 10^6$	0.5×10^6 to 5×10^6 (0.5×10^6 to 10×10^6)	$\sim 10^4$ to 10^5 (?)	$\sim 10^2$ to 10^4 (?)
Rate of segment lengthening (long-term migration)	0 to 50 mm/year (0 to 30 mm/year?)	0 to 100 mm/year (0 to 30 mm/year)	Indeterminate — no off-axis trace	Indeterminate — no off-axis trace
Rate of segment lengthening (short-term propagation)	0 to 100 mm/year (?)	0 to 500 mm/year (0 to 50 mm/year)	Indeterminate — no off-axis trace	Indeterminate — no off-axis trace
		Discontinuities		
Type	Transform, large propagating rifts	Overlapping spreading centers (oblique shear zones, rift valley jogs)	Overlapping spreading centers (intervolcano gaps)	Devals, offsets of axial summit caldera (intravolcano gaps)
Offset (km)	>30 km	2 to 30 km	0.5 to 2.0 km	<1 km
Offset age (years)[c]	$>0.5 \times 10^6$ ($>2 \times 10^6$)	$<0.5 \times 10^6$ ($<2 \times 10^6$)	~0	~0
Depth anomaly	300 to 600 m (500 to 2000 m)	100 to 300 m (300 to 1000 m)	30 to 100 m (50 to 300 m)	0 to 50 m (0 to 100 m?)
Off-axis trace	Fracture zone, pseudo-fault	V-shaped discordant zone	None	None
High amplitude magnetization?	Yes	Yes	Rarely (?)	No? (?)
Breaks in axial magma chamber?	Always	Yes, except during OSC[d] linkage? (N.A.)	Yes, except during OSC linkage? (N.A.)	Rarely, 4 of 21 for data through '90 (N.A.)
Break in axial low-velocity zone?	Yes (N.A.)	No, but reduction in volume (N.A.)	Small reduction in volume (N.A.)	Small reduction in volume? (N.A.)
Geochemical anomaly?	Yes	Yes	Usually	30 to 50%
Break in high-temperature venting?	Yes	Yes	Yes (N.A.)	Often (N.A.)

[a] $\pm 1\sigma$.

[b] Where information differs for slow- vs. fast-spreading ridges (>60 mm/year), it is placed in parentheses. N.A. means nonapplicable (that is, a magma chamber has not been detected at slow-spreading centers yet, so a break in the chamber does not apply). A question mark means not presently known or poorly constrained.

[c] Offset age refers to the age of the seafloor, which is juxtaposed to the spreading axis at a discontinuity.

[d] Overlapping spreading center.

From Macdonald, K. C., Scheirer, D. S., and Carbotte, S. M., Mid-ocean ridges: discontinuities, segments, and giant cracks, *Science,* 253, 986, 1991. With permission.

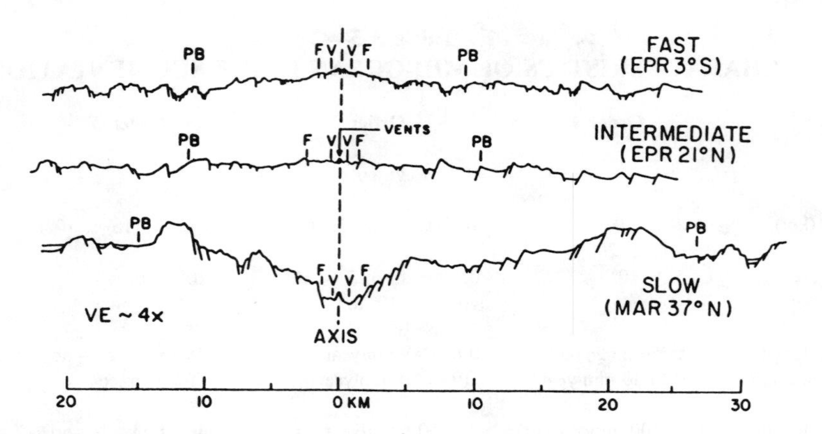

FIGURE 3.5—2. High-resolution deep-tow profiles of mid-ocean ridges at fast, intermediate, and slow rates. Neovolcanic zone bracketed by Vs, zone of fissuring by Fs, plate boundary zone (width of active faulting) by PBs. The neovolcanic zone is generally very narrow (~1 km). Active faulting occurs up to 10 to 30 km off axis. (From Macdonald, K. C., Mid-ocean ridges: fine scale tectonic, volcanic, and hydrothermal processes within the plate boundary zone, *Annu. Rev. Earth Planet. Sci.*, 10, 155, 1982. With permission of Annual Reviews Inc., © 1982.)

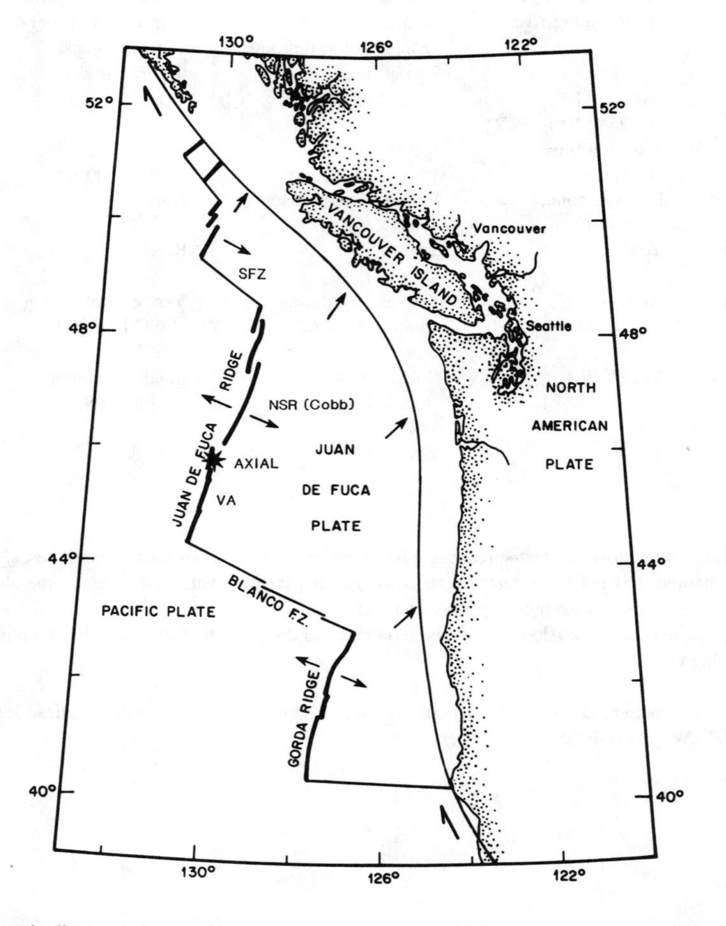

FIGURE 3.5—3. Schematic diagram of the ridge segments of the Juan de Fuca Ridge, in the northeast Pacific. Thick lines indicate the spreading center segments; thinner lines indicate fracture zones and the location of the continental margin. The following abbreviations are used: SFZ, Sovanco Fracture Zone; NSR, Northern Symmetrical Ridge; AXIAL, Axial Seamount Segment; VA, Vance Segment (or segment B in the older literature). (From Johnson, H. P. and Embley, R. W., Axial Seamount: an active ridge axis volcano on the central Juan de Fuca Ridge, *J. Geophys. Res.*, 95, 12, 689, 1990. With permission.)

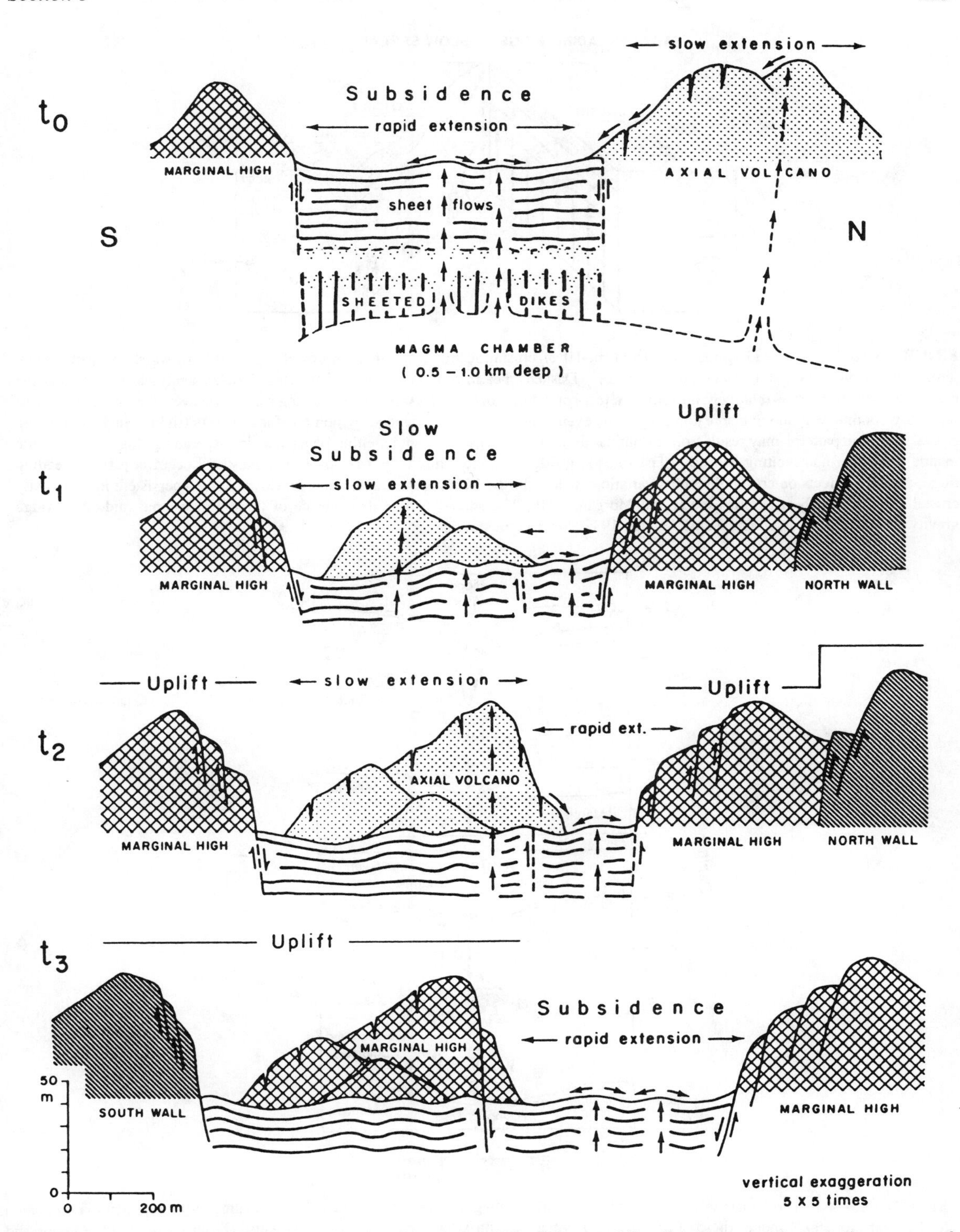

FIGURE 3.5—4. A model of volcanic and structural evolution of the Galapagos Rift valley illustrating the periodic sequence of eruption, subsidence, and extension that occurs at this seafloor spreading center. The sequence t_0 to t_3 is continually repeated for millions of years, with each step probably completed in less than 1000 years. Terrain outlines: dotted, pillow basalts; horizontal lines, sheet flows; drawn vertical arrows, rapid lava transport; dashed arrows, slow lava transport. (From van Andel, T. H. and Ballard, R. D., The Galapagos Rift at 86°W. 2. Volcanism, structure, and evolution of the rift valley, *J. Geophys. Res.*, 84, 5390, 1979. With permission.)

FAST SPREADING RIDGE SLOW SPREADING RIDGE

DISCONT.
DISCONT.
DISCONT.
CRUST
DEPLETED MANTLE

FIGURE 3.5—5. A spreading-rate dependent model of crustal accretion and mantle upwelling. Solid arrows show mantle flow directions. Open arrows show plate motion vectors. Dashed lines in the mantle show isotherms. Gravity analyses indicate that the crustal density structure is relatively uniform at a fast-spreading ridge (left). At a slow-spreading ridge, however, the crustal thickness may vary continuously along a spreading segment, even if the segment is bounded by nontransform offsets (right). Such contrasting crustal accretion patterns may result from a dominantly plumelike upwelling and melting beneath a slow-spreading ridge and sheetlike mantle upwelling and melting beneath a fast-spreading ridge. Smaller amplitude 3-D upwellings may still occur at a fast-spreading ridge, but their effects on crustal thickness variations will be further reduced by along-axis melt flows along a persistent low-viscosity crustal magma chamber. (From Lin, J. and Morgan, J. P., The spreading rate dependence of three-dimensional mid-ocean ridge gravity structure, *Geophys. Res. Lett.*, 19, 13, 1992. With permission.)

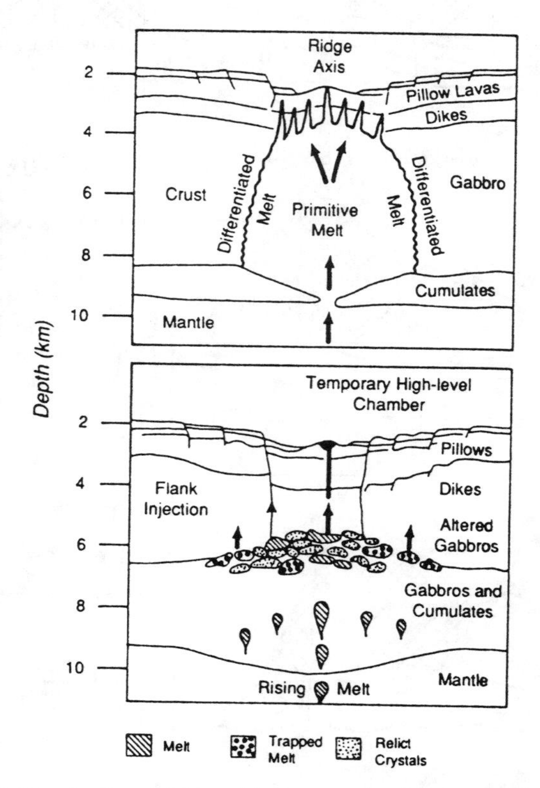

FIGURE 3.5—6. The "infinite onion" model (above) for magma chambers beneath fast-spreading ridge segments is compared to the "infinite leak" model (below) for magma storage beneath slow-spreading ridge segments. (From Cann, J. R., Onions and leaks: magma at mid-ocean ridges, *Oceanus,* 34, 36, 1991. Courtesy of *Oceanus* Magazine/Woods Hole Oceanographic Institution.)

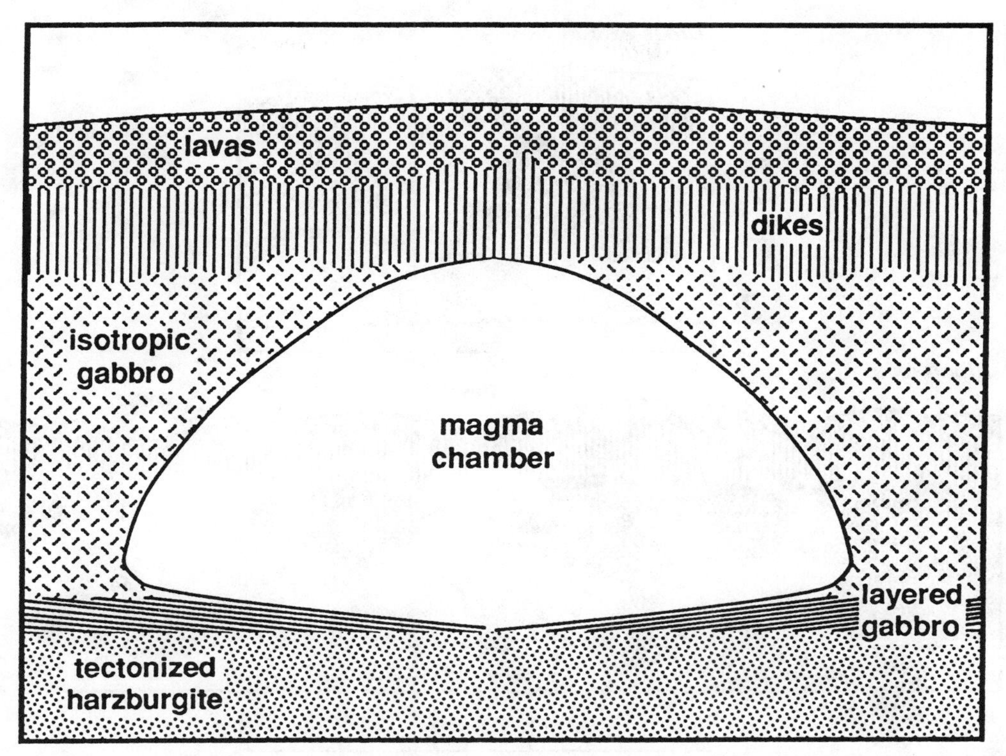

FIGURE 3.5—7. Early models depicted subaxial magma chambers as large, molten features. (From Cann, J. R., A model for oceanic crustal structure developed, *Geophys. J. R. Astron. Soc.*, 39, 169, 1974; copyright by the American Geophysical Union.)

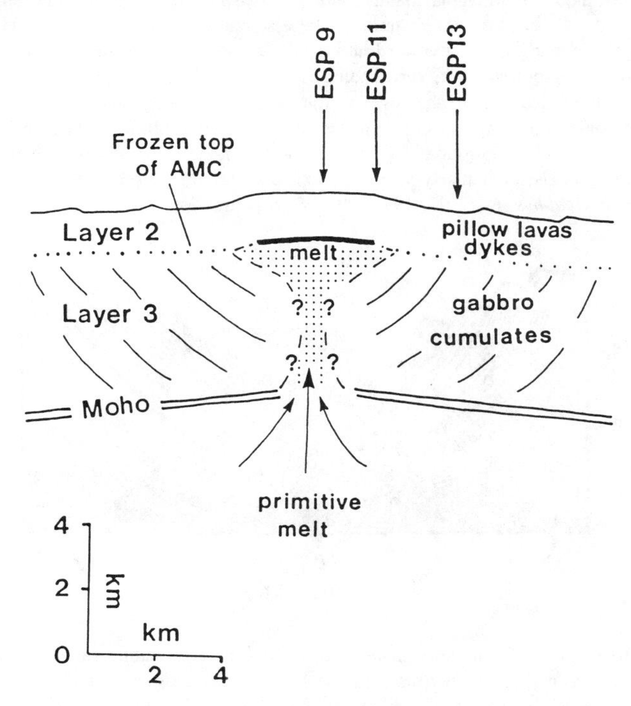

FIGURE 3.5—8. Schematic model of a magma chamber along the fast-spreading East Pacific Rise inferred from multichannel seismic studies. ESP, extended spread profile, AMC, axial magma chamber. (From Detrick, R. S., Buhl, P., Vera, E., Mutter, J., Orcutt, J., Madsen, J., and Brocher, T., Multichannel seismic imaging of a crustal magma chamber along the East Pacific Rise, *Nature*, 326, 35, 1987. With permission.)

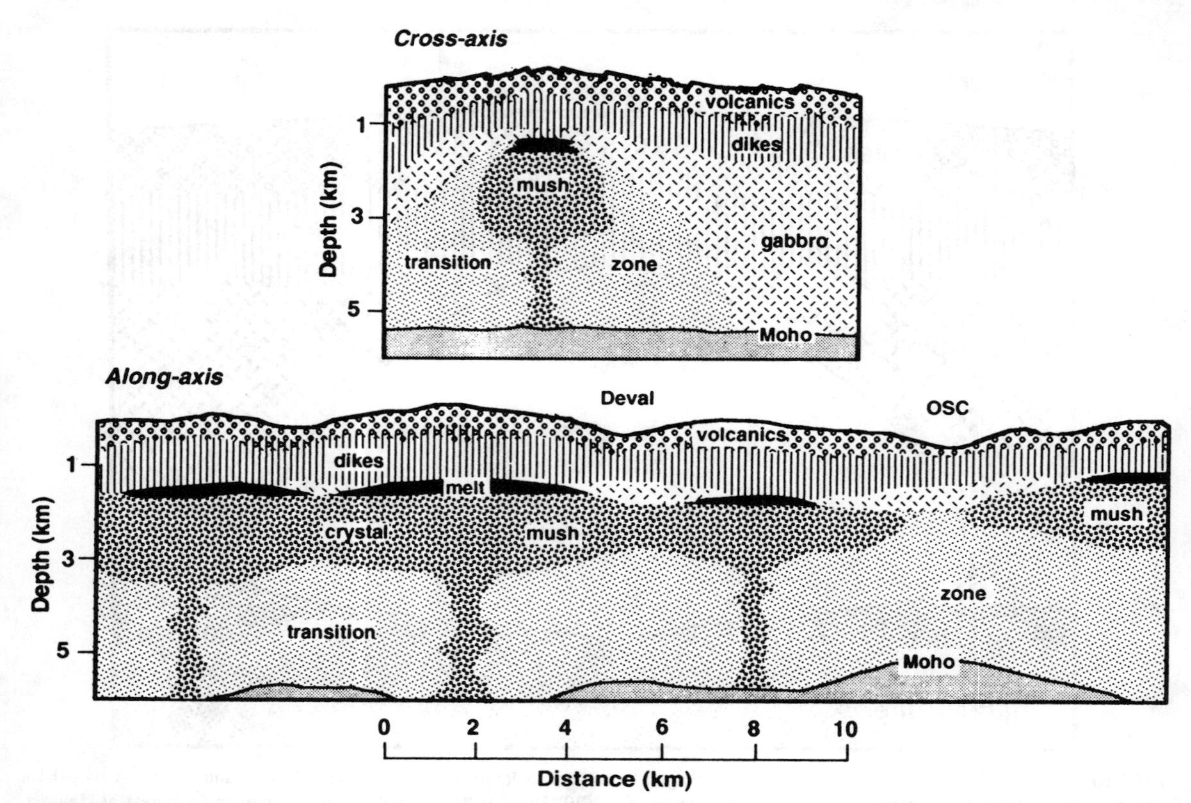

FIGURE 3.5—9. Interpretive model of a magma chamber along a fast-spreading (high magma supply) ridge like the East Pacific Rise based on recent geophysical and petrological constraints. The essential elements of this model are a narrow, sill-like body of melt 1 to 2 km below the ridge axis that grades downward into a partially solidified crystal mush zone, which is in turn surrounded by a transition zone to the solidified, but still hot, surrounding rock. The solidus, which defines the limit of magma, can occur anywhere from the boundary of the mush zone to the edges of the axial low-velocity zone (LVZ). Because the solidus may not be isothermal and significant lithologic variations can occur in the lower layered gabbros, isolated pockets of magma with low melt percentages can occur throughout the LVZ. Eruptions will mainly tap the molten, low-viscosity melt lens. The relative volumes of melt and mush vary along the ridge axis, particularly near ridge axis discontinuities. (From Sinton, J. M. and Detrick, R. S., Mid-ocean ridge magma chambers, *J. Geophys. Res.*, 97, 197, 1992. With permission.)

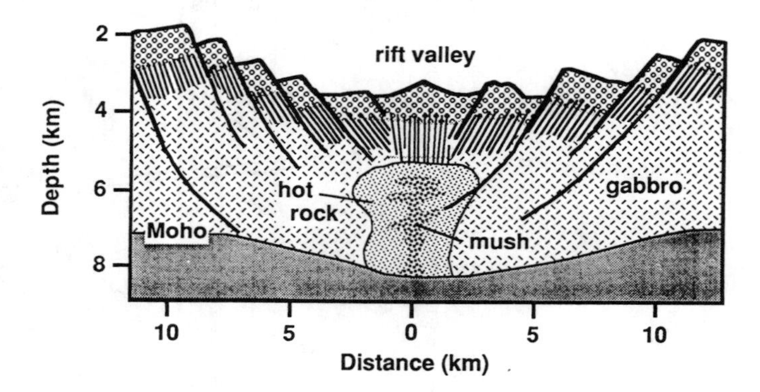

FIGURE 3.5—10. Interpretive model of a magma chamber beneath a slow-spreading (low magma supply) ridge like the Mid-Atlantic Ridge based on recent geophysical and petrological constraints. Such ridges are unlikely to be underlain by an eruptable magma lens in any steady-state sense. A dike-like mush zone is envisioned below the rift valley, forming small sill-like intrusive bodies which progressively crystallize to form oceanic crust. Eruptions will be closely coupled in time to injection events of new magma from the mantle. Faults bordering the rift valley may root in the brittle-ductile transition within the partially molten magma chamber. (From Sinton, J. M. and Detrick, R. S., Mid-ocean ridge magma chambers, *J. Geophys. Res.*, 97, 197, 1992. With permission.)

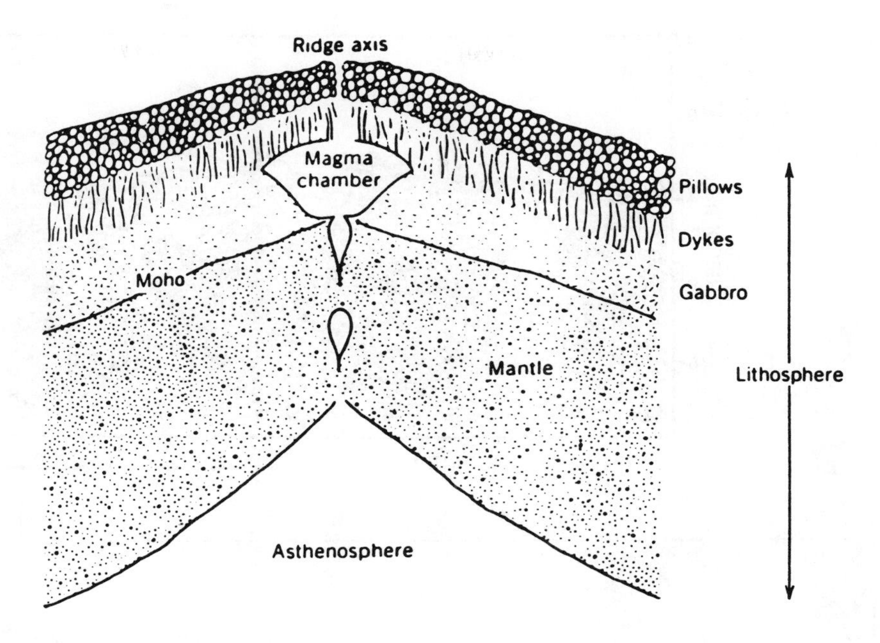

FIGURE 3.5—11. Section of ocean crust at the ridge axis showing the location of the subaxial magma chamber. (From Anderson, R. N., *Marine Geology: A Planet Earth Perspective,* John Wiley & Sons, New York, 1986, 71. Reproduced by permission of copyright © owner, John Wiley & Sons, Inc.)

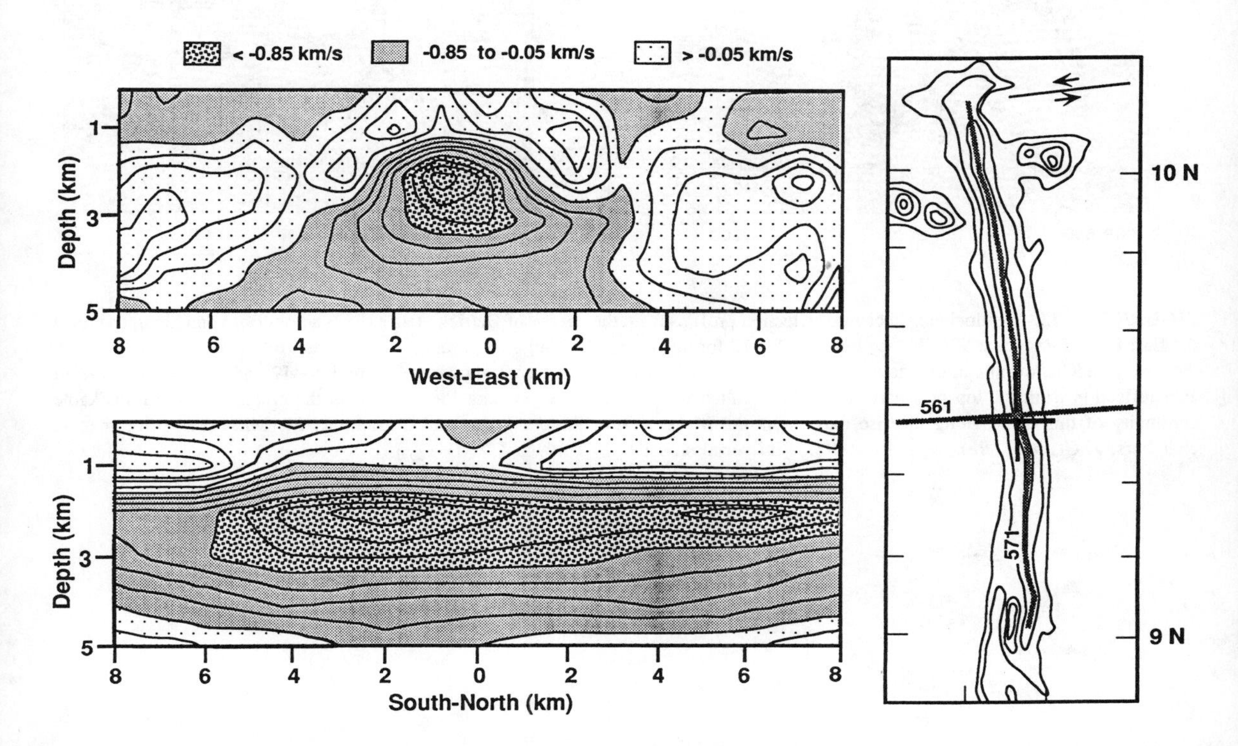

FIGURE 3.5—12. Vertical sections displaying the anomalous P-wave velocity structure across (top) and along (bottom) the East Pacific Rise crest near 9°30′N from the seismic tomography study of Toomey et al. (1990). The location of these sections is depicted by the solid lines in the map on the right. The largest velocity anomaly (−0.85 km/s) is confined to a narrow (<2 km-wide), thin (<1 to 1.5-km thick) zone in the mid-crust beneath the rise axis. This body is surrounded by a broader seismic low-velocity zone (LVZ), shaded gray, which has a width of 10 to 12 km and extends to the base of the crust. The relatively small velocity anomaly associated with the bulk of this LVZ precludes the existence of a large molten magma chamber as envisioned in the model shown in Figure 3.5—8. (From Sinton, J. M. and Detrick, R. S., Mid-ocean ridge magma chambers, *J. Geophys. Res.,* 97, 197, 1992. With permission.)

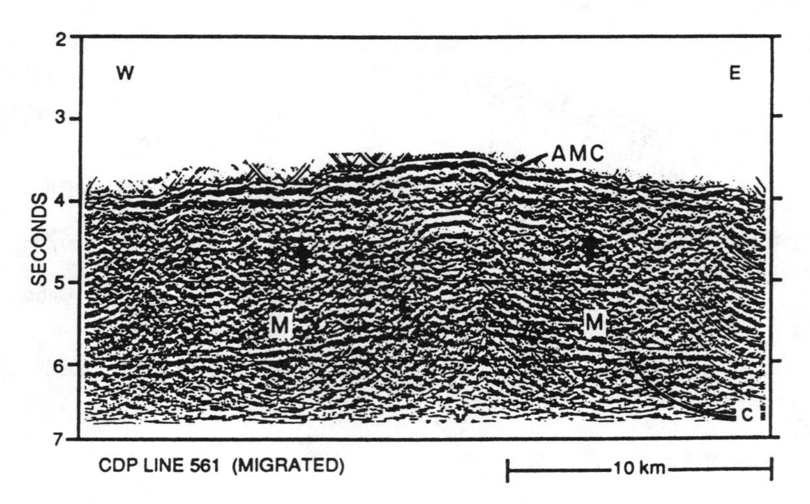

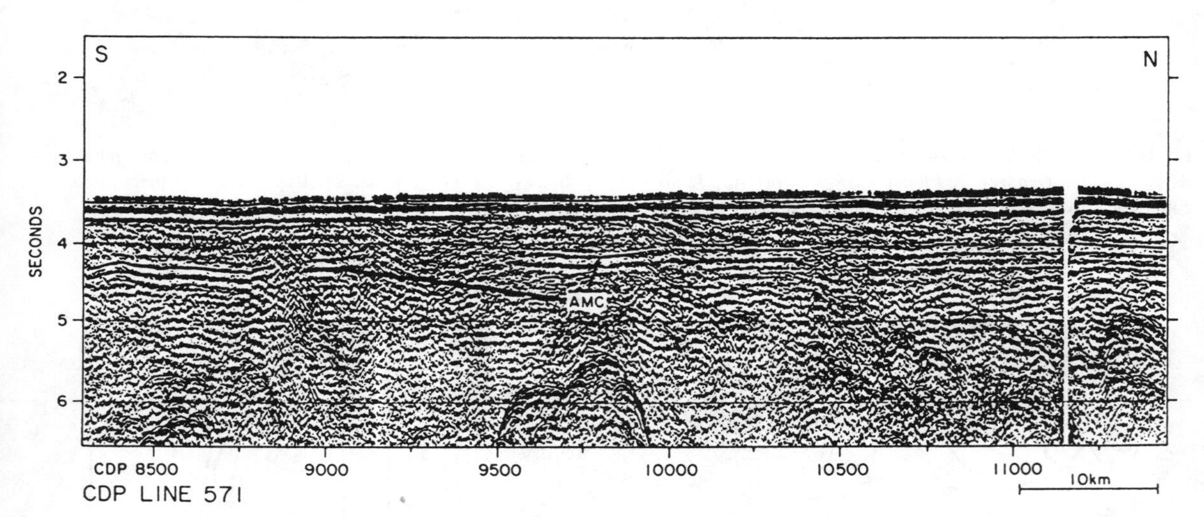

FIGURE 3.5—13. Multichannel seismic reflection profiles from the survey of Detrick et al. (1987) across (top) and along (bottom) the East Pacific Rise near 9°30′N (see Figure 3.5—12 for the location). A high-amplitude, subhorizontal reflector is present beneath the rise axis. This event can be unequivocally tied to the top of the seismic LVZ shown in Figure 3.5—12 and has been interpreted as a reflection from the top of a narrow, sill-like molten body or axial magma chamber (AMC) in the crust. Note the remarkable continuity of this event along the rise axis. M is the Moho. (From Sinton, J. M. and Detrick, R. S., Mid-ocean ridge magma chambers, *J. Geophys. Res.,* 97, 197, 1992. With permission.)

3.6 SEISMIC AND MAGNETIC DATA

Table 3.6—1
MODELS OF OCEANIC SEISMIC STRUCTURE

Three-layer model			Multiple-layer model*		
Layer	Velocity V_p, km/s	Thickness, km	Layer	Velocity V_p, km/s	Thickness, km
1	~2.0	~0.5	1	1.7–2.0	0.5
2	5.07 ± 0.63	1.71 ± 0.75	2A	2.5–3.8	0.5–1.5
			2B	4.0–6.0	0.5–1.5
3	6.69 ± 0.26	4.86 ± 1.42	3A	6.5–6.8	2.0–3.0
			3B	7.0–7.7	2.0–5.0
Mantle	8.13 ± 0.24	—	Mantle	8.1	—

* Houtz and Ewing (1976) proposed dividing layer 2 into layers 2A, B, and C having average velocities of 3.64, 5.19, and 6.09 km/s, respectively.

From Salisbury, M. H., Stephen, R., Christensen, N. I., Francheteau, J., Hamans, Y., Hobart, M., and Johnson, D., The physical state of the upper levels of Cretaceous oceanic crust from the results of logging, laboratory studies, and the oblique seismic experiment at DSDP sites 417 and 418, in *Deep Drilling Results in the Atlantic Ocean: Ocean Crust,* Talwani, M., Harrison, C. G., and Hayes, D. E., Eds., Technical Volume, American Geophysical Union, Washington, D.C., 1979, 113. With permission.

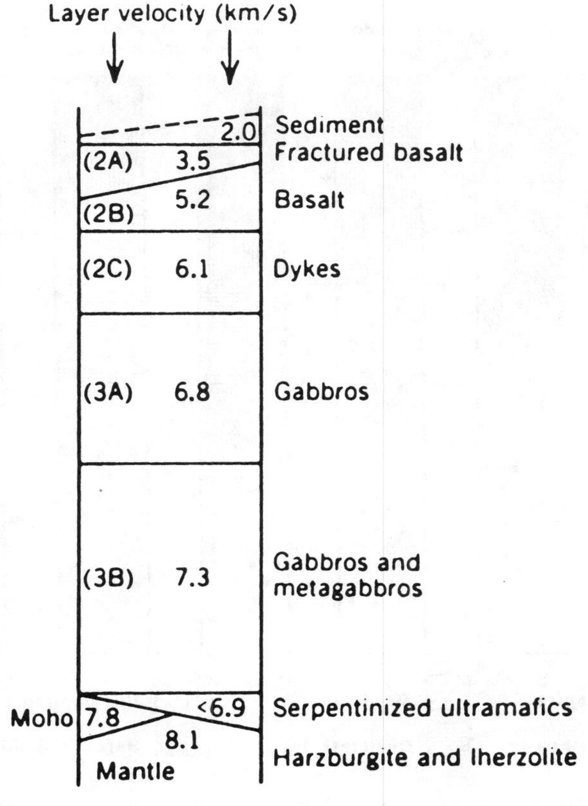

FIGURE 3.6—1. Seismic velocities recorded in the oceanic crust and upper mantle. (From Anderson, R. N., *Marine Geology: a Planet Earth Perspective,* John Wiley & Sons, New York, 1986, 68. Reproduced by permission of copyright © owner, John Wiley & Sons, Inc.)

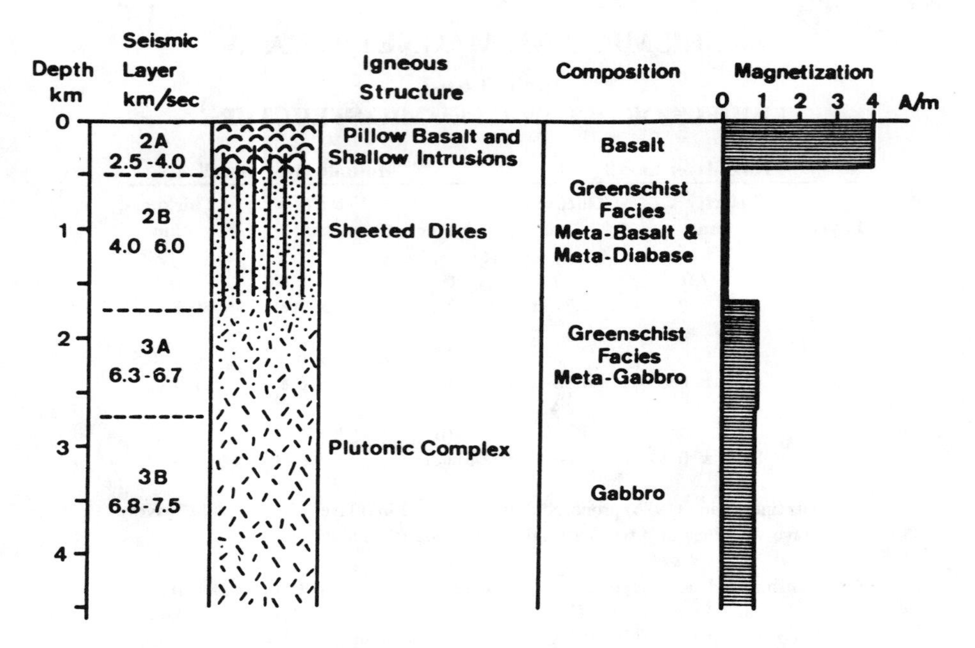

FIGURE 3.6—2. Model of the magnetization structure of the oceanic crust. (From Lowrie, W., Geomagnetic reversals and ocean crust magnetization, in *Deep Drilling Results in the Atlantic Ocean: Ocean Crust,* Talwani, M., Harrison, C. G., and Hayes, D. E., Eds., Technical Volume, American Geophysical Union, Washington, D.C., 1979, 135. With permission.)

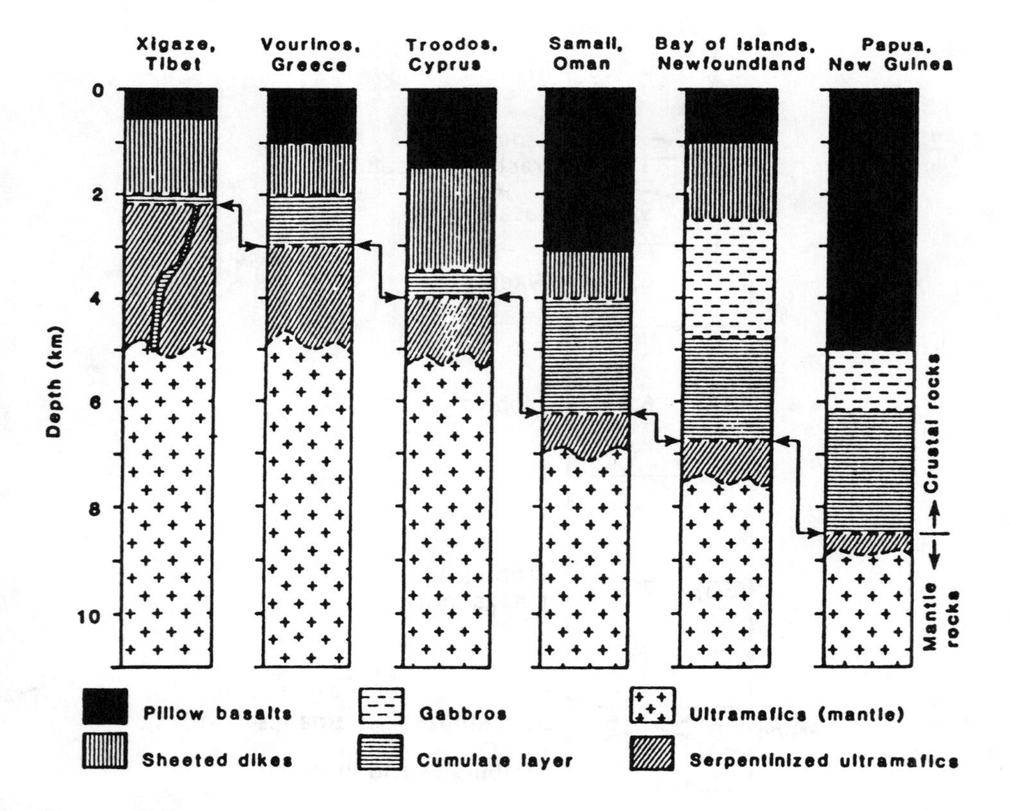

FIGURE 3.6—3. Diagrammatic structure of six ophiolites (obducted oceanic crust) typified by the sequence of pillow basalts, sheeted dikes, gabbros, and ultramafics. The layered sequence of the ophiolites correlates closely with the vertical structure of the oceanic lithosphere. (Modified from Lewis, B. T. R., The process of formation of ocean crust, *Science,* 220, 151, 1983. With permission.)

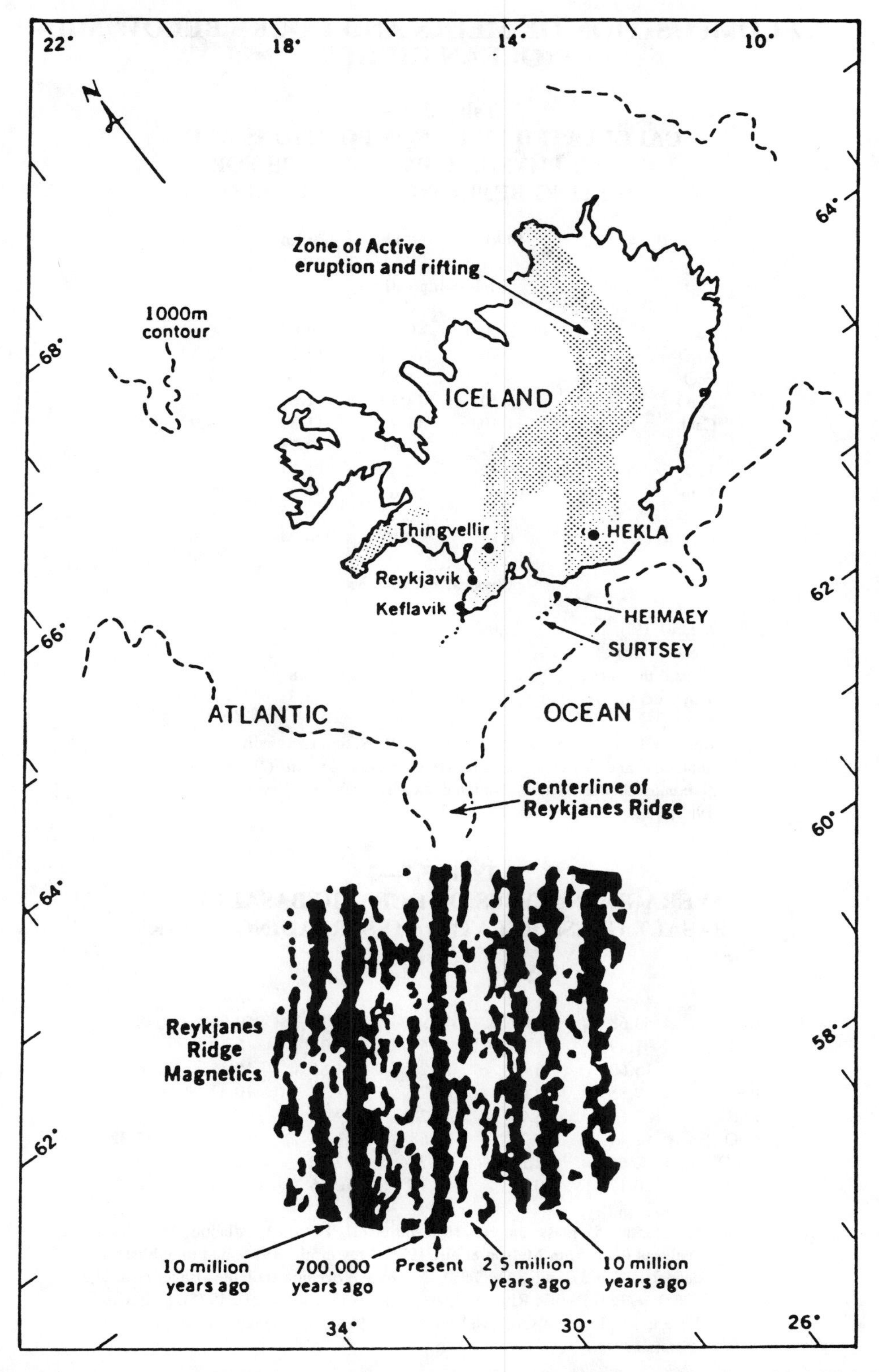

FIGURE 3.6—4. Magnetic striped patterns of the seafloor. The black stripes represent bands of seafloor with the rock magnetized as it is today, with the north pole pointing north. White stripes represent rock magnetized so that the north pole of the rock points in the reverse direction, to the south. This pattern can be used to age-date the seafloor. (From Anderson, R. N., *Marine Geology: a Planet Earth Perspective,* John Wiley & Sons, New York, 1986. Reproduced by permission of copyright © owner, John Wiley & Sons, Inc.)

3.7 COMPOSITION OF MELTS AND ROCKS BELOW MID-OCEAN RIDGES

Table 3.7—1
CALCULATED MELT COMPOSITIONS AND OTHER PHYSICAL PARAMETERS FOR MELTING BENEATH OCEAN RIDGES

P_o =	40 kb	30 kb	20 kb	14 kb
	Melt compositions			
SiO_2	48.50	49.03	49.85	50.40
Al_2O_3	12.70	14.20	15.20	16.40
FeO	9.53	8.80	8.22	7.70
MgO	15.10	13.60	11.80	10.20
CaO	11.10	11.90	11.30	9.90
Na_2O	1.31	1.61	2.13	2.70
TiO_2	0.72	0.88	1.14	1.40
Sum	99.0	100.0	99.6	98.7
Mg#	0.74	0.73	0.72	0.70
CaO/Al_2O_3	0.87	0.83	0.74	0.60
	Physical parameters			
Mean F (%)	20.0	15.6	10.8	7.8
Mean P (kb)	16.4	11.7	7.3	4.9
Crustal thickness	22.5	13.8	6.8	3.6
Water depth	0	1.7	3.3	4.1

Reprinted with permission from Detrick, R. S. and Langmuir, C. H., The geometry and dynamics of magma chambers, in *Mid-Ocean Ridge: A Dynamic Global System,* National Academy Press, Washington, D.C., 1988, 123.

Table 3.7—2
AVERAGE ANALYSES OF OCEANIC BASALTS AND BASALT GLASS RELATED TO SPREADING RIDGES

	1	2	3	4	5	6
SiO_2	50.68	50.67	49.94	50.19	50.93	49.61
TiO_2	1.49	1.28	1.51	1.77	1.19	1.43
Al_2O_3	15.60	15.45	17.25	14.86	15.15	16.01
FeO*	9.85	9.67	8.71	11.33	10.32	11.49
MgO	7.69	8.05	7.28	7.10	7.69	7.84
CaO	11.44	11.72	11.68	11.44	11.84	11.32
Na_2O	2.66	2.51	2.76	2.66	2.32	2.76
K_2O	0.17	0.15	0.16	0.16	0.14	0.22

Note: **1**, Atlantic, 51 glass analyses (Melson et al., 1975); **2**, Atlantic, 155 glass analyses (data from Melson et al., 1975, Frey et al., 1974, and unpublished data from leg 37 and FAMOUS); **3**, average oceanic tholeiite (Engel et al., 1965); **4**, East Pacific Rise, 38 glass analyses (Melson et al., 1975); **5**, Indian Ocean, 12 glass analyses (Melson et al., 1975); **6**, average oceanic tholeiite (Cann, 1971).

From Bryan, W. B., Thompson, G., Frey, F. A., and Dickey, J. S., Inferred geologic settings and differentiation in basalts from the deep-sea drilling project, *J. Geophys. Res.*, 81, 4285, 1976. With permission.

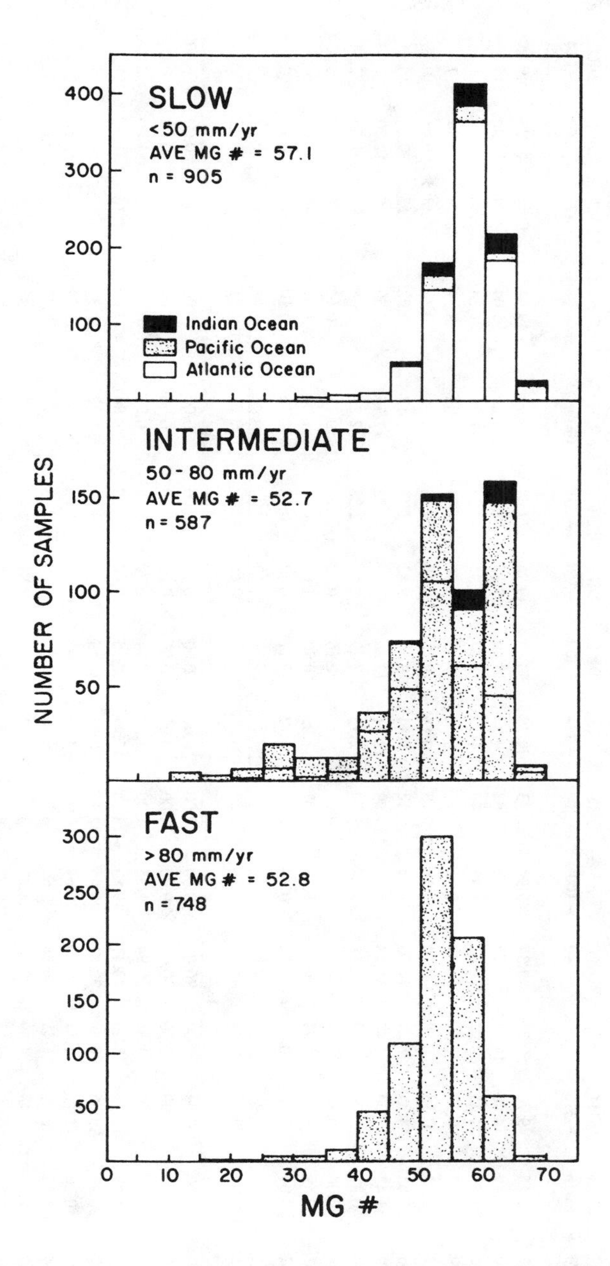

FIGURE 3.7—1. Histograms of over 2200 glass compositions from mid-ocean ridges at three different spreading rate ranges, keyed according to ocean. Slow-spreading centers include the Mid-Atlantic Ridge, Cayman Trough, Galapagos Rift spreading center west of about 96°W, Southwest Indian Ridge, Central Indian Ridge, and Carlsbad Ridge. Intermediate-spreading centers include the Galapagos spreading center east of 96°W, East Pacific Rise north of the Rivera Transform and west of 145°W (Pacific Antarctic plate boundary), Juan de Fuca spreading center, and Southeast Indian Ridge. All fast-spreading ridge lavas derive from the East Pacific Rise. (From Sinton, J. M. and Detrick, R. S., Mid-ocean ridge magma chambers, *J. Geophys. Res.*, 97, 197, 1992. With permission.)

Table 3.7—3
CHEMICAL ANALYSES (wt %) OF BASALTIC ROCKS FROM THE RIFT VALLEY IN THE ATLANTIC OCEAN NEAR 36°N

	Picritic basalts		Olivine basalts													
	DR12 −320	DR12 −316	DR11 −315	DR11 −315A	DR4 −303	DR5 −100D	DR5 −304	DR8 −314	DR1 −122	DR9 −308	DR9 −319	DR9 −322 (inter.)	DR9 −322 (margin)	DR2 −174	DR9 −309	DR1 −124
SiO_2	46.27	47.45	49.70	50.24	49.34	49.93	50.33	49.79	50.45	49.96	49.83	49.92	49.94	50.62	49.28	49.55
Al_2O_3	14.20	13.49	14.86	14.72	14.75	14.53	14.92	14.86	15.40	14.83	14.82	15.29	15.04	14.84	15.30	15.19
Fe_2O_3	1.20	1.04	1.65	0.98	0.91	1.45	1.43	1.39	1.37	1.46	1.35	1.66	1.34	1.56	1.68	1.41
FeO	7.69	7.72	9.28	9.50	7.42	8.62	7.71	8.32	7.41	8.06	9.37	8.03	8.24	8.76	1.44	7.48
MnO	0.17	0.17	0.18	0.18	0.18	0.18	0.18	0.18	0.18	0.18	0.18	0.18	0.18	0.18	0.18	0.18
MgO	16.06	16.65	8.23	8.24	11.00	9.22	10.12	9.54	9.33	9.82	8.44	9.12	9.02	8.12	9.20	10.53
CaO	11.89	11.42	11.90	11.82	11.48	12.08	11.89	12.51	12.26	12.43	11.82	12.55	12.59	11.95	12.70	12.18
Na_2O	1.98	0.37	2.28	2.14	1.55	2.34	1.79	2.18	2.23	2.01	2.28	2.23	2.08	1.91	2.55	2.51
K_2O	0.09	0.08	0.22	0.23	0.16	0.23	0.17	0.22	0.16	0.12	0.23	0.13	0.13	0.24	0.17	0.15
TiO_2	0.62	0.59	1.45	1.44	0.86	1.27	1.07	1.09	0.90	1.02	1.44	1.04	1.04	1.41	1.02	0.91
P_2O_5	0.07	0.09	0.18	0.18	0.10	0.18	0.10	0.12	0.11	0.12	0.18	0.10	0.10	0.18	0.13	0.12
PF	0.48	0.53	0.62	0.55	0.84	0.72	0.69	0.52	0.51	0.65	0.57	0.63	0.62	0.70	0.79	0.51
Total	100.74	100.59	100.54	100.21	98.58	100.74	100.39	100.71	100.30	100.65	100.50	100.87	100.31	100.46	100.43	100.71
Norm																
Qtz	0.00	0.00	0.00	0.00	0.00	0.00	0.00	0.00	0.00	0.00	0.00	0.00	0.00	1.36	0.00	0.00
Ne	2.41	0.00	0.00	0.00	0.00	0.00	0.00	0.00	0.00	0.00	0.00	0.00	0.00	0.00	0.00	0.00
Or	0.53	0.47	1.30	1.35	0.94	1.35	1.00	1.30	0.94	0.70	1.35	0.76	0.76	1.41	1.00	0.88
Ab	12.29	11.59	19.29	18.10	13.11	19.80	15.14	18.44	18.86	17.00	19.29	18.86	17.60	16.16	21.57	21.23
An	29.59	30.42	29.66	29.88	32.81	28.46	32.17	30.11	31.53	31.08	29.52	31.32	31.31	31.21	29.79	29.73
Di wo	12.08	10.70	11.77	11.51	9.80	12.64	10.92	13.01	11.92	12.43	11.66	12.64	12.72	11.23	13.50	12.48
Di en	8.44	7.50	6.75	6.45	6.38	7.66	6.99	8.00	7.51	7.78	6.68	7.79	7.70	6.55	8.56	8.10
Di fs	2.62	2.29	4.48	4.59	2.74	4.29	3.20	4.26	3.66	3.89	4.46	4.11	4.33	4.14	4.08	3.52
Hy en	0.00	6.60	10.90	12.15	17.74	8.27	16.66	7.53	10.86	11.18	9.73	8.54	10.15	13.67	2.69	4.77
Hy fs	0.00	2.01	7.24	8.64	7.63	4.63	7.63	4.01	5.29	5.60	6.50	4.51	5.71	8.65	1.28	2.07
Fo	22.11	19.16	1.98	1.33	2.29	4.92	1.07	5.75	3.39	3.84	3.22	4.46	3.23	0.00	8.16	9.35
Fa	7.56	6.44	1.45	1.04	1.08	3.04	0.54	3.38	1.82	2.12	2.37	2.59	2.00	0.00	4.28	4.48
Mt	1.73	1.50	2.39	1.42	1.31	2.10	2.07	2.01	1.98	2.11	1.95	2.40	1.94	2.26	2.43	2.04
Hm	1.17	1.12	2.75	2.73	1.63	2.41	2.03	2.07	1.70	1.93	2.73	1.97	1.97	2.67	1.93	1.72
Ap	0.16	0.21	0.42	0.42	0.23	0.42	0.23	0.28	0.25	0.28	0.42	0.23	0.23	0.42	0.30	0.28

Table 3.7—4
CHEMICAL CHARACTERISTICS OF SOME POTENTIAL PRIMARY OCEAN-FLOOR BASALTIC MAGMAS (RECALCULATED TO 100% ON AN ANHYDROUS BASIS)

	1	2	3	4	5	6
SiO_2	49.7	49.5	49.1	50.8	50.0	50.5
TiO_2	0.72	0.81	0.62	0.62	0.65	0.76
Al_2O_3	16.4	15.7	16.5	14.4	14.5	15.6
FeO	7.89	7.45	8.78	6.89	8.26	7.3
MnO	0.12	0.15	0.15	0.10	n.d.	0.2
MgO	10.1	10.0	10.3	12.0	11.7	9.4
CaO	13.2	13.0	12.4	13.6	13.0	13.7
Na_2O	2.00	1.95	1.92	1.43	1.71	2.1
K_2O	0.01	0.17	0.07	0.07	0.03	0.4
P_2O_5	—	0.08	0.06	—	—	—
Cr_2O_3	0.07	0.14	0.06	0.10	—	—
Mg′-value	0.72	0.73	0.70	0.77	(0.72)	0.72
CaO/Al_2O_3	0.81	0.82	0.75	0.94	0.90	0.88
Ni ppm	320	249	232	—	—	—
Sm ppm	1.60	1.95	1.37	—	—	—
La/Sm	0.63	1.81	1.57	—	—	—

1. **Frey et al.,** *J. Geophys. Res.*, 79, 5507, 1974; 3–14.
2. **Rhodes** (unpublished data); Chain 43 #23, 45°N.
3. **Langmuir et al.,** *Earth Planet. Sci. Lett.*, 36, 133, 1977; FAMOUS, 527–1–1.
4. **Donaldson and Brown,** *Earth Planet. Sci. Lett.*, 37, 81, 1977; Avg. melt inclusion in spinel.
5. **Dungan and Rhodes,** *Contr. Mineral. and Petrol.*, 79, 1979; interpolated from melt inclusion data.
6. **Watson,** *J. Volcanol. Geotherm. Res.*, 1, 73, 1976; interpolated from melt inclusion data.

From Rhodes, J. M. and Dungan, M. A., The evolution of ocean-floor basaltic magmas, in *Deep Drilling Results in the Atlantic Ocean: Ocean Crust,* Talwani, M., Harrison, C. G., and Hayes, D. E., Eds., Technical Volume, American Geophysical Union, Washington, D.C., 1979, 262. With permission.

Table 3.7—5
OCEANIC ULTRAMAFIC ROCKS (wt. %)

	1	2	3	4
SiO_2	45.4	45.9	46.07	45.95
TiO_2	0.1	0.2	0.24	0.17
Al_2O_3	1.7	3.7	3.60	3.69
(FeO)	8.3	8.3	8.35	8.29
MnO	0.1	0.1	0.11	0.11
MgO	42.9	38.6	38.42	38.53
CaO	0.7	2.3	2.17	2.34
Na_2O	0.2	0.3	0.49	0.37
K_2O	0.05	0.05	0.05	0.05
Cr_2O_3	0.3	0.3	0.3	0.3
NiO	0.2	0.2	0.2	0.2

1. Average oceanic harzburgite (75 analyses).
2. Average oceanic lherzolite (64 analyses).
3. Calculated lherzolite as the sum of 87% average harzburgite + 13% of primary tholeiitic melt under 8–10 kbar.
4. Calculated lherzolite as the sum of 86.6% average harzburgite + 13.4% of primary tholeiitic melt under 3–5 kbar.

From Dmitriev, L. V., Sobolex, A. V., and Suschenskaja, N. M., The primary melt of the oceanic tholeiite and the upper mantle composition, in *Deep Drilling Results in the Atlantic Ocean: Ocean Crust,* Talwani, M., Harrison, C. G., and Hayes, D. E., Eds., Technical Volume, American Geophysical Union, Washington, D.C., 1979, 302. With permission.

Table 3.7—6
PETROGRAPHIC CHARACTERISTICS OF UNALTERED SUBMARINE BASALT AND BASALT GLASS

	Group I		Group II	
	Unfractionated	**Fractionated**	**Unfractionated**	**Fractionated**
Plagioclase	An_{90-80}; molecular Mg > Fe; K_2O < 0.02; zoning inconspicuous	An_{80-60}; molecular Mg ≤ Fe; K_2O < 0.04; minor zoning	An_{85-60}; molecular Mg < Fe; K_2O > 0.05; marginal zoning conspicuous	An_{60}; alkali feldspar; molecular Mg < Fe; K_2O > 0.10; prominent marginal zoning or alkali feldspar overgrowths
Olivine	Fo_{90-85}; present as phenocrysts and groundmass phase and as skeletal crystals in glass	Fo_{85-80}; present as phenocrysts, absent in groundmass or as quench crystals in glass	Fo_{85-80}; present as phenocrysts, rare in groundmass or as quenched skeletal crystals in glass	Absent
Pyroxene	Diopsidic augite (?), detectable only in groundmass, usually as skeletal crystals	Aluminous diopsidic phenocrysts and microphenocrysts, usually optically strained and sector-zoned; may form skeletal quench crystals in glass	Titaniferous augite, rarely as phenocrysts, and prominent in groundmass	Titaniferous augite or ferroaugite as phenocrysts; distinct Ti and Al enrichment
Spinel	Aluminous, magnesian spinel; microphenocrysts associated with olivine	Absent or with distinct marginal reaction rims or magnetite overgrowths	Rare or absent	Absent
Magnetite	Absent in glass; moderately abundant as skeletal groundmass crystals	Absent in glass except as overgrowths on spinel; skeletal groundmass crystals may be abundant	Rare as phenocrysts, common to abundant in groundmass	May form euhedral phenocrysts, prominent in groundmass
Ilmenite	Absent	Absent	May be associated with magnetite in groundmass	May be associated with magnetite in groundmass; rare as phenocrysts

From Bryan, W. B., Thompson, G., Frey, F. A., and Dickey, J. S., Inferred geologic settings and differentiation in basalts from the deep-sea drilling project, *J. Geophys. Res.*, 81, 4285, 1976. With permission.

Table 3.7—7
COMPARISON OF ESTIMATES OF MAJOR ELEMENT FLUXES DURING HYDROTHERMAL (≥200°C) ALTERATION OF BASALTS (UNITS = g/cm³)

From oceanic metabasalts	From experimental studies	From heat flow estimates
MgO −1.8 to +11.4 (−54.5 to +.345 × 10^{12} g/yr	+4.12 (125 × 10^{12} g/yr)	+5.74 to +11.5 (174 to 348 × 10^{12} g/yr)
Na_2O −2.6 to +3.1	0 to +2.81	
SiO_2 +3 to −24		−0.99 to −2.96
CaO −0.01 to −17.5	−3.39 to −9.31	−2.15 to −4.44
K_2O −0.46 to +0.59	−0.35	

Note: + taken up by rock from seawater; − lost from rock to seawater.

From Donnelly, T. W., Thompson, G., and Robinson, P. T., Very low temperature hydrothermal alteration of the oceanic crust and the problem of fluxes of potassium and magnesium, in *Deep Drilling Results in the Atlantic Ocean: Ocean Crust*, Talwani, M., Harrison, C. G., and Hayes, D. E., Eds., Technical Volume, American Geophysical Union, Washington, D.C., 1979, 369. With permission.

3.8 DEEP-SEA SEDIMENTS

Table 3.8—1
RADIOMETRIC SEDIMENT ACCUMULATION RATES ON THE NORTHERN MID-ATLANTIC RIDGE

Core and location		Sedimentation rate, cm/10^3 y	
		Bulk	Carbonate free
	Near 42°N		
CH 82–24	Ridge flank, on foothill	2.7	0.95
CH 82–26	Ridge flank on plain among low hills	2.1	0.53
CH 82–30(1)	Ridge flank on small rise	1.5	0.27
CH 82–41	Foothills west of rift valley	1.3	0.31
	Near 37°N, FAMOUS Area		
Core 52*	Interaction of FZB and RV3	0.98	0.19
Core 124*	Intersection of FZB and RV2	0.94	0.17
Core 125*	Intersection of FZB and RV2	0.50	0.10
Core 118*	Near FZA on scarp	0.35	0.04
Core 527–3*	Floor of inner valley, near center of RV2	2.9	0.78
	TAG Area 26°N		
Core 4A*	Median valley floor	1.8	0.74
	24°N		
ZEP 12	Ridge flank	—	0.45
ZEP 13*	Ridge crest valleys	~0.4	~0.2
ZEP 15*	Ridge crest valleys	~0.22	~0.05
ZEP 18*	Ridge crest valleys	~0.25	~0.11
ZEP 22	Ridge flank	0.25	—
	20°N		
AII 42–13	Ridge flank	0.5	0.17
AII 42–17*	Ridge crest; hills adjacent to median valley	0.73	0.13
AII 42–33	Ridge flank	0.47	0.31
AII 42–41	Abyssal hill west of ridge	1.3	1.26

* Median valley cores.

From Scott, M. R., Salter, P. F., and Barnard, L. A., Chemistry of ridgecrest sediments from the North Atlantic Ocean, in *Deep Drilling Results in the Atlantic Ocean: Ocean Crust,* Talwani, M., Harrison, C. G., and Hayes, D. E., Eds., Technical Volume, American Geophysical Union, Washington, D.C., 1979, 403. With permission.

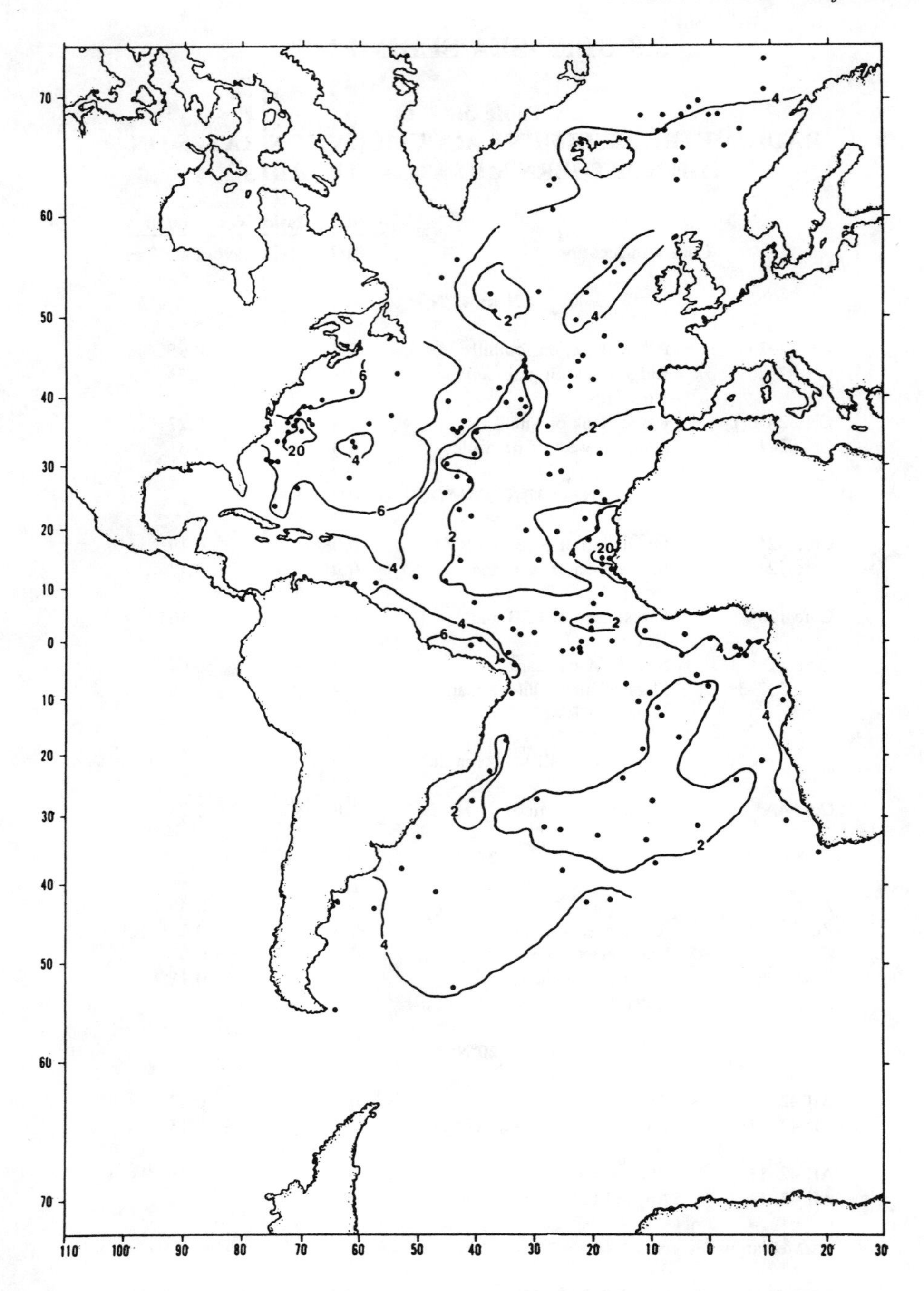

FIGURE 3.8—1. Sedimentation accumulation rates in the Atlantic Ocean in centimeters per 1000 years. (From Balsam, W. L. and Deaton, B. C., Sediment dispersal in the Atlantic Ocean: evaluation by visible light spectra, *Rev. Aquat. Sci.*, 4, 411, 1991.)

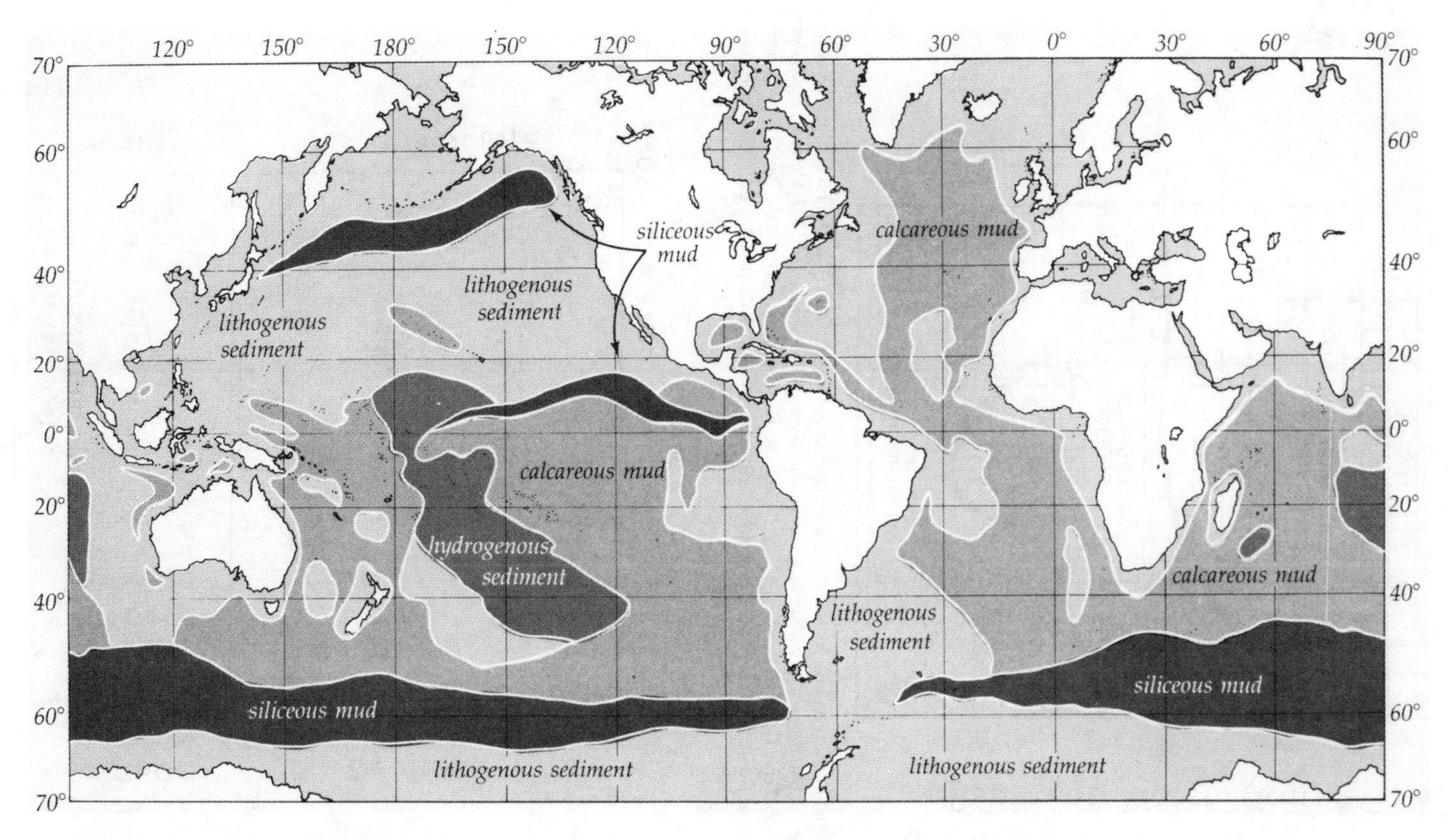

FIGURE 3.8—2. Distribution of deep-sea sediments. (From Ingmanson, D. A. and Wallace, W. J., *Oceanology: an Introduction,* 2nd ed., Wadsworth Publishing Company, Belmont, CA, 1979. With permission.)

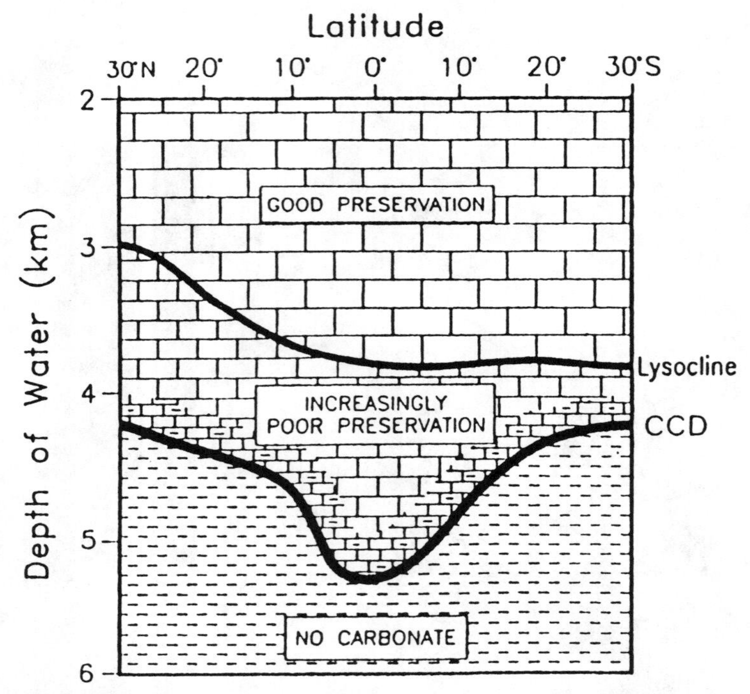

FIGURE 3.8—3. Depth of the lysocline and calcium carbonate compensation depth (CCD) for a N-S profile of the tropical central Pacific Ocean showing relative degrees of carbonate preservation. Note depression of the CCD due to enhanced productivity in the equatorial upwelling area. (From Grotsch, J., Wu, G., and Berger, W. H., Carbonate cycles in the Pacific: reconstruction of saturation fluctuations, in *Cycles and Events in Stratigraphy,* Einsele, G., Ricken, W., and Seilacher, A., Eds., Springer-Verlag, Berlin, 1991, 110. With permission.)

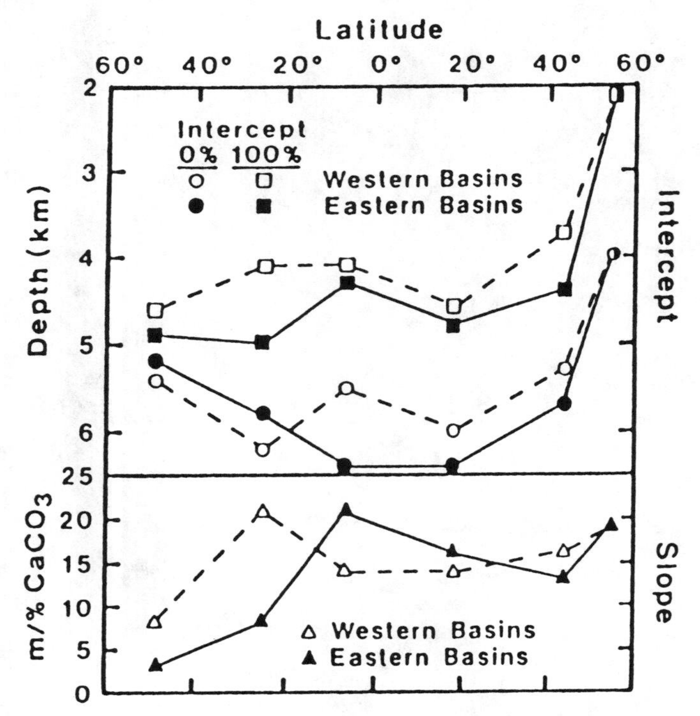

FIGURE 3.8—4. Variations in carbonate content vs. depth characteristics as functions of latitude in different regions of the Atlantic Ocean basin. (From Biscaye, P. E., Kolla, V., and Turekian, K. K., Distribution of calcium carbonate in surface sediments of the Atlantic Ocean, *J. Geophys. Res.,* 81, 2595, 1976. With permission.)

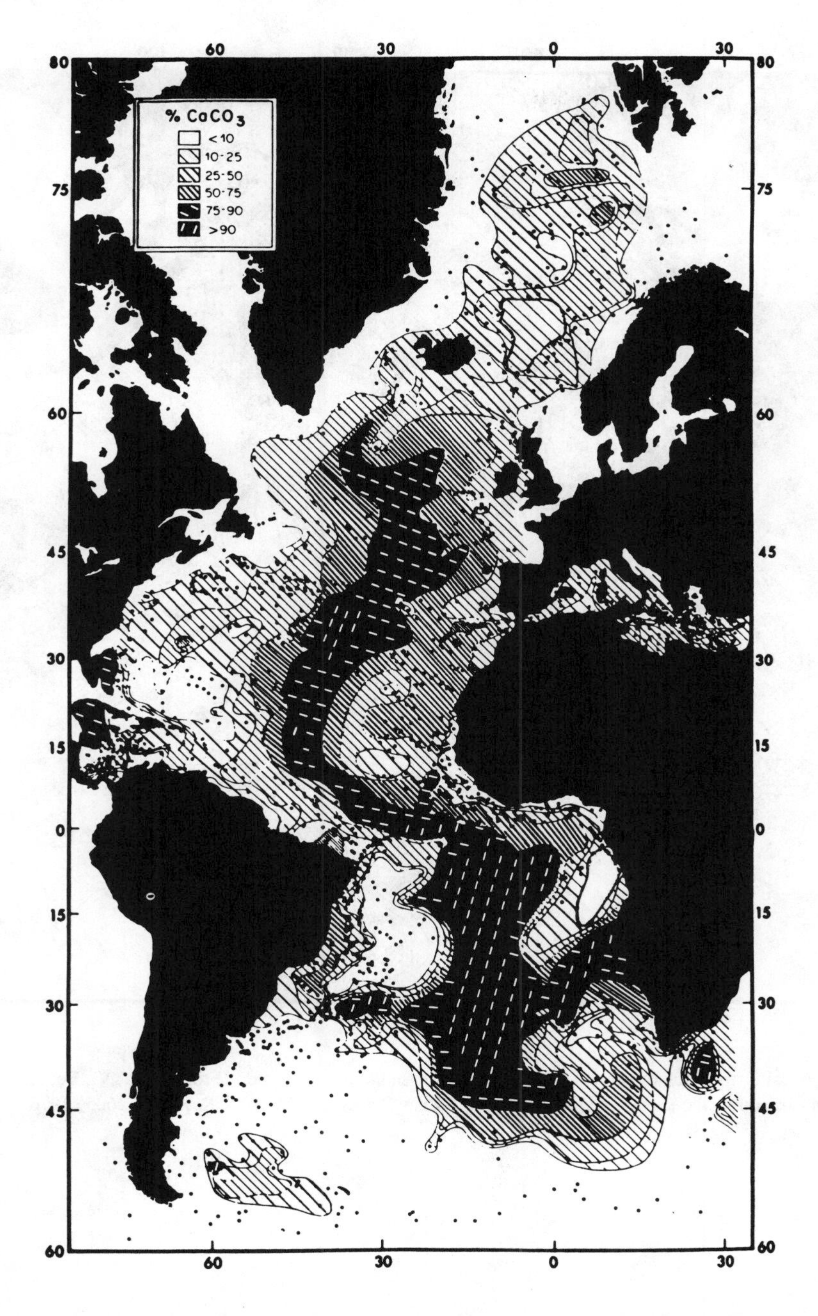

FIGURE 3.8—5. Distribution of percent calcium carbonate in surface sediments of the Atlantic Ocean and adjacent seas and oceans. (From Biscaye, P. E., Kolla, V., and Turekian, K. K., Distribution of calcium carbonate in surface sediments of the Atlantic Ocean, *J. Geophys. Res.*, 81, 2595, 1976. With permission.)

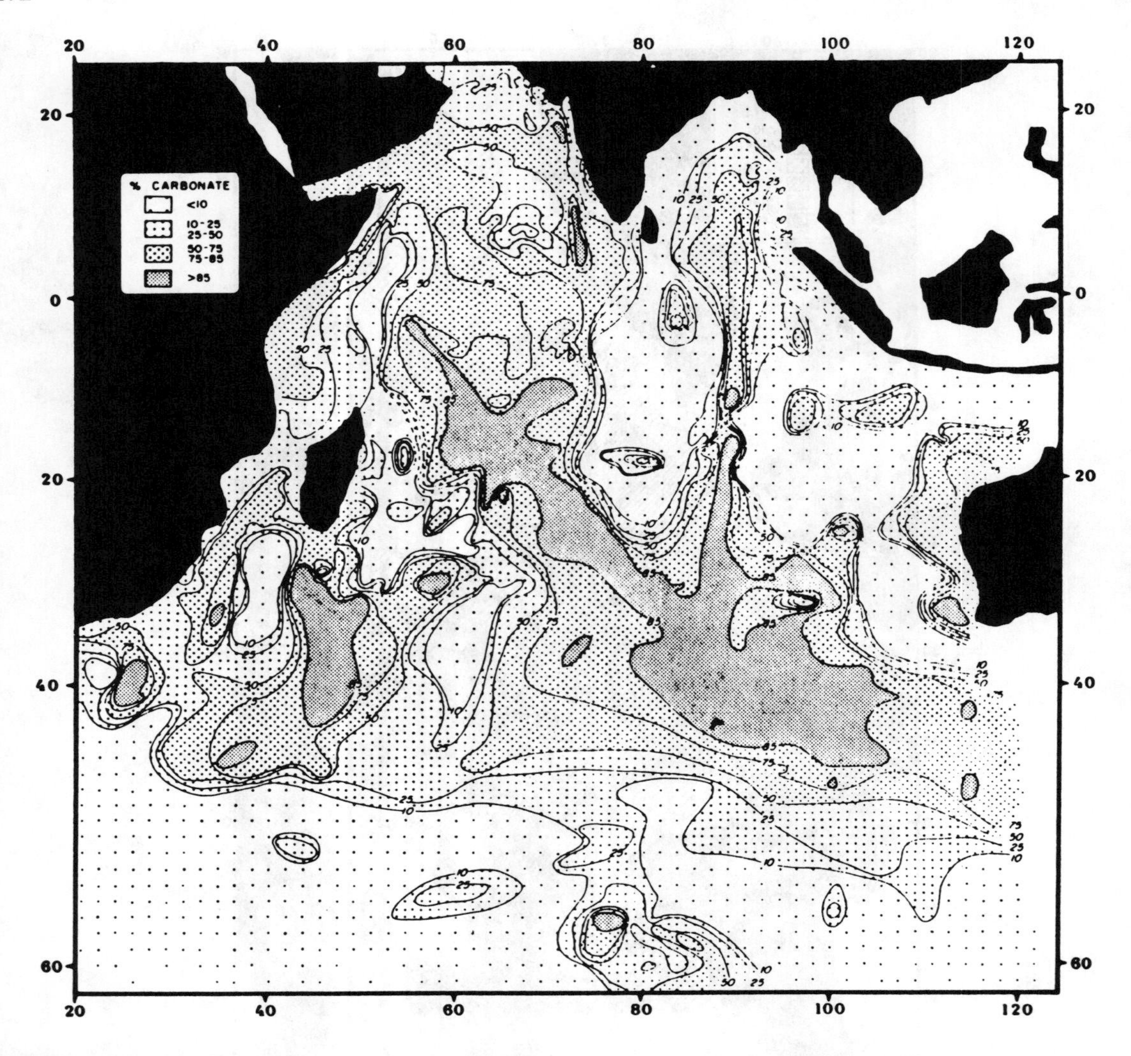

FIGURE 3.8—6. Areal distribution of percent calcium carbonate in the Indian Ocean. (From Kolla, V., Be, A. W. H., and Biscaye, P. E., Carbonate distribution in the surface sediments of the Indian Ocean, *J. Geophys. Res.*, 81, 2605, 1976. With permission.)

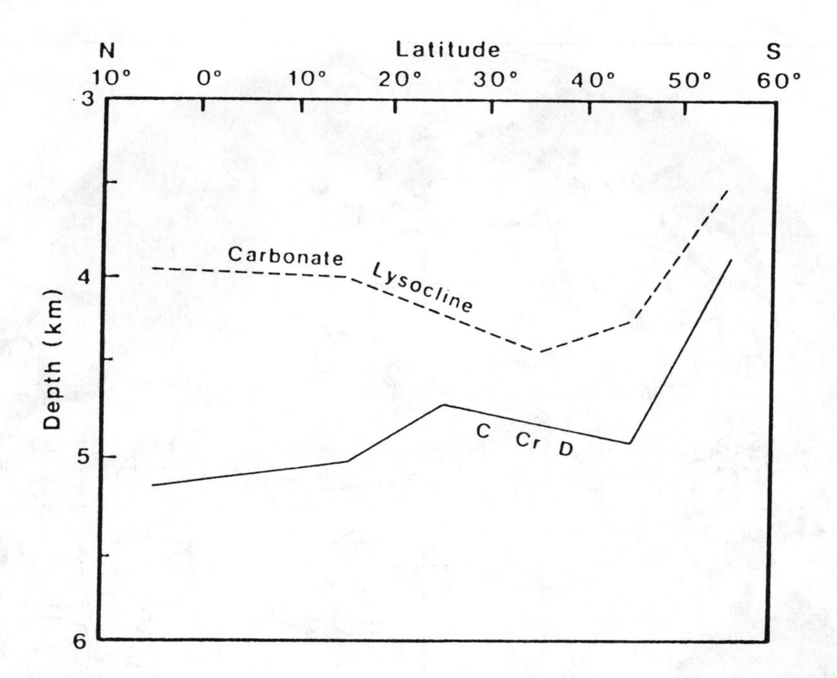

FIGURE 3.8—7. Carbonate lysocline and carbonate critical depth (CCrD) variations with latitude in the Indian Ocean. (From Kolla, V., Be, A. W. H., and Biscaye, P. E., Carbonate distribution in the surface sediments of the Indian Ocean, *J. Geophys. Res.*, 81, 2605, 1976. With permission.)

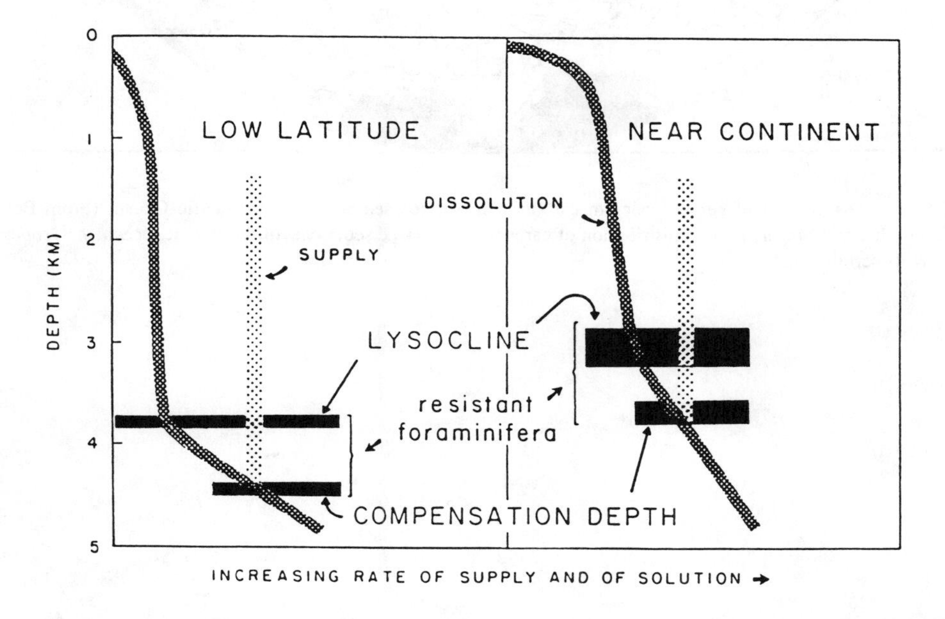

FIGURE 3.8—8. Model of the relationship between the lysocline and calcium carbonate compensation depth in the southeast Pacific. (From Berger, W. H., Planktonic foraminifera: selective solution and the lysocline, *Mar. Geol.*, 8, 111, 1970. With permission.)

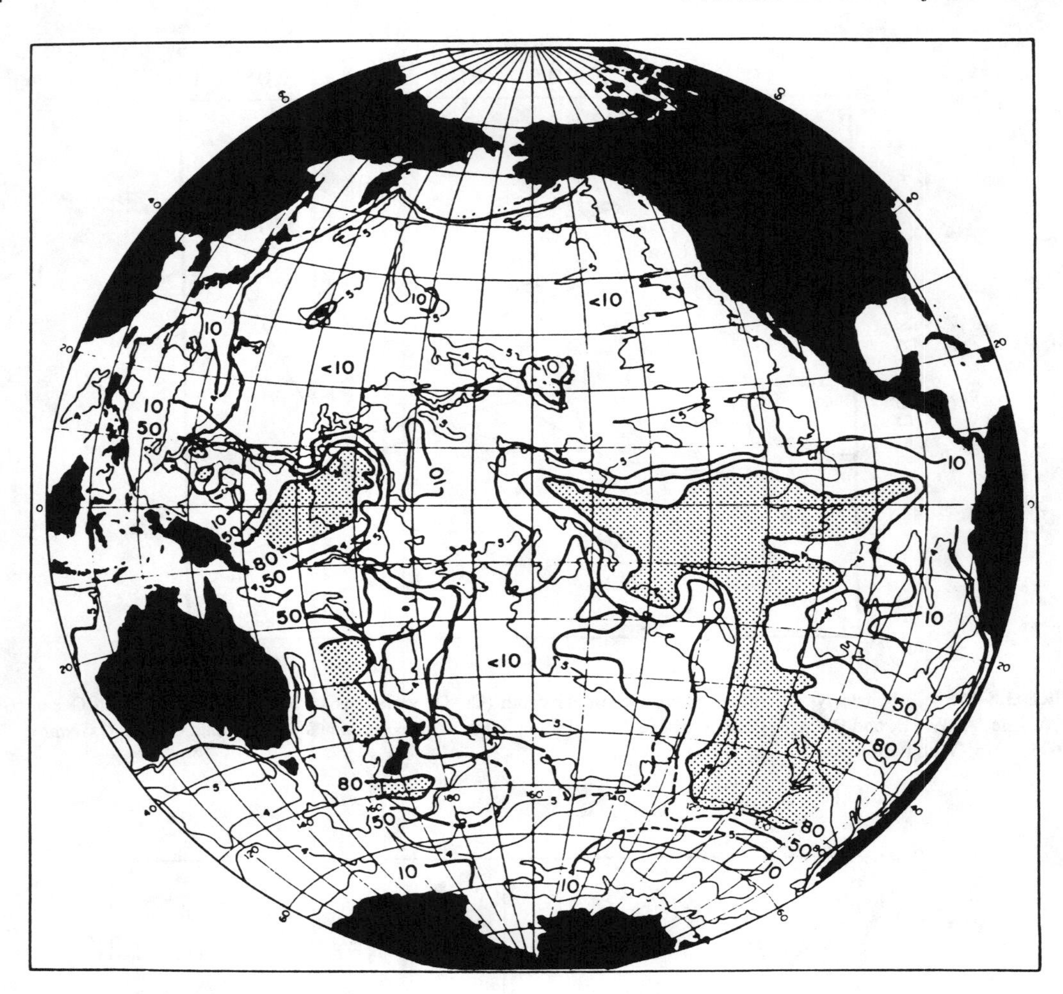

FIGURE 3.8—9. Distribution of percent calcium carbonate in surface sediments of the Pacific Ocean. (From Berger, W. H., Adelseck, C. G., Jr., and Mayer, L. A., Distribution of carbonate in surface sediments of the Pacific Ocean, *J. Geophys. Res.*, 81, 2617, 1976. With permission.)

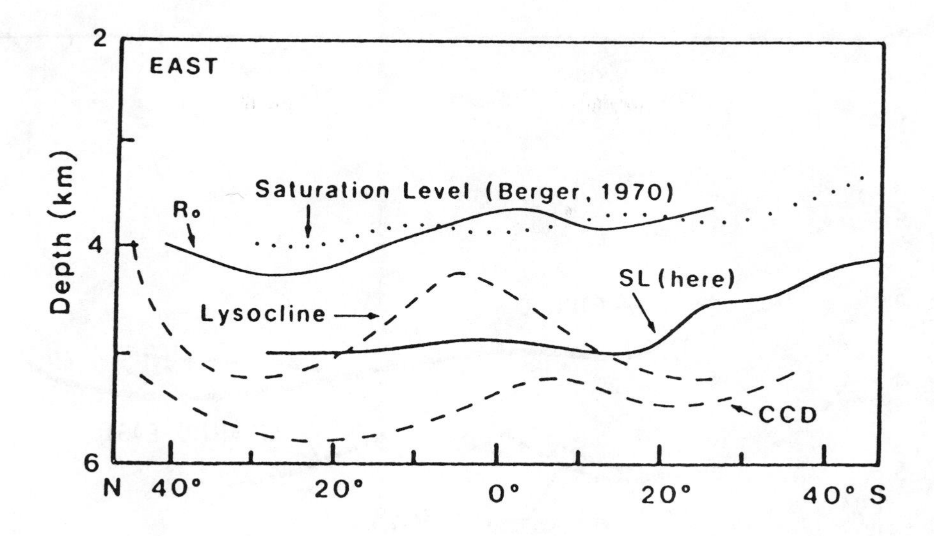

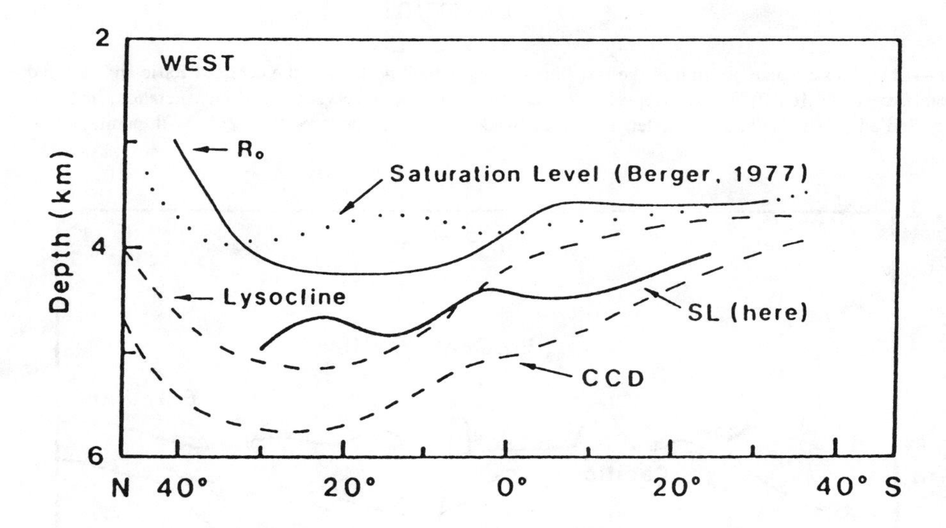

FIGURE 3.8—10. Depth distribution of the R_o, calcite saturation level, lysocline, and CCD as a function of latitude in the eastern and western Atlantic Ocean. (From Morse, J. W. and Mackenzie, F. T., *Geochemistry of Sedimentary Carbonates,* Elsevier, Amsterdam, 1990. With permission.)

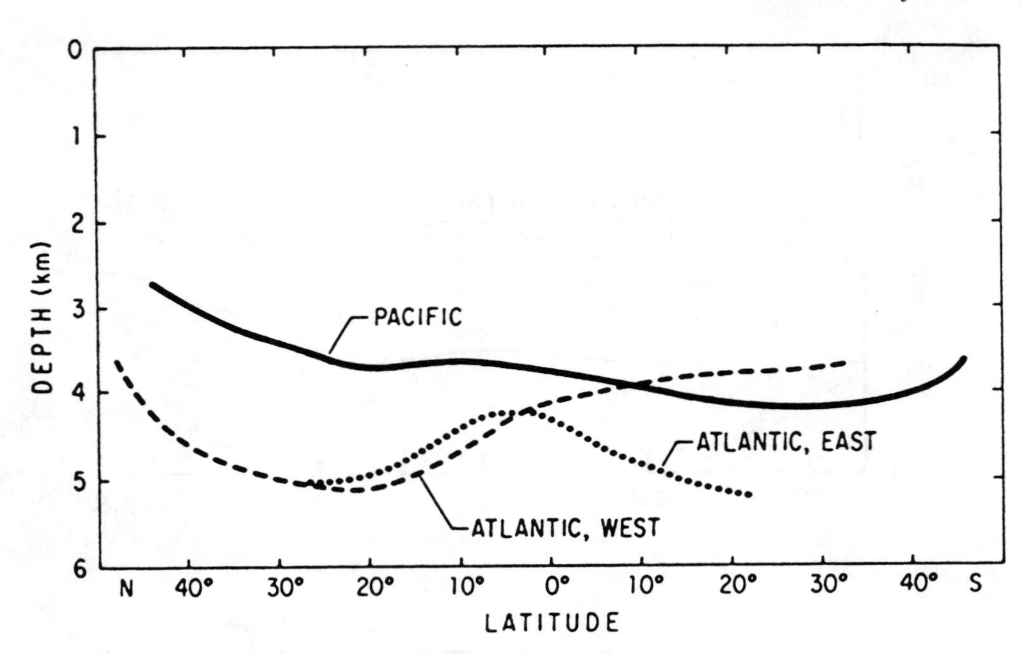

FIGURE 3.8—11. Approximate position of the lysocline in longitudinal profiles of the central Pacific and east and west Atlantic troughs. (From Berger, W. H., Biogenous deep-sea sediments: production, preservation, and interpretation, in *Chemical Oceanography,* Vol. 5, 2nd ed., Riley, J. P. and Chester, R., Eds., Academic Press, London, 1976, 265. With permission.)

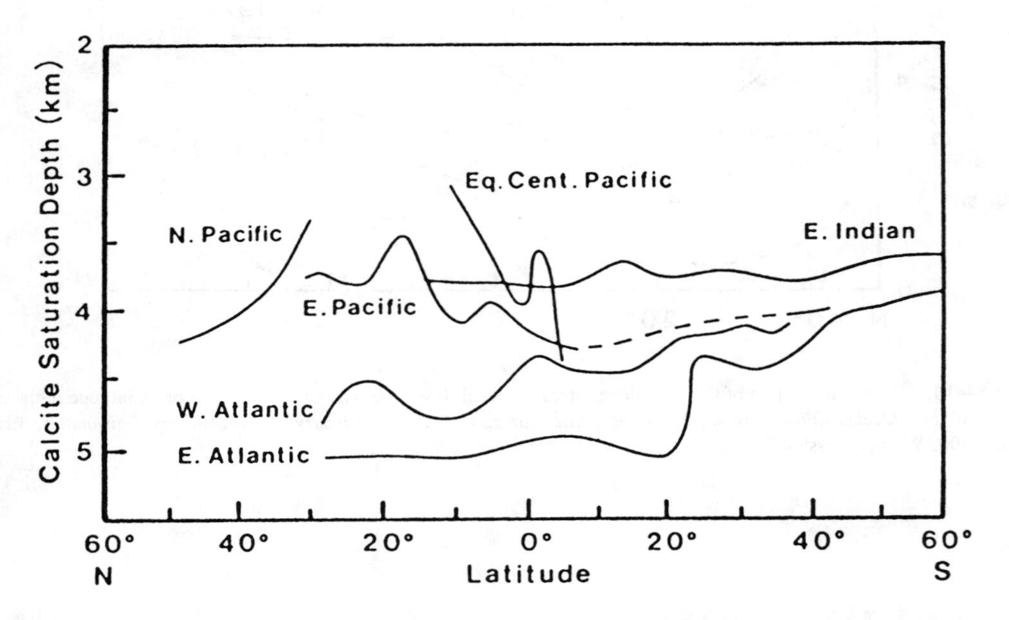

FIGURE 3.8—12. Saturation depths with respect to calcite in different areas of the Atlantic, Pacific, and Indian Oceans. (From Morse, J. W. and Mackenzie, F. T., *Geochemistry of Sedimentary Carbonates,* Elsevier, Amsterdam, 1990. With permission.)

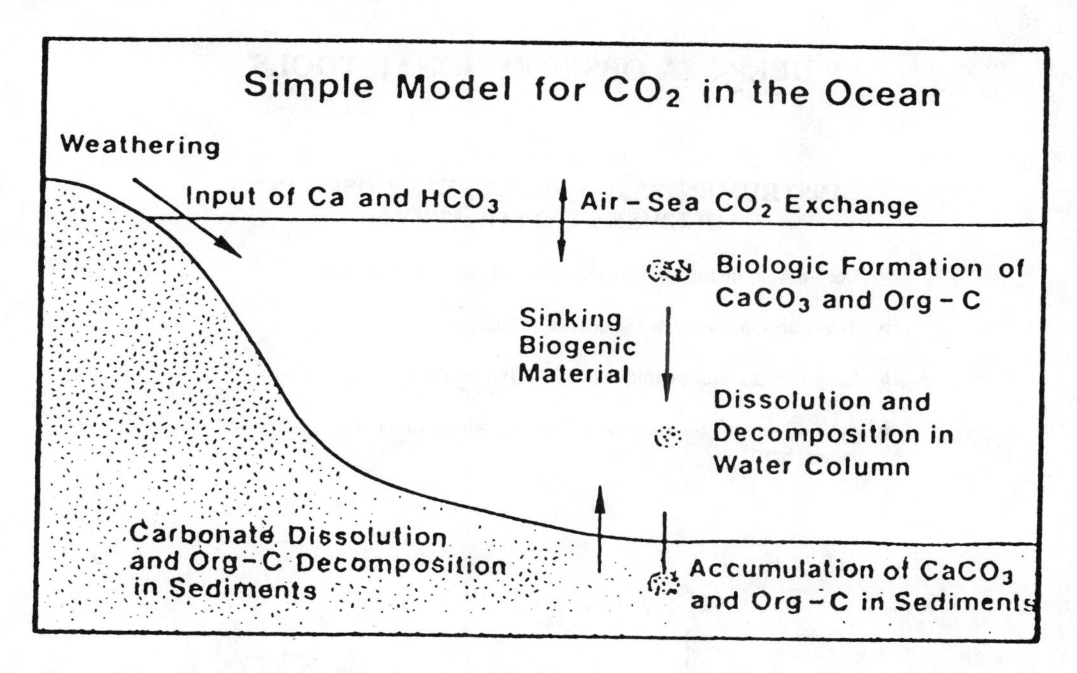

FIGURE 3.8—13. Simple model of the major components of the oceanic carbonate system. Pelagic organisms in the upper ocean represent the principal source of calcium carbonate in seafloor sediments. (From Morse, J. W. and Mackenzie, F. T., *Geochemistry of Sedimentary Carbonates,* Elsevier, Amsterdam, 1990. With permission.)

3.9 DEEP-SEA OBSERVATIONAL TOOLS

Table 3.9—1
INSTRUMENTATION FREQUENTLY USED IN DEEP-SEA EXPLORATION

- Deep-towed, multibeam sonar systems (SEABEAM and SASS)
- Side-scan sonar systems (SeaMARC, GLORIA)
- Deep-towed sensor platforms (DEEPTOW, ANGUS, ARGO, RAIE)
- Subseafloor surveys (seismic reflection, seismic refraction, seismic tomography)

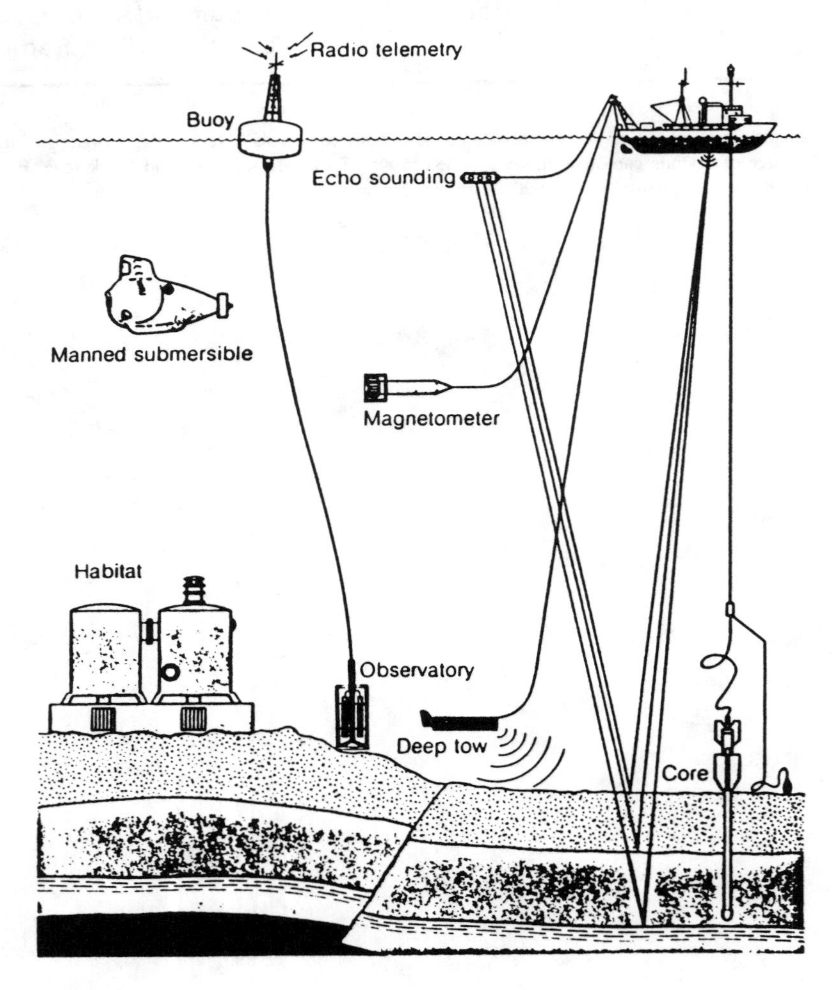

FIGURE 3.9—1. Schematic representation of the diverse deep-sea observations made during cruises of oceanographic research vessels.

3.10 SHALLOW-WATER ENVIRONMENTS, BEACHES, AND SEDIMENTS

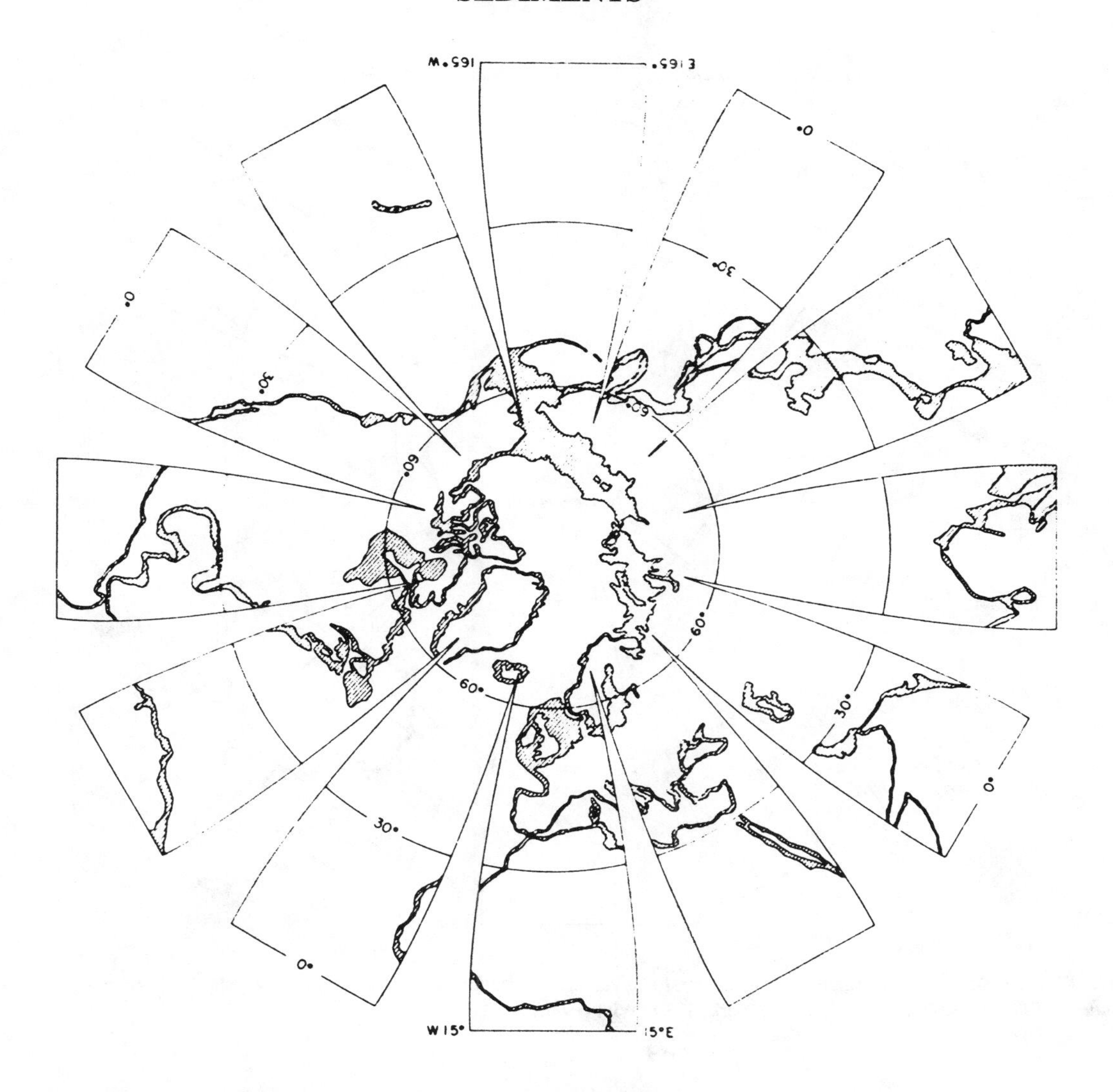

FIGURE 3.10—1a. The distribution of continental shelves in the northern hemisphere. (From Walsh, J. J., *On the Nature of Continental Shelves,* Academic Press, New York, 1989. With permission.)

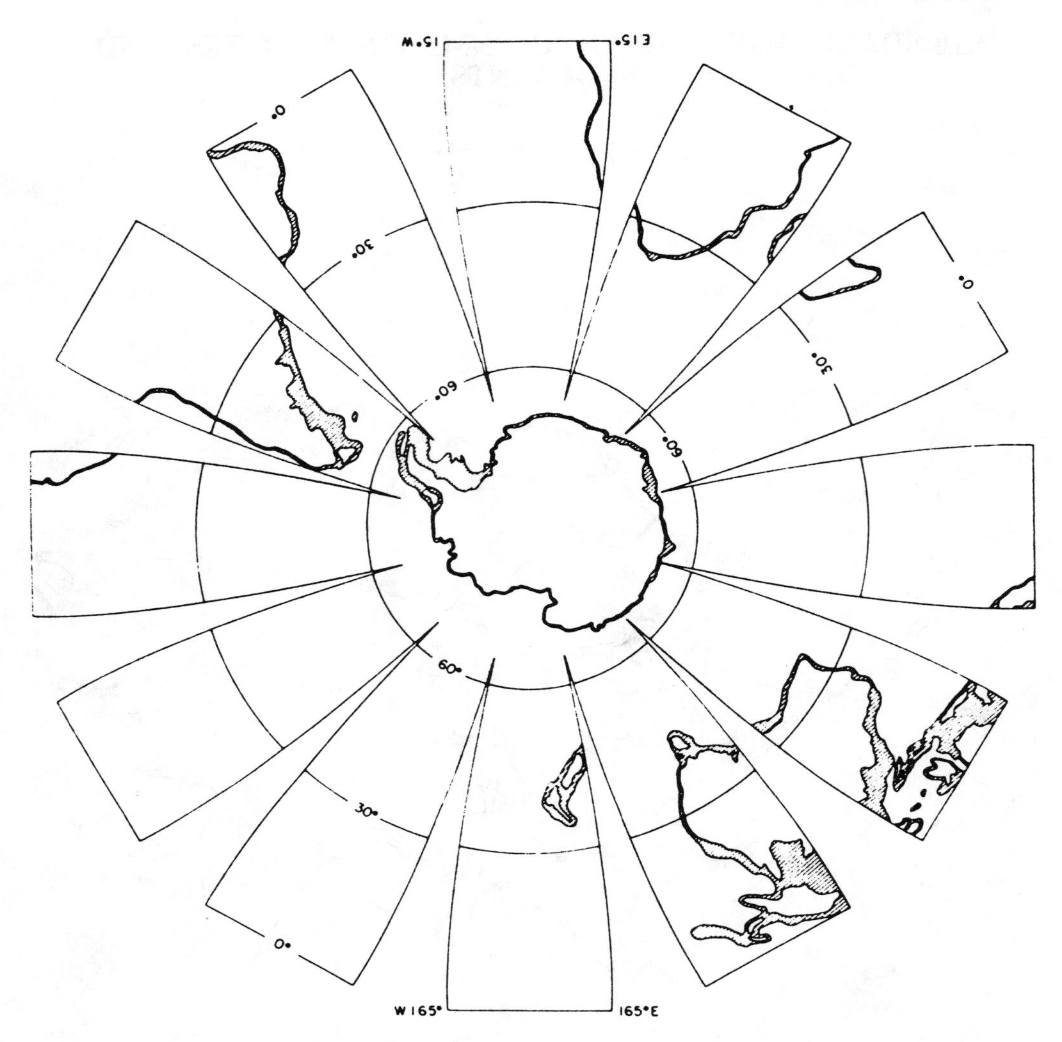

FIGURE 3.10—1b. The distribution of continental shelves in the southern hemisphere. (From Walsh, J. J., *On the Nature of Continental Shelves,* Academic Press, New York, 1989. With permission.)

Table 3.10—1
SEDIMENT DEPTH COLLECTED BY DIFFERENT SAMPLERS UNDER OPTIMAL CONDITIONS (~2 m FINE-GRAINED SEDIMENT)

Sediment depth sampled	Sampling equipment
0–10 cm	Lightweight, small-volume grabs (for example, Birge-Ekman, Ponar and mini-Ponar, mini-Shipek)
0–30 cm	Heavy, large-volume grabs (for example, Van Veen, Smith-McIntyre, Petersen)
0–50 cm	Single gravity corers (for example, Kajak-Brinkhurst and Phleger corers)
	Box corers
	Multiple corers
0–2 m	Single gravity corer (for example, Benthos and Alpine corers)
Deeper than 2 m	Piston corers

From Murdoch, A. and MacKnight, S. D., Eds., *CRC Handbook of Techniques for Aquatic Sediments Sampling,* CRC Press, Boca Raton, FL, 1991.

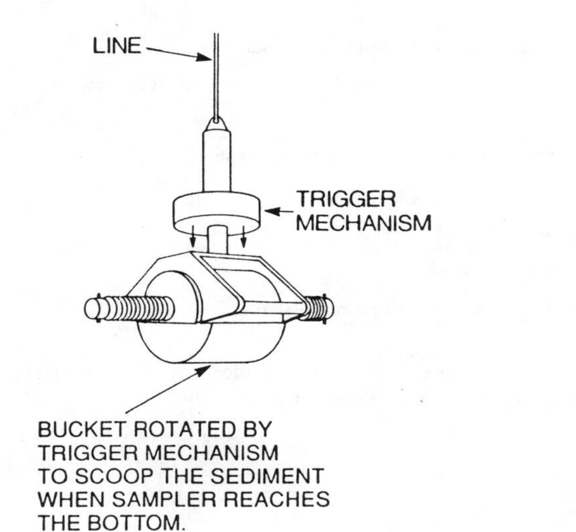

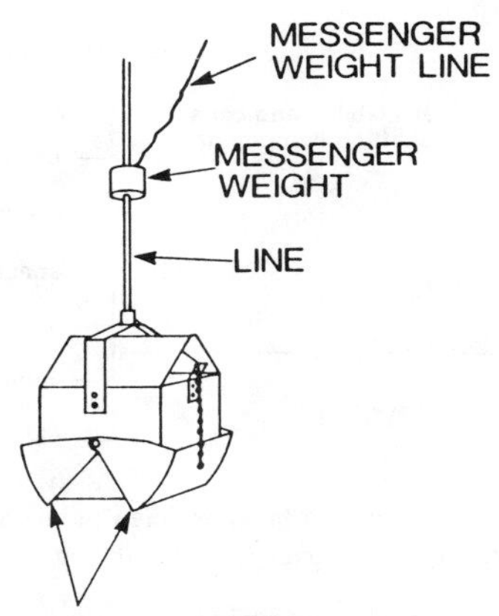

FIGURE 3.10—2. Sediment grab samplers with their essential parts. (From Murdoch, A. and MacKnight, S. D., Eds., *CRC Handbook of Techniques for Aquatic Sediments Sampling,* CRC Press, Boca Raton, FL, 1991.)

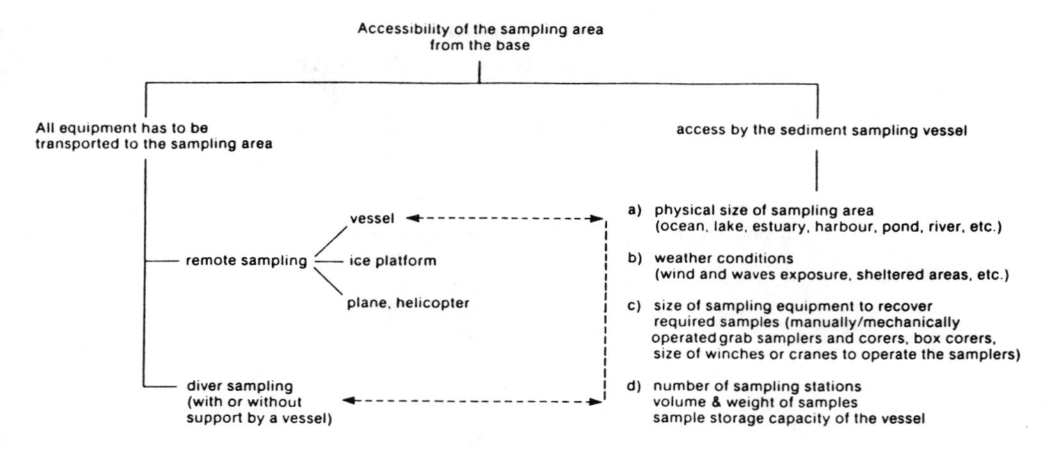

FIGURE 3.10—3. Factors to be considered in selection of sediment sampling equipment. (From Murdoch, A. and MacKnight, S. D., Eds., *CRC Handbook of Techniques for Aquatic Sediments Sampling,* CRC Press, Boca Raton, FL, 1991.)

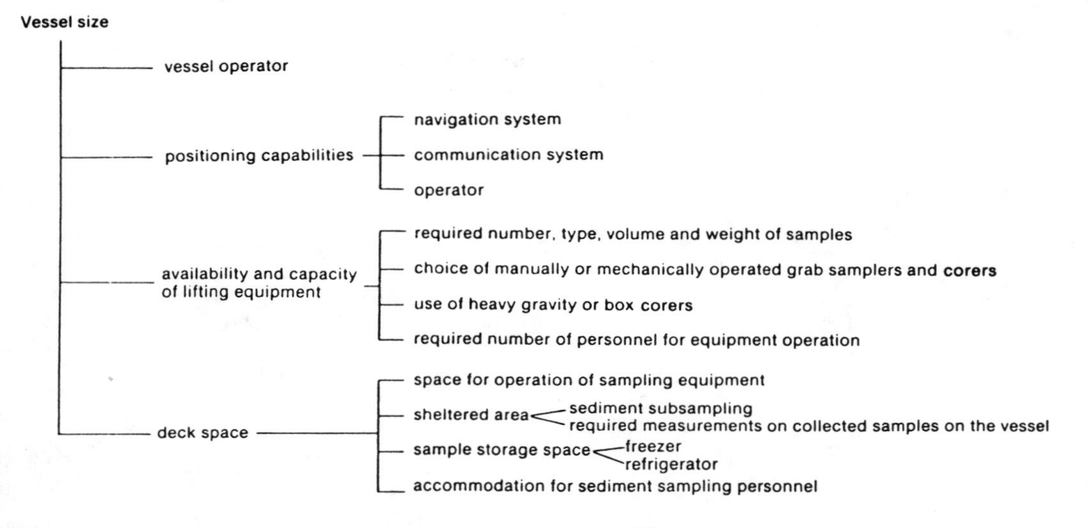

FIGURE 3.10—4. Elements to be considered in sediment sampling from a vessel. (From Murdoch, A. and MacKnight, S. D., Eds., *CRC Handbook of Techniques for Aquatic Sediments Sampling,* CRC Press, Boca Raton, FL, 1991.)

Table 3.10—2
THE UDDEN-WENTWORTH PARTICLE SIZE CLASSIFICATION

Grade limits (diameter in mm)	Grade limits (diameter in phi units)	Name
>256	≤ −8	Boulder
256–128	−8– −7	Large cobble
128–64	−7– −6	Small cobble
64–32	−6– −5	Very large pebble
32–16	−5– −4	Large pebble
16–8	−4– −3	Medium pebble
8–4	−3– −2	Small pebble
4–2	−2– −1	Granule
2–1	−1–0	Very coarse sand
1–1/2	0–1	Coarse sand
1/2–1/4	1–2	Medium sand
1/4–1/8	2–3	Fine sand
1/8–1/16	3–4	Very fine sand
1/16–1/32	4–5	Coarse silt
1/32–1/64	5–6	Medium silt
1/64–1/128	6–7	Fine silt
1/128–1/256	7–8	Very fine silt
1/256–1/512	8–9	Coarse clay
1/512–1/1024	9–10	Medium clay
1/1024–1/2048	10–11	Fine clay

Table 3.10—3
SETTLING VELOCITIES OF SEDIMENTARY PARTICLES

Sediment	Median diameter (μm)	Settling velocity (m/day)
Fine sand	250—125	1040
Very fine sand	125—62	301
Silt	31.2	75.2
Silt	15.6	18.8
Silt	7.8	4.7
Silt	3.9	1.2
Clay	1.95	0.3
Clay	0.98	0.074
Clay	0.49	0.018
Clay	0.25	0.004
Clay	0.12	0.001

From King, C. A. M., *Introduction to Marine Geology and Geomorphology*, Edward Arnold, London, 1975. With permission.

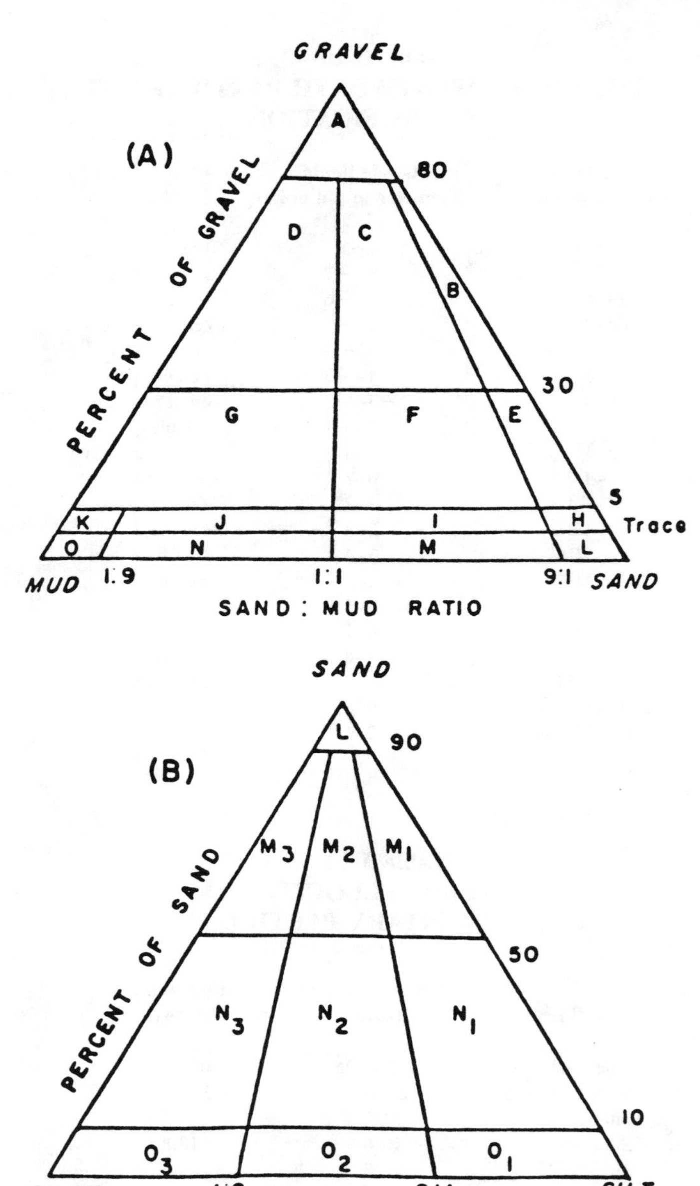

FIGURE 3.10—5. Ternary diagrams used to classify sediments. Letters in ternary diagrams relate to the following terminology for mixtures of gravel, sand, silt, and clay: (A) gravel, (B) sandy gravel, (C) muddy sandy gravel, (D) muddy gravel, (E) gravelly sand, (F) gravelly muddy sand, (G) gravelly mud, (H) slightly gravelly mud, (I) slightly gravelly muddy sand, (J) slightly gravelly sand mud, (K) lightly gravelly mud, (L) sand, (M_1) silty sand, (M_2) muddy sand, (M_3) clayey sand, (N_1) sandy silt, (N_2) sandy mud, (N_3) sandy clay, (O_1) silt, (O_2) mud, and (O_3) clay. (From Folk, R. L., The distinction between grain size and mineral composition in sedimentary rock nomenclature, *J. Geol.*, 1954, 344. Copyright 1954 by the University of Chicago. With permission.)

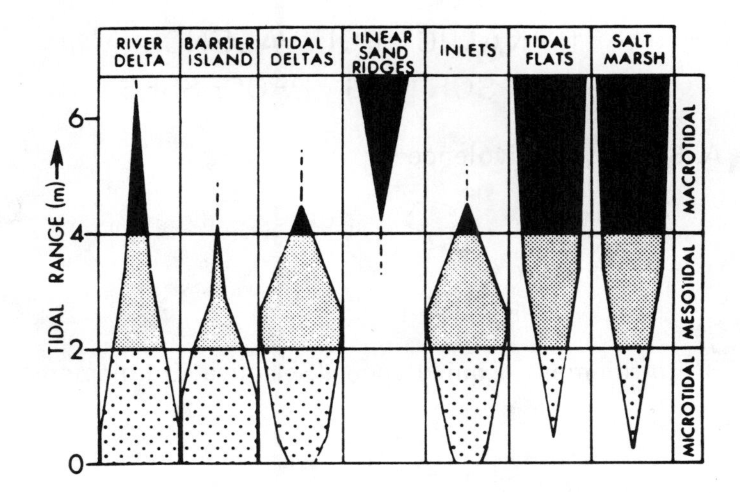

FIGURE 3.10—6. Variation of morphology of coastal plain shorelines as a function of tidal range. Note that tidal deltas and inlets are most common in mesotidal areas. (From Hayes, M. O., in *Estuarine Research,* Vol. 2, Cronin, L. E., Ed., Academic Press, New York, 1975, 3. With permission.)

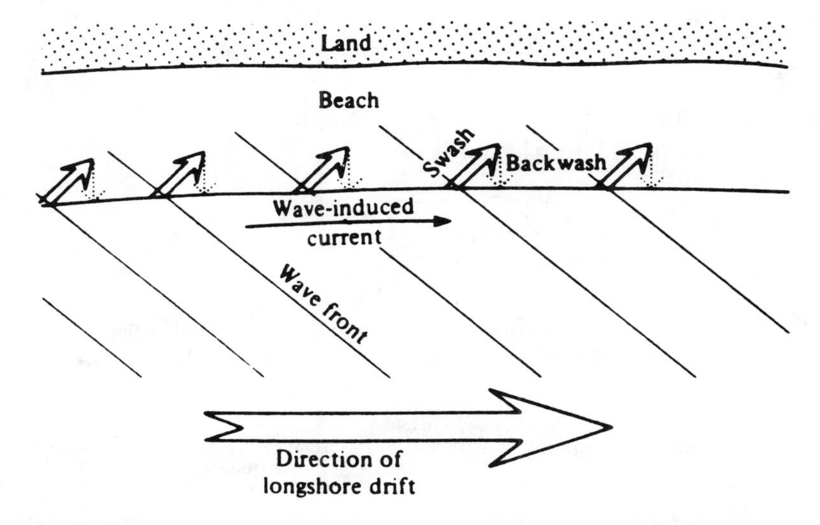

FIGURE 3.10—7. Diagram demonstrating the process of longshore drift. (From Barnes, R. S. K., *Coastal Lagoons: The Natural History of a Neglected Habitat,* Cambridge University Press, Cambridge, 1980. With permission.)

DYNAMIC EQUIVALENCE
AND SORTING PROCESSES

A. Settling Equivalence

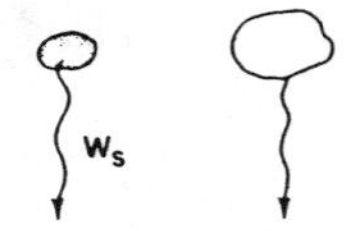

B. Entrainment Equivalence (Selective Entrainment)

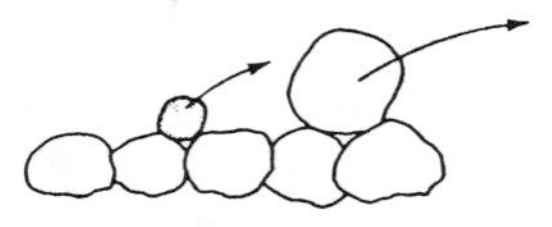

C. Transport Equivalence (Transport Sorting)

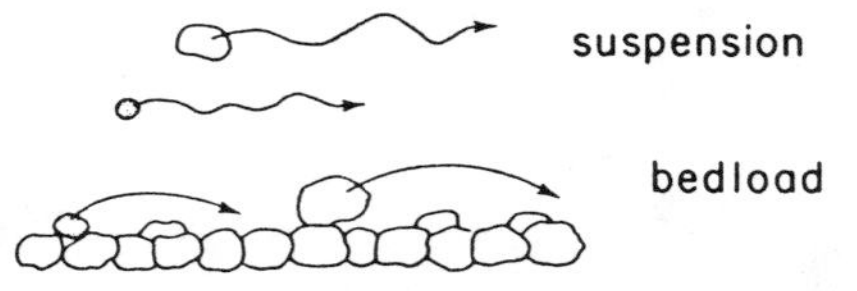

D. Dispersive – Pressure Equivalence
(Shear Sorting)

FIGURE 3.10—8. Schematic illustration of sorting processes and possible dynamic equilibria acting on sediments. (From Komar, P. D., Physical processes of waves and currents and the formation of marine placers, *Rev. Aquat. Sci.*, 1, 393, 1989.)

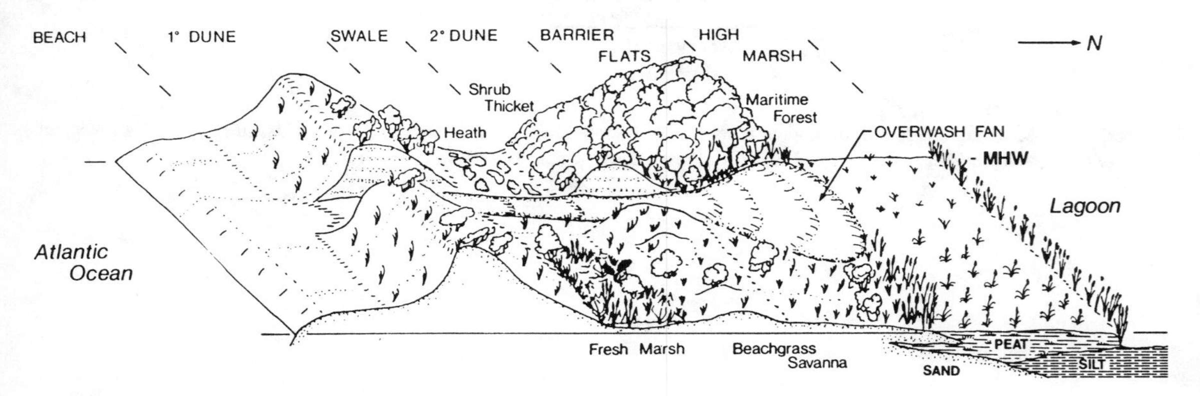

FIGURE 3.10—9. Barrier-beach environments, vegetation, and sediments. Sand deposits from overwash and aeolian activity produce new substrates for colonization by plants. These deposits may then be buried by peat as marshes develop, forming discrete lenses. (From Clark, J. S., Population and evolutionary implications of being a coastal plant: long-term evidence from the North Atlantic coasts, *Rev. Aquat. Sci.*, 2, 509, 1990.)

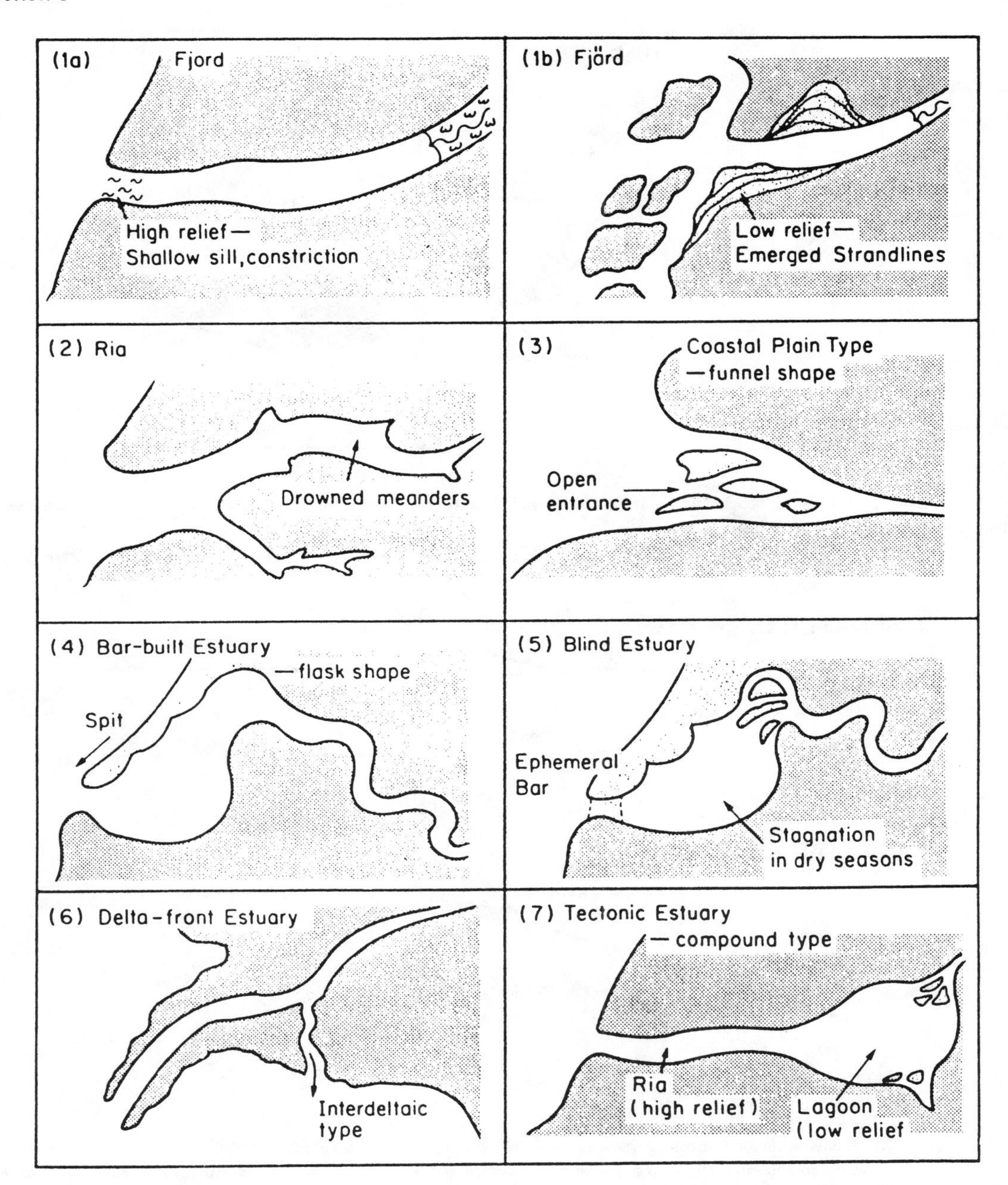

FIGURE 3.10—10. Illustrations of the seven basic physiographic types of estuaries. (From Fairbridge, R. W., in *Chemistry and Biogeochemistry of Estuaries,* Olausson, E. and Cato, I., Eds., John Wiley & Sons, Chichester, 1980, 1. Reproduced by permission of copyright © owner, John Wiley & Sons, Ltd.)

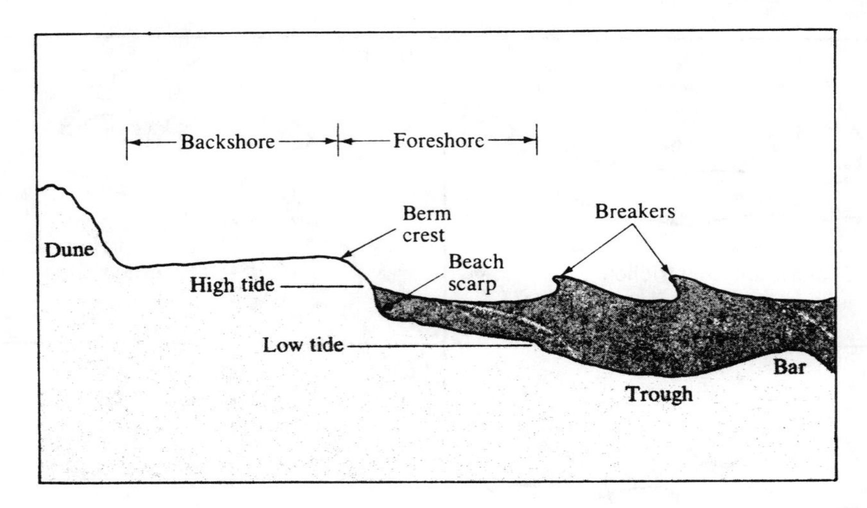

FIGURE 3.10—11. Profile of beach and adjacent coastal features. (From Gross, M. G., *Oceanography,* 3rd ed., Charles E. Merrill, Columbus, OH, 1976. With permission.)

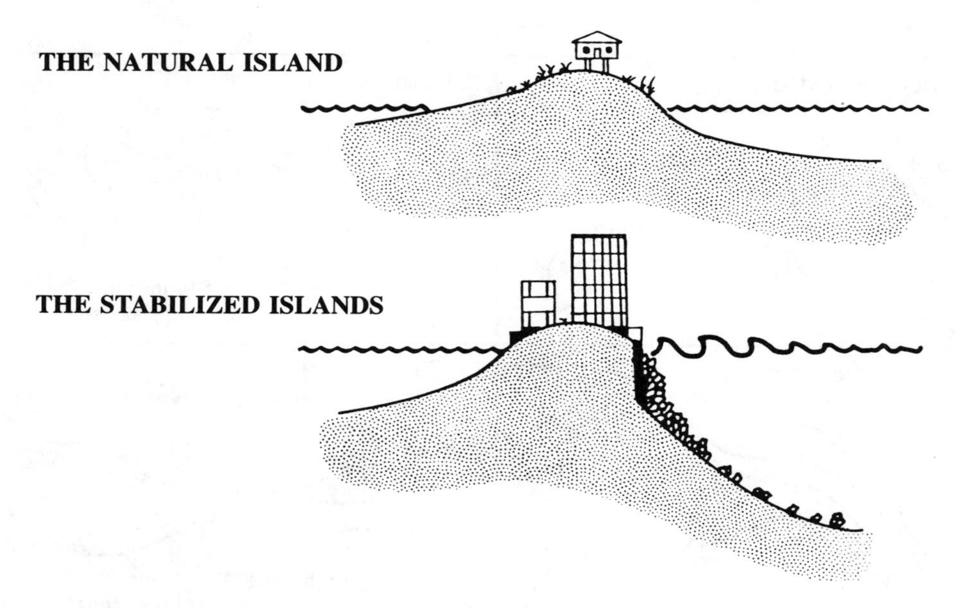

FIGURE 3.10—12. A diagrammatic illustration showing the long-range (10 to 50 year) effect of stabilization on a shoreline system. In the final stage the armored shoreline sits atop a steepened shoreface and all waves strike the seawall. (From Ragotzkie, R. A., Ed., *Man and the Marine Environment,* CRC Press, Boca Raton, FL, 1983.)

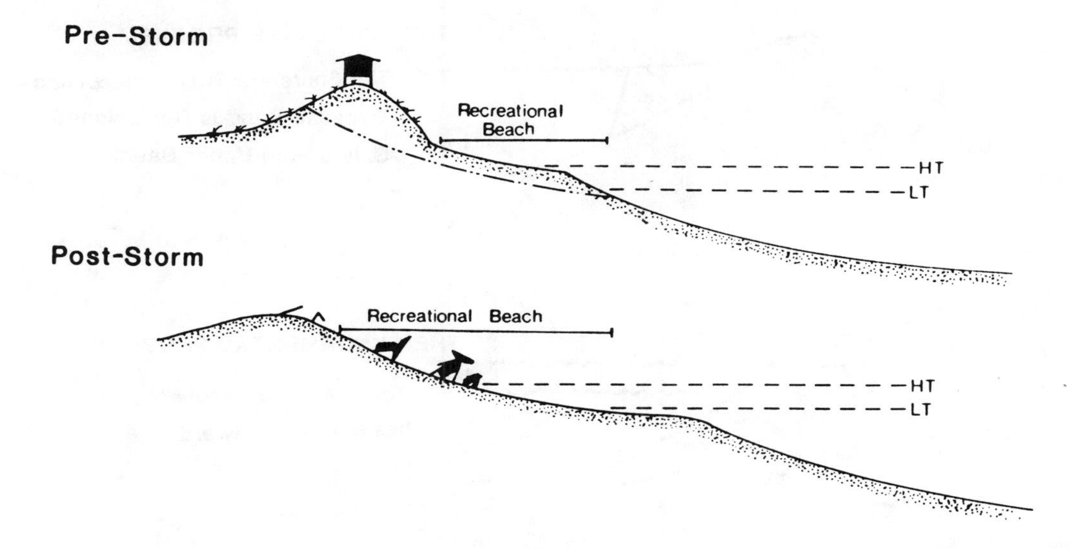

FIGURE 3.10—13. The storm response at a natural beach system. The flattened beach is capable of dissipating wave energy over a broadened surface area relative to the prestorm beach. (From Ragotzkie, R. A., Ed., *Man and the Marine Environment,* CRC Press, Boca Raton, FL, 1983.)

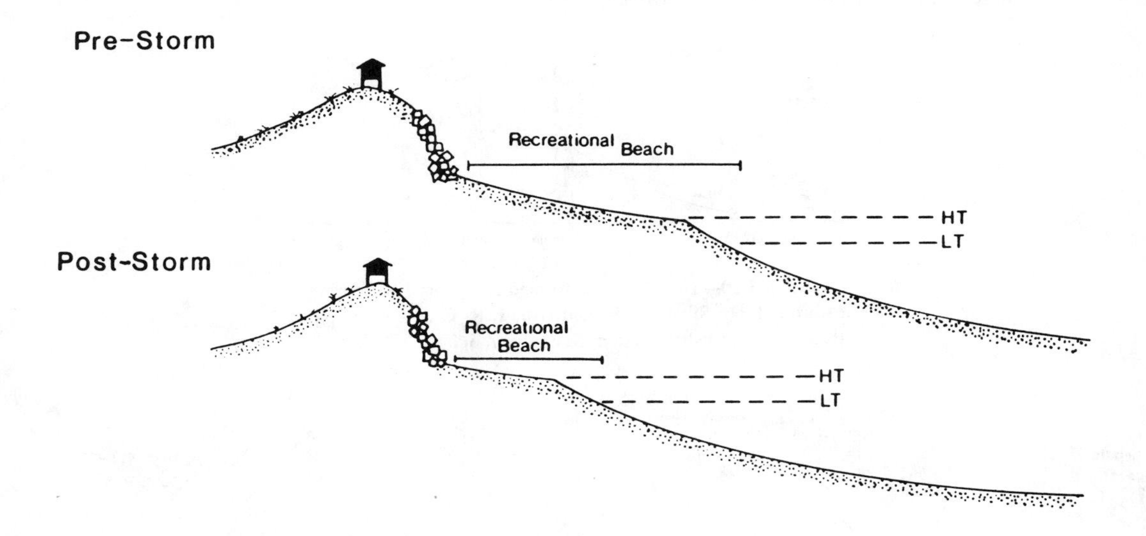

FIGURE 3.10—14. The revetted-beach storm response is a steepened shoreface. Compare with Figure 3.10—11. (From Ragotzkie, R. A., Ed., *Man and the Marine Environment,* CRC Press, Boca Raton, FL, 1983.)

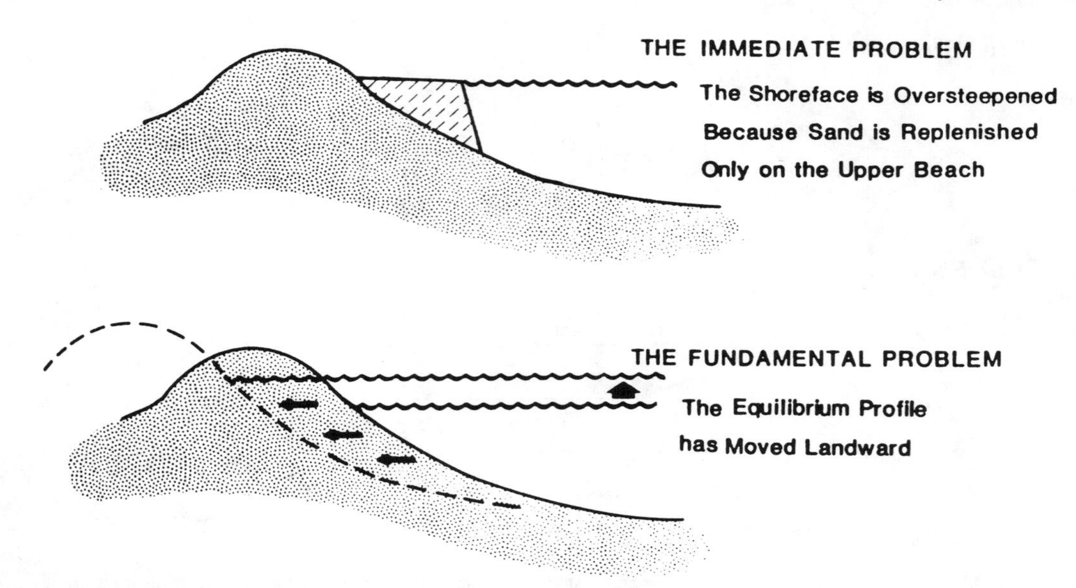

FIGURE 3.10—15. A diagrammatic illustration showing why a replenished beach disappears more rapidly than its natural predecessor. The first problem is that the sand is piled atop the upper beach which steepens the whole system. In order to get back to a profile in equilibrium with local sand supply and wave climate conditions, the sand rapidly disappears across the entire shoreface. (From Ragotzkie, R. A., Ed., *Man and the Marine Environment,* CRC Press, Boca Raton, FL, 1983.)

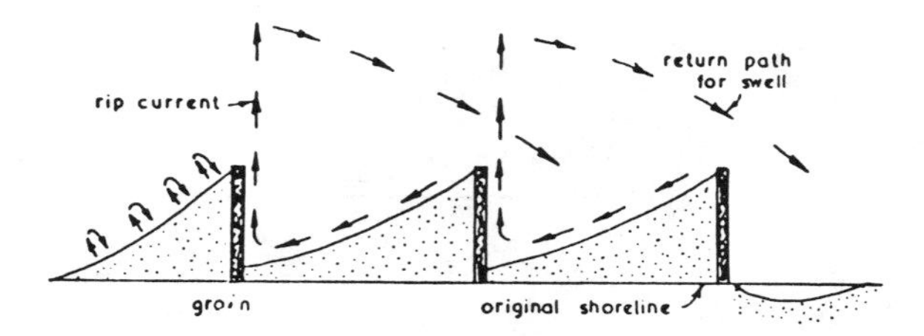

FIGURE 3.10—16. Currents formed at a groin field during a storm. (From Silvester, R. and Hsu, J. R. C., New and old ideas in coastal sedimentation, *Rev. Aquat. Sci.,* 4, 375, 1991.)

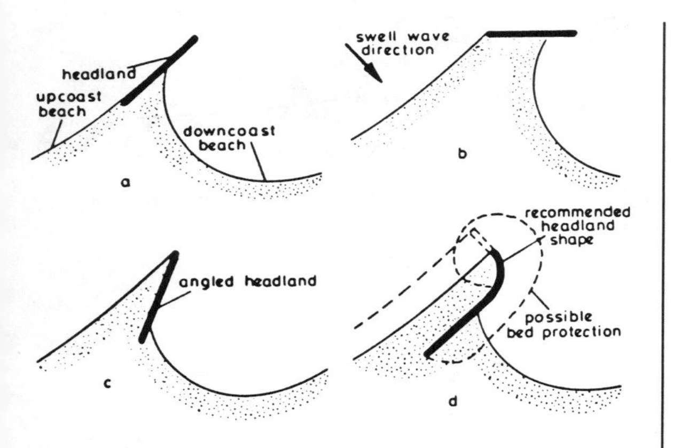

FIGURE 3.10—17. Headland shapes and orientations to produce bays. (From Silvester, R. and Hsu, J. R. C., New and old ideas in coastal sedimentation, *Rev. Aquat. Sci.,* 4, 375, 1991.)

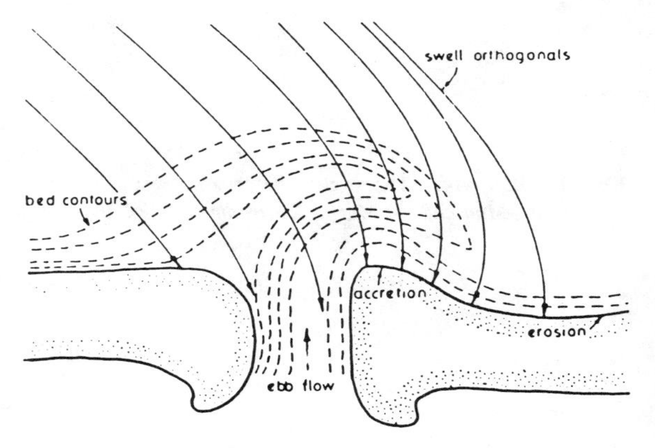

FIGURE 3.10—18. Arcuate-shaped bar at inlet showing wave refraction. (From Silvester, R. and Hsu, J. R. C., New and old ideas in coastal sedimentation, *Rev. Aquat. Sci.,* 4, 375, 1991.)

Table 3.10—4
WORLDWIDE COASTAL CLIFF EROSION RATES

Location	Lithology	Erosion rate (m/year)	Interval	Method
Australia				
Point Peron near Perth	Limestone	0.0002—0.001	1953—1962	Steel pegs
Warrnambool, Victoria	Aeolianite	0.014	(130 yr)	Surveys
Barbados				
Mullins Bay	Coral rock	0.002	(4.5 yr)	Pegs
Paynes Bay	Coral rock	0.0005	—	Pegs
Canada				
Port Bruce, Ontario	Glacial deposits	2.2	1966—1976	Maps
Central part, Lake Erie, Ontario	Glacial deposits	0.25—2.75	1810—1964	Surveys, air photos
Scarborough, Ontario	Glacial deposits	0.76	1952—1976	Surveys, air photos
France				
Ault, Somme	Chalk	0.08—0.37	(100 yr)	—
Dieppe, Seine Maritime	Chalk	0.4	—	—
St. Jouin, Seine Maritime	Chalk	0	1834—1894	Maps
Germany				
North of Kiel, Baltic Sea	Glacial clay	0.6	1873—1934	—
		0.8	1876—1938	—
Heligoland, North Sea	Mesozoic sandstone	1	—	—
Cape Arkona, Rugen	Chalk	3—4	(100 yr)	—
Iceland				
Surtsey Island	Lava	25—37	1967—1975	Air-photogrammetric surveys
Indian Ocean				
Aldabra Atoll	Limestone	0.0010—0.0013	1969—1971	Micro-erosion meter
Indonesia				
Krakatoa Island	Volcanic ash	33	1883—1928	—
Japan				
Haranomachi, Fukushima	Pliocene sandstone, mudstone	0.3—0.7	1912—1959	Maps
Iwaki, Fukushima	Pliocene mudstone	0.13—2.05	1974—1978	Surveys
Okuma, Fukushima	Pliocene mudstone, sandstone	0.62	1947—1961	Air-photogrammetric maps, surveys
		0.31	1961—1963	
		1.08	1963—1965	
Suetsugi to Hirono, Fukushima	Pliocene siltstone	0.80—1.4	1947—1973	Air-photogrammetric maps
	Pliocene sandstone	0.74	1947—1973	
Byobugaura, Chiba	Pliocene mudstone	0.4—1.1	1888—1950	Maps, surveys
Byobugaura, Chiba	Pliocene mudstone	0.79	1946—1960	Air-photogrammetric maps
		0.91	1960—1965	
		1.47	1965—1967	

Table 3.10—4 (continued)
WORLDWIDE COASTAL CLIFF EROSION RATES

Location	Lithology	Erosion rate (m/year)	Interval	Method
Byobugaura, Chiba	Pliocene mudstone	0.73	1884—1969	Maps
Taito-misaki, Chiba	Pliocene mudstone	1.11	1960—1966	Air-photogrammetric maps
		0.70	1966—1970	
Toban, Hyogo	Pleistocene deposits	1.0—1.5	1893—1955	Maps
Southern coast, Atsumi peninsula	Pleistocene deposits	0.03—0.6	1888—1959	Maps
Toyama coast	Alluvial fan deposits	0.5—1.0	—	—
West coast, Kanazawa city	Alluvial clay	1—2	—	Surveys
Nozuka, Tokachi, Hokkaido	Alluvium	2.3	1964—1974	Surveys
Habuseura, Niijima Island	Volcani sand	5.5	1961—1965	Maps, surveys
Nishinoshima Island	Volcanic ejecta	80	1974—1977	Air-photogrammetric surveys
New Zealand				
Kai-iwi Beach, Wellington	Pliocene siltstone	1.5	1876—1893	Surveys
Point Kean, Kaikoura	Tertiary mudstone	0.24	1942—1974	Air photos
Sharks Tooth Point, Kaikoura	Teritary mudstone	0.33	1942—1974	Air photos
Fifth Bay, Kaikoura	Tertiary mudstone	0.08—0.10	1942—1974	Air photos
Mudstone Bay, Kaikoura	Tertiary mudstone	0.06—0.13	1942—1974	Air photos
Poland				
Baltic Sea coast	Quaternary deposits	1	—	—
Sweden				
Hallshuk, Gotland	Silurian limestone	0.018	1899—1955	Photos
North of Ygne, Gotland	Silurian marl	0.025	1950—1962	Photos
U.K.				
Fourth Bight, Whitby, Yorkshire	Upper Lias shale	0.023	1971—1972	Micro-erosion meter
Tees estuary to Ravenscar, Yorkshire	Upper Lias shale, sandstone	0.036	1891— ?	Maps, surveys
Flamborough Head, Humberside	Chalk	0.3	—	—
Bridlington to Sewerby, Humberside	Glacial deposits	1.8	—	—
Holderness, Humberside	Glacial deposits	3.3	—	—
Holderness, Humberside	Glacial deposits	0.29—1.75	1852—1952	Surveys
Norfolk	Glacial deposits	0.9	1880—1967	Maps, air photos
West Runton to East Runton, Norfolk	Glacial deposits	0.2	1971—1973	Cliff-top stakes
Cromer to Mudesley, Norfolk	Glacial deposits	4.2	1838—1861	—
		5.7	1861—1905	—
Suffolk	Glacial deposits	0.8	1880—1950	Maps
Hopton, Suffolk	Glacial deposits	0.8—0.9+	1926—1950	Maps
Pakefield, Suffolk	Glacial deposits	0.6—0.9	1926—1950	Maps
Covehithe, Suffolk	Glacial deposits	5.1	1925—1950	Maps
Southwold, Suffolk	Glacial deposits	3—3.3	1925—1950	Maps

Southwold, Suffolk	Glacial deposits	4.5—13.5	—	—
Dunwich, Suffolk	Glacial deposits	1.6	1589—1753	Maps
		0.9	1753—1824	
		1.5	1824—1884	
		1.1	1884—1925	
		0.15	1925—1977	Maps, surveys
Studd Hill, Kent	London clay	1.5	1872—1898	Maps
		2	1898—1931	
		3.4	1931—1939	
		1.2	1939—1961	
Beltinge, Kent	London Clay	0.7	1872—1907	Maps
		0.9	1907—1933	
		1.1	1933—1939	
		0.9	1939—1959	
Grenham Bay, Birchington, Kent	Chalk	0.08—0.4	1872—1932	Maps
South Foreland, Kent	Chalk	0.19	1878—1962	—
Seven Sisters, Sussex	Chalk	0.51	1873—1962	—
Birling Gap, Seven Sisters, Sussex	Chalk	0.91	1875—1916	—
		0.99	1950—1962	Surveys
Middle Bottom, Dorset	Chalk	0.05	1882—1962	—
White Nothe to Hamburg Tout, Dorset	Chalk	0.22	1882—1962	—
Ballard Down, Dorset	Chalk	0.23	1882—1962	—
Fairy Dell, Dorset	Marls	0.4—0.5	1887—1969	Maps, air photos
Vale of Glamorgan, Wales	Lower Lias limestone, shale	0.008—0.099	1967—1969	Cliff-top stakes
Southeast County Down, N. Ireland	Glacial deposits	0.21—0.84	1834—1962	Maps, air photos
U.S.A.				
Martha's Vineyard, Mass.	Glacial deposits	1.7	1846—1886	Surveys
Boston Harbor, Mass.	Glacial deposits	0.23	1860—1908	Surveys
		0.31	1903—1905	Maps
Boston Harbor, Mass.	Glacial deposits	0.1	—	—
Outer Cape Cod, Mass.	Glacial deposits	1	1848—1888	—
Outer Cape Cod, Mass.	Glacial deposits	0.96	1848—1888	Surveys
Outer Cape Cod, Mass.	Glacial deposits	0.9—1.5+	(Long continued observations)	
Outer Cape Cod, Mass.	Glacial deposits	0.8	1879—1959	Surveys
Outer Cape Cod, Mass.	Glacial deposits	0.12—2.2	1938—1974	Air photos
Highland, Outer Cape Cod, Mass.	Glacial deposits	0.3	1953—1958	Surveys

Table 3.10—4 (continued)
WORLDWIDE COASTAL CLIFF EROSION RATES

Location	Lithology	Erosion rate (m/year)	Interval	Method
North Shore, Long Island, N.Y.	Glacial depsoits	0—0.5	1933—1966	—
North Shore, Long Island, N.Y.	Glacial deposits	0.5	(80 yr)	Charts, air photos
East Fort Point, Long Island, N.Y.	Glacial deposits	0.9+	1833—1883	—
Lake County, Ohio	Glacial deposits	0.3—0.9	1876—1937	Maps, air photos
		0.3 (max.)	1937—1973	Air photos
Berrien County, Mich.	Glacial deposits	3.8	1971—1974	Air photos
Eastern shore, Lake Michigan, Mich.	Glacial deposits	2.5	1970—1974	Surveys
Shannon Point, Wash.	Glacial deposits	0.2	1893—1969	Maps
North of Yaquina Bay, Ore.	Miocene sandstone	0.6	1880—1963	Photos
Rockaway Beach, Calif.	Franciscan volcanic and metavolcanic rocks	0.15	—	Maps
Montara, Calif.	Miocene conglomerate	0.26—0.29	1912—1965	Maps
Santa Cruz, Calif.	Mio-Pliocene mudstone	0—0.2	1943—1963	Air photos
Santa Cruz, Calif.	Miocene mudstone with shale	0.4	(80 yr)	Maps, surveys
Santa Barbara, Calif.	Miocene shale	0.2	1951—1965	Surveys
	Mio-Pliocene shale	0.08	1923—1967	
	Pliocene siltstone	0.3	1927—1947	
Isla Vista, Calif.	Mio-Pliocene shale	0.05—0.4	1966—1973	Maps, surveys
Leucadia to Pacific Beach, Calif.	Sandstone, claystone	0—0.5	1970—1975	Nails
La Jolla, Calif.	Cretaceous-Eocene sandstone	0.0003—0.0006	—	Dated inscriptions and graffiti
La Jolla, Calif.	Cretaceous sand-stone, Eocene shale	0.01—0.2	1940—1979	Photos
La Jolla, Calif.	Alluvium	0.5	(11 yr)	Photos
La Jolla, Calif.	Alluvium	0.28—0.50	1918—1930	Measurements
La Jolla, Calif.	Alluvium	0.09—0.3	1912—1975	Maps
Sunset Cliffs, San Diego, Calif.	Upper Cretaceous sandstone, siltstone	0.012	(75 yr)	Photos, observations
Sunset Cliffs, San Diego, Calif.	Pre-Miocene sand-stone, shale	0.01—0.3	—	Photos, maps

U.S.S.R.				
Barrents Sea coast	Granitic rocks	0.001—0.002	—	—
Laptev Sea coast	Glacial clay	4—6	—	—
Laptev Sea coast	Glacial clay	4	—	—
Laptev Sea coast	Glacial clay	50 (max.)	—	—
Laptev Sea coast	Glacial clay	3.5—4	—	—
Bering Sea coast	Moraine deposits	1—2 (max.)	—	—
Okhotsk Sea coast	Volcanic ash	50 (max.)	—	—
Okhotsk Sea coast	Quaternary brown loam and clay	40 (max.)	—	—
Baltic Sea coast	Granitic rocks	0.001—0.002 (max.)	—	—
Baltic Sea coast	Sandstone	3 (max.)	—	—
Baltic Sea coast	Sandy deposits	20 (max.)	—	—
Baltic Sea coast	Sandy deposits	2	—	—
Baltic Sea coast	Sandy deposits	1—1.5	—	—
Azov Sea coast	Quaternary brown loan and clay	3	—	—
Black Sea coast	Flysh	0.02—0.03	(20 yr)	Surveys, photos
Black Sea coast	Flysh, shale	0.01—0.02	—	—
Black Sea coast	Coquinite	0.002—0.005	—	—
Black Sea coast	Crystallized limestone	0.003	—	—
Black Sea coast	Massive limestone	0.3—0.5 (max.)	—	—
Black Sea coast	Limestone with loess	2—3	—	—
Black Sea coast	Limestone with loess	0.61	—	—
Black Sea coast	Quaternary loess	0.5—1.0	—	—
Black Sea coast	Quaternary conglomerate	12	—	—
Black Sea coast	Quaternary brown loam and clay	1 (max.)	—	—
Black Sea coast	Quaternary clay	2—3	—	—
Black Sea coast	Diluvial deposits	0.11	—	—

From Komar, P. D., Ed., *CRC Handbook of Coastal Processes and Erosion,* CRC Press, Boca Raton, FL, 1983.

Section 4

Marine Biology

Section 4

MARINE BIOLOGY

INTRODUCTION

Marine biology is the branch of marine science concerned with the study of organisms and their habitats in the oceans and estuaries of the world. It also deals with the physical and chemical factors that affect them. Over the years, marine organisms have been investigated for a variety of reasons, most notably for basic scientific research, for environmental impact assessment, for exploitation of natural resources, for purposes of management of amenity, and for identification of sites for conservation priority.[1] Because of marine pollution problems, numerous biotic monitoring programs have assessed anthropogenic impacts on aquatic populations, particularly during the last 2 decades (see Section 5). In many of these programs, data have been collected on the abundance, distribution, and diversity of marine organisms, together with information on physical and chemical parameters.[2] These data have resulted in the compilation of additional information on the structure of entire communities, as well as the biotic interactions and responses of organisms taking place within the systems under investigation. It is the objective of this section to examine the organisms, biological processes, and habitats commonly found in oceanic and estuarine environments.

BACTERIA

The microbiology of oceanic and estuarine environments is a rapidly evolving area of scientific inquiry. Marine bacteria play a critical role in these environments as decomposers of organic matter, as transformers of organic substrates into inorganic compounds, and as agents influencing the physical-chemical properties of these shallow coastal systems. Only relatively recently has the role of bacteria and their grazers in marine food webs become a focal point of research efforts. The concept of microbial loops and the view of bacteria as competitors for food and nutrients of other estuarine organisms has stimulated much interest in the interplay of microbial groups, material cycles, and the factors affecting them. The overwhelming significance of bacteria and protozoa in the heterotrophic activities of estuarine and marine ecosystems has precipitated investigations assessing the controls on microheterotrophic processes. Workers continue to evaluate microbial (carbon) production from different habitats and the significance of this carbon as an energy source that passes up the food chain to microheterotrophs (e.g., flagellates and ciliates) and detritivores.

Marine bacteria may be classified according to their habitat. Microbial habitats include planktonic, neustonic, epibiotic, benthic, and endobiotic types. Based on their life forms, these microbes may be broadly subdivided into autotrophic and heterotrophic organisms. In terms of their ecological roles, marine bacteria are essential to food chains and the cycling of elements (e.g., nitrogen and sulfur cycles).

The abundance of bacteria peaks in estuaries and coastal waters and gradually declines into offshore areas and the open ocean. Estimates of bacterial numbers in estuarine waters are in the range of 10^6 to 10^7 cells per milliliter; in the coastal ocean, in the range of 1 to 3 $\times$ 10^6 cells per milliliter; and in neritic waters of the ocean, in the range of 10^4 to 10^6 cells per milliliter. Similar spatial trends have been reported for bacterial biomass, with highest values occurring in estuarine waters and exceeding 10 μg C per liter. In coastal oceanic waters, biomass measurements commonly are between 5 and 10 μg C per liter and in offshore waters between 1 and 5 μg C per liter. Bacterial biomass may be as high as 50 to 100 μg C per liter in localized areas.[3–8]

The greatest number of microbes inhabit bottom sediments.[8,9] Abundances of bacteria in salt marsh and mudflat sediments often surpass 10^9 cells per cubic centimeter.[6,8] Bacterial counts drop when proceeding from intertidal to subtidal sediments in concert with the diminution in the concentration of organic matter. Cell counts typically decrease with increasing depth below the sediment-water interface within the top 20 cm of the sediment profile. A strong positive correlation exists here between the amount of organic matter and the abundance of bacteria; for example, the density of bacteria rises in deposits containing detritus. A correlation also holds between the number of bacteria and the size of sedimentary grains. Finer particles (clay and silt) have a larger microbial population attached to them than coarser particles (sand), as a consequence of the greater volume-specific surface area among the smaller grain sizes. Only a small portion of a sedimentary particle is actually colonized by the bacteria. Estimates of the density of bacteria on sediment surfaces range from one cell per 0.3 μm^2 to one cell per 400 μm^2. The microbes concentrate in cracks, crevices, and depressions of the grains where they are protected from mechanical damage. Extracellular polysaccharides produced by bacteria in contact with sediment surfaces serve as a food source for deposit feeders.

Bacteria in the ocean are most often small in size (less than 0.4 μm in diameter) and unattached,[10–12] with the larger cells mainly occurring in enriched waters at physical interfaces or attached to particles.[13] Kemp[8] specifies that bacteria inhabiting sediments generally average 0.6 μm in diameter. Wiebe and Pomeroy[14] detected few attached bacteria on particles in most coastal and open oceanic environments. Much (greater than 90%) of the heterotrophic activity in oceanic samples processed by Azam and Hodson[10] was apparent in the water fraction passing through 1-μm Nucleopore® filters. This finding differed from that of Hanson and Wiebe[15] who, investigating bacterial processes in estuarine waters, observed the bulk of heterotrophic activity associated with large particles. A similar result was reported by Goulder.[16]

The abundance of free-living bacteria relative to attached bacteria increases when progressing from estuarine waters to the open ocean.[17] For instance, Bell and Albright[18] documented declining attachment of bacteria downestuary in the Frazer River estuary in British Columbia, Canada. Sieburth and Davis[19] likewise confirmed diminishing numbers of attached bacteria downestuary in Narragansett Bay, Rhode Island.

Various techniques of study have facilitated research on bacterial growth and production. Chief among these methods are the use of epifluorescence microscopy, stains (e.g., acridine orange), and radioisotopic applications. A major breakthrough has been the use of tritiated thymidine incorporation for determining bacterial production in estuarine and marine environments.

Prior to the use of direct counting by epifluorescence microscopy in the mid-1970s, estimates of bacterial abundance in marine sediments were obtained primarily via a variety of indirect most-probable-number techniques.[20] The inefficiency of these techniques, however, resulted in severe underestimates of bacterial density.[21] With the advent of measuring rates of nucleic acid synthesis, especially DNA, microbiologists have gathered much more meaningful *in situ* productivity values on natural bacterial assemblages.

Moriarty[22] and Karl[21,23] explored the strengths and weaknesses of the nucleic acid synthesis techniques. The basic methodology for calculating growth rates by means of nucleic acid synthesis is, in practice, relatively simple, involving the measurement of the rate of incorporation of radioactive precursor in a macromolecule. Moriarty and Pollard[24] developed the use of tritiated thymidine incorporation into DNA for measuring the growth rate of bacteria in marine sediments, whereas Fuhrman and Azam[25] perfected the technique for application in seawater samples. While some microbiologists advocate radioactive thymidine for labeling DNA, others employ adenine as a labeling agent.[26] Adenine labels RNA as well as DNA; hence, estimates of microbial growth rates can also be derived from rates of RNA synthesis.[27] Current radioisotopic methods for determining bacterial production, although holding great promise in microbial investigations of oceans and estuaries, are expected to undergo refinement as their applications accelerate. This refinement should enable improvements to be made in the interpretation of results.

Bacteria are the main processors of organic matter in aerobic and anaerobic environments. Together with fungi, bacteria decompose substantial quantities of organic matter in aerobic zones of estuarine sediments. Perhaps equally important are bacteria inhabiting anaerobic zones. These organisms have more specific roles in the transformation of organic matter; while aerobic forms utilize a diversity of natural substrates, the anaerobes use a more restricted group of compounds. On the basis of their specialized biochemistry, anaerobic forms have been differentiated into four broad types: (1) fermenting bacteria, (2) dissimilatory sulfate-reducing bacteria, (3) dissimilatory nitrogenous oxide-reducing bacteria, and (4) methanogenic bacteria. Anaerobic microbial metabolism is, in some respects, enigmatic and far more complex than heterotrophic aerobic microbial metabolism which principally involves the transformation of dissolved organic matter (DOM) or particulate organic matter (POM) to biomass and carbon dioxide.

Aerobic heterotrophic bacteria can be found in three distinct habitats, that is, in the water column either suspended or attached to detrital particles, in the top layer of seafloor sediments, and in living and dead tissues of plants and animals. Free-living bacteria rely on dissolved organic carbon as an energy source, whereas attached bacteria not only use dissolved organic carbon as an energy source, but also the attached particle as a substrate and habitat. Fundamental differences between free-living and attached bacteria, such as the inequality in cell size and rate of assimilation of organic substrates, have been established. Metabolic activities — for example, nutrient uptake rates, utilization, and release — probably differ as well. The abundance of the attached microbes fluctuates markedly with the amount of particulate organic carbon; however, the number of free-living bacteria is more constant. As the density of attached bacteria rises in the presence of detritus, microbial metabolism seems to be altered.

Bacterioplankton play a vital role in the trophodynamic pathways of pelagic food webs. Of the total phytoplankton production converted to DOM, 10 to 50% is consumed by bacterioplankton.[28] Ingestion of the bacteria by phagotrophic protists which are consumed, in turn, by microzooplankton or larger zooplankton creates a microbial loop that salvages much of the DOM typically lost through remineralization. Microflagellates constitute the principal grazers of bacteria in the water column. The importance of this trophic link in marine and estuarine food chains remains uncertain.

Bacteria are essential links in detritus-based food webs, too, converting POM into DOM to meet their nutritional requirements and serving as a source of food for microfauna (e.g., ciliates and flagellates) and macrofauna (e.g., detritivores).[29–31] Detritivores depend on microbial enrichment of the more refractory detritus prior to assimilation, but may be capable of directly assimilating the more readily decomposable detritus (e.g., macroalgae) with little or no microbial enrichment.[32] Most detritus ingested by detritus feeders is voided relatively unchanged, albeit free of attached bacteria. The detritus may be consumed repeatedly by detritivores, therefore, with the attached bacteria being continually removed from the detritus for nutrition.[30] Although multiple investigations of detritus feeders in estuaries corroborate the significance of attached bacteria as a food source, recent evidence indicates that many deposit feeders obtain at least part of their energy from the detritus itself because the biomass of the bacteria consumed is thought to be insufficient to support the larger bacterivores.[33]

A successional sequence of organisms colonizes detrital particles. Bacteria probably act as the initial colonizers followed by fungi, algae, protozoans, other microfauna, and larger grazers.[34] Grazing on bacteria serves to stimulate their growth, promoting detrital mineralization. The chemical composition of the detritus also changes through time, becoming higher in nitrogen content with the increase in microbial biomass.[30,31] Extracellular excretion products (mucopolysaccharides) are responsible for much of the protein enrichment of detritus as it ages.[35,36]

During the decomposition of POM, three processes are discernible: (1) leaching of soluble compounds, (2) microbial decay, and (3) heterotrophic consumption.[30] An initial leaching phase during decomposition removes either soluble or autolyzed substances, and it is rather short-lived, lasting for several minutes to a few weeks, depending on the kind of detritus. A microbial decay phase, mediated by hydrolysis via enzymes released from microbial cells, occurs subsequent to the leaching phase. Compounds more resistant to decay (e.g., cellulose and lignin) predominate as decomposition proceeds, while the labile components, such as certain proteins and sugars, are depleted. The detritus eventually becomes more recalcitrant to decay, being composed of fulvic and humic acids consisting of refractory phenolic polymers and complexes that form marine humus. This refractory material concentrates in bottom sediments of estuarine and coastal marine systems. Heterotrophic consumption of bacteria takes place at any time during the process. Mechanical fragmentation of the detritus by physical and biological activity reduces the size of the particulate material, exposing new surface area to microbial attack.

As in the case of bacterioplankton, attached bacteria utilize DOM in the water column that originates from allochthonous (e.g., atmospheric and tributary systems) and autochthonous (e.g., exudation products of plants and animals and microbial decomposition products) sources. Research on the uptake of DOM by bacteria has often employed ^{14}C-labeled substrates (e.g., ^{14}C-glucose).[37] Bacteria have been shown to outcompete all other organisms for DOM in marine and estuarine environments.[38,39] Thus, the flux of DOM is closely tied to microbial metabolism. However, other factors contribute to the flux of DOM in these systems as well, such as the conversion of DOM to particulate form during the formation of amorphous organic particles.

Because fungi are rare or inactive in anaerobic environments, bacteria represent the principal transformers of organic matter in anoxic seas. Anaerobic conditions characteristically exist in deeper sediment layers, in systems having very poor water circulation resulting from physical restrictions of stratification (e.g., bottom waters of fjord-type estuaries), and some polluted regions. Much microbial decomposition and mineralization of organic matter take place in salt marsh and estuarine bottom sediments only a few millimeters or centimeters below the sediment-water interface. The depth to the anaerobic zone in seafloor sediments is a function of physical-chemical properties and biological processes. For example, light intensity at the sediment surface, the degree of turbulence on the seafloor, permeability of the sediment, bioturbation, and content of organic matter affect the thickness of the anaerobic zone.[40]

Anaerobic microbial metabolism yields a number of important substances — biomass, carbon dioxide, methane, soluble low-molecular weight compounds — potentially assimilable by aerobic organisms. Two pathways of anaerobic decomposition in estuaries include fermentation and dissimilatory sulfate reduction. Fermenting bacteria utilize various substrates, for instance, alcohols, amino acids, cellulose, pectin, purines, and sugars. Hydrogen, carbon dioxide, ammonia, and a group of organic compounds (e.g., alcohols and short-chain fatty acids) are derived from this type of anaerobic metabolism. Sulfate-reducing bacteria use SO_4^{2-} as a terminal electron acceptor during the decomposition of organic matter, with hydrogen sulfide generated in the process. The production of sulfides creates a black coloration in anoxic sediments. Sulfur reduction usually accounts for a significant amount of oxidation of organic matter in estuarine sediments; in some systems greater than 50% of the oxidation of carbon is attributable to this pathway.[41]

Dissimilatory nitrogenous oxide reduction occurs via several specific pathways: (1) denitrification, (2) dissimilatory reduction (terminating at NO_2), (3) dissimilatory ammonia production, and (4) nitrification N_2O pathway: ammonia to nitrous oxide. Nitrogenous oxides act as terminal electron acceptors. These pathways influence the flux of nitrogen in bottom sediments. They also govern, in part, nitrogen cycling in these complex environments. Dissimilatory nitrogenous oxide reduction pathways generally are responsible for less mineralization of organic matter than oxygen uptake and sulfate reduction.

Methanogenic bacteria differ from other anaerobes by using carbon dioxide or a methyl group as an electron acceptor while supplying methane.[42] The oxidation of organic matter by means of methanogenesis can be quantitatively significant in areas where sulfate concentrations remain low. Anaerobic zones of estuarine mudflats and salt marsh sediments that have high organic concentrations are likely sites for methanogenesis.[30,43]

PHYTOPLANKTON

Phytoplankton are free-floating, microscopic plants — unicellular, filamentous, or chain-forming species — inhabiting surface waters (photic zone) of open oceanic and coastal environments. Encompassing a variety of algal groups, these predominantly autotrophic organisms consist of single cells or relatively simply organized filamentous or chain-forming cell systems. They have been subdivided on the basis of size into ultraplankton (less than 5 μm in diameter), nanoplankton (5 to 70 μm), microphytoplankton (70 to 100 μm), and macrophytoplankton (greater than 100 μm). However, two broad fractions, the net plankton and nanoplankton, often are differentiated by means of the nominal aperture size of the plankton net deployed during field sampling. All phytoplankton retained by the net (approximately 64-μm apertures) comprise the net plankton, and those passing through the net constitute the nanoplankton.[44,45]

Diatoms (class Bacillariophyceae), dinoflagellates (class Dinophyceae), coccolithophores (class Prymnesiophyceae), silicoflagellates (class Chrysophyceae), and blue-green algae (class Cyanophyceae) are the principal phytoplankton taxa in the ocean,[30] although other taxonomic groups, including the green-colored algae (class Chlorophyceae), brown-colored phytoflagellates (class Haptophyceae), and euglenoid flagellates (class Euglenophyceae) also may be locally important in estuarine and lagoonal systems.[46] Algologists classify phytoplankton by using various criteria, notably biochemistry, life history, and fine cellular structure. Characteristics of the cell have greatly aided workers in the identification of phytoplankton. Such features as the cell shape, cell dimensions, cell wall, mucilage layers, chloroplasts, flagella, reserve substance (e.g., starch, oil, and leucosin), cell vacuoles, and trichocysts have facilitated taxonomic work.[46,47]

Diatoms, forming a frustule or external skeleton composed primarily of silica, frequently dominate phytoplankton communities in high-latitude regions, temperate inshore waters, and upwelling systems. Most marine diatoms are planktonic and nonmotile, ranging in size from 10 to 200 μm.[30,48] They tend to sink in nonturbulent waters, although a number of morphological, physiological, and physical adaptations promote their suspension. Storage products of photosynthesis include chrysolaminarin (a polysaccharide) and lipids, making these microflora a high-energy ration for zooplankton. Most diatoms reproduce by vegetative cell division, and under optimal conditions, cells double at a rate of 0.5 to 6 doublings per day, which enables blooms to develop.[49] During unfavorable conditions, some diatoms (e.g., numerous centric diatoms) produce resting cells or dormant stages that may be important in their occurrence, distribution, succession, and survival.[50,51]

Dinoflagellates are also widely distributed in oceanic and estuarine environments, being dominant in many subtropical and tropical regions and abundant in temperate and boreal autumnal assemblages.[29] Typically unicellular, biflagellated, and autotrophic, these plankters often compose a major fraction of the phytoplankton communities of estuaries. Two groups can be distinguished by the presence or absence of thecal plates: (1) the Peridinales (thecate or armored types) and (2) the

Gymnodinales (nonthecate or naked forms). While some dinoflagellates reproduce sexually, most undergo asexual reproduction by binary cell division. Rates of reproduction are variable, depending on environmental conditions, but approach those of the diatoms. The vast majority of dinoflagellates consist of autotrophic forms; however, heterotrophic (e.g., *Kofoidinium* and *Polykrikos*) and parasitic (e.g., *Blastodinium* and *Oodinium*) types also exist. A number of species produce toxins which, when liberated, occasionally cause mass mortality of fish, shellfish, and other organisms. These effects become most pronounced during toxic, "red tide" blooms in estuaries associated with the proliferation (densities from 5×10^5 to 2×10^6 cells per liter)[45] of such genera as *Gonyaulax* and *Ptychodiscus* (= *Gymnodinium*). Paralytic shellfish poisoning, a neurological disorder in humans, is ascribable to the toxic agent saxitoxin (produced by dinoflagellates) that accumulates in shellfish tissues.

Coccolithophores are small, biflagellate, unicellular algae characteristically covered by coccoliths or small calcareous plates embedded in a gelatinous sheath surrounding the cell.[52] These small phytoplankters (5 to 50 μm) reach peak abundance in tropical and subtropical, open oceanic waters, but sometimes proliferate in coastal environments. Although most species inhabit warmer seas, a few (e.g., *Pontosphaera huxleyi* and *Syracophsaera* spp.) attain maximum abundance in colder regions. The major fraction of photosynthesis in certain areas is ascribable to coccolithophores, which periodically comprise a significant portion of calcareous sediment (i.e., Globigerina ooze) on the seafloor of the open oceans.[53] Despite the overwhelmingly autotrophic existence of plankters in this group, a few taxa obtain energy heterotrophically below the photic zone.

Plankton nets, plankton pumps, and water bottles are three types of gear typically employed during phytoplankton field sampling. Once samples are obtained, the standing crop or biomass of phytoplankton can be determined by direct cell counts, chlorophyll *a* determinations, or ATP content. The standing crop of phytoplankton varies substantially both in space and time owing to the patchy distribution of the microflora. The patchiness of phytoplankton populations develops from organismal responses to hydrographic, light, and nutrient distributions, to predation and symbiosis, and to mechanical aggregation by physical processes.[54,55]

Phytoplankton productivity pertains to the rate of carbon fixation (organic synthesis), and is determined by measuring rates of photosynthesis or respiration. Methods used to assess primary productivity of phytoplankton are the measurements of: (1) oxygen released during photosynthesis, (2) carbon dioxide uptake, (3) pH, (4) the rate of appearance of new algal biomass over time, and (5) uptake of the radioactive isotope, ^{14}C. The ^{14}C method represents the most widely adapted technique for estimating primary planktonic productivity in the sea.

The annual production of phytoplankton averages about 50 g C/m^2/year in the open ocean, 100 g C/m^2/year in coastal marine waters (ranging from 50 to 250 g C/m^2/year in areas removed from upwelling effects), and 6.8 to 530 g C/m^2/year in estuaries.[31,56–59] Lower values are encountered in the centers of subtropical gyres (e.g., Sargasso Sea) and in turbid coastal and estuarine waters. Similar to phytoplankton biomass, phytoplankton productivity varies widely both in space and time.

The primary production of phytoplankton is a function of the interaction of a number of physical, chemical, and biological factors, the most important of which are light, temperature, water circulation, salinity, nutrients, and grazing.[29,30,44,45] Light energy has been deemed to be a limiting factor controlling the distribution of phytoplankton.[31,44] Seasonal variations in illumination with latitude result in different seasonal production patterns in tropical, temperate, boreal, and polar regions.

The amount of illumination in the sea depends on the angle of the sun during the day, the season of the year, the latitude, and the local climatic conditions (e.g., the percent of cloud cover). A portion of the solar radiation incident at the sea surface is lost by scattering and reflection. In the water column, absorption and scattering of light by water molecules, suspended particles, and dissolved matter further attenuate light. Light attenuation is greater in estuaries than in coastal or open oceanic waters where turbidity tends to be lower. In addition, different spectral distributions characterize estuarine, coastal, and oceanic waters. In turbid coastal systems, the maximum transmittance occurs at approximately 575 nm compared to 465 nm in clearer oceanic systems.[60,61]

Within the range of photosynthetically active radiation, approximately 400 to 700 nm, phytoplankton utilize chlorophyll pigments as well as accessory pigments (e.g., fucoxanthin and peridinin) to absorb all wavelengths of light.[62] Photosynthesis increases logarithmically with increasing light intensity until a maximum value is reached. Near the sea surface, where light saturation exists, however, photosynthesis remains inhibited. With increasing depth below the sea surface, photosynthesis rises to a maximum amount. At greater depths, photosynthesis diminishes; at the compensation depth, it just balances the metabolic losses due to respiration. Because of high turbidity in estuaries, the compensation depth may be very shallow, in some cases 1 m or less. Photosynthesis-irradiance curves have proven to be useful in studies of phytoplankton production in relationship to changing light intensity and quality.[63]

Light has been implicated as a major factor triggering winter-spring phytoplankton blooms in mid- and high-latitude systems. Nutrients that accumulate during the winter in estuarine and coastal embayments are assimilated by phytoplankton when light increases, thereby accelerating growth and initiating the algal blooms. Phytoplankton pulses commonly appear in estuaries at various times during the spring and summer months.[29,44–46]

Temperature affects enzymatic activities (respiration and photosynthesis) and growth rate processes of phytoplankton. Laboratory cultures indicate that the cell division rates of phytoplankton generally increase by two to four times with a 10°C increase in temperature, as long as the temperatures lie within a range favorable to growth.[64] The development of a thermocline in mid- and high-latitude waters due to seasonal temperature changes has been related to improved growth of phytoplankton and seasonal production cycles.

Phytoplankton employ several osmotic mechanisms to ameliorate the effects of highly variable salinity levels. Because of their ability to osmoregulate, phytoplankton typically have euryhaline tolerances. Experimental work demonstrates that estuarine phytoplankton are more euryhaline than coastal and oceanic forms.[65] The rate of cell division of these microflora, as well as their occurrence, distribution, and productivity, is influenced by salinity.

Nutrients are necessary for adequate growth and production of phytoplankton. The major nutrient elements include nitrogen, phosphorus, and silicon. Trace elements, such as iron, manganese, zinc, copper, cobalt, and molybdenum, may limit phytoplankton growth if present in insufficient concentrations.

However, some trace metals (e.g., copper and zinc) can be toxic to phytoplankton even at low concentrations and, consequently, may hinder their productivity. Some species require the vitamins cobalamine, thiamin, and biotin, as well as other organic compounds.[30,44–46]

ZOOPLANKTON

Many scientists usually classify zooplankton according to their size or the length of their planktonic life. Three size categories (i.e., micro-, meso-, and macrozooplankton) are typically used in the classification of zooplankton. Those forms passing through a plankton net with a mesh size of 202 μm comprise the microzooplankton, and individuals retained by this net constitute the mesozooplankton. Still larger members of the zooplankton community collected with plankton nets constructed of 505-μm mesh compose the macrozooplankton.[66,67] Among the microzooplankton less than about 60 μm in size, protozoans and tintinnids, together with meroplanktonic larvae of benthic invertebrates and copepod nauplii, are most numerous. Important groups of the mesozooplankton include cladocerans, copepods, rotifers, and larger meroplankton. In terms of absolute abundance and biomass, copepods, in particular, may completely overshadow other mesozooplankton in some oceanic and estuarine regions. Three major macrozooplankton groups are commonly observed: (1) the jellyfish group (hydromedusae, comb jellies, true jellyfishes); (2) crustaceans (amphipods, isopods, mysid shrimp, true shrimp); and (3) polychaete worms.

Zooplankton may be classified into three classes based on the duration of planktonic life, namely holoplankton, meroplankton, and tychoplankton.[29,44,48,67] Holoplankton are those organisms which remain in the plankton their entire lives. Copepods, cladocerans, and rotifers exemplify a holoplanktonic existence. Meroplankton refer to those animals which live only a part of their life cycle in the plankton and consist largely of the larvae of benthic invertebrates, benthic chordates, and fishes. Tychoplankton are demersal zooplankton which are periodically inoculated into the plankton by bottom currents, wave action, and bioturbation.[68–70] Some migrate into bottom waters to feed. These organisms do not normally constitute a quantitatively significant fraction of the zooplankton community. Amphipods, isopods, cumaceans, and mysids are common members of the tychoplankton, and all represent potential food sources for omnivorous and carnivorous zooplankton as well as planktivorous fish.

A number of biotic and abiotic factors affect zooplankton dynamics and community structure. Light is a major environmental factor regulating diel vertical migration of these organisms.[71] Acting as an environmental cue which triggers the migration process, illumination changes at sunrise and sunset elicit vertical movements of zooplankton populations.[72,73] Three general vertical migration patterns are recognized, specifically nocturnal, twilight, and reverse migration. The following light cues may precipitate a vertical migration: (1) a change in depth of a specific light intensity, (2) a change in underwater spectra, (3) a change in the polarized light pattern, (4) an absolute amount of change in light intensity, and (5) a relative rate of intensity change.[74–76]

Metabolic rates of zooplankton are a function of temperature, which influences the growth, fecundity, longevity, and other life processes of zooplankton.[44,77] Because higher temperatures accelerate molting with less growth occurring during intermolt periods, susceptible zooplankton inhabiting warmer waters usually grow to a smaller size than individuals living in cooler regions. Fecundity is lower in adults of smaller size; therefore, lower temperatures favor zooplankton with greater fecundity.[78] In regard to ecological effects attributable to temperature changes, seasonal variations in the species composition and abundance of zooplankton are well documented, especially in temperate waters. The community of zooplankton in mid-Atlantic Bight estuaries, for example, undergoes a succession of species populations in which winter-spring dominants (e.g., *Acartia hudsonica*) are replaced by summer-fall dominants (e.g., *A. tonsa*).[79] From spring to fall, increased spawning of benthic invertebrates triggered by temperature changes occasionally results in pulses of meroplankton that often contribute to peak concentrations of zooplankton during these seasons.

Zooplankton are responsive to salinity levels encountered along the longitudinal and vertical axes of estuaries. The salinity tolerance of zooplankton varies among species and may change among ontogenetic stages of a single species.[67] As a result, the patterns of species succession and dominance along the longitudinal axis of an estuary are contingent upon salinity levels. In the vertical plane, salinity limitations may restrain the vertical migration of zooplankton populations. Salinity extremes often arrest growth and increase mortality of susceptible forms.

Zooplankton have an essential role in marine and estuarine food chains as an intermediate link between primary producers (i.e., phytoplankton) and secondary consumers. The grazing of zooplankton provides energy for higher-trophic-level organisms; however, many zooplankters are omnivorous, which complicates the structure of food webs (see below). Observations of copepods indicate that they feed by means of a passive filtration process of small particles or by active capture of large particles.[80,81] In the active capture of prey, copepods initially detect a food particle by chemo- or mechanoreception, and, subsequently, they seize it. Feeding appears to be much more sophisticated among copepods than once believed; until recently, consumption was considered to entail indiscriminate filter feeding or simple sieving of phytoplankton. Some copepods clearly display selective feeding behavior. A number of investigators have advocated a coupling between diel vertical migration behaviour and feeding activity of these organisms.

The rate of ingestion of zooplankton increases with increasing prey density up to a limit above which the ingestion rate is nearly constant. This relationship can be described by using a number of models, such as the Ivlev equation and Michaelis-Menten equation. A feeding threshold of zooplankton also exists below which phytoplankton are no longer exploited.[82] Feeding thresholds preclude the elimination of phytoplankton populations via grazing pressure. While zooplankton ingest large amounts of food, only a portion is assimilated, with the remainder being egested. Assimilation efficiencies for zooplankton measured by several different methods often exceed 50% and may equal 99% in some cases.

BENTHOS

BENTHIC FLORA

The benthos consists of all organisms, both plant and animal, living on or in various bottom substrates (e.g., sediments, rocks, man-made structures) of estuarine and oceanic environments. In coastal and estuarine systems, benthic flora and fauna are important to energy budgets and may serve as habitat formers. Macrophytes (algae and vascular plants) constitute the major fraction of benthic biomass in these coastal systems, with emergent and submergent vascular plants providing the most organic carbon. Bottom sediments are often devoid of benthic macroalgae, which typically attach to hard surfaces, including man-made structures, shells, stones, and rocky shores. They often drift passively over the estuarine and coastal seafloor. Dense meadows of vascular plants (e.g., seagrasses) commonly grow in shallow subtidal areas. Salt marsh grasses and mangroves are conspicuous features of the intertidal zone of temperate and tropical regions, respectively.

The benthic flora of nearshore habitats can be separated into micro- and macrofloral components. The microflora are most extensively developed in intertidal habitats where single-cell algae (i.e., diatoms, blue-green algae, and dinoflagellates) frequently discolor the sediments due to their large numbers. Single cells or filamentous colonies of microflora adhere to sediments and also attach to the surfaces of rocks, other plants, animals, and man-made materials. Motile forms migrate vertically in sediments in response to changing light intensities. Mats of green and blue-green algae underlie salt marshes and tidal flats. Blue-green algal mats are prominent in warm-water systems, such as those located along the Gulf of Mexico.

In certain habitats (e.g., tidal flats), benthic microalgae supply a considerable amount of primary production. Annual production estimates typically range from about 25 to 2000 g C/m^2/year.[30] Benthic microalgae, which form mats on mudflats, bind sediment via the secretion of extracellular mucous films, thereby stabilizing the substrate and mitigating erosion. Many microalgal species grow as epiphytes on the leaves and stems of macrophytes where they form a furlike covering termed periphyton or Aufwuchs. Grazers utilize the periphyton as food.[83] Grazing by mud snails, fiddler crabs, herbivorous fish, and other fauna controls the growth of epibenthic algae, limiting biomass and production. Epiphytic felt flora are habitation sites of micro- and meiofauna (i.e., protozoans, nematodes, rotifers, tardigrades, copepods, and ostracods) that consume associated microorganisms (e.g., bacteria and diatoms) and, in turn, supply ration for macrofauna. Species of *Cladophora, Enteromorpha, Hypnea,* and *Polysiphonia* are common components of the algal felt flora.

The attachment of epiphytes to submerged macroflora is beneficial for several reasons. First, the submerged plants act as support or anchorage for the epiphytes. Second, the greater proximity of the microflora to light enhances epiphytic growth, and water currents remove growth-inhibiting substances and transport nutrients.

Benthic macroalgae (seaweeds) cover only small areas of estuarine and coastal marine substrates, usually attaining peak densities on hard surfaces. Because of strong currents and wave action, they generally require attachment to a solid substrate. Adhesive, rootlike holdfasts or basal disks provide effective means of attachment for these plants.

Environmental conditions in estuaries are not conducive to luxuriant growth of benthic algae. High turbidity and sedimentation, for instance, limit light penetration. These factors account for the species-poor condition of benthic macroalgae in many of these complex systems. Most of the species belong to the Chlorophyta (green algae), Phaeophyta (brown algae), and Rhodophyta (red algae).[67] Representative genera of the Chlorophyta include *Cladophora, Enteromorpha,* and *Ulva.* Among the Phaeophyta, species of *Ascophyllum, Fucus,* and *Pelvetia* may be common, especially on rocky shores. *Ceramium, Gracilaris,* and *Polsiphonia* are genera of the Rhodophyta reported in many estuaries.[83] Some macroalgal taxa (e.g., *Blidingia minima* var. *subsalsa, E. clathrata,* and *Vaucheria* spp.) reach maximum abundances in estuaries. The distribution of benthic macroalgae in these systems is distinctive along the longitudinal axis. Species diversity decreases when proceeding from the estuarine mouth to its head. Annual species, particularly green algae, extend farther upestuary compared to perennial red or brown algae, which are more prevalent downestuary.

Seaweeds often display a zonation pattern, notably on rocky shores, due to differences in dominance among populations.[84] Species zonation may be more obvious on artificial embankments and other man-made structures than on natural banks.[85] Factors contributing to zonation of macrofloral populations are species competition, grazing, and physiological stresses associated with emersion.

Seaweeds serve as an important source of food for herbivores. Live plant matter not ingested by benthic grazers ultimately enters the detritus food web where microorganisms colonize and decompose the material. Macroalgal detritus breaks down rapidly relative to vascular plant detritus, which is much more refractory. It may be degraded in only a few weeks in estuaries and nearshore habitats, but the more recalcitrant vascular plant detritus requires several months or even years to decompose.[30,86–88]

The benthic macroflora are best represented in three major plant communities: salt marshes, seagrasses, and mangroves. Rooted vascular plants predominate in these communities, with benthic macroalgae being less important in terms of biomass and production. Globally, salt marshes range from mid-temperate to high latitudes and are replaced from suitable sites in the tropics by mangroves, which generally lie between 25°N and 25°S latitude. Seagrasses have a very wide distribution; they can be found in shallow waters of all latitudes except the most polar.[89]

Salt-marsh plants dominate the vegetation of intertidal zones in mid- and high-latitude regions.[89–91] Consisting of halophytic grasses, sedges, and succulents, which develop on muddy sediments at and above the mid-tide level, salt marshes comprise much of the 2 × 10^2 ha of estuarine wetlands in the lower 48 contiguous states. Most of the flora in salt marshes belong to a few cosmopolitan genera (e.g., *Spartina, Juncus,* and *Salicornia*) that are broadly distributed.

Six types of salt marshes have been distinguished (i.e., estuarine, Wadden, lagoonal beach plain, bog, and polderland types).[90] The formation of each type depends on salinity and tidal range. Three geographical units of salt marshes are recognized in North America: (1) Arctic, (2) eastern North America, and (3) western North America marshes. Salt marshes are well developed on the southern shore of Hudson Bay, the East Coast of Canada and the U.S., and the Gulf Coast of the U.S.[92] Discernible differences in species composition occur among the salt marshes of these three geographical units, although the predominance of *Spartina alterniflora* and *S. patens* has been noted in mesohaline and polyhaline marshes. The salt-marsh grasses bordering estuaries frequently exhibit a well-defined zonation. For example, the coastal plain marshes along the Atlantic Coast south of Chesapeake Bay are characterized by a lower marsh of tall form *S. alterniflora* and an upper marsh of *S. patens, Distichlis spicata, Juncus roemerianus,* and *Sali-*

cornia sp. Along the Atlantic Coast of North America, salt marsh communities tend to be less diverse and their zonation and succession patterns less complex than those along the Pacific Coast.

Sheltered coastal areas, where sedimentation is ensured and erosion remains slight, are ideal sites for salt marsh development. The halophytes initially establish populations on mudflat surfaces between the levels of mean high water neap (MHWN) and MHW.[90] Growth of the salt marsh proceeds as sedimentation produces a surface above the MHWN, which forms the limit for the establishment of the halophytes. Maturation of the marsh is promoted by gradual accretion of sediments. Sedimentation rates of 3 to 10 mm/year enhance salt marsh maturation. Higher rates of sedimentation, however, may arrest successional development and restrict species richness. Sediment progradation plays a crucial role in the state of marsh maturation (i.e., youth, maturity, and old age), which has been related to the stages of plant and animal succession.[92]

Salt-marsh systems serve at least five significant ecological functions. They are highly productive; estimates of aboveground production of *Spartina* salt marshes range from 200 to 3000 g C/m^2/year, and below-ground production may be even greater.[31] Much of this primary production enters the detritus food web upon the death of the plants, and forms an essential energy source for many estuarine organisms. Many animals utilize the salt-marsh habitat for food, shelter, and reproduction. The rooted vegetation anchors the sediment and reduces erosion. The salt marshes also behave as sources or sinks of trace metals and nutrients.

Tidal freshwater marshes replace tidal salt marshes as salinity diminishes inland. They develop in areas where the annual salinity averages 0.5‰ or less. Farther upstream, these marshes grade into nontidal freshwater wetlands.

A recurrent biotope of exceptional value in estuaries worldwide is the seagrass community composed of monocotyledonous angiosperms that grow in shallow subtidal waters. Twelve genera of seagrasses (i.e., *Amphibolis, Cymodocea, Enhalus, Halodule, Halophila, Heterozostera, Posidonia, Phyllospadix, Syringodium, Thalassia, Thalassodendron,*. and *Zostera*), incorporating about 50 species, have been identified worldwide.[93,94] *Zostera* (eelgrass) is the dominant genus in temperate waters of North America, and *Thalassia* (turtlegrass) predominates in subtropical and tropical regions. Other abundant genera in North America include *Halodule* and *Phyllospadix*. Seven genera of seagrasses occur in tropical latitudes compared to five in temperate latitudes. *Zostera* has the broadest distribution of all seagrass genera in the temperate and boreal waters of the Atlantic and Pacific Oceans. It also has been the most intensely studied seagrass.

Locally, seagrasses are restricted to lower intertidal and subtidal zones in estuarine and coastal oceanic environments. Anchored to bottom sediments by a network of roots and rhizomes and structured with a dense arrangement of stems and leaves, seagrasses create a habitat for many other aquatic organisms. Sediment fauna, rhizome and stem biota, leaf epiphyton, and nekton form four productive subhabitats within seagrass systems. The thick foliage affords prey protection from predators and shelters organisms from strong currents and wave action.

Salinity, light, and turbidity strongly affect the distribution and growth of seagrasses. In turbid estuaries, where high seston concentrations restrict light transmission, seagrasses often are limited to waters less than 1 m deep. However, in regions of greater water clarity, seagrasses may extend to depths of 30 m.

Seagrasses are ecologically significant because of their high primary production and value as habitat formers. Estimates of their primary production range from 58 to 1500 g C/m^2/year, surpassing maximum biomass values by about 2 to 2.5 times.[31,94] Numerous invertebrate and fish populations utilize seagrass habitats as nursery, feeding, and reproductive grounds. Large quantities of detritus generated from seagrasses also are important in the energy flow of many estuarine ecosystems. Additionally, these rooted vascular plants play a role in the cycling of certain critical nutrient elements.

Mangroves or mangals are assemblages of halophytic trees, shrubs, palms, and creepers that form dense thickets or forests in intertidal and shallow subtidal zones of subtropical and tropical waters.[59,89,95] They thrive in protected embayment areas, tidal lagoons, and estuaries between 25°N and 25°S latitudes. Florida has the most elaborate mangrove communities in the U.S., dominated by the black mangrove (*Avicennia germinans*), red mangrove (*Rhizophora mangle*), and white mangrove (*Laguncularis racemosa*). Mangroves rim more than 170,000 ha of the Florida coastline; these plant associations are prominent along the Gulf Coast as well. Although extending northward as far as 32°N in the U.S., mangroves have established most successful communities south of 28 to 29°N latitude.[89,96]

Worldwide, about 80 species of monocots and dicots belonging to 16 genera have been described in mangrove communities, with at least 34 species in 9 genera believed to be true mangroves.[89,95] Based on hydrodynamics (i.e., tidal action and freshwater hydrology) and topography, mangroves can be classified into six types of wetland systems: (1) overwash mangrove islands, (2) fringe mangrove wetlands, (3) riverine mangrove wetlands, (4) basin mangrove wetlands, (5) hammock mangrove wetlands, and (6) dwarf mangrove wetlands. Mangroves exhibit a zonal pattern related to the salinity tolerances of the plant populations, tidal inundation, and other factors. In South Florida, for example, an *R. mangle* fringe builds into shallow subtidal waters. *A. germinans* occurs immediately landward of *R. mangle* from the lower to the upper intertidal zone. Upland of *A. germinans, L. racemosa* occupies sites in the middle to upper intertidal zone. Farther upland, *Conocarpus erecta* or buttonwood replaces *L. racemosa,* and, although occasionally inhabiting the upper intertidal zone, usually is part of the sand/strand vegetation. Some workers consider mangroves to be pioneer or seral successional stages evolving toward a freshwater, terrestrial plant community. According to this postulation, an estuarine/mudflat community can be transformed into a tropical rain forest.

The prop or drop-type roots of mangrove trees are shallow, and cable roots projecting horizontally from the stem base provide additional support. High rates of sedimentation typify mangrove systems. Submerged portions of mangrove roots frequently have dense growths of benthic algae attached to them (e.g., species of the genera *Bostrychia, Caloglossa, Catonella,* and *Murrayella*). The subtidal root systems of mangroves protect the shoreline, enhance bank stabilization, and promote land accretion, while mitigating erosion.

As in the case of salt marshes, tidal freshwater marshes, and seagrass systems, mangroves are highly productive, having primary production estimates ranging from 350 to 500 g C/m^2/year.[59] A substantial portion of this production supports detritus-based food webs. A highly diverse assemblage of vagile and sessile faunal populations also inhabits mangroves. Insects, reptiles, and terrestrial mammals have also successfully adapted to mangrove forests.

BENTHIC FAUNA

Classification

Benthic fauna can be subdivided according to size into micro-, meio-, macro-, and megafaunal components.[29,31] Individuals which pass through sieves of 0.04- to 0.1-mm mesh constitute the microfauna. The meiofauna, metazoans weighing less than 10^{-4} g (wet weight), are retained on sieves of 0.04- to 0.1-mm mesh.[97] The macrofauna are larger animals captured on sieves of 0.5- to 2.0-mm mesh. Still larger benthic fauna visible to the naked eye are categorized as megafauna.

Benthic microfauna consist essentially of protozoans. The meiofauna can be subdivided into: (1) temporary meiofauna which are juvenile stages of the meiofauna, and (2) permanent meiofauna (i.e., gastrotrichs, kinorhynchs, nematodes, rotifers, archiannelids, halacarines, harpacticoid copepods, ostracods, mystacocarids, and tardigrades as well as representatives of the bryozoans, gastropods, holothurians, hydrozoans, oligochaetes, polychaetes, turbellarians, nemertines, and tunicates).[98] Whereas the absolute abundance of micro- and meiofauna far exceeds the benthic macrofauna, the biomass of the macrobenthos is much greater.

The species composition of meiofauna in a given location is a function in part of the sediment type. For instance, the meiofauna of sand flats are vermiform, interstitial species that negotiate crevices between sedimentary grains. In muddy sediments, burrowing forms predominate. The distribution of meiofauna is patchy when proceeding from intertidal to subtidal zones due to variability of the physical environment (e.g., temperature, salinity, desiccation, and sediment grain size) and biological interactions (e.g., predation, competition, and bioturbation). Horizontal gradients in salinity strongly modulate the species composition of meiofauna. The species richness of this group generally increases with rising salinity from the head to the mouth of estuaries. Higher salinities also correspond to greater meiofaunal densities. Changes in the species composition and abundance of meiofauna have likewise been correlated with seasonal temperature changes.

Chemical zonation in bottom sediments exerts some control on vertical distribution of meiofauna. Vertical zonal patterns in subtidal sediments, for example, reflect the ability of the meiofauna to tolerate low oxygen concentrations. Other factors influencing the vertical zonation of the fauna include food availability, amount of organic matter, sediment grain size, and selective predation.[99]

In sediments on the continental shelf, meiofaunal densities range from about 10^5 to $10^7/m^2$,[100,101] and standing crop dry weight biomass figures range from approximately 1 to 2 g/m^2.[99] Annual meiofaunal production exceeds 20 g C/m^2/year in some estuaries. The fecundity of the meiofauna is less than that of most macrofaunal species. These diminutive organisms usually release from 1 to 10 eggs at a time, although members of the archiannelid genus *Polygordius* and several species of *Protodrilus* produce up to 200 eggs per reproductive event. Most estuarine meiofauna undergo direct development; brood protection and viviparity are common. The meiofauna are a highly successful group of benthic organisms because of several reproductive adaptations, such as the year-round reproduction of some of the taxa, relatively short generation times, and the delay of life-history development via resting eggs or larval stage delay.

The species composition and abundance of benthic macrofauna have wide temporal and spatial variations.[102–104] Fluctuations in abundance are produced among opportunistic species (e.g., *Mulinia lateralis, Capitella capitata, Cistenides* (= *Pectinaria*) *gouldii,* and *Polydora ligni*). Large changes in abundance of the macrofauna during the year may be ascribable to normal periodicities of reproduction, recruitment, and mortality. However, they may also be attributable to random environmental perturbations where vagaries in physical or chemical conditions can precipitate large aperiodic density changes.

The benthic fauna can also be classified on the basis of life habits and adaptations into epifauna, which live on the seafloor or attached to a firm substrate, and infauna, which live in bottom sediments. Other (nonparasitic) species, however, are best classified as interstitial, boring, swimming, and commensal-mutualistic forms. Using feeding habits as criteria, five types of benthic fauna have been identified: (1) suspension feeders, (2) deposit feeders, (3) herbivores, (4) carnivores, and (5) scavengers.[29,105]

Spatial Distribution

The spatial distribution of benthic macroinvertebrates has been assessed on three levels: (1) small-scale distributions (e.g., locations within an estuary), (2) regional distributions (e.g., on the scale of the estuary), and (3) global distributions.[106] Physical, chemical, and biological factors clearly influence the functional morphology, behavior, and distribution of benthic fauna as alluded to above.[98] The species composition and local distributions of the benthic macrofauna have been related to various physical factors, notably waves and currents, sediment characteristics, and depth. The species composition of the benthic macrofauna within habitats is strongly dependent on sediment type. The polychaete *P. ligni* and the bivalve *Scrobicularia plana,* for example, prefer muddy sediments, while the amphipods *Bathyporeia* spp. and *Haustorius* spp. and the polychaetes *Ophelia* spp. live in sandy bottoms.[106] Biological factors (e.g., predation and competition) and chemical factors (e.g., oxygen concentrations) also may greatly affect the local distribution of the benthos. Biotic factors can act as limiting factors;[99] hence, the mere tolerance of a species to physical and chemical conditions oftentimes does not provide sufficient explanation of an observed distribution pattern. The occurrence of a species depends on biological adaptation as well. Similarities in reproductive season, modes of feeding, size, and other biotic factors, for instance, may enable cohabitation of several species in the same general environment.[100,101]

As the concentration of dissolved oxygen diminishes with increasing depth in bottom sediments, the absolute abundance of the macrofauna also drops. The abundance of benthic macroinvertebrates peaks in the oxygenated surface layers. This relationship may not hold for various micro- and meiofauna, however, which may attain maximum numbers in deeper sediments (e.g., near the chemocline).[107]

The distribution of the macrobenthos in the sediment column has also been related to the stage of successional development of the benthic community in addition to local physical disturbances.[108–110] In habitats subjected to frequent physical disturbances, pioneering species of infauna — tubicolous or otherwise sedentary invertebrates living in near-surface sediments — tend to dominate the community. The pioneering forms feed near the sediment surface or from the water column. Habitats devoid of physical disturbances harbor higher-order successional stages or equilibrium stages of benthos dominated by bioturbating infauna which feed at greater depths within bottom sediments.

Aside from physical disturbances caused by wave and current action, sediment deposition and erosion, as well as other factors, biological disturbances due to grazing, predation, and competition control the vertical distribution of the infauna. The proximity of epifaunal predators elicits responses from infaunal prey that leads to their repositioning within the sediment. Disturbances of the seafloor have been enlisted by benthic ecologists as principal causative agents of the spatial mosaic pattern observed in bottom communities. Sediment profile imaging has proven to be a viable method of assessing vertical profiles of the seafloor and of determining the stage of successional development of the benthos in shallow water environments. The sediment-profile camera has been a valuable tool in ascertaining the structure of the benthic community.[111]

Larger-scale distributional patterns of the benthic macrofauna may be evident when examining population responses to gradients in environmental conditions. For example, changes in salinity along the length of an estuary have a profound effect on the species distribution of the benthic fauna. Abrupt changes in biota have been discerned at salinities of 0.5, 5, 18, and 30‰. Along the homoiohaline gradient of the Chesapeake Bay, gradual and relatively uniform benthic macrofaunal changes occur except within the salinity ranges of 3 to 8 and 15 to 20‰ where the changes in the macrobenthos are more accelerated.[112]

The distribution of benthic macroinvertebrates is a function of larval dispersal and success of recruitment. Many benthic faunal populations have a planktonic larval stage early in ontogeny, and at this time, the larvae experience high mortality due to predation and the vagaries of environmental conditions. Nevertheless, currents can transport the larvae long distances to populate favorable habitats. The length of the planktonic larval stage of benthic invertebrates varies, depending on the mode of development, environmental factors, and chemical or physical cues which induce the larvae to settle and metamorphose. Three major categories of planktonic larvae are recognized: (1) long-life planktotrophic forms, (2) short-life planktotrophic forms, and (3) lecithotrophic forms.[113,114] Two other modes of development of benthic macroinvertebrate larvae are direct development, omitting a pelagic phase, and viviparity in which the larvae develop within parental organisms. Some workers recognize only two primary categories of larval development, namely planktotrophic and nonplanktotrophic development.[115,116]

The degree of larval dispersal is not only contingent upon the duration of the pelagic stages, but also upon the behavior of the larvae and, perhaps most importantly, upon the hydrodynamic processes of the water body.[117,118] While horizontal advective processes and the longevity of the pelagic larval stage primarily determine the potential for dispersal, larval behavior is responsible in many cases for the actual degree of dispersal.[119] Hence, swimming behavior of the larvae in the water column can maximize or minimize horizontal advective transport. Strong swimmers (e.g., decapod crustacean larvae) maintain greater control over larval dispersal by the strength of their movements in the water column. Larval swimming behavior is responsive to environmental factors. Directional swimming (e.g., geotaxis and phototaxis) and changes in swimming velocity (e.g., barokinesis) of larvae have been linked to light, gravity, and hydrostatic pressure. Planktotrophic larvae, because of their higher abundance and longer pelagic existence, have greater dispersal capability than lecithotrophic larvae.[115,116] Predation and harsh environmental conditions, however, cause elevated mortality rates that may limit their spatial distribution.

Meroplankton of many benthic macroinvertebrates delay metamorphosis until a favorable substrate is located. Certain chemical or physical cues trigger metamorphosis and settlement of these larvae. Chemicals released by adults of the same species purportedly induce larvae to metamorphose, which partially dictates population distributions along the seafloor. Additional factors possibly involved in regulating larval settlement are sediment size distributions and the presence of algal substrates, bacterial coatings, and organic matter.[119]

Larval recruitment (settlement, attachment, and metamorphosis) of certain benthic populations (e.g., clams and oysters) reflects gregarious behavioral patterns mediated by adult-derived chemical cues. Pheromones, for example, elicit behavioral responses in the larvae that foster gregarious settlement. Biochemical control of larval recruitment to the benthos has been demonstrated among annelids, arthropods, bryozoans, ascidian chordates, coelenterates, echinoderms, and mollusks. Clumped distributions of benthic macroinvertebrates may arise from factors other than gregarious settlement of larvae to chemical cues. For instance, suitable substrates for larval settlement may themselves have a patchy distribution.[120]

On a global scale, relatively few macrobenthic species have a cosmopolitan distribution. Few benthic species which inhabit estuaries nearly exclusively display a worldwide distribution. Fouling species living in marine and brackish water systems (e.g., *Balanus amphitrite* and *B. improvisus*) tend to have a wide distribution. The distribution of the benthos on a world scale can be increased by means of adult migration, transport by rafting, and transport by human activities.

Biomass and Diversity

Significant differences are apparent between benthic fauna of shallow continental shelf environments and those of the deep sea. When proceeding from benthic habitats on the shelf (i.e., supratidal, intertidal, subtidal zones) to those along the continental slope in the deep sea (bathyal zone), the biomass of benthic fauna (e.g., bivalves, polychaetes, gastropods, benthic foraminifera) decreases while the species diversity increases.[29,121] However, from continental rise depths to the abyssal plain (abyssal zone), species diversity decreases again.[104,122,123] An exceptionally high biomass of benthic organisms occurs at deep-sea hydrothermal vent communities along mid-ocean ridges; these communities exhibit biomass and diversity characteristics that are comparable to those of shallow-water communities.[124–126]

The high chemosynthetic primary production at hydrothermal vents in the deep sea provides nutrition directly to luxuriant populations of clams, mussels, and tube worms while supporting indirectly a substantial number of grazers, scavengers, detritus feeders, and predators.[124,125] The clams, mussels, and tube worms attain surprisingly elevated biomasses and densities at active vent sites. It has been estimated, for example, that the combined wet weight biomass of the two most common vent species (i.e., the tube worm *Riftia pachyptila* and the mussel *Bathymodiolus thermophilus*) along the Galapagos Rift exceeds 20 to 30 kg/m^2 in certain areas of the active vent fields.[127] Since 1977, when deep-sea hydrothermal vents were initially discovered at a depth of 2500 m along the Galapagos Rift spreading center, other hydrothermal vent communities have been found along the East Pacific Rise (e.g., 9°N, 11°N, 13°N, and 21°N), in Guaymas Basin, along the Gorda, Juan de Fuca, and Explorer Ridges, and in the Mariana back-arc spreading center, along the Mid-Atlantic Ridge, and elsewhere.[126] Moreover, comparable fauna have been collected at cold seep localities at the base of the Florida Escarpment, the Gulf of Mexico slope off Louisiana, Alaminos Canyon, and along the Oregon subduction zone. Biological processes (e.g., growth rates) of some vent organisms (e.g., *Calyptogena magnifica*) have been shown to proceed at rates that are extremely rapid for a deep-sea environment and comparable to those from some shallow water temperate environments.

Diversity is a useful parameter for assessing the stability of macrobenthic communities. According to the stability-time hypothesis, which Sanders[128] postulated, the species diversity of benthic communities is dependent on the stability of the environment as well as its geological history. For low-diversity communities in fluctuating environments, he coined the words "physically controlled", and for high diversity communities in stable or predictable environments, he fashioned the term "biologically accommodated". Diversity tends to be greater in geologically ancient environments that are physically stable than in recent, variable environments. Because they are geologically young, unstable or unpredictable, and ephemeral, estuaries have a lower probability of speciation and a greater likelihood of extinction than more constant environments such as the deep sea. Organisms inhabiting estuaries must adapt to a highly fluctuating environment where perturbations in physical, chemical, and biological conditions are extreme. Thus, specialized species in estuaries typically do not survive as well as generalists with broad adaptability.

Feeding Strategies

As noted above, five types of benthic macrofaunal feeders exist: (1) suspension feeders, (2) deposit feeders, (3) herbivores, (4) carnivores-scavengers, and (5) parasites.[105] Most benthic macrofauna are suspension and deposit feeders which inhabit sandy and muddy sediments, respectively. Environments with high concentrations of suspended solids cannot support suspension feeders because the particles clog their filtering structures. Suspension feeders mainly consume phytoplankton, but also remove some detritus, although the nutritional value of the detritus remains unclear. Deposit feeders are either selective or nonselective. Selective deposit feeders separate their food from sedimentary particles; nonselective deposit feeders ingest sediment and food particles together, digest the organic matter therein, and excrete the sediment and other nondigestible components. Bacteria, benthic microalgae, meiofauna, and microfauna comprise important food sources for nonselective deposit feeders, and bacteria and microalgae are excellent foods for selective deposit feeders. As a group, deposit feeders are ecologically significant because they continually rework bottom sediments, and in this process impact the biological, chemical, and physical characteristics of the seafloor.

Benthic fauna modify seafloor sediments via feeding, burrowing, and physiological activities.[109] Animal tubes, pits and depressions, excavation and fecal mounds, crawling trails or burrows, and organisms protruding above the sediment-water interface exacerbate bed roughness, thereby affecting fluid motion and sediment erosion and transport in the benthic boundary layer.[129] Bioturbation by benthic macrofauna (i.e., biogenic particle manipulation and pore water exchange) influences interparticle adhesion, water content of sediments, bed roughness, and geochemistry of interstitial waters. Both the stabilization and destabilization of bottom sediments have been attributed to biogenic activity of the benthos. These organisms also enhance nutrient mixing in sediments and the exchange of gases across the sediment-water interface. Among bioturbating organisms, pioneering species consisting principally of tubicolous or sedentary forms rework sediments most intensely in the upper 2 cm. In contrast, benthic communities dominated by high-order successional stages rework sediments at greater depths to several centimeters. Conveyor-belt, deposit-feeding species, for example, rework sediments well below the sediment-water interface.[109]

Biological Interactions

Investigations of rocky intertidal zones have supplied much information on the biological interactions of benthic communities. In intertidal and subtidal, soft-bottom environments, exclusion experiments using cages and other devices to preclude predators from killing prey organisms have been invaluable in evaluating the significance of predation on the structure of benthic communities.[130–132] Results of exclusion studies on unvegetated soft bottoms of lagoons and estuaries indicate a two- to threefold increase in the density of macroinvertebrate prey in areas protected from predators.[133] Caging experiments on vegetated soft bottoms reveal less predation and only slight effects on the benthic community. The vegetation serves as protective coverage for the prey organisms. Vegetated soft bottoms generally harbor communities with greater abundance and species richness than unvegetated bottoms. Predation effects are pronounced in sediments of unvegetated tidal flats (e.g., Wadden Sea) where the density of benthic populations in protected enclosures increases by as much as tenfold that of unprotected areas.[134,135]

Competition for space may be intense in certain communities (e.g., fouling communities). In many estuaries, competitive interactions control the abundance and distribution of fouling populations. Competitive strategies of fouling organisms in hard-substrate communities include overgrowth caused by relatively rapid growth rate, aggressive behavior, allelochemicals, and preemption of larval settlement by established individuals.[136,137] Solitary forms appear to be superior competitors for space in rocky intertidal zones, but the overgrowth of colonial fauna dominates in many other communities.

Strategies of Study

Studies of bottom-dwelling invertebrates have progressed on two basic levels: (1) the species level, whereby data are gathered on life history and population dynamics of the fauna; and (2) the community level, whereby an assemblage of organisms is assessed in terms of multiple variables, such as species composition, abundance, diversity, and productivity.[138] According to the community concept as recounted by Erwin,[139] the patterns of faunal and floral assemblages can be investigated using two approaches: (1) the level bottom approach in which assemblages are largely constant and related to the physical environment, primarily the substrate; and (2) the zonation approach, where assemblages are related to a physical gradient or a combination of physical gradients. The zonation approach has been successfully applied to intertidal and shallow subtidal zones, whereas the level-bottom community concept has been most fruitful in investigations of subtidal zones of estuaries and shallow coastal marine environments.

Benthic research has been greatly accelerated in recent years by the use of newly developed acoustic and photographic techniques in the field. Side-scan sonography facilitates data collection on bed roughness, sediment type, and morphology. The

REMOTS® sediment profile camera is especially valuable in studying the species composition of the benthic community and the geochemistry, bottom micromorphology, and microstratigraphy of the upper part of the sediment column in shallow-water environments.[111] Electromagnetic current meters, acoustic current meters, thermistors and heat transfer-type current meters, and impellors and laser-doppler current meters have supplied impressive amounts of data on current profiles and turbulence in the benthic boundary layer. Tripods have mollified problems in data acquisition on the bottom or benthic boundary layer over a long time series of several days or weeks. Types of tripods currently employed in field research are the U.S. Geological Survey GEOPROBE, the University of Connecticut systems, University of Washington, the Virginia Institute of Marine Science tripod, and the BASS® tripod.[129] In the deep sea, the use of submersible research vehicles (e.g., *Alvin*) fitted with advanced instrumentation and sampling devices have been extremely valuable in the study of deep-sea benthic communities, such as those currently under investigation in hydrothermal vent fields along mid-ocean ridge systems.

FISHES

CLASSIFICATION

Fish populations are a major component of the nektonic community of estuarine and marine waters, and they constitute some of the largest and most motile faunal elements of these ecosystems. The great mobility of fish confers significant advantages over sessile organisms, enabling them to avoid unfavorable environmental conditions and predatory attacks. Because of their value as a source of food, many finfishes are of great recreational and commercial importance to man.

The classification of Moyle and Cech[140] categorizes fish populations into five broad classes: (1) freshwater, (2) diadromous, (3) true estuarine, (4) nondependent marine, and (5) dependent marine fishes. A similar classification, involving six rather than five categories, has been proposed by McHugh.[141] His scheme focuses on breeding, migratory, and ecological criteria as follows: (1) freshwater fishes that occasionally enter brackish waters; (2) truly estuarine species which spend their entire lives in the estuary; (3) anadromous and catadromous species; (4) marine species which pay regular visits to the estuary, usually as adults; (5) marine species which use the estuary primarily as a nursery ground, spawning and spending much of their adult life at sea, but often returning seasonally to the estuary; and (6) adventitous visitors which appear irregularly and have no apparent estuarine requirements.

Estuarine fish faunas are characterized by the numerical dominance of a few species which tend to be widespread, reflecting the broad tolerances and range of adaptations of these fishes. Juveniles, which use the estuarine environment as a nursery area, comprise the most abundant forms in these coastal ecotones. Many of these young fishes enter estuaries from coastal populations. Estuaries harbor numerous migratory finfish species that seasonally move into and out of these systems. A smaller number of populations are permanent residents.

Estuarine nursery areas play a vital role in maintaining commercial offshore stocks of fish.[142] In the U.S., this relationship holds particularly strong for estuaries along the Atlantic and Gulf Coasts, but less so for those along the Pacific Coast.[141] McHugh[143] reported that more than 60% of the U.S. domestic commercial fish landings in 1980 by weight and more than 45% by dollar value occurred within 4.8 km of the coast. More recent statistics confirm his findings.[144] He also specified that species which spend at least part of their lives within estuaries compose nearly two thirds of the U.S. commercial catch by dollar value and much of the marine recreational catch as well.[145,146] Seven of the ten species groups most important to commercial fisheries in the U.S. are typically marine, constituting 58% of the commercial catch. The economic value of estuarine-dependent fisheries raises concern over the potential impact of anthropogenic stresses on natural populations of fish (see Section 5).[147,148]

The structure of fish communities depends on abiotic and biotic factors.[149,150] Among abiotic factors, both chemical and physical conditions — especially salinity, temperature, and dissolved oxygen — strongly influence the species composition, abundance, and distribution of fishes in estuarine waters. Fishes that dominate the community in estuaries tend to have a broad salinity tolerance.[151] The truly estuarine forms typically adapt readily to salinity changes. The adaptability of some marine fishes to lower salinities seems remarkable. However, fewer osmotic problems are encountered by marine teleosts at the mouth of estuaries; as a result, the species richness of the fish community generally peaks in this region. Progressing upestuary, osmotic problems arise for various populations, and at a salinity of approximately one third that of seawater, major osmotic difficulties may develop.

In the estuarine environment, fishes must deal with temperature as well as salinity gradients.[151–153] In temperate and boreal regions, in particular, seasonal temperature changes have a marked effect on the community structure. The migratory patterns of many species, for example, are strongly coupled to seasonal temperature levels. Changing thermal gradients can act as barriers to certain species (e.g., bluefish), thereby affecting their migratory behavior. In order to maximize survivorship in estuarine habitats, fishes thermoregulate behaviorally, avoiding or selecting environmental temperatures. However, the observed distribution of a species in an estuary reflects its response to other factors as well, such as food availability, nutritional state, competition, predation, and habitat requirements.

When dissolved oxygen concentrations drop below 4 ml/l and approach 0 ml/l in estuaries, ichthyofauna are impacted. In severe cases, migration routes can be effectively blocked by oxygen-depleted water masses that spread over broad areas, persisting for months.[141] Fishes entering waters devoid of oxygen can be trapped, increasing their risk of death. Mass mortality of fishes ascribable to low dissolved oxygen levels has been periodically documented. Other factors shown to occasionally inflict heavy losses on fish populations are large quantities of suspended sediment and red tides.

REPRESENTATIVE FISH FAUNAS[140,148–154]

Estuaries

Common fishes of estuaries include the anchovies (Engraulidae), killifish (Cyprinodontidae), silversides (Atherinidae), herrings (Clupeidae), mullet (Mugilidae), pipefish (Syngnathidae), drums (Sciaenidae), flounders (Bothidae, Pleuronectidae), eels (Anguillidae), and gobies (Gobiidae).

Pelagic Environment

Neritic zone — Characteristic fishes in the neritic zone are herrings (Clupeidae), eels (*Anguilla*), mackerels (Scombridae), bluefish (*Pomatomus*), butterfishes (Stromateidae), tunas

(*Thunnus*), marlin (*Makaira*), snappers (Lutjanidae), grunts (Pomadasyidae), porgies (Sparidae), barracudas (Sphyraenidae), and sharks.

Epipelagic zone — Occupants of the epipelagic zone are some albacores, bonitos, and tunas (Scombridae), dolphins (*Coryphaena*), mantas (Mobulidae), marlin (*Makaira*), sailfish (*Istiophorus*), molas (*Mola*), and lanternfish (Myctophidae).

Mesopelagic zone — In the mesopelagic zone, fish examples include deep-sea eels (*Synaphobranchus*), the deep-sea swallower (*Chiasmodus*), lanternfishes (Myctophidae), stalkeyed fish (*Idiacanthus*), and stomiatoids.

Bathypelagic zone — Examples of fishes found in the bathypelagic zone are the deep-sea swallower (*Chiasmodus*), deep-water eels (e.g., *Cyema*), stomiatoids (e.g., *Chauliodus* and *Malacosteus*), scorpionfishes (Scorpaenidae), dories (Zeidae), gulpers (*Eurypharynx*), and swallowers (*Saccopharynx*).

Abyssopelagic zone — Some of the fishes encountered in the abyssopelagic zone are deep-water eels (*Cyema*), deep-sea anglers (*Borophryne* and *Melanocetus*), gulpers (*Eurypharynx*), and stomiatoids (*Chauliodus*).

Benthic Environment

Supratidal zone — Only a few species have established niches in this environment. Some of these include gobies (Gobiidae), eels (Anguilliformes), and clingfishes (Gobiesocidae).

Intertidal zone — Representative fishes of the intertidal zone include stingrays (Dasyatidae), flounders (Bothidae, Pleuronectidae), soles (Soleidae), eels (*Anguilla*), morays (Muraenidae), clingfishes (Gobiesocidae), sculpins (Cottidae), searobins (Triglidae), blennies (Blenniidae), gobies (Gobiidae), pipefishes and seahorses (Syngnathidae), and cusk-eels (Ophidiidae).

Subtidal zone — Fishes commonly occurring in the inner subtidal zone of the continental shelf (to a depth of about 50 m) are skates (Rajidae), stingrays (Dasyatidae), flounders and soles (Pleuronectiformes), searobins (Triglidae), dogfish sharks (Squalidae), bonefish (Albulidae), eels (Anguillidae), morays (Muraenidae), seahorses and pipefishes (Syngnathidae), croakers, kingfish, and drums (Sciaenidae), hakes (Gadidae), wrasses (Labridae), butterflyfishes and angelfishes (Chaetodontidae), parrotfishes (Scaridae), trunkfishes (Ostraciidae), puffers (Tetraodontidae), and blennies (Blenniidae). In the outer subtidal zone of the continental shelf (from 50 to ~200 m depth) common fishes include cod (*Gadus*), haddock (*Melanogrammus*), hakes (*Merluccius* and *Urophycis*), halibuts (*Hippoglossus*), chimaeras (*Chimaera*), hagfishes (Myxinidae), eels (*Anguilla*), and pollock (*Pollachius*).

Bathyal zone — Some typical bathyal fishes are halibuts (*Hippoglossus*), chimaeras (*Chimaera*), cods (*Gadus*), and hagfishes (Myxinidae).

Abyssal zone — Abyssal fishes include rat-tails or grenadiers (Macrouridae) eels (*Synaphobranchus*), brotulas (Brotulidae), and relatives of the lanternfish (e.g., *Bathypterois* and *Ipnops*).

Hadal zone — Examples of fishes inhabiting the hadal region are rat-tails (Macrouridae), deep-water eels (*Synaphobranchus*), and brotulids (*Bassogigas*).

RESEARCH METHODS

Research on the population dynamics of fishes involves collecting data on growth, mortality, and recruitment.[141,148,155,156] Age determinations are essential to this work as well as to the proper management of fisheries resources. Estimates on age can be obtained by three methods: (1) mark-and-recapture experiments, (2) length-frequency distribution, and (3) interpretations of layers recorded in hard parts (i.e., scales and otoliths).

The von Bertalanffy, logistic, and Gompertz equations are three of the most frequently used relationships in fisheries research for assessing the growth of fish populations. In models applying the von Bertalanffy equation, fish grow toward a theoretical length or weight; the rate of change of size declines as the length approaches a maximum limit. The logistic curve also reveals size tending to an asymptote. The Gompertz model, while transcribing a curve whereby length or weight at age tends to an asymptote, differs from the von Bertalanffy and logistic models in that the curve generated has an asymmetrically placed inflection point, at an age less than halfway up the curve. A valuable method of estimating growth parameters is the Walford plot.

Investigations of ichthyofaunal mortality consider both natural and fishing death. Fishery management programs evaluate both types of mortality. The death rates of fish populations in the absence of fishing is not completely understood.

Density-dependent and density-independent types of mortality have been invoked as key factors regulating the size of fish populations.[140,155] Density-dependent mortality (e.g., disease and predation) arises from actions of the fish populations, whereas density-independent mortality is unrelated to any population action (e.g., temperature extremes and storms). Compensatory responses in mortality of fishes due to density-dependent factors normally originate during the critical early life-history stages.

Recruitment variability has been related to the variability in larval survival.[157] Major sources of recruitment losses are predation, low food availability, and feeding effects. The spatial clustering of eggs and larvae and their subsequent removal by predators impose significant limitations on recruitment success.

The management of fisheries employs the application of general stock-recruitment models. The models of Beverton and Holt[158] and Ricker[159] are widely used in recruitment studies. The Ricker model delineates a series of curves with low recruitment at high stock levels. The Beverton and Holt recruitment model in turn entails a group of asymptotic curves that depict constant recruitment beyond a certain stock density.

TROPHIC RELATIONSHIPS

The food webs of estuarine and marine ecosystems consist of a complex interrelationship of producer, consumer, and decomposer organisms. The movement of nutrient materials through the biota of estuarine and marine communities occurs in a predictable sequence of trophic (eating) relationships.[160] The trophic structure of a community centers around the concept of the food chain, which groups organisms into categories or trophic levels (i.e., producers, consumers, and decomposers) for the delineation of food energy transfer within the system.[89,161] Hence, the base of the food chain consists of autotrophs or chemoautotrophs that fix carbon via photosynthesis or chemosynthesis and provide energy for consumer organisms (i.e., heterotrophs). At successively higher trophic levels, primary consumers serve as a food source for secondary consumers which in turn are consumed by tertiary consumers (i.e., carnivores). Decomposers (i.e., saprophagic bacteria and fungi) assimilate dead plant or animal matter, transforming it into dissolved organic matter to meet their energy requirements, while releasing mineral nutrients. At each trophic level, approximately 80 to 90% of the potential energy is lost as heat, thereby limiting the length of food chains to three or four trophic levels.[162,163]

The multiple feeding strategies of many estuarine and marine organisms result in complex food webs structured by two major interlocking components of energy flow, that is, grazing and detrital pathways. The distinction between grazing and detritus food webs occurs at the primary producer-primary consumer levels.[164] At secondary consumer levels and above, a network of heterotrophs derives energy and nutrients from both primary pathways; consequently, the separation between grazing and detrital food webs becomes obscured among upper-trophic-level organisms.

Phytoplankton and other live plant communities form the base of grazing food webs.[29,59] This type of food web is most evident in deeper, clearer estuarine and nearshore systems and in neritic and epipelagic waters. Zooplankton and zooplankton grazers (e.g., bay anchovies) are critical primary and secondary consumers of grazing food webs.

The base of detrital food webs is detritus — a general term denoting both particulate and dissolved nonliving organic matter. The bulk of detritus in shallow water systems results from the accumulation of plant material, primarily benthic macrophytes (i.e., macroalgae, seagrasses, salt marsh grasses, and mangroves), that largely remain ungrazed while alive and enter the detritus pool upon death.[31] Because of low exploitation rates by grazing herbivores, more than 90% of the primary production by benthic macrophytes passes to detritus food webs in estuaries.[59] Another potentially significant source of detritus consists of biodeposits, feces, and pseudofeces of animals. Based principally upon vascular plant detritus, detritus food webs are common features of shallow estuaries, particularly those characterized by extensive communities of seagrass, salt marsh, and mangroves.

Much estuarine detritus is consumed directly by detritivores living on or in bottom sediments. The detritus forms a substrate for the growth of bacteria, fungi, and microalgae which provide a rich food supply for primary consumers. Bacteria and fungi are the decomposers that attack the detrital substrate, regenerating nutrients and transforming the POM into DOM which they assimilate.[30] Detritivores that consume the bacteria are essential foods for secondary consumers (larger invertebrates, fishes, birds), many of which have omnivorous feeding habits. Thus, they generally ingest organisms from more than one trophic level, accounting for the complex network of feeding relationships manifested as links or arrows in detrital food web diagrams.

In the open ocean, grazing (phytoplankton)-based food webs control the flow of energy through biotic communities. Recently, however, the concept of a microbial loop has emerged for pelagic food webs in which bacteria process nonliving organic matter while being grazed by protozoans; the microbial food web is the ultimate food resource for metazooplankton.[165] The microbial loop has stirred a "sink or link" debate among marine scientists concerning the relative importance of microbes as a sink for fixed carbon via respiratory losses or as a source of food for the metazoan food web.[166,167] Sherr and Sherr[168] consider this controversy to be a nonissue, however, viewing the microbial loop as " . . . a component (and integral) part of a larger microbial food web, which includes all pro- and eukaryotic unicellular organisms, both autotrophic and heterotrophic. From this standpoint, the entire microbial food web, not simply phytoplankton, supports the metazoan food web."

At deep-sea hydrothermal vent environments, the reaction of seawater with crustal rocks at high temperatures produces reduced organic compounds that discharge from hot springs on the seafloor and provide energy for free-living and/or chemosynthetic bacteria (alternatively referred to as chemoautotrophs, chemolithotrophs, or chemolithoautotrophs) which form the base of the food web in these unique habitats.[169,170] By oxidizing sulfides, especially hydrogen sulfide (H_2S), as well as other reduced substrates (lithotrophy) such as hydrogen (H_2), iron (Fe^{2+}), or manganese (Mn^{2+}), released from vents or cold seeps, the microbes obtain energy to synthesize organic compounds from carbon dioxide (CO_2) in seawater (autotrophy).[171,172] In so doing, the bacteria support copious populations of specifically adapted invertebrates living in close proximity to the vents.[170] Jannasch[169] and Jannasch and Mottl[170] summarize the types of chemoautotrophic bacteria capable of supporting benthic invertebrate communities in these environments.

SECTION PLAN

The purpose of this section is to provide useful data on the major biotic groups (i.e., bacteria, phytoplankton, zooplankton, benthos, and nekton) of oceanic and estuarine ecosystems. Information contained in tables and figures assesses the productivity of the oceans and estuaries of the world. The biomass, density, and geographical distribution of characteristic flora and fauna are also treated. In addition, data are presented on key habitats (e.g., salt marshes, seagrasses, mangroves, deep-sea hydrothermal vents) within these ecosystems.

REFERENCES

1. **Baker, J. M., Hartley, J. P., and Kicks, B.,** Planning biological surveys, in *Biological Surveys of Estuaries and Coasts,* Baker, J. M. and Wolff, W. J., Eds., Cambridge University Press, Cambridge, 1987.
2. **Baker, J. M. and Wolff, W. J., Eds.,** *Biological Surveys of Estuaries and Coasts,* Cambridge University Press, Cambridge, 1987.
3. **Sieburth, J. M.,** Bacterial substrates and productivity in marine ecosystems, *Annu. Rev. Ecol. Syst.,* 7, 259, 1976.
4. **Morita, R. Y.,** The role of microorganisms in the environment, in *Oceanic Sound Scattering Prediction,* Anderson, N. R. and Zahuranec, B., Eds., Plenum Press, New York, 1977, 445.
5. **Fenchel, T.,** Ecology of heterotrophic microflagellates. IV. Quantitative occurrence and importance as consumers of bacteria, *Mar. Ecol. Prog. Ser.,* 9, 35, 1982.
6. **Rublee, P. A.,** Bacteria and microbial distribution in estuarine sediments, in *Estuarine Comparisons,* Kennedy, V. S., Ed., Academic Press, New York, 1982, 159.

7. **Fenchel, T.,** Suspended marine bacteria as a food source, in *Flows of Energy and Materials in Marine Ecosystems: Theory and Practice,* Fasham, M. J. R., Ed., Plenum Press, New York, 1984, 301.
8. **Kemp, P. F.,** The fate of benthic bacterial production, *Rev. Aquat. Sci.,* 2, 109, 1990.
9. **Rublee, P. A.,** Seasonal distribution of bacteria in salt marsh sediments, North Carolina, *Estuarine Coastal Shelf Sci.,* 15, 676, 1982.
10. **Azam, F. and Hodson, R. E.,** Size distribution and activity of marine microheterotrophs, *Limnol. Oceanogr.,* 22, 492, 1977.
11. **Watson, S. W. and Hobbie, J. E.,** Measurement of bacterial biomass as lipopolysaccharide, in *Native Aquatic Bacteria: Enumeration, Activity, and Ecology,* Costerton, J. W. and Colwell, R. R., Eds., American Society for Testing Materials, Philadelphia, 1979, 82.
12. **Sleigh, M. A., Ed.,** *Microbes in the Sea,* Ellis Horwood, Chichester, England, 1987.
13. **Wiebe, W. J.,** Physiological and biochemical aspects of marine bacteria, in *Heterotrophic Activity in the Sea,* Hobbie, J. E. and Williams, P. J. le B., Eds., Plenum Press, New York, 1984, 55.
14. **Wiebe, W. J. and Pomeroy, L. R.,** Microorganisms and their association with aggregates and detritus in the sea: a microscopic study, *Mem. Ist. Ital. Idrobiol. Dott Marco de Marchi Pallanza Italy,* 29 (Suppl.), 325, 1972.
15. **Hanson, R. B. and Wiebe, W. J.,** Heterotrophic activity associated with particulate size fractions in a *Spartina alterniflora* salt-marsh estuary, Sapelo Island, Georgia, U.S.A. and the continental shelf waters, *Mar. Biol.,* 42, 321, 1977.
16. **Goulder, R.,** Attached and free bacteria in an estuary with abundant suspended solids, *J. Appl. Bacteriol.,* 43, 399, 1977.
17. **Sieburth, J. M.,** Protozoa bacterivory in pelagic marine waters, in *Heterotrophic Activity in the Sea,* Hobbie, J. E. and Williams, P. J. le B., Eds., Plenum Press, New York, 1984, 405.
18. **Bell, C. R. and Albright, L. J.,** Attached and free-floating bacteria in the Frazer River estuary, British Columbia, Canada, *Mar. Ecol. Prog. Ser.,* 6, 317, 1981.
19. **Sieburth, J. M. and Davis, P. G.,** The role of heterotrophic nanoplankton in the grazing and nurturing of planktonic bacteria in the Sargasso and Caribbean Seas, *Ann. Inst. Oceanogr. (Paris),* 58, 285, 1982.
20. **Alongi, D. M.,** The role of soft-bottom benthic communities in tropical mangrove and coral reef ecosystems, *Rev. Aquat. Sci.,* 1, 243, 1989.
21. **Karl, D. M.,** Determination in *in situ* microbial biomass, viability, metabolism and growth, in *Bacteria in Nature,* Poindexter, J. S. and Leadbetter, E. R., Eds., Plenum Press, New York, 1986, chap. 3.
22. **Moriarty, D. J. W.,** Measurement of microbial growth rates in aquatic systems using rates of nucleic acid synthesis, *Adv. Microb. Ecol.,* 9, 245, 1986.
23. **Karl, D. M.,** Selected nucleic acid precursors in studies of aquatic microbial ecology, *Appl. Environ. Microbiol.,* 44, 891, 1982.
24. **Moriarty, D. J. W. and Pollard, P. C.,** DNA synthesis as a measure of bacterial productivity in seagrass sediments, *Mar. Ecol. Prog. Ser.,* 5, 151, 1981.
25. **Fuhrman, J. A. and Azam, F.,** Bacterioplankton secondary production estimates for coastal waters of British Columbia, Antarctica, and California, *Appl. Environ. Microbiol.,* 39, 1085, 1980.
26. **Karl, D. M.,** Simultaneous rates of RNA and DNA synthesis for estimating growth and cell division of aquatic microbial communities, *Appl. Environ. Microbiol.,* 42, 802, 1981.
27. **Karl, D. M., Winn, C. D., and Wong, D. C. L.,** RNA synthesis as a measure of microbial growth in aquatic environments. I. Evaluation, verification and optimization of methods, *Mar. Biol.,* 64, 1, 1981.
28. **Turner, J. T., Tester, P. A., and Ferguson, R. L.,** The marine cladoceran *Penilia avirostris* and the "microbial loop" of pelagic food webs, *Limnol. Oceanogr.,* 33, 245, 1988.
29. **Levinton, J. S.,** *Marine Ecology,* Prentice-Hall, Englewood Cliffs, NJ, 1982.
30. **Valiela, I.,** *Marine Ecological Processes,* Springer-Verlag, New York, 1984.
31. **Kennish, M. J.,** *Ecology of Estuaries,* Vol. 1, CRC Press, Boca Raton, FL, 1986.
32. **Newell, R. C.,** The energetics of detritus utilization in coastal lagoons and nearshore waters, *Oceanol. Acta,* 32, 347, 1982.
33. **Lopez, G. R. and Levinton, J. S.,** Ecology of deposit-feeding animals in marine sediments, *Q. Rev. Biol.,* 62, 235, 1987.
34. **Morrison, S. J., King, J. D., Bobbie, R. J., Bechtold, R. E., and White, D. C.,** Evidence for microfloral succession on allochthonous plant litter in Apalachicola Bay, Florida, U.S.A., *Mar. Biol.,* 41, 229, 1977.
35. **Cagle, G. C.,** Fine structure and distribution of extracellular polymer surrounding selected aerobic bacteria, *Can. J. Microbiol.,* 21, 395, 1975.
36. **Tenore, K. R., Cammen, L., Findlay, S. E. G., and Phillips, N.,** Perspectives of research on detritus: do factors controlling the availability of detritus to macroconsumers depend on its source?, *J. Mar. Res.,* 40, 473, 1982.
37. **Meyer-Reil, L.-A.,** Bacterial biomass and heterotrophic activity in sediments and overlying waters, in *Heterotrophic Activity in the Sea,* Hobbie, J. E. and Williams, P. J. le B., Eds., Plenum Press, New York, 1984, 523.
38. **Meyer-Reil, L.-A.,** Autoradiography and epifluorescence microscopy combined for the determination of number and spectrum of actively metabolizing bacteria in natural waters, *Appl. Environ. Microbiol.,* 36, 506, 1978.
39. **Kirchman, D. L., Mazella, L., Alberte, R. S., and Mitchell, R.,** Epiphytic bacterial production on *Zostera marina, Mar. Ecol. Prog. Ser.,* 15, 17, 1984.
40. **Fenchel, T. M.,** The ecology of micro- and meiobenthos, *Annu. Rev. Ecol. Syst.,* 9, 99, 1978.
41. **Jorgensen, B. B.,** The sulfur cycle of a coastal marine sediment (Limfjorden, Denmark), *Limnol. Oceanogr.,* 22, 814, 1977.

42. **Wiebe, W. J., Christian, R. R., Hansen, J. A., King, G., Sherr, B., and Skyring, A.,** Anaerobic respiration and fermentation, in *The Ecology of a Salt Marsh,* Pomeroy, L. R. and Wiegert, R. G., Eds., Springer-Verlag, New York, 1981, 137.
43. **Lipschultz, F.,** Methane release from a brackish intertidal salt-marsh embayment of Chesapeake Bay, Maryland, *Estuaries,* 4, 143, 1981.
44. **Raymont, J. E. G.,** *Plankton and Productivity in the Oceans,* Vol. 1, 2nd ed., Pergamon Press, Oxford, 1980.
45. **Dawes, C. J.,** *Marine Botany,* John Wiley & Sons, New York, 1981.
46. **Boney, A. D.,** *Phytoplankton,* Edward Arnold, London, 1975.
47. **Dodge, J. D.,** *The Fine Structure of Algal Cells,* Academic Press, New York, 1973.
48. **McConnaughey, B. H. and Zottoli, R.,** *Introduction to Marine Biology,* C. V. Mosby, St. Louis, 1983.
49. **Eppley, R. W.,** The growth and culture of diatoms, in *The Biology of Diatoms,* Werner, D., Ed., Blackwell Scientific, Oxford, 1977, 24.
50. **Hargraves, P. E.,** Studies on marine planktonic diatoms. II. Resting spore morphology, *J. Phycol.,* 12, 118, 1976.
51. **Steidinger, K. A. and Walker, L. M.,** Introduction, in *Marine Plankton Life Cycle Strategies,* Steidinger, K. A. and Walker, L. M., Eds., CRC Press, Boca Raton, FL, 1984.
52. **Gross, M. G.,** *Oceanography: A View of the Earth,* 5th ed., Prentice-Hall, Englewood Cliffs, NJ, 1990.
53. **Kennish, M. J. and Lutz, R. A.,** Calcium carbonate in surface sediments of the deep ocean floor, *Rev. Aquat. Sci.,* 6, 183, 1992.
54. **McCarthy, J. J. and Altabet, M. A.,** Patchiness in nutrient supply: implications for phytoplankton ecology, in *Trophic Interactions within Aquatic Ecosystems,* Selected Symp. 85, Meyers, D. G. and Strickler, J. R., Eds., American Association for the Advancement of Science, Washington, D.C., 1984, 29.
55. **Bennett, A. F. and Denman, K. L.,** Phytoplankton patchiness: inferences from particle statistics, *J. Mar. Res.,* 43, 307, 1985.
56. **Ryther, J. H.,** Geographic variations in productivity, in *The Sea,* Vol. 2, Hill, M. N., Ed., Wiley-Interscience, New York, 1963, 347.
57. **Ryther, J. H.,** Photosynthesis and fish production, *Science,* 166, 72, 1969.
58. **Mann, K. H.,** Macrophyte production and detritus food chains in coastal waters, *Mem. Ist. Ital. Idrobiol. Dott Marco de Marchi Pallgaza Italy,* 29 (Suppl.), 353, 1972.
59. **Mann, K. H.,** *Ecology of Coastal Waters: a Systems Approach,* University of California Press, Berkeley, 1982.
60. **Jerlov, N. G.,** *Marine Optics,* Elsevier Oceanography Series, Vol. 14, Elsevier, Amsterdam, 1976.
61. **Smith, R. C. and Baker, K. S.,** Optical properties of the clearest natural waters (200–800 nm), *Appl. Opt.,* 20, 177, 1981.
62. **Steemann Nielsen, E.,** *Marine Photosynthesis: with Special Emphasis on the Ecological Aspects,* Elsevier Oceanography Series, Vol. 13, Elsevier Scientific, Amsterdam, 1975.
63. **Lewis, M. R., Warnock, R. E., Irwin, B., and Platt, T.,** Measuring photosynthetic action spectra of natural phytoplankton populations, *J. Phycol.,* 21, 310, 1985.
64. **Rice, T. R. and Ferguson, R. L.,** Response of estuarine phytoplankton to environmental conditions, in *Physiological Ecology of Estuarine Organisms,* Vernberg, F. J., Ed., University of South Carolina Press, Columbia, 1975, 1.
65. **Brand, L. E.,** The salinity tolerance of forty-six marine phytoplankton isolates, *Estuarine Coastal Shelf Sci.,* 18, 543, 1984.
66. Biological Methods Panel Committee on Oceanography, Recommended Procedures for Measuring the Productivity of Plankton Standing Stock and Related Oceanic Properties, National Academy of Sciences, Washington, D.C., 1969.
67. **Lippson, A. J., Haire, M. S., Holland, A. F., Jacobs, F., Jensen, J., Moran-Johnson, R. L., Polgar, T. T., and Richkus, W. A.,** *Environmental Atlas of the Potomac Estuary,* The Johns Hopkins University Press, Baltimore, 1981.
68. **Marcus, N. H.,** Recruitment of copepod nauplii into the plankton: importance of diapause eggs and benthic processes, *Mar. Ecol. Prog. Ser.,* 15, 47, 1984.
69. **Marcus, N. H. and Schmidt-Gengenbach, J.,** Recruitment of individuals into the plankton: the importance of bioturbation, *Limnol. Oceanogr.,* 31, 206, 1986.
70. **Alldredge, A. L. and King, J. M.,** The distance demersal zooplankton migrate above the benthos: implications for predation, *Mar. Biol.,* 84, 253, 1985.
71. **Forward, R. B., Jr.,** Diel vertical migration: zooplankton photobiology and behaviour, *Oceanogr. Mar. Biol. Annu. Rev.,* 26, 361, 1988.
72. **Forward, R. B., Jr.,** A consideration of the shadow response of a larval crustacean, *Mar. Behav. Physiol.,* 12, 99, 1986.
73. **Stearns, D. E.,** Copepod grazing behavior in simulated natural light and its relation to nocturnal feeding, *Mar. Ecol. Prog. Ser.,* 30, 65, 1986.
74. **Stearns, D. E. and Forward, R. B., Jr.,** Copepod photobehavior in a simulated natural light environment and its relation to nocturnal vertical migration, *Mar. Biol.,* 82, 91, 1984.
75. **Swift, M. C. and Forward, R. B., Jr.,** Absolute light intensity vs. rate of relative change in light intensity: the role of light in the vertical migration of *Chaoborus punctipennis* larvae, *Bull. Mar. Sci.,* 43, 604, 1988.
76. **Forward, R. B., Jr.,** Behavioral responses of larvae of the crab *Rhithropanopeus harrisii* (Brachyura: Xanthidae) during diel vertical migration, *Mar. Biol.,* 90, 9, 1985.
77. **Heinle, D. R.,** Temperature and zooplankton, *Chespeake Sci.,* 10, 186, 1969.
78. **McLaren, I. A.,** Demographic strategy of vertical migration by a marine copepod, *Am. Nat.,* 108, 91, 1974.

79. **Jeffries, H. P. and Johnson, W. C.,** Distribution and abundance of zooplankton, in *Coastal and Offshore Environmental Inventory: Cape Hatteras to Nantucket Shoals,* Marine Publ. Ser. No. 2, University of Rhode Island, Kingston, 1973, 4–1.
80. **Miller, C. B.,** The zooplankton of estuaries, in *Estuaries and Enclosed Seas,* Ketchum, B. H., Ed., Elsevier, Amsterdam, 1983, 103.
81. **Price, H. J.,** Feeding mechanisms in marine and freshwater zooplankton, *Bull. Mar. Sci.,* 43, 327, 1988.
82. **McAllister, C. D.,** Zooplankton rations, phytoplankton mortality, and the estimation of marine production, in *Marine Food Chains,* Steele, J. H., Ed., University of California Press, Berkeley, 1970, 419.
83. **Day, J. H.,** The estuarine fauna, in *Estuarine Ecology: with Particular Reference to Southern Africa,* Day, J. H., Ed., A. A. Balkema, Rotterdam, 1981, 147.
84. **den Hartog, C.,** Brackish water as an environment for algae, *Blumea,* 15, 31, 1967.
85. **Tittley, I.,** Zonation and seasonality of estuarine benthic algae: artificial embankments in the River Thames, *Bot. Mar.,* 28, 1, 1985.
86. **Sibert, J. R. and Naiman, R. J.,** The role of detritus and the nature of estuarine ecosystems, in *Marine Benthic Dynamics,* Tenore, K. R. and Coull, B. C., Eds., University of South Carolina Press, Columbia, 1980, 311.
87. **Valiela, I., Teal, J. M., Allen, S. D., Van Etten, R., Goehringer, D., and Volkmann, S.,** Decomposition in salt marsh ecosystems: the phases and major factors affecting disappearance of above-ground organic matter, *J. Exp. Mar. Biol. Ecol.,* 89, 29, 1985.
88. **Wilson, J. O., Buchsbaum, R., Valiela, I., and Swain, T.,** Decomposition in salt marsh ecosystems: phenolic dynamics during decay of litter of *Spartina alterniflora, Mar. Ecol. Prog. Ser.,* 29, 177, 1986.
89. **Boaden, P. J. S. and Seed, R.,** *An Introduction to Coastal Ecology,* Blackie and Son, Glasgow, 1985.
90. **Beeftink, W. G.,** Salt-marshes, in *The Coastline,* Barnes, R. S. K., Ed., John Wiley & Sons, Chichester, England, 1977.
91. **Long, S. P. and Mason, C. F.,** *Saltmarsh Ecology,* Blackie and Son, Glasgow, 1983.
92. **Frey, R. W. and Basan, P. B.,** *Coastal Sedimentary Environments,* 2nd ed., Davis, R. A., Ed., Springer-Verlag, New York, 1985, 225.
93. **den Hartog, C.,** *The Seagrasses of the World,* North Holland, Amsterdam, 1970.
94. **Phillips, R. C. and McRoy, C. P., Eds.,** *Handbook of Seagrass Biology: an Ecosystem Perspective,* Garland STPM Press, New York, 1980.
95. **Lin, P.,** *Mangrove Vegetation,* Springer-Verlag, New York, 1988.
96. **Rutzler, K. and Feller, C.,** Mangrove swamp communities, *Oceanus,* 30, 16, 1987.
97. **Fenchel, T. M.,** The ecology of micro- and meiobenthos, *Annu. Rev. Ecol. Syst.,* 9, 99, 1978.
98. **Perkins, E. J.,** *The Biology of Estuaries and Coastal Waters,* Academic Press, London, 1974.
99. **Coull, B. C. and Bell, S. S.,** Perspectives of marine meiofaunal ecology, in *Ecological Processes in Coastal Marine Systems,* Livingston, R. J., Ed., Plenum Press, New York, 1979, 189.
100. **Tietjen, J. H.,** The ecology of shallow water meiofauna in two New England estuaries, *Oecologia (Berlin),* 2, 252, 1969.
101. **Tietjen, J. H.,** Potential roles of nematodes in polluted ecosystems and the impact of pollution on meiofauna, in *Ecological Stress and the New York Bight: Science and Management,* Mayer, G. F., Ed., Estuarine Research Federation, Columbia, SC, 1982, 225.
102. **Holland, A. F.,** Long-term variation sof macrobenthos in a mesohaline region of Chesapeake Bay, *Estuaries,* 8, 93, 1985.
103. **Holland, A. F., Shaughnessy, A. T., and Hiegel, M. H.,** Long-term variation in mesohaline Chesapeake Bay macrobenthos: spatial and temporal patterns, *Estuaries,* 10, 227, 1987.
104. **Gage, J. D. and Tyler, P. A.,** *Deep-Sea Biology: a Natural History of Organisms at the Deep-Sea Floor,* Cambridge University Press, Cambridge, 1991.
105. **Rhoads, D. C.,** Organisms-sediment relations on the muddy sea floor, *Oceanogr. Mar. Biol. Annu. Rev.,* 12, 263, 1974.
106. **Wolff, W. J.,** Estuarine benthos, in *Estuaries and Enclosed Seas,* Ketchum, B. H., Ed., Elsevier, Amsterdam, 1983, 151.
107. **Fenchel, T. and Riedl, R. J.,** The sulfide system: a new biotic community underneath the oxidized layer of marine sand bottoms, *Mar. Biol.,* 7, 255, 1970.
108. **Rhoads, D. C., McCall, P. L., and Yingst, J. Y.,** Disturbance and production on the estuarine seafloor, *Am. Sci.,* 66, 577, 1978.
109. **Rhoads, D. C. and Boyer, L. F.,** The effects of marine benthos on physical properties of sediments: a successional perspective, in *Animal-Sediment Relations: the Biogenic Alteration of Sediments,* McCall, P. L. and Tevesz, M. J. S., Eds., Plenum Press, New York, 1982, 3.
110. **Rhoads, D. C. and Germano, J. D.,** Interpreting long-term changes in benthic community structure: a new protocol, *Hydrobiologia,* 142, 291, 1986.
111. **Rhoads, D. C. and Germano, J. D.,** Characterization of organism-sediment relations using sediment profile imaging: an efficient method of remote ecological monitoring of the seafloor (REMOTS® System), *Mar. Ecol. Prog. Ser.,* 8, 115, 1982.
112. **Boesch, D. F.,** A new look at the zonation of benthos along the estuarine gradient, in *Ecology of Marine Benthos,* Coull, B. C., Ed., University of South Carolina Press, Columbia, 1977, 245.
113. **Thorson, G.,** Reproduction and larval development of Danish marine bottom invertebrates with special reference to the plankton larvae in the sound (Oresund), *Medd. Komm. Dan. Fis., Havunders. Ser. Plankton,* 4, 1, 1946.
114. **Thorson, G.,** Reproductive and larval ecology of marine bottom invertebrates, *Biol. Rev.,* 25, 1, 1950.
115. **Jablonski, D.,** Larval ecology and macroevolution in marine invertebrates, *Bull. Mar. Sci.,* 39, 565, 1986.

116. **Jablonski, D. and Lutz, R. A.,** Larval ecology of marine benthic invertebrates: paleobiological implications, *Biol. Rev.,* 58, 21, 1983.
117. **Day, R. and McEdward, L.,** Aspects of the physiology and ecology of pelagic larvae of marine benthic invertebrates, in *Marine Plankton Life Cycle Strategies,* Steidinger, K. A. and Walker, L. M., Eds., CRC Press, Boca Raton, FL, 1984, 93.
118. **Scheltema, R. S.,** On dispersal and planktonic larvae of benthic invertebrates: an eclectic overview and summary of problems, *Bull. Mar. Sci.,* 39, 290, 1986.
119. **Crisp, D. J.,** Overview of research on marine invertebrate larvae, 1940–1980, in *Marine Biodeterioration: an Interdisciplinary Study,* Costlow, J. D. and Tipper, R. C., Eds., Naval Institute Press, Annapolis, MD, 1984, 103.
120. **Chia, F. S. and Rice, M. E., Eds.,** *Settlement and Metamorphosis of Marine Invertebrates,* Elsevier/North Holland, New York, 1978.
121. **Sanders, H. L. and Hessler, R. R.,** Ecology of the deep-sea benthos, *Science,* 163, 1419, 1969.
122. **Rex, M. A.,** Community structure in the deep-sea benthos, *Annu. Rev. Ecol. Syst.,* 12, 331, 1981.
123. **Rex, M. A.,** Geographical patterns of species diversity in the deep-sea benthos, in *The Sea,* Vol. 8, Rowe, G. T., Ed., John Wiley & Sons, New York, 1983.
124. **Tunnicliffe, V.,** The biology of hydrothermal vents: ecology and evolution, *Oceanogr. Mar. Biol. Annu. Rev.,* 29, 319, 1991.
125. **Tunnicliffe, V.,** Hydrothermal-vent communities of the deep sea, *Am. Sci.,* 80, 336, 1992.
126. **Lutz, R. A.,** The biology of deep-sea vents and seeps, *Oceanus,* 34, 75, 1991.
127. **Grassle, J. F.,** The ecology of deep-sea hydrothermal vent communities, *Adv. Mar. Biol.,* 23, 301, 1986.
128. **Sanders, H. L.,** Marine benthic diversity: a comparative study, *Am. Nat.,* 102, 243, 1968.
129. **Wright, L. D.,** Benthic boundary layers of estuarine and coastal environments, *Rev. Aquat. Sci.,* 1, 75, 1989.
130. **Reise, K.,** Predator exclusion experiments in an intertidal mud flat, *Helgol. Wiss. Meeresunters.,* 30, 263, 1977.
131. **Reise, K.,** *Tidal Flat Ecology: an Experimental Approach to Species Interactions, Ecological Studies,* Vol. 54, Springer-Verlag, Berlin, 1985.
132. **Virnstein, R. W.,** Predation on estuarine infauna: patterns of component species, *Estuaries,* 2, 69, 1979.
133. **Peterson, C. H.,** Predation, competitive exclusion and diversity in the soft-bottom communities of estuaries and lagoons, in *Ecological Processes in Coastal and Marine Systems,* Livingston, R. J., Ed., Plenum Press, New York, 1979, 233.
134. **Reise, K.,** Experiments on epibenthic predation in the Wadden Sea, *Helgol. Wiss. Meeresunters.,* 31, 51, 1978.
135. **Reise, K.,** Moderate predation on meiofauna by the macrobenthos of the Wadden Sea, *Helgol. Wiss. Meerestunters.,* 32, 453, 1979.
136. **Smedes, G. W.,** Seasonal changes and fouling community interactions, in *Marine Biodeterioration: an Interdisciplinary Study,* Costlow, J. D. and Tipper, R. C., Eds., Naval Institute Press, Annapolis, MD, 1984, 155.
137. **Lopez Gappa, J. J.,** Overgrowth competition in an assemblage of encrusting bryozoans settled on artificial substrata, *Mar. Ecol. Prog. Ser.,* 51, 121, 1989.
138. **Pratt, S. D.,** Benthic fauna, in *Coastal and Offshore Environmental Inventory, Cape Hatteras to Nantucket Shoals,* Mar. Publ. Ser. No. 2, University of Rhode Island, Kingston, 1973.
139. **Erwin, D. G.,** The community concept, in *Sublitoral Ecology: the Ecology of the Shallow Sublittoral Benthos,* Earll, R. and Erwin, D. G., Eds., Clarendon Press, Oxford, 1983, 144.
140. **Moyle, P. B. and Cech, J. J., Jr.,** *Fishes: an Introduction to Ichthyology,* Prentice-Hall, Englewood Cliffs, NJ, 1982.
141. **McHugh, J. L.,** Estuarine nekton, in *Estuaries,* Lauff, G. H., Ed., Publ. 83, American Association for the Advancement of Science, Washington, D.C., 1967, 581.
142. **Gunter, G.,** Some relationships of estuaries to the fisheries of the Gulf of Mexico, in *Estuaries,* Lauff, G. H., Ed., Publ. 83, American Association for the Advancement of Science, Washington, D.C., 1967, 621.
143. **McHugh, J. L.,** *Fishery Management,* Springer-Verlag, New York, 1984.
144. NOAA, Fisheries of the United States, 1987, Current Fishery Statistics 8700, U.S. Department of Commerce, NOAA, NMFS, 1988.
145. **McHugh, J. L.,** Management of estuarine fisheries, in *A Symposium on Estuarine Fisheries,* American Fisheries Society, Spec. Publ. 3, Bethesda, MD, 1966, 133.
146. **McHugh, J. L.,** Estuarine fisheries: are they doomed?, in *Estuarine Processes,* Vol. 1, Wiley, M., Ed., Academic Press, New York, 1976, 15.
147. **McHugh, J. L.,** Are estuaries necessary?, *Commer. Fish. Rev.,* 30, 37, 1968.
148. **Stickney, R. R.,** *Estuarine Ecology of the Southeastern United States and Gulf of Mexico,* Texas A & M University Press, College Station, TX, 1984.
149. **Evans, D. H., Ed.,** *The Physiology of Fishes,* CRC Press, FL, 1993.
150. **Pitcher, T. J., Ed.,** *The Behavior of Teleost Fishes,* The Johns Hopkins University Press, Baltimore, 1986.
151. **Day, J. H., Blaber, S. J. M., and Wallace, J. H.,** Estuarine fishes, in *Estuarine Ecology: with Particular Reference to Southern Africa,* Day, J. H., Ed., A. A. Balkema, Rotterdam, 1981, 197.
152. **Haedrich, R. L.,** Estuarine fishes, in *Estuaries and Enclosed Seas,* Ketchum, B. H., Ed., Elsevier Scientific, Amsterdam, 1983, 183.
153. **Yanez-Arancibia, A., Linares, F. A., and Day, J. W., Jr.,** Fish community structure and function in Terminos Lagoon, a tropical estuary in the southern Gulf of Mexico, in *Estuarine Perspectives,* Kennedy, V. S., Ed., Academic Press, New York, 1980, 465.

154. **Lager, K. F., Bardach, J. E., and Miller, R. R.,** *Ichthyology,* John Wiley & Sons, New York, 1962.
155. **Everhart, W. H., Eipper, A. W., and Youngs, W. D.,** *Principles of Fishery Science,* Cornell University Press, Ithaca, NY, 1975.
156. **Weatherley, A. H. and Gill, H. S.,** *The Biology of Fish Growth,* Academic Press, Orlando, FL, 1987.
157. **Rothschild, B. J. and DiNardo, G. T.,** Comparison of recruitment variability and life history data among marine and anadromous fishes, *Am. Fish. Soc. Symp.,* 1, 531, 1987.
158. **Beverton, R. J. H. and Holt, S. J.,** On the dynamics of exploited fish populations, *Min. Agric. Fish Food (U.K.) Fish. Invest. London,* Ser. 2, 19, 1957.
159. **Ricker, W. E.,** Stock and recruitment, *J. Fish. Res. Board Can.,* 11, 559, 1954.
160. **Odum, H. T.,** *Systems Ecology,* John Wiley & Sons, New York, 1983.
161. **McLusky, D. S.,** *The Estuarine Ecosystem,* Halsted Press, New York, 1981.
162. **Pimm, S. L. and Kitching, R. L.,** The determinants of food chain lengths, *Oikos,* 50, 302, 1987.
163. **Pimm, S. L.,** Energy flow and trophic structure, in *Concepts of Ecosystem Ecology: a Comparative View,* Pomeroy, L. R. and Alberts, J. J., Eds., Springer-Verlag, New York, 1988, 263.
164. **Odum, E. P. and Biever, L. J.,** Resource quality, mutualism, and energy partitioning in food chains, *Am. Nat.,* 124, 360, 1984.
165. **Azam, F.,** The ecological role of water-column microbes in the sea, *Mar. Ecol. Prog. Ser.,* 10, 257, 1983.
166. **Ducklow, H. W., Purdie, D. A., Williams, P. J. le B., and Davis, J. M.,** Bacterioplankton: a sink for carbon in a coastal marine plankton community, *Science,* 232, 865, 1986.
167. **Sherr, E. B., Sherr, B. F., and Albright, L. J.,** Bacteria: link or sink, *Science,* 235, 88, 1987.
168. **Sherr, E. B. and Sherr, B. F.,** Role of microbes in pelagic food webs: a revised concept, *Limnol. Oceanogr.,* 33, 1225, 1988.
169. **Jannasch, H. W.,** The chemosynthetic support of life and the microbial diversity at deep-sea hydrothermal vents, *Proc. R. Soc. London,* B225, 277, 1985.
170. **Jannasch, H. W. and Mottl, M. J.,** Geomicrobiology of deep-sea hydrothermal vents, *Science,* 229, 717, 1985.
171. **Cavanaugh, C. M.,** Symbiosis of chemoautotrophic bacteria and marine invertebrates from hydrothermal vents and reducing sediments, in *Hydrothermal Vents of the Eastern Pacific: An Overview,* Jones, M. L., Ed., Bull. Biol. Soc. Wash., D.C., 6, 373, 1985.
172. **Lutz, R. A. and Kennish, M. J.,** Deep-sea hydrothermal vent communities: a review, *Rev. Geophys.,* 31, 211, 1993.

4.1 PLANKTON

Table 4.1—1 CLASSIFICATION OF PLANKTON

Size category	Size range
Megaplankton	20–200 cm
Macroplankton	2–20 cm
Mesoplankton	0.2–20 mm
Microplankton	20–200 μm
Nanoplankton	2–20 μm
Picoplankton	0.2–2 μm
Femtoplankton	0.02–0.2 μm

From Wotton, R. S., Ed., *The Biology of Particles in Aquatic Systems,* CRC Press, Boca Raton, FL, 1990, 4.

Table 4.1—2
AMOUNT OF PLANKTON ($cm^3/1,000\ m^3$) AT VARIOUS DEPTHS AND IN DIFFERENT PARTS OF THE NORTH ATLANTIC

Depth (m)	Gulf Stream 40–43°N	Continental slope of North America 38–41°N	Gulf Stream 35–37°N	Sargasso Sea 20–37°N	North equatorial current 16–19°N	African littoral 15–22°N	Canaries current 27–36°N	32° 29′N 20° 0.9′W	32° 34′N 16° 19′W
0–50	74.5	199.0	54.9	64.7	89.6		78.6		
50–100	42.3	94.0	35.9	59.9	80.7	214.0	55.0		
100–200	26.2	35.1	23.3	32.1	30.6	94.7	27.0	26.0	28.0
200–500	25.4	30.8	11.1	10.2	15.9	59.9	15.6		
500–1,000	15.7	19.0	6.0	4.1	5.4	27.2	6.5		
1,000–2,000	5.0	7.8	2.8	1.2	1.6	–	2.1	5.0	4.0
2,000–3,000	–	–	2.1	0.6	1.0	–	0.8	1.0	0.7
3,000–4,000	–	–	–	0.3	0.3	–	0.2	0.6	–
4,000–5,000	–	–	–	0.1	–	–	–	–	–

Table 4.1—3
CHANGES IN THE RELATIVE NUMBERS OF CALANOIDA WITH DEPTH IN VARIOUS REGIONS OF THE OCEAN

	Open ocean			Isolated basin				
						Mediterranean Sea (Ionian Sea)		
Depth (m)	Northwestern Pacific Ocean (spring)	Tropical zone Pacific Ocean	Tropical zone Indian Ocean	Norwegian Sea (annual average)	Central North Polar Basin	Winter	Summer	Sea of Japan (winter)
0–50								
50–100	63.5	93.1	—	29.2	92.5	91.5	53.0	58.4
100–200	30.6	67.0	64.5	33.6	105	71.5	49.3	35.0
200–500	80.5	52.3	31.0	33.6	27.1	37.1	26.1	137.0
500–1000	16.2	40.0	61.0	160	27.1	8.8	27.0	56.4
1000–2000	59.0	28.5	18.7	5.1	5.8	3.5	4.7	3.1

Note: Numbers in each layer are expressed in % of numbers in the overlying layer.

4.2 CHEMICAL COMPOSITIONS

Table 4.2—1
TOTAL CARBON, CARBONATE CARBON, AND ORGANIC CARBON CONCENTRATION IN PLANKTONIC MARINE ORGANISMS

Organism	Total carbon (% DW)	Carbonate carbon (% DW)	Organic carbon (% DW)	Organic carbon (% af DW)
Cnidaria				
Cyanea capillata	13.8	–	13.8	36.0
Physalia physalis	31.4	–	31.4	62.8
Pelagia noctiluca	12.9	–	12.9	26.0
Pelagia noctiluca	15.9	–	15.9	31.2
Aequorea vitrina	26.8	–	26.8	52.5
Average	17.5	–	17.5	41.7
Ctenophora				
Mnemiopsis sp.	6.4	–	6.4	20.6
Arthropoda				
Euphausia krohnii	35.8	–	35.8	43.9
Centropages hamatus	36.3	–	36.3	46.2
C. typicus 1:1				
Calanus finmarchicus	41.7	–	41.7	50.5
Meganyctiphanes norvegica	42.0	–	42.0	51.6
Lophogaster sp.	46.8	–	46.8	57.4
Centropages sp.	38.5	–	38.5	49.7
Centropages sp.	38.7	–	38.7	50.0
Sagitta elegans 1:1				
Idotea metallica	33.2	2.36	30.8	48.0
Calanus finmarchicus	39.8	–	39.8	48.3
Mixed copepods	35.6	–	35.6	46.0
Calanus finmarchicus (a small admixture of euphausiids and shell-less pteropods)	37.8	–	37.8	46.0
Average	38.3	–	38.0	48.9
Mollusca				
Limacina sp.	28.3	2.74	25.6	56.0
Ommastrephes sp.	45.1	–	45.1	48.8
Sthenoteuthis sp.	37.2	–	37.2	40.4
Clione limacina	26.3	–	26.3	39.4
Illex illecebrosus	39.2	–	39.2	42.6
Squid eggs (*Loligo*)	21.7	–	21.7	45.0
Average	33.1	–	32.7	45.4
Chordata				
Salpa sp.	10.6	–	10.6	46.1
Salpa sp.	9.6	–	3.6	39.0
Salpa fusiformis	7.8	–	7.8	33.9
Pyrosoma sp.	9.4	–	9.4	41.0
Average	9.4	–	9.4	40.0

Table 4.2—1 (continued)
TOTAL CARBON, CARBONATE CARBON, AND ORGANIC CARBON CONCENTRATION IN PLANKTONIC MARINE ORGANISMS

Organism	Total carbon (% DW)	Carbonate carbon (% DW)	Organic carbon	
			(% DW)	(% af DW)
Mixed Samples				
Mixed copepods and phytoplankton	29.8	–	29.8	38.5
Copepods and phytoplankton	25.2	–	25.2	48.0
Phytoplankton and fish	4.8	1.54	3.3	32.7
Phytoplankton and copepods	6.6	–	6.6	56.0
Copepods and phytoplankton	14.3	–	14.3	48.0
Mixed zooplankton	28.4	–	28.4	48.6
Average	18.2	–	17.9	38.8

Note: All entries are % of dry weight (DW) except last column, which is ash-free (af) dry weight. Where no results are listed, the inorganic carbonate was not detectable.

From Curl, H. S., Jr., Analyses of carbon in marine plankton organisms, *J. Mar. Res.*, 20(3), 185, 1962. With permission.

Table 4.2—2
ORGANIC CONTENT OF COPEPODS AND SAGITTAE BASED ON DRY WEIGHTS

	Protein %	Fat %	Carbohydrate %	Ash %	P_2O_5 %	Nitrogen %
Copepods	70.9–77.0	4.6–19.2	0–4.4	4.2–6.4	0.9–2.6	11.1–12.0
Sagittae	69.6	1.9	13.9	16.3	3.6	10.9

Table 4.2—3
CLASSIFICATION OF ORGANIC MATERIAL IN AQUATIC SYSTEMS

Size category	Size range (diameter)
CPOM — Coarse particulate organic material (or matter)	>1 mm
FPOM — Fine particulate organic material (or matter)	<1 mm but >0.45 μm
DOM — Dissolved organic material (or matter)	<0.45 μm

From Wotton, R. S., Ed., *The Biology of Particles in Aquatic Systems,* CRC Press, Boca Raton, FL, 1990, 4.

4.3 GEOGRAPHICAL DISTRIBUTION OF CHARACTERISTIC FAUNAL GROUPS

Table 4.3—1
AREAS OF NORTH PACIFIC IN WHICH LISTED SPECIES HAVE BEEN SHOWN TO OCCUR

Organism	Subarctic	Transitional	Central	Equatorial	Eastern Tropic Pacific	Warm-water cosmopolites	Comments
PROTOZOA							
Foraminifera							
Globigerina quinqueloba	+						
Globigerinoides minuta	+						
Globigerina pachyderma	+						
Globorotalia truncatulinoides			+				
Pulleniatina obliquiloculata				+			
Sphaeroidinella dehiscens				+			
Globigerina conglomerata				+			
Globorotalia tumida				+			
Globorotalia hirsuta				+			
Globigerinella aequilateralis						+	Pure
Globigerinoides conglobata						+	Pure
Globigerinoides rubra						+	Pure
Orbulina universa						+	Pure
Globigerinoides sacculifera						+	Peak at equator
Globorotalia menardii						+	Peak at equator
Globigerina eggeri						+	Edge effect
Hastigerina pelagica						+	Edge effect
Radiolaria							
Castanidium apsteini	+						
Castanidium variabile	+						Doubtful; may also be deep central
Haeckeliana porcellana	+						Doubtful; may also be deep central
Castanea amphora			+				
Castanissa brevidentata			+				
Castanella thomsoni			+				T. zone w/upwelled water?
Castanea henseni			+				T. zone w/upwelled water?
Castanea globosa			+				T. zone w/upwelled water?
Castanidium longispinum				+			
Castanella aculeata				+			

CHAETOGNATHA							
Sagitta elegans	+						
Eukrohnia hamata	+						
Sagitta scrippsae		+					
Sagitta pseudoserratodentata			+				
Sagitta californica			+				Crossing W.T.P.
Sagitta ferox				+			
Sagitta robusta				+			Patchy
Sagitta regularis				+			
Sagitta hexaptera						+	
Sagitta enflata						+	Peak at equator
Pterosagitta draco						+	Peak at equator
Sagitta pacifica						+	Peak at equator
Sagitta minima						+	Edge effect
ANNELIDA							
Tomopteris septentrionalis						+	
Tomopteris pacifica	+						
Poeobius meseres	+						
ARTHROPODA							
Copepoda							
Calanus pacificus	+						May be T. zone
Calanus plumchrus	+						
Calanus tonsus	+						
Calanus cristatus	+						
Eucalanus bungii bungii	+						
Eucalanus elongatus hyalinus		+					South Pacific also
Eucalanus bungii californicus		+					
Clausocalanus pergens		+					
Clausocalanus lividus			+				
Eucalanus subcrassus				+			
Rhincalanus cornutus				+			
Eucalanus inermis					+		
Eucalanus crassus						+	Patchy, 'pure'
Rhincalanus nasutus						+	Very patchy, almost pure equatorial
Eucalanus attenuatus						+	Peak at equator; some edge effect

Table 4.3—1 (continued)
AREAS OF NORTH PACIFIC IN WHICH LISTED SPECIES HAVE BEEN SHOWN TO OCCUR

Organism	Subarctic	Transitional	Central	Equatorial	Eastern Tropic Pacific	Warm-water cosmopolites	Comments
Eucalanus subtenuis				+			Patchy, peak at equator; some edge effect
Clausocalanus arcuicornis							
Eucalanus longiceps				+			
Rhincalanus gigas							
Clausocalanus laticeps							
Euphausiacea							
Thysanoessa longipes	+						
Euphausia pacifica	+						
Thysanopoda acutifrons		+					
Thysanoessa gregaria		+					
Euphausia gibboides		+					
Nematoscelis difficilismegalops		+					
Nematoscelis atlantica			+				
Euphausia brevis			+				
Euphausia hemigibba			+				
Euphausia gibba			+				
Euphausia mutica			+				
Stylocheiron suhmii			+				Crossing in W.T.P.
Euphausia diomediae				+			
Euphausia distinguenda				+			
Nematoscelis gracilis				+			
Euphausia distinguenda					+		
Euphausia eximia					+		
Euphausia lamelligera					+		
Euphausia tenera						+	Peak at equator
Stylocheiron abbreviatum						+	Avoids E.T.P.
Euphausia superba							
Amphipoda							
Parathimisto pacifica	+						
MOLLUSCA							
Pteropoda							
Limacina helicina	+						
Clio polita	+						
Corolla pacifica		+					

Clio balantium		+					
Cavolinia inflexa			+				
Clio pyramidata			+				Crossing in W.T.P.
Styliola subula			+				Crossing in W.T.P.
Limacina lesueuri			+				Crossing in W.T.P.
Clio n.sp.				+			
Cavolinia uncinata				+			Very patchy
Limacina trochiformis					+		
Limacina inflata						+	
Cavolinia longirostris						+	Very patchy; almost pure equatorial
Cavolinia gibbosa						+	Very patchy, avoids E.T.P.
Hyalocylix striata						+	
Creseis virgula						+	Edge effect
Creseis acicula						+	
Cavolinia tridentata						+	Peak at equator
Diacria trispinosa						+	Peak at equator
Limacina bulimoides						+	Avoids E.T.P.
Clio antarctica						+	
Heteropoda							
Caranaria japonica		+					
Gymnosomata							
Clione limacina	+						

Note: E.T.P., Eastern Tropical Pacific; W.T.P., Western Tropical Pacific.

From McGowan, J. A., Oceanic Biogeography of the Pacific, in *The Micropalaeontology of Oceans,* Cambridge University Press, Cambridge, 1971, 14. With permission.

Table 4.3—2
TYPICAL COSMOPOLITAN OCEANIC SPECIES

Siphonophora	**Mollusca**	**Copepoda**
Physophora hydrostatica	*Euclio pyramidata*	*Rhincalanus nasutus*
Agalma elegans	*Euclio cuspidata*	*Eucalanus elongatus*
Dimophyes arctica	*Diacria trispinosa*	*Pleuromamma robusta*
Lensia conoidea	*Pneumodermopsis ciliata*	*Euchirella rostrata*
Chelopheys appendiculata	*Taonidium pfefferi*	*Euchirella curticaudata*
Sulculeolaria biloba	*Tracheloteuthis risei*	*Oithona spinirostris*
Medusae	**Polychaeta**	**Other Crustacea**
Cosmetira pilosella	*Travisiopsis lanceolata*	*Lepas* sp.
Laodicea undulata	*Vanadis formosa*	*Munnopsis murrayi*
Halicreas sp.	*Rhynchonerella angelini*	*Brachyscelus crusulum*
Periphylla periphylla	*Tomopteris septentrionalis*	*Meganyctiphane norvegica*
	Chaetognatha	
	Sagitta serratodentata f. tasmanica	*Euphausia krohni*
	Sagitta hexaptera	*Anchialus agilis*
	Thaliacea	
	Salpa fusiformis	
	Dolioletta gegenbauri	

Table 4.3—3
SOME PLANKTONIC SPECIES TYPICAL OF DEEP WATER

Gaetaenus pileatus	*Amalopenaeus elegans*	*Sagitta macrocephala*
Arietellus plumifer	*Hymenodora elegans*	*S. zetesios*
Pontoptilus muticus	*Boreomysis microps*	*Eukrohnia fowleri*
Centraugaptilus rattrayi	*Eucopia unguiculata*	*Nectonemertes miriabilis*
Augaptilus megalaurus and many other copepods	*Cyphocaris anonyx*	*Spiratella helicoides*
	Scina sp.	*Histioteuthis boneltiana*

Table 4.3—4
SPECIES COMPOSITION OF THE FIVE WORLD DISTRIBUTIONAL ZONES OF PLANKTONIC FORAMINIFERA

Northern and Southern Cold-Water Regions

1. Artic and antarctic zones
   Globigerina pachyderma (Ehrenberg): Left-coiling variety; right-coiling in subarctic and subantarctic zones
2. Subarctic and subantarctic zones
   Globigerina quinqueloba (Natland)
   Globigerina bulloides (d'Orbigny)
   Globigerinita bradyi (Wiesner)
   Globorotalia scitula (Brady)

Transition Zones

3. Northern and south transition zones between cold-water and warm-water regions
   Globorotalia inflata (d'Orbigny): With mixed occurrences of subpolar and tropical-subtropical species

Warm-Water Region

4. Northern and southern subtropical zones

Globigerinoides ruber (d'Orbigny): Pink variety in Atlantic Ocean only	
Globigerinoides conglobatus (Brady): Autumn species	
Hastigerina pelagica (d'Orbigny)	
Globigerinita glutinata (Egger)	
Globorotalia truncatulinoides (d'Orbigny)	
Globorotalia hirsuta (d'Orbigny)	Winter species
Globigerina rubescens (Hofker)	Winter species
Globigerinella aequilateralis (Brady)	Prefer outer margins of subtropical central water masses and into transitional zone
Orbulina universa (d'Orbigny)	
Globoquadrina dutertrei (d'Orbigny)	
Globigerina falconensis (Blow)	
Globorotalia crassaformis (Galloway and Wissler)	

5. Tropical Zone

Globigerinoides sacculifer (Brady): Including *Sphaeroidinella dehiscens* (Parker and Jones)	
Globorotalia menardii (d'Orbigny)	
Globorotalia tumida (Brady)	
Pulleniatina obliquiloculata (Parker and Jones)	
Candeina nitida (d'Orbigny)	
Hastigerinella digitata (Rhumbler)	
Globoquadrina conglomerata (Schwager)	Restricted to Indo-Pacific
Globigerinella adamsi (Banner and Blow)	Restricted to Indo-Pacific
Globoquadrina hexagona (Natland)	Restricted to Indo-Pacific

Note: The species are listed under the zone where their highest concentrations are observed, but they are not necessarily limited to these areas. Most species listed under the Subtropical Zones are also common in the tropical waters.

* Usually located in central water masses between 20°N and 40°N, or between 20°S and 40°S latitude.

4.4 DEEP-SEA BIOLOGY INCLUDING HYDROTHERMAL VENT AND COLD SEEP COMMUNITIES

Table 4.4—1
MAJOR DEEP-SEA HYDROTHERMAL VENT SITES WHERE BIOLOGICAL INVESTIGATIONS HAVE BEEN CONDUCTED

Vent site	Coordinates	Depth (m)	Major vent fields
Galapagos	0°47′ 48′N: 86°01′ 14′W	2480	Rose Garden, Garden of Eden, Musselbed-Clambake
East Pacific Rise: 17 south	17°30′ 21°26′S: 113–114°W	2600–2825	
East Pacific Rise: 13 EPR	12°38′–54′N: 103°54′–58′W	~2600	Parigo, Pogosud, Actinoir Pogonord, Genesis
East Pacific Rise: 21 EPR	20°51′N:109°05′W	2620	Clam Acres, Hanging Gardens
Guaymas Basin (EPR)	27°01′-03′N: 111°24′W	2050	
Northern Gorda Ridge	42°45′N:126°43′W	~2700	G14
Southern Gorda Ridge	41°00′N:127°29′W	3240	Nesca
Cleft Segment (Juan de Fuca)	44°38′–58′N: 130°15′–22′W	2210–2280	Megaplume, South Cleft
Axial Seamount (Juan de Fuca)	45°57′N:130°01′W	1560	Ashes, Casm
Endeavour Segment (Juan de Fuca)	47°56′–58′N: 129°08′W	2250	
Middle Valley (Juan de Fuca)	48°30′N:128°41′W	~2400	
Explorer Ridge	49°46′N:139′16′W	1870	Magic Mountain
Lau Basin	21°25′–22°40′S: 176°30′W	1800–1960	Hine Hina, Vai Lili
Fiji Basin	17°00′S:173°55′E	2000	White Lady, Mussel Valley
Manus Basin	3°10′S:150°17′E	~2500	
Mariana Trough	18°13′N:144°42′E	3550–3750	Snail Pits
Okinawa Trough	27°16′N:127°05′E	1350–1550	JADE
Mid-Atlantic Ridge	23°22′N:44°57′W	3460	Snakepit
Mid-Atlantic Ridge	26°08′N:44°50′W	3620–3700	TAG

From Tunnicliffe, V., The biology of hydrothermal vents: ecology and evolution, *Oceanogr. Mar. Biol. Annu. Rev.*, 29, 319, 1991. With permission.

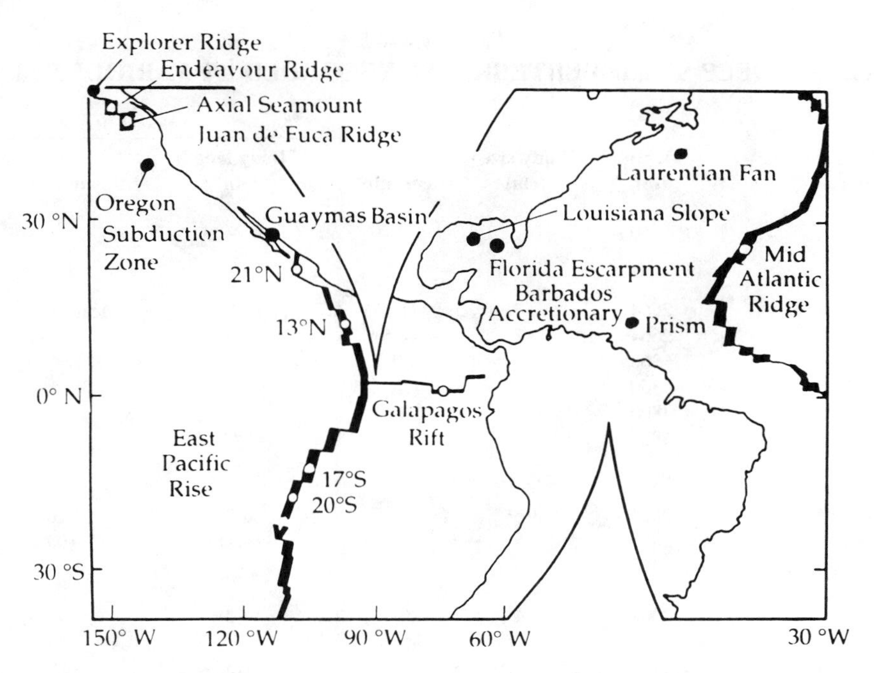

FIGURE 4.4—1. Locations of deep-sea hydrothermal vent and seep communities discovered in the eastern Pacific and Atlantic. (From Gage, J. D. and Tyler, P. A., *Deep-Sea Biology: a Natural History of Organisms at the Deep-Sea Floor,* Cambridge University Press, Cambridge, 1991. With permission.)

Table 4.4—2
RATES OF DEEP-SEA INVERTEBRATES RECORDED AT VARIOUS DEPTHS

Taxon	Depth (m)	Body size (cm)	Speed: cm min^{-1}	Body length min^{-1}	cm h^{-1}	Body length h^{-1}
Vermes	360		2–9	~1		
"Green worm"						
Crustacea						
Glyphocrangon sculpta	2664	7	(1.67)[a]	(0.24)[a]	Mean 100	Mean 14.3
	2664	7	(4.5)[a]	(0.65)[a]	(270 max)[a]	39
Cancer borealis	360		7.7			
Sclerocrangon sp.	360		1.7–4.5			
"Decapod"	4800		0.4			
Hemichordata sp.	4873		0.2			
Echinodermata						
Asteroidea						
Hymenaster membranaceus	2664	4–9	(5.17)[a]	(0.83)[a]	310	50
Bathybiaster vexillifer	2008	16	(1.67)[a]	(0.1)	100	6
Henricia sp.	360					
Echinoidea						
Echinus affinis	2008	4	(0.78)[a]	(0.2)[a]	47	12
Ophiuroidea						
?Ophiomyxa tumida	1180	~4 (max)	6[b]	1.5[b]		
Ophiomusium lymani	1200	20 (max)	0.7–3.4	0.004–0.17		
"Ophiuroid"	4800					
Holothurioidea						
Benthogone rosea	2008	17	(1.48)[a]	(0.09)[a]	89	5
Scotoplanes sp.	1200	ca. 10	0.6	0.06		
Gastropoda						
Bathybembix bairdi	1300		1.5			
Neptunia amianta	1300		1.4			
Gastropod sp. "A"	4800		1.0			
Gastropod sp. "A"	4800		0.2			
Gastropod sp. "B"	4800		0.1			
Gastropod sp.	2664	4	(2.57)[a]	(0.64)	154	38

[a] Time-lapse rate, 64 min.
[b] Time-lapse rate, 36 s.

From Gage, J. D., Biological rates in the deep sea: a perspective from studies on processes in the benthic boundary layer, *Rev. Aquat. Sci.*, 5, 49, 1991. With permission.

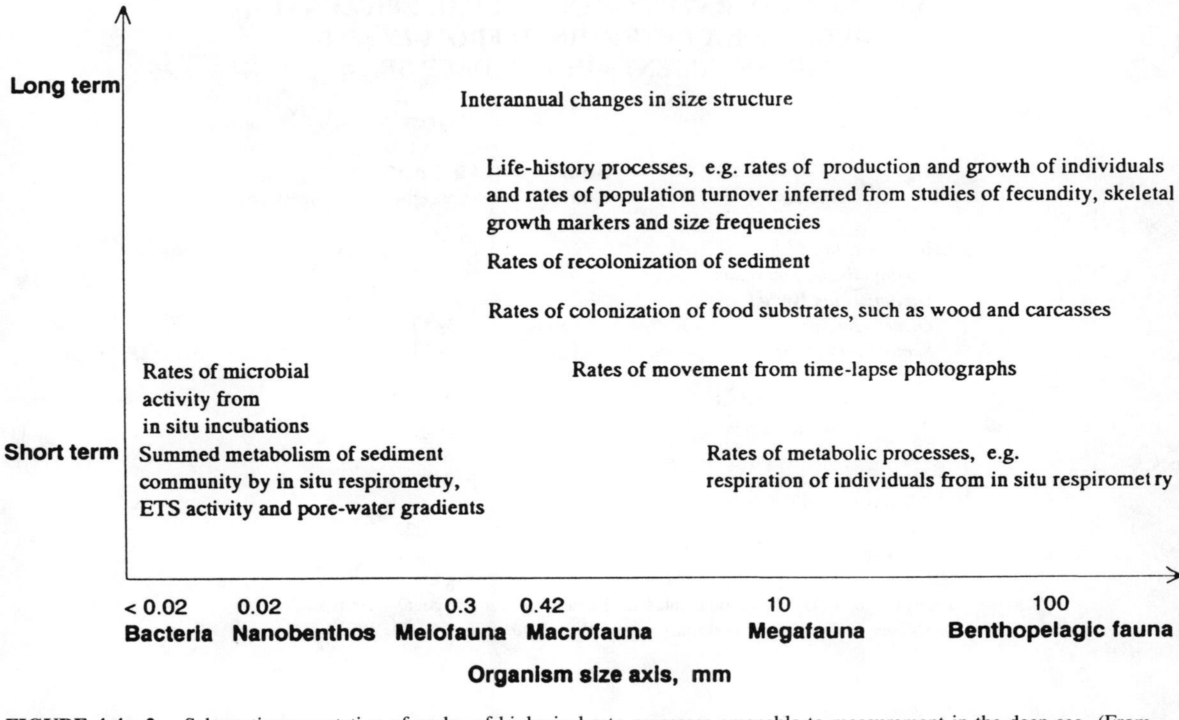

FIGURE 4.4—2. Schematic presentation of scales of biological rate processes amenable to measurement in the deep sea. (From Gage, J. D., Biological rates in the deep sea: a perspective from studies on processes in the benthic boundary layer, *Rev. Aquat. Sci.*, 5, 49, 1991.)

Table 4.4—3
RESPIRATION RATES OF INDIVIDUAL EPIFAUNAL MEGAFAUNA DETERMINED FROM *IN SITU* MEASUREMENTS IN THE DEEP SEA

Species	Depth (km)	Oxygen consumption (μl h^{-1} mg^{-1} wet weight)	No. measured
Echinodermata			
Ophiophthalmus normani	1.3	26–43	9
Ophiomusium lymani	1.23	52	1
O. armigerum	3.65	23–30	2
Scotoplanes globosa		4–13	5
Fish			
Coryphaenoides acrolepis	1.23	2.4	1
C. (Nematonurus) armatus	3.63	2.7–3.7	2
	2.75	3.1	1
Eptratrelas deani	1.23	2.2	1
Sebastolobus altivelis			
Adult	1.3	2.7–3.22	4
Juvenile	1.3	26.6–34.4	2

From Gage, J. D., Biological rates in the deep sea: a perspective from studies on processes in the benthic boundary layer, *Rev. Aquat. Sci.*, 5, 49, 1991.

4.5 BIOLOGICAL PRODUCTION IN THE OCEAN

Table 4.5—1
GLOBAL PRIMARY PRODUCTION IN MARINE ECOSYSTEMS

Ecosystem	Proportion of ocean	Area (10^6 km^2)	Mean net productivity (g C_{org} m^{-2} $year^{-1}$)	Total net production (10^9 t C $year^{-1}$)
Algal bed and reef	0.0016	0.6	1166.7	0.7
Estuary	0.0039	1.4	714.3	1.0
Upwelling zone	0.0011	0.4	250.0	0.1
Continental shelf	0.0737	26.6	161.6	4.3
Open ocean	0.9197	332.0	56.3	18.7
Total	1.0000	361.0	68.7	24.7

From Berkman, P. A., The Antarctic marine ecosystem and humankind, *Rev. Aquat. Sci.*, 6, 295, 1992.

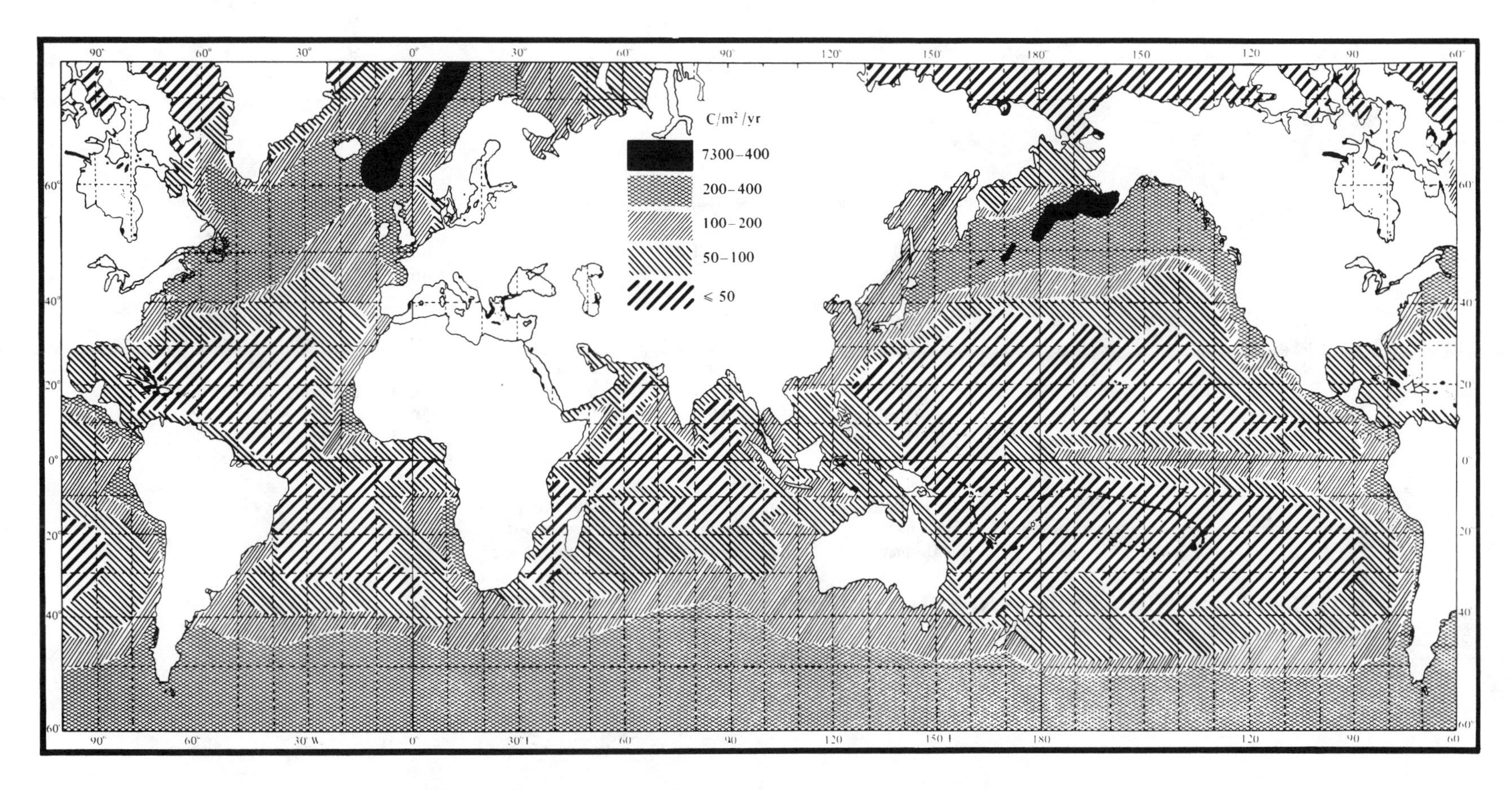

FIGURE 4.5—1. Estimation of organic production in the oceans. Contours are in grams of carbon biologically fixed under each square meter of sea surface per year.

Table 4.5—2
PRIMARY PRODUCTION AND NITROGEN FIXATION IN RELATION TO LOSSES OF SEDIMENT STORAGE OF ORGANIC CARBON, DENITRIFICATION, AND METHANE PRODUCTION IN VARIOUS AQUATIC ECOSYSTEMS

Region	Area (km^2)	Net primary production ($\times 10^9$ tons C yr^{-1})	Nitrogen fixation ($\times 10^7$ tons N_2 yr^{-1})	Sediment organic carbon sink ($\times 10^9$ tons C yr^{-1})	Denitrification loss ($\times 10^7$ tons N_2 yr^{-1})	Methane emission ($\times 10^7$ tons CH_4 yr^{-1})
Open ocean	3.1×10^8	18.60	0.43	0.20	0	0.36
Continental shelf	2.6×10^7	5.20	0.27	0	2.97	0.04
Continental slope	3.2×10^7	2.24	0.06	0.50	5.50	0.03
Freshwater marshes	1.6×10^6	1.51	2.21	0.15	6.40	3.10
Estuaries/deltas	1.4×10^6	0.92	0.06	0.20	1.04	0.60
Salt marshes	3.5×10^5	0.49	0.48	0.05	1.40	0.80
Rivers/lakes	2.0×10^6	0.40	1.88	0.13	0.26	5.10
Coral reefs	1.1×10^5	0.30	0.28	0.01	0	0.32
Seaweed beds	2.0×10^4	0.03	0	0	0	0.08
Total aquatic area	3.8×10^8	C INPUT: 29.7	N_2 INPUT: 5.7	C OUTPUT: 1.2	N_2 OUTPUT: 17.6	CH_4 OUTPUT: 10.4

From Walsh, J. J., *On the Nature of Continental Shelves*, Academic Press, New York, 1989. With permission.

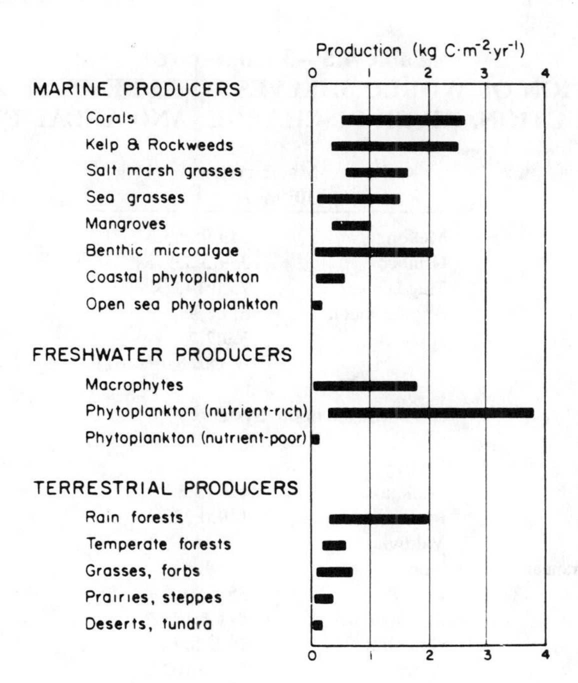

FIGURE 4.5—2. Annual net production rates of marine, freshwater, and terrestrial producers. (From Valiela, I., *Marine Ecological Processes,* Springer-Verlag, New York, 1984. With permission.)

Table 4.5—3
GEOGRAPHIC LOCATION OF WORLD SHELVES WITH RESPECT TO ANNUAL PRIMARY PRODUCTION, RIVER DISCHARGE, AND AREAL EXTENT

Latitude	Region	Major rivers (10^2 m^3 sec^{-1})		Area (10^5 km^2)	Unit production (g C m^{-2} yr^{-1})
		Eastern Boundary Currents			
0–30	Ecuador-Chile		—	2.7	1000–2000
	Southwest Africa		—	1.7	1000–2000
	Northwest Africa		—	2.6	200–500
	Baja California		—	1.1	600
	Somali Coast	Juba	5.5:5.5	0.6	175
	Arabian Sea	Indus	75.5:86.0	3.8	200
30–60	California-Washington	Columbia	79.6:97.4	1.6	150–200
	Portugal-Morocco	Tagus	3.1:8.9	0.8	60–290
			% Discharge (3)[a]	14.9	Mean: 644.2
		Western Boundary Currents			
0–30	Brazil	Amazon	1750.0:2005.1	6.0	90
	Gulf of Guinea	Congo	396.4:589.9	3.5	130
	Oman/Persian Gulfs	Tigris	14.5:14.5	4.3	—
	Bay of Bengal	Ganges	116.0:371.8	3.0	110
	Andaman Sea	Irrawaddy	135.6:150.6	4.0	50
	Java/Banda Seas	Brantas	4.0:8.6	6.7	—
	Timor Sea	Fitzroy	1.8:1.8	4.0	100
	Coral Sea	Fly	24.5:62.5	3.2	20–175
	Arafura Sea	Mitchell	3.6:40.3	14.2	150
	Red Sea	Awash	<0.4	1.6	34
	Mozambique Channel	Zambezi	70.8:93.3	2.6	100–150

Table 4.5—3 (continued)
GEOGRAPHIC LOCATION OF WORLD SHELVES WITH RESPECT TO ANNUAL PRIMARY PRODUCTION, RIVER DISCHARGE, AND AREAL EXTENT

Latitude	Region	Major rivers (10^2 m^3 sec^{-1})		Area (10^5 km^2)	Unit production (g C m^{-2} yr^{-1})
	South China Sea	Mekong	149.0:363.3	15.7	215–317
	Caribbean Sea	Orinoco	339.3:339.3	3.7	66–139
	Central America	Magdalena	75.0:143.3	4.5	180
	West Florida Shelf	Appalachicola	6.9:6.9	2.1	30
	South Atlantic Bight	Altamaha	3.9:7.2	1.4	130–350
			% Discharge (62)[a]	80.5	Mean: 137.4
		Mesotrophic Systems			
30–60	Australian Bight	Murray	7.4:7.7	8.2	50–70
	New Zealand	Waikato	4.1:25.8	2.9	—
	Argentina-Uruguay	Parana	149.0:220.1	10.4	—
	Southern Chile	Valdivia	4.5:44.3	4.0	—
	Southern Mediterranean	Nile	9.5:9.5	2.9	30–45
	Gulf of Alaska	Fraser	35.4:80.8	3.1	50
	Nova Scotia-Maine	St. Lawrence	141.6:178.7	6.7	130
	Labrador Sea	Churchill	15.8:147.5	21.2	24–100
	Okhotsk Sea	Amur	103.0:103.0	7.1	—
	Bering Sea	Kuskokwim	12.8:39.3	11.7	170
			% Discharge (12)[a]	78.2	Mean: 81.9
		Phototrophic Systems			
60–90	Beaufort Sea	Mackenzie	97.2:97.2	2.6	10–20
	Chukchi Sea	Yukon	62.0:62.0	6.1	40–180
	East Siberian Sea	Kolyma	22.4:39.8	7.8	—
	Laptev Sea	Lena	163.09:180.2	6.9	—
	Kara Sea	Ob	122.0:300.0	10.1	—
	Barents Sea	Pechora	33.6:78.7	7.3	25–96
	Greenland-Norwegian Seas	Tjorsa	3.6:13.4	2.8	40–80
	Weddell-Ross Seas	—		3.9	12–86
			% Discharge (11)[a]	47.5	Mean: 58.9
		Eutrophic Systems			
30–60	Mid-Atlantic Bight	Hudson	3.7:24.6	1.3	300–380
	Baltic Sea	Vistula	10.1:71.1	3.9	75–150
	East China Sea	Yangtze	220.0:240.3	10.7	—
	Sea of Japan	Ishikari	5.0:22.8	1.7	100–200
	North-Irish Seas	Rhine	25.4:47.8	8.7	100–250
	Northern Mediterranean	Po	14.7:49.0	2.9	68–85
	Caspian Sea	Volga	75.8:75.8	1.5	—
	Black Sea	Danube	65.3:97.9	1.6	50–150
	Bay of Biscay	Loire	8.3:16.7	2.7	—
	Texas/Louisiana	Mississippi	178.0:200.1	2.1	100
			% Discharge (12)[a]	37.1	Mean: 154.2

[a] Total percentage of the freshwater input from the 215 largest rivers of the earth (>30 m^3 sec^{-1}) within each of these shelf regions.

From Walsh, J. J., *On the Nature of Continental Shelves,* Academic Press, New York, 1989. With permission.

Table 4.5—4
ANNUAL PRODUCTION ESTIMATES IN THE ANTARCTIC MARINE ECOSYSTEM[a]

Species group	Minimum production		Maximum production	
	Carbon (10^8 tons)	Total weight[b] (10^6 tons)	Carbon (10^6 tons)	Total weight[b] (10^6 tons)
Primary production	661	6610	4450	44500
Microbial loop	215	2150	1584	15840
Zooplankton	209	2090	1540	15400
Fish + squid	58	580	428	4280
Birds + mammals	1	10	8	80
Benthic flux	82	820	602	6020

[a] Calculations based on the ecological efficiency of energy transfer (Table 4.5—5) from the next lower trophic level as described.[1]

[b] Assumed relation between carbon and total weight is 1:10.[2]

From Berkman, P. A., The Antarctic marine ecosystem and humankind, *Rev. Aquat. Sci.*, 6, 306, 1992.

REFERENCES

1. **Huntley, M. E., Lopez, M. D. G., and Karl, D. M.,** Top predators in the Southern Ocean: a major leak in the biological carbon pump, *Science,* 253, 64, 1991.
2. **Ryther, J. H.,** Photosynthesis and fish production in the sea, *Science,* 166, 72, 1969.

Table 4.5—5
ECOLOGICAL EFFICIENCIES IN THE ANTARCTIC MARINE ECOSYSTEM

Species group	Relative ingestion (I)	Relative growth (G)	Ecological efficiency[a] (G_n/I_n)
Microbial loop (M)			
Phytoplankton (P)	0.125		
Total M	0.125	0.044	0.352
Zooplankton (Z)			
P	0.875		
M[b]	0.075		
Total Z	0.950	0.325	0.342
Fish + squid (F)			
Z	0.234		
Total F	0.234	0.065	0.278
Bird + mammals (B)			
Z	0.091		
F	0.065		
Total B	0.156	0.003	0.019
Organic flux to the benthos (f)[c]			
Z			0.285
F			0.023
B			0.034
Total f			0.342

Note: Mean ingestion (I) and growth (G) data calculated relative to one unit of fixed carbon.[1]

[a] Values correspond to gross growth efficiencies in Reference 1.
[b] Microbial growth plus egestion were ingested by the zooplankton.
[c] Based on estimates of egestion from Z, F, and B.

From Berkman, P. A., The Antarctic marine ecosystem and humankind, *Rev. Aquat. Sci.*, 6, 306, 1992.

REFERENCES

1. **Huntley, M. E., Lopez, M. D. G., and Karl, D. M.,** Top predators in the Southern Ocean: a major leak in the biological carbon pump, *Science*, 253, 64, 1991.
2. **Ryther, J. H.,** Photosynthesis and fish production in the sea, *Science*, 166, 72, 1969.

Table 4.5—6
PRIMARY PRODUCTION OF ESTUARINE HABITATS (VALUES EXPRESSED IN g C/m²/year OR g dry wt/m²/year)

Plant type	Location	g C/m²/year	g dry wt/m²/year
Phytoplankton	Baltic Sea	48—94	
	St. Margaret's Bay, Canada	190	
	Cochin Backwater, India	124	
	Ems estuary, Netherlands	13—55	
	Grevelingen, Netherlands	130	
	Wadden Sea, Netherlands	100—200	
	Loch Etive, Scotland	70	
	Lynher, U.K.	81.7	
	Alewife Cove, U.S.	162	
	Barataria Bay, U.S.	210	
	Beaufort, U.S.	52.5	
	Bissel Cove, U.S.	56	
	Charlestown River, U.S.	42	
	Core Sound, U.S.	67	
	Duplin River, U.S.	248	
	Flax Pond, U.S.	60	
	Hempstead Bay, U.S.	177	
	Jordan Cove, U.S.	66	
	Long Island Sound, U.S.	205	
		308	
	Narragansett Bay, U.S.	242	
	Niantic River, U.S.	72	
	North Inlet, U.S.	346	
Microbenthic algae	Danish fjords	116	
	Wadden Sea, Netherlands	115—178	
		101 ± 39	
	Grevelingen, Netherlands	25—37	
	Ythan estuary, U.K.	31	
	Lynher, U.K.	143	
	Alewife Cove, U.S.	45	
	Barataria Bay, U.S.	240	
	Bissel Cove, U.S.	52	
	Charlestown River, U.S.	41	
	Delaware, U.S.	160	
	False Bay, U.S.	143—226	
	Flax Pond, U.S.	52	
	Hempstead Bay, U.S.	62	
	Jordan Cove, U.S.	41	
	Niantic River, U.S.	32	
	Sapelo Island, U.S.	180	
Seagrasses			
Halodule wrightii	North Carolina, U.S. (intertidal)	70—240	
	South Florida — Gulf Coast, U.S. (intertidal)	70—240	
Thalassia testudinum	South Florida — Gulf Coast — Texas, U.S. (subtidal)	580—900	
Zostera marina	Denmark		
	Leaves[a]		856
	Roots, rhizomes[b]		241
	Alaska, U.S.	19—552	
	Beaufort, U.S.	350	
	North Carolina, U.S. (intertidal)	330	

Table 4.5—6 (continued)
PRIMARY PRODUCTION OF ESTUARINE HABITATS (VALUES EXPRESSED IN g C/m²/year OR g dry wt/m²/year)

Plant type	Location	g C/m²/year	g dry wt/m²/year
Zostera marina (continued)	Little Egg Harbor, U.S.		466
	Pacific Coast, U.S. (subtidal)	90—540	
	Puget Sound, U.S.	58—330	
		58—1500	116—680
			10—1200
Salt marsh grasses			
Carex spp.	Arctic Waters	10—120[a]	
Distichlis spicata	Atlantic Coast, U.S.		1070—3400[b]
	Gulf Coast, U.S. (Louisiana)	1600[a]	
	Pacific Coast, U.S.	300—600[a]	
Juncus gerardi	Maine, U.S.		485[a]
	Atlantic Coast, U.S.		1620—4290[b]
Juncus roemerianus	North Carolina, U.S.		754[a]
	Georgia, U.S.		2200[a]
	Gulf Coast, U.S. (Louisiana)	1700[a]	
	Gulf Coast, U.S.		1360—7600[b]
Puccinellia phrygondes	Arctic waters	25—70[a]	
Salicornia virginica	Atlantic Coast, U.S.		430—1430[b]
Salicornia spp.	Norfolk, U.K.		867[a]
	Pacific Coast, U.S.	325—1000[a]	
Spartina alterniflora	Massachusetts, U.S.		3500[b]
	Atlantic Coast, U.S.	200—800[a]	
	Atlantic Coast, U.S.	220—1680[b]	
	Atlantic Coast, U.S.		220—3500[b]
	Atlantic and Gulf Coasts, U.S.	200—2000[a]	
	Gulf Coast, U.S. (Louisiana)	1300[a]	279—6000[a]
Spartina foliosa	Pacific Coast, U.S.	400—850[a]	
Spartina patens	Maine, U.S.		5163[a]
	Atlantic Coast, U.S.		310—3270[b]
	Atlantic and Gulf Coasts, U.S.	500—700[a]	
	Gulf Coast, U.S. (Louisiana)	3000[a]	
Mixed species	Netherlands	100—500[a]	200—1000[a]
	Europe		11—1100[a]
	Barataria Bay, U.S.	590[a]	1175[a]
	Gulf Coast, U.S.	500[a]	
	North America	100—1700[b]	
	North America		300—4000[a]
Rhizophora mangle	South Florida, U.S.	400	
		300	
			100
			470—730 (leaf fall)

[a] Above ground.
[b] Below ground.

From Kennish, M., *Ecology of Estuaries,* Vol. 1, CRC Press, Boca Raton, FL, 1986, 163.

Table 4.5—7
ANNUAL PRODUCTION RATES OF BENTHIC MICROALGAE IN SELECTED ESTUARINE ECOSYSTEMS

Location	Production[a]	Ref.
Danish fjords	116	7
Wadden Sea, Netherlands	115–178	1
	101 ± 39	3
Grevelingen, Netherlands	25–37	4
Ythan estuary, U.K.	31	2
Lynher, U.K.	143	5
Alewife Cove	45	6
Barataria Bay	240	6
Bissel Cove	52	6
Charlestown River	41	6
Delaware	160	8
False Bay	143–226	9
Flax Pond	52	6
Hempstead Bay	62	6
Jordan Cove	41	6
Niantic River	32	6
Sapelo Island	180	10

[a] Production values in g C per square meter per year.

From Kennish M. J., *Ecology of Estuaries,* Vol. 1, CRC Press, Boca Raton, FL, 1986.

REFERENCES

1. **Cadée, G. C. and Hegeman, J.,** *Neth. J. Sea Res.,* 8, 260, 1974.
2. **Sibert, J. R. and Naiman, R. J.,** *Marine Benthic Dynamics,* Tenore, K. R. and Coull, B. C., Eds., University of South Carolina Press, Columbia, 1980, 311.
3. **McLusky, D. S.,** *The Estuarine Ecosystem,* Halsted Press, New York, 1981.
4. **Wolff, W. J.,** *Ecology of Marine Benthos,* Coull, B. C., Ed., University of South Carolina Press, Columbia, 1977, 267.
5. **Joint, I. R.,** *Estuarine Coastal Mar. Sci.,* 7, 185, 1978.
6. **Welsh, B. L., Whitlatch, R. B., and Bohlen, W. F.,** *Estuarine Comparisons,* Kennedy, V. S., Ed., Academic Press, New York, 1982, 53.
7. **Grøntved, J.,** *Medd. Dan. Fisk. Havunders.,* 3, 55, 1960.
8. **Gallagher, J. L. and Daiber, F. C.,** *Limnol. Oceanogr.,* 19, 390, 1974.
9. **Pamatmat, M. M.,** *Int. Rev. Gesamten Hydrobiol.,* 53, 211, 1968.
10. **Pomeroy, L. R.,** *Limnol. Oceanogr.,* 4, 386, 1959.

Table 4.5—8
TROPHIC TYPES OF TIDAL FLATS INDICATING BENTHIC MICROALGAL PRODUCTION

Trophic type	Primary production	Pool of detritus; decomposition	Consumption biomass
Exposed sandy shore	Inhibited by unstable sediment	Small, irregular import; oxic	Interstitial fauna, some large scavengers
Beds of suspension feeders	High in phytoplankton, variable in benthic micro- and macroalgae	Moderate to large, accumulation of fecal material, variable import; oxic and anoxic	Large suspension feeders, particularly bivalves; few grazers and deposit feeders
Mudflat	High in benthic microalgae,[a] occasionally drifting green algae	Accumulation of allochthonous detritus; primarily anoxic	Deposit feeders and algal grazers dominate; often small (nematodes)
Sandflat	High in benthic microalgae[a]	Variable, small import; oxic and anoxic	Grazers dominate; large deposit feeders
Seagrass bed	High in aquatic phanerogams, benthic and epiphytic microalgae[a]	Large, variable import; oxic and anoxic	Trophically very diverse, many grazers on epiphytes
Saltmarsh and mangroves	High in halophytes, variable in benthic microalgae[a]	Very large, high import; primarily anoxic	Trophically diverse; very often small

[a] Includes photosynthetic bacteria.

From Reise, K., *Tidal Flat Ecology: an Experimental Approach to Species Interactions, Ecological Studies,* Vol. 54, Springer-Verlag, New York, 1985. With permission.

Table 4.5—9
NET ANNUAL AERIAL PRIMARY PRODUCTION ESTIMATES FOR *SPARTINA ALTERNIFLORA* AND *JUNCUS ROEMERIANUS*

g dry weight per square meter			
Streamside	*S. alterniflora* high marsh	Whole marsh	*J. roemerianus*
3300	2200	2288	1500
3700	1350	1538	2200
2000	400	528	

From Pomeroy, L. R., Darley, W. M., Dunn, E. L., Gallagher, J. L., Haines, E. B., and Whitney, D. M., *The Ecology of a Salt Marsh, Ecological Studies,* Vol. 38, Pomeroy, L. R. and Wiegert, R. G., Eds., Springer-Verlag, New York, 1981, 39. With permission.

Table 4.5—10
SPARTINA ALTERNIFLORA ABOVE- AND BELOW-GROUND PRIMARY PRODUCTION ALONG THE ATLANTIC COAST OF NORTH AMERICA (kg/m²/year)

Area	Form	Above	Below	Total
Nova Scotia	NR	0.8	1.1	1.9
Maine	Short	0.8	0.2	1.0
Massachusetts	Tall	0.4	3.5	3.9
New York	NR	0.6	0.9	1.5
New Jersey	Short	0.5	2.3–3.2	2.8–3.7
North Carolina	Short	0.6	0.5–0.6	1.1–1.2
	Tall	1.3	0.5	1.8
South Carolina	Short	1.3	5.4	6.7
	Tall	2.5	2.4	4.9
Georgia	Short	1.4	2.0	3.4
	Medium	2.8	4.8	7.6
	Tall	3.7	2.1	5.8

Note: NR, not reported.

From Dame, R. F., The importance of *Spartina alterniflora* to Atlantic Coast estuaries, *Rev. Aquat. Sci.,* 1, 639, 1989.

Table 4.5—11
RATES OF CARBONATE PRODUCTION

Geographic location	Environment	Substrate	Depth	Raw production (per m^{-2} $year^{-1}$)	Carbonate production (g $CaCO_3 \cdot m^{-2}$ $year^{-1}$)	Comments
Lynher Estuary, Tamar River, England	Mudflat	Mud	Intertidal	3.65 g AFDW	47.5	
Pelican Point, Corona Del Mar, California	Beach	Rocks with sand, gravel	Intertidal	3,036 g $CaCO_3$	3,036	Average of 2 stations
Georgia coast	Salt marsh	Sand and mud	<2.5 m	25 kcal	68.0	
San Francisco Bay, California	Mudflat	Sandy mud	0–1 m	66.5 g AFDW	864.0	Average of 3 stations
Lynher Estuary, England	Mudflat	Mud	Intertidal	1.5 g org-C	103.1	
Seafield Bay, England	Salt marsh	Mud	Intertidal	0.15 g org-C	10.5	
Seafield Bay, England	Salt marsh	Mud	Intertidal	0.29 g org-C	20.4	
Hamble Spit (Southampton Water), England	Mudflat	Mud	Intertidal	62.7 g AFDW	815.2	Average groups II, III, IV
Balgzand, Dutch Wadden Sea	Mudflat	Mud	Intertidal	12 g $CaCO_3$	12	*Macoma balthica* only
Petpeswick Inlet, eastern Canada	Tidal flat	Mud, sand	Intertidal	24.4 kcal	6.5	*Spartina* habitat
Petpeswick Inlet, eastern Canada	Tidal flat	Mud, sand	Intertidal	83.6 kcal	35.5	*Zostera* habitat
North Inlet, South Carolina	Oyster reef		Intertidal	3,461 kcal	92,991	
Coral reefs, worldwide					1,400–35,000; median = 5,400	
Prévost Lagoon, Mediterranean Sea, France	Bay	Sand, muddy sand		15–888 g; average = 161.3; median = 75	195–11,544; average = 2,097; median = 975	Mollusks only
Kiel Bay, Baltic Sea	Bay	Mud	>10 m	31.1 g shelled wet wt	1.9	*Abra alba* only
Ythan Estuary, Scotland	Bay	Sandy mud	2.5 m	25.6 g org-C	1,798.5	
Copano Bay, Texas	Bay	Muddy sand	1 m	43.7 g $CaCO_3$	43.7	
Dublin Bay, Ireland	Bay	Sand		0.01 g shell-free wet wt	0.04	Bivalves only

Carmarthen Bay, Wales	Bay	Sand	10–17 m	19.1 g AFDW	248.4	
Grovelingen Estuary, Netherlands	Bay	Sand, mud	6–10 m	42.5 g AFDW	552.9 range = 1.2–1,558.7	Average of 6 stations
Gunnamatta Bay, Australia	Bay	Sand, silt	11 m	67.4 kJ	113.7	
Biscayne Bay, Florida	Bay	Muddy sand	3 m	18.2 g $CaCO_3$	18.2	Area A
Biscayne Bay, Florida	Bay	Muddy sand	3 m	29.6 g $CaCO_3$	29.6	Area B
Blackhall, Saltburn, Longscar, Robin Hoods Bays	Bay	Sand	10–18 m	0.45 AFDW	5.8: range = 4.0–28.6	Mean bivalve production, 4 bays
Delaware Bay	Bay	Sand	21 m	21,775 mg AFDW	283.3	
Delaware Bay	Bay mouth	Muddy sand	20 m	19,206 mg AFDW	249.9	
Laguna Madre, Texas	Hypersaline lagoon	Sand	1 m	15.0 g $CaCO_3$ (without *Ensis*); 520.3 with *Ensis*	15.0 (without *Ensis*); 520.3 with *Ensis*	No *Ensis* were preserved
Las Maritas, Venezuela	Shallow subtidal	Mud	2 m	8.5 kcal	24.8	
San Luis, Venezuela	Shallow subtidal	Sand	4–11 m	21.0 kcal	61.3	
Bothnian Sea, Baltic Sea	Marginal sea	Various	Various	2.1 g org-C	150.6	
Gulf of Finland, Baltic Sea	Marginal sea	Various	Various	4.1 g org-C	285.0	
Gulf of Riga, Baltic Sea	Marginal sea	Various	Various	1.7 g org-C	117.3	
Baltic Sea proper	Marginal sea	Various	Various	2.4 g org-C	170.6	
Delaware shelf	Continental shelf	Sand	15 m	1.571 mg AFDW	20.4	
Bird Rock, San Diego, California	Continental shelf	Rocks with sand, gravel	8–27 m	381 g $CaCO_3$	381	Average of 15 stations
Pelican Point, Corona Del Mar, California	Continental shelf	Rocks with sand, gravel	14–22 m	400 g $CaCO_3$	400	Average of 4 stations
Naples Reef, Santa Barbara, California	Continental shelf	Rocks with sand, gravel	9 m	373 g $CaCO_3$	373	
Southern California	Continental shelf	Rocks with sand, gravel	<30 m	400 g $CaCO_3$	400	
Southern California	Continental shelf		30–100 m	6 g $CaCO_3$	6	
Northumberland coast, England	Continental shelf	Sandy mud	80 m	0.23 g AFDW	2.94	
Northumberland coast, England	Continental shelf	Sandy mud	80 m	0.14 g AFDW	1.78	
New Zealand	Continental shelf	Various	Various	8–40 g $CaCO_3$	8–40, 16 most likely	

Table 4.5—11 (continued)
RATES OF CARBONATE PRODUCTION

Geographic location	Environment	Substrate	Depth	Raw production (per m^{-2} $year^{-1}$)	Carbonate production (g $CaCO_3 \cdot m^{-2}$ $year^{-1}$)	Comments
Orkney Islands, Scotland	Continental shelf	Silt to gravel	20–100 m	18–312 g $CaCO_3$	18–312; 77 most likely	Average over 6,000 year
North Ronaldsay, North Bank, Scotland	Continental shelf	Sand to gravel	27–50 m	541 g $CaCO_3$	541	Average over 6,000 year
East Orkney Platform edge, Scotland	Continental shelf	Silt to gravel	30–90 m	114–646 g $CaCO_3$	114–646; 248 most likely	Average over 6,000 year
Stronsay Firth Banks, Scotland	Continental shelf	Silt to gravel	20–40 m	250–400 g $CaCO_3$	250–400; 400 most likely	Average over 6,000 year
North Orkney Platform, Scotland	Continental shelf	Silt to gravel	30–70 m	129 g $CaCO_3$	129	Average over 6,000 year
East Orkney shelf, Scotland	Continental shelf	Silt to gravel	30–90 m	123 g $CaCO_3$	123	Average over 6,000 year
Orkney Sounds	Continental shelf	Various	0–50 m	65 g $CaCO_3$	65	Average over 6,000 year
Long Island Sound, New York	Continental shelf	Various	8–35 m	70 g $CaCO_3$	70	Based on Sanders
Southern California	Continental slope		100–1000 m	5 g $CaCO_3$	5	
Southern California	Continental slope basin		>1000 m	<1 g $CaCO_3$	<1	

From Powell, E. N., Staff, G. M., Davies, D. J., and Callender, W. R., Macrobenthic death assemblages in modern marine environments: formation, interpretation, and application, *Rev. Aquat. Sci.*, 1, 555, 1989.

Table 4.5—12
PERCENTAGE OF PRIMARY PRODUCTION CONSUMED BY HERBIVORES IN MARINE AND TERRESTRIAL ENVIRONMENTS[a]

Coastal environments	Percentage of production eaten by herbivores	Number of trophic steps	Ref. (sources from Valiela, 1984)
Vascular plants			
Eelgrass, North Sea	4	3	Nienhuis and van Ierland (1978)
Salt marsh, Georgia	4.6	3—4	Teal (1962)
Salt marsh, North Carolina	58		Smith and Odum (1981)
Mangrove swamp, Florida	9—27[b]		Onuf et al. (1977)
Phytoplankton			
Long Island Sound	73[c]	4	Riley (1956)
Narragansett Bay	0—30[d]	4	Martin (1970)
Cochin Backwater, India	10—40		Qasim and Odum (1981)
Beaufort Sound	1.9—8.9		Williams et al. (1968)
Offshore California	7—52 (average 23)		Beers and Stewart (1971)
Peruvian upwelling	92, 54—61	3	Walsh (1975), Whiteledge (1978)
Open seas (all phytoplankton)			
Georges Bank	50—54	4	Riley (1963), Cohen et al. (1981)
North Sea	75—80	4—6	Crisp (1975)
Sargasso Sea	100	5	Menzel and Ryther (1971)
Eastern Tropical Pacific	39—140 (average 70)[e]	5	Beers and Stewart (1971)

[a] Annual consumption except where indicated otherwise. These values are rough, but best possible estimates based on many assumptions and extrapolations.
[b] Leaves and buds only.
[c] This is an estimate of consumption of organic matter in the water column. Larger zooplankton consume about 20%, microzooplankton and bacteria an additional 43%. In the bottom, benthic animals use an estimated 31% of net primary production.
[d] Of standing stock of algae.
[e] Includes only microzooplankton that passed through a 202-μm mesh. The biomass of these small species was about 24% of that of the larger zooplankton. Total consumption could easily be larger than reported if any of the larger species are herbivorous.

From Valiela, I., *Marine Ecological Processes*, Springer-Verlag, New York, 1984. With permission.

Table 4.5—13
ESTIMATES OF ZOOPLANKTON PRODUCTION

Organism or group	Area and period	Production mgC/m³/day	Daily production ratios x/y standing crop	Production net primary production ratios x/y
Calanus finmarchicus	E. Barents Sea; year	7.8	0.002	0.03
Calanus cristatus	N.W. Pacific; summer	5.6	0.012	—
Calanus plumchrus	N.W. Pacific; summer	4.6	0.010	—
Eucalanus bungii	N.W. Pacific; summer	3.5	0.014	—
Calanus glacialis	N. Bering Sea; year	0.7	—	0.005
Calanus plumchrus	W. Bering Sea; year	3.1	—	0.012
Calanus cristatus	W. Bering Sea; year	3.8	—	0.015
Eucalanus bungii	W. Bering Sea; year	7.3	—	0.03
Diaptomus salinus	Aral Sea; year	0.66	0.007	—
Acartia clausi	Black Sea; year	0.38	0.035	0.001
Centropages kröyeri	Black Sea; summer	0.19	0.077	0.0002
Euphasia pacifica	N.E. Pacific; year	0.9	0.008	0.0048

Table 4.5—13 (continued)
ESTIMATES OF ZOOPLANKTON PRODUCTION

Organism or group	Area and period	Production mgC/m³/day	Daily production ratios x/y standing crop	Production net primary production ratios x/y
Acartia tonsa	Chesapeake Bay estuary; summer	77	0.50	0.05
Acartia clausi	Black Sea bay; June	15	0.17	—
Acartia clausi	Black Sea, open sea; June	6.6	0.23	0.08
Calanus helgolandicus	Black Sea, open sea; June	28	0.15	0.07
Zooplankton	Georges Bank; year	200	0.03	0.25
Zooplankton	English Channel; year	75	0.10	0.30
Zooplankton	Long Island Sound; year	166	0.17	0.30
Zooplankton	N. North Sea; April–Sept.	180	0.048	0.58
Herbivorous copepods	North Sea; Jan.–June	4.9	0.08	0.14
Copepods (mainly *Calanus*)	North Sea; March–June	46 (author's calculation)	0.10 (author's calculation)	0.20 (author's calculation)
Zooplankton	Gulf of Panama; Jan.–April	70 or 234	0.29 or 0.98	0.09 or 0.31

From Mullin, M., Production of zooplankton in the ocean: the present status and problems, *Oceanogr. Mar. Biol. Annu. Rev.*, 7, 308, 1970. With permission by George Allen and Unwin Ltd., London.

Table 4.5—14
BACTERIAL PRODUCTION IN SEDIMENT (mg C/m/day) TO 1 cm DEPTH, AS MEASURED BY TRITIATED THYMIDINE INCORPORATION

Production	Total measured depth	Habitat description
2.7	1	Freshwater river sand
3.7	15	Beach sand
2.5	0.4	Beach sand
40.5	25	Nearshore sand
16.5	25	Offshore sand
800	2	Mangrove intertidal
200–700	1	Mangrove creek bank
230	1	Mangrove creek bank
300–925	0.4	Coral reef sand
20.1	20	Salt marsh
450	0.5	Salt marsh
85	2	Seagrass bed
40	0.3	Seagrass bed
60	0.5	Seagrass bed
1200	1	Saline pond
250–500	1	Aquaculture ponds

From Kemp, P. F., The fate of benthic bacterial production, *Rev. Aquat. Sci.*, 2, 109, 1990.

4.6 BACTERIA AND PROTOZOA

Table 4.6—1
BACTERIAL ABUNDANCE IN SOME INTERTIDAL AND SUBTIDAL SEDIMENTS[a]

Location/site description	Number of samples	Depth (cm)	Cell numbers × 10^9 Per g dry weight	Cell numbers × 10^9 Per cm^3	Ref. (sources from Rublee, 1982)
Marsh Sediments					
Newport River Estuary, NC; *Spartina alterniflora* marsh, yearly mean	13	0—1	13.8 (8.5—22.0)	8.5 (5.3—13.7)	This report
	13	5—6	5.7 (3.2—7.7)	5.4 (3.3—8.1)	
	13	10—11	2.9 (1.8—4.4)	3.0 (1.7—4.2)	
	13	20—21	1.8 (1.1—30)	1.9 (1.0—2.9)	
Newport River Estuary, NC; transect across *Spartina* marsh	4	0—1	19.7 (12.0—34.1)	9.4 (8.4—10.9)	Rublee and Dornseif (1978)
	4	5—6	7.2 (4.6—9.8)	6.1 (5.1—7.1)	
	4	10—11	3.2 (2.3—5.0)	3.5 (2.4—4.4)	
	4	20—21	2.1 (1.6—2.3)	2.3 (2.2—2.6)	
Rhode River Estuary, MD; *Typha angustifolia/Scirpus* spp. low marsh	3	0—1	66.3 (57.2—84.6)	16.0 (13.8—20.4)	This report
	3	5—6	35.6 (24.2—48.0)	9.7 (6.6—13.1)	
	3	10—11	29.9 (27.9—31.9)	3.0 (7.6—8.1)	
	2	20—21	8.8 (7.4—10.3)	3.0 (2.5—3.5)	
Rhode River Estuary, MD; three high marsh sites, *Spartina cynusoroides, S. patens, Distichlis spicata, Scirpus* spp., *Hibiscus* spp.	3	0—1	60.1 (37.5—76.6)	12.0 (11.3—12.4)	This report
	3	5—6	51.2 (36.9—69.1)	9.5 (7.2—11.0)	
	3	10—11	37.0 (30.8—43.2)	6.1 (4.3—9.3)	
	3	20—21	23.8 (19.9—25.9)	3.9 (2.8—5.1)	
	1	30—31	31.7	3.6	
Great Sippewissett marsh, MA; tall *Spartina alterniflora,* annual mean	12	0—1	38.0 (10.0—64.0)		J. E. Hobbie and J. Helfrich (personal communication)
	12	2—3	28.0		
	12	5—6	35.0		
	12	9—10	42.0		
	12	19—20	24.0		
	12	29—30	16.0		
Great Sippewissett marsh, MA; short *S. alterniflora,* annual mean	12	0—1	54.0 (24.0—80.0)		J. E. Hobbie and J. Helfrich (personal communication)
	12	2—3	29.0		
	12	5—6	29.0		
	12	9—10	28.0		
	12	19—20	18.0		
	12	29—30	12.0		
Great Sippewissett marsh, MA; High marsh, *S. patens,* annual mean	12	0—1	49.0		J. E. Hobbie and J. Helfrich (personal communication)
	12	2—3		(16.0—120.0)	
	12	5—6	35.0		
	12	9—10	30.0		
	12	19—20	29.0		
	12	29—30	18.0		
			19.0		
Intertidal/Subtidal Mud and Sand Flats					
Sapelo Island, GA; intertidal sand flat adjacent to *Spartina* marsh, winter samples	3	0—1		1.0 (0.6—1.2)	S. Y. Newell (personal communication)
Newport River Estuary, NC; subtidal sand, June—November	18	0—1	3.3 (1.6—5.9)	2.7 (1.2—4.8)	Shelton (1979)
	18	5—6	2.4 (1.0—5.9)	1.9 (0.8—4.8)	
Newport River Estuary, NC; subtidal mud, June—November — "sulfuretum"	18	0—1	8.8 (3.1—22.9)	4.2 (2.0—8.3)	Shelton (1979)
	18	5—6	5.8 (1.3—20.5)	2.7 (1.1—7.4)	

Table 4.6—1 (continued)
BACTERIAL ABUNDANCE IN SOME INTERTIDAL AND SUBTIDAL SEDIMENTS[a]

Location/site description	Number of samples	Depth (cm)	Cell numbers × 10^9		Ref. (sources from Rublee, 1982)
			Per g dry weight	Per cm^3	
Intertidal/Subtidal Mud and Sand Flats (continued)					
Rhode River, MD; subtidal mud flat	3	0—1	17.4 (9.6—26.8)	9.1 (5.0—14.0)	This report
	2	2—3	16.0 (11.7—19.7)	8.4 (6.1—10.3)	
	3	5—6	13.0 (11.8—14.2)	8.7 (7.9—9.5)	
	3	10—11	8.0 (6.7—9.1)	6.2 (5.2—7.1)	
	2	20—21	4.7 (3.3—6.2)	3.6 (2.5—4.7)	
Lowe's Cove Damariscotta Estuary, ME; March—August	6	sfc	2.8 (1.2—4.7)		M. DeFlaun (personal communication)
Halifax, Nova Scotia; high intertidal mud flats, April—November	13	sfc	6.0 (2.2—8.6)	3.5 (2.2—4.8)	L. M. Cammen (personal communication)
	13	1	5.7 (3.7—7.9)	4.5 (3.5—5.5)	
	13	5	3.5 (2.3—5.2)	3.6 (2.9—4.3)	
Halifax, Nova Scotia; low intertidal mudflat, April—November	13	sfc	4.2 (1.5—7.0)	2.8 (1.4—4.3)	L. M. Cammen (personal communication)
	13	1	4.0 (2.8—5.2)	3.5 (2.1—4.1)	
	13	5	4.0 (2.7—5.2)	3.8 (2.2—5.1)	
Petpeswick Inlet, Nova Scotia, seven intertidal mud and sandflat stations, May—September	17	sfc	2.8 (0.3—10.0)		Dale (1974)
	2	1	2.9 (2.1—3.6)		
	1	2	6.3		
	16	5	2.5 (0.1—7.0)		
	14	10	1.8 (0.1—4.7)		

[a] Values given as mean (and range) of observations. sfc = surface.

From Rublee, P. A., *Estuarine Comparisons*, Kennedy, V. S., Ed., Academic Press, New York, 1982, 159. With permission.

Table 4.6—2
COMPARISON OF MEAN BACTERIAL NUMBERS IN RELATIONSHIP TO ORGANIC MATTER AND GRAIN SIZE IN SEDIMENTS

Location	Date	Temperature (°C)	Sediment		Bacterial number (10^8 g^{-1})	Organic matter (mg g^{-1})	Grain size (μm)	Remarks	Ref. (sources from Meyer-Reil, 1984)
			Type	Depth (cm)					
Petpeswick Inlet, Nova Scotia	6/3/72	—	Intertidal flats	0	99.70	3.8[a]	19.1	7 stations, different horizons, maximum/minimum values	Dale (1974)
	8/30/72		Intertidal flats	5	1.17	0.1[a]	132.5		
Beaufort Sea, AK	Summer 1975/1976	1.2/−0.1	Arctic sediment	Surface	6.6/106	—	—	33/11 stations	Griffiths et al. (1978)
	Winter 1976	−1.9	Arctic sediment	Surface	10	—	—	13 stations	
Kampinge, Baltic Sea	6/1—5/77	15	Sandy beach	0	14.2	2.7	277	4 stations	Meyer-Reil et al. (1978)
Kiel Bight, Baltic Sea	11/21/74	—	Sandy (14 m)	0	14.02	5.4	410	1 station	Weise and Rheinheimer (1978)
Halifax Harbor, Nova Scotia	—	2	Muddy, aerobic/anaerobic (24 m)	0—100	5—30	2—3[a]	—	1 station, different horizons, different sampling dates	Kepkay et al. (1979)
Windermere, South Basin	4/79	—	Littoral sediment	2.9	290	—	—	Organic matter and particle size given for different sediment fractions	Jones (1980)
			Profundal sediment	2.1	340	—	—		
Blelham Tarn	4/79	—	Littoral sediment	3.6	280	—	—		
			Profundal sediment	1.2	119	—	—		
Kiel Fjord/Kiel Bight, Baltic Sea	7/4—12/77	21.0	Sandy beach	0	10.3	5.9	293	12 stations, 3 sampling dates	Meyer-Reil et al. (1980) and Meyer-Reil (unpublished)
	3/6—11/78	3.0	Sandy beach	0	4.5	5.1	355		
	11/6—11/78	9.0	Sandy beach	0	15.3	11.7	282		
Mactan, Philippines	4/2/80	27	Coral beach	0	14.7	—	—	3 stations	Meyer-Reil (1981)

Table 4.6—2 (continued)
COMPARISON OF MEAN BACTERIAL NUMBERS IN RELATIONSHIP TO ORGANIC MATTER AND GRAIN SIZE IN SEDIMENTS

Location	Date	Temperature (°C)	Sediment Type	Sediment Depth (cm)	Bacterial number (10^8 g^{-1})	Organic matter (mg g^{-1})	Grain size (μm)	Remarks	Ref. (sources from Meyer-Reil, 1984)
Kiel Bight, Baltic Sea	5/30/80	8	Sandy (10 m)	0—2	8.8	12.7	—		Meyer-Reil (unpublished)
				2—4	6.7	8.1	—		
				4—6	0.7	7.2	—		
	5/28/80		Muddy, anaerobic (28 m)	0—2	84.7	164.4	—		
				2—4	69.0	157.7	—		
				4—6	86.8	160.3	—		
Antarctica	1/26/81	2	Muddy (1951 m)	0	14.2	109.4	—	1 station, different horizons	Meyer-Reil (unpublished)
				100	28.6	105.8	—		
				400	6.5	39.8	—		
				800	14.6	98.6	—		
				1100	11.1	82.5	—		

[a] Organic carbon (% by weight).

From Meyer-Reil, L.-A., *Heterotrophic Activity in the Sea,* Hobbie, J. E. and Williams, P. J. le B., Eds., Plenum Press, New York, 1984, 523. With permission.

Table 4.6—3
DIRECT COUNTS OF BACTERIA IN SOME TROPICAL MANGROVE AND CORAL REEF SEDIMENTS

Location	Habitat and season	Depth (cm)	Cell numbers × 10^9
Kaneohe Bay, Hawaii	Fine coral sand	0–1	3.7(2.4–5.0)
Kanoehe Bay, Hawaii	Coral rubble	0–1	0.5(0.3–0.7)
	Coarse coral sand	0–1	0.5(0.3–0.7)
Majuro atoll, Marshall Islands	Lagoon sands	0–1	2.2(0.8–4.3)
Majuro atoll, Marshall Islands	Dead coral	Surface	3.9(1.9–7.2)
Heron Island, GBR, Australia	Reef flat, summer	0–1	2.8(2.3–3.2)
Lizard Island, GBR, Australia	Reef flat, summer	0–0.5	0.5(0.36–0.60)[a]
	Reef flat, winter	0–0.5	0.72
Hamilton Harbor, Bermuda	Carbonate silt	0–1	2.6[b]
	Carbonate silt	3	2.0
	Carbonate silt		1.5
	Carbonate silt	7	1.3
	Carbonate silt	9	1.4
	Carbonate silt	11	1.1
Davies Reef, Australia	Reef front, winter	0–1	0.9(0.3–1.5)
	Reef flat, winter	0–1	1.5(1.4–1.6)
	Shallow lagoon, winter	0–1	1.2(0.7–1.9)
	Shallow lagoon, summer	0–1	0.6(0.2–1.0)[c]
	Deep lagoon, winter	0–1	1.7(1.2–2.4)
	Deep lagoon, summer	0–1	1.0(0.7–2.1)[c]
Umtata, Southern Africa	Mangrove mud, winter	0–1	0.9(0.8–1.0)
Cienaga Grande, Columbia	Tropical lagoon, mangrove muds, winter	Surface	0.5(0.4–1.3)[d]
Hinchinbrook Island, Australia	Low intertidal mangroves	0–0.1	1.30 (±70)[e]
Bowling Green Bay, Australia	Mangrove creek bank	0–1	2.4(2.33–2.46)
Cape York peninsula, Australia	Mangroves, winter	0–2	150(11–359)
	Mangroves, summer	0–2	77(20–344)
Hinchinbrook Island, Australia	Mangroves, autumn	0–2	23(20–25)
	Mangroves, winter	0–2	153(51–225)
	Mangroves, spring	0–2	170(105–297)
	Mangroves, summer	0–2	163(125–225)
Hinchinbrook Island, Australia	Low intertidal mangroves, autumn	0–2	16.4(8.7–24.0)
	Low intertidal mangroves, autumn	2–4	29.8(12.9–46.7)
	Low intertidal mangroves, autumn	4–6	7.7(5.7–9.6)
	Low intertidal mangroves, autumn	6–8	7.0(5.9–8.1)
	Low intertidal mangroves, autumn	8–10	3.1(2.3–3.8)
	High intertidal mangroves, autumn	0–2	16.3(14.6–18.0)
	High intertidal mangroves, autumn	2–4	10.5(10.2–10.7)
	High intertidal mangroves, autumn	4–6	2.6(2.4–2.7)
	High intertidal mangroves, autumn	6–8	3.0(2.3–3.6)
	High intertidal mangroves, autumn	8–10	4.6(4.1–5.0)
Chunda Bay, Australia	Low intertidal mangroves	0–5	11.3(0.7–34.3)
	High intertidal mangroves	0–5	8.1(0.1–30.1)

Note: Values are given as mean (and range) of cell numbers per gram weight of sediment.

[a] Estimated from biomass measurements using 20 fg cell^{-1}.
[b] Mean values only, as estimated from graphs in Figure 2 in Hines, M. E., *Proc. 5th Intl. Coral Reef Congr.*, Vol. 3, Moorea, French Polynesia, 1985, 427.
[c] Unpublished data (Alongi).
[d] Converted from cells ml^{-1} using dry wt conversion; standard deviation in parentheses.
[e] Converted from fatty acids assuming 10 fg carbon cell^{-1}.

From Alongi, D. M., The role of soft-bottom benthic communities in tropical mangrove and coral reef ecosystems, *Rev. Aquat. Sci.*, 1, 243, 1989.

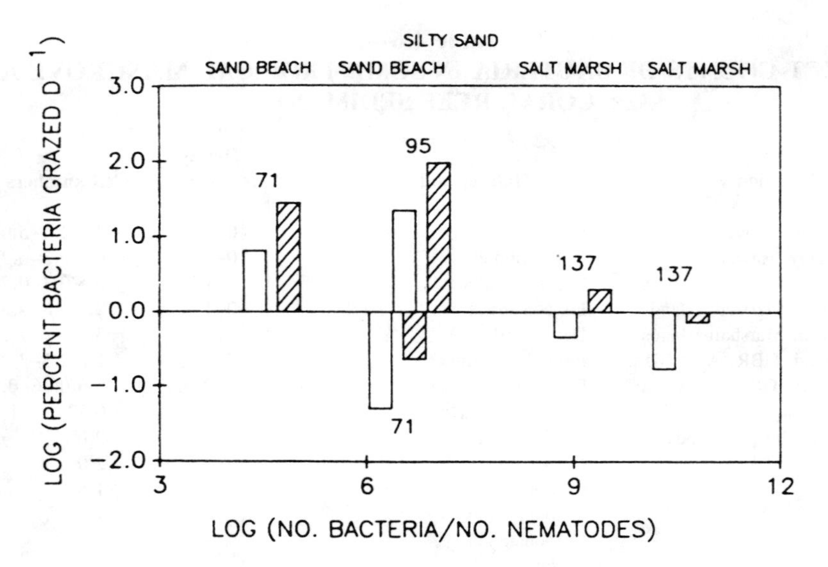

FIGURE 4.6—1. Predicted percent of bacterial abundance grazed by nematodes per day, assuming all nematodes were bacterivorous. Open bars are based on minimum grazing rates (0.07 μg C per individual per day), filled bars on maximum grazing rates (0.3 μg C per individual per day). Adjacent numbers are references from which nematode and bacterial abundances were taken. (From Kemp, P. F., The fate of benthic bacterial production, *Rev. Aquat. Sci.*, 2, 109, 1990.)

Table 4.6—4
COMPARISON OF THE SUITABILITY OF SOME CURRENT METHODS FOR MEASURING BACTERIAL GROWTH RATES IN SEAWATER

Criterion of suitability[a]	1	2	3	4
FDC	+ +	+	+ +	+ +
^{35}S-SO_4 assimilation	− −	+	+	+
^{3}H-Adenine incorporation into RNA	− −	− −	+	+ +
^{3}H-Thymidine incorporation into RNA	+ +	+	+	+ +
Increase in cell number	+ +	+ +	−	−
Increase in cell ATP	−	+	−	+

Note: + +, Meets criterion; +, probably meets criterion; −, probably does not meet criterion; − −, does not meet criterion.

[a] Criteria:
1. Method should be specific for bacteria.
2. Conversion factor should not be growth-rate dependent.
3. Experimental manipulation should not change growth rate.
4. Should be sensitive enough to allow short incubations (i.e., min-h).

From Azam, F. and Fuhrman, J. A., *Heterotrophic Activity in the Sea,* Hobbie, J. E. and Williams, P. J. le B., Eds., Plenum Press, New York, 1984, 179. With permission.

Table 4.6—5
SOME INGESTION RATES (BACTERIA CELLS/HOUR) OF CILIATES AND FLAGELLATES

Organism	Ingestion rate	Method used
Benthic ciliate sp.	180–600	Emptying of vacuoles
Benthic ciliates	59–410	FL beads, bacteria
Benthic ciliates	128	FL bacteria
Benthic ciliates	414–1881	Decreasing numbers
Pelagic ciliate sp.	100–800	Latex beads
Pelagic ciliate sp.	$4.3–237 \times 10^3$	FL beads
Pelagic ciliates	33–76	FL beads
Pelagic ciliates	60–120	FL bacteria
Pelagic ciliates	731–2150	FL bacteria
Benthic flagellates	9–12	Decreasing numbers
Pelagic flagellates	9–35	FL bacteria
Pelagic flagellates	23,117	Decreasing numbers
Pelagic flagellates	0–144	Decreasing numbers
Pelagic flagellates	50–210	Decreasing numbers
Pelagic flagellates	27–254	Decreasing numbers
Pelagic flagellates	10–75	Decreasing numbers
Pelagic flagellates	20–80	Decreasing numbers

Note: Flagellate ingestion rates based on microbead ingestion are not included as many are potentially biased by preservation artifacts. FL = Fluorescently labeled.

From Kemp, P. F., The fate of benthic bacterial production, *Rev. Aquat. Sci.*, 2, 109, 1990. With permission.

Table 4.6—6
PROTOZOAN ABUNDANCES IN SEDIMENT WHERE ABUNDANCE WAS REPORTED PER UNIT VOLUME

Number/ml		
Ciliates	**Flagellates**	**Habitat**
6–50	75	Mangrove
120	90	Beach sand
4.6–49	19–43	Subtidal
9	700	Lagoon
7–260	6–180	Mangrove
3340	18500	River mud
~50	10^6	Tundra pond
187–637[a]		Sheltered beach
448–1056	6880[c]	Salt marsh
7616	1.2×10^6	Saline pond
300–6200[b]		Sand beach

[a] Reported as cm^{-2}; most ciliates were at 0–1 cm depth.
[b] Total protozoa reported were ''mostly ciliates''.
[c] Kemp, unpublished data.

From Kemp, P. F., The fate of benthic bacterial production, *Rev. Aquat. Sci.*, 2, 109, 1990.

Table 4.6—7
ESTIMATES OF FREE-LIVING PROTOZOANS (EXCLUDING FORAMINIFERA) IN SOME TROPICAL MANGROVE AND CORAL REEF SEDIMENTS

Location	Habitat and season	Depth (cm)	Cell numbers × 10^5
Kaneohe Bay, Hawaii	Reef rubble	0–1	24.0(14–37)[a]
Isla Providencia, Colombia	Reef sand	0–1	9.0–23.0[a]
Moorea Island, French Polynesia	Trahura reef	0–3	0.2(0–2)[a]
Umtata, South Africa	Mngazana River mangrove muds		
	High intertidal sites	0–30	3.0(0.6–7.8)[a]
	Mid-intertidal sites	0–15	2.0(0.3–6.2)[a]
	Low intertidal sites	0–15	0.30(0.1–0.5)[a]
Davies Reef, Great Barrier Reef	Shallow lagoon, autumn	0–2	Ciliates: 1.8(±1.6) Flagellates: 14.0(±4.0)[b]
Davies Reef, Great Barrier Reef	Reef front, winter	0–5	Ciliates: 8.0(3.0–13.1) Flagellates: 97.1(44.1–147.2)
	Reef flat, winter	0–5	Ciliates: 2.0(2.0–4.0) Flagellates: 13.1(3.2–21.0)
	Shallow lagoon, winter	0–5	Ciliates: 3.1(1.1–4.0) Flagellates: 94.0(5.0–131.3)
	Deep lagoon, winter	0–5	Ciliates: 3.3(3.0–4.4) Flagellates: 151.3(100.0–202.4)
Cape York peninsula, Qld. Australia	Mangroves, winter	0–5	Ciliates: 3.8(0.6–11.0) Flagellates: 3.4(0.6–10.0)
	Mangroves, summer	0–5	Ciliates: 15.0(2.8–26.0) Flagellates: 11.4(6.4–16.0)
Hinchinbrook Island, Australia	Mangroves, autumn	0–5	Ciliates: 2.3(2.1–24) Flagellates: 15.5(12.3–19.0)
	Mangroves, winter	0–5	Ciliates: 2.3(1.8–3.0) Flagellates: 3.2(3.1–3.3)
	Mangroves, spring	0–5	Ciliates: 7.5(6.2–10.4) Flagellates: 4.9(4.7–5.2)
	Mangroves, summer	0–5	Ciliates: 4.0(2.1–5.7) Flagellates: 2.3(2.0–3.1)
Chunda Bay, Qld. Australia	Low intertidal mangroves	0–5	Ciliates: 11.9(0.1–32.3) Flagellates: 10.3(0.2–28.7)
	High intertidal mangroves	0–5	Ciliates: 5.2(0.0–15.5) Flagellates: 7.0(0.0–30.2)

Note: Values are given as mean and range of cell numbers per square meter.

[a] Ciliates enumerated only.
[b] Standard deviations in parentheses.

From Alongi, D. M., The role of soft-bottom benthic communities in tropical mangrove and coral reef ecosystems, *Rev. Aquat. Sci.*, 1, 250, 1989.

Table 4.6—8
RELATIVE BIOMASSES OF BACTERIA, MACROFAUNA, MEIOFAUNA, AND PROTOZOA IN g C/m^2 USING THE CONVERSION g C = 50% g DRY WEIGHT WHERE NECESSARY

Bacteria						
To 1 cm	Total	Macrofauna	Meiofauna	Protozoa	Habitat description	Ref. (sources from Kemp, 1990)
0.06					Intertidal sand	136
	0.04	0.2	0.5	0.12	Exposed beach sand	21
	0.04	0.9[a]	0.8	0.17	Exposed beach sand	21
	2.5	5.0	0.5		Subtidal sand	95
0.036	0.54	1.25			Exposed beach sand	52
0.6	1.2			0.002[b]	Salt marsh	64
0.8	1.6			0.02[b]	Saline pond	64
	5.25	1.9	1.6		Sand beach	71
	18.0	3.2	2.2	0.2	Exposed sand beach	72
	3.6	0	0.3	0.08	Very exposed beach	72
0.14	0.28		0.073		Intertidal sand	89
1.2–2.3			0.7–0.8		Salt marsh	57
31.84			1.24		Salt marsh	137
0.025	0.067				Subtidal, oil seep	90
0.013	0.035				Subtidal, normal	90
	2.8		1.0		Coral sand	36
	1.03	6.6	0.05		Cont. shelf sand	74
	0.93	0.6	0.036		Cont. slope	74
	0.12	0.014	0.0033		Abyssal plain	74
1.5	14.0				Salt marsh	24
	0.52	1.5	1.2		Intertidal mud	138

[a] Excluding filter-feeding bivalves.
[b] Protozoan biomass includes ciliates only.

From Kemp, P. F., The fate of benthic bacterial production, *Rev. Aquat. Sci.*, 2, 109, 1990.

4.7 MEIOFAUNAL DENSITY

Table 4.7—1
DENSITY OF INTERTIDAL MEIOFAUNA[a]

Deposit	Location	Nematoda	Copepoda	Ostracoda	Oligochaeta	Total Meiofauna
Sand	S. of Stockholm, Locality "A", Sweden	38—169	31—833	0	214—498	391—1,529
	S. of Stockholm, Locality "B", Sweden	48—88	24—129	0	92—109	173—402
	Gothland, Sweden	8—14	0	0	17—112	225—1,048
	Scanian East Coast, Sweden	1	2—45	0	16—160	49—318
	Kattegat, Sweden	6—106	1—52	0	1—243	11—845
	Øresund, Denmark	115—497	1—10	0	1—375	249—764
	Hard Wadden, W. Denmark	10—1,050	<1—840	0—61	+	13—1,914
	Soft Wadden, W. Denmark	31—367	13—389	4—12	+	50—768
	Boulogne, France	3—367	3—192	0—<1	0	19—389
	Acachon, France	67—204	5—331	<1—18	0—12	83—591
	Eden estuary, St. Andrews, Scotland[x]	3,163	17	13	+	3,193
	Blyth estuary, England	300—1,320	*	*	*	—
	Whitstable, England	1,136—5,220	13—486	21—742	0—7	1,264—5,817
	Miami, FL	0—110	0—260	0	0	14—872
	Porto Novo, S. India	594—1,150	186—448	0	0—2	968—1,960
	E. Coast of Malaya	346—8,068	86—2,416	0	0	700—10,212
Coral sand	E. Coast of Malaya	32	144	0	12	244
Mud	Bristol Channel (Wales)	70—10,440	0—500	0—790	0—780	90—11,820
	Southampton Water, England	*	81—1,021	*	*	—
	Blyth estuary, England	228—2,830	*	*	*	—
	Salt marsh, Massachusetts	1,440—2,130	*	*	*	—
	Salt marsh, Georgia	460—16,300	*	*	*	—
	Vellar estuary, S. India	307—3,240	5—490	0—63	0	420—3,815

Note: * = not studied.

[a] Numbers below 10 cm^2 of surface.

From McIntyre, A. D., *Biol. Rev.*, 44, 245, 1969. With permission.

Table 4.7—2
DENSITIES OF MEIOBENTHOS (NUMBER OF INDIVIDUALS 10 cm^{-2}) IN SOME TROPICAL MANGROVE AND CORAL REEF SEDIMENTS

Location	Habitat and season	Total faunal densities	Sediment depth (cm)	Percent of dominant taxon	Ref. (sources from Alongi, 1989)
Tuamotu, French Polynesia	Mururoa Lagoon, summer	440(390—1293)	0—1	Nematodes, 52—56%	96
Tuamotu, French Polynesia	Maturei Vavao				55
	Atoll lagoon, autumn	111(36—240)	0—5	Nematodes, 16—61%	
Bermuda	Beach and lagoon, carbonate sands (all seasons)	522(122—1333)	0—10	Nematodes, $\bar{x}$ = 73%	103
Moorea Island, French Polynesia	Reef flat, winter	79(63—107)	0—3	Polychaetes, $\bar{x}$ = 42%	98
	Lagoon channel	60			
	Reef flat	57(39—93)			
	Deep outer reef slope	120			
Grand Recif, Madagascar	Inner reef flat, winter	327(263—610)	0—1.5	Copepods, $\bar{x}$ = 61%	98
Godavari River, India	Low intertidal, all seasons mangroves	2131	0—20	Nematodes, $\bar{x}$ = 86%	105
Lakshadweep, S. Arabian Sea	Carbonate beach sand, winter				
	High beach	25(7—48)	0—5	Nematodes, $\bar{x}$ = 31%	100
	Mid-beach	35(3—60)			
	Low beach	28(3—48)			
Bay of Bengal, India	Mangroves, low intertidal		0—9	Nematodes, 50—67%	110
	Winter	156(35—270)			
	Spring	193(90—280)			70
	Summer	173(80—280)			
	Autumn, monsoon	88(60—160)			
Cebu, Philippines	Beach-reef carbonates autumn	3139(744—8769)	0—5	Nematodes, $\bar{x}$ = 39%	117
Great Barrier Reef, Australia	Lagoon, winter		0—5		118
	Lizard Island	1182		Nematodes, 85%	
	Orpheus Island	445		Copepods, 47%	
	Davies Reef	223		Copepods, 82%	
Mangroves, South Cuba	Tolete lagoon		0—5		
	Dry season	196		Nematodes, $\bar{x}$ = 54%	119
	Wet season	115			
	Basto lagoon				
	Dry season	245			
	Wet season	36			
Davies Reef, Great Barrier Reef	Reef front, winter	477(280—511)	0—5	Copepods, $\bar{x}$ = 86%	33
	Reef flat, winter	615(415—1010)		Copepods, $\bar{x}$ = 51%	
	Shallow lagoon, winter	631(622—640)		Copepods, $\bar{x}$ = 48%	
	Deep lagoon, winter	231(220—255)		Nematodes, $\bar{x}$ = 50%	
Isla del Caro Reefs, Costa Rica	Reef flat				
	Winter	248(99—330)	0—6	Foraminfera, $\bar{x}$ = 21%	232
	Spring	317(270—363)			
	Summer	221(117—325)		Copepods, $\bar{x}$ = 20%	
	Autumn	501(427—575)			

Table 4.7—2 (continued)
DENSITIES OF MEIOBENTHOS (NUMBER OF INDIVIDUALS 10 cm^{-2}) IN SOME TROPICAL MANGROVE AND CORAL REEF SEDIMENTS

Location	Habitat and season	Total faunal densities	Sediment depth (cm)	Percent of dominant taxon	Ref. (sources from Alongi, 1989)
Cape York peninsula, Australia	Mangrove estuaries		0—5		
	Summer wet	776(217—2454)		Turbellaria, $\bar{x}=70\%$	19
	Winter dry	330(66—1660)		Turbellaria, $\bar{x}=46\%$	
Hinchinbrook Is., Australia	Mangroves, all seasons				
	Low intertidal	767(347—1840)	0—5	Turbellaria, $\bar{x}=68\%$	19
	Mid-intertidal	94(57—187)		Nematoda, $\bar{x}=86\%$	
	High intertidal	42(14—85)		Nematoda, $\bar{x}=67\%$	

Note: Values depict mean (and range) except where noted.

From Alongi, D. M., The role of soft-bottom benthic communities in tropical mangrove and coral reef ecosystems, *Rev. Aquat. Sci.*, 1(2), 255, 1989.

4.8 MACROBENTHOS

Table 4.8—1
ANNUAL PRODUCTION OF MACROBENTHOS (PRODUCTION IN DRY WEIGHT UNLESS OTHERWISE STATED)

Species	Production (m^2/year)	$P/\bar{B}$	Maximum age (years)	Locality	Ref. (sources from Kennish, 1986)
Nephtys incisa	9.34 g	2.16	3	Long Island Sound, U.S., 4–30 m	190
Cistenoides gouldii	1.70 g	1.94	2	Long Island Sound, U.S., 4–30 m	190
Yoldia limatula	3.21 g	2.28	2	Long Island Sound, U.S., 4–30 m	190
Pandora gouldiana	6.13 g	1.99	2	Long Island Sound, U.S., 4–30 m	190
Moira atropos	2.52 g	0.70	6	Biscayne Bay, Florida, U.S., 3 m	191
Tagelus divisus	21.0 g	1.78	2	Biscayne Bay, Florida, U.S., L.W.S.	192
Ampharete acutifrons	0.719 g (wet)	4.58	1	Long Island Sound, U.S., 917 m	193
Neomysis americana	36.2 mg	3.66	1?	Long Island Sound, U.S., 917 m	193
Crangon septemspinosa	0.519 g	3.82	3	Long Island Sound, U.S., 917 m	193
Asterias forbesi	4.52 g	2.64	3	Long Island Sound, U.S., 917 m	193
Tellina martinicensis	0.23 g	2.4	2	Biscayne Bay, Florida, U.S., 3 m	194
Chione cancellata	8.9 g	0.42	7	Biscayne Bay, Florida, U.S., M.L.W.S.	195
Dosinia elegans	0.13 g	1.25	2	Biscayne Bay, Florida, U.S., 3 m	196

Table 4.8—1 (continued)
ANNUAL PRODUCTION OF MACROBENTHOS (PRODUCTION IN DRY WEIGHT UNLESS OTHERWISE STATED)

Species	Production (m²/year)	P/B̄	Maximum age (years)	Locality	Ref. (sources from Kennish, 1986)
Pectinaria hyperborea	10.6 g	4.6	2	St. Margaret's Bay, Nova Scotia, 60 m	197
Scrobicularia plana	60 kcal	0.29	7?	North Wales, lower shore	198
	13.3 kcal	0.67	4	North Wales, upper shore	198
Anodontia alba	14.09 g	1.43	4 + (?)	Biscayne Bay, Florida, U.S., low water	199
Strongylocentrotus droebachiensis	401.0 kcal	0.80	6	St. Margaret's Bay, Nova Scotia, intertidal	200
Neanthes virens	45.2 kcal	1.62	3	Thames estuary, U.K., intertidal	201
Ammotrypane aulogaster	359 mg	2.08	?	Northumberland, U.K., 80 m	202
Heteromastus filiformis	297 mg	1.01	2	Northumberland, U.K., 80 m	202
Spiophanes kroyeri	196 mg	1.40	3	Northumberland, U.K., 80 m	202
Glycera rouxi	192 mg	0.37	5	Northumberland, U.K., 80 m	202
Calocaris macandreae	142 mg	0.12	9.5	Northumberland, U.K., 80 m	202
Abra nitida	118 mg	1.11	3	Northumberland, U.K., 80 m	202
Lumbrineris fragilis	78 mg	1.34	3	Northumberland, U.K., 80 m	202
Chaetozone setosa	50 mg	1.28	3	Northumberland, U.K., 80 m	202
Brissopsis lyrifera	108 mg	0.30	4	Northumberland, U.K., 80 m	202
Mya arenaria	11.6 g	2.54	3	Petpeswick Inlet, E. Canada, intertidal	203
Macoma balthica	1.93 g	1.53	3	Petpeswick Inlet, E. Canada, intertidal	203
Littorina saxatilis	3.25 g	4.11	1	Petpeswick Inlet, E. Canada, Intertidal	203
Macoma balthica	10.07 g	2.07	6	Ythan Estuary, Scotland, intertidal	204
Nephtys hombergi	7.34 g	1.9	3	Lynher Estuary, U.K., Intertidal	205
Ampharete acutifrons	2.32 g	5.5	1	Lynher Estuary, U.K., intertidal	205
Mya arenaria	2.66 g	0.5	8	Lynher Estuary, U.K., intertidal	205
Scrobicularia plana	0.48 g	0.2	9	Lynher Estuary, U.K., intertidal	205
Macoma balthica	0.31 g	0.9	6	Lynher Estuary, U.K., intertidal	205
Cerastoderma edule	0.21 g	0.2	7	Lynher Estuary, U.K., intertidal	205
Ampelisca brevicornis	4.26 g (wet)	3.95	1.25	Helgoland Bight, 28 m	206
	2.43 g (wet)	3.68	1.25		
Pectinaria californiensis	2.02 g C	5.3	1.2	Puget Sound, Washington, U.S., 34 m	207
	2.798 g C	3.3	2.1	Puget Sound, Washington, U.S., 203 m	207
	3.471 g C	4.1	1.8	Puget Sound, Washington, U.S., 254 m	207
	1.386 g C	5.5	1.9	Puget Sound, Washington, U.S., 207 m	207
	4.816 g C	3.4	2.4	Puget Sound, Washington, U.S., 71 m	207
Cerastoderma edule	29.25 g	1.59	5	Southampton Water, U.K., intertidal	208
	71.36 g	1.10	5		
	46.44 g	2.61	5		

Table 4.8—1 (continued)
ANNUAL PRODUCTION OF MACROBENTHOS (PRODUCTION IN DRY WEIGHT UNLESS OTHERWISE STATED)

Species	Production (m²/year)	P/B̄	Maximum age (years)	Locality	Ref. (sources from Kennish, 1986)
Mercenaria mercenaria	3.99 g	0.52	8	Southampton Water, U.K., intertidal	208
	14.00 g	0.28	8		
	6.19 g	0.17	9		
Venerupis aurea	0.70 g	1.11	5	Southampton Water, U.K., intertidal	208
	1.25 g	1.10	5		
Crassostrea virginica	3828 kcal	1.87	?	South Carolina, U.S., intertidal	209
Littorina littorea	6.13 g	0.61	?	Grevelingen Estuary, Netherlands, intertidal	210
Hydrobia ulvae	7.23 g	1.78	1	Grevelingen Estuary, Netherlands, intertidal	210
	8.80 g	1.24	1		
	12.79 g	1.36	2		
Cardium edule	10.21 g	0.69	3.5	Grevelingen Estuary, Netherlands, intertidal	210
	119.82 g	2.56	3.5		
	51.76 g	1.13	3.5		
Macoma balthica	3.40 g	1.93	8.10	Grevelingen Estuary, Netherlands, intertidal	210
	0.95 g	1.00	8.10		
	0.07 g	0.30	8.10		
	−0.74 g	−0.25	8.10		
Arenicola marina	3.79 g	1.14	3	Grevelingen Estuary, Netherlands, intertidal	210
	6.26 g	0.72	3		
	3.32 g	0.99	3		
Pontoporeia affinis	3.17 g	1.90	3	North Baltic, 64 m	211
Pontoporeia femorata	3.03 g	1.43	3	North Baltic, 64 m	211
Harmothoe sarsi	0.23 g	1.99	3	North Baltic, 64 m	211
Pharus legumen	16.12 g	0.56	6	Carmarthen Bay, South Wales, 13.5 m	212
Spiophanes bombyx	3.35 g	4.86	?	Carmarthen Bay, South Wales, 13.5 m	212
Ensis siliqua	1.37 g	0.27	10	Carmarthen Bay, South Wales, 13.5 m	212
Donax vittatus	0.72 g	2.10	2.5	Carmarthen Bay, South Wales, 13.5 m	212
Magelona papillicornis	0.69 g	1.10	3	Carmarthen Bay, South Wales, 13.5 m	212
Venus striatula	0.62 g	0.41	10	Carmarthen Bay, South Wales, 13.5 m	212
Ophiura texturata	0.46 g	0.68	3	Carmarthen Bay, South Wales, 13.5 m	212
Tellina fabula	0.29 g	0.90	6	Carmarthen Bay, South Wales, 13.5 m	212
Glycera alba	0.28 g	0.97	3	Carmarthen Bay, South Wales, 13.5 m	212
Sigalion mathildae	0.17 g	0.44	?	Carmarthen Bay, South Wales, 13.5 m	212
Tharyx marioni	0.015 g	0.79	2	Carmarthen Bay, South Wales, 13.5 m	212
Astropecten irregularis	0.0004 g	0.005	?	Carmarthen Bay, South Wales, 13.5 m	212
Echinocardium cordatum	−0.012 g	−0.02	3	Carmarthen Bay, South Wales, 13.5 m	212

Modified from Warwick, R. M., in *Marine Benthic Dynamics*, Tenore, K. R. and Coull, B. C., Eds., University of South Carolina Press, Columbia, 1980, 1. With permission.

Table 4.8—2
PRODUCTION ESTIMATES FOR ESTUARINE MACROBENTHIC ASSEMBLAGES IN SELECTED ESTUARIES OF EUROPE AND THE U.S.[a]

Area	Depth	Production	Ref. (sources from Wolff, 1983)
Long Island Sound	Subtidal	21.4 (infauna)	Sanders (1956)
	Subtidal	5.8 (epifauna)	Richards and Riley (1967)
Firemore Bay, Scotland	Whole bay	2.5 (g carbon)	McIntyre and Eleftheriou (1968)
Kiel Bight, Germany	15 m	17.9	Arntz and Brunswig (1975)
Tamar Estuary, England	Intertidal	13.3	Warwick and Price (1975)
Southampton Water, England	Intertidal	152—225 (?)	Hibbert (1976)
Grevelingen Estuary, The Netherlands	Whole estuary	50.3—57.4	Wolff and De Wolf (1977)

[a] Values expressed as g ash-free dry weight per square meter per year.

From Wolff, W. J., *Estuaries and Enclosed Seas*, Ketchum, B. H., Ed., Elsevier, Amsterdam, 1983, 151.

Table 4.8—3
BIOMASS VALUES OF MACROBENTHIC SPECIES ASSEMBLAGES IN SELECTED EUROPEAN AND U.S. ESTUARIES

Area	Depth	Biomass (g m^2)	Ref. (sources from Wolff, 1983)
Wadden Sea, Denmark	Intertidal	174—497 (wet)	Smidt (1951)
Ringkøbing Fjord, Denmark	Intertidal	267—450 (wet)	Smidt (1951)
Long Island Sound	Subtidal	54.6 (dry)	Sanders (1956)
Barnstable Harbor	Intertidal	38—40 (dry)	Sanders et al. (1962)
North Carolina Estuary		8.1 (dry tissue)	Williams and Thomas (1967)
Puget Sound	12—195 m	8—9 (ash-free)	Lie (1968)
Firemore Bay, Scotland	Beach	1.3 (dry)	McIntyre and Eleftheriou (1968)
	Subtidal	3.7 (dry)	McIntyre and Eleftheriou (1968)
Biscayne Bay	Intertidal	30 (dry)	Moore et al. (1968)
Narragansett Bay	Intertidal musselbed	1852 (ash-free)	Nixon et al. (1971)
Lower Mystic River	15 m	1.0—50 (wet)	Rowe et al. (1972)
Balgzand, The Netherlands	Intertidal	9.6—25.1 (ash-free)	Beukema (1974)
Kiel Bay, Germany	>15 m	26.3 (ash-free)	Arntz and Brunswig (1975)
Orwell Estuary, England	Intertidal	46.0 (ash-free)	Kay and Knights (1975)
Stour Estuary, England	Intertidal	28.1—38.7 (ash-free)	Kay and Knights (1975)
Colne Estuary, England	Intertidal	7.8 (ash-free)	Kay and Knights (1975)
Dengie Flats, England	Intertidal	18.5 (ash-free)	Kay and Knights (1975)
Crouch Estuary, England	Intertidal	8.7 (ash-free)	Kay and Knights (1975)
Roach Estuary, England	Intertidal	8.7 (ash-free)	Kay and Knights (1975)
Foulness Flats, England	Intertidal	19.6 (ash-free)	Kay and Knights (1975)
Thames Estuary, England	Intertidal	18.8 (ash-free)	Kay and Knights (1975)
Medway Estuary, England	Intertidal	24.1 (ash-free)	Kay and Knights (1975)
Swale Estuary, England	Intertidal	32.8 (ash-free)	Kay and Knights (1975)
Tamar Estuary, England	Intertidal	13 (ash-free)	Warwick and Price (1975)
Southampton Water, England	Intertidal	90—190 (ash-free)	Hibbert (1976)
Wadden Sea, The Netherlands	Intertidal	27 (ash-free)	Beukema (1976)
Grevelingen Estuary, The Netherlands	Whole estuary	20.8 (ash-free)	Wolff and De Wolf (1977)
Byfjord, Sweden	Subtidal	4.5 (dry)	Rosenberg et al. (1977)
Delaware River	Intertidal fresh water	7.0 (dry)	Crumb (1977)
San Francisco Bay	Intertidal	10.1—30.5 (ash-free)	Nichols (1977)

From Wolff, W. J., *Estuaries and Enclosed Seas*, Ketchum, B. H., Ed., Elsevier, Amsterdam, 1983, 151. With permission.

Table 4.8—4
DENSITIES OF MACROBENTHOS (NUMBER OF INDIVIDUALS m^{-2}) IN SOME TROPICAL MANGROVE, CORAL REEF, AND ADJACENT MUD- AND SANDFLAT SEDIMENTS

Location	Habitat	Season	Total densities	Species richness (no. site)	Ref. (sources from Alongi, 1989)
Selangor, Malaysia	Mangrove, low intertidal	All	70(±102)	0–11	122
	Epifauna				
	Infauna		137(±91)	7–24	
Morrumbere Estuary	Mangroves, low intertidal	All	170(±152)	31–74	124
	Sandflat		242(±235)	42–103	
Phuket Island, Thailand	Low intertidal mangroves	Autumn	80(±28)	26	125
	Mid-intertidal mangroves		218(±34)	92	
	High intertidal mangroves		129(±65)	60	
	Mudflat		52	36	
	Sandflat		147(±20)	47	
Surin Island, Thailand	Low intertidal mangroves	Spring	4	5	126
	Mid-intertidal mangroves		10	8	
	High intertidal mangroves		28	34	
	Mudflat		26	11	
	Sandflat		43	11	
Kuala Lumpur, Malaysia	Mudflat near mangroves	All	304(±247)	22	127
Northwest Cape, W. Australia	Mudflat	Spring	992(±722)	122	128
	Avicennia forest		257(±390)	59	
	Rhizophora forest		473(±319)	31	
	High intertidal flat		1(±3)		
Cochin Estuary, India	Low intertidal mangroves	Premonsoon	5,872	N.A.	123
		Monsoon	420		

	Mudflat	Premonsoon	16,000		
		Monsoon	1,036		
Ka Yao Yai, Thailand	Low intertidal mangroves	All	49	43	129
	Mid-intertidal mangroves		107	55	
	High intertidal mangroves		142–178	70	
	Mudflat		247	109	
	Sandflat		190	48	
Maturei Vavao, Polynesia	Coral lagoon	Autumn	1,138(249–3,264)	N.A.	55
Discovery Bay, Jamaica	Coral lagoon	Spring	1,610(918–2,660)	38(26–50)	130
Grand Recif, Madagascar	Coral lagoon	Winter	632(84–1,752)	57(27–160)	97
Florida Key, U.S.	Carbonate lagoon	Spring	5,528(275–12,825)	55(33–76)	120
		Summer	7,181(725–27,020)	58(24–76)	
Red Sea, Sudan	Fore reef	Autumn	30,599–42,840	35–79	131
	Reef lagoon		5,981–12,125	18–21	
Gulf of Aquaba, Red Sea	Fringing reef	Winter	673	57	132
	Lagoon	Summer	449	41	
		Autumn	307	49	
Moorea, Polynesia	Fringing reef	Summer	2,242–4,866		
	Reef flat		496–1,824	N.A.	
	Reef slope		632		98
	Lagoon		1,335		
Carrie Bow Cay, Belize	Carbonate Sandflat	Spring	16,750	32–133	121

Note: Values depict means and SD or ranges in parentheses; N.A. = information not provided.

From Alongi, D. M., The role of soft-bottom benthic communities in tropical mangrove and coral reef ecosystems, *Rev. Aquat. Sci.*, 1(2), 259, 1989.

4.9 BIOTIC STRUCTURING OF THE HABITAT

Table 4.9—1
BIOTIC EFFECTS ON HABITAT CHARACTERISTICS PROPOSED BY FUNCTIONAL-GROUP HYPOTHESES

Habitat alteration	Functional-group hypothesis proposing effect	Examples[a]
Bioturbation of sediments (sediment reworking and destabilization)	Adult-larval interactions, trophic-group amensalism, mobility-mode interactions	*Callianassa californiensis* (50 ml/ind./d)
		Callianassa spp. (up to 1300 ml/ind./d or 3.9 kg/m^2/d)
		Saccoglossus kowalevskii (36 ml/m^2/hr)
		Enteropneust sp. (up to 700 ml/ind./d or 600 ml/m^2/d)
		Arenicola marina (2500 ml/m^2/d)
Resuspension of sediments	Trophic-group amensalism	*Callianassa californiensis* (150g/m^2/d)
		Callianassa spp. (up to 1300 ml/ind./d or 3.9 kg/m^2/d)
		Upogebia pugettensis (6 g/m^2/d)
		Enteropneust sp. (600 ml/m^2/d)
		Arenicola marina (400 l/m^2/yr)
Sediment stabilization	Mobility-mode interactions	*Zostera marina* (reduced erosion in experimental aquaria)
		Tube mat (reduced erosion)
		Salt marsh plants (inhibition of bioturbator species)
		Diopatra cuprea tubes (inhibition of bioturbator species)

[a] Sediment reworking and resuspension rates given in parentheses.

From Posey, M. H., Functional approaches to soft-substrate communities: how useful are they?, *Rev. Aquat. Sci.*, 2(3,4), 345, 1990.

Table 4.9—2
MARINE INVERTEBRATES IN WHICH THE PRESENCE OF CONSPECIFICS INFLUENCES SETTLEMENT AND METAMORPHOSIS OF LARVAE

Phylum	Species	Evidence for chemical cue
Cnidaria		
Anthozoa	*Ptilosarcus guerneyi*	Yes
Hydrozoa	*Nemertesia antennina*	Yes
	Clava squamata	Yes
	Kirchenpaueria pinnata	Yes
	Tubularia larynx	
Mollusca		
Bivalvia	*Crassostrea virginica*	Yes
	Ostrea edulis	Yes
	Mercenaria mercenaria	Yes
Gastropoda	*Haliotis disces*	Yes
	Bittium reticulatum	
	Collisella strigatella	
	Rossoa splendida	

Table 4.9—2 (continued)
MARINE INVERTEBRATES IN WHICH THE PRESENCE OF CONSPECIFICS INFLUENCES SETTLEMENT AND METAMORPHOSIS OF LARVAE

Phylum	Species	Evidence for chemical cue
Annelida	*Sabellaria alveolata*	Yes
	S. vulgaris	Yes
	Spirorbis borealis	
	Phragmatopoma californica	Yes
	Pomatoleios kraussi	
	Hydroides dianthus	
	Spirorbis pagenstecheri	
	Sabellaria spinulosa	Yes
	Polydora ligni	
Arthropoda	*Aerceriella enigmata*	
	Balanus balanoides	
	Elminius modestus	Yes
	Balanus crenatus	Yes
	B. amphitrite	
	B. tintinnabulum	
	Chthamalus stellatus	
Sipuncula	*Golfingia misakiana*	Yes
Echiura	*Urechis caupo*	Yes
Echinodermata		
Crinoidea	*Florometra serratissima*	
Holothuroidea	*Psolus chitonoides*	
Echinoidea	*Dendraster excentricus*	Yes
Chordata		
Urochordata	*Chelyosoma productum*	
	Pyura haustor	Yes

From Burke, R. D., *Bull. Mar. Sci.*, 39, 323, 1986. With permission.

4.10 ASSIMILATION EFFICIENCY

Table 4.10—1
EFFICIENCY WITH WHICH PARTICLES OF VARIOUS DIAMETERS ARE CAPTURED BY ANIMALS THAT USE DIFFERENT CAPTURE METHODS

Taxon	Capture method	Particle size (μm)	Efficiency (%)
Demosponges	Choanocyte collar	0.4—0.6	96
	+ phagocytosis	2	20—45
Haliclona (sponge)	Choanocyte collar	0.6	85—99
		2—4	<20
		5—7	>50
Terebratulina (brachipod)	Cilia	2.7	50
		~6	90
Crepidula (gastropod)	Mucous net	1.2	90
		1.5—2.0	100
Mytilus (bivalve)	Cirri on gills	2	80
Crassostrea (bivalve)	Cirri on gills	2	50
		5—6	100
Solen (bivalve)	Cirri on gills	1.5	0
		1.5—2.0	15—35

Table 4.10—1 (continued)
EFFICIENCY WITH WHICH PARTICLES OF VARIOUS DIAMETERS ARE CAPTURED BY ANIMALS USING DIFFERENT CAPTURE METHODS

Taxon	Capture method	Particle size (μm)	Efficiency (%)
		2.0—2.5	40—60
		2.5—3.0	70—90
Geukensia (bivalve)	Cirri on gills	0.2—0.4	30[a]
		0.4—0.6	86[a]
Chlamys islandica (bivalve)	Cirri on gills	1.2	4
		2	~25
		2.5	57
		3	75
		4	92
		6	96
Perna (bivalve)	Cirri on gills	0.6	18
		2	80
Aulacomya (bivalve)	Cirri on gills	0.6	18
		2	86
Geukensia (bivalve)	Cirri on gills	2	35—70
Spisula (bivalve)	Cirri on gills	>4	100
Brachidontes (bivalve)	Cirri on gills		
Mercenaria (bivalve)	Cirri on gills		
Argopecten (bivalve)	Cirri on gills	2	15
		5—6	100
Pecten (bivalve)	Cirri on gills	1	20
		6—7	100
Arctica (bivalve)	Cirri on gills	1	70
Cultellus (bivalve)	Cirri on gills	4	100
Mya (bivalve)	Cirri on gills	1	25
		4	100
Ostrea (bivalve)	Cirri on gills	~1	0
		4	100
Monia (bivalve)	Cirri on gills	3.2	50
		5.2	90
		6—7	100
Cardium (bivalve)	Cirri on gills	1.1	50
		2.8	90
		4	100
Anodonta (bivalve)	Cirri on gills	1.6	50
		4	90
		6—7	100
Chaetopterus (polychaete)	Mucous net	~0.5	50
		1.2—1.7	90
		2—3	100
Sabella (polychaete)	Cilia	1.5	50
		3	90
Ascidia (ascidian)	Mucous net	0.9—1.1	50
		1.1—2.0	90
Styela (ascidian)	Mucuous net	1	90
Ascidians	Mucous mesh	1	70
		2—3	100
Ascidians	Mucous mesh	>0.6	100

[a] Efficiency expressed in terms of retention of particles of diameter 0.6 μm.

From Wotton, R., *The Biology of Particles in Aquatic Systems*, CRC Press, Boca Raton, FL, 1990, 173.

Table 4.10—2
ASSIMILATION OF PHYTOPLANKTON BY ZOOPLANKTON

Species	Food	Method	Assimilation (%)	Reference (sources from Heinle, 1981)
Calanus finmarchicus	Various diatoms and flagellates	Tracer-isotope [^{32}P]	15–99	Marshall and Orr (1961)
	Skeletonema costatum	Tracer-isotope [^{14}C]	60–78	Marshall and Orr (1955b)
Temora longicornis	*S. costatum*	Tracer-isotope [^{32}P]	50–98	Berner (1962)
Euphausia pacifica	*Dunaniella primolecta*	Tracer-isotope [^{14}C]	85–99	Lasker (1960)
Ostrea edulis (larvae)	*Isochrysis galbana*	Tracer-isotope [^{32}P]	13–50	Walne (1965)
Calanus helgolandicus	Natural particulate matter	Chemical analyses	74–91	Corner (1961)
Metridia lucens	*T. nordenskioldii Ditylum* spp. *Artemia* nauplii	"Ratio method"	50–84 35 59	Haq (1967)
C. hyperboreus	*Exuviella* spp.	"Ratio method" Chemical analyses	39.0–85.6 54.6–84.6	
Natural zooplankton	Natural particulate material	"Ratio method"	32.5–92.1	Conover (1966a)
C. finmarchicus	*Skeletonema costatum*	"Ratio method"	53.8–64.4	Corner et al. (1967)
	S. costatum	Chemical analyses	57.5–67.5	

From Heinle, D. R., *Functional Adaptations of Marine Organisms,* Vernberg, F. J. and Vernberg, W. B., Eds., Academic Press, New York, 1981, 85. With permission.

Table 4.10—3
ASSIMILATION EFFICIENCY (AE) OF MARINE COLLECTORS FEEDING ON VARIOUS FOODS

Taxon	AE (%)	Food
Microheterotrophs	63	Amino acids
Protozoa		
Favella	67	Natural food
Favella	71	Dinoflagellates
Balanion	82	Dinoflagellates
Coelenterata		
Gorgonian coral	22	Coral mucus
Mollusca		
Geukensia	14	Refractory carbon
Crassostrea	1	Refractory carbon
Oysters	29	Natural food
Spisula	33	Algae (low conc.)
Spisula	76	Algae (high conc.)
Bivalves	Up to 94	P in *Phaeodactylum*
Crustacea		
Acartia	47	Mucus + microorganisms
Acartia	50	Coral mucus organics
Calanus	19–88	Different diatoms
Calanus	69–85	Diatom carbon
Calanus	74–93	Diatom nitrogen
Mysidium	44	Mucus + microorganisms
Corophium	82–92	Surface of particles

Table 4.10—3 (continued)
ASSIMILATION EFFICIENCY (AE) OF MARINE COLLECTORS FEEDING ON VARIOUS FOODS

Taxon	AE (%)	Food
Echinodermata		
Stichopus	~30	Organic sediments
Parastichopus	17	Organic sediments
Chordata		
Pyura	42	Kelp detritus
Pyura	75	Algae
Pyura	34	Natural food

From Wotton, R., *The Biology of Particles in Aquatic Systems,* CRC Press, Boca Raton, FL, 220, 1990.

Table 4.10—4
ASSIMILATION EFFICIENCY (AE) OF FRESHWATER AND MARINE SHREDDERS FEEDING ON VARIOUS FOODS

Freshwater shredders

Taxon	AE(%)	Food
Insecta: Plecoptera		
Pteronarcys	13—19	Leaves
Pteronarcys	9—16	Detritus
Insecta: Trichoptera		
Zelandopsyche	8—24	Beech leaves
Crustacea: Isopoda		
Asellus	23	Decayed leaves
Asellus	26—44	Decomposing leaves
Crustacea: Amphipoda		
Gammarus	10	Decaying leaves
Gammarus	30—40	Alder leaves
Gammarus	0—35	Beech leaves
Gammarus	42—76	Fungal mycelium
Gammarus	73—96	Fungal protein
Gammarus	10	Elm/maple leaves
Gammarus	14—18	Leaf protein
Hyalella	9	Elm leaves
Hyalella	22	Elm leaf protein
Hyalella	73	Epiphytes
Hyalella	80	Epiphyte protein
Hyalella	7—15	Lake sediments
Hyalella	14—23	Sediment protein

Marine shredders

Taxon	AE(%)	Food
Crustacea		
Palaeomonetes	73	Sterile detritus
Palaeomonetes	91	Unsterilized detritus

From Wotton, R., *The Biology of Particles in Aquatic Systems,* CRC Press, Boca Raton, FL 1990, 173.

Table 4.10—5
SOME INDICATIVE Q_{10} DETERMINATIONS FOR VARIOUS PHYSIOLOGICAL PROCESSES IN A RANGE OF TAXA

Metabolic process	Q_{10}	Temperature range of determination	Species
Oxygen uptake	1.3–2.2	5.5–27.5	*Nereis* spp.
	2.98	30	*Palaemon pacificus*
	4.69	15	*P. pacificus*
	1.5	9–16	*Carcinus maenas*
	1.4–1.6	na	Intertidal pagurids
	2.1–2.4	na	Subtidal pagurids
	2.6–2.7	na	Supratidal pagurids
	1.96–2.40	2–15	*Mytilus edulis*
	1.44	20–29	*Macrognathus aculeatum*
	1.27	20–29	*Anabas testudineus*
	1.30	20–29	*Channa punctatus*
	2.91	29–32	*C. punctatus*
	3.05	29–32	*Macrognathus aculeatum*
Gill clearance	1.11	10–20	*Branchionus plicatilis*
Food ingestion	1.75	10–20	*B. plicatilis*
Ventilation/heart beat	1.9–3.6	na	Decapod crustaceans
Total sediment oxygen	6.5	na	Benthic community
Zinc accumulation	1.8	10–20	*Palaemon elegans*

Note: na, not available; see original source for references.

From Saenger, P. and Holmes, N., *Pollution in Tropical Aquatic Systems,* Connell, D. W. and Hawker, D. W., Eds., CRC Press, Boca Raton, FL, 75, 1992.

4.11 PHYTOPLANKTON

Table 4.11—1
NATURAL AND MAN-IMPOSED CONDITIONS WHICH DETERMINE LEVELS AND RATES OF CHANGE OF FACTORS AFFECTING ABUNDANCE AND SUCCESSION OF ESTUARINE PHYTOPLANKTON

Factors	Natural conditions	Man-imposed conditions
Salinity	Precipitation, runoff, evaporation, circulation of water	Water impoundment, channelization, dredge and fill, mosquito ditching
Temperature	Latitude, season, weather, time of day, circulation of water	Heated effluent, dams, canals and waterways, stream channelization
Light intensity		
At surface	Latitude, season, weather, time of day	Air pollution — smog
Below surface	Reflection, absorption, scattering	Dredging, waste dumping, erosion
Nutrients	Drainage, runoff, circulation of water, sediments	Sewage and industrial wastes, urban and agricultural drainage, erosion
Metabolites	Living and dead plants and animals	Sewage, urban and agricultural drainage, erosion
Toxic substances		
Petroleum	Deposits	Leaks and spills during drilling, transport, storage, use or disposal
Radionuclides	Primordial deposits, cosmic ray produced	Fallout, nuclear power reactors, other releases
Heavy metals	Terrestrial deposits, sediments, land drainage	Industrial and domestic wastes, mining, erosion
Synthetic toxicants		Industrial, agricultural, and domestic use

From Rice, T. R. and Ferguson, R. L., *Physiological Ecology of Estuarine Organisms,* Vernberg, F. J., Ed., University of South Carolina Press, Columbia, 1975, 1. With permission.

Table 4.11—2
DISTRIBUTION OF PIGMENTS IN ALGAL CLASSES[a]

Pigments	Chlorophyceae	Prasinophyceae	Euglenophyceae	Chrysophyceae	Haptophyceae	Xanthophyceae[b]	Bacillariophyceae	Dinophyceae	Cyanophyceae		Cryptophyceae
Chlorophyll *a*	+	+	+	+	+	+	+	+	+		+
Chlorophyll *b*	+	+	+	—	—	—	—	—	—		—
Chlorophyll *c*	—	—	—	+	+	—	+	+	—		—
Chlorophyll *e*	—	—	—	—	—	+	—	—	—		—
α-Carotene	(+)	(+)	—	—	+	—	(+)	(+)	—		+
β-Carotene	+	+	+	+	+	+	+	+	+		—
γ-Carotene	+	(+)	—	—	—	—	—	—	—		—
ϵ-Carotene	—	—	—	—	—	—	(+)	—	—		—
Lutein	+	+	+	+	—	+	(+)	—	(+)		—
Zeaxanthin	+	+	?	—	—	—	—	—	+		+
Violaxanthin	+	+	—	—	—	+ ?	—	—	—		
Neoxanthin	+	+	+	—	—	+ ?	—	—	—		
Fucoxanthin	—	—	—	+	+	+	+	—	—		
Diatoxanthin	—	—	—	—	—	—	+	—	—		+
Diadinoxanthin	—	—	—	—	—	—	+	+	—		—
Neodiadinoxanthin	—	—	—	—	—	—	—	+	—		—
Dinoxanthin	—	—	—	—	—	—	—	+	—		—
Neodinoxanthin	—	—	—	—	—	—	—	+	—		—
Peridinin	—	—	—	—	—	—	—	+	—		—
Neoperidinin	—	—	—	—	—	—	—	+	—		—
Myxoxanthin	—	—	—	—	—	—	—	—	+		—
Oscilloxanthin	—	—	—	—	—	—	—	—	+		—
C-Phycoerythrin	—	—	—	—	—	—	—	—	+	Phycoerythrin }	found in some genera
C-Phycocyanin	—	—	—	—	—	—	—	—	+	Phycocyanin }	

[a] Classes with planktonic representatives
[b] Representative genera planktonic, but not extensively reported on in plankton studies.

From Boney, A. D., *Phytoplankton*, Edward Arnold, London, 1975. With permission.

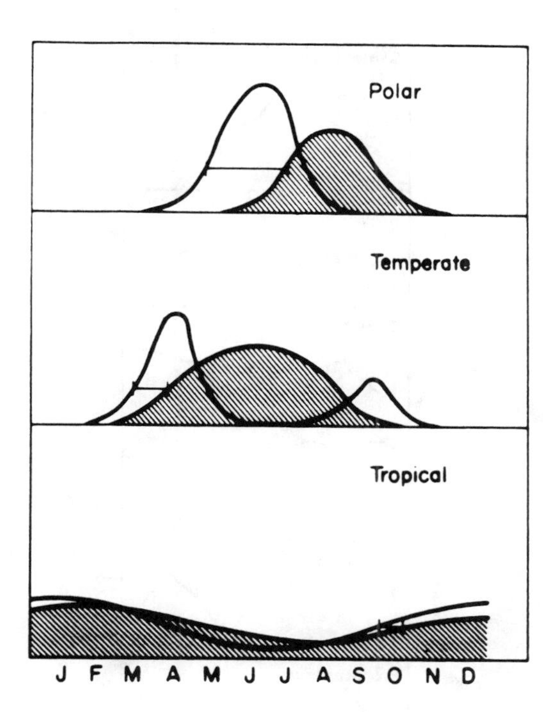

FIGURE 4.11—1. Seasonal cycles of phytoplankton in Arctic temperate-boreal, and tropical waters. Stippled area delineates zooplankton seasonal cycles in these regions. The horizontal bar indicates the lag period between the increases of phytoplankton and zooplankton. (From Dring, M. J., *The Biology of Marine Plants,* Edward Arnold, London, 1982. With permission.)

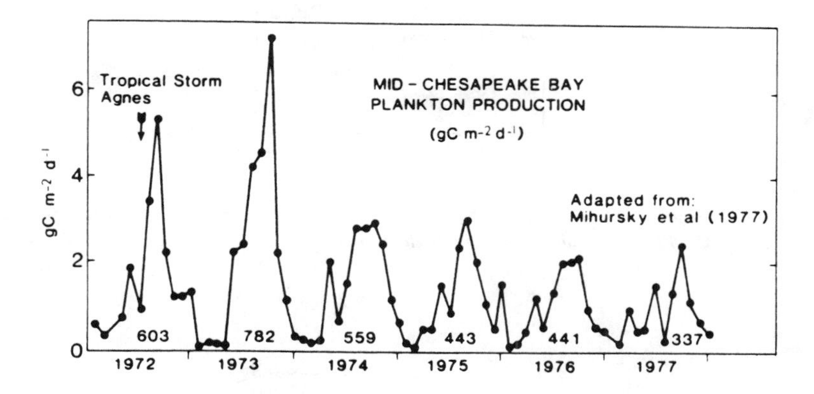

FIGURE 4.11–2. Mean monthly phytoplankton production rates in central Chesapeake Bay from January 1972 through December 1977. Values below peaks are estimates of annual phytoplankton production. (From Mihursky, J. A., Heinle, D. R., and Boynton, W. R., Unpubl. Tech. Rep., UMCEES Ref. No. 77–28-CBL, Chesapeake Biological Laboratory, Solomons, MD, 1977. With permission.)

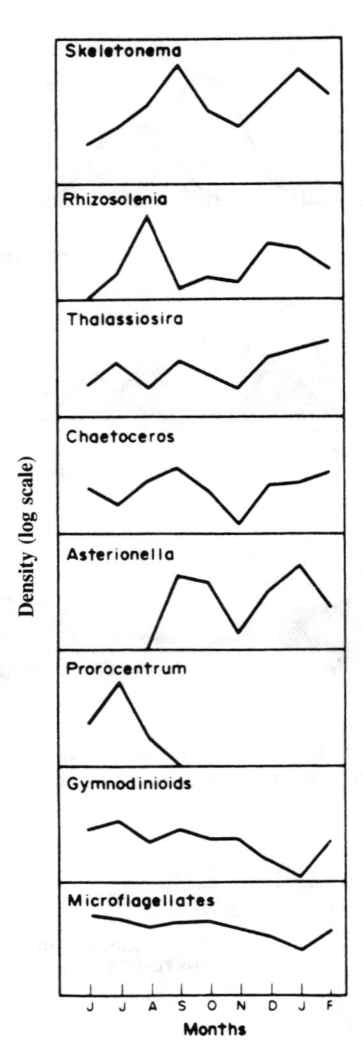

FIGURE 4.11—3. Seasonal abundance of some major phytoplankton species in Narragansett Bay, depicting general successional patterns. (From Raymont, J. E. G., *Plankton and Productivity in the Oceans,* Vol. 1, 2nd ed., Pergamon Press, Oxford, 1980. With permission.)

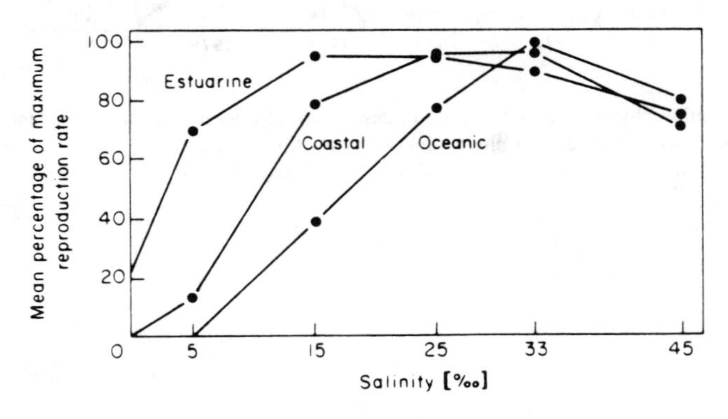

FIGURE 4.11—4. Mean percentage of the maximum reproduction rates of estuarine, coastal, and oceanic groups of phytoplankton species. (From Brand, L. E., *Estuarine Coastal Shelf Sci.*, 18, 543, 1984. With permission.)

4.12 ZOOPLANKTON

Table 4.12—1
SALINITY RANGES OF VARIOUS COPEPOD SPECIES OBSERVED IN NATURE

Copepod species	0	10	20	30	40	50	60	70	80
Acartia (Paracartia) africana									
A. (Paracartia) longipatella									
A. (Arcartiella) natalensis									
Calanoides carinatus									
Centropages brachiatus									
C. chierchiae									
C. furcatus									
Clausidium sp.									
Clausocalanus furcatus									
Corycaeus spp.									
Ctenocalanus vanus									
Euterpina acutifrons									
Halicyclops spp.									
Harpacticus ? gracilis									
Hemicyclops sp.									
Nannocalanus minor									
Oithona brevicornis/nana									
O. plumifera									
O. similis									
Paracalanus aculeatus									
P. crassirostris									
P. parvus									
Porcellidium sp.									
Pseudodiaptomus stuhlmanni									
P. hessei									
P. nudus									
Rhincalanus nasutus									
Saphirella stages									
Temora turbinata									
Tortanus capensis									
Tegastes sp.									

From Grindley, J. R., *Estuarine Ecology: With Particular Reference to Southern Africa,* Day, J. H., Ed., A. A. Balkema, Rotterdam, 1981, 117. With permission.

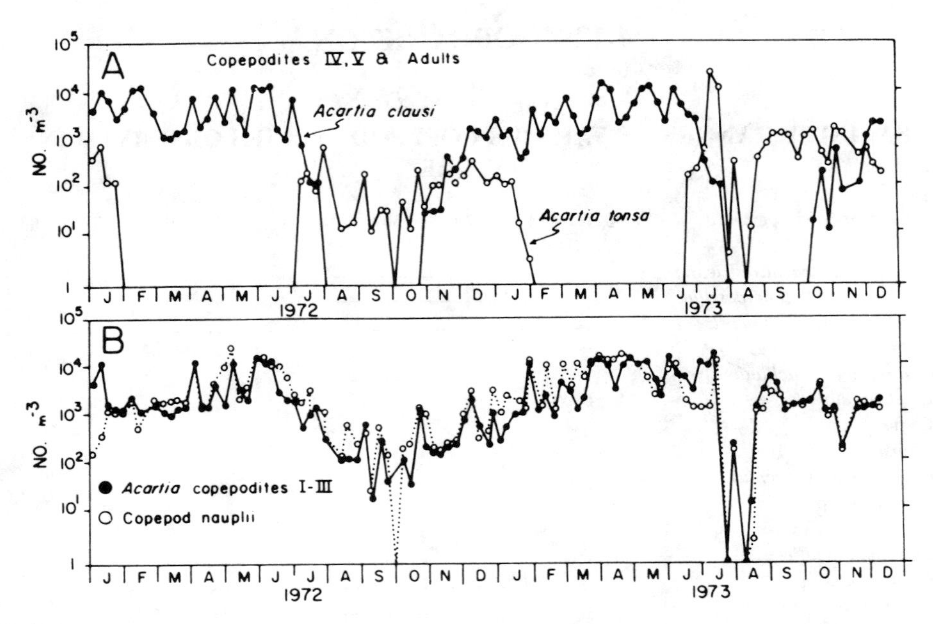

FIGURE 4.12—1. Abundance cycles of *Acartia hudsonica* (=*A. clausi*) and *A. tonsa* in Narragansett Bay. (A) Densities of older copepodites (summed) and adults; (B) densities of nauplii and younger copepodites. Scales are linear between each logarithmic interval. (From Miller, C. B., *Estuaries and Enclosed Seas,* Ketchum, B. H., Ed., Elsevier, Amsterdam, 1983, 103. With permission.)

Table 4.12—2
DISTRIBUTION AND OCCURRENCE OF MAJOR HOLOPLANKTONIC SPECIES WITHIN THE MID-ATLANTIC BIGHT AREA

Species	Distribution	Occurrence
Dominant		
Acartia hudsonica	Estuarine and marine[a]	Winter-spring
A. tonsa	Estuarine and marine[a]	Summer-fall
Calanus finmarchicus	Stenohaline marine[b]	Year-round
Centropages hamatus	Euryhaline marine[c]	Year-round
C. typicus	Stenohaline marine[b]	Summer-fall
Pseudocalanus minutus	Euryhaline marine[c]	Winter-spring
Temora longicornis	Euryhaline marine[c]	Winter-spring
Common		
Labidocera aestiva	Euryhaline marine[c]	Summer-fall
Tortanus discaudatus	Estuarine and marine[a]	Spring-summer
Sagitta elegans	Stenohaline marine[b]	Winter-spring
Frequent		
Oithona brevicornis	Estuarine and marine[a]	Summer-fall
O. similis	Euryhaline marine[c]	Year-round

[a] Estuarine and marine — widely distributed in an estuary and reproducing to a limited extent in open coastal waters.
[b] Stenohaline marine — reproducing in open neritic waters.
[c] Euryhaline marine — reaching maximum population densities near the coast, close to the mouths of estuaries.

From Jeffries, H. P. and Johnson, W. C., *Coastal and Offshore Environmental Inventory: Cape Hatteras to Nantucket Shoals,* Marine Publ. Ser. No. 2, University of Rhode Island, Kingston, 1973, 4—1. With permission.

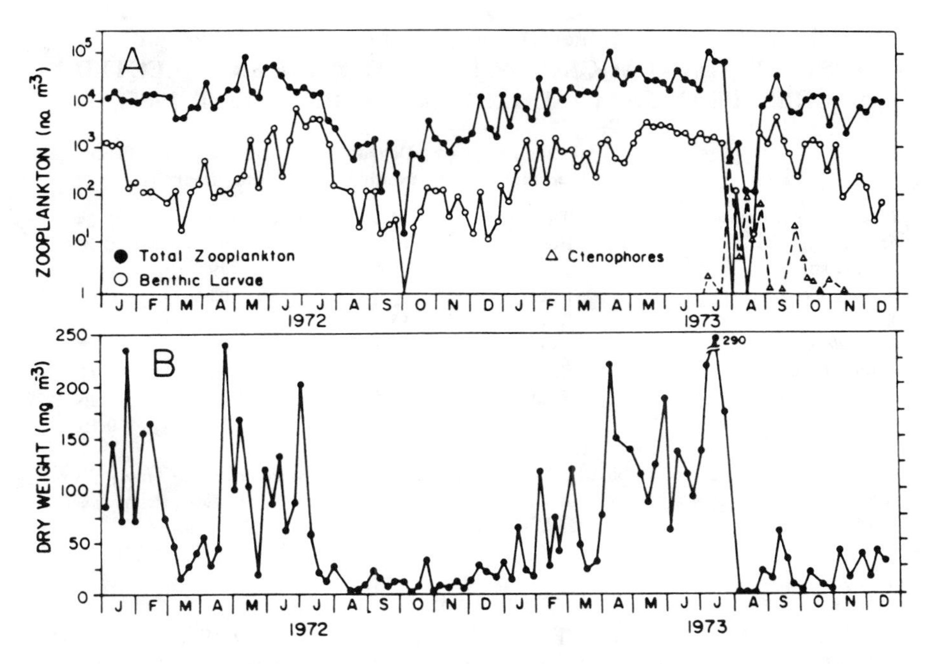

FIGURE 4.12—2. (A) Abundance cycles of holoplankton in Narragansett Bay with comparisons to abundance of benthic larvae and ctenophores. Scale is linear between each logarithmic interval. (B) Seasonal cycles of dry weight biomass of zooplankton in Narragansett Bay. Note linear scale. (From Miller, C. B., *Estuaries and Enclosed Seas,* Ketchum, B. H., Ed., Elsevier, Amsterdam, 1983, 103. With permission.)

Table 4.12—3
30 MOST ABUNDANT MICROZOOPLANKTON TAXA COLLECTED IN CHESAPEAKE BAY (AUGUST 1984 TO DECEMBER 1985)

Taxon	Taxonomic group	Density (no./liter)	Max. density (no./liter)	Percent of total
Synchaeta spp.	Rotifera	117.1	2513	29.38
Tintinnopsis fimbriata	Tintinnina	106.0	4334	20.59
Copepod nauplii	Copepoda	39.2	1006	19.45
Synchaeta sp. A	Rotifera	34.2	609	2.89
Tintinnopsis dadayi	Tintinnina	17.6	486	2.89
Keratella cochlearis cochlearis	Rotifera	27.2	481	2.80
Polyarthra sp.	Rotifera	36.8	378	2.38
Difflugiidae	Sarcodina	14.3	270	2.06
Synchaeta sp. B	Rotifera	65.2	296	1.80
Tintinnopsis subactua	Tintinnina	11.7	442	1.76
T. radix	Tintinnina	22.1	439	1.37
Brachionus angularis	Rotifera	24.9	205	1.21
Trichocerca sp.	Rotifera	14.5	264	1.11
B. calyciflorus	Rotifera	16.2	170	1.05
Pelecypoda larvae	Pelecypoda	7.4	63	1.02
Keratella cochlearis tecta	Rotifera	16.0	232	0.89
Filinia sp.	Rotifera	21.0	223	0.66
Polyarthra vulgaris	Rotifera	17.4	101	0.62
Notholca acuminata	Rotifera	12.3	93	0.48
B. plicatilus	Rotifera	11.4	66	0.33
Hexarthra mira	Rotifera	16.3	132	0.30

Table 4.12—3 (continued)
30 MOST ABUNDANT MICROZOOPLANKTON TAXA COLLECTED IN CHESAPEAKE BAY (AUGUST 1984 TO DECEMBER 1985)

Taxon	Taxonomic group	Density (no./liter)	Max. density (no./liter)	Percent of total
Tintinnopsis subacuta	Tintinnina	4.5	28	0.24
Filinia longiseta	Rotifera	10.9	47	0.24
Acineta sp.	Ciliophora	15.6	149	0.22
B. caudatus	Rotifera	10.0	63	0.22
Keratella sp.	Rotifera	15.2	132	0.22
Cyphoderiidae	Sarcodina	1.8	20	0.17
Centrophyxidae	Sarcodina	2.7	30	0.17
Conochilus unicornis	Rotifera	15.6	68	0.16
Arcella sp.	Sarcodina	2.9	46	0.15

From Brownlee, D. C. and Jacobs, F., *Contaminant Problems and Management of Living Chesapeake Bay Resources,* Majumdar, S. K., Hall, L. W., Jr., and Austin, H. M., Eds., Pennsylvania Academy of Science, Easton, 1987, 217. With permission.

Table 4.12—4
MEAN ABUNDANCE, PERCENT COMPOSITION, AND UBIQUITY OF MESOZOOPLANKTON COLLECTED IN CHESAPEAKE BAY FROM AUGUST 1984 TO DECEMBER 1985[a]

Taxon	Density (5 s/m²)	Percent of total	Ubiquity (% of total samples in which taxon occurred)
Acartia tonsa	5503.82	54.13	91.25
Eurytemora affinis	2032.20	19.99	53.61
Bosmina longirostris	715.45	7.04	32.89
Polychaete larvae	506.94	4.99	59.51
Barnacle nauplii	336.12	3.31	73.57
A. clausi	254.32	2.50	9.70
Diaphanosoma leuchtenbergianum	153.73	1.51	14.07
Moina micrura	86.48	0.85	13.88
Podon polyphemoides	82.21	0.81	24.71
Acartia sp.	81.54	0.80	7.22
Mesocyclops edax	80.57	0.79	14.26
Copepod nauplii	76.93	0.76	61.60
Cyclops vernalis	43.18	0.42	25.10
Ostracoda	27.71	0.27	57.03
Oithona colcarva	26.03	0.26	28.71
Daphnia retrocurva	23.06	0.23	18.63
Scapholeberis kingi	17.80	0.18	6.46
Gastropod larvae	16.41	0.16	12.93
Centropages hamatus	16.13	0.16	9.32
Pseudodiaptomus coronatus	13.41	0.13	22.62
Harpacticoida	10.99	0.11	30.80
Halicyclops magnaceps	9.82	0.10	4.75
Mollusca	8.49	0.08	13.12
Alonella sp.	4.68	0.05	8.17
Alona spp.	4.33	0.04	1.33
Cyclops bicuspidatus	4.29	0.04	10.08
Diaptomus sp.	3.97	0.04	15.59
Mysid	3.17	0.03	46.20
Ergasilus sp.	2.98	0.03	13.31
Polychaeta	2.96	0.03	3.04
Unid. fish eggs	2.69	0.03	11.79

Table 4.12—4 (continued)
MEAN ABUNDANCE, PERCENT COMPOSITION, AND UBIQUITY OF MESOZOOPLANKTON COLLECTED IN CHESAPEAKE BAY FROM AUGUST 1984 TO DECEMBER 1985[a]

Taxon	Density (5 s/m²)	Percent of total	Ubiquity (% of total samples in which taxon occurred)
Ilyocryptus spinifer	1.76	0.02	10.65
Sida crystallina	1.37	0.01	6.84
Brachyurian zoea	1.33	0.01	12.17
Paracyclops fimbriatus poppei	1.09	0.01	5.32
Sagitta sp.	0.97	0.01	14.83
Chydorius sp.	0.84	0.01	9.89
Camptocercus rectirostris	0.82	0.01	3.61
Hydracarina	0.79	0.01	4.75
Leptodora kindtii	0.75	0.01	2.66
Ilyocryptus sp.	0.74	0.01	2.28
Eucyclops agilis	0.70	0.01	9.89
Palaemonetes sp.	0.62	0.01	12.36
Unid. crab zoea	0.49	0.00	1.71
Unid. fish larvae	0.46	0.00	18.25
Alona affinis	0.43	0.00	8.56
Temora turbinata	0.41	0.00	1.33
Gammarus fasciatus	0.40	0.00	17.30
Centropages furcatus	0.38	0.00	0.57
Paracalanus crassirostris	0.32	0.00	1.14
Cyclopoida	0.16	0.00	0.76
Oligochaeta	0.13	0.00	0.19
Morone americana	0.13	0.00	1.14
Corophium lacustre	0.07	0.00	8.17
Alona costata	0.06	0.00	3.42
Argulus sp.	0.06	0.00	13.50
Morone sp.	0.04	0.00	0.57
Alona sp.	0.04	0.00	0.57
Sapherella sp.	0.04	0.00	0.76
Bosmina sp.	0.03	0.00	0.38
Clupeidae	0.02	0.00	0.95
M. saxatilis	0.02	0.00	0.76
Monoculodes edwardsi	0.02	0.00	4.94
Eubosmina coregoni	0.01	0.00	0.95
Leptocheirus plumulosus	0.01	0.00	1.33
Chaoborus sp.	0.01	0.00	3.23
Chironomid larvae	0.01	0.00	6.46
Isopoda	0.01	0.00	7.22
Parathemisto compressa	0.01	0.00	1.90
Micropogon undulatus	0.00	0.00	2.28
Anchoa mitchilli	0.00	0.00	1.33
Brachyurian megalops	0.00	0.00	0.95
Pseudopleuronectes americanus	0.00	0.00	0.38
Menidia	0.00	0.00	0.76
Gobiosoma bosci	0.00	0.00	0.76
Piscicolidae	0.00	0.00	0.38
Lucifer faxoni	0.00	0.00	0.57
Dipteran larvae	0.00	0.00	0.38
Euceramus praelongus	0.00	0.00	0.19

[a] Mean, maximum, and percent of total abundance over all stations and dates.

From Brownlee, D. C. and Jacobs, F., *Contaminant Problems and Management of Living Chesapeake Bay Resources,* Majumdar, S. K., Hall, L. W., Jr., and Austin, H. M., Eds., Pennsylvania Academy of Science, Easton, 1987, 217. With permission.

4.13 SEAGRASSES, SALT MARSHES, AND MANGROVES

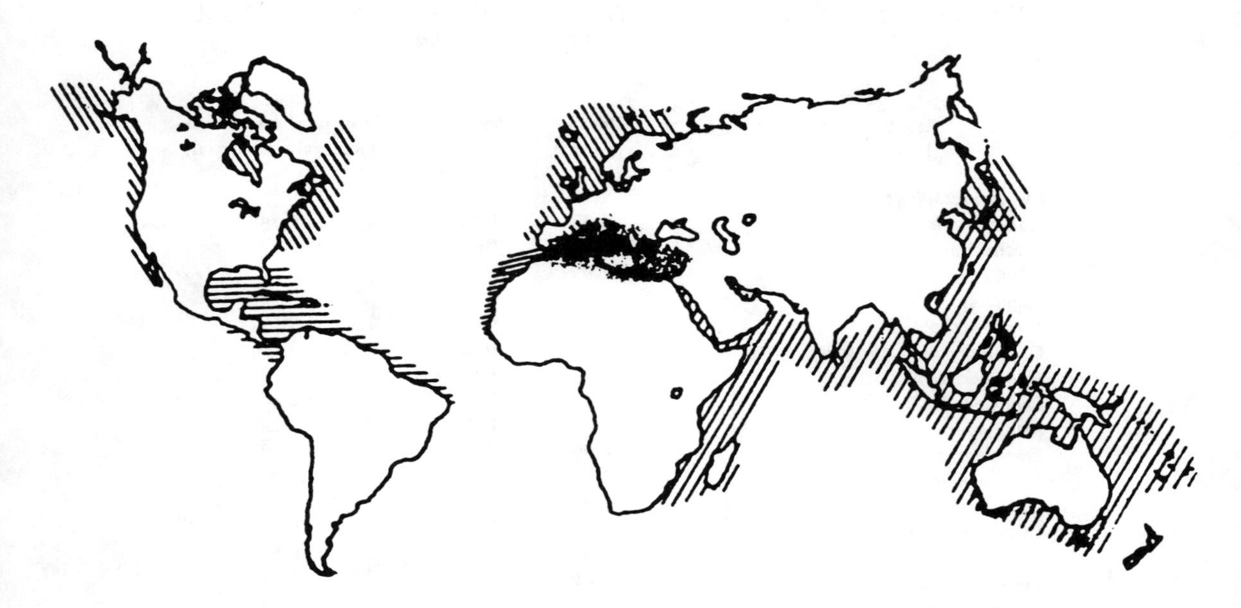

FIGURE 4.13—1. Geographic distribution of genera of marine seagrasses. *Zostera* (\\\\\\\\\\). *Posidonia* (⁘), *Thalassia* and *Halophila* (≡). *Cymodocea* (//////), *Syringodium, Thalassia, Enhalus, Halodule,* and *Cymodocea* (//////). (From Thayer, G. W. et al., in *Wetland Functions and Values,* Greeson, P., Clark, J. R., and Clark, J. E., Eds., Proc. Natl. Symp. Wetlands, American Water Resources Association, Minneapolis, 1979, 235. With permission.)

FIGURE 4.13—2. Geographic distribution of salt marshes (bold, stippled area) and mangroves (cross-hatched area). (From Ferguson, R. L., Thayer, G. W., and Rice, T. R., *Functional Adaptation of Marine Organisms,* Vernberg, F. J. and Vernberg, W. B., Eds., Academic Press, New York, 1981, 9. With permission.)

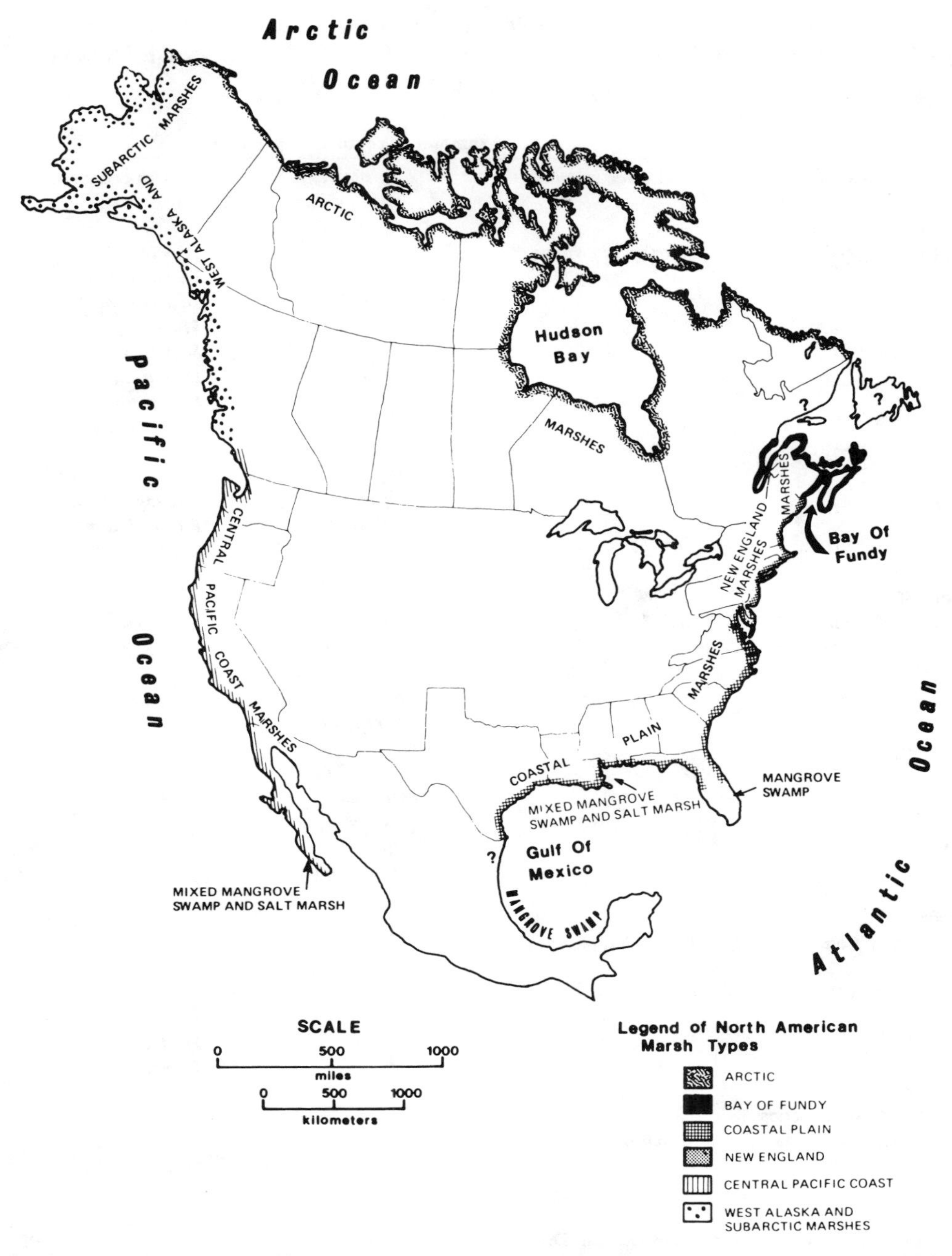

FIGURE 4.13—3. Schematic distribution of the three major groups of salt marshes in North America: (1) Bay of Fundy and New England marshes, (2) Atlantic and Gulf coastal plain marshes, and (3) Pacific marshes. (From Frey, R. W. and Basan, P. B., *Coastal Sedimentary Environments,* 2nd ed., Davis, R. A., Ed., Springer-Verlag, New York, 1985, 225. With permission.)

Table 4.13—1
COMMON SALT MARSH PLANTS FROM VARIOUS REGIONS OF NORTH AMERICA

Location	Examples of common vegetation	
	Lower marsh	Upper marsh
Eastern North America		
New England	*Spartina alterniflora*	*S. patens* *Distichlis spicata* *Juncus gerardi* *S. alterniflora* (dwarf)
Coastal Plain	*S. alterniflora*	*S. patens* *D. spicata* *Salicornia* sp. *J. roemerianus*
Bay of Fundy	*S. alterniflora*	*S. patens* *Limonium nashii* *Plantago oliganthos* *Puccinellia maritima* *J. gerardi*
Gulf of Mexico		
North Florida/South Alabama and Mississippi	Dominant *J. roemerianus* Subdominant *S. patens* *S. alterniflora*	
Louisiana	Dominant *S. alterniflora* Subdominant *S. patens* *D. spicata* *J. roemerianus*	
Arctic		
Northern Canada/Europe	*Puccinellia phryganodes*	*Carex subspathacea*
Western Alaska	*P. phryganodes*	*Puccinellia triflora* *Plantago maritima* *Triglochin* sp. *P. maritima*
Western North America		
Southern California	*S. foliosa*	*Salicornia pacifica* *Suaeda californica* *Batis maritima*

From Chapman, V. J., Salt Marshes and Salt Deserts of the World, Wiley-Interscience, New York, 1960, 392.

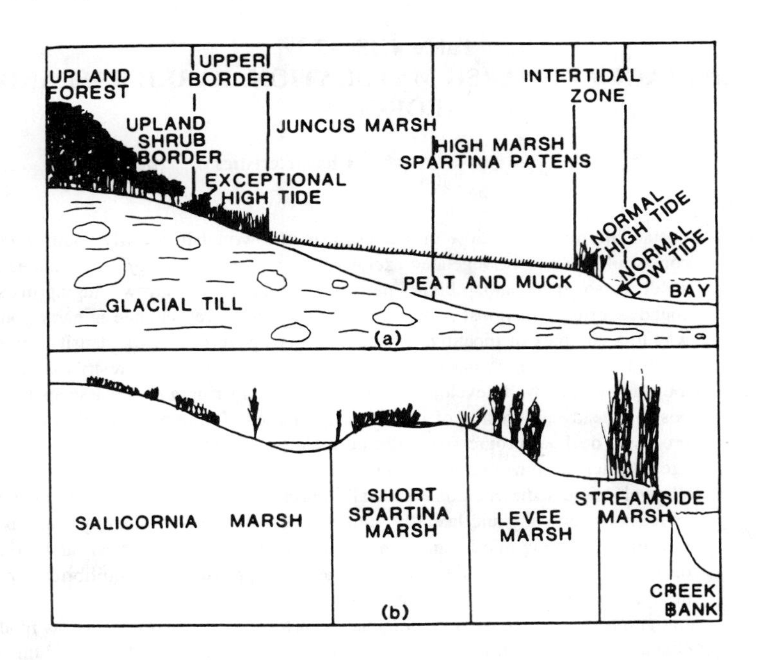

FIGURE 4.13—4. Plant zonation in typical salt marshes of North America. (a) New England salt marsh. (b) Georgia salt marsh. (a: From Teal, J. M., *Ecology*, 43, 614, 1962. With permission. b: From Miller, W. R. and Egler, F. E., *Ecol. Monogr.*, 20, 143, 1950. With permission.)

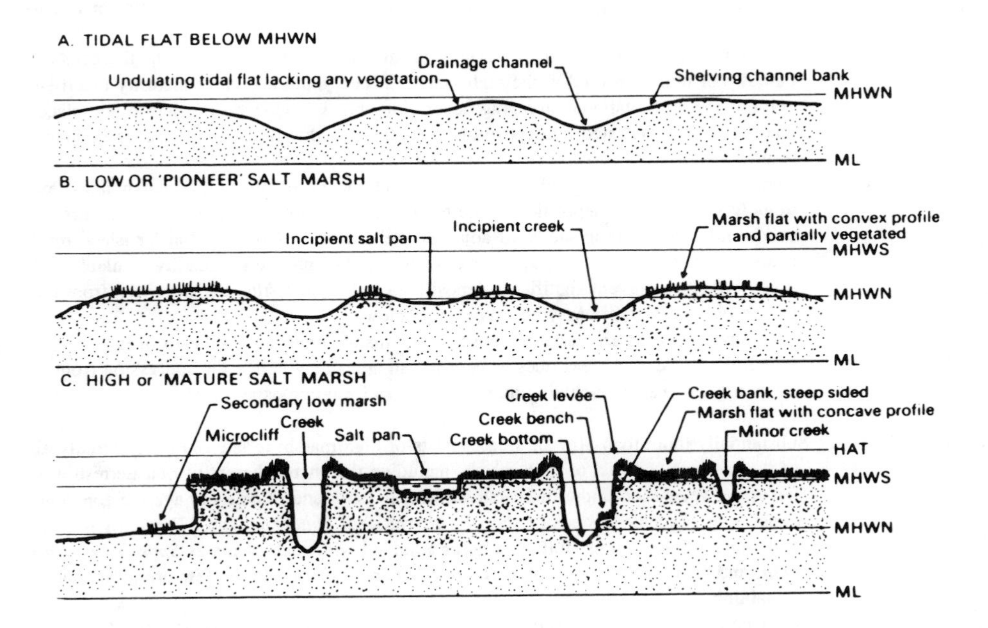

FIGURE 4.13—5. Profiles of tidal flat and salt marsh habitats depicting physiographical features and a hypothetical sequence of salt marsh development. (A) High level tide flat; (B) low level marsh: vascular plants colonize the higher points of the tidal flat; and (C) high marsh: surface completely vegetated, creeks and pans sharply defined. Secondary low marsh: develop where the surface has been eroded due to creek movement or the formation of an erosion microcliff. (From Long, S. P. and Mason, C. F., *Saltmarsh Ecology*, Blackie & Son, Glasgow, 1983. With permission.)

Table 4.13—2
IDEALIZED STAGES IN MARSH MATURATION, BARRIER ISLANDS OF GEORGIA[a]

Stage		Characteristics
Youth	A	Substantially more than 50% of area consists of low marsh; in early youth, total area may consist of low marsh, vegetated exclusively by *Spartina alterniflora*, initiated either as small marsh islands along higher parts of tidal flats or as narrow fringing marshes around sound or estuary margins; zonation patterns absent to very simple in early youth, giving way to a complex of repetitive low-marsh zones governed by the density and distribution of tidal drainages; high-marsh vegetation restricted mainly to terrestrial fringes
	B	Drainage systems well developed; pronounced meandering and crevasse splaying, with possible headward erosion of individual tributaries, during early youth; intensity of these processes declines progressively during middle and late youth and the positions of drainage channels become correspondingly more stable
	C	Relatively rapid sedimentation, especially during early youth; marsh substrates actively accrete, both vertically and laterally, until lateral growth is inhibited by margins of sounds or estuaries; during middle and late youth, accretion is most vertical, and the rate of deposition decelerates as the marsh increasingly approaches an equilibrium among topography, tidal hydraulics, and sediment supply
	D	Stratigraphic record consists predominantly of low-marsh environments, as modified by channel migrations; in many places the vertical sequence consists only of thin veneers of Holocene sediment spread over shallow basements of Pleistocene sand, rather than the fill of open lagoons or estuaries; these sand platforms are remnants of old barrier islands
Maturity	A	Low- and high-marsh areas approximately equal in size; low marsh and lower edge of high marsh vegetated by ecophenotypes of *S. alterniflora*; in early maturity the remaining high-marsh zones may consist of mixtures of *S. alterniflora, Salicornia* spp., or *Distichlis spicata*, followed by isolated stands of *Juncus roemerianus*, as is true in late youth, whereas in later maturity these plants typically occur either in mosaic clumps or in narrow, concentric zones
	B	Good drainage system remains, especially in low-marsh areas; but many tidal creeks are partially or totally infilled in the high marsh; in late youth and early maturity much lateral erosion and rotational slumping occur along creek banks; yet erosion in one place tends to be compensated by deposition in another, so that little net difference in channels results
	C	Relatively slow deposition; tidal sedimentation restricted mainly to low-marsh areas; extremely slow rates of deposition in the high marsh, except where barrier washovers occur or torrential rains sweep sands off adjacent Pleistocene or Holocene barrier island remnants; where supplies of washover or terrestrial sediments are not readily available and tidal hydraulics prevent significant deposition of clays and silts, the transition from mature to old-age marshes may be exceedingly slow
	D	Stratigraphic records are variable; those in low-marsh areas are similar to ones from youthful marshes, whereas ones in high-marsh areas may depict the succession from lowest to highest marsh, including numerous channel migrations and fills
Old age	A	Substantially more than 50% of area consists of high marsh; in late old age, virtually all of the area may consist of high marsh, including the encroachment of quasiterrestrial or terrestrial vegetation upon the marsh surface; in early and middle old age the zonation consists of concentric bands of short *S. alterniflora, Salicornia* spp., *Distichlis spicata*, and *Juncus roemerianus*, followed by or admixed with such plants as *Sporobolus, Borrichia*, and *Batis*; taller forms of *S. alterniflora* are restricted to the few drainage channels remaining
	B	Drainage mostly by surface runoff; most tidal channels are filled, and the marsh substrate is more or less uniform in elevation; aeolian processes are correspondingly more important in the distribution and reworking of sediments
	C	Extremely slow rates of deposition; tidal processes are largely ineffectual, and the transition to a terrestrial environment depends upon the availability of terrestrial sediments and mechanisms for their dispersal

Table 4.13—2 (continued)
IDEALIZED STAGES IN MARSH MATURATION, BARRIER ISLANDS OF GEORGIA[a]

Stage	Characteristics
D	Stratigraphic records should show complete sequences, from underlying subtidal or low intertidal sediments, through the earliest marsh stages, to the oldest marsh or quasiterrestrial environments; old marshes probably are not obtained without the deposition of washover or terrestrial sands; hence, the final record would be a coarsening-upward sequence

[a] This model, which requires considerable testing, applies only to the more seaward marshes in Georgia, such as those associated with dissected barrier islands. More landward marshes, particularly those adjoining freshwater drainages, differ not only with respect to discrete modes of growth, but also by plant zonations. In brackish marshes, for example, *S. alterniflora* may be largely or totally replaced by *S. cynosuroides,* and *J. roemerianus* may occupy most of the marsh area.

From Frey, R. W. and Basan, P. B., *Coastal Sedimentary Environments,* 2nd ed., Davis, R. A., Ed., Springer-Verlag, New York, 1985, 225. With permission.

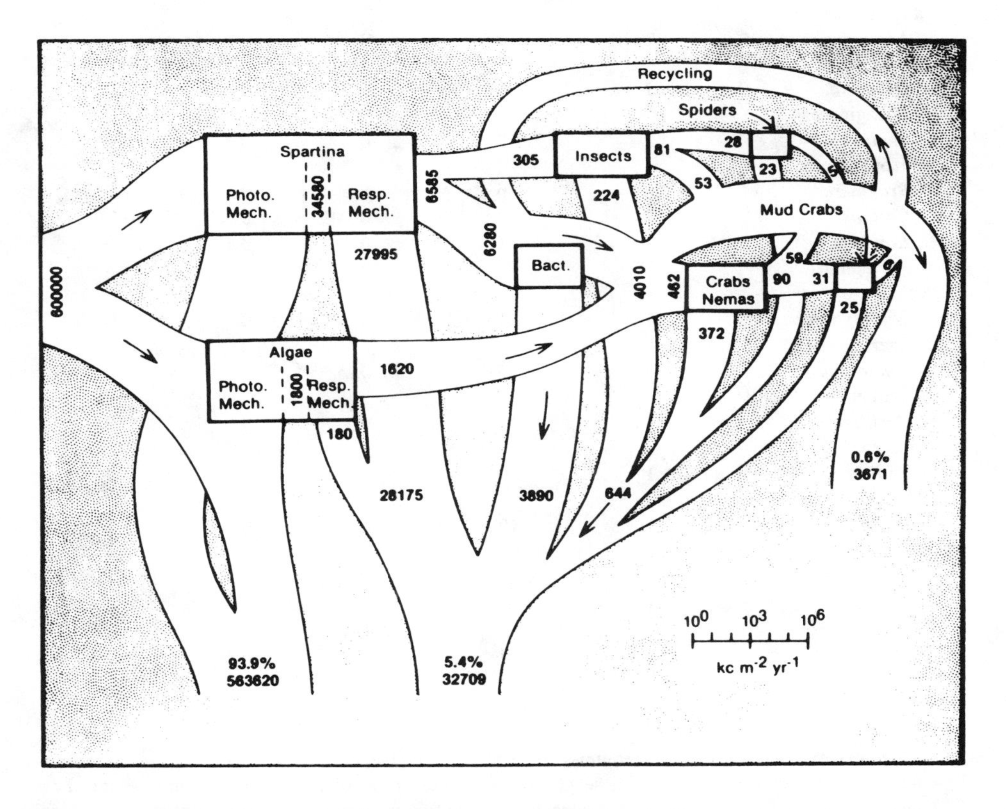

FIGURE 4.13—6. Energy flow in Georgia salt marshes depicted in a box model. The two outward flows at the bottom delineate heat losses. The outward flow at the right shows net secondary production exported from the marsh. (From Montague, C. L., Bunker, S. M., Haines, E. B., Pace, M. L., and Wetzel, R. L., *The Ecology of a Salt Marsh, Ecological Studies,* Vol. 38, Pomeroy, L. R. and Wiegert, R. G., Eds., Springer-Verlag, New York, 1981, 69. With permission. [Adapted from Teal, J. M., *Ecology,* 43, 614, 1962.])

Table 4.13—3
A LIST OF FAMILIES AND GENERA OF MANGROVES

Families and genera	Number of species	Indian Ocean West Pacific	Pacific America	Atlantic America	West Africa
Dicots					
Avicenniaceae					
Avicennia	11	6	3	2	1
Bombacaceae					
Camptostemon	2	2	0	0	0
Chenopodiaceae					
Suaeda[a]	2	0	0	1	1
Combretaceae					
Conocarpus[b]	1	0	1	1	1
Languncularia	1	0	1	1	1
Lumnitzera	2	2	0	0	0
Euphorbiaceae					
Exoecaria[b]	1	1	0	0	0
Leguminosae					
Machaerium[b]	1	0	1	1	1
Meliaceae					
Xylocarpus	10	8	?	2	1
Myrsinaceae					
Aegiceras	2	2	0	0	0
Myrtaceae					
Osbornia	1	1	0	0	0
Plumbaginaceae					
Aegiatilis	2	2	0	0	0
Rhizophoraceae					
Bruguiera	6	6	0	0	0
Ceriops	2	2	0	0	0
Kandelia	1	1	0	0	0
Rhizophora	7	5	2	3	3
Rubiaceae					
Scyphiphora	1	1	0	0	0
Sonneratiaceae					
Sonneratia	5	5	0	0	0
Sterculiaceae					
Heritiera[b]	2	2	0	0	0
Theaceae					
Pellicera	1	0	1	0	0
Tiliaceae					
Brownlowia[b]	17	17	0	0	0
Monocots					
Arecaceae					
Nypha[a]	1	1	0	0	0
Pandanaceae					
Pandanus[b]	1	1	0	0	0
Total	80	65	9	11	9

[a] *Suaeda* typically is a small- to medium-sized bush, but can become a small tree.

[b] At least some of the species in these genera are more typical of freshwater swamps behind the mangrove coastal swamp and are not considered to be true mangroves by some botanists.

From Dawes, C. J., *Marine Botany*, John Wiley & Sons, New York, 1981. With permission.

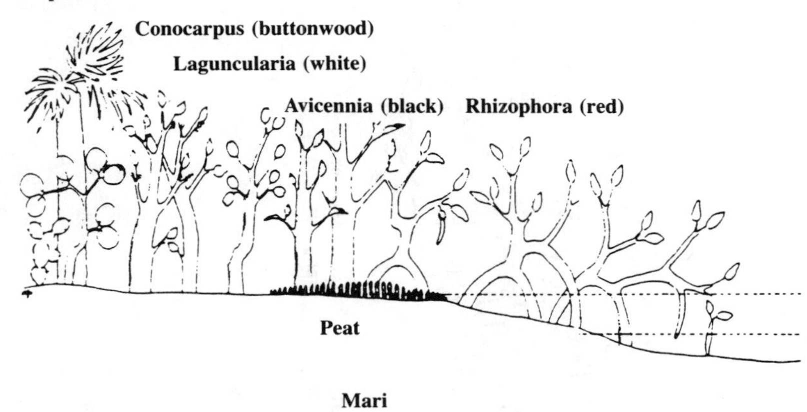

FIGURE 4.13—7. Mangrove swamp in southern Florida showing a typical zonation pattern. Red mangroves form the outer fringe in shallow subtidal waters and are replaced landward by black mangroves, white mangroves, buttonwoods, and a tropical rain forest. (From Dawes, C. J., *Marine Botany,* John Wiley & Sons, New York, 1981. Reproduced by permission of copyright © owner, John Wiley & Sons, Inc.)

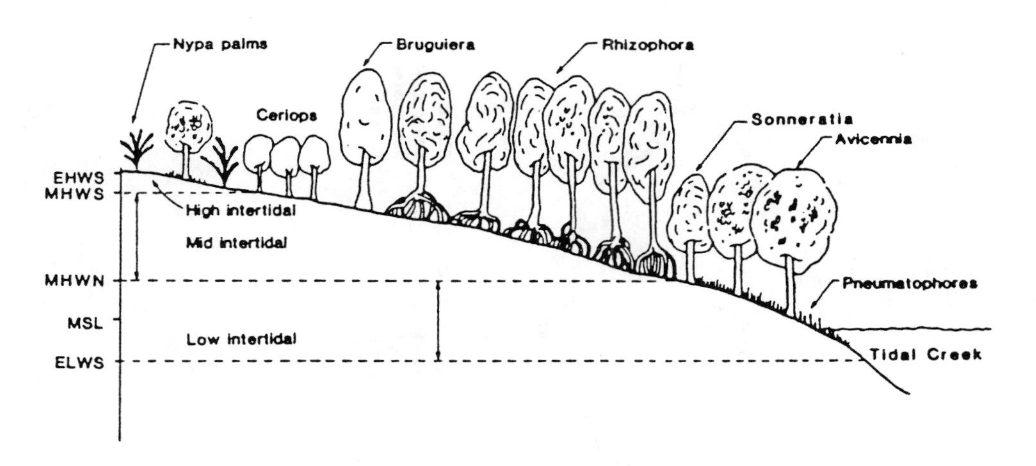

FIGURE 4.13—8. Diagram of an idealized mangrove forest showing the three intertidal zones. EHWS, extreme high water spring; MHWS, mean high water spring; MHWN, mean high water neap; MSL, mean sea level; ELWS, extreme low water spring. (From Alongi, D. M., The role of soft-bottom benthic communities in tropical mangrove and coral reef ecosystems, *Rev. Aquat. Sci.,* 1, 243, 1989.)

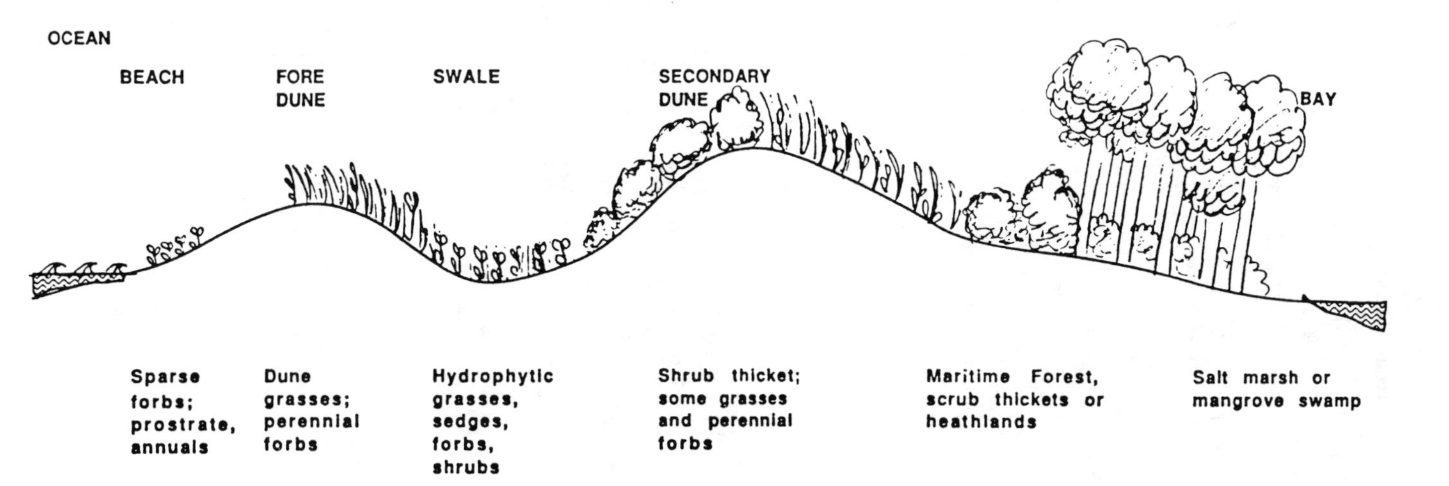

FIGURE 4.13—9. Typical array of plant communities on a transect across a hypothetical barrier island. (From Ehrenfeld, J. G., Dynamics and processes of barrier island vegetation, *Rev. Aquat. Sci.*, 2, 437, 1990.)

4.14 DISSOLVED AND PARTICULATE ORGANIC CARBON (DOC; POC)

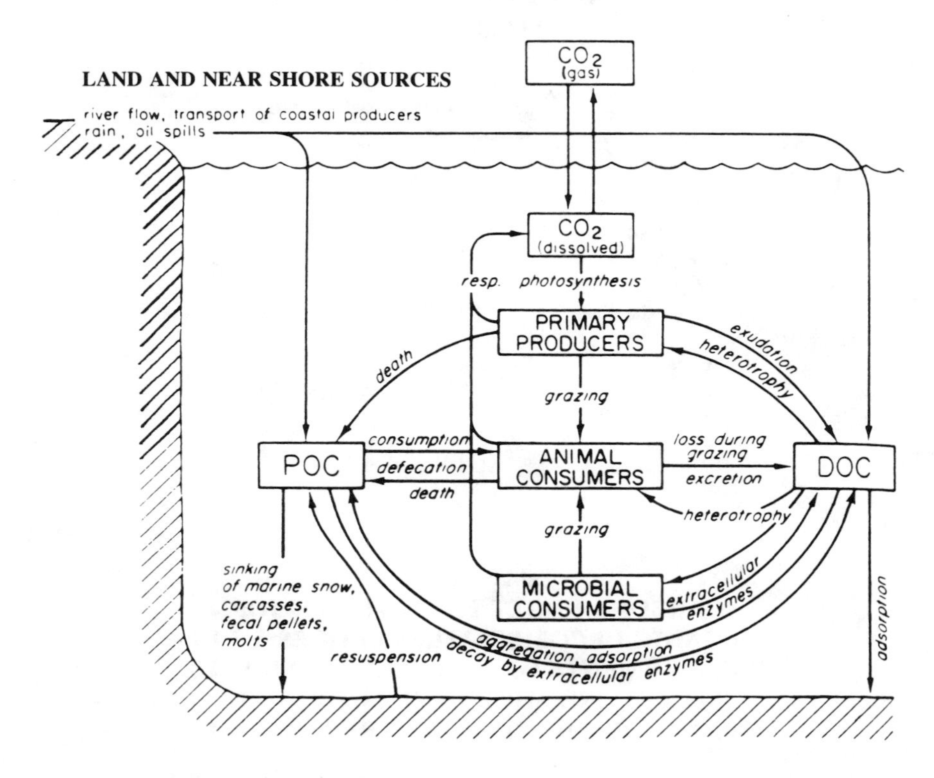

FIGURE 4.14—1. Carbon transfer in aerobic marine environments. The boxes represent pools and the arrows, processes. The inorganic parts of the cycle are simplified. Organic aggregates and debris comprise marine snow. Some resuspension of dissolved organic carbon from sediments into the overlying water is not shown in the diagram. (From Valiela, I., *Marine Ecological Processes,* Springer-Verlag, New York, 1984. With permission.)

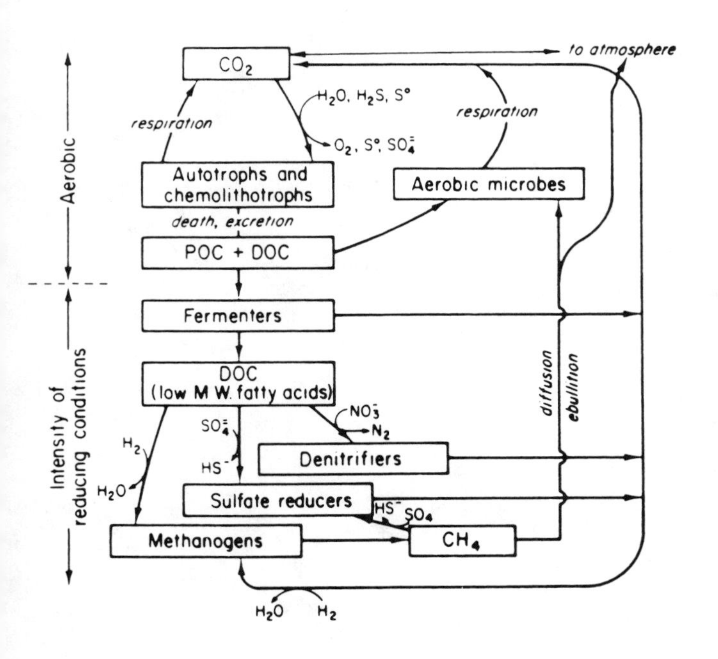

FIGURE 4.14—2. Carbon transformations in the transition from aerobic to anaerobic situations. The gradient from aerobic to anaerobic can be thought of as representing a sediment profile, with increased reduction and different microbial processes deeper in the sediment. Boxes represent pools or operations that carry out processes; arrows are processes that can be biochemical transformations or physical transport. Elements other than carbon are shown, where relevent, to indicate the couplings to other nutrient cycles. Some arrows indicate oxidizing and some reducing pathways. (From Valiela, I., *Marine Ecological Processes,* Springer-Verlag, New York, 1984. With permission.)

Table 4.14—1
ORGANIC MATTER ACCUMULATION IN *SPARTINA* MARSHES

Area	TNPP accumulating in sediments (%)	Sedimentation rate (mm year^{-1})
Massachusetts	5.3	1
New York	37	2–6.3
New York	0	−9.5–37
North Carolina	75	>1.2
South Carolina	0	1.3

From Dame, R. F., The importance of *Spartina alterniflora* to Atlantic Coast estuaries, *Rev. Aquat. Sci.*, 1, 639, 1989.

Table 4.14—2
CARBON BUDGET FROM THE MAJOR ORGANIC DECOMPOSITION PROCESSES IN THE SEDIMENTS OF A NEW ENGLAND SALT MARSH AND SHORT BELOWGROUND *SPARTINA* PRODUCTION AT THE SAME SITE

Process	Carbon fluxes (g C m^{-2} year^{-1})	BNPP (%)
Decomposition		
Aerobic respiration	361	31.7
Fermentation + sulfate reduction	432	37.9
Nitrate reduction	5	0.4
Methanogenesis	6	0.5
Burial	89	7.8
Export (as DOC)	36	3.2
Total losses	929	
Below-ground production	1140	

Note: Decomposition was measured in terms of carbon dioxide.

From Dame, R. F., The importance of *Spartina alterniflora* to Atlantic Coast estuaries, *Rev. Aquat. Sci.*, 1, 639, 1989. With permission.

Table 4.14—3
FLUX ESTIMATES FROM A NUMBER OF MARSH-ESTUARINE SYSTEMS DOMINATED BY *SPARTINA ALTERNIFLORA*

System	Type	POC	DOC	TOC	NH_4	NN	DON	PN	TN	PO_4	TP
Crommet Creek	I				2.1	0.3				0.6	3.2
Flax Pond	I	68	−8	53	−2.0	1.0				−1.4	−0.3
Canary Creek	I	−62	−38	−100	0.7	1.9	−0.9	−2.9	−1.2	−0.1	
Bly Creek	I	38	−260	−222	−0.8	0.3	−14.6	2.8	−12.5	−0.2	−0.7
Sippewisset	II	−76			−4.2	−3.9	−9.8	−6.7	−24.6	−0.6	
Gott's Marsh	II	−7			−0.4	−0.9	−2.1	−0.3	−3.7		−0.3
Rhode River	II			31		−3800			−8400		−2200
Ware Creek	II	−35	−80	−115	−2.9	2.2	−2.3		−2.8	−0.1	0.7
North Inlet	II	−128	−328	−456	−6.3	−0.9			−42.7		
Dill Creek	II	−303									−6.4
Sapelo Island	II	−208	−108	−316							
Carter Creek	III	−116	−25	−141	−0.3	0.3	−9.2	4.6	−4.0	−0.6	

Note: All values are in grams per square meter per year of marsh.

From Dame, R. F., The importance of *Spartina alterniflora* to Atlantic Coast estuaries, *Rev. Aquat. Sci.*, 1, 639, 1989.

4.15 TIDAL FRESHWATER MARSHES

Table 4.15—1
COMMON SPECIES OF VASCULAR PLANTS OCCURRING IN THE TIDAL FRESHWATER HABITAT

Species	General characteristics	Habitat preference	Salinity tolerance	Associated species
Acorus calamus (sweet-flag)	Grows in dense colonies propagating mainly by rhizome; stemless plants up to 1.5 m with stiff, narrow basal leaves; cylindrical inflorescence emerges from side of stem (open spadix); aromatic	Shallow water or wet soil; channel margins	Fresh	*Peltandra virginica* *Polygonum* spp. *Impatiens capensis*
Alternanthera philoxeroides (alligatorweed)	Hollow stems with simple branches bearing opposite, lance-shaped leaves; forms dense mats; flowers on long panicles; perennial	Extremely adaptable; often emersed	Fresh to oligohaline	—
Amaranthus cannabinus (water-hemp)	Erect, fleshy and stout; up to 2 m; leaves lanceolate with blades as long as 20 cm; not conspicuous until midsummer when it towers above other marsh forbs	Common to levee sections of the tidal marsh habitat; tolerates periodic inundation	Fresh to mesohaline	*Peltandra virginica* *Polygonum* spp. *Bidens* spp.
Asclepias incarnata (swamp milkweed)	Tall, leafy, pink-flowered herb growing solitary or in small, loose groups; lance-shaped, opposite leaves; reproduces via seeds or rhizomes	Cosmopolitan; grows in many wetland situations; high marsh species	Fresh to oligohaline	High marsh herbs
Bidens coronata B. laevis (burmarigold)	Annual plants up to 1.5 m tall, solitary or in small scattered groups; loosely branched above with opposite leaves; leaf shape variable but generally toothed or lanceolate; impressive yellow bloom late in the growing season	Cosmopolitan, growing in the upper two thirds of the intertidal zone on wet mud or in shallow water	Fresh	*Polygonum* spp. *Amaranthus cannabinus* Other *Bidens* spp.
Calamagrostis canadensis (reed-bentgrass)	Slender grass up to 1.5 m, generally forming dense colonies; long, flat leaves; loose, ovoid panicle with purplish color; perennial	Wet meadows and thickets	Fresh?	*Typha* spp. *Acorus calamus*
Carex spp. (sedges)	Grasslike sedges, culms mostly three-angled, bearing several leaves with rough margins; up to 2 m tall and usually in groups; perennial from long, stout rhizomes	Low areas with frequent flooding or damp soil	Fresh	—

Cephalanthus occidentalis (buttonbush)	Branched shrub up to 1.5 m tall with leathery smooth opposite leaves and white flowers crowded into dense, spherical, stalked heads; flowers June through August; leaf petioles reddish	Upland margins and raised hummocks of tidal freshwater marshes; wet soil	Fresh to oligohaline	*Hibiscus* spp. *Cornus amommum*
Echinochloa walteri (water's millet)	Grass up to 2 m, solitary or in small groups; long, moderately wide leaf blades; flowers in a terminal panicle which is ovoid; greenish purple, and appears in July/August	Shallow water; moist areas, disturbed sites	Fresh to oligohaline	—
Hibiscus spp. *Kosteletzkya virginica* (mallows)	Shrubform herbs up to 2 m, scattered or in large colonies; leaves wedge-shaped or rounded and alternate; large, showy pink or white flowers appearing in midsummer; perennial	Freshwater marshes or the upland margin of saline marshes with freshwater seepage	Fresh to mesohaline	*Typha* spp. *Spartina cynosuroides* *Polygonum* spp. *Impatiens capensis*
Eleocharis palustris E. quadrangulata (spikerushes)	Perennials with horizontal rootstocks; culms stout, slender, and cylindrical or squarish with a basal sheath; flowers crowded onto terminus of spikelet; between 0.5 and 1.5 m	Channel margins or stream banks in shallow water; muddy, organic substrates	Fresh to oligohaline	*Pontederia cordata* *Scirpus* spp. *Juncus* spp. *Leersia oryzoides*
Impatiens capensis (jewelweed)	Annual plants up to 2 m with succulent, branched stems with swelling at the joints; colonial; leaves alternate and ovate or elliptic with toothed margins; flowers orange and funnel-like, appearing in July/August	Same as *Bidens* spp.; also grows in shaded portions of marshes	Fresh	*Bidens* spp. *Typha* spp. *Polygonum* spp.
Iris versicolor (blue flag)	Flat, swordlike leaves arising from a stout creeping rhizome; large, purplish-blue flowers emerge in spring from a stiff upright stem; perennial	High, shaded portions of the intertidal zone in damp soil; will not tolerate long inundations	Fresh	None in particular
Leersia oryzoides (rice cutgrass)	Weak slender grass growing in dense, matted colonies; leaf sheaths and blades very rough; emerges from creeping rhizomes and often sprawls on other vegetation	Midintertidal zones of marshes; high diversity vegetation patches	Fresh to oligohaline	Many; none in particular

Table 4.15—1 (continued)
COMMON SPECIES OF VASCULAR PLANTS OCCURRING IN THE TIDAL FRESHWATER HABITAT

Species	General characteristics	Habitat preference	Salinity tolerance	Associated species
Lythrum salicaria Decodon verticillatus (loosestrife)	Shrubform herbs forming large, dense colonies; aggressive; up to 1.5 m in height with lanceolate leaves opposite or whorled; upper axils branched with small purplish-pink flowers; terminal spikes pubescent; annual	Moist portions of marshes; high intertidal or upland areas	Fresh to oligohaline	*Hibiscus* spp. *Convolvulus* spp.
Mikania scandens (climbing hempweed)	Long, herbaceous vine forming matted tangles over other emergent plants; heart-shaped leaves; dense, pinkish flower clusters; slender stem; propagates by both seed and rhizome; perennial	Open, wooded swamps and marshes; shrub thickets	Fresh to oligohaline	—
Myrica cerifera (wax-myrtle)	Compact, tall, evergreen shrub with leathery alternate leaves; spicy aroma; waxy, berry-like fruits; forms extensive thickets	Most all coastal habitats; border between intertidal zone and uplands	Fresh	*Acer rubrum* *Nyssa* spp. *Taxodium distichum*
Nuphar lutecum (*N. advena*) (spatterdock)	Plant with floating or emergent leaves and flowers attached to flexible underwater stalks; rises from thick rhizomes imbedded in benthic muds; flowers deep yellow, appearing throughout the summer	Constantly submerged areas up to 1.5 m depth, or, if tidal, near or below mean low water in deep organic muds	Fresh	Usually in pure stand
Nyssa sylvantica *N. aquatica* (gum)	Medium-sized tree (10 m) with numerous horizontal, crooked branches; leaves crowded at twig ends turning scarlet in fall; flowers appear in April/May	Marsh/upland borders	Fresh	*Acer rubrum* *Myrica cerifera* *Alnus* spp.
Panicum virgatum (switchgrass)	Perennial grass 1—2 m in height in large bunches with partially woody stems; nest of hairs where leaf blade attaches to sheath; large, open, delicately branched seed head produced in late summer; rhizomatous	Dry to moist sandy soils or the midintertidal portions of tidal freshwater marshes; disturbed areas	Fresh to mesohaline	*Hibiscus* spp. *Scirpus* spp. *Eleocharis palustris*

Peltandra virginica (arrow-arum)	Stemless plants, 1—1.5 m tall, growing in loose colonies; several arrowhead-shaped leaves on long stalks; emerge in rather dense clumps from a thick subsurface tuber; flowers from May to June	Grows predominantly as an emergent on stream margins or intertidal marsh zones on rich, loose silt	Fresh to oligohaline	*Pontederia cordata* *Zizania aquatica* Many other species
Phragmites australis (common reed)	Tall, coarse grass with a feathery seed head; 1—4 m in height; grows aggressively from long, creeping rhizomes; perennial; flowers from July to September	Extremely cosmopolitan, growing in tidal and nontidal marshes and often associated with disturbed areas	Fresh to mesohaline	*Spartina cynosuroides* *Zizania aquatica*
Polygonum arifolium *P. sagittatum* (tearthumbs)	Plants with long, weak stems up to 2 m tall, usually leaning on other vegetation; leaves sagitate in shape and alternate; leaf midribs and stems armed with recurved barbs; flowers small and appearing in late summer; annual	Shallow water or damp soil; middle to upper intertidal zone	Fresh to oligohaline	*Bidens* spp. *Hibiscus* spp. *Impatiens capensis*
Polygonum punctatum *P. densiflorum* *P. hydropiperoides* (smartweeds)	Upright plants growing from a fibrous tuft of roots; narrowly to widely lanceolate leaves with stalks basally enclosed within a membranous sheath; up to 1 m; flowers at spike at end of stalk	Upper three quarters of intertidal zone in freshwater marshes on wet or damp soil	Fresh to oligohaline	Many species
Pontederia cordata (pickerelweed)	Rhizomatous perennial growing in dense or loose colonies; plants up to 2 m tall; fleshy, heart-shaped leaves with parallel veins and emerging from spongy stalks; flowers dark violet-blue, appearing June to August	Lower intertidal zone of tidal freshwater marshes	Fresh to oligohaline	*Nuphar luteum* *Peltandra virginica* *Sagittaria latifolia*
Rosa palustris (swamp rose)	Shrub up to 2 m growing in loose colonies; stems lack prickles except for those occurring at bases of leaf stalks; pinnately compound leaves with fine serrate margins; showy, pink flowers appearing July/August	High intertidal zones or wet meadows	Fresh to oligohaline	*Cephalanthus occidentalis*
Rumex verticillatus (water dock)	Erect, robust annual with dark-green, lance-shaped leaves; stem swollen at nodes; attains heights over 1.5 m and grows solitary or in loose colonies; flower head is	Wet meadows or pond margins on mud or in shallow water	Fresh to oligohaline	—

Table 4.15—1 (continued)
COMMON SPECIES OF VASCULAR PLANTS OCCURRING IN THE TIDAL FRESHWATER HABITAT

Species	General characteristics	Habitat preference	Salinity tolerance	Associated species
	evident in late spring and can be 50 cm in length			
Sagittaria latifolia (duck-potato) *S. falcata* (bultongue)	Perennial herbs; stemless, up to 2 m in height and emerging from fibrous tubers; leaves arrowhead-shaped or lanceolate with white flowers in whorls appearing on a naked stalk in July/August	Borders of rivers or marshes in low intertidal zones on organic, silty mud	Fresh to oligohaline	*Peltandra virginica* *Pontederia cordata*
Scirpus validus (soft-stem bulrush) *S. cyperinus* (woolgrass) *S. americanus* (common three square)	Medium to large rushes with cylindrical or triangular stems; inconspicuous leaf sheaths; usually grow in small groups; bear seed clusters on end or side of stem; perennial	Brackish to fresh shallow water or low to middle intertidal zones on organic clay substrates	Fresh to mesohaline	Other rushes *Typha* spp.
Sparganium eurycarpum (great burread)	Stout upright forbs up to 1 m with limp, underwater, emergent leaves attached basally and alternating up the stem; toward the terminus, stems zig-zag bearing sphere-like clusters of pistillate and staminate flowers	Partially submerged, shallow water marsh areas; lower to middle intertidal zones	Fresh	*Zizania aquatica* *Leersia oryzoides* *Polygonum* spp.
Spartina cynosuroides (big cordgrass)	Perennial grass attaining heights in excess of 3 m, having long, tapering leaves and growing from vigorous underground rhizomes; found in dense monospecific or mixed stands	Channel and creek margins in tidal oligohaline marshes	Fresh to mesohaline	*Phragmites australis* *Typha* spp.
Taxodium distichum (bald cypress)	Tall tree with straight trunk (40 m), conifer-like but deciduous; light porous wood covered by stringy bark; unbranched shoots originating from roots as knees	Marsh/upland borders	Fresh?	*Nyssa* spp. *Acer rubrum*

Typha latifolia *T. domingensis* *T. angustifolia* (cattails)	Stout, upright reeds up to 3 m forming dense colonies; basal leaves, long and sword-like, appearing before stems; yellowish male flower disintegrates leaving a thick, velvety-brown swelling on the spike; rhizomatous; perennial	Very cosmopolitan, occurring in shallow water or upper intertidal zones; some disturbed areas	Fresh to mesohaline	Many associates
Zizania aquatica (wild rice)	Annual or perennial aquatic grass, 1—4 m tall, usually found in colonies; short underground roots, stiff hollow stalk, and long, flat, wide leaves with rough edges; male and female flowers separate along a large terminal panicle in late summer	Fresh to slightly brackish marshes and slow streams, usually in shallow water; requires soft mud and slowly circulating water	Fresh to oligohaline	*Peltandra virginica* Many other species
Zizaniopsis mileacea (giant cutgrass)	Perennial by creeping rhizome; culms 1—4 m high; long, rough-edged leaves geniculate at lower nodes; large, loose terminal panicles appearing in midsummer; aggressive	Swamps and margins of tidal streams	Fresh?	—

From Odum, W. E., Smith, T. J., III, Hoover, J. K., and McIvor, C. C., The Ecology of Tidal Freshwater Marshes of the United States East Coast: A Community Profile, FWS/OBS-83/17, U.S. Fish and Wildlife Service, Washington, D.C., 1984.

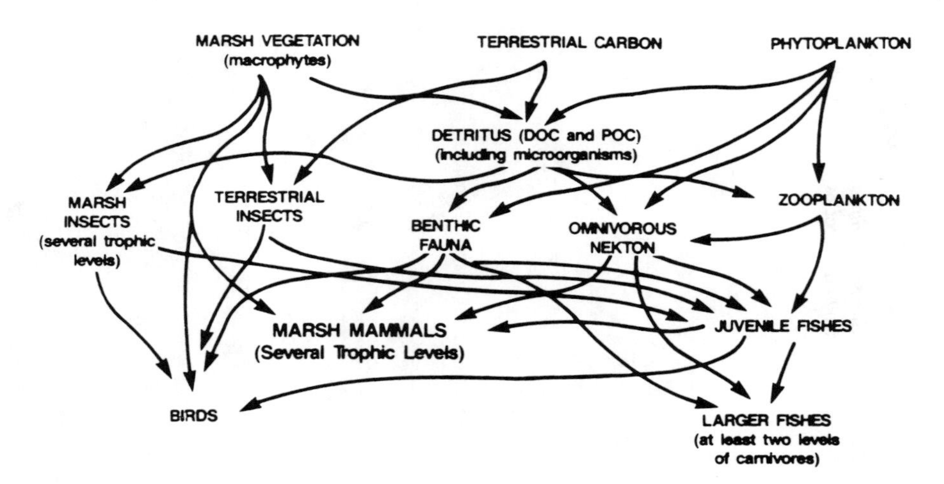

FIGURE 4.15—1. Hypothetical energy flow diagram for a tidal freshwater marsh. (From Odum, W. E., Smith, T. J., III, Hoover, J. K., and McIvor, C. C., The Ecology of Tidal Freshwater Marshes of the United States East Coast: A Community Profile, FWS/OBS-83/17, U.S. Fish and Wildlife Service, 1984. With permission.)

Table 4.15—2
REPRESENTATIVE BENTHIC MACROINVERTEBRATES OF TIDAL FRESHWATER MARSHES OF MID-ATLANTIC REGION (U.S.)

Sponges
- *Spongilla lacustris* and other species

Hydra
- *Hydra americana*
- *Protohydra* spp.

Bryozoans
- *Barentsia gracilus*
- *Lophopodella* sp.
- *Pectinatella magnitica*

Leeches
- Families Glossiphoniidae, Piscicolidae

Oligochaetes
- Families Tubificidae, Naididae

Insects
- Dipteran larvae (especially family Chironomidae)
- Larvae of Ephemeroptera, Odonata, Trichoptera, and Coleoptera

Amphipods
- *Hyallela azteca*
- *Gammarus fasciatus*
- *Lepidactylus dytiscus* (southeastern states)

Crustaceans
- Crayfish
- Blue crab, *Callinectes sapidus*
- Caridean shrimp, *Palaemonetes paladosus*

Mollusks
- Fingernail clam, *Pisidium* spp.
- Asiatic clam, *Corbicula fluminea* (formerly *C. manilensis*)
- Brackishwater clam, *Rangia cuneata*
- Pulmonate snails (at least six families)

From Odum, W. E., Smith, T. J., III, Hoover, J. K., and McIvor, C. C., The Ecology of Tidal Freshwater Marshes of the United States East Coast: A Community Profile, FWS/OBS-83/17, U.S. Fish and Wildlife Service, Washington, D.C., 1984.

4.16 FISHES

Table 4.16—1
ABUNDANCES OF FISHES COLLECTED AT THREE STATIONS IN THE LOWER CHESAPEAKE BAY[a]

Species	Station 1	Station 2	Station 3	Total
Leiostomus xanthurus	3,925	3,666	7,683	15,274
Syngnathus fuscus	1,489	1,024	795	3,308
Anchoa mitchilli	493	571	949	2,013
Bairdiella chrysoura	175	393	721	1,289
Menidia menidia	95	317	65	475
Syngnathus floridae	121	93	79	293
Apeltes quadracus		8	236	244
Hypsoblennius hentzi	30	51	105	186
Fundulus heteroclitus	8	113	5	126
Orthropristis chrysoptera	12	7	104	123
Tautoga onitis	29	19	51	99
Lagodon rhomboides	34	44	18	96
Paralichthys dentatus	35	20	33	88
Centropristis striata	11	15	57	83
Chasmodes bosquianus	19	32	23	74
Gobiosoma bosci	18	16	11	45
Cyprinodon variegatus		44		44
Morone saxatilis	13	8	21	42
Pseudopleuronectes americanus	27	7	2	36
Micropogonias undulatus	20	8	5	33
Opsanus tau	1	15	15	31
Cynoscion nebulosus	6	9	10	25
Anguilla rostrata	6	15		21
Sphoeroides maculatus	1	2	15	18
Mycteroperca microlepis		2	13	15
Lucania parva		14		14
Gobiesox strumosus	3	6	3	12
Urophycis regius	3	7		10
Sciaenops ocellata		8	1	9
Trinectes maculatus	5		3	8
Chilomycterus schoepfi	1	2	4	7
Menidia beryllina		6		6
Synodus foetens		4	2	6
Hippocampus erectus	2	3		5
Brevoortia tyrannus	2		1	3
Pomatomus saltatrix		3		3
Prionotus carolinus	3			3
Monacanthus hispidus			3	3
Alosa pseudoharengus		2		2
Chaetodon ocellatus	1		1	2
Aluterus schoepfi			1	1
Fistularia tabacaria			1	1
Fundulus majalis		1		1
Peprilus paru	1			1
P. triacanthus			1	1
Scophthalmus aquosus			1	1
Stenotomus chrysops			1	1
Rissola marginata			1	1
Total number of species	31	36	37	48
Total number of individuals	6,589	6,555	11,040	24,182

[a] Data collected from unvegetated areas and night trawls not included.

From Orth, R. J. and Heck, K. L., Jr., *Estuaries*, 3, 278, 1980. Reprinted with permission of Estuarine Research Federation, copyright 1980.

Table 4.16—2
RANK ABUNDANCE OF TEN MOST COMMON SPECIES IN TRAWL SAMPLES IN VICINITY OF CALVERT CLIFFS, MD[a,b]

Species	1969	1970	1971	1972	1973	1974	1975	1976	1977	1978	1979	1980	1981	Total
Anchoa mitchilli	1	1	1	1	1	1	2	2	2	1	1	1	1	16
Leiostomus xanthurus	2	2	2	2	2	2	1	1	1	2	2	2	2	23
Micropogonias undulatus	3	8	4	7	5	3	3	3	8	3	8	3	5	63
Morone americanus	—	6	3	4	3	5	7	5	7	—	6	7	—	86
Pseudopleuronectes americanus	10	—	10	5	—	9	6	4	4	5	3	6	3	87
Trinectes maculatus	4	5	6	3	4	7	5	—	9	10	4	—	8	87
Brevoortia tyrannus	—	—	—	—	8	9	4	7	5	4	5	3	4	93
Cynoscion regalis	6	9	7	6	—	6	—	9	3	7	7	—	6	99
Menidia menidia	5	7	5	10	7	—	—	6	—	—	—	—	—	117
Alosa aestivalis	—	10	—	—	6	—	—	6	—	—	—	5	7	122
A. pseudoharengus	7	4	—	—	10	—	—	—	—	6	—	9	—	124
Paralichthys dentatus	—	—	—	—	—	8	8	10	10	8	9	8	10	126
Morone saxatilis	—	3	8	9	9	—	—	—	—	—	—	—	—	128
Urophycis regius	—	—	—	—	—	10	9	8	—	9	—	—	—	135
Peprilus alepidotus	—	—	9	—	—	—	—	—	—	—	10	10	—	139
Bairdiella chrysoura	8	—	—	—	—	—	—	—	—	—	—	—	—	140
Anguilla rostrata	—	—	—	8	—	—	—	—	—	—	—	—	—	140
Apeltes quadracus	9	—	—	—	—	—	—	—	—	—	—	—	—	141
Anchoa hepsetus	—	—	—	—	—	—	—	—	—	—	—	—	9	141
Proportion of total CPUE in three most abundant species	89	94	88	88	98	96	99	93	95	96	97	97	97	

[a] Ranks are based on the average monthly catch-per-unit-effort.
[b] — equals rank greater than 10. Total of ranks computed with rank greater than 10 counted as 11.

From Horwitz, R. J., *Ecological Studies in the Middle Reach of Chesapeake Bay: Calvert Cliffs,* Heck, K. L., Jr., Ed., Springer-Verlag, Berlin, 1987, 167. With permission.

Table 4.16—3
FEEDING STRATEGIES OF REPRESENTATIVE FISHES OF BARNEGAT BAY, NJ

Species	Feeding strategy
Alosa aestivalis (blue herring)	Primarily planktivorous; consumes copepods, mysids, ostracods, small fishes
Ammodytes americanus (sand lance)	Principally feeds on small crustaceans
Anchoa mitchilli (bay anchovy)	Primarily planktivorous; consumes mysids, copepods, organic detritus, small fishes
Anguilla rostrata (American eel)	Omnivorous; consumes annelids, crustaceans, echinoderms, mollusks, eelgrass, small fishes
Bairdiella chrysoura (silver perch)	Omnivorous; consumes calanoid copepods, annelids, fishes, organic detritus
Brevoortia tyrannus (Atlantic menhaden)	Planktivorous; detritivorous; principally filter-feeds on phytoplankton; also feeds on zooplankton and organic detritus
Caranx hippos (crevalle jack)	Fish predator
Centropristis striata (black sea bass)	Omnivorous; benthophagous; feeds on crustaceans, mollusks, fishes, plants
Cynoscion regalis (weakfish)	Primarily a fish predator; mostly consumes anchovies and silversides; also feeds on small crustaceans
Fundulus heteroclitus (mummichog)	Omnivorous; consumes small animals and plants
Gasterosteus aculeatus (threespine stickleback)	Omnivorous; feeds on algae, copepods, fish eggs
Gobiosoma spp. (gobies)	Consume small invertebrates and fishes
Leiostomus xanthurus (spot)	Detritivore; benthophagous; consumes organic detritus and microbenthos
Menidia menidia (Atlantic silverside)	Omnivorous; feeds on small crustaceans, annelids, algae, insects
Morone americana (White perch)	Omnivorous; consumes annelids, crustaceans, fishes, organic detritus
M. saxatilis (striped bass)	Primarily a fish predator; also feeds on annelids, crustaceans, mollusks
Opsanus tau (oyster toadfish)	Omnivorous; principally consumes annelids, crustaceans, fishes
Paralichthys dentatus (summer flounder)	Predator; feeds on fishes, crustaceans, mollusks
Pomatomus saltatrix (bluefish)	Fish predator; consumes large numbers of Atlantic menhaden and bay anchovy
Pseudopleuronectes americanus (winter flounder)	Benthophagous; feeds on annelids, crustaceans, mollusks
Sphoeroides maculatus (northern puffer)	Primarily consumes small crustaceans
Stenotomus chrysops (scup)	Benthophagous; feeds on annelids, crustaceans, small fishes
Syngnathus sp. (pipefish)	Primarily feeds on amphipods and copepods
Tautog onitis (tautog)	Consumes crustaceans, mollusks, and other invertebrates

From Kennish, M. J. and Loveland, R. E., *Ecology of Barnegat Bay, New Jersey,* Kennish, M. J. and Lutz, R. A., Eds., Springer-Verlag, New York, 1984, 302. With permission.

Table 4.16—4
PRIMARY FOODS OF SOME ESTUARINE FISHES ALONG THE U.S. ATLANTIC AND GULF COASTS

Scientific name	Common name	Food habits
Megalops atlanticus	Tarpon	Juveniles consume copepods, ostracods, grass shrimp, and fish
Brevoortia tyrannus	Atlantic menhaden	Algae, planktonic crustacea
Harengula jaguana	Scaled sardine	Harpacticoid copepods, amphipods, mysids, isopods, and chironomid larvae
Anchoa hepsetus	Striped anchovy	Copepods, mysids, isopods, mollusks, fish, zooplankton
A. mitchilli	Bay anchovy	Zooplankton, fish, decapods, amphipods, mysids, detritus
Synodus foetens	Inshore lizardfish	Fish
Arius felis	Sea catfish	Amphipods, decapods, insects, mollusks, copepods, schizopods, isopods, hydroids
Opsanus tau	Oyster toadfish	Crustaceans, mollusks, polychaetes
Fundulus majalis	Striped killifish	Mollusks, crustaceans, insects, fish
F. pulvereus	Bayou killifish	Insects, isopods
F. similis	Longnose killifish	Harpacticoid copepods, ostracods, barnacle larvae, insects, isopods, amphipods
Cyprinodon variegatus	Sheepshead minnow	Plant detritus, small crustaceans, nematodes, diatoms, blue-green algae, filamentous algae, formas, insects
Adinia xenica	Diamond killifish	Plant detritus, filamentous algae, amphipods, insects, small copepods, diatoms
Lucania parva	Rainwater killifish	Insects, crustacean larvae, annelids, mysids, amphipods, cumaceans, copepods, plant detritus, small mollusks
Gambusia affinis	Mosquito fish	Amphipods, chironomids, insects, algae
Poecilia latipinna	Sailfin molly	Algae, diatoms, vascular plant detritus, inorganic matter
Menidia beryllina	Tidewater silverside	Isopods, amphipods, copepods, mysids, detritus, algae, insects, barnacle larvae
Membras martinica	Rough silverside	Copepods, barnacle larvae, amphipods, insects, shrimp fish
Morone saxatilis	Striped bass	Fish, crustaceans
Lutjanus griseus	Gray snapper	Fish, crustaceans
Diapterus plumieri	Striped mojarra	Mysids, amphipods, mollusks, ostracods, detritus, copepods
Eucinostomus gula	Silver jenny	Copepods, amphipods, mollusks, detritus, mysids
E. argenteus	Spotfin mojarra	Amphipods, copepods, mysids, mollusks, detritus
Archosargus probatocephalus	Sheepshead	Shrimp, mollusks, small fish, crabs, other crustaceans, algae, plant detritus
Lagodon rhomboides	Pinfish	Fish, crustaceans, vascular plants, algae, detritus, copepods, mysids, mollusks
Bairdiella chrysoura	Silver perch	Decapods, schizopods, copepods, mysids, amphipods, polychaetes, ectoprocts, fish, detritus
Cynoscion nebulosus	Spotted seatrout	Copepods, decapods, mysids, carideans, fish, mollusks
C. regalis	Weakfish	Polychaetes, copepods, amphipods, mysids, stomatopods, decapods, fishes
Leiostomus xanthurus	Spot	Polychaetes, copepods, isopods, amphipods, mysids, cumacea, fishes
Micropogonias undulatus	Atlantic croaker	Polychaetes, mollusks, amphipods, isopods, copepods, decapods, stomatopods, mysids, cumacea, ascidians, fish
Stellifer lanceolatus	Star drum	Amphipods, isopods, copepods, cumaceans, mysids, stomatopods, decapods, fish
Chaetodipterus faber	Atlantic spadefish	Small crustaceans, annelids, detritus, ctenophores
Mugil cephalus	Striped mullet	Algae, detritus, vascular plants, crustaceans, bacteria, diatoms

Table 4.16—4 (continued)
PRIMARY FOODS OF SOME ESTUARINE FISHES ALONG THE U.S. ATLANTIC AND GULF COASTS

Scientific name	Common name	Food habits
Gobiosoma robustum	Code goby	Amphipods, mysids, insect larvae, cladocerans, algae, detritus, mollusks
Ancylopsetta quadrocellata	Ocellated flounder	Mysids, copepods, other crustaceans, polychaetes, fish
Citharichthys spilopterus	Bay whiff	Mysids, other crustaceans, fish
Etropus crossotus	Fringed flounder	Polychaetes, mollusks, copepods, isopods
Scophthalmus aquosus	Windowpane	Mysids, other crustaceans, fish
Trinectes maculatus	Hogchoker	Annelids, algae, amphipods, detritus, foraminifera, plant seeds, copepods, insect larvae, mollusks, cumaceans
Symphurus plagiusa	Blackcheek tonguefish	Mollusks and crustaceans

From Stickney, R. R., *Estuarine Ecology of the Southeastern United States and Gulf of Mexico,* Texas A & M University, College Station, 1984. With permission.

Table 4.16—5
17 MOST ABUNDANT FISH SPECIES TAKEN IN TRAWL SAMPLES IN THE TERMINOS LAGOON (JULY 1976 TO MARCH 1979)

Species	Total number
Anchoa mitchilli	3895
Eucinostomus gula	3673
Arius melanopus	1734
Archosargus unimaculatus	1656
Sardinella macrophthalmus	1358
Bairdiella chrysoura	1307
Anchoa lamprotaenia	857
Cichlasoma urophthalmus	609
Sphoeroides testudineus	598
Arius felis	570
Diapterus rhombeus	534
Cetengraulis edentulus	311
Eucinostomus argenteus	279
Cytharichthys spilopterus	163
Opsanus beta	124
Chaetodipterus faber	121
Cynoscion nebulosus	115

From Yanez-Arancibia, A., Linares, F. A., and Day, J. W., Jr., *Estuarine Perspectives,* Kennedy, V. S., Ed., Academic Press, New York, 1980, 465. With permission.

Table 4.16—6
U.S. DOMESTIC COMMERCIAL FISH LANDINGS IN 1987, BY DISTANCE FROM SHORE OR OFF FOREIGN COASTS[a]

	Fishes	
Distance from shore	Weight	Dollar value
0 to 4.8 km	3431	897
4.8 to 320 km	5748	897
International waters (including foreign coasts)	604	276
Totals	9773	2070

[a] In millions of pounds and millions of dollars.

From Fisheries of the United States, 1987, Current Fishery Statistics 8700, U.S. Department of Commerce, NOAA, NMFS, May 1988.

Table 4.16—7
BIOMASS IN CERTAIN ECOSYSTEMS

Group	Area	g/m²
Birds	New Hampshire forest	0.04
Moose	Isle Royale, MI	0.7
Humans	U.S.	0.9
Fishes	Unpolluted rivers	1—5
	Georges Bank	1.6—7.4
	Atlantic salmon river, Matamek, Quebec	2.1—17.8
	Narragansett Bay	3.2
Large mammals	Central and East African grasslands	3.5—23.6
Fishes	Gulf of Mexico	5.6—31.6
	Flax Pond (Long Island) Estuary (annual average)	24.0
	California kelp bed	33.2—37.6
	Bermuda coral reef in summer	59.3
	Narrangansett Bay salt marsh embayment	69.2
Anchovy	Peruvian upwelling in autumn	216.7

From Haedrich, R. L. and Hall, C. A. S., *Oceanus*, 19, 55, 1976. With permission.

4.17 TROPHIC RELATIONSHIPS

Table 4.17—1
DETRITUS-GENERATED AND PHYTOPLANKTON-GENERATED FOOD WEBS IN ESTUARIES

Detritus-Generated Food Webs

Detritus — benthos (epifauna) — benthophagous fishes
Detritus — benthos (infauna) — benthophagous fishes (rays)
Detritus — benthos — benthophagous fishes — large fish predators (sharks)
Detritus — small benthos — larger invertebrates and small benthic fishes — large fishes
Detritus — large detritivorous fishes (mullet): "telescoping" of food chain
Detritus — benthos — large predators (rays)
Detritus — micronekton — intermediate predators (snappers, croaker)
Detritus — zooplankton — small fishes and invertebrates
Detritus — zooplankton — small fishes and invertebrates — larger fishes

Phytoplankton-Generated Food Webs

Phytoplankton — zooplankton — planktivorous, pelagic, and benthopelagic fishes
Phytoplankton — zooplankton — planktivorous fishes — large fish predators
Phytoplankton — phytoplanktonic fishes (menhaden): summer
Phytoplankton — zooplankton — menhaden: winter
Phytoplankton — zooplankton — large carnivores
Phytoplankton — (dinoflagellates) — mullet: alteration of usual feeding habits

Modified from de Sylva, D. P., *Estuarine Research,* Vol. 1, Cronin, L. E., Ed., Academic Press, New York, 1975, 420. With permission.

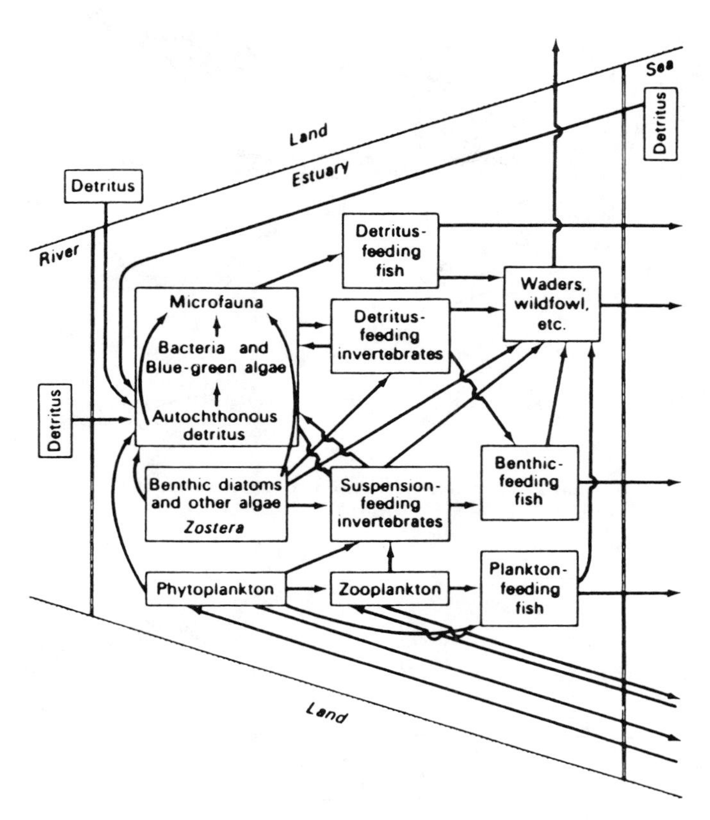

FIGURE 4.17—1. A simplified estuarine food web. (From Barnes, R. S. K., *Estuarine Biology,* Edward Arnold, London, 1974. With permission.)

Table 4.17—2
TROPHIC COMPOSITION OF TRAWL SAMPLES AT FOUR STATIONS IN MID-CHESAPEAKE BAY, 1969–1981; SOME SPECIES MAY COMPRISE MORE THAN ONE GROUP

	Proportion of annual CPUE[a]		CPUE[a] (catch per 30 min)	
Predominant food type(s)	Median	Range	Median	Range
Zooplankton-benthos-detritus	93.2	77.6—99.0	1121	469—3144
Zooplankton-benthos-detritus[b,c]	0.019	0.002—0.34	0.12	0.021—4.1
Zooplankton[b]	0.57	0.069—3.7	7.9	2.1—43.7
Benthos[c]	2.9	0.48—19.9	30.6	5.6—155.3
Polychaetes	2.4	0.36—18.5	25.5	4.2—144.1
Mysids	0.38	0.017—4.2	7.1	0.19—17.2
Mysids, mysids + fish	1.6	0.10—6.3	17.6	2.0—112.0
Variety of small benthos epifauna	0.30	0.026—3.2	3.6	0.30—38.8
Fish	0.60	0.95—4.2	11.4	2.1—106.3
Phytoplankton-detritus	0.28	0.002—5.4	4.5	0.028—63.6
Other	0.045	0.006—0.11	0.52	0.094—8.6

[a] Catch-per-unit effort.
[b] Excluding *Anchoa mitchilli.*
[c] Excluding *Leiostomus xanthurus* and *Micropogonias undulatus.*

From Horwitz, R. J., *Ecological Studies in the Middle Reach of Chesapeake Bay: Calvert Cliffs,* Heck, K. L., Jr., Ed., Springer-Verlag, Berlin, 1987, 167. With permission.

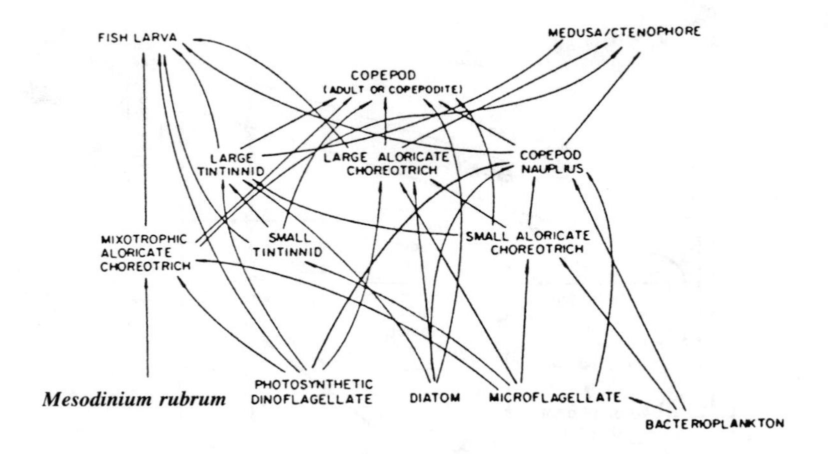

FIGURE 4.17—2. Trophic pathways involving marine ciliates. (From Pierce, R. W. and Turner, J. T., Ecology of planktonic ciliates in marine food webs, *Rev. Aquat. Sci.*, 6, 139, 1992.)

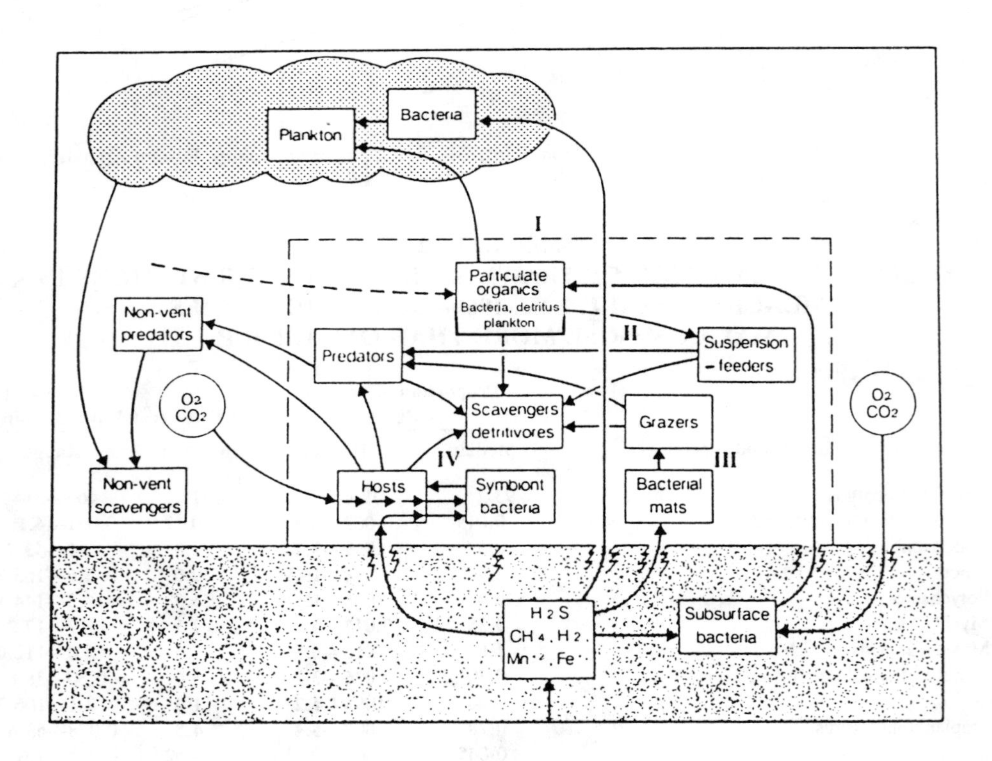

FIGURE 4.17—3. The trophic interactions of a hydrothermal vent community. Four distinct carbon-energy pathways are recognizable: I, immediate exit of either reduced compounds or chemosynthetic bacteria such that primary production is concentrated in the hydrothermal plume; II, the suspension-feeder loop that uses particulate organics from both *in situ* bacterial production and external input; III, the localized grazer-scavenger interaction dependent upon chemosynthetic bacterial mats; and IV, the greatest biomass generation in the form of symbiotic chemosynthetic microbes inside invertebrate hosts. (From Tunnicliffe, V., The biology of hydrothermal vents: ecology and evolution, *Oceanogr. Mar. Biol. Annu. Rev.*, 29, 319, 1991. With permission.)

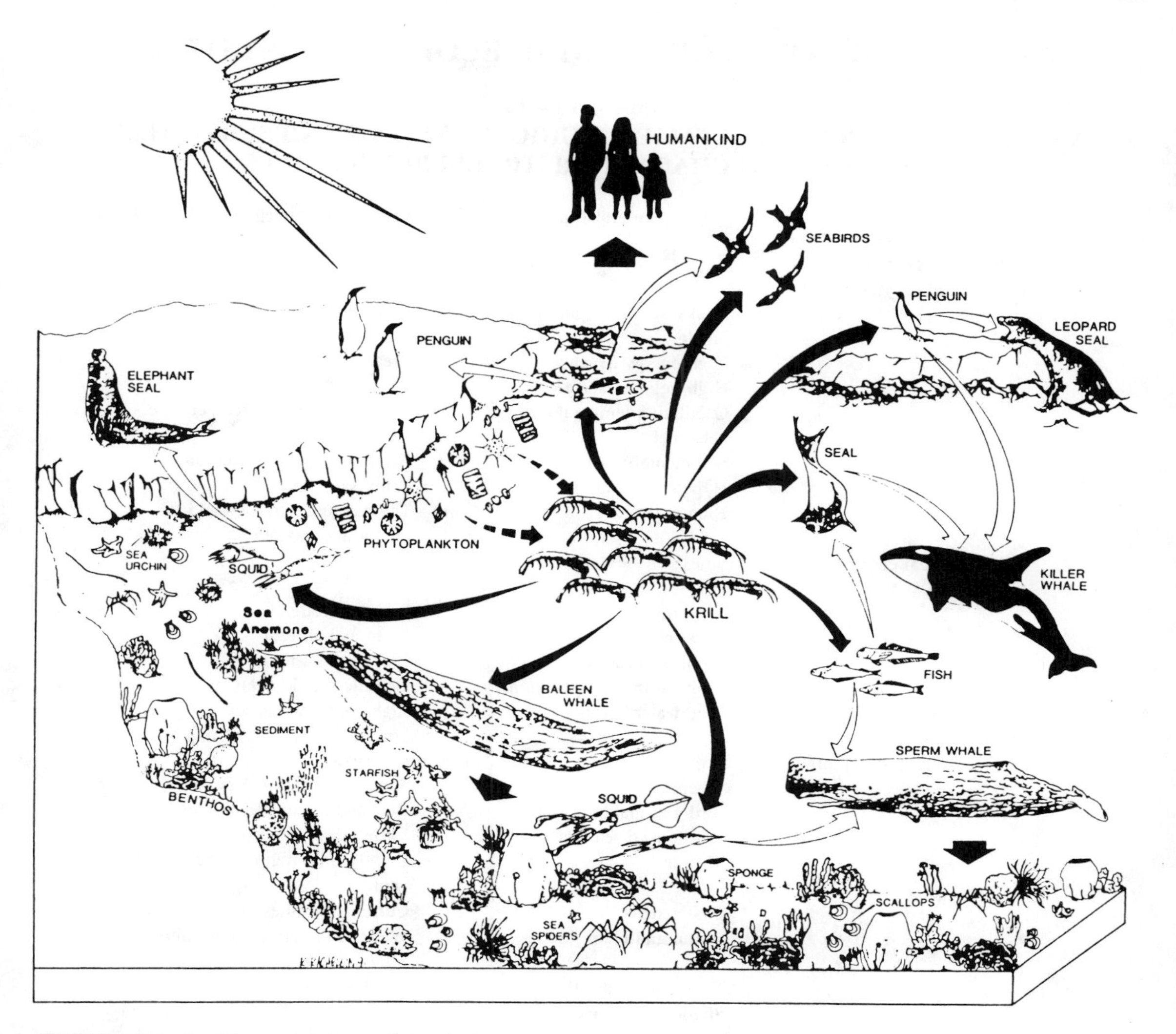

FIGURE 4.17—4. Diagram of the sea-ice, pelagic, and benthic components of the Antarctic marine ecosystem. Representative trophic interactions are shown: from the primary producers at the marginal ice zone to the krill (dashed open arrows), to the krill/zooplankton predators (solid arrows), to the top consumers (open arrows). Ultimately, the organic materials in the Antarctic marine ecosystem flow into the benthos or are consumed by man (large solid arrow). (From Berkman, P. A., The Antarctic marine ecosystem and humankind, *Rev. Aquat. Sci.*, 6, 295, 1992.)

4.18 CLASSIFICATION OF COASTAL ECOLOGICAL SYSTEMS

Table 4.18—1
A CLASSIFICATION OF COASTAL ECOLOGICAL SYSTEMS AND SUBSYSTEMS ACCORDING TO CHARACTERISTIC ENERGY SOURCES

Category	Name of type	Characteristic energy source or stress
1. Naturally stressed systems of wide latitudinal range		High stress energies
	Rocky sea fronts and intertidal rocks	Breaking waves
	High energy beaches	Breaking waves
	High velocity surfaces	Strong tidal currents
	Oscillating temperature channels	Shocks of extreme temperature range
	Sedimentary deltas	High rate of sedimentation
	Hypersaline lagoons	Briny salinities
	Blue-green algal mats	Temperature variation and low nighttime oxygen
2. Natural tropical ecosystems of high diversity		Light and little stress
	Mangroves	Light and tide
	Coral reefs	Light and current
	Tropical meadows	Light and current
	Tropical inshore plankton	Organic supplements
	Blue water coasts	Light and low nutrient
3. Natural temperate ecosystems with seasonal programming		Sharp seasonal programming and migrant stocks
	Tidepools	Spray in rocks, winter cold
	Bird and mammal islands	Bird and mammal colonies
	Landlocked sea waters	Little tide, migrations
	Marshes	Lightly tidal regimes and winter cold
	Oyster reefs	Current and tide
	Worm and clam flats	Waves and current, intermittent flow
	Temperate grass flats	Light and current
	Shallow salt ponds	Small waves; light energy concentrated in shallow zone
	Oligohaline systems	Saltwater shock zone, winter cold
	Medium salinity plankton estuary	Mixing intermediate salinities with some stratification
	Sheltered and stratified estuary	Geomorphological isolation by sill
	Kelp beds	Swells, light and high salinity
	Neutral embayments	Shelfwaters at the shore
	Coastal plankton	Eddies of larger oceanic systems
4. Natural Arctic ecosystems with ice stress		Winter ice, sharp migrations and seasonal programming
	Glacial fjords	Icebergs
	Turbid outwash fjords	Outflow of turbid icewater lens
	Ice stressed coasts	Winter exposure to freezing
	Inshore Arctic ecosystems with ice stress	Ice, low light
	Sea- and under-ice plankton	Low light
5. Emerging new systems associated with man		New but characteristic manmade energy sources and/or stresses
	Sewage wastes	Organic and inorganic enrichment
	Seafood wastes	Organic and inorganic enrichment
	Pesticides	Organic poison
	Dredging spoils	Heavy sedimentation by man

Table 4.18—1 (continued)
A CLASSIFICATION OF COASTAL ECOLOGICAL SYSTEMS AND SUBSYSTEMS ACCORDING TO CHARACTERISTIC ENERGY SOURCES

Category	Name of type	Characteristic energy source or stress
5. Emerging new systems associated with man (continued)		
	Impoundments	Blocking of current
	Thermal pollution	High and variable temperature discharges
	Pulp mill wastes	Wastes of wood processing
	Sugar cane wastes	Organics, fibers, soils of sugar industry wastes
	Phosphate wastes	Wastes of phosphate mining
	Acid waters	Release or generation of low pH
	Oil shores	Petroleum spills
	Pilings	Treated wood substrates
	Salina	Brine complex of salt manufacture
	Brine pollution	Stress of high salt wastes and odd element ratios
	Petrochemicals	Refinery and petrochemical manufacturing wastes
	Radioactive stress	Radioactivity
	Multiple stress	Alternating stress of many kinds of wastes in drifting patches
	Artificial reef	Strong currents
6. Migrating subsystems that organize areas		Some energies taxed from each system

From Odum, H. T. and Copeland, B. J., in *Coastal Ecological Systems of The United States*, Vol. 1, Odum, H. T., Copeland, B. J., and McMahan, E. A., Eds., The Conservation Foundation, Washington, D.C., 1974, 5. With permission.

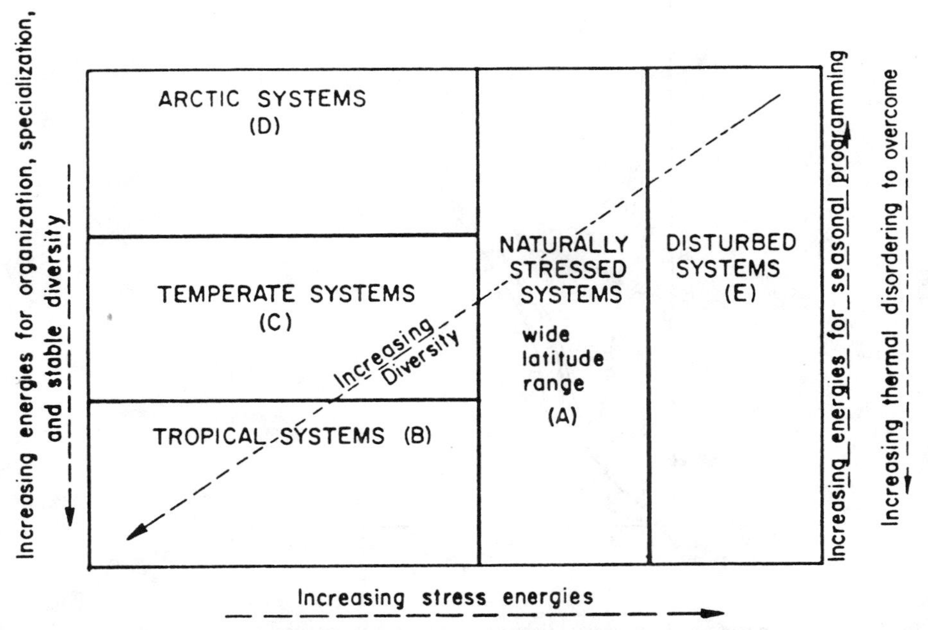

FIGURE 4.18—1. Systems analysis diagram applied to the classification of estuaries. (From Odum, H. T. and Copeland, B. J., in *Coastal Ecological Systems of the United States*, Vol. 1, Odum, H. T., Copeland, B. J., and McMahan, E. A., Eds., The Conservation Foundation, Washington, D.C., 1974, 5. With permission.)

4.19 DISTRIBUTIONAL CLASSES OF SPECIES WITH RESPECT TO SALINITY

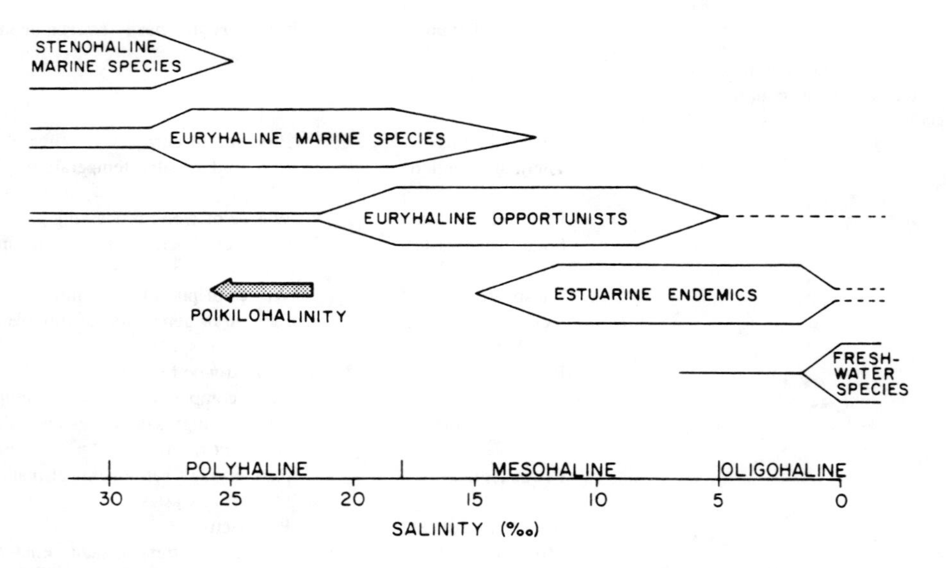

FIGURE 4.19—1. General model of the distributional classes of species in a homoiohaline estuary. (From Boesch, D. F., *Ecology of Marine Benthos,* Coull, B. C., Ed., University of South Carolina Press, Columbia, 1977, 245. With permission.)

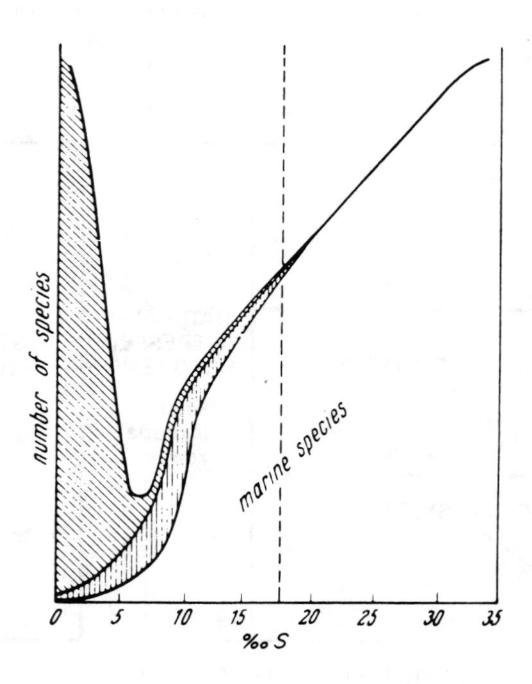

FIGURE 4.19—2. Relationship of salinity to species richness. Obliquely hatched area: freshwater species. Vertically hatched area: brackish-water species. White: marine species. (From Wolff, W. J., *Estuaries and Enclosed Seas,* Ketchum, B. H., Ed., Elsevier, Amsterdam, 1983, 151. With permission.)

Section 5
Marine Pollution

Section 5

MARINE POLLUTION

INTRODUCTION

Estuarine and coastal marine environments in particular are affected by a wide range of pollutants owing to increased anthropogenic activity from a burgeoning population in the coastal zone. Estuaries have long served as major repositories for the disposal of industrial and municipal wastes, sewage sludge, and dredged material. Contaminants associated with these wastes (e.g., heavy metals, chlorinated hydrocarbons, polynuclear aromatic hydrocarbons, petroleum hydrocarbons, and radioactivity) have impacted biotic communities and sensitive habitat areas. Pollutants primarily enter estuaries and nearshore oceanic waters via pipeline discharges, disposal from vessels, riverine input, atmospheric deposition, and nonpoint source runoff from land. Ocean dumping of anthropogenic wastes, employed increasingly in the U.S. until 1972, has been gradually reduced since that time because of the enactment of federal regulations designed to minimize environmental impacts on aquatic ecosystems.

In addition to the planned disposal of wastes at sea, accidental spills are responsible for occasional, yet substantial, quantities of pollutants in estuarine and shallow marine waters. For example, large oil spills at times have eradicated entire communities and destroyed many hectares of adjoining wetlands area. More insidious pollution problems (e.g., organic loading), while commonly perceived to be less damaging to the environment, often have a greater long-term impact on these coastal systems. Concomitant with the growing societal concern of marine pollution has been the development of various ocean monitoring programs. Most important among them is National Oceanic and Atmospheric Administration's (NOAA's) National Status and Trends (NS&T) Program which has measured toxic contamination at almost 300 estuarine and coastal locations throughout the U.S. since 1984 to determine trends of environmental quality in space and time and to discern biological responses to the chemical contamination. On a national scale, the highest and most biologically significant concentrations of contaminants in the NS&T Program occur predominantly in urbanized estuaries.

While some pollutants in estuarine and coastal marine waters are relatively innocuous, others significantly impact sensitive habitat areas and aquatic communities. Many of the anthropogenic problems encountered in the coastal zone have been attributed to specific waste disposal activities.[1,2] However, nonpoint sources of pollutants may be equally deleterious to nearshore marine environments.[3] These problems originate largely from the generation and release of waste products by a burgeoning human population in coastal regions. According to government projections, 200 million Americans representing 80% of the total U.S. population will reside in coastal areas by the turn of the century.[4] These demographic statistics raise concern over the potential stress that the populace will place on nearby estuarine and marine resources in future years.

A waste management problem clearly exists in the U.S. today. To set this problem in proper perspective, the U.S. annually disposes of 1.3 billion tonnes of waste, with nearly two thirds of this total being industrial refuse and dredged material. While approximately 80% of the waste generated in the U.S. is currently dumped in some 6500 landfills, more than 2000 of these facilities are scheduled to close within 5 years.[5] The reduction in the number of landfills nationwide will shift a greater burden of the responsibility for processing the waste on incineration plants and recycling programs, since the U.S. Congress enacted legislation prohibiting the disposal of industrial waste and sewage sludge in the ocean after December 31, 1991. Dredged spoils are presently the only solid material in the U.S. that can be legally deposited in the marine environment.

The disposal of pollutants and release of chemical contaminants in coastal marine environments have been coupled to various ecological and human health effects.[6–8] Because of the seriousness of these effects, it has been necessary to differentiate pollution from contamination. GESAMP[9] defined pollution as "the introduction by man, directly or indirectly, of substances or energy into the marine environment, resulting in such deleterious effects as harm to living resources, hazards to human health, hindrance to marine activities including fisheries, impairment of quality for use of seawater, and reduction of amenities." Contamination is defined as "the introduction of substances into the marine environment which alters the concentration and distribution of substances within the ocean."

Contaminants enter estuarine and oceanic waters via five principal pathways: (1) direct pipeline discharges; (2) discharges and dumping from ships; (3) riverine input; (4) atmospheric deposition; and (5) nonpoint source runoff from land.[10–12] Industrial and municipal wastes, sewage sludge, and dredged material are the most common anthropogenic wastes found in these coastal environments.[13] Contaminants typically associated with these pollutants include heavy metals, toxic nonmetals, chlorinated hydrocarbons, petroleum hydrocarbons, polynuclear aromatic hydrocarbons, artificially produced radionuclides, and pathogens.

Despite its vastness, its unique physical properties, and its importance as a prime dilutor of deposited waste, the ocean has a limited assimilative capacity for pollutants. Given sufficient time and dilution, the ocean can assimilate degradable organic and inorganic substances,[14–16] but unassimilated materials, such as long-life radioactive wastes and synthetic toxic organic compounds, will accumulate and remain essentially unaltered in the marine environment.[17–20] The persistence of these hazardous wastes in oceanic environments may pose a long-term danger to marine food webs. The utilization of the ocean as a repository of anthropogenic waste, therefore, is in direct conflict to its immense value as a resource to man.

The primary strategies for controlling anthropogenic wastes in estuarine and coastal marine systems are containment and dispersal.[1,21] Containment is the strategy of choice when dealing with the disposal of very hazardous refuse (e.g., high-level radioactive waste) or when attempting to contain dredged materials by employing submarine pits and caps. Dispersal relies on natural mechanisms (i.e., currents) at sea to transport and distribute materials away from dumpsites. One of the reasons for discharging and dumping anthropogenic wastes in estuaries and coastal oceanic waters is an expected dilution response, with transport and dispersion of contaminants to the offshore presumably mitigating impacts on nearshore habitats. However, hydrographic processes in coastal zones generally produce more complex circulation patterns than in offshore waters,[22] as evidenced by coastal boundary layer effects, broad spectra of turbulent eddies, and flow modulated by local bathymetry and shoreline configurations.[23] These factors make accurate prediction of pollutant dispersal extremely tenuous. Estuarine circulation appears to be even more problematical.[24–27]

Physical processes (e.g., advection, diffusion, and sedimentation) affect the distribution and fate of anthropogenic wastes in the water column. Chemical processes (e.g., adsorption, desorption, dissolution, oxidation, reduction, flocculation, volatilization, neutralization, and precipitation) influence the availability, persistence, and degradation of the wastes in water and sediments. Biological processes (e.g., incorporation/accumulation, toxicity response, degradation) may modulate food chain contamination and the long-term stability of populations.[1,28] Physical, chemical, and biological processes taken together, therefore, not only govern the distribution of anthropogenic wastes, but also their fate and overall effect on the marine environment by altering the concentration, chemical form, bioavailability, or toxicological strength of contaminants.[11,12]

While the public commonly perceives ocean dumping and major accidents of supertankers offshore as the primary sources of pollution in marine waters, 80 to 90% of the pollutants entering the marine environment actually derive from land-based sources through pipeline discharges, runoff into estuaries and coastal oceanic waters, and atmospheric influx.[29] The largest volume of anthropogenic wastes dumped at sea in past years has been dredged material, industrial products (i.e., acid-iron and alkaline waste, scrap metal, fish by-products, coal ash, and flue-gas desulfurization sludges), and sewage sludge. Most typically, ships or barges have dumped these wastes directly into the water, but some vessels are fitted with specially constructed incinerators for burning liquid organic wastes (e.g., polychlorinated biphenyls [PCBs] and organohalogens).[28] According to Capuzzo et al.,[13] the principal envionmental concerns of waste disposal in the sea are the following:

1. The accumulation and transfer of metals and xenobiotic compounds in marine food webs, including accumulation in commercial resources
2. The toxic effects of such contaminants on the survival and reproduction of marine organisms and the resulting impact on marine ecosystems
3. The uptake and accumulation of pathogenic organisms in commercially harvested species destined for human consumption
4. The release of degradable organic matter and nutrients to the ocean resulting in localized eutrophication and organic enrichment

Marine pollution monitoring and survey programs, such as the NOAA Mussel Watch and National Status and Trends Programs, have been developed to detect changes in estuarine and coastal marine populations attributable to human impacts. Short- and long-term monitoring of biotic communities is necessary for decision makers to devise sound solutions to acute and insidious aquatic pollution. However, biotic monitoring programs have been carried out in coastal environments for reasons other than strictly environmental impact assessment. For example, surveys are commonly executed for purposes of management of amenity, for exploitation of natural resources, for identification of sites for conservation priority, and for basic scientific research.[30]

This section provides an overview of pollution problems encountered in estuarine and coastal marine environments. It examines the deleterious and pervasive effects that human activities have on biotic communities and sensitive habitat areas in these ecosystems. Also addressed are the strategies employed by man to control anthropogenic wastes in the coastal zone.

ENVIRONMENTAL LEGISLATION

The many, varied, and increasingly complex problems arising from man's use of estuaries and coastal marine waters have focused attention on the pressing need to assess anthropogenic effects and to protect estuarine and marine resources. As a result of the heightened concern in the U.S., numerous detailed and quantitative monitoring and research programs were undertaken in the 1960s, 1970s, and 1980s, which provided a basic understanding of the problems and impacts plaguing these ecosystems. It became increasingly evident from this work that pollution studies of estuarine and marine environments involved both ecological and societal issues. Many of these problems and impacts stemmed from overpopulation and development in the coastal zone. The U.S. population living within 80 km of the ocean more than doubled between 1940 (42 million) and 1980 (89 million). Recent demographic statistics indicate that the population continued to mount in this region throughout the 1980s, exacerbating the potential environmental dangers in coastal waters.

Growing societal awareness of the adverse effects of pollution on aquatic habitats led to the passage of significant legislation by Congress between 1969 and 1988 which has ameliorated the degradation of the estuarine and oceanic environments. Some of the more important Acts of Congress promulgated during this period included:

1. National Environmental Policy Act of 1969
2. Federal Water Pollution Control Act Amendments of 1972
3. Marine Protection, Research, and Sanctuaries Act of 1972 (commonly called the Ocean Dumping Act) which regulated the dumping of all types of materials into marine waters and prevented or strictly limited any material that would adversely affect human health, welfare, the marine environment, ecosystems, or economic potentialities[31]
4. Toxic Substances Act of 1976 which enabled the U.S. Environmental Protection Agency to ban the production of PCB in the U.S. in 1977
5. Clean Water Act of 1977
6. Ocean Dumping Act Amendments of 1977 which called for an end to ocean dumping of sewage sludge and industrial waste as soon as possible
7. Low-Level Radioactive Waste Policy Act of 1980
8. Nuclear Waste Policy Act of 1982 which provided a framework for resolving many of the management problems associated with low-level and high-level radioactive waste
9. Water Resources Development Act of 1986 which moved the site for dumping of sewage sludge from the New York Bight Apex to the edge of the continental shelf (190 km east of Atlantic City, NJ)
10. Ocean Dumping Ban Act of 1988 which stressed that all dumping of sewage sludge and industrial waste would cease after December 31, 1991

In addition to this natural environmental legislation, several key international treaties for marine pollution control were enacted during the above mentioned 2-decade period. Of these treaties, the most important was the Convention on the Prevention of Marine Pollution by Dumping of Wastes and Other Matter of 1972 (commonly known as the London Dumping Convention). This international agreement to regulate dumping at sea entered into force on August 20, 1975, and as of January 1, 1990, 64 countries had become "contracting states". The London Dumping Convention controls waste dumping in the ocean principally through the national legislation and regulations of states contracting to the Convention. A second global treaty of major relevance was the International Convention for the Prevention of Pollution from Ships of 1973. The main part of this Convention, amended by a Protocol in 1978 (MARPOL of 1973/78), entered into force in 1983. It has dealt primarily with oil pollution from ships. Annex V of this Convention placed a ban on the dumping of all plastics at sea; it entered into force in 1988.[32] Table 5.10–1 provides a list of substances controlled by the London Dumping Convention.

MAGNITUDE OF THE PROBLEM

The waste disposal problem in coastal waters is inextricably linked to an ever expanding global population as noted previously. To set this problem in proper context, the world population has increased from slightly less than a billion in 1800 to about 5.5 billion today. By the year 2000, the world population is projected to approach 7 billion people. Projections into the 21st century appear to be even less promising, since more than 40% of all people now inhabiting the earth are less than 15 years old.[5]

Equally disturbing is the rate of change of the global population since the 17th century. The world population grew at a rate of about 0.3% per year between 1650 and 1750, 0.5% per year between 1750 and 1850, and 0.8% per year between 1850 and 1950. The death rate, meanwhile, declined in the 1800s and 1900s, largely due to the industrial revolution and the accompanying advances in agricultural production and medicine.[33] As a consequence, taking the total number of all people born from 1500 to 1993, more than two thirds of that total number are alive today. Concomitant with this explosive world population has been an increase in destruction of land, depletion of natural resources, production of all forms of waste, and pollution of aquatic ecosystems on earth.[2,7,8,34] The U.S. alone generates more than a billion metric tons (MT) of solid waste each year, amounting to 2.2 kg of waste for each man, woman, and child. This far exceeds the solid waste produced by any other country of the world.

One of the most dramatic demographic trends in industrialized nations, as well as some undeveloped ones, has been rapid urbanization. Along with the migration of people to large cities has been a rise in insidious and acute pollution problems in nearby estuaries and coastal marine waters, as reflected by the data of comprehensive aquatic monitoring projects in the U.S., such as the National Status and Trends Program of the NOAA (see below).[35,36] Anthropogenic degradation of various major coastal systems, including (but not limited to) the New York Bight, Boston Harbor, Puget Sound, San Francisco Bay, and Santa Monica Bay, provides examples. Multiple sources of pollution affect these systems: for example, (1) chemical contaminants derived from industrial processes, domestic activities, and the widespread use of pesticides, herbicides, and fertilizers in agriculture; (2) organic carbon enrichment related to elevated inorganic nutrient input; (3) polynuclear aromatic hydrocarbon (PAH) compounds originating from industrial effluents, the pyrolysis of organic matter, and other sources; and (4) petroleum hydrocarbons. In many impacted coastal waters, especially those in proximity to major metropolitan centers, nonpoint source runoff contributes significant concentrations of pollutants.

By the late 1970s, the decimation of many hectares of estuarine and coastal marine habitat and contiguous wetlands area became a serious concern in the U.S. Many people targeted ocean dumping of sewage sludge, industrial wastes, and dredged material for the observed deterioration of these habitats. In 1978, vessels in U.S. waters discharged a total of 65 million MT of dredged material, 5 million MT of solid sewage waste, and 2.4 million MT of industrial and construction waste into marine waters. The monetary cost of this disposal was high. In 1980, industry spent an estimated $100 million for the disposal of more than 2 million MT of industrial waste, whereas the disposal of dredged spoils cost the U.S. Army Corps of Engineers approximately $75 million. In addition, $30 million of an estimated $186 million Federal budget targeted for pollution-related work was allotted for research on marine waste disposal at this time.[37] Despite these expenditures, significant environmental impacts in the coastal zone continued unabated in some regions.

Environmental groups directed their attention to terminating ocean dumping of all toxic wastes in the early 1980s. In part because of their efforts, a global moratorium on radioactive waste dumping at sea was enacted in 1983. They also helped to accelerate the drafting of laws in the U.S. which prohibit the ocean dumping of industrial wastes and sewage sludge, as well as ocean incineration.[29]

While ocean dumping has clearly contributed to the degradation of coastal marine systems, by all accounts 80 to 90% of the polluting substances in the sea derive from land-based sources — pipeline discharges, nonpoint source runoff, and atmospheric input. Hence, ocean dumping comprises a small fraction of the total pollution loadings in coastal marine waters. Solid waste originating from operational and accidental disposal from ships or landfills nearby the shoreline is a primary culprit of marine pollution. For instance, repeated incidences of solid waste washing ashore on New Jersey beaches in the summer of 1988 raised the consciousness and ire of people across the country regarding the damaging effects of coastal pollution on the environment and tourism industry. Pictures of hospital wastes (i.e., bandages, hypodermic needles, syringes, and plastic bags of blood products, some tainted with hepatitis B and AIDS viruses) littering sandy beaches incited a public outcry demanding greater surveillance of coastal waters and increased enforcement of environmental regulations. Although health officials deemed these incidents to be nonsignificant health threats, they resulted in a tightening up of regulations to avoid such events in the future and to control the input of all types of wastes from municipalities, industries, and other sources.[32]

MAJOR POLLUTANTS

The primary pollutants of estuarine and coastal marine waters include the following:

1. Sewage and other oxygen-demanding wastes (principally carbonaceous organic matter) which may lead to anoxia or hypoxia of coastal waters subsequent to microbial degradation
2. Progressive enrichment of estuarine and coastal marine

waters due to the addition of organic and inorganic nutrients in wastewater discharges
3. Pathogens and other infectious agents often associated with sewage wastes
4. Oil originating from tanker accidents and other major spillages, routine operations during oil transportation, and influx from river runoff, urban runoff, municipal wastes, and effluents from nonpetroleum industries
5. Heavy metals derived from a variety of anthropogenic activities such as the burning of fossil fuels, smelting, sewage-sludge disposal, ash disposal, dredged material disposal, antifouling paints, seed dressings and slimicides (Hg), power station corrosion products (Cu, Cr, and Zn), automobile emissions (Pb), oil refinery effluent, and other industrial processes
6. Polynuclear aromatic hydrocarbons entering coastal ecosystems from sewage and industrial effluents, petroleum spills, creosote oil, combustion of fossil fuels, and forest and brush fires
7. Chlorinated hydrocarbon compounds (e.g., organochlorine pesticides and polychlorinated biphenyls) input largely from agricultural and industrial sources)
8. Radioactive wastes generated by uranium mining and milling, nuclear power plants, and industrial, medical, and scientific uses of radioactive materials
9. Calefaction of natural waters owing largely to the discharge of condenser cooling waters from electric generating stations

In recent years, the amount of wastes discharged into U.S. marine waters has approached 50 million MT per year. Dredging activities have accounted for about 80% of this waste, industrial products approximately 10%, and sewage sludge about 9%. Although the volume of industrial and sewage wastes is small relative to that of dredged spoils, these wastes have posed much more serious problems for these environments because of their potentially hazardous chemicals, notably toxic elements and compounds, and radioactive materials.

Agricultural, industrial, and municipal wastes are the main sources of the aforementioned pollutants. Agricultural wastes contribute a wide range of pollutants to coastal marine ecosystems, including pesticides, insecticides, high oxygen-demanding materials, nutrient elements (i.e., nitrogen and phosphorus), and other substances. Industrial wastes are also highly variable in content and may be assessed on the basis of their biological oxygen demand, concentration of suspended solids, and specific toxicity of inorganic and organic constituents. The volume of these wastes can be strictly controlled at the point of generation within a facility or by applying various levels of pretreatment prior to their discharge. Municipal facilities process wastewaters from homes and commercial establishments by employing primary (grit removal, screening, grinding, flocculation, sedimentation), secondary (oxidation of dissolved organic matter), and tertiary (nutrient removal) methods of treatment. The ultimate goal is to reduce or eliminate suspended solids, oxygen-demanding substances, dissolved inorganic (nitrogen and phosphorus) compounds, and bacteria that may impact receiving waters.

The Federal Water Pollution Control Act Amendments of 1972 established stringent controls and cleanup deadlines for industrial and municipal polluters. These amendments set rigorous standards for wastewater discharges designed to mollify environmental impacts in receiving waters. The Clean Water Act of 1977, in particular, targeted toxic pollutants in natural waters by adding provisions to mitigate potentially adverse effects of these contaminants.

ORGANIC LOADING

Many estuaries, particularly those nearby metropolitan centers, are burdened with nutrient enrichment and organic loading problems which in many cases have a dramatic effect on the dissolved oxygen levels and biotic communities found there. When nutrient enrichment exceeds the assimilative capacity of a system, eutrophic conditions may develop which can foster changes in the structure of the communities. Numerous coastal environments worldwide receive large concentrations of nutrient elements and carbon from industrial and municipal wastewaters, dredged material, and sewage sludge. Nonpoint sources of enrichment, such as agricultural and atmospheric addition, as well as urban runoff, exacerbate these pollution effects.

The oxidation of pollutant organic matter may produce acute oxygen deficiency in susceptible bodies of water. Two types of oxygen depletion zones have been differentiated in estuaries: (1) anoxic zones (bottom waters with <0.1 mg/l dissolved oxygen); and (2) hypoxic zones (bottom waters <2.0 mg/l dissolved oxygen).[38] The most sensitive systems appear to be those characterized by poor circulation where oxygen-depleted waters cannot be effectively reoxygenated. These affected bodies often are poorly flushed shallow coastal bays characterized by low freshwater inflow and restricted tidal ranges.[39] While enrichment with inorganic nutrients may in fact maintain healthy and productive biotic communities in some of these shallow coastal bays, excess nitrogen input has, in more than one case, generated phytoplankton blooms that have led to oxygen deficits in bottom waters, ultimately causing the death of much aquatic life. Even larger estuaries (e.g., Chesapeake Bay and Long Island Sound) have fallen victim to hypoxia and anoxia, resulting in a series of undesirable effects and culminating in multiple fishkills and the loss of valuable shellfish beds.[40–44]

Broad expanses of the Chesapeake Bay experience anoxia in late spring and summer.[40,41] The intensity of anoxia has been related to the degree of stratification of the water column and to the freshwater flow into the bay.[45] Seasonal increase in density stratification of the estuary, ascribable to the spring freshet, increases water column stability and minimizes advective transport of oxygen from the surface to the deep layer. These hydrographic conditions, together with the decomposition of phytoplankton and organic matter derived from sewage inputs and other sources (e.g., industrial outfalls), cause a decrease in dissolved oxygen of bottom waters. During the months from June to September, anoxic conditions purportedly spread along the bottom of the channel of the bay in waters greater than 10 m.[46] Eutrophic conditions are common in tributaries of the Chesapeake Bay (e.g., Patuxent River), owing to increases in nitrate and phosphate concentrations and abundance of phytoplankton.[47,48]

Parker and O'Reilly,[43] evaluating historical data, documented a general east-to-west gradient of decreasing dissolved oxygen in Long Island Sound. While these workers found no evidence of hypoxia in the eastern basin and only moderate hypoxia in the central basin, they uncovered repeated episodes of hypoxia in the western basin and severe, persistent, and recurrent hypoxia in the East River and Eastern Narrows. Koppelman et al.[49] linked accelerated eutrophication in the western perimeter of the sound in recent years to heavy urbanization and associated nutrient enrichment that spurs phytoplankton growth in these waters. Sewage effluent from the East River has unquestionably influenced nutrient distributions and contributed to the east-to-west gradient of increasing nutrients in the sound.[43,50] Focusing on the physical and biological mechanisms which create and maintain summertime hypoxia in Long Island Sound, Welsh and Eller[44] have envisaged oxygen depletion as a cumulative process through the summer months corresponding to a period of thermally controlled stratification in the estuary.

Sewage effluent and sludge not only contribute to the degradation of estuarine and marine environments worldwide, but also threaten human health because of pathogens they contain. Bacteria, viruses, protozoa, and helminths typically occur in these pollutants. They represent a possible health hazard for people who either ingest contaminated shellfish or swim in contaminated water. Among bacteria, the genus *Salmonella* includes the organism responsible for typhoid; *Shigella* spp. cause dysentery; and some species of the *Clostridia* produce exotoxins pathogenic to man.[51] Certain viruses in contaminated areas give rise to infectious hepatitis. Sewage treatment can lower the number of pathogens in wastewater effluent, specifically during the sludge-forming process.[13] Thermophilic digestion and other sewage treatment techniques have the potential to further reduce the population size of these pathogenic organisms.[2]

Biggs et al.[52] described the susceptibility of estuaries to eutrophication and pollution. They developed a classification scheme of estuarine pollution susceptibility based on physical and hydrologic data paired with a watershed classification based on anthropogenic activity and physical characteristics of the watershed.

OIL POLLUTION

Oil introduced into estuarine and coastal marine waters originates from a diversity of sources. Of all the oil reaching the sea, approximately 45% is derived from river runoff, urban runoff, municipal wastes, and effluents from nonpetroleum industries. Activities related to oil transportation account for another 33% of the polluting oil, with only about 25% of this total ascribable to accidents and major oil spills and 8% to normal operational losses. Natural oil seeps release 20 to 23% of the oil found in the marine hydrosphere. Roughly 75% of the accidental oil spills in the U.S. occur in coastal waters, primarily estuaries, enclosed bays, and wetlands, and most of the chronic oil pollution is associated with routine operations of oil refineries and oil installations, as well as discharges of industrial and municipal wastes into these habitats.[4,53,54] Because these systems constitute some of the most productive areas on earth, the effect of acute and chronic oil pollution on biota inhabiting them raises concern among estuarine and marine scientists.

The input of petroleum hydrocarbons to the sea from transportation, fixed installations, and other anthropogenic sources amounts to 4.94 mta. Most of the input to estuaries results from river runoff (1.40 mta), urban runoff (0.40 mta), and municipal and industrial waste (0.45 mta). Tanker accidents and routine transportation activities may supply more than 0.50 mta to these coastal habitats as well.[8]

According to Clark,[8] oil input originating from the users of petroleum products far exceeds that from the extraction and transport industries which are responsible for little more than 25% of the total input of oil to the sea. However, estimates by the National Research Council[55] on the global influx of petroleum hydrocarbons into the marine environment indicate that transportation losses may account for more than 40% of the polluting oil. Most of the losses are from routine operational discharges, such as tanker balast wastewater, and not from accidents, which generate only about 12% of the oil entering the ocean each year.

The degradation of spreading oil on the sea surface proceeds via several abiotic and biotic processes. Important physical-chemical processes that alter the oil include evaporation, photochemical oxidation, and dissolution. Microbial degradation of petroleum hydrocarbons, principally by marine bacteria, is the chief biological method by which the oil is broken down. Through time, the polluting oil commonly converts to stable water-in-oil emulsions that promote the formation of viscous masses known as chocolate mouse, which are extremely persistent and can exacerbate the overall impact of the oil. As a consequence of these physical-chemical and microbial processes, the density of the oil increases, thereby enhancing its settlement to the seafloor. The sorption of hydrocarbons to particulate matter (e.g., clay, silt, sand, shell fragments, and organic material) also facilitates sedimentation of the oil. Once the oil accumulates in seafloor sediments, the rate of microbial degradation of the oil decreases appreciably. The application of chemical dispersants to an oil slick soon produces dispersed oil droplets in the water column; this process decreases the potential for sedimentation or beaching of the oil.

Elevated petroleum hydrocarbon concentrations have been observed in most estuarine and coastal sediments around the world.[56–61] Petroleum hydrocarbon compounds are hydrophobic contaminants. They typically sorb to fine particulate material suspended in estuarine waters and become trapped in sheltered inshore areas.[62]

In terms of habitat damage, estuaries and coastal marine waters rank high among the most sensitive ecosystems to ecological degradation from the effects of polluting oil. Salt marshes, mangroves, mudflats, and other low-energy habitats tend to trap oil. The accumulation of hydrocarbon contaminants easily degrades these sensitive habitats, and the toxicity of the oil can decimate entire communities of organisms. Unlike high-energy environments (e.g., open stretches of shoreline exposed to strong wave action that physically breaks up and disperses oil relatively quickly), embayments and other sheltered coastal areas often sequester petroleum compounds in bottom sediments for years, re-releasing them to the surrounding environment, especially at times of major meteorologic events (e.g., hurricanes, storm surges, etc.). This effect creates an inhospitable environment for organisms that may persist for a decade or more.

Estuarine and marine biota are impacted by polluting oil either directly via the toxic effects of water-soluble components in crude oils and refined products, or indirectly via the degradation of critical habitat areas. In addition, chemical dispersants, solvents, and agents that reduce surface tension and hasten the removal of oil slicks from the water surface may also be toxic to estuarine and marine life.[55] While some researchers point to the toxicity of oil dispersants as the primary factor in ecological damage from oil spills,[7] others allude to laboratory findings that reveal lower acute lethal toxicities of dispersants currently in use than crude oils and their refined products.

Once a hydrocarbon compound is taken up by an estuarine or marine organism, it follows one of three general pathways. The compound may be metabolized, stored with possible elimination at a later time, or excreted unchanged. The hydrocarbons enter the organism in several ways: by active uptake of dissolved or dispersed substances; by ingestion of petroleum-sorbed particles, including live or dead organic matter; and by drinking or gulping of water containing the chemicals, as in the case of fish.[63]

The effects of polluting oil on aquatic communities can vary greatly because the diverse number of chemical compounds present in various proportions typically have different degrees of toxicity to the flora and fauna exposed to them. Moreover, the susceptibility of the organisms to the damage of polluting oil depends on the life history stages exposed to the compounds and the length of time of exposure. Both short- and long-term effects have been discerned on impacted biota. Short-term effects may be manifested in high mortality of individuals immediately subsequent to the release or discharge of the polluting oil. Long-term effects, such as lower population abundance through time, may develop from sublethal doses of toxins to egg, larval, and juvenile life stages.

Long-term biological impacts are exemplified by the spill of oil from the barge *Florida* into Buzzards Bay in September 1969, which caused a great loss of marine life for years in Wild Harbor River and contiguous waters. The grounding of the tanker *Arrow* in Chedabucto Bay, Nova Scotia, in February 1970, impacted biota of the bay for years. The rupturing of the tanker *Exxon Valdez* on a reef in Prince William Sound, Alaska, in March 1989, and the release of large volumes of oil into coastal waters have adversely affected biota and sensitive habitat areas of the sound to this day.

POLYNUCLEAR AROMATIC HYDROCARBONS (PAH)

Among the most widespread chemical contaminants in estuarine and nearshore environments are PAH, a group of ubiquitous compounds commonly occurring in bottom sediments, overlying waters, and biota, especially those in proximity to urban industrialized areas. These compounds consist of hydrogen and carbon arranged in the form of two or more fused benzene rings in linear, angular, or cluster arrangements with unsubstituted groups possibly attached to one or more rings.[64] Two general types of PAH have been characterized: (1) fused-ring compounds (e.g., alkyl naphthalenes, phenanthrene, and benzo(a)pyrene); and (2) linked-ring compounds (e.g., biphenyls). Together, PAH comprise a homologous series of fused aromatic ring compounds of increasing environmental concern.[65]

PAH enter estuarine and nearshore marine environments via several routes, most notably sewage and industrial effluents, oil spills, creosote oil, combustion of fossil fuels, and forest and brush fires.[66–68] Urban and agricultural runoff, groundwater inflow, and direct discharges transport large amounts of these hydrophobic contaminants to coastal systems. Pyrolysis of organic matter appears to be a primary source of PAH in aquatic habitats.[69] Natural sources of PAH, such as direct biosynthesis and diagenesis of sediment, are small relative to anthropogenic inputs. Oil spills and atmospheric deposition account for large quantities of PAH. Eisler[64] compiled statistics of the major sources of PAH compounds in aquatic environments.

Because they are potentially carcinogenic, mutagenic, and teratogenic to aquatic organisms, PAH have been the focus of numerous studies (for a review see Neff[66] and Kennish[2]). Not all PAH are potent carcinogens and mutagens, but some of them can initiate carcinogenesis and mutagenesis after metabolic activation.[70,71] The ability of aquatic organisms to take up PAH from contaminated environments is well chronicled.[72] Coupled to this uptake is the potential for trophic transfer of PAH from aquatic and terrestrial organisms. Hence, it is necessary to assess the availability of PAH to aquatic organisms, their mobilization in various compartments of contaminated systems, and the capability of exposed organisms to metabolize and excrete the contaminants.[73]

Owing to their relative insolubility in water and strong adsorption to particulate matter, PAH tend to concentrate in bottom sediments. Benthic invertebrates inhabiting polluted environments, therefore, are often continuously exposed to PAH. Yet, less information is available on the ability of these animals to metabolize and excrete PAH compounds[74,75] than is available on fish.[76,77] Bivalves have low or undetectable (MFO) enzyme activity for metabolizing PAH and typically bioaccumulate them with little alteration.[78] Consequently, certain shellfish populations (e.g., mussels [*Mytilus* sp.] and oysters [*Crassostrea* and *Ostrea*]) have been successfully employed as sentinel organisms to monitor PAH contamination in coastal waters.[79,80] PAH have been implicated in impacts on shellfish and finfish in Chesapeake Bay, Oregon Bay, and Puget Sound.[68]

Estuaries and nearshore coastal waters located near urban and industrial centers are major repositories of PAH.[62,81,82] For example, estuarine or coastal marine environments in proximity to Boston, Providence, and Baltimore exhibit elevated PAH concentrations. In an assessment of the environmental quality in the Gulf of Maine, Larsen[83] reported high PAH values in Casco, Penobscot, and Massachusetts Bays, averaging 4300, 2600, and 1600 ppb, respectively. Even higher PAH levels were recorded in Boston Harbor and Salem Harbor, two heavily populated areas. Boston and the industrialized southern New England region were deemed to be the main source of the PAH.

Urban runoff, sewage discharges, and fossil fuel combustion account for a significant influx of PAH in Narragansett Bay.[62] High concentrations occur in the northern perimeter of the estuary at a sewage outfall site near Providence, RI. Petroleum oils and coal tar also contribute PAH compounds to the system.

Peak levels of PAH in the Chesapeake Bay are observed in bottom sediments in the northern reaches near areas of greatest human population densities. Concentrations gradually decline downestuary. The Elizabeth River, a polluted subestuary of the lower Chesapeake Bay, is known to be heavily contaminated with PAH.[84] Anthropogenic sources, primarily combustion or high-temperature pyrolysis of carbonaceous fuels, are responsible for most of the PAH.[85]

CHLORINATED HYDROCARBONS

A large group of chemical contaminants — the chlorinated hydrocarbons — resists breakdown in aquatic environments, creating acute as well as insidious pollution problems. Included in the group of chlorinated hydrocarbons or organochlorine compounds encountered in estuarine and nearshore oceanic waters is a wide variety of biocides (insecticides, herbicides, and fungicides) and other substances that may persist for decades. Many chlorinated hydrocarbons are potentially toxic to estuarine and coastal marine organisms and, because of their fat solubility, tend to accumulate in the lipid tissues of animals. They also undergo biomagnification in food chains, thereby posing a hazard to man.

In addition to DDT, an extensive list of chlorinated hydrocarbon compounds has been compiled over the years for use as commercial pesticides, including (but not limited to) aldrin, chlordane, endrin, heptachlor, dieldrin, perthane, and toxaphene. A second group of pesticides, organophosphates, while exhibiting greater overall toxicity than chlorinated hydrocarbon pesticides, degrades much more rapidly. These pesticides are usually broken down into harmless compounds within 1 week.[86] Little is known regarding the impact of phosphorus-containing biocides (among them dipterex, fenthion, gluthion, malathion, naled, parathion, and ronnel) on estuarine and marine organisms. Wetlands fringing estuarine water bodies often represent ideal breeding sites for mosquitos and other insects.[87] In the past, some of these habitats have received considerable quantities of insecticides. Furthermore, the runoff of pesticides and herbicides (especially the stable, long-lived organochlorine compounds) from farmlands upstream of estuaries has been detrimental to aquatic biota.

PCBs are another class of chlorinated hydrocarbon compounds that have been linked to various environmental and public health concerns. These synthetic chemical compounds, consisting of a mixture of chlorinated biphenyls which contain varying numbers of substituted chlorine atoms on aromatic rings,[88] have been scrutinized extensively since the 1970s when their presence in the aquatic environment was initially perceived as a serious contamination problem.[89] In 1976, the Toxic Substances Control Act was passed, enabling the U.S. Environmental Protection Agency to ban the production of PCBs in the U.S. and to regulate their disposal. This banning occurred because PCBs were coupled to acute and chronic health effects in man (e.g., cancer, liver disease, skin lesions, and reproductive disorders). In addition, these compounds concentrate at higher trophic levels in organisms (e.g., finfish) that serve as a food source for humans. Despite the ban on production of PCBs in the U.S. in 1977, PCB contamination in estuaries and nearshore oceanic waters continues to be a universal problem.[2,90]

PCBs, like DDT, are highly stable in marine waters. They have a high specific gravity and a high affinity for solids. With a low solubility in water and high solubility in fats and oils, PCBs tend to partition out of the aquatic compartment into organismal tissue. PCBs bioaccumulate in the food chain principally because of their great stability and persistence and because they are poorly metabolized by biological systems.[90]

DDT and PCBs are probably the best known chlorinated hydrocarbon contaminants in the marine biosphere. These persistent, stable, and toxic substances have been recorded in the tissues of many estuarine organisms.[91] Their accumulation in bottom sediments is largely responsible for the continued contamination of biota inhabiting these coastal ecosystems.[92] Because of their potential carcinogenicity and mutagenicity to man, DDT and PCBs have been monitored intensively during the past 2 decades.[93]

In a synthesis of the trends of chlorinated pesticide (including DDT) and PCB contamination in U.S. coastal fish and shellfish, Mearns et al.[93] utilized more than 35,000 biotic samples from at least 300 separate surveys conducted between 1940 and 1985. Based on this database, regional and nationwide patterns of fish and shellfish contamination with DDT, dieldrin, and to a lesser extent, other pesticides, as well as PCBs were mapped for three heavily sampled periods (i.e., 1965 to 1972, 1976 to 1977, and 1984). Chemical residues targeted by this survey were analyzed in biotic samples collected in more than 200 individual bays, estuaries, and other distinct embayments.

Results of the survey by Mearns et al.[93] clearly show that DDT concentrations have decreased dramatically in U.S. coastal fish and shellfish since the early 1970s. On a national level, the decline may approach 100-fold. Other organochlorine compounds — PCB, dieldrin, and toxaphene — by comparison exhibited decreasing contamination over the same period for selected species and sites. DDT was a most widespread and prevalent organochlorine contaminant in estuarine invertebrates and fish in the late 1960s and early 1970s. During this time period, DDT concentrations averaged about 0.024 ppm (wet weight) in bivalves[94] and 0.7 to 1.1 ppm in whole fish.[95] Areas having the highest levels of DDT contamination were southern California, central California, Mobile Bay, and San Francisco Bay. One short-lived "hot-spot" region existed in Florida. A broad, low-level gradient of DDT contamination also occurred along the East Coast of the U.S. radiating north and south of Delaware Bay and possibly secondarily from Long Island, NY.

HEAVY METALS

Estuaries and coastal waters receive heavy metals from three principal sources, namely freshwater runoff, the atmosphere, and point sources of anthropogenic activity. While the weathering of rocks and leaching of soils are responsible for a portion of the heavy metal concentration in river water, anthropogenic input from domestic or industrial sewage, landfill leachates, and boating and shipping activities may far exceed the heavy metal fraction derived from weathering processes. Estuaries represent sites of accumulation of heavy metals.[24,96] In these systems, bottom sediments act as a sink for river-borne heavy metals[97,98] and therefore serve as a source of these elements for the benthos and overlying waters.[99–103] Among the major estuarine compartments of heavy metals (i.e., biota, water, or sediment), sediments are quantitatively most important.[24] The estuarine seafloor is a repository for heavy metals transported from land to sea; estuarine circulation tends to trap these pollutants within the estuary, thereby reducing their export to coastal waters.

As defined by Viarengo,[104] heavy metals comprise a group of elements with atomic weights ranging from 63.546 to 200.590, and are characterized by similar electron distribution in the external shell (e.g., Cu, Zn, Cd, and Hg). These exclude the alkaline earth metals, alkali metals, lanthanides, and actinides. Important anthropogenic sources of heavy metals, in addition to those alluded to above, include smelting operations, ash disposal, dredged-spoil dumping, and the burning of fossil fuels. Substantial volumes of copper, lead, and zinc released from pipes and tanks in domestic systems commonly enter natural waters that discharge to estuaries and coastal waters. Industrial sources — the mining of metal ores, finishing and plating of metals, and manufacture of metal objects, dyes, paints, and textiles — add significant amounts of heavy metals to coastal ecosystems as well. Municipal and industrial discharges in urbanized/industrialized regions generally account for heavy, localized contamination in impacted systems. Examples are Commencement Bay and Elliott Bay on Puget Sound, Boston Harbor, and Newark Bay. Historically, sewage sludge and dredged-spoil disposal have contaminated significant volumes of bottom sediments. Disposal sites in the coastal waters of the New York Bight and Southern California Bight contain tons of heavy metals. In Europe, heavy metal enrichment has been chronicled in Liverpool Bay, the Thames estuary, the Rhine-Waal/Meuse/Scheldt estuaries, the German Bight, and the Baltic Sea.

Heavy metals as a group are potentially toxic to aquatic organisms above a threshold availability.[105] However, many of these elements have been determined to be essential for proper metabolism of biota at lower concentrations. Cobalt, copper, iron, manganese, molybdenum, vanadium, strontium, and zinc, for example, are trace metals necessary for normal life processes, but they can be toxic to organisms at elevated concentrations by acting as enzyme inhibitors. For this reason, as well as others, toxicity levels have been established for heavy metals in aquatic environments. Water quality standards have also been developed on the basis of toxicity data, and extensive tables of heavy metals for freshwater fish, invertebrates, and marine organisms are now available.[106] Abel[7] listed heavy metals in the approximate order of their decreasing toxicity: Hg, Cd, Cu, Zn, Ni, Pb, Cr, Al, and Co. However, this sequence is tentative and subject to change, contingent upon physical-chemical conditions in the environment that affect chemical speciation of the metals. Moreover, the toxicity of a given metal will vary with the species investigated. The biochemistry of an organism plays a vital role in its susceptibility to metal toxicity. Metallothioneins and lysosomes, for instance, bind metal contaminants within the organism, thus enhancing cellular detoxification.[104,107,108] Engel and Brouwer[107] confirm the involvement of metallothioneins in the sequestration of elevated levels of trace metals, but hypothesize that their primary function is regulating normal metal metabolism of marine and estuarine organisms. The toxic effects of heavy metals in these organisms typically are manifested in aberrant physiology, reproduction, feeding, and development.

Estuarine organisms, particularly mollusks, concentrate heavy metals; consequently, they have been employed as bioindicators of heavy metal pollution. Mussels (e.g., the blue mussel, *Mytilus edulis*) are especially useful in this regard and have been utilized in many pollution studies of estuaries.[109] While bioaccumulation of trace metals is conspicuous among estuarine organisms, evidence of a general biomagnification effect is lacking.

Among the major estuaries in the U.S., Chesapeake Bay and San Francisco Bay exhibit local areas of heavy metal enrichment.[102,110,111] Most heavy metals enter these two estuaries in large influent systems. Anthropogenic input of metals is responsible for significant localized contamination of areas in proximity to urbanized/industrialized zones. San Francisco Bay may be more heavily impacted by anthropogenic activity because of the long history of mining and the urban and industrial development along its shores and in watershed areas upstream of river mouths. Metal-induced stress in aquatic communities of both estuaries is most conspicuous within benthic habitats subjected to high trace metal enrichment.

RADIOACTIVITY

The nuclear fuel cycle and the detonation of nuclear explosives have been the principal anthropogenic sources of radioactivity in estuarine and marine environments.[112] Human-generated radioactivity in the form of fallout from the detonation of nuclear explosives in the atmosphere contributed significant concentrations of radionuclides to these environments between 1945 and 1980. During this period, for example, more than 1200 nuclear weapons tests were conducted worldwide. The global stock of nuclear power plants grew from 66 to 398 from 1970 to 1986, and these facilities became a greater source of anthropogenic radioactivity in estuaries and oceans than nuclear detonations after the termination of atmospheric nuclear testing in 1980. Agricultural, medical, and scientific uses of radioisotopes have also contributed to the pool of artificial radionuclides in these environments, especially since 1975. While the radioactivity from all natural radionuclides in seawater amounts to about 750 dpm/l (97% derived from ^{40}K), the oceanic input from human activity has varied substantially since 1960, because of reductions in fallout from nuclear explosions after the second nuclear test ban treaty in 1963, the increase in the number of nuclear power plants on line after 1970, and accelerated usage of radionuclides in agriculture, industry, and medicine in the last 2 decades.

Radioactive waste is classified under six categories: (1) high-level wastes; (2) transuranic wastes; (3) low-level wastes; (4) uranium and mill tailings; (5) decontamination and decommissioning wastes from nuclear reactors; and (6) gaseous effluents. The 1972 International Convention on the Prevention of Marine Pollution by Dumping of Wastes and Other Matter and the U.S. Marine Protection, Research and Sanctuaries Act of 1972 prohibit the dumping of high-level radioactive wastes in the sea.[113] The disposal of low-level radioactive wastes is regulated by the Low-Level Radioactive Waste Policy Act of 1980 and its amendments (1985). The Nuclear Waste Policy Act of 1982 provides a framework for resolving many of the management problems associated with low- and high-level radioactive waste in the U.S.

Several waste management strategies minimize the environmental impact of radioactive waste. These include dilution and dispersion, isolation and containment, and isotopic dilution with stable isotopes. In the case of estuaries, which most commonly receive artificial radionuclides in low-level wastes discharged in liquid effluents from nuclear power plants, the preferred management option is one of dilution and dispersion.

Radionuclides enter biotic compartments of estuaries by organismal uptake from water, sediments, or other organisms via ingestion. The uptake of radionuclides by aquatic organisms and their potential biomagnification through food chains is a major human health concern because radiation damages reproductive and somatic cells leading to chromosome aberrations that may result in cancer or other malfunctions.[112,114] Hence, estuarine and marine organisms of direct dietary importance to man (e.g., crabs, clams, mussels, oysters, and fish) have been monitored continuously for radionuclides in many regions of the world.[115–117] The consumption of seafood, however, does not represent the major source of mankind's exposure to radioactivity. Most of the annual average radiation dose in man originates from natural radiation sources on land, with inhalation (e.g., radon) being the most important pathway, followed by external irradiation and ingestion of radioactive substances.[118]

Once radioactive substances enter the estuarine environment, various biological, chemical, and physical processes act to concentrate, disperse, or dilute it. Many estuarine and marine organisms concentrate ^{210}Po.[119] Naturally occurring ^{210}Po represents the principal radiation dose to these organisms,[120] and it ultimately returns to man via seafood.[121] Natural levels of ^{210}Po in surface seawater range from 0.6 to 4.2×10^{-2} pCi/l, which are greater than the concentrations of thorium nuclides, but less than those of ^{40}K, tritium, isotopes of uranium, and a number of other radionuclides. The total background radioactivity in surface seawater ($\sim$340 pCi/l) is less than that in marine muds (20,000 to 30,000 pCi/kg) and sands (5,000 to 10,000 pCi/kg). Since fine sediments, with their large surface area, adsorb more radioactivity than coarse sediments and have higher concentrations of organic matter and benthic populations, they may provide a more important pathway for the movement of radionuclides, particularly heavier species, through food chains.

The sediment column is the ultimate repository for most radionuclides in estuaries.[122–124] However, once deposited, the nuclides can be resuspended along with sediments by turbulence from bottom currents or storms which roil grains. Furthermore, the bioturbating activity of benthic fauna displaces the sediment and alters the erodibility of the substrate, often facilitating particle resuspension. As radionuclides migrate downward from sediment layers, they eventually become isolated from contact with biological processes. At this point, the estuary effectively becomes a sink for the removal of the substances from biotic compartments in the system.

Many estuarine organisms quickly accumulate radionuclides. Most conspicuous in this regard are smaller forms (e.g., phytoplankton and zooplankton) which have large surface area-to-volume ratios. The degree of uptake by larger biota (e.g., macroalgae, macroinvertebrates, and fish), while usually much less than that of smaller biota, nevertheless can be significant in those individuals that ineffectively regulate the radionuclides in their tissues. For large numbers of marine species, the uptake of radionuclides is proportional, or nearly proportional, to the elemental concentration in the seawater medium.[116] Small plankton and macroalgae exemplify this phenomenon. Elemental concentration factors among classes of marine organisms vary widely, due largely to species differences. When the radionuclide concentration is assessed in food chains, little evidence exists for biomagnification at high trophic levels, although exceptions can be found, such as ^{137}Cs, which preferentially accumulate in higher trophic level fish.[115] Biomagnification of radionuclides by estuarine and marine organisms is manifested most clearly in areas near nuclear fuel processing plants and industries producing nuclear weapons.[125,126]

Estuarine plants assimilate radionuclides directly from the water as well as from sediments. Some of these substances may even be assimilated from surface-deposited material that enters plant tissue through stomates or epidermal tissues. In grazing food webs, herbivores accumulate radioactive materials primarily by ingesting primary producers. One direct pathway of radionuclides to man involves the ingestion of marine algae containing natural or synthetic radioactivity.

Because of the accumulation of radionuclides in bottom sediments and associated organic detritus, the manipulation and processing of radionuclide-sorbed organic detritus by detritus feeders is a key linkage to the recycling of radioactive material through biotic compartments. The radionuclides directly attach to the detritus via deposition and sorption phenomena, and subsequent mineralization of the detritus by microbes releases the radioactive substances to sediments or interstitial waters. Alternatively, radionuclides enter the organic detritus reservoir after the death of plants and animals that sequestered radionuclides in their tissues when alive. The solubilization or resuspension of the nuclides in the detritus enables them to reenter estuarine food chains or to remobilize to other areas of the system. Detritus food chains represent a source of radionuclides for grazing food chains and vice versa.

Carnivores accumulate radionuclides by consuming herbivores and other prey that store radionuclides. They also receive radionuclides from environmental compartments such as water and sediments. The propensity of these organisms to accumulate radioactive substances from the environment is related to their behavioral patterns, physical attributes, and other inherent characteristics. The external morphology of the animals, for example, influences surface adsorption of radionuclides, with smaller individuals adsorbing more of the substances per unit mass than larger forms. In addition, benthic feeders coming in contact with contaminated sediments may be expected to have higher radionuclide concentrations than pelagic feeders inhabiting pristine waters. Finally, the physiology and metabolism of the organism greatly affect the intake, assimilation, tissue distribution, and retention of radionuclides by the carnivores.

DREDGING AND DREDGED-SPOIL DISPOSAL

Dredging is undertaken principally to maintain the navigability of harbors and waterways, although nonmaintenance activities (e.g., construction of port facilities and power plants) also require dredging operations that ultimately translocate significant volumes of sediment. Dredged-spoil disposal involves the discharge, release, or dumping of this sediment at subaerial or subaqueous sites. The environmental impacts of dredging and dredged-spoil disposal in estuaries mainly concern direct effects on habitat and organisms or indirect effects attributable to changes in water quality. For example, dredging causes damage directly by the physical disturbance associated with the removal and relocation of sediment.[127] The removal of bottom sediment destroys the benthic habitat, and operation of the dredge increases mortality of the benthos by mechanical injury during dredging, as well as by smothering of the organisms after they are picked up. A by-product of this process — the resuspension of bottom sediment — releases nutrients and remobilizes contaminants, thus affecting water quality and overall estuarine chemistry.[86,128] Dredged-spoil disposal in open water may adversely affect the benthic habitat by blanketing the substrate with potentially contaminated sediment.[129]

Dredged materials comprise about 80% by weight of all materials dumped in the nearshore waters of the U.S.[21] An average of 300 million m^3 of sediment are removed each year for the maintenance of existing waterways in the U.S., with an additional 80 million m^3 extracted for new dredging projects.[2] A substantial volume of this material is disposed of in the ocean, although incomplete estimates have been compiled. More than 100 coastal disposal sites in the U.S. receive dredged material. Nearly all open ocean dumping of dredged spoils takes place on the continental shelf. A total of 28 million metric tons of dredged spoils are dumped annually at 60 licensed offshore sites in England and Wales.[8]

In the past, coastal marshes and other wetlands have been utilized as dredged-spoil repositories. However, the use of these habitats has been severely curtailed since the 1960s because of the recognition that they are important nursery and breeding grounds for many estuarine organisms and zones of high biological productivity for estuaries and nearshore oceanic regions. Hence, depending on the conditions, open water release of dredged spoils may be the only viable option of disposal.

Two types of sites are generally considered in open water disposal of dredged spoils, namely retentive and dispersive sites.[130] Retentive sites ensure that disposed sediments remain within each site; low-energy hydrodynamic environments typically characterize these disposal localities. However, when a low-energy hydrodynamic environment is lacking, an inerodible cap may be placed over the disposed sediments to prevent their movement and the mobilization of contaminants from the dredged material. Despite the design effectiveness of a sheltered site or a protective cap of clean sediments for retentive purposes, both approaches can fail. Major storms roil bottom sediments even in the most protected embayments, and the construction of an impenetrable protective cap over disposal sediments may not be achievable. Still, the monitoring of capped disposal sites in ocean waters for extended periods of time has demonstrated that capping is technically feasible under normal tidal and wave conditions.[131,132]

Dredges are either mechanical or hydraulic devices. Mechanical dredges (i.e., bucket, dipper, and ladder dredges) lift sediment from the seafloor and transport it to a disposal site. Hydraulic dredges (i.e., agitation, dustpan, cutterhead, hopper, sidecasting, and suction dredges) loosen sediment by the action of cutterheads or the agitation of water jets, creating a slurry that can be easily transported through a pipeline to a dredged-spoil disposal area. Alternatively, the slurry is stored in hoppers, moved to disposal locations, and dumped. Hydraulic dredges offer a number of advantages over mechanical dredges. In particular, they are more efficient and economical to operate.[133]

As conveyed by Nichols,[134] dredging and disposal of spoils upset the hydraulic and sediment regime of an estuary. The deepening of an estuarine channel, for example, alters the stability it has attained by the adjustment of bed geometry and hydraulic regime over time. This modification reduces tidal currents which accelerates sediment deposition in the channel. Consequently, dredging is partly a self-perpetrating activity. The disposal of sediment also impairs natural processes, and it usually results in misplaced sediment in a foreign environment of deposition. An anomalous bed topography is often produced by the uneven or heterogeneous distribution of the sediment, typically in disequilibrium with the prevailing energy regime. If disposed alongside a dredged channel, therefore, the misplaced sediment gradually returns to the channel and eventually must be dredged again.

The most acute biotic impacts of dredging and dredged-spoil disposal involve the benthos. The immediate effect of dredging is the physical removal of benthic organisms and their transfer to spoil disposal areas. Mortality results from mechanical damage or smothering and approaches 100% when the organisms are dumped at terrestrial disposal sites. Less mortality originates from subaqueous disposal, but injuries and death can still be quite high, depending on the types of organisms affected and their new habitat. Benthic flora seem to be more easily impacted by dredging activities than benthic fauna. They are directly killed by the action of the dredge itself, and secondary effects, such as high turbidity levels, can limit their production. The dumping of spoils over plant communities inhabiting subaqueous or subaerial environments destroys many hectares of plant communities each year.

Through recolonization and succession, the benthic communities slowly recover from dredging effects. Recovery of a dredged site by benthic communities is temporally and spatially variable and site specific, but usually requires at least 6 months to 1 year or more to complete. Opportunistic, pioneering invertebrate species typically comprise the initial colonizers of a dredged-spoil dump site. Equilibrium assemblages of benthos succeed these opportunistic forms as time passes. Gradually, the same species that inhabited the dredged-spoil site prior to the disturbance return to repopulate the area. The more motile invertebrates generally incur less mortality than sedentary types. Two factors that can increase the probability of death of benthic fauna are the release of large concentrations of toxic contaminants from the sediments or acute changes in sediment type due to dredging activities.[135]

The release of toxic substances from dredged materials and the dispersal of pollutants in the water column can further imperil estuarine biota. The adverse effects of these contaminants may be exacerbated by increased nutrient and turbidity loads that can be detrimental, especially to planktonic organisms. Capuzzo et al.[21] identify four general categories of chemical contamination associated with dredging and disposal of sediments: (1) high concentrations of organic matter fostering anoxia and the presence of hydrogen sulfide; (2) transition and heavy metal contamination; (3) petroleum hydrocarbons; and (4) synthetic organic chemicals.

The principal effect of dredging on the water quality of estuaries relates to chemical exchanges between the dispersed sediment and water. As the dredged sediments disperse, several events occur that affect water quality. Initially, heavy metals in the water sorb to the sedimentary particles in suspension, thereby lowering their concentrations in the water. Subsequently, the sediments may release some of their heavy metal load, raising the metal concentration in the water. Second, estuarine sediments release ammonia when dispersed which stimulates phytoplankton production; this activity is followed by higher values of pH, dissolved oxygen, and biological oxygen demand (BOD)[2].

When dredging or dredged-spoil disposal operations are conducted in heavily utilized harbors and waterways containing large quantities of inorganic and organic pollutants, it is imperative to chemically characterize the dredged material. This goal can be achieved most expeditiously via four approaches: (1) bulk chemical analysis; (2) the elutriated test; (3) selective chemical leaching; and (4) bioassay tests.[129] Bulk chemical analysis provides a basis for estimating mass loads of wastes to estuarine and marine environments, and therefore, a basis for assessing the amount of substances that possibly can alter the biology and chemistry of a disposal site. The elutriate test measures the quantity of a substance exchanged between the sediment and the aqueous phase during dredging and disposal. The chemical state of pollutants associated with dredged sediments is determined by means of selective chemical leaching. Bioassay tests yield an operational measure of the toxicity of dredged material on estuarine organisms.

To accurately delineate the effects of suspended dredged material on water quality, quantitative chemical analyses of the sediments should be conducted. Examples of specific chemical analyses of the sediments include the determination of sediment sorption capacities, ionic and cationic sediment exchange capacities, and sediment elemental partitioning.[136] The partitioning of contaminants between solid and solution phases, the persistence of the pollutants sorbed to sediments in the estuarine environment, and the effects of contaminants on biota are all important in the impact assessment of dredged material. Processes such as dissolution, diagenesis, resuspension, and microbial degradation influence the flux and residence times of contaminants released with sediments into the water column during dredging and dredged-spoil disposal operations and, consequently, their distribution and fate in the estuary.

EFFECTS OF ELECTRIC GENERATING STATIONS

The siting of large (>500 MW) electric generating stations along estuaries causes a wide range of ecological impacts on aquatic communities. The construction and operation of both open-cycle, once-through cooling, and closed-cycle systems have been linked to a variety of adverse effects on estuarine organisms. The construction of docks and other structures, as well as the dredging of intake and discharge canals, promotes sediment erosion and increases sediment loads in adjacent waterways which may be dangerous to the benthos. They can also decimate or eliminate habitat areas. Man-made changes in bathymetry and the diversion of fluid flow oftentimes alter local hydrography and water quality. The discharge of condenser cooling water results in the calefaction or thermal loading of receiving waters. The elevated water temperatures interfere with the physiological processes of the biota (e.g., enzyme activity, feeding, reproduction, respiration, and photosynthesis) and frequently contribute to their death. Other impacts are associated with the release of chemical substances (e.g., biocides), the impingement of organisms on intake screens, and the entrainment of various life forms in cooling water systems.

Much of the effect of large, open-cycle, once-through power plants centers around their demands for large volumes of condenser cooling water which requires that they be built along estuaries and other coastal bodies of water.[4,125] For nuclear power plants which are thermally less efficient (~33% efficiency) than their fossil-fueled counterparts (~40% efficiency), a greater amount of waste heat is released to the environment. Furthermore, the larger quantity of cooling water drawn into nuclear power plants usually translates into higher impingement and entrainment mortality.

Most studies of power plant impacts on aquatic organisms have focused on the effects of thermal discharges, chemical releases, and impingement and entrainment.[2,137–140] Thermal discharges influence both the water quality and aquatic populations of impacted estuaries, with the most acute effects found in near-field regions in close proximity to the power plants. Increasing temperature lowers the dissolved oxygen content of receiving waters while raising the metabolic rate of its faunal inhabitants. These changes cause some individuals to become more susceptible to chemical toxins or disease and others to die forthwith. Bacterial decomposition of organic matter also accelerates at higher temperatures, further depressing dissolved oxygen concentrations. Moreover, the density, viscosity, surface tension, and nitrogen solubility of estuarine waters diminish with increasing temperature.

The calefaction of estuarine waters elicits physiological and behavioral responses in organisms, including the alteration of metabolic rates leading to a diminution of growth and behavioral adjustment manifested in avoidance or attraction reactions to the heated effluent. Even though an organism may not succumb directly to the elevated temperatures, it can experience sublethal physiological effects that can be detrimental, such as a decrease in reproduction or hatching success of eggs, an inhibition in development of larvae, and variations in respiratory rates. The avoidance or attraction responses of finfish to thermal plumes have been intensely studied by ichthyologists. Heat and (especially) cold shock mortality of finfishes are related to the attraction of the poikilotherms to the thermal plumes.

The ultimate consequences of artificial warming on estuarine communities are shifts in their structure. Changes in species composition, species diversity, and population density are commonly observed in the near-field regions of power plants. In some cases, opportunistic or nuisance species may replace equilibrium populations at heavily impacted sites.

Some of the chemicals released by electric generating stations pose a potential hazard to estuarine organisms owing to their toxicity. Most important in this respect are chlorine, which is injected into cooling water to control biofouling on heat exchanger surfaces, trace metals originating from the dissolution of piping within the condenser cooling system, and low-level radioactive wastes produced by nuclear reactors. Among these chemicals, chlorine has received much attention because at times it accounts for massive mortality of susceptible organisms in discharge waters (e.g., bacteria and phytoplankton).

Perhaps the greatest biological impact of power plants is ascribable to impingement of large organisms on intake screens and entrainment of plankton, microinvertebrates, and small juvenile fish passively drawn into the cooling water systems. The numerical losses of estuarine organisms from impingement and entrainment at power plants can be extremely high. For example, at the P. H. Robinson Generating Plant in Galveston Bay, TX, more than 7 million fish were impinged in 1 year (1969 to 1970). Impingement mortality of Atlantic menhaden (*Brevoortia tyrannus*) at the Millstone Nuclear Power Station on Niantic Bay, CT, exceeded 2 million individuals in the late summer and early fall of 1971.[4] Approximately 13 million fish and macroinvertebrates were impinged at the Oyster Creek Nuclear Generating Station on Barnegat Bay, NJ, from September 1975 through August 1977.[141] During the period January to December 1986, impingement of fish and invertebrates at the Pilgrim Nuclear Power Station on Cape Cod Bay, MA, averaged 1.26 and 1.91 individuals per hour, respectively.[142] Entrainment losses of Atlantic menhaden and river herring at the Brayton Point Power Plant on Mount Hope Bay, MA, during the summer of 1971 ranged from 7 to more than 160 million individuals. Effects of entrainment at the Millstone Nuclear Power Station killed some 36 million young fish in 16 d in November 1971.[143] An estimated 5.16×10^{13} microzooplankton were entrained in the Oyster Creek Nuclear Generating Station from September 1975 through August 1976 and 4.03×10^{13} microzooplankton from September 1976 through August 1977. Macrozooplankton losses during these intervals amounted to an estimated 4.25×10^{11} and 9.98×10^{10} individuals from 1975 to 1976 and 1976 to 1977, respectively.[141] The number of fish eggs entrained through the Pilgrim Nuclear Power Station in 1986 equalled 1.70×10^{9} and the number of larval fish, 2.76×10^{8}.[142]

Although the absolute number of estuarine organisms impinged or entrained by power plants commonly ranges into the millions over an annual cycle, the actual impact of these losses on resident populations of phytoplankton, zooplankton, benthos, and fish is difficult to ascertain. Clearly, the need exists for comparison studies of power plant-induced mortality of populations to the resident population sizes in waters not impacted by impingement and entrainment. For instance, Marcy et al.[144] aptly summarized the percentages of larval fish populations entrained by various power plants in the U.S.; the numbers entrained ranged from about 2 to 12% of the estimated population sizes.

Much speculation and controversy surround the total effect of impingement and entrainment mortality. Mathematical population models have been formulated to assess these impacts on resident species.[145,146] Major limitations to the use of the models include inaccurate estimates of population sizes and the inability to accurately account for marked fluctuations in reproductive success and survival of organisms from year to year.[125,145] In addition, biological compensation, a basic tenet of fisheries management, may operate to mollify or eliminate impingement and entrainment effects.[147] For instance, high egg and larval mortalities can be compensated by increases in fecundity and survival of resident populations, together with predator mortality, unless the populations diminish to very low levels.[125,148]

A number of abiotic and biotic factors affect the impingement and entrainment of estuarine organisms. Among the most important abiotic factors are the siting and cooling water system design of the power plant and the volume and ambient conditions of the cooling water used. Biotic factors of significance relate to the abundance, survival, ecological roles, and reproductive strategies of the affected organisms.[149] Considerable variation in impingement or entrainment impact may be rooted to a multitude of factors such as seasonal fluctuations, organismal migrations, climatic changes, tides, storms, and unscheduled changes in plant operations.[125]

Seasonal variations in the number of impinged organisms can be substantial due to seasonal migration or emigration of populations and seasonal changes in water temperature that control the ability of the organisms to withstand currents and hence impingement on intake screens. Unequivocally, the swimming performance of finfishes affects the probability of their impingement. Factors influencing swimming performance include the type of species, its developmental stage, parasite infestation, nutritional condition, presence of predators or prey, salinity, temperature, oxygen concentration, suspended matter, tides, time of day, and season.

Various modifications of intake systems, incorporating deflection methods and exclusion structures, have been designed to mitigate impingement by attracting or repelling fish from intake areas. The main deflection techniques developed to ameliorate impingement are electrical barriers, air bubble curtains, artificial lights, acoustic barriers, louvers, and velocity caps. Among these methods, louvers and velocity caps appear to offer the most success in directing fish away from intake screens and guiding them safely into areas for removal or bypass. In regard to exclusion devices, fine screens or clinker bunds may be effective, although they clog easily and consequently require much backwashing and general maintenance. The greatest decreases in impingement mortality are documented at power plants where intake velocities have been substantially reduced. Other reductions in impingement losses have been recorded at power plants where conventional vertical traveling screens have been retrofitted with a more advanced technology, such as Ristroph screens comprised of a continuously rotating traveling design modified with a low-pressure spray wash and fish recovery system.[139]

The annual entrainment estimates for eggs or plankton at large, once-through, open-cycle power plants located on estuaries may exceed 1 billion individuals. Entrainment mortality results from thermal stress, mechanical effects (e.g., turbulence, pressure changes, abrasion of suspended organic particles, and impacts against piping), and chemical toxicity (e.g., biocides such as chlorine). Although no consensus exists regarding the relative significance of thermal, mechanical, and chemical stresses on entrainment mortality of biota, when detailed studies have been conducted, mechanical damage appears to be of overriding importance. Nevertheless, the total entrainment mortality is the cumulative effect of these three types of stress. It can be reduced most effectively by lowering the volume of cooling water pumped through plant condensers.

Entrainment mortality is also a function of biotic factors. For example, observed entrainment mortalities are closely coupled to the seasonal densities of entrainable organisms in the impacted systems. In addition, the sizes, life stages, and relative susceptibility to injury of the entrained forms play a critical role in the total mortality of estuarine organisms in power plants.

NATIONAL STATUS AND TRENDS PROGRAM

A prime objective of the NS&T Program of the NOAA is to assess the effects of anthropogenic activities on environmental quality in coastal and estuarine areas of the U.S. Initiated in 1984, the NS&T Program monitors the concentrations of more than 70 chemical contaminants in livers of benthic fish, whole soft parts of mussels and oysters, and sediments of nearly 300 coastal and estuarine sites nationwide. Samples collected annually at these locations provide an effective measure of the trends of chemical contamination in space and time and limited measures of the biological significance of that contamination. To this end, bioassay tests are conducted on water or sediment at a sampling site, and indigenous species are examined for evidence of response to chemical contaminants. Because the presence of tumors in fish is usually interpreted as a response to contamination,[150] the livers of fish at NS&T sites are carefully examined for them.[36]

The NS&T Program of the NOAA is designed to describe both regional and national distributions of chemical contamination. The average distance between NS&T sampling sites is 20 km within bays and estuaries and 100 km along the open coast.[151] Almost half of all sampling sites occur in urban estuaries within 15 km of population centers in excess of 100,000 people because the contaminant levels tend to be high and more spatially variable in urban, rather than rural, systems. Similarly, since high and biologically significant concentrations of contaminants are limited primarily to urbanized estuaries, the sampling sites lie closer together in estuaries and bays than along open coasts.[36]

The NS&T Program consists of two principal components: (1) the Benthic Surveillance Project and (2) the Mussel Watch Project. The Benthic Surveillance Project determines the concentrations of toxic chemicals in bottom-dwelling fish and sediments. The Mussel Watch Project delineates levels of chemical contamination in sediments, as well as in bivalve mollusks (i.e., mussels and oysters). These shellfish are analyzed for radionuclides in addition to the same contaminants as analyzed in the bottomfish of the Benthic Surveillance Project. Bottom sediments are monitored for pollutants in both the Benthic Surveillance and Mussel Watch Projects because they are known repositories of chemical contaminants. The exposure of bottom-dwelling fish to chemical pollutants is closely linked to the sediments, and hence these organisms are considered to be reliable indicators of local pollution. Thus, they are important target organisms in the Benthic Surveillance Project. Since bivalve mollusks have been successfully used as sentinel organisms to detect aquatic contamination, mussels and oysters have been selected in the Mussel Watch Project as indicators of environmental quality.

Samples from the Benthic Surveillance Project have been analyzed in five laboratories: (1) NOAA National Marine Fisheries laboratories in Woods Hole, MA; (2) Gloucester, MA; (3) Beaufort, NC; (4) Charleston, SC; and (5) Seattle, WA. The Mussel Watch samples are analyzed at the following locations: (1) Battelle Laboratories in Duxbury, MA; (2) Battelle Laboratories in Sequim, WA; (3) Texas A & M University Geochemical and Environmental Research Group in College Station, TX; and (4) La Jolla, CA Laboratory of Scientific Applications International Corporation. Data generated in these laboratories identify the areas around the U.S. coast having the highest levels of contamination. Once discovered, these highly contaminated areas become the subject of more intensive surveys of biological and chemical conditions to discern whether or not the observed levels of contamination are degrading biological communities and habitats. These more intensive studies are performed by the NOAA National Marine Fisheries Laboratories in Beaufort, NC and Seattle, WA. Gradients of contamination and measures of biological properties (e.g., reproductive impairment, genetic damage, and sediment toxicity) are typically carried out in these laboratories.

Results of 8 years of NS&T sampling throughout the coastal U.S. reveal that higher levels of chemical contamination of surface sediments are characteristic of urbanized estuaries. This is clearly evident for urbanized areas of the northeastern states; near San Diego, Los Angeles, and Seattle on the West Coast; and, except at a few sites, relatively rare in the southeastern states and along the Gulf of Mexico coast. The high levels of contaminants in urbanized estuaries, however, are lower than those expected to cause sediment toxicity.[36] Liver tumors in fish and sediment toxicity, both extreme responses to contamination, have been found infrequently.[151] Analyses of mussels and oysters are beginning to uncover temporal trends in contaminant levels at NS&T sites. Trends observed over longer time scales (>5 years) indicate that the concentrations of most contaminants measured in the NS&T Program may be declining.[36]

SUMMARY AND CONCLUSIONS

Estuarine and coastal marine environments are susceptible to a multitude of human wastes from a burgeoning population in the coastal zone. These highly sensitive ecosystems serve as repositories for dredged spoils, sewage sludge, industrial and municipal effluents, and other types of pollution. Areas with the highest levels of pollution border metropolitan centers, where industrial, municipal, and domestic contaminants have accumulated for years. Boston Harbor and Raritan Bay on the East Coast of the U.S. and San Diego Harbor and Puget Sound on the West Coast provide examples. Others are San Francisco Bay, contaminated with cadmium, copper, mercury, nickel, and other heavy metals; Elliott Bay nearby Seattle, WA, contaminated with arsenic, cadmium, copper, lead, zinc, and PCB; and Commencement Bay, contaminated with pollutants from Tacoma, WA. The U.S. Environmental Protection Agency has designated Commencement Bay as a Superfund site, underscoring the severity of the dilemma in this estuary. Most of the aforementioned problems stem from overpopulation and development in the coastal zone. For example, the U.S. population living within 80 km of the ocean more than doubled between 1940 (42 million) and 1980 (89 million). Recent demographic statistics indicate that the population continued to mount in this region throughout the 1980s, exacerbating the potential environmental hazards in coastal waters. As many as 200 million Americans, representing approximately 80% of the total U.S. population, will reside in coastal areas by the year 2000 according to government projections.

Growing public awareness of pollution effects in the marine hydrosphere in the 1960s and 1970s led to a reassessment of the value of these critical regions. Their high biotic productivity, rivaling the most intensively cultivated farmlands, was emphasized. Even though they comprise only 8% of the total area of the oceanic realm, estuarine and nearshore marine environments account for 50% of the world fisheries harvest. Only the highly productive upwelling areas, essentially responsible for the remaining 50% of the world fisheries harvest, have yielded comparable statistics. Despite their acknowledged productivity, estuaries and shallow oceanic waters are by far the most heavily affected by pollution.

While man is the source of severe and persistent environmental degradation of estuarine and marine waters, natural processes interacting with anthropogenic wastes may, in some cases, contribute to habitat destruction. Strong currents associated with storm surges, hurricanes, and other meteorological events roil bottom sediments and disperse pollutants from impacted sites to originally unaffected areas; this has been demonstrated for lipophilic organic substances (PAH, PCB, and other chlorinated hydrocarbons), oil-derived hydrocarbons, trace metals, radionuclides, and pathogens. Although nutrients from sewage effluent are readily biodegraded or assimilated (but in excessive amounts promote eutrophic conditions), the lipophilic organic compounds and pathogens often concentrate in biota and sediments, adversely influencing food webs and the functioning of the ecosystems. In this respect, toxic chemicals discharged to estuaries and coastal marine waters create more insidious problems than some visible pollutants (e.g., oil spills or sewage). Nevertheless, when nutrients overwhelm the capacity of a system to assimilate them, red tides, brown tides, and eutrophic conditions commonly arise which frequently alter its trophic structure through the loss of large numbers of heterotrophs. The uncontrolled growth of certain phytoplankton species (e.g., *Ptychodiscus brevis*) has periodically inflicted mass mortality on finfish and shellfish populations. Neurotoxin released by these dinoflagellate cells accumulates in the siphons and hepatopancreas of shellfish, causing paralytic shellfish poisoning (a neurological disorder) in humans who consume them. Saxitoxin is the toxic agent responsible for this acute illness.

Short- and long-term monitoring of biotic communities is necessary for decision makers to devise sound solutions to acute and insidious aquatic pollution. Ideally, both pre- and post-impact surveys of biota should be conducted to accurately predict pollution effects and to formulate essential remedial action. An ultimate goal of marine science is to develop quantitative and predictive models (relying on field and laboratory data) not only of physical, chemical, and biological processes, but also of the anthropogenic stresses imposed on natural systems.

The literature on the biological aspects of marine ecology is replete with references to pollution impacts on populations attributable to organochlorine insecticides, PCB, organotin compounds, heavy metals, toxic metals, oil spills, and nutrient loadings. Although nearly all phyla have been subjected at some time to contaminants in coastal and offshore waters, the most adversely affected forms are benthic populations with limited or no mobility; consequently, they have been the focal point of environmental impact studies. Since 1984, the NOAA has conducted the NS&T Program to assess the effects of anthropogenic activities on the environmental quality of estuarine and coastal marine waters.

Biotic monitoring programs have been carried out in estuaries and marine waters for reasons other than strictly environmental impact assessment. For instance, surveys are commonly executed for purposes of management of amenity, for exploitation of natural resources, for identification of sites for conservation priority, and for basic scientific research. In nearly all cases, the surveys collect data on the abundance and distribution of organisms, together with information on physical and chemical parameters.

It is evident that much ecological research on marine environments requires at least a fundamental knowledge of the biology of the constituent populations. Considering the pervasive and enigmatic anthropogenic problems plaguing these environments, it behooves aquatic scientists from all disciplines to make a concerted effort to elucidate the population dynamics of the biota and the effects of pollutants on them. Unequivocally, in order to ameliorate or solve these ecological problems and to rehabilitate or protect sensitive habitat areas, it will be necessary to compile additional data on the structure of the communities and the biotic interactions and responses of the organisms within the systems under investigation. Perhaps more importantly, an integral part of evaluating the health of these waterways entails the delineation of organism-environment interactions, which in many respects is much more problematical. *In situ* observations and laboratory experimentation of pollutant impacts on estuarine and marine organisms must proceed if the estuaries and oceans are to continue to be a source of aesthetic, commercial, and recreational resources for mankind into the 21st century.

REFERENCES

1. **McDowell-Capuzzo, J. E.,** Effects of wastes on the ocean: the coastal example, *Oceanus,* 33, 39, 1990.
2. **Kennish, M. J.,** *Ecology of Estuaries: Anthropogenic Effects,* CRC Press, Boca Raton, FL, 1992.
3. **Alm, A. L.,** Nonpoint sources of water pollution, *Environ. Sci. Technol.,* 24, 967, 1989.
4. **Hall, C. A. S., Howarth, R., Moore, B., III, and Vorosmarty, C. J.,** Environmental impacts of industrial energy systems in the coastal zone, *Annu. Rev. Energy,* 3, 395, 1978.
5. **Spencer, D. W.,** Introduction: the ocean and waste management, *Oceanus,* 33, 5, 1990.
6. **Wolfe, D. A. and O'Connor, T. P., Eds.,** *Oceanic Processes in Marine Pollution,* Vol. 5, Urban Wastes in Coastal Marine Environments, Robert E. Krieger Publishing, Malabar, FL, 1988.
7. **Abel, P. D.,** *Water Pollution Biology,* Ellis Horwood, Chichester, England, 1989.
8. **Clark, R. B.,** *Marine Pollution,* 2nd ed., Clarendon Press, Oxford, 1989.
9. GESAMP (IMCO/FAO/UNESCO/WMO/WHO/IAEA/UN/UNEP Joint Group of Experts on Scientific Aspects of Marine Pollution), Scientific Criteria for the Selection of Waste Disposal Sites at Sea. Reports and Studies No. 16, Inter-Governmental Maritime Consultative Organization, London, 1982.
10. **Bourdeau, P. and Barth, H.,** Estuarine, coastal, and ocean pollution: EEC policy and research, in *Estuarine and Coastal Pollution: Detection, Research, and Control,* Moulder, D. S. and Williamson, P., Eds., Pergamon Press, New York, 1986, 1.
11. **Capuzzo, J. M. and Kester, D. R., Eds.,** *Oceanic Processes in Marine Pollution,* Vol. 1, Biological Processes and Wastes in the Ocean, Robert E. Krieger Publishing, Malabar, FL, 1987.
12. **Capuzzo, J. M. and Kester, D. R., Eds.,** Biological effects of waste disposal: experimental results and predictive assessments, in *Oceanic Processes in Marine Pollution,* Vol. 1, Biological Processes and Wastes in the Ocean, Capuzzo, J. M. and Kester, D. R., Eds., Robert E. Krieger Publishing, Malabar, FL, 1987, 3.
13. **Capuzzo, J. M., Burt, W. V., Duedall, I. W., Park, P. K., and Kester, D. R.,** Future strategies for nearshore waste disposal, in *Wastes in the Ocean,* Vol. 6, Nearshore Waste Disposal, Ketchum, B. H., Capuzzo, J. M., Burt, W. V., Duedall, I. W., Park, P. K., and Kester, D. R., Eds., John Wiley & Sons, New York, 1985, 491.
14. **Topping, G.,** Sewage and the sea, in *Marine Pollution,* Johnston, R., Ed., Academic Press, London, 1976, 303.
15. **Kennish, M. J.,** *Ecology of Estuaries,* Vol. 2, Biological Aspects, CRC Press, Boca Raton, FL, 1990.
16. **Nedwell, D. B. and Lawson, P. A.,** Degradation of digested sewage sludge in marine sediment-water model systems and fate of metals, *Mar. Pollut. Bull.,* 21, 87, 1990.
17. **Park, P. K. and O'Connor, T. P.,** Ocean dumping research: historical and international development, in *Ocean Dumping of Industrial Wastes,* Ketchum, B. H., Kester, K. R., and Park, P. K., Eds., Plenum Press, New York, 1981, 3.
18. **Park, P. K., Kester, D. R., and Duedall, I. W., Eds.,** *Radioactive Wastes and the Ocean,* John Wiley & Sons, New York, 1983.
19. **Eisenbud, M.,** *Environmental Radioactivity,* 4th ed., McGraw-Hill, New York, 1988.
20. **Gage, J. D. and Tyler, P. A.,** *Deep-Sea Biology: a Natural History of Organisms at the Deep Seafloor,* Cambridge University Press, Cambridge, 1991.
21. **Capuzzo, J. M., Burt, W. V., Duedall, I. W., Park, P. K., and Kester, D. R.,** The impact of waste disposal in nearshore environments, in *Wastes in the Ocean,* Vol. 6, Nearshore Waste Disposal, Ketchum, B. H., Capuzzo, J. M., Burt, W. V., Duedall, I. W., Park, P. K., and Kester, D. R., Eds., John Wiley & Sons, New York, 1985, 3.
22. **Dyke, P. P. G., Moscardini, A. O., and Robson, E. H., Eds.,** *Offshore and Coastal Modelling,* Springer-Verlag, New York, 1985.
23. **Lam, D. C. L., Murthy, C. R., and Simpson, R. B.,** *Effluent Transport and Diffusion Models for the Coastal Zone,* Springer-Verlag, New York, 1984.
24. **Kennish, M. J.,** *Ecology of Estuaries,* Vol. 1, Physical and Chemical Aspects. CRC Press, Boca Raton, FL, 1986.
25. **van der Kreeke, J., Ed.,** *Physics of Shallow Estuaries and Bays,* Springer-Verlag, New York, 1986.
26. **Kjerfve, B.,** *Hydrodynamics of Estuaries,* Vol. 1, CRC Press, Boca Raton, FL, 1988.
27. **Kjerfve, B.,** *Hydrodynamics of Estuaries,* Vol. 2, CRC Press, Boca Raton, FL, 1988.
28. **Duedall, I. W.,** A brief history of ocean disposal, *Oceanus,* 33, 29, 1990.
29. **Curtis, C. E.,** Protecting the oceans, *Oceanus,* 33, 19, 1990.
30. **Baker, J. M. and Wolff, W. J., Eds.,** *Biological Surveys of Estuaries and Coasts,* Cambridge University Press, Cambridge, 1987.
31. **Kitsos, T. R. and Bondareff, J. M.,** Congress and waste disposal at sea, *Oceanus,* 33, 23, 1990.
32. **Waldichuk, M.,** The state of pollution in the marine environment, *Mar. Pollut. Bull.,* 20, 598, 1989.
33. **Ehrlich, P. R. and Ehrlich, A. H.,** *Population, Resources, Environment,* 2nd ed., W. H. Freeman, San Francisco, 1972.

34. **Turk, A., Turk, J., and Wittes, J. T.,** *Ecology, Pollution, Environment,* W. B. Saunders, Philadelphia, 1972.
35. NOAA, National Status and Trends Program for Marine Environmental Quality: a Summary of Selected Data on Chemical Contaminants in Sediments Collected during 1984, 1985, 1986, and 1987, Technical Report, NOAA Office of Oceanography and Marine Assessment, Ocean Assessments Division, Rockville, MD, 1988.
36. **O'Connor, T.,** Coastal Environmental Quality in the United States, 1990: Chemical Contamination in Sediment and Tissues, Special NOAA 20th Anniversary Report, NOAA Ocean Assessments Division, Rockville, MD, 1990.
37. **Manheim, F. T.,** Who is doing what in marine dumping?, in *Wastes in the Ocean,* Vol. 1, Industrial and Sewage Wastes in the Ocean, Duedall, I. W., Ketchum, B. H., Park, P. K., and Kester, D. R., Eds., John Wiley & Sons, New York, 1983, 47.
38. **Pokryfki, L. and Randall, R. E.,** Nearshore hypoxia in the bottom water of the northwestern Gulf of Mexico from 1981 to 1984, *Mar. Environ. Res.,* 22, 75, 1987.
39. **Reyes, E. and Merino, M.,** Diel dissolved oxygen dynamics and eutrophication in a shallow, well-mixed tropical lagoon (Cancun, Mexico), *Estuaries,* 14, 372, 1991.
40. **Officer, C. B., Biggs, R. B., Taft, J. L., Cronin, L. E., Tyler, M. A., and Boynton, W. R.,** Chesapeake Bay anoxia: origin, development, and significance, *Science,* 223, 22, 1984.
41. **Seliger, H. H., Boggs, J. A. and Biggley, W. H.,** Catastrophic anoxia in the Chesapeake Bay in 1984, *Science,* 228, 70, 1985.
42. **Correll, D. L.,** Nutrients in Chesapeake Bay, in *Contaminant Problems and Management of Living Chesapeake Bay Resources,* Majumdar, S. K., Hall, L. W., Jr., and Austin, H. M., Eds., Pennsylvania Academy of Science, Easton, PA, 1987, 298.
43. **Parker, C. A. and O'Reilly, J. E.,** Diel dissolved oxygen dynamics and eutrophication in a shallow, well-mixed tropical lagoon (Cancun, Mexico), *Estuaries,* 14, 372, 1991.
44. **Welsh, B. L. and Eller, F. C.,** Mechanisms controlling summertime oxygen depletion in western Long Island Sound, *Estuaries,* 14, 265, 1991.
45. **Tyler, M. A.,** Flow-induced variation in transport and deposition pathways in the Chesapeake Bay: the effect on phytoplankton dominance and anoxia, in *Estuarine Variability,* Wolfe, D. A., Ed., Academic Press, New York, 1986, 161.
46. **Taft, J. L., Taylor, W. R., Hartwig, E. O., and Loftus, R.,** Seasonal oxygen depletion in Chesapeake Bay, *Estuaries,* 3, 242, 1980.
47. **Day, J. W., Jr., Hall, C. A. S., Kemp, W. M., and Yanez Arancibia, A.,** *Estuarine Ecology,* John Wiley & Sons, New York, 1989.
48. **Kuo, A. Y., Park, K., and Moustafa, M. Z.,** Spatial and temporal variabilities of hypoxia in the Rappahannock River, Virginia, *Estuaries,* 14, 113, 1991.
49. **Koppelman, L. E., Weyl, P. K., Gross, M. G., and Davies, D. S.,** *Urban Sea: Long Island Sound,* Praeger Publishers, New York, 1976, 223.
50. **Bowman, M. J.,** Nutrient distributions and transport in Long Island Sound, *Estuarine Coastal Mar. Sci.,* 5, 531, 1977.
51. **Perkins, E. J.,** *The Biology of Estuaries and Coastal Waters,* Academic Press, New York, 1974.
52. **Biggs, R. B., DeMoss, T. B., Carter, M. M., and Beasley, E. L.,** Susceptibility of U.S. estuaries to pollution, *Rev. Aquat. Sci.,* 1, 189, 1989.
53. National Academy of Sciences, *Petroleum in the Marine Environment,* National Academy Press, Washington, D.C., 1975.
54. National Academy of Sciences, *Oil in the Sea: Inputs, Fates, and Effects,* National Academy Press, Washington, D.C., 1985.
55. National Research Council, *Using Oil Spill Dispersants on the Sea,* National Academy Press, Washington, D.C., 1989.
56. **Farrington, J. W., Frew, M. N., and Gschwend, P. M.,** Hydrocarbons in cores of northwestern Atlantic coastal and continental marine sediments, *Estuarine Coastal Mar. Sci.,* 5, 793, 1977.
57. **Laflamme, R. E. and Hites, R. A.,** The global distribution of polyaromatic hydrocarbons in recent sediments, *Geochim. Cosmochim. Acta,* 42, 289, 1978.
58. **DeLaune, R. D., Gambrell, R. P., Pardue, J. H., and Patrick, W. H., Jr.,** Fate of petroleum hydrocarbons and toxic organics in Louisiana coastal environments, *Estuaries,* 13, 72, 1990.
59. **Fang, C. S.,** Petroleum drilling and production operations in the Gulf of Mexico, *Estuaries,* 13, 89, 1990.
60. **Lytle, T. F. and Lytle, J. S.,** Contaminants in sediments from the central Gulf of Mexico, *Estuaries,* 13, 98, 1990.
61. **Pelletier, E., Ouellet, S., and Paquet, M.,** Long-term chemical and cytochemical assessment of oil contamination in estuarine intertidal sediments, *Mar. Pollut. Bull.,* 22, 273, 1991.
62. **Lake, J. L., Norwood, C., Dimock, C., and Bowen, R.,** Origins of polycyclic aromatic hydrocarbons in estuarine sediments, *Geochim. Cosmochim. Acta,* 43, 1847, 1979.
63. **Lee, R. F. and Benson, A. A.,** Fates of petroleum in the sea: biological aspects, in *Proceedings of the Workshop on Inputs, Fates and Effects of Petroleum in the Marine Environment,* Vol. 2, National Academy of Sciences, Washington, D.C., 1973, 541.
64. **Eisler, R.,** Polycyclic Aromatic Hydrocarbon Hazards to Fish, Wildlife, and Invertebrates: A Synoptic Review, Biol. Rept. 85 (1.11), U.S. Fish and Wildlife Service, Washington, D.C., 1987.
65. **Wijayaratne, R. D. and Means, J. C.,** Sorption of polycyclic aromatic hydrocarbons by natural estuarine colloids, *Mar. Environ. Res.,* 11, 77, 1984.
66. **Neff, J. M.,** *Polycyclic Aromatic Hydrocarbons in the Aquatic Environment: Sources, Fates, and Biological Effects,* Applied Science Publishers, London, 1979.
67. **Neff, J. M.,** Polycyclic aromatic hydrocarbons, in *Fundamentals of Aquatic Toxicology,* Rand, G. M. and Petrocelli, S. M. Eds., Hemisphere, New York, 1985.

68. **Klauda, R. J. and Bender, M. E.,** Contaminant effects on Chesapeake Bay finfishes, in *Contaminant Problems and Management of Living Chesapeake Bay Resources,* Majumdar, S. K., Hall, L. W., Jr., and Austin, H. M., Eds., Pennsylvania Academy of Science, Easton, PA, 1987, 321.
69. **Jackim, E. and Lake, C.,** Polynuclear aromatic hydrocarbons in estuarine and nearshore environments, in *Estuarine Interactions,* Wiley, M. L., Ed., Academic Press, New York, 1978, 415.
70. **Richards, N. L. and Jackson, B. L.,** Symposium: Carcinogenic Polynuclear Aromatic Hydrocarbons in the Marine Environment, U.S. Environmental Protection Agency, EPA-600/9–82–013, Gulf Breeze, FL, 1982.
71. **Mix, M. C.,** Polycyclic aromatic hydrocarbons in the aquatic environment: occurrence and biological monitoring, in *Reviews in Environmental Toxicology,* Vol. 1, Hodgson, E., Ed., Elsevier, Amsterdam, 1984, 51.
72. **McElroy, A. E., Farrington, J. W., and Teal, J. M.,** Bioavailability of polycyclic aromatic hydrocarbons in the aquatic environment, in *Metabolism of Polycyclic Aromatic Hydrocarbons in the Aquatic Environment,* Varanasi, U., Ed., CRC Press, Boca Raton, FL, 1989, 1.
73. **McElroy, A. E.,** Polycyclic aromatic hydrocarbon metabolism in the polychaete *Nereis virens. Aquat. Toxicol.,* 18, 35, 1990.
74. **Lee, R. F.,** Mixed function oxygenases (MFO) in marine invertebrates, *Mar. Biol. Lett.,* 2, 87, 1981.
75. **James, M. O.,** Biotransformation and deposition of PAH in aquatic invertebrates, in *Metabolism of Polycyclic Aromatic Hydrocarbons in the Aquatic Environment,* Varanasi, U., Ed., CRC Press, Boca Raton, FL, 1989, 69.
76. **Stegeman, J. J.,** Polynuclear aromatic hydrocarbons and their metabolism in the marine environment, in *Polycyclic Hydrocarbons and Cancer,* Vol. 3, Gelboin, H. V. and Ts'O, P. O. P., Eds., Academic Press, New York, 1981, 1.
77. **Varanasi, U., Stein, J. S., and Nishimoto, M.,** Biotransformation and disposition of PAH in fish, in *Metabolism of Polycyclic Aromatic Hydrocarbons in the Aquatic Environment,* Varanasi, U., Ed., CRC Press, Boca Raton, FL, 1989, 93.
78. **Wade, T. L., Kennicutt, M. C., II, and Brooks, J. M.,** Gulf of Mexico hydrocarbon seep communities. Part III. Aromatic hydrocarbon concentrations in organisms, sediments, and water, *Mar. Environ. Res.,* 27, 19, 1989.
79. **Lauenstein, G. G. and O'Connor, T. P.,** Measuring the health of the U.S. coastal waters, *Sea Technol.,* 29, 29, 1988.
80. **Wade, T. L., Atlas, E. L., Brooks, J. M., Kennicutt, M. C., II, Fox, R. G., Sericano, J., Garcia-Romerio, B., and DeFreitas, D.,** NOAA Gulf of Mexico Status and Trends Program: trace organic contaminant distribution in sediments and oysters, *Estuaries,* 11, 171, 1988.
81. **Olsen, C. R., Cutshall, N. H., and Larsen, I. L.,** Pollutant-particle association and dynamics in coastal marine environments: a review, *Mar. Chem.,* 11, 501, 1982.
82. **Guerin, W. F. and Jones, G. E.,** Estuarine ecology of phenanthrene-degrading bacteria, *Estuarine Coastal Shelf Sci.,* 29, 115, 1989.
83. **Larsen, P. F.,** Marine environmental quality in the Gulf of Maine, *Rev. Aquat. Sci.,* 6, 67, 1992.
84. **Roberts, M. H., Jr., Sved, D. W., and Felton, S. P.,** Temporal changes in AHH and SOD activities in feral spot from the Elizabeth River, a polluted sub-estuary, *Mar. Environ. Res.,* 23, 89, 1987.
85. **Huggett, R. J., de Fur, P. O., and Bieri, R. H.,** Organic compounds in Chesapeake Bay, *Mar. Pollut. Bull.,* 19, 454, 1988.
86. **Stickney, R. R.,** *Estuarine Ecology of the Southeastern United States and Gulf of Mexico,* Texas A & M University Press, College Station, TX, 1984.
87. **Dale, P. E. R. and Hulsman, K.,** A critical review of salt marsh management methods for mosquito control, *Rev. Aquat. Sci.,* 3, 281, 1990.
88. U.S. Environmental Protection Agency, Ambient Water Quality Criteria for Polychlorinated Biphenyls, U.S. Environmental Protection Agency, Washington, D.C., 1980.
89. **Brown, M. P., Werner, M. B., Sloan, R. J., and Simpson, K. W.,** Polychlorinated biphenyls in the Hudson River, *Environ. Sci. Technol.,* 19, 656, 1985.
90. **Waid, J. S., Eds.,** *PCBs and the Environment,* Vol. 1, CRC Press, Boca Raton, FL, 1986.
91. **O'Connor, J. M., Klotz, J. B., and Kneip, T. J.,** Sources, sinks and distribution of organic contaminants in the New York Bight ecosystem, in *Ecological Stress and the New York Bight: Science and Management,* Estuarine Research Federation, Columbia, SC, 1982, 631.
92. **Young, D. R., Gossett, R. W., and Heesen, T. C.,** Persistence of chlorinated hydrocarbon contamination in a California marine ecosystem, in *Oceanic Processes in Marine Pollution,* Vol. 5, Urban Wastes in Coastal Marine Environments, Wolfe, D. A. and O'Connor, T. P., Eds., Robert E. Krieger, Malabar, FL, 1988, 33.
93. **Mearns, A. J., Matta, M. B., Simecek-Beatty, D., Buchman, M. F., Shigenaka, G., and Wert, W. A.,** PCB and Chlorinated Pesticide Contamination in U.S. Fish and Shellfish: a Historical Assessment Report, NOAA Technical Memorandum NOS OMA 39, National Oceanic and Atmospheric Administration, Seattle, WA, 1988.
94. **Butler, P. A.,** Organochlorine residues in estuarine mollusks, 1965–72. National Pesticide Monitoring Program, *Pest. Monit. J.,* 6, 238, 1973.
95. **Schmitt, C. J., Ribick, M. A., Ludke, J. L., and May, T. W.,** National Pesticide Monitoring Program: Organochlorine Residues in Freshwater Fish, 1976–79, Resource Publication 152, U.S. Department of the Interior, Fish and Wildlife Service, Washington, D.C., 1983.
96. **Cutter, G. A.,** Trace elements in estuarine and coastal waters — U.S. studies from 1986–1990, *Rev. Geophys., Suppl.,* 1991, 639.
97. **Turkekian, K. K.,** The fate of metals in the oceans, *Geochim. Cosmochim. Acta,* 41, 1139, 1977.
98. **Duinker, J. C. and Nolting, R. F.,** Mixing, removal, and mobilization of trace metals in the Rhine estuary, *Neth. J. Sea Res.,* 12, 205, 1978.

99. **Muller, G. and Forstner, U.,** Heavy metals in sediments of the Rhine and Elbe estuaries: mobilization or mixing effect?, *Environ. Geol.*, 1, 33, 1975.
100. **Langston, W. J.,** Metals in sediments and benthic organisms in the Mersey estuary, *Estuarine Coastal Shelf Sci.*, 23, 239, 1986.
101. **Lyngby, J. E. and Brix, H.,** Monitoring of heavy metal contamination in the Limfjord, Denmark, using biological indicators and sediment, *Sci. Total Environ.*, 64, 239, 1987.
102. **Sinex, S. A. and Wright, D. A.,** Distribution of trace metals in the sediments and biota of Chesapeake Bay, *Mar. Pollut. Bull.*, 19, 425, 1988.
103. **Panutrakul, S. and Baeyens, W.,** Behaviour of heavy metals in a mud flat of the Scheldt estuary, Belgium, *Mar. Pollut. Bull.*, 22, 128, 1991.
104. **Viarengo, A.,** Heavy metals in marine invertebrates: mechanisms of regulation and toxicity at the cellular level, *Rev. Aquat. Sci.*, 1, 295, 1989.
105. **Nemerow, N. L.,** *Stream, Lake, Estuary, and Ocean Pollution,* Van Nostrand Reinhold, New York, 1985.
106. **Mance, G.,** *Pollution Threat of Heavy Metals in Aquatic Environments,* Elsevier, New York, 1987.
107. **Engel, D. W. and Brouwer, M.,** Trace metal-binding proteins in marine molluscs and crustaceans, *Mar. Environ. Res.*, 13, 177, 1984.
108. **Viarengo, A., Moore, M. N., Mancinelli, G., Mazzucotelli, A., and Pipe, R. K.,** Significance of metallothioneins and lysosomes in cadmium toxicity and homeostasis in the digestive gland cells of mussels exposed to the metal in presence or absence of phenanthrene, *Mar. Environ. Res.*, 17, 184, 1985.
109. **Lauenstein, G. G., Robertson, A., and O'Connor, T. P.,** Comparison of trace metal data in mussels and oysters from a mussel watch programme of the 1970s with those from a 1980s programme, *Mar. Pollut. Bull.*, 21, 440, 1990.
110. **Nichols, F. H., Cloern, J. E., Luoma, S. N., and Peterson, D. H.,** The modification of an estuary, *Science,* 231, 567, 1984.
111. **Wright, D. A. and Phillips, D. J. H.,** Chesapeake and San Francisco Bays: a study in contrast and parallels, *Mar. Pollut. Bull.*, 19, 405, 1988.
112. **Broaden, P. J. S. and Seed, R.,** *An Introduction to Coastal Ecology,* Blackie & Sons, London, 1985.
113. **Schell, W. R. and Nevissi, A. E.,** Radionuclides at the Hudson Canyon disposal site, in *Wastes in the Ocean,* Vol. 3, Radioactive Wastes and the Ocean, Park, P. K., Kester, D. R., Duedall, I. W., and Ketchum, B. H., Eds., John Wiley & Sons, New York, 1983, 183.
114. **Kiefer, J.,** *Biological Radiation Effects,* Springer-Verlag, New York, 1990.
115. **Pentreath, R. J.,** Radionuclides in marine fish, *Mar. Biol. Annu. Rev.*, 15, 365, 1977.
116. **Fowler, S. W.,** Biological transfer and transport processes, in *Pollutant Transfer and Transport in the Sea,* Kullenberg, G., Ed., CRC Press, Boca Raton, FL, 1982, 1.
117. **Martin, J. M. and Thomas, A. J.,** Origins, concentrations, and distributions of artificial radionuclides discharged by the Rhone River to the Mediterranean Sea, *J. Environ. Radioact.*, 11, 105, 1990.
118. United Nations, United Nations Scientific Committee on the Effects of Atomic Radiation, Ionizing Radiation: Sources and Biological Effects, Technical Report, United Nations, New York, 1982.
119. **Cherry, R. D. and Shannon, L. V.,** The alpha radioactivity of marine organisms, *Atomic Energy Rev.*, 12, 3, 1974.
120. **Cherry, R. D. and Heyraud, M.,** Evidence of high natural radiation dose domains in mid-water oceanic organisms, *Science,* 218, 54, 1982.
121. **Osterberg, C. L.,** Nuclear power wastes and the ocean, in *Wastes in the Ocean,* Vol. 4, Energy Wastes in the Ocean, Duedall, I. W., Kester, D. R., Park, P. K., and Ketchum, B. H., John Wiley & Sons, New York, 1985, 127.
122. **Hayes, D. W. and Sackett, W. M.,** Plutonium and cesium radionuclides in sediments of the Savannah River estuary, *Estuarine Coastal Shelf Sci.*, 25, 169, 1987.
123. **Donoghue, J. F., Bricker, O. P., and Olsen, C. R.,** Particle-borne radionuclides as tracers for sediment in the Susquehanna River and Chesapeake Bay, *Estuarine Coastal Shelf Sci.*, 29, 341, 1989.
124. **Hamilton, E. I.,** Radionuclides and large particles in estuarine sediments, *Mar. Pollut. Bull.*, 20, 603, 1989.
125. **Landford, T. E.,** *Electricity Generation and the Ecology of Natural Waters,* Liverpool University Press, Liverpool, England, 1983.
126. **Eisenbud, M.,** *Environmental Radioactivity,* 3rd ed., Academic Press, New York, 1987.
127. **Wilson, J. G.,** *The Biology of Estuarine Management,* Croom Helm, London, 1988.
128. **Ketchum, B. H., Capuzzo, J. M., Burt, W. V., Duedall, I. W., Park, P. K., and Kester, D. R., Eds.,** *Wastes in the Ocean,* Vol. 6, Nearshore Waste Disposal, John Wiley & Sons, New York, 1985.
129. **Kester, D. R., Ketchum, B. H., Duedall, I. W., and Park, P. K., Eds.,** The problem of dredged-material disposal, in *Wastes in the Ocean,* Vol. 2, Dredged-Material Disposal in the Ocean, Kester, D. R., Ketchum, B. H., Duedall, I. W., and Park, P. K., Eds., John Wiley & Sons, New York, 1983, 3.
130. **McAnally, W. H. and Adamec, S. A., Jr.,** Designing Open Water Disposal for Dredged Muddy Sediments. Technical Report, U.S. Army Engineer Waterways Experiment Station, Vicksburg, Mississippi, 1987.
131. **O'Connor, J. M. and O'Connor, S. G.,** Evaluation of the 1980 Capping Operations at the Experimental Mud Dump Site, New York Bight Apex, Technical Report D-83–3, U.S. Army Engineer Waterways Experiment Station, Vicksburg, Mississippi, 1983.
132. **Parker, J. H. and Valente, R. M.,** Long-Term Sand Cap Stability: New York Dredged-Material Disposal Site, Contract Report CERC-88–2, U.S. Army Engineer Waterways Experiment Station, Vicksburg, Mississippi, 1988.

133. **Herbich, J. B. and Haney, J. P.,** Dredging, in *The Encyclopedia of Beaches and Coastal Environments,* Schwartz, M. L., Ed., Dowden, Hutchinson & Ross, Stroudsburg, PA, 1982, 379.
134. **Nichols, M. M.,** The problem of misplaced sediment, in *Ocean Dumping and Marine Pollution,* Palmer, H. D. and Gross, M. G., Eds., Dowden, Hutchinson & Ross, Stroudsburg, PA, 1979, 147.
135. **Rhoads, D. C., McCall, P. L., and Yingst, J. Y.,** Disturbance and production on the estuarine seafloor, *Am. Sci.,* 66, 577, 1978.
136. **Kirby, C. J., Keeley, J. W., and Harrison, J.,** An overview of the technical aspects of the Corps of Engineers national dredged-material research program, in *Estuarine Research,* Vol. 2, Geology and Engineering, Cronin, L. E., Ed., Academic Press, New York, 1975, 523.
137. **Morgan, R. P., II and Carpenter, E. J.,** Biocides, in *Power Plant Entrainment: a Biological Assessment,* Schubel, J. R. and Marcy, B. C., Academic Press, New York, 1978, 95.
138. **Hocutt, C. H., Stauffer, J. R., Jr., Edinger, J. E., Hall, L. W., Jr., and Morgan, R. P., II, Eds.,** *Power Plants: Effects on Fish and Shellfish Behaviour,* Academic Press, New York, 1980.
139. **Kennish, M. J., Roche, M. B., and Tatham, T. R.,** Anthropogenic effects on aquatic communities, in *Ecology of Barnegat Bay, New Jersey,* Kennish, M. J. and Lutz, R. A., Eds., Springer-Verlag, New York, 1984, 318.
140. **Abbe, C. R.,** Thermal and other discharge-related effects on the bay ecosystem, in *Ecological Studies in the Middle Reach of Chesapeake Bay: Calvert Cliffs,* Heck, K. L., Jr., Ed., Springer-Verlag, Berlin, 1987, 270.
141. Jersey Central Power and Light Company, Oyster Creek and Forked River Nuclear Generating Stations 316(a) and (b) Demonstration, Volume 1. Technical Report, Jersey Central Power and Light Company, Morristown, NJ, 1978.
142. Boston Edison Company, Marine Ecology Studies related to Operation of Pilgrim Station, Semi-Annual Report No. 29, Boston Edison Company, Braintree, MA, 1987.
143. **Clark, J. R. and Brownell, W.,** Electric Power Plants in the Coastal Zone: Environmental Issues, Spec. Publ. 7, American Littoral Society, Highlands, NJ.
144. **Marcy, B. C., Jr., Beck, A. D., and Ulanowics, R. E.,** Effect and impacts of physical stress on entrained organisms, in *Power Plant Entrainment: a Biological Assessment,* Schubel, J. and Marcy, B. C., Jr., Eds., Academic Press, New York, 1978, 135.
145. **Van Winkle, W., Ed.,** *Proceedings of the Conference for Assessing the Effects of Power-Plant-Induced Mortality on Fish Populations,* Pergamon Press, New York, 1977.
146. **Ogawa, H.,** Modelling of power plant impacts on fish populations, *Environ. Manag.,* 3, 321, 1979.
147. **O'Connor, S. G. and McErlean, A. J.,** The effects of power plants on productivity of the nekton, in *Estuarine Research,* Vol. 1, Cronin, L. E., Ed., Academic Press, New York, 1975, 494.
148. **Goodyear, C. P.,** Assessing the impact of power plant mortality on the compensatory reserve of fish populations, in *Proceedings of the Conference for Assessing the Effects of Power-Plant-Induced Mortality on Fish Populations,* Van Winkle, V., Ed., Pergamon Press, New York, 1977, 198.
149. **Marcy, B. C., Jr., Kranz, V. R., and Barr, R. P.,** Ecological and behavioral characteristics of fish eggs and young influencing their entrainment, in *Power Plants: Effects on Fish and Shellfish Behavior,* Hocutt, C. H., Stauffer, J. R., Jr., Edinger, J., Hall, L. W., Jr., and Morgan, R. P., II, Eds., Academic Press, New York, 1980, 29.
150. **Susani, L.,** Liver Lesions in Feral Fish: a Discussion of Their Relationship to Environmental Pollutants, NOAA Technical Memorandum NOS OMA 27, NOAA Office of Oceanography and Marine Assessment, Rockville, MD, 1986.
151. **O'Connor, T. P., Price, J. E., and Parker, C. A.,** Results from NOAA's national status and trends program on distributions, effects, and trends of chemical contamination in the coastal and estuarine United States, *Oceans 89 Conf. Proc.,* 2, 569, 1989.

5.1 ORGANIC LOADING, SEWAGE TREATMENT, AND EUTROPHICATION

Table 5.1—1
SITE SELECTION CONSIDERATIONS FOR WASTE DISPOSAL IN THE OCEAN

Physical	Sedimentary	Biological
Ocean flow	Physical and chemical properties of wastes and sediment	Fishing grounds
Surface waves	Sorption capacity	Aquaculture sites
Wind-driven surface currents	Distribution coefficients	Breeding and nursery grounds
Interior circulation	Sedimentation	Migration routes
Turbulent diffusion	Sedimentation disperson	Productivity
Shear diffusion	Bioturbation	Recreational areas
Vertical mixing	Sediment stability	
Modeling advection and diffusion		

From GESAMP (IMCO/FAO/UNESCO/WMO/WHO/IAEA/UN/UNEP Joint Group of Experts on Scientific Aspects of Marine Pollution), Scientific Criteria for the Selection of Waste Disposal Sites at Sea, Reports and Studies No. 16, Inter-Governmental Maritime Consultative Organization, London, 1982. With permission.

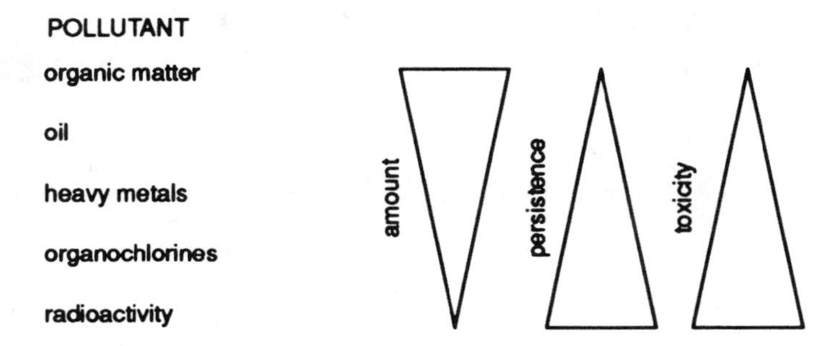

FIGURE 5.1—1. Major categories of contaminants commonly discharged into estuarine and coastal marine waters, together with the properties which make them an environmental concern. (From Wilson, J. G., *The Biology of Estuarine Management,* Croom Helm, London, 1988. With permission.)

Table 5.1—2
PATHOGENS LIKELY TO BE ASSOCIATED WITH SEWAGE SLUDGE

Bacteria	Viruses	Protozoa	Helminths
Salmonella spp.	Poliovirus	*Entamoeba histolytica*	*Echinococcus granulosus*
Shigella spp.	Coxsackie A and B	*Acanthamoeba* spp.	*Hymenolepis nana*
Vibrio spp.	Echovirus	*Giardia* spp.	*Taenia saginata*
Mycobacterium spp.	Adenovirus		*Fasciola hepatica*
Bacillus anthracis	Reovirus		*Ascaris lumbricoides*
Clostridium perfringens	Parvovirus		*Enterobius vermicularis*
Yersinia spp.	Rotavirus		*Strongyloides* spp.
Campylobacter spp.	Hepatitis A		*Trichuris trichiura*
Pseudomonas spp.	Norwalk and related gastroenteric viruses		*Toxocara canis*
			Trichostrongylus spp.
Leptospira spp.			
Listeria monocytogenes			
Escherichia coli			
Clostridium botulinum			

From Alderslade, R., The problems of assessing possible hazards to the public health associated with the disposal of sewage sludge to land, in *Recent Experience in the United Kingdom: Characterization, Treatment, and Use of Sewage Sludge,* Hermite, P. L. and Ott, H., Eds., Reidel Publishing, Dordrecht, 1981, 372. With permission.

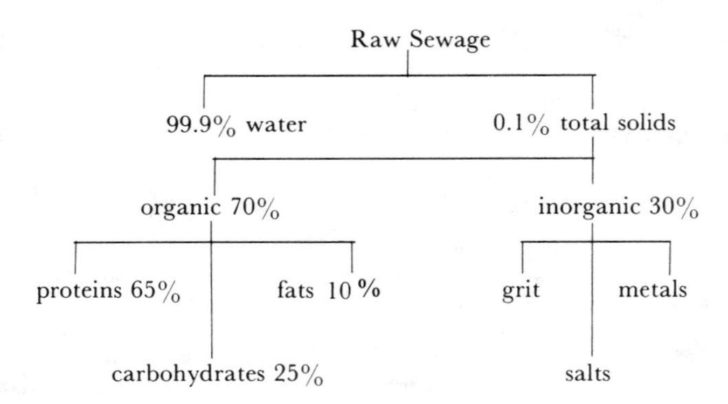

FIGURE 5.1—2. Composition of a typical raw sewage. (From Harrison, R. M., Ed., *Pollution: Causes, Effects, and Control,* 2nd ed., Royal Society of Chemistry, Cambridge, 1990. With permission.)

Table 5.1—3
PHYSICAL METHODS OF PRETREATMENT

Process	Aim	Examples
Screening	Removal of coarse solids	Vegetable canneries, paper mills
Centrifuging	Concentration of solids	Sludge dewatering in chemical industry
Filtration	Concentration of fine solids	Final polishing and sludge dewatering in chemical and metal processing
Sedimentation	Removal of settleable solids	Separation of inorganic solids in ore extraction, coal, and clay production
Flotation	Removal of low-specific gravity solids and liquids	Separation of oil, grease, and solids in chemical and food industry
Freezing	Concentration of liquids and sludges	Recovery of pickle liquor and nonferrous metals
Solvent extraction	Recovery of valuable materials	Coal carbonizing and plastics manufacture
Ion exchange	Separation and concentration	Metal processing
Reverse osmosis	Separation of dissolved solids	Desalination of process and wash water
Adsorption	Concentration and removal of trace impurities	Pesticide manufacture, dyestuffs removal

From Harrison, R. M., Ed., *Pollution: Causes, Effects, and Control,* 2nd ed., Royal Society of Chemistry, Cambridge, 1990. With permission.

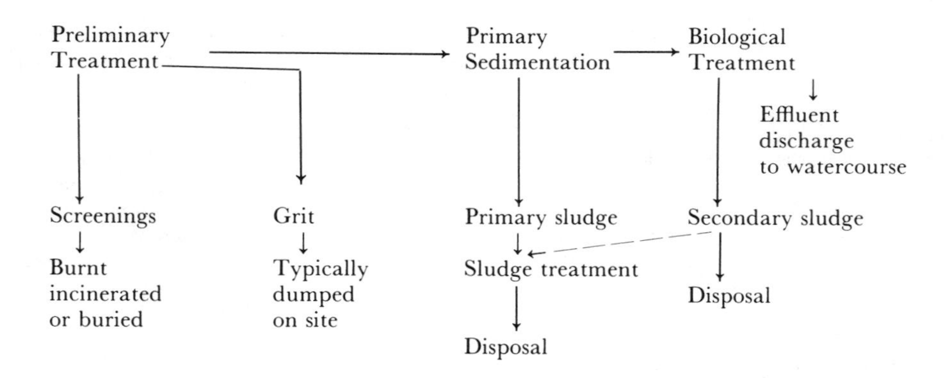

FIGURE 5.1—3. Flow diagram of a conventional sewage treatment works. (From Harrison, R. M., Ed., *Pollution: Causes, Effects, and Control,* 2nd ed., Royal Society of Chemistry, Cambridge, 1990. With permission.)

Table 5.1—4
AMOUNTS OF HEAVY METAL IONS REMOVED FROM SEWAGE BY SLUDGES

	Primary sedimentation		Percolating filter treatment		Activated-sludge process	
Heavy metal ion	Metal concentration in crude sewage (mg l^{-1})	Proportion removed by treatment (%)	Metal concentration in settled sewage (mg l^{-1})	Proportion removed by treatment (%)	Metal concentration in settled sewage (mg l^{-1})	Proportion removed by treatment (%)
Copper	Up to 0.8	45	Up to 0.44	20	0.4	54
					Up to 0.44	80
					0.4–25	50–79
	Up to 5	12			28	90–93
Dichromate	(as Cr)		(as Cr)		(as Cr)	
	Up to 1.2	28	Up to 0.86	32	Up to 0.86	67–70
					4.0	6.3
					0.5–2	~100
					5	50
					50	10
Iron (Ferric)	3–9	40	1.8–5.4	Nil	1.8–5.4	80
Lead	0.3–0.9	40	0.18–0.54	30	0.18–0.54	90
Nickel	0.1–0.3	20	0.08–10	40	0.08–0.24	30
					2.0	31
					2.5–10	30
Zinc	0.7–1.6	40	0.4–1.0	30	0.4–1.0	60
					2.5	90
					2.5	95
	Up to 5	12			7.5	100
					15	78
					20	74

From Harrison, R. M., Ed., *Pollution: Causes, Effects, and Control*, 2nd ed., Royal Society of Chemistry, Cambridge, 1990. With permission.

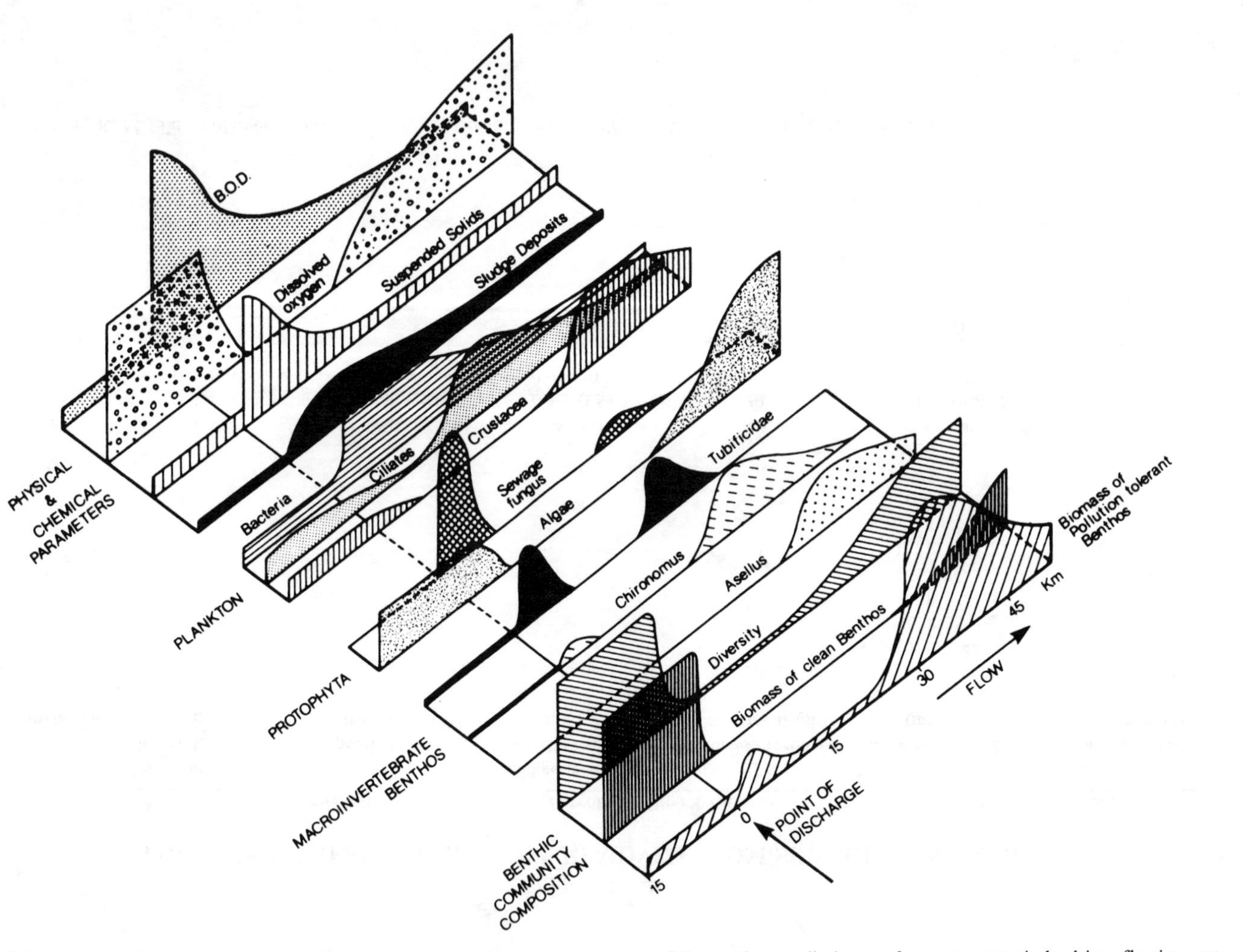

FIGURE 5.1—4. Spatial variation of physical, chemical, and biological consequences of the continuous discharge of a severe organic load into flowing water. (From Boudou, A. and Ribeyre, F., Eds., *Aquatic Ecotoxicology: Fundamental Concepts and Methodologies,* CRC Press, Boca Raton, FL, 1989.)

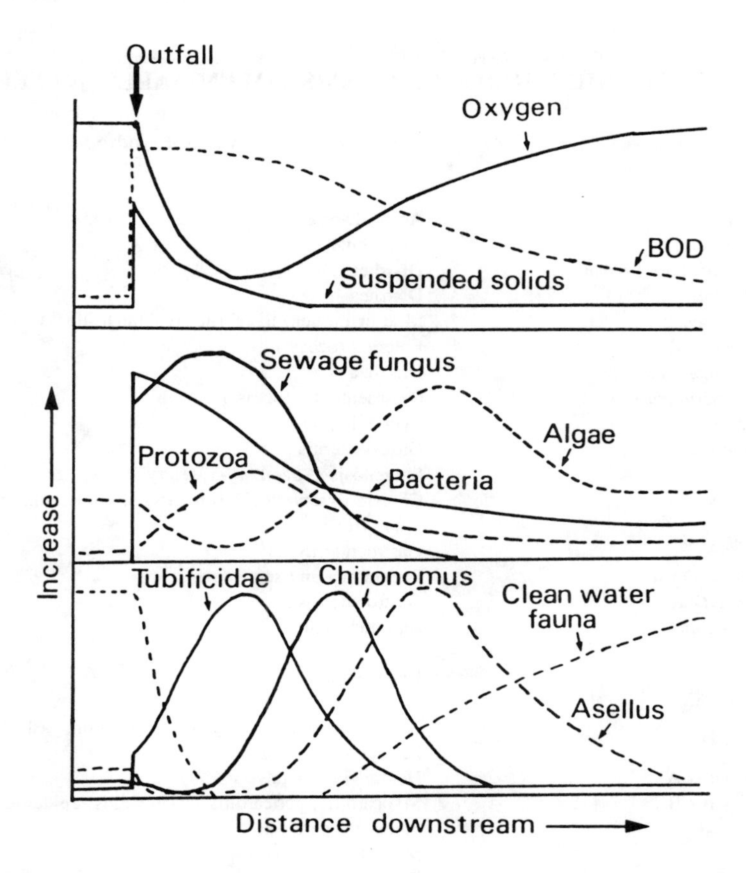

FIGURE 5.1—5. Changes in water quality and populations of organisms in a river below a discharge of an organic effluent. (From Harrison, R. M., Ed., *Pollution: Causes, Effects, and Control,* 2nd ed., Royal Society of Chemistry, Cambridge, 1990. With permission.)

Table 5.1—5
HUMAN PATHOGENIC MICROORGANISMS POTENTIALLY WATERBORNE

Pathogen	Clinical syndrome
Bacteria	
Aeromonas hydrophila	Acute diarrhea
Campylobacter spp.	Acute enteritis
Enterotox. *Clostridium perfringens*	Diarrhea
Enterotox. *E. coli*	Diarrhea
Francisella tularensis	Mild or influenzal, febrile, typhoidal illness
Klebsiella pneumoniae	Enteritis (occas.)
Plesiomonas shigelloides	Diarrhea
Pseudomonas aeruginosa	Gastroenteritis (occas.)
Salmonella typhi	Typhoid fever
Other salmonellae	Gastroenteritis
Shigella spp.	Shigellosis (''bacillary dysentry'')
Vibrio cholerae	Cholera dysentery (01 serovars) or choleralike infection (non-01)
V. fluvialis	Gastroenteritis
Lactose-positive *Vibrio*	Pneumonia and septicemia
V. parahaemolyticus	Gastroenteritis
Yersinia enterocolitica	Enteritis, ileitis
Cyanobacteria	
Cylindrospermopsis spp.	Hepatoenteritis
Viruses	
Enteroviruses	Aseptic meningitis, respiratory infection, rash, fever
Poliovirus	Paralysis, encephalitis
Coxsackie virus A	Herpangina, paralysis
Coxsackie virus B	Myocarditis, pericarditis, encephalitis, epidemic pleurodynia, transient paralysis
Echovirus	Meningitis, enteritis
Types 68—71	Encephalitis, acute hemorrhagic conjunctivitis
Hepatitis A	Infectious hepatitis type A
Hepatitis non-A, non-B	Hepatitis type non-A, non-B
Influenza A	Influenza
Norwalk and other parvoviruslike agents	Epidemic, acute nonbacterial gastroenteritis
Rotavirus	Nonbacterial, endemic, infantile gastroenteritis; epidemic vomiting and diarrhea
Protozoa	
Balantidium coli	Balantidiasis (balantidial dysentery)
Cryptosporidium	Cryptosporidiosis
Entamoeba histolytica	Amoebiasis (amoebic dysentery)
Giardia lamblia	Giardiasis — mild, acute, or chronic diarrhea
Helminths	
Ascaris	Ascariasis (roundworm infection)
Ancylostoma	Hookworm infection
Clonorchis	Clonorchiasis (Chinese liver fluke infection)
Diphyllobothrium	Diphyllobothriasis (broadfish tapeworm infection)
Dracunculus mediensis	Dracontiasis (Guinea worm infection)
Fasciola	Fascioliasis (sheep liver fluke infection)
Fasciolopsis	Fasciolopsiasis (giant intestinal fluke infection)
Paragonimus	Paragonimiasis (Oriental lung fluke infection)
Spirometra mansoni	Sparganosis (plerocercoid tapeworm larvae infection)
Taenia	Taeniasis (tapeworm infection)
Trichostrongylus	Trichostrongyliasis
Trichuris	Trichuriasis (whipworm infection)

From Harrison, R. M., Ed., *Pollution: Causes, Effects, and Control,* 2nd ed., Royal Society of Chemistry, Cambridge, 1990. With permission.

Table 5.1—6
HUMAN PATHOGENIC MICROORGANISMS POTENTIALLY WATER-CONTACT TRANSMITTED

Pathogen	Clinical syndrome
Bacteria	
Aeromonas hydrophila	Wound and ear infections, septicemia, meningitis, endocarditis, corneal ulcers
A. sobria	Wound and ear infections
Chromobacterium violaceum	Septicemia
Clostridium perfringens	Wound infection — gas gangrene
Klebsiella pneumoniae	Pneumonia, bacteremia
Legionella spp.	Legionellosis (Legionnaires' disease)
Leptospira spp.	Leptospirosis (Weil's disease — jaundice, hemorrhages, aseptic meningitis)
Mycobacterium marinum	Skin infection (''swimming pool granuloma'')
M. ulcerans	Skin infection (progressive subcutaneous ulceration)
Pseudomonas aeruginosa	Otitis externa and media; follicular dermatitis (pruritic pustular rash)
P. pseudomallei	Meliodosis (glanderslike infection)
Staphylococcus aureus	Wound and skin infections
Halophilic vibrios (incl. *Vibrio parahaemolyticus, Vibrio alginolyticus,* lactose positive *Vibrio*)	Wound and ear infections, conjunctivitis, salpingitis, pneumonia, septicemia
Viruses	
Adenovirus	Pharyngoconjunctivitis (swimmning pool conjunctivitis), respiratory infection
Adenosatellovirus	Associated with adenovirus type 3 conjunctivitis and respiratory infection in children but etiology not clearly established
Protozoa	
Naegleria fowleri	Primary amoebic meningoencephalitis (PAME)
Helminths	
Schistosoma spp.	Schistosomiasis (bilharzia)
Avian schistosomes *(Trichobilharzia, Austrobilharzia)*	Schistosome dermatitis (swimmer's itch)
Ancylostoma duodenale	Hookworm infection
Necator americanus	Hookwork infection

Compiled from McNeill, A. R., Australian Water Resources Tech. Pap. No. 85, Australian Government Publishing Service, Canberra, 1985, 561.

Table 5.1—7
CHEMICALS USED IN INDUSTRIAL WASTE TREATMENT

Chemical	Purpose
Calcium hydroxide	pH adjustment, precipitation of metals, and assisting sedimentation
Sodium hydroxide	Used mainly for pH adjustment in place of lime
Sodium carbonate	pH adjustment and precipitation of metals with soluble hydroxide
Carbon dioxide	pH adjustment
Aluminum sulfate	Solids separation
Ferrous sulfate	Solids separation
Chlorine	Oxidation

From Harrison, R. M., Ed., *Pollution: Causes, Effects, and Control,* 2nd ed., Royal Society of Chemistry, Cambridge, 1990. With permission.

Table 5.1—8
SCALE OF EUTROPHICATION IN CHESAPEAKE BAY AND ITS TRIBUTARY ESTUARIES

Ecosystem		Ecological description	Surface area (m^2) (10^6)	Vol. (m^3) (10^6)	Avg. depth (m)	Retention time (years)
Patuxent	1963	Noneutrophic	137	660	4.8	1.70
	1969–71	Eutrophic	137	660	4.8	1.70
	1978	Eutrophic	137	660	4.8	1.70
York	1969–71	Noneutrophic	210	910	4.3	0.72
Rappahannock	1969–71	Noneutrophic	400	1,780	4.5	1.27
James	1969–71	Eutrophic	600	2,400	3.6	0.39
Potomac	1913	Noneutrophic	1,250	7,150	5.8	1.07
	1954	Eutrophic	1,250	7,150	5.8	1.07
	1969–71	Hypereutrophic	1,250	7,150	5.8	1.07
	1977–78	Eutrophic	1,250	7,150	5.8	1.07
Chesapeake Bay	1969–71	Localized				
(Including Tribs)		eutrophic	11,500	74,000	6.5	1.16
(Excluding Tribs)		conditions	6,500	52,000	8.4	1.32

From Jaworski, N. A., in *Estuaries and Nutrients,* Neilson, B. J. and Cronin, L. E., Eds., Humana Press, Clifton, NJ, 1981, 37. With permission.

Table 5.1—9
EXTERNAL NUTRIENT LOADINGS FOR CHESAPEAKE BAY AND ITS TRIBUTARY ESTUARIES

		External phosphorus			External nitrogen			
		(g/yr) (10^6)	(g/m^2/yr)	(g/m^3/yr)	(g/yr) (10^6)	(g/m^2/yr)	(g/m^3/yr)	At. N:P ratio of load
Patuxent	1963	170	1.24	0.26	930	6.7	1.4	12
	1969–71	250	1.82	0.38	1,110	8.1	1.7	10
	1978	420	3.06	0.64	1,500	11.4	2.4	8
York	1969–71	160	0.76	0.18	1,190	5.6	1.3	17
Rappahannock	1969–71	180	0.45	0.10	1,500	3.8	0.8	19
James	1969–71	1,780	2.70	0.70	10,300	15.6	4.2	13
Potomac	1913	910	0.73	0.13	18,600	14.8	2.6	46
	1954	2,000	1.63	0.28	22,600	18.1	3.1	26
	1969–71	5,380	4.30	0.80	25,200	20.2	3.5	11
	1977–78	2,520	2.01	0.35	32,800	26.2	4.6	30
Chesapeake Bay	1969–71							
(Including Tribs)		15,000	1.30	0.20	109,100	9.5	1.5	16
(Excluding Tribs)		7,350	1.10	0.10	70,160	10.8	1.3	22

From Jaworski, N. A., in *Estuaries and Nutrients,* Neilson, B. J. and Cronin, L. E., Eds., Humana Press, Clifton, NJ, 1981, 37. With permission.

Table 5.1—10
CONCENTRATIONS OF NITROGEN AND PHOSPHORUS IN 28 ESTUARINE ECOSYSTEMS[a]

Estuary	Nutrient concentration	
	Nitrogen	**Phosphorus**
River dominated		
Pamlico River, North Carolina	1.5	8.0
Narragansett Bay, Rhode Island	0.6	1.6
Western Wadden Sea, Netherlands	3.0	2.0
Eastern Wadden Sea, Netherlands	4.0	2.5
Mid-Patuxent River, Maryland	4.2	2.3
Upper Patuxent River, Maryland	10.0	2.0
Long Island Sound, New York	1.5	0.5
Lower San Francisco Bay, California	20.6	3.8
Upper San Francisco Bay, California	11.5	2.0
Barataria Bay, Louisiana	4.6	0.8
Victoria Harbor, British Columbia	11.5	2.0
Mid-Chesapeake Bay, Maryland	4.5	0.6
Upper Chesapeake Bay, Maryland	5.0	6.0
Duwamish River, Washington	60.0	3.0
Hudson River, New York	5.0	0.16
Apalachicola Bay, Florida	10.0	0.1
Embayments		
Roskeeda Bay, Ireland	0.4	2.2
Bedford Basin, Nova Scotia	0.6	0.5
Central Kaneohe Bay, Hawaii	0.8	0.3
S.E. Kaneohe Bay, Hawaii	1.0	0.5
St. Margarets Bay, Nova Scotia	1.1	0.5
Vostok Bay, Russia	1.0	0.05
Lagoons		
Beaufort Sound, North Carolina	0.5	0.5
Chincoteague Bay, Maryland	3.2	2.5
Peconic Bay, New York	1.9	1.3
High Venice Lagoon, Italy	2.4	0.05
Fjords		
Baltic Sea	1.3	0.1
Loch Etive, Scotland	1.1	0.06

[a] Nutrient concentrations in μg-at/l.

From Boynton, W. R., Kemp, W. M., and Keefe, C. W., in *Estuarine Comparison,* Kennedy, V. S., Ed., Academic Press, New York, 1982, 69. With permission.

Table 5.1—11
SUMMARY OF POLLUTANT CONCENTRATION SUSCEPTIBILITY IN ESTUARIES

Most susceptible systems	Least susceptible systems
General Population	General Population
Brazos River**	Willapa Bay**
Ten Thousand Islands	St. Catherines/Sapelo Sound**
San Pedro Bay**	Penobscot Bay**
North/South Santee Rivers**	Humboldt Bay**
Galveston Bay**	Broad River**
Suisun Bay**	Hood Canal**
Sabine Lake**	Coos Bay**
St. Johns River*	Casco Bay**
Apalachicola Bay	Grays Harbor
San Antonio Bay**	Chincoteague Bay*
Connecticut River*	Bogue Sound**
Great South Bay*	St. Andrew/St. Simons Sound*
Merrimack River	Sheepscot Bay*
Atchafalaya/Vermillion Bays*	Apalachee Bay
Matagorda Bay*	Rappahannock River
Heavy Industry	St. Helena Sound
Brazos River**	Puget Sound*
North/South Santee Rivers**	Heavy Industry
Galveston Bay**	St. Catherines/Sapelo Sound**
Sabine Lake**	Hood Canal**
San Pedro Bay**	Penobscot Bay**
Connecticut River*	Casco Bay**
Calcasieu Lake	Humboldt Bay**
Hudson River/Raritan Bay	Buzzards Bay
Charleston Harbor	Boston Bay
Perdido Bay	Coos Bay**
Potomac River	Broad River**
San Antonio Bay**	Willapa Bay**
Mobile Bay	Bogue Sound**
Suisun Bay**	Puget Sound*
Great South Bay*	Narragansett Bay
Baffin Bay	Santa Monica Bay
Chesapeake Bay	Saco Bay
Agricultural Activities	St. Andrew/St. Simons Sound*
Brazos River**	Agricultural Workers
Suisun Bay**	Humboldt Bay**
North/South Santee Rivers**	Hood Canal**
St. Johns River*	Penobscot Bay**
Matagorda Bay*	Coos Bay**
Atchafalaya/Vermillion Bays*	St. Catherines/Sapelo Sound**
San Pedro Bay**	Chincoteague Bay*
Sabine Lake**	Bogue Sound**
Corpus Christi Bay	Long Island Sound
Galveston Bay**	Casco Bay**
San Antonio Bay**	Willapa Bay**
Winyah Bay	Broad River**
Albemarle Sound	Sheepscot Bay*
Neuse River	Klamath River
Laguna Madre	

Note: Systems that are present in all three categories are marked with two asterisks; systems present in two categories are marked by one asterisk.

From Biggs, R. B., DeMoss, T. B., Carter, M. M., and Beasley, E. L., Susceptibility of U.S. estuaries to pollution, *Rev. Aquat. Sci.*, 1, 203, 1989.

5.2 OIL POLLUTION

Table 5.2—1
ESTIMATED WORLD INPUT OF PETROLEUM HYDROCARBONS TO THE SEA[a]

	Oil industry	Other	Total
Transportation			
Tanker operations	0.60		
Tanker accidents	0.30		
Dry docking	0.25		
Other shipping operations		0.12	
Other shipping accidents		0.10	
	1.15	0.22	1.37
Fixed installations			
Offshore oil production	0.06		
Coastal oil refineries	0.06		
Terminal loading	0.001		
	0.12		0.12
Other sources			
Industrial waste		0.15	
Municipal waste		0.30	
Urban runoff		0.40	
River runoff		1.40	
Atmospheric fallout		0.60	
Natural seeps		0.60	
		3.45	3.45
	1.27	3.67	4.94
Biosynthesis of hydrocarbons			
Production by marine phytoplankton		26,000	
Atmospheric fallout		100–4,000	

[a] Millions of tons per year.

From Clark, R. B., *Marine Pollution,* 2nd ed., Clarendon Press, Oxford, 1989. With permission of Oxford University Press.

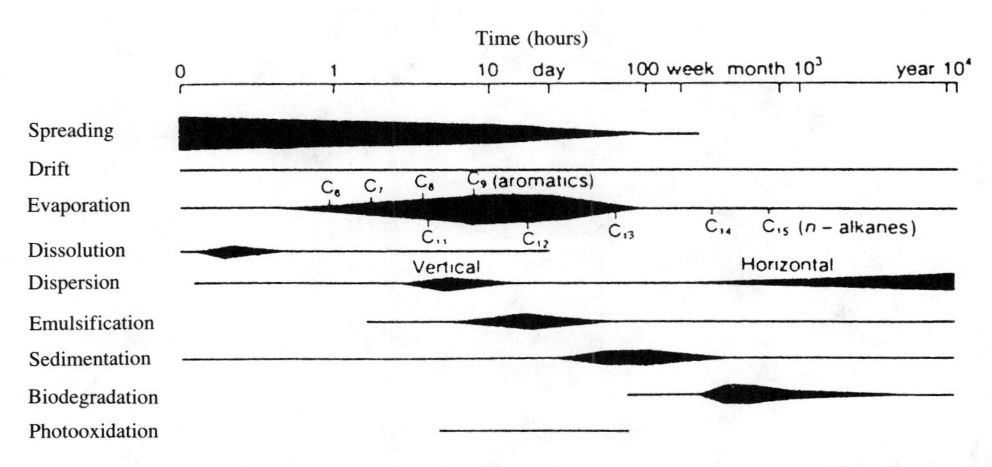

FIGURE 5.2—1. Time-course of factors affecting an oil spill at sea. (From Clark, R. B., *Marine Pollution,* 2nd ed., Clarendon Press, Oxford, 1989. With permission of Oxford University Press.)

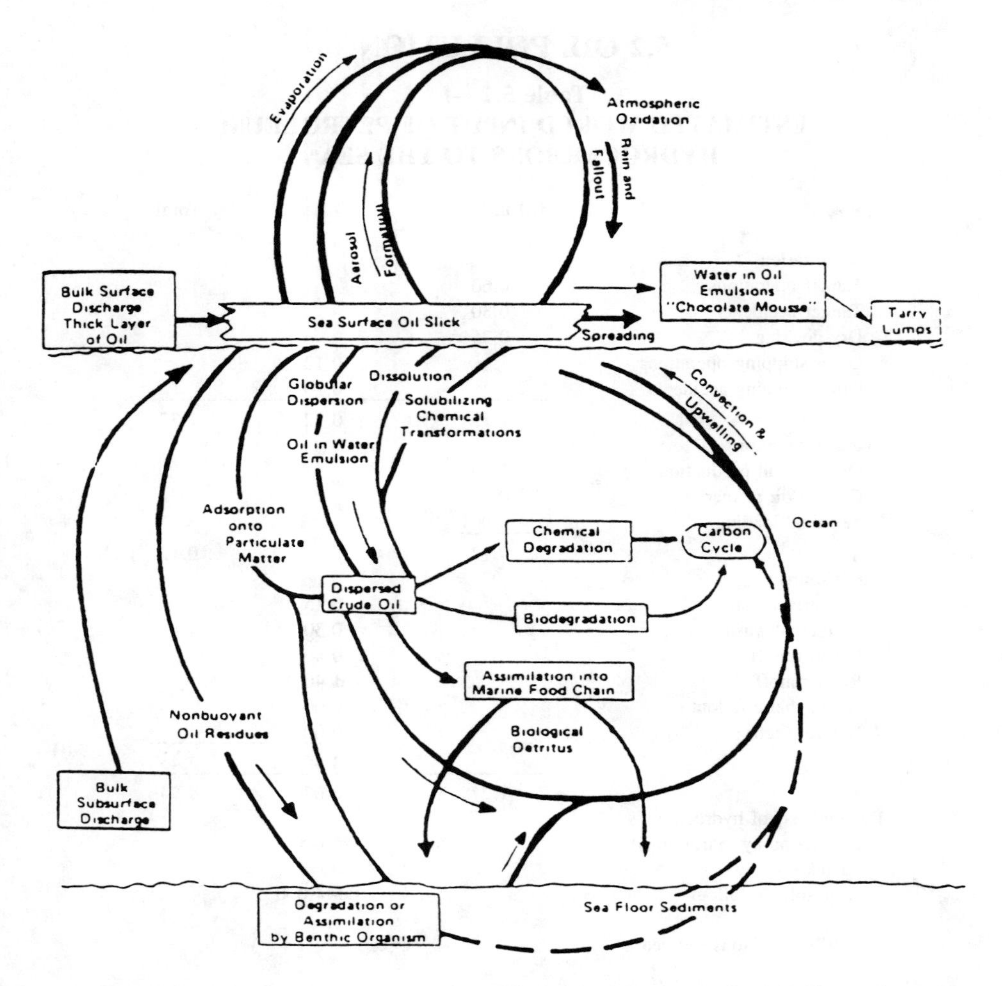

FIGURE 5.2—2. Fate and weathering of polluting oil in estuarine and marine waters showing various abiotic and biotic processes that act to alter the oil. (From Burwood, R. and Speers, G. C., *Estuarine Coastal Mar. Sci.*, 2, 117, 1974. With permission.)

Table 5.2—2
RECOVERY OF VARIOUS ECOSYSTEMS SUBJECTED TO CATASTROPHIC OIL SPILLS

	Time between major stresses (in years)				
Ecosystem type	**3**	**5**	**10**	**20**	**100**
River					
Headwaters	50—70% of species recovered	Recovery less than 95% of species	Recovered	Recovered	Recovered
Middle reach	50—75% of species recovered	State of constant recovery less than 95 % of species	Recovered	Recovered	Recovered
Slow	50—75% of species recovered	Recovery less than 95% of species	Recovered	Recovered	Recovered
Lakes	Most species would not be recovered	Biological integrity not maintained	State of constant recovery	Final state of recovery	Recovered
Estuaries	Principally clams and mollusks are recovered	Clam and mollusk populations still reduced	Recovered	Recovered	Recovered
Marine					
Beaches	Beaches are in state of final repopulation	Repopulated and probably recovered	Recovered	Recovered	Recovered
Rock Shore	Colony communities not recovered	Colony communities generally recovered	Recovered	Recovered	Recovered
Tidal Flat	Principally bivalves not recovered	Bivalves still reduced	Recovered	Recovered	Recovered
Marshes	Annual plants and short life span	Long-lived plants not reestablished; most organisms recovered	Final stages of recovery	Recovered for very large systems	Recovery depends upon size of area affected
Open Water	Very small area repopulated	Long life-span organisms in recovery	Most species present	Recovered except for very large systems	Recovery depends upon size of area affected

From Cairns, J., Jr., Ed., *Rehabilitating Damaged Ecosystems*, Vol. 2, CRC Press, Boca Raton, FL, 1988.

Table 5.2—3
DISPERSED OIL AND OIL EFFECTS ON MANGROVES

Location	Type	Dispersant used & dilution	Type of oil	Amount of spill	Date	Resource affected	Impact	Dispersant effect
Panama	Field	Corexit 9527, 24 h, 1:20	50 ppm Prudhoe Bay crude	Exp.[a]	1984	Mangroves	Defoliation, death	Dispersed oil before it reached mangroves
Coast on Caribbean side of Panama	Accidental	Corexit 9527, ca. 21,000 l, 1:20	Medium weight crude	55,000—60,000	Apr 27, 1986	*R. mangle*	Defoliation, death	
Coast on Caribbean side of Panama	Experimental	Corexit 9527, 1:20	Prudhoe Bay crude 50 ppm	Exp.	1985	*R. mangle*	28% trees defoliated	No defoliation at sites with dispersant
South Florida Turkey Pt., Biscayne Bay, FL	Field	Corexit 9527, 1:20	LA Crude concentrated	Exp.	1982—1986	*R. mangle*		
Panama	Field spill	Corexit 9527, 1:20	Medium weight crude		Fall, 1986	Mangroves *R. mangle*	Observed mangrove death	If dispersed before oil on mangroves, less mortality
Jamaica	Lab	11 dispersants	Venezuelan	Exp.	1988—1989	*Rhizophora Avicennia Laguncularia*	Defoliations, death of root	Various at 1250 ppm not low

[a] Exp. = experimental.

From Connell, D. W. and Hawker, D. W., Eds., *Pollution in Tropical Aquatic Systems,* CRC Press, Boca Raton, FL, 1992, 106.

Table 5.2—4
TROPICAL AND SUBTROPICAL SEAGRASS DISPERSANT OIL AND OIL EFFECTS ON SEAGRASSES

Location	Type	Dispersant used (dilution)	Type and conc of dispersed oil	Amount of spill	Date	Resource affected	Impact	Dispersant effect
	Lab outdoors	Corexit 9527 (1:20)	50 ppm oil, 1:20, 24 h	50 ppm oil, lab	1984	*Thalassia testudinum*	LD_{50} 12 & 96 h bioassays, oil & dispersed oil	Oil with dispersant has lower toxicity than without dispersant
Miami, FL	Lab outdoors	Corexit 9527, (1:20)	LA crude, Murban	Lab	1983–1984	*Thalassia* *Halodule* *Syringudium*	LD_{50} vs. time & conc at 5 to 100 h	At medium conc High High
Miami, FL	Lab outdoors	Arco D-609, (1:10)	LA crude, Murban	Lab		*Thalassia* *Halodule* *Syringudium*	LD_{50} 5 h 100 h	Low to medium Low to medium Low to medium at 75 & 125 ml
Miami, FL	Lab outdoors	Conco K (K), 1:10)	LA crude, Murban	Lab		*Thalassia* *Halodule* *Syringudium*	LD_{50} at 5 & 100 h	Medium to high High High
Panama	Field	Corexit 9527	Prudhoe Bay crude, 50 ppm, 24 h	Lab	1985	*T. testudinum*	None to *Thalassia*	No effect on *Thalassia*
Miami, FL	Lab outdoors	Corexit 9550 (1:20)	LA crude 125 and 75 ml oil 1:20 disp. in 100,000 cc SW	Lab	1986	*Thalassia* *Halodule* *Syringudium*	LD_{50} at 100 h	Low Medium Low to medium
Miami, FL	Lab outdoors	OFC-D-607 (1:10)	LA crude 75 and 125 ml	Lab	1986	*Thalassia* *Halodule* *Syringudium*	LD_{50} at 100 h	Low Low Medium
Miami, FL	Lab outdoors	Cold Clean 500 (1:10)	LA crude 75 and 125 ml in 100,000 cc SW	Lab	1986	*Thalassia* *Halodule* *Syringudium*	LD_{50} 100 h	Low Low to medium Low
Miami, FL	Lab	Finsol OSP-7 (1:10)	LA crude 75 and 125 ml in 100,000 cc SW	Lab	1986	*Thalassia* *Hodule* *Syringudium*	LD_{50} 100 h	Medium Low Low to medium low

From Thorhaug et al., Lab 10 dispersants in Venezuelan lab, 1988–89, Thai, Syr., Hal LD_{50} 3 low, 3 med., 4 high; Connell, D. W. and Hawker, D. W., Eds., *Pollution in Tropical Aquatic Systems,* CRC Press, Boca Raton, FL, 1992, 110.

Table 5.2—5
DISPERSED OIL AND OIL EFFECTS ON CORALS

Location	Type	Dispersant used & dilution	Conc. of Dispersant	Amount of spill	Date	Resource affected	Impact	Dispersant effect
Bermuda	Field & lab	Corexit 9527 BP 1100 WD	1:20 1:10	Arabian light crude	1981—1986	Corals	6—24 h after, 1—50 ppm on *Diploria strigosa*	No effect to brief exposures; when oil dispersed 20 ppm polychaetes, bivalves crustacea intolerant unclear after 9 mo. whether dispersant had effect or not
Arabian Gulf	Field	Corexit 9527 20:1		Arabian light crude experi- ment	1980	Corals	No impact imme- diately, some death after 6 mo. during winter cold	
Panama	Field	Corexit 9527	50 ppm 20:1	Prudhoe Bay crude experi- ment	1985	Corals, seagrasses mangroves	No coral death at 24 h exposure	No death of corals with dispersant
Panama	Spill	Corexit 9527	20:1	50,000 med. wt. crude	1986	Corals, seagrasses mangroves	Coral death	Reports intertidal reefs extensive mortality, subtidal to 2 m mortality
Jamaica	Lab	10 dispersants	1:10	Venezuela light	1988—1989	Corals	Various	3 nontoxic, 5 hightly toxic

From Connell, D. W. and Hawker, D. W., Eds., *Pollution in Tropical Aquatic Systems,* CRC Press, Boca Raton, FL, 1992.

Table 5.2—6
COSTS OF OIL SPILL CLEANUP TECHNIQUES

Type	Cost per barrel
Mechanical	$65–5000
Dispersant	$15–65
Shoreline	$650–7000
Gels	Above $7000

From Connell, D. W. and Hawker, D. W., Eds., *Pollution in Tropical Aquatic Systems,* CRC Press, Boca Raton, FL, 1992, 122.

Table 5.2—7
SHORT-TERM OIL SPILL CLEANUP AND INDIRECT LONG-TERM COSTS

	Capitalization costs	Direct costs	Indirect waste removal	Indirect environmental costs	Socioeconomic costs
Mechanical	High	$1.50—0.60/gal	$2/gal		None if successful, high if not successful
Beach cleanup	Low	$8—17/gal	$4/gal	Very high in productive ecosystems (mangroves, reefs)	High if tourism is involved or if mangroves/coral dies; low if rock or sand not used
No action	0	0	0	About $100,000/acre if mangroves, seagrasses corals are killed	Very high if fisheries or tourism is affected or if mangroves or corals die
Dispersant	Medium to low	$0.06—0.30/gal	0	None if nontoxic dispersants used, some if toxic dispesants used	Low

From Connell, D. W. and Hawker, D. W., Eds., *Pollution in Tropical Aquatic Systems,* CRC Press, Boca Raton, FL, 1992, 122.

5.3 POLYNUCLEAR AROMATIC HYDROCARBONS (PAH)

Table 5.3—1
PAH AND OTHER AROMATIC HYDROCARBONS

Structure	1957 I.U.P.A.C. name	Other names	Mol wt	Relative carcinogenicity[a]	Common liter. abbrev. (if any)
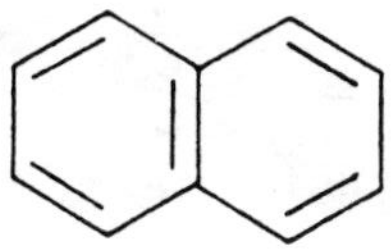	Naphthalene	—	128	—	—
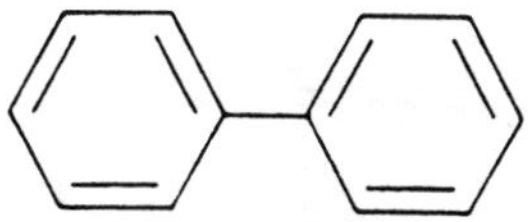	Biphenyl	—	154	—	—
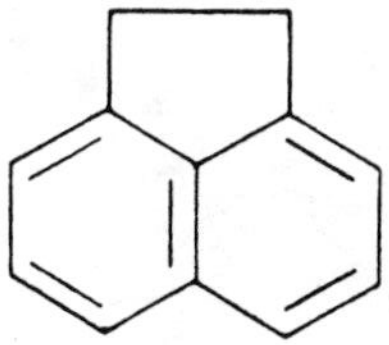	Acenaphthene	—	154	—	—
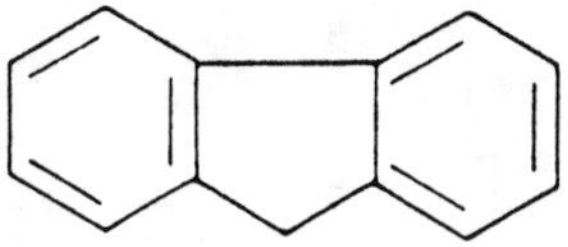	Fluorene	—	166	—	—
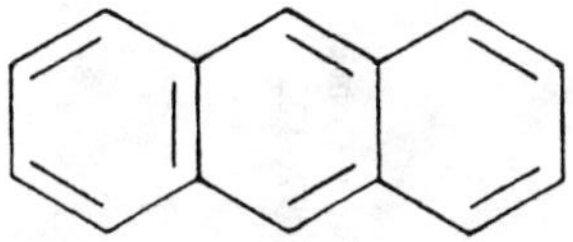	Anthracene	—	178	—	—

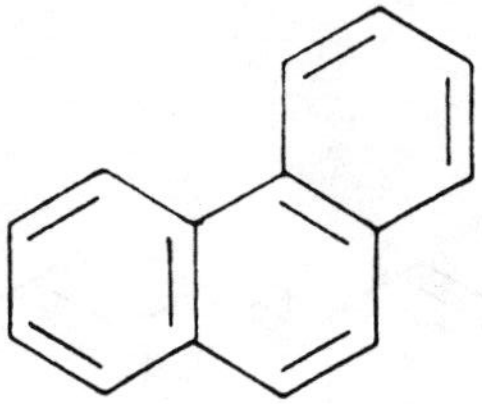	Phenanthrene	—	178	—	—
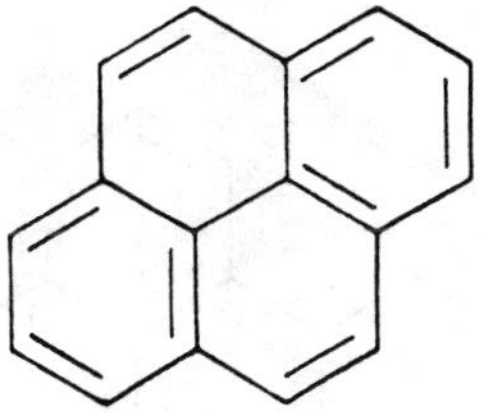	Pyrene	—	202	—	—
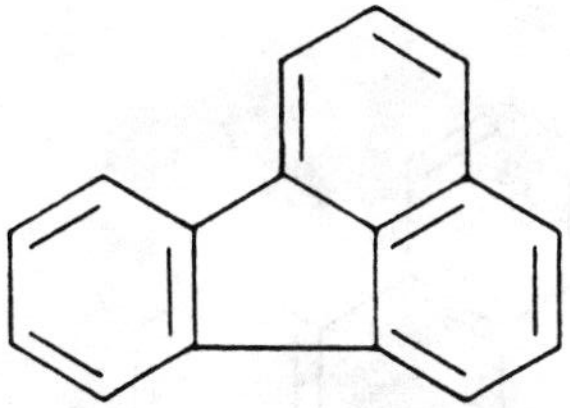	Fluoranthene	—	202	—	—
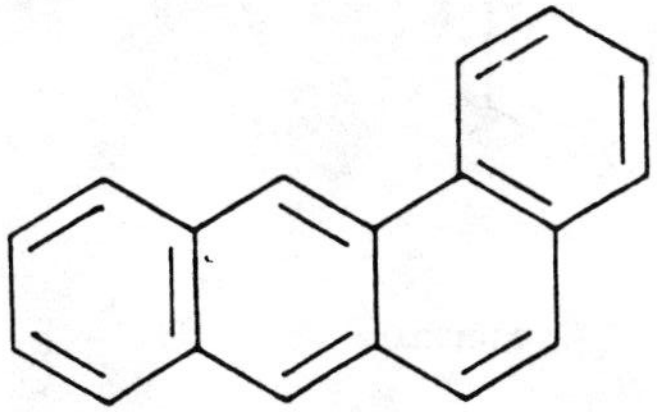	Benzo(a)anthracene	1,2 Benzanthracene	228	<+	B(a)A

Table 5.3—1 (continued)
PAH AND OTHER AROMATIC HYDROCARBONS

Structure	1957 I.U.P.A.C. name	Other names	Mol wt	Relative carcinogenicity[a]	Common liter. abbrev. (if any)
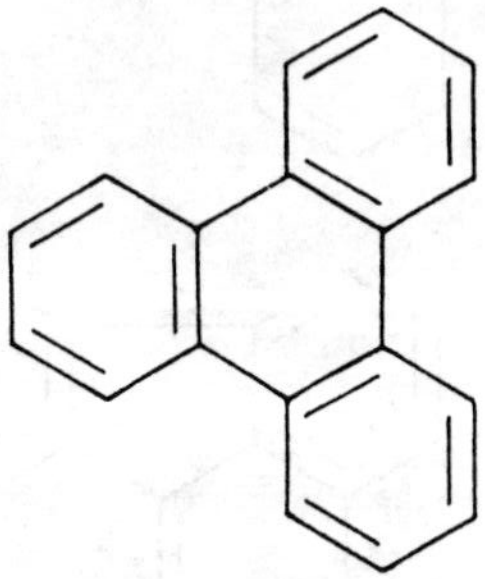	Triphenylene	—	228	—	—
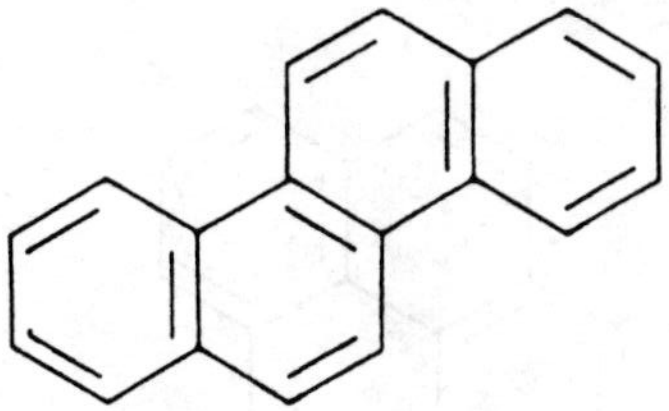	Chrysene	—	228	<+	—
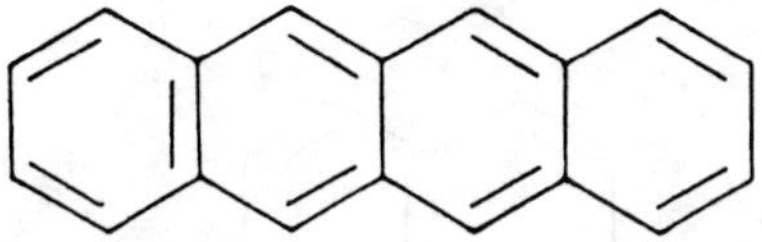	Naphthacene	Tetracene	228	—	—

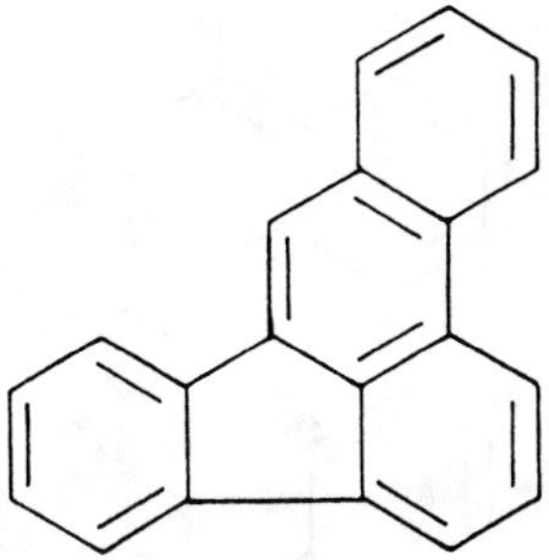 Benzo(b)fluoranthene	3,4 Benzfluoranthene	252	+ +	B(b)F
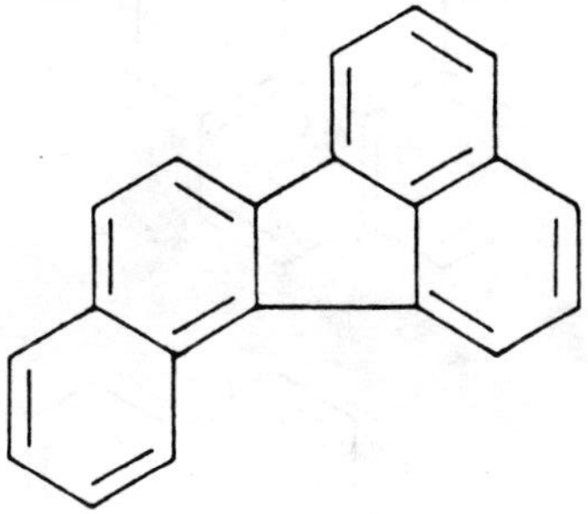 Benzo(j)fluoranthene	10,11 Benzfluoranthene	252	+ +	B(j)F
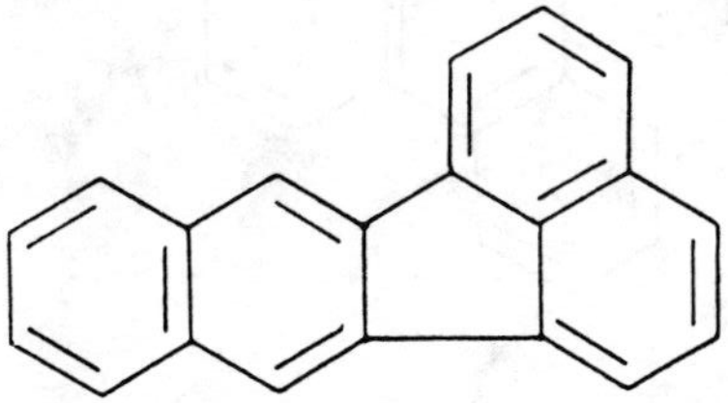 Benzo(k)fluoranthene	11,12 Benzfluoranthene	252	—	B(k)F

Table 5.3—1 (continued)
PAH AND OTHER AROMATIC HYDROCARBONS

Structure	1957 I.U.P.A.C. name	Other names	Mol wt	Relative carcinogenicity[a]	Common liter. abbrev. (if any)
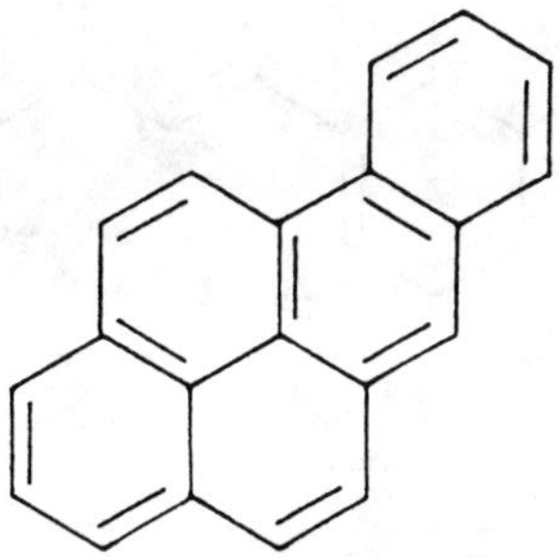	Benzo(a)pyrene	3,4 Benzopyrene	252	+ + + +	B(a)P
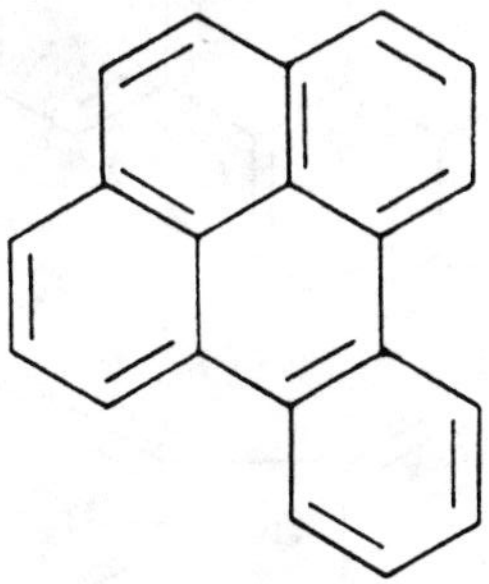	Benzo(e)pyrene	1,2 Benzopyrene	252	< +	B(e)P
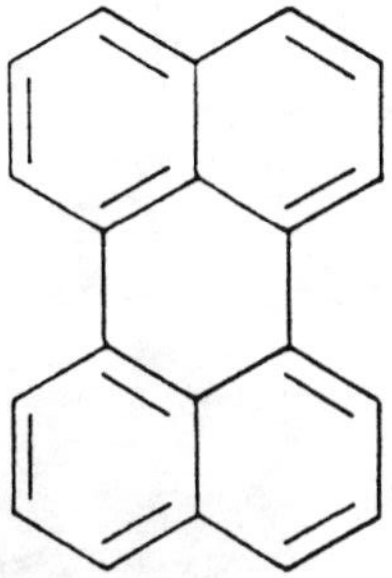	Perylene	—	252	—	—

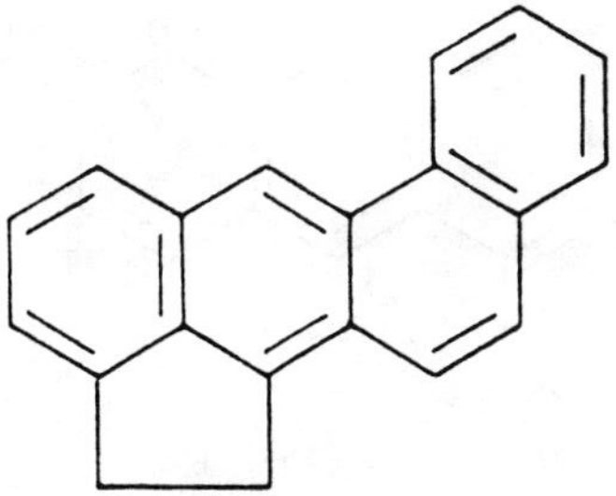	Cholanthrene	—	254	—	—
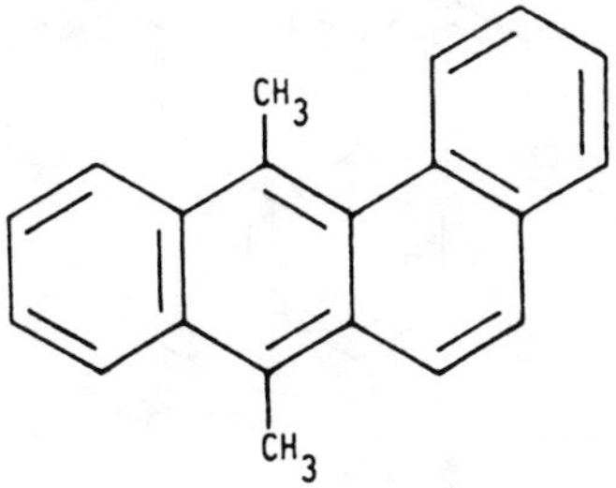	7,12 Dimethylbenz-(a)anthracene	7,12 Dimethyl-1,2-benzan-thracene	256	+ + + + +	—
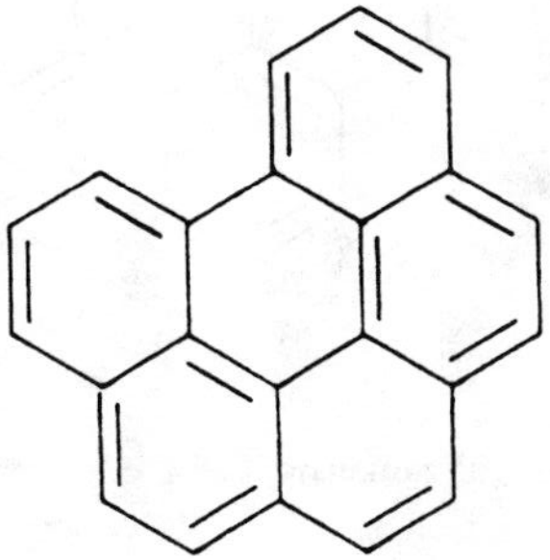	Benzo(ghi)perylene	1,12 Benzperylene	276	—	—

Table 5.3—1 (continued)
PAH AND OTHER AROMATIC HYDROCARBONS

Structure	1957 I.U.P.A.C. name	Other names	Mol wt	Relative carcinogenicity[a]	Common liter. abbrev. (if any)
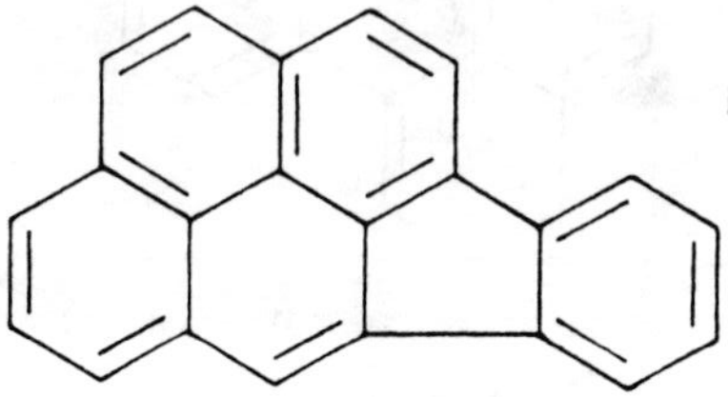	Indeno(1,2,3-cd)pyrene	o-Phenylenepyrene	276	+	IP
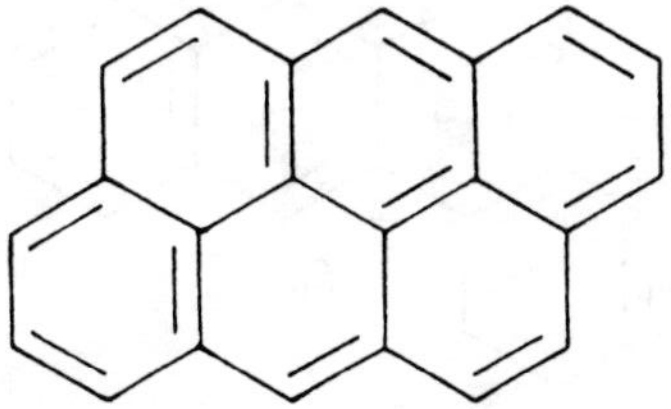	Anthanthrene	—	276	<+	—
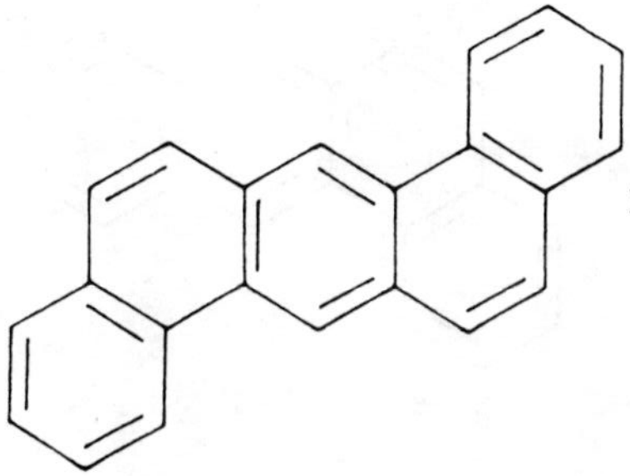	Dibenz(a,h)anthracene	1,2,5,6 Dibenzanthracene	278	+ + +	—

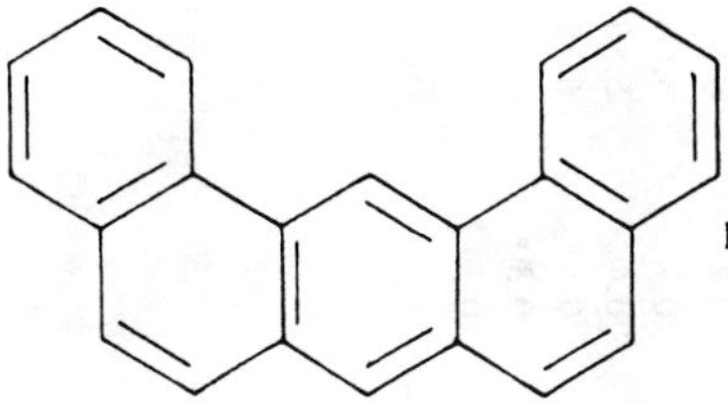 Dibenz(a,j)anthracene	1,2,7,8 Dibenzanthracene	278	—	—
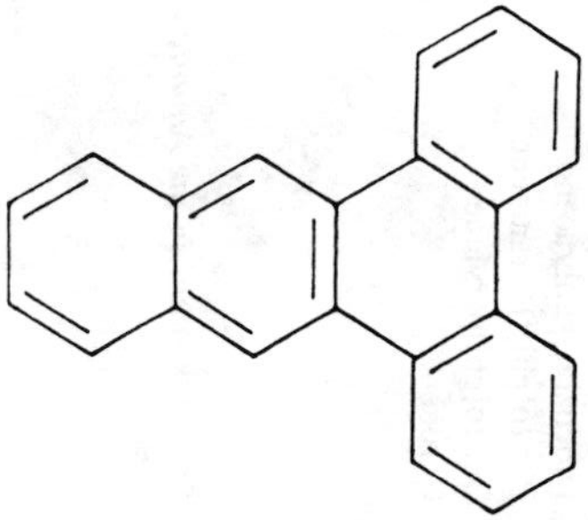 Dibenz(a,c)anthracene	1,2,3,4, Dibenzanthracene	278	—	—
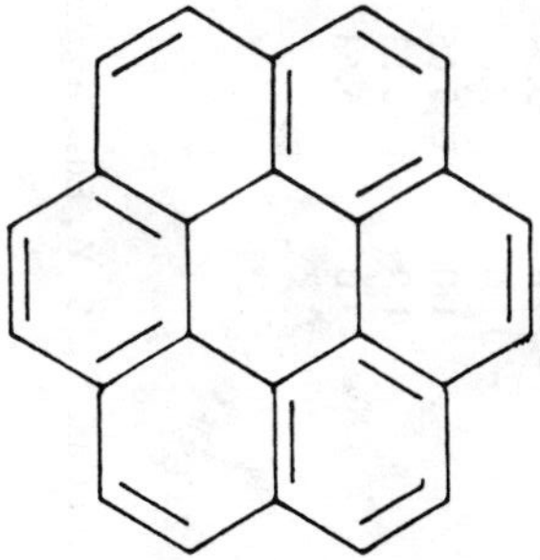 Coronene	—	300	—	—

[a] + + + + + = extremely active; + + + + = very active; + + + = active; + + = moderately active; + = weakly active; < = less than; — = inactive or unknown.

From Futoma, D. J., Smith, S. R., Smith, T. E., and Tanaka, J., *Polycyclic Aromatic Hydrocarbons in Water Systems,* CRC Press, Boca Raton, FL, 1981, 2.

Table 5.3—2
PHYSICAL PROPERTIES OF POLYCYCLIC AROMATIC HYDROCARBONS

Compound	Density	Vapor pressure at 25°C	Sat. conc (ng/m^3)	Vapor equilibrium −10°C	Vapor 30°C
Fluorene	1.203	—	—	—	—
Anthracene	1.25	$2.6.10^{-5}$	$1.9.10^{7}$	—	—
Phenanthrene	1.79	$9.1.10^{-5}$	$6.5.10^{7}$	—	—
Fluoranthene	1.252	—	—	—	—
Pyrene	1.271	$9.1.10^{-8}$	$7.4.10^{4}$	5.810^{2}	$1.4.10^{5}$
Benzo(a)anthracene	—	$1.5.10^{-8}$	$1.3.10^{3}$	3.4	$2.8.10^{3}$
Chrysene	1.274	—	—	—	—
Benzo(k)fluoranthene	—	$1.3.10^{-11}$	$1.3.10^{1}$	$1.3.10^{-2}$	$3.0.10^{1}$
Benzo(a)pyrene	1.351	$7.3.10^{-10}$	$7.5.10^{1}$	$1.5.10^{-1}$	$1.6.10^{2}$
Benzo(e)pyrene	—	$7.4.10^{-10}$	$7.5.10^{1}$	$1.5.10^{-1}$	$1.6.10^{2}$
Perylene	1.35	—	—	—	—
Benzo(ghi)perylene	—	$1.3.10^{-11}$	1.5	$1.8.10^{-3}$	3.4
Dibenz(ghi,pqr)chrysene	1.377	$2.0.10^{-13}$	$2.0.10^{-2}$	$1.8.10^{-6}$	$5.8.10^{-2}$

From Afghan, B. K. and Chau, A. S. Y., Eds., *Analysis of Trace Organics in the Aquatic Environment,* CRC Press, Boca Raton, FL, 1989, 220.

Table 5.3—3
SOLUBILITY AND OCTANOL-WATER PARTITION COEFFICIENTS OF PAH AT 25°C

Compound	Solubility (μg/l)	K_{ow}	Compound	Solubility (μg/l)	K_{ow}
Fluorene	800	—	7,12-Dimethylbenz[a]anthracene	1.5	6.36
Anthracene	59	4.5	Benzo[b]fluoranthene	2.4	6.21
Phenanthrene	435	4.46	Benzo[j]fluoranthene	2.4	6.21
2-Methylanthracene	21.3	4.77	Cholanthene	2.0	6.28
9-Methylphenanthrene	261.0	4.77	Benzo[a]pyrene	3.8	6.04
1-Methylphenanthrene	269.0	4.77	Benzo[e]pyrene	2.4	6.21
Fluoranthene	260.0	5.03	Perylene	2.4	6.21
Pyrene	133.0	4.98	Dibenzo[a,h]fluorene	0.8	6.57
9,10-Dimethylanthracene	56.0	5.13	Dibenzo[a,g]fluorene	0.8	6.57
Benzo[a]fluorene	45.0	5.34	Dibenzo[a,c]fluorene	0.8	6.57
Benzo[b]fluorene	29.6	5.34	3-Methylcholanthene	0.7	6.64
Benzo[a]anthracene	11.0	5.63	Dibenz[a,j]anthracene	0.4	6.86
Naphthacene	1.0	5.65	Benzo[ghi]fluoranthene	0.5	6.78
Chrysene	1.9	5.63	Benzo[ghi]perylene	0.3	6.78
Triphenylene	43.0	5.63	Coronene	0.14	7.36

Note: Salinity 32% @22°C.

From Afghan, B. K. and Chau, A. S. Y., Eds., *Analysis of Trace Organics in the Aquatic Environment,* CRC Press, Boca Raton, FL, 1989, 220.

Table 5.3—4
RELATIVE CARCINOGENICITY INDEX OF SOME PAH

Compound	Carcinogenicity index
Benzo[a]anthracene	+
7,12-Dimethylbenz[a]anthracene	+ + + +
Dibenz[a,j]anthracene	+
Dibenz[a,h]anthracene	+ + +
Dibenz[a,c]anthracene	+
Benzo[c]phenanthrene	+ + +
Dibenzo[a,g]fluorene	+
Dibenzo[a,h]fluorene	UC
Dibenzo[a,c]fluorene	+
Benzo[b]fluoranthene	+ +
Benzo[j]fluoranthene	+ +
Benzo[j]aceanthrylene	+ +
3-Methylcholanthrene	+ + + +
Benzo[a]pyrene	+ + +
Dibenzo[a,l]pyrene	UC
Dibenzo[a,h]pyrene	+ + +
Dibenzo[a,i]pyrene	+ + +
Indeno(1,2,3-cd)pyrene	+
Chrysene	UC
Dibenzo[b,def]chrysene	+ +
Dibenzo[def,p]chrysene	+

Note: UC, unknown.

From Afghan, B. K. and Chau, A. S. Y., Eds., *Analysis of Trace Organics in the Aquatic Environment,* CRC Press, Boca Raton, FL, 1989, 221.

Table 5.3—5
MAJOR SOURCES OF PAH IN ATMOSPHERIC AND AQUATIC ENVIRONMENTS

Ecosystem and sources	Annual input (metric tons)
Atmosphere	
Total PAH	
Forest and prairie fires	19,513
Agricultural burning	13,009
Refuse burning	4,769
Enclosed incineration	3,902
Heating and power	2,168
Benzo(a)pyrene	
Heating and power	
Worldwide	2,604
U.S. only	475
Industrial processes (mostly coke production)	
Worldwide	1,045
U.S. only	198
Refuse and open burning	
Worldwide	1,350
U.S. only	588
Motor vehicles	
Worldwide	45
U.S. only	22
Aquatic environments	
Total PAH	
Petroleum spillage	170,000
Atmospheric deposition	50,000
Wastewaters	4,400
Surface land runoff	2,940
Biosynthesis	2,700
Total benzo(a)pyrene	700

From Eisler, R., Polycyclic Aromatic Hydrocarbon Hazards to Fish, Wildlife, and Invertebrates: A Synoptic Review, Biol. Rep. 85 (1.11), U.S. Fish and Wildlife Service, Washington, D.C., 1987.

Table 5.3—6
COMPARISON OF TOTAL PAH CONCENTRATIONS IN MARINE AND FRESHWATER (FW) SURFICIAL SEDIMENTS

Location	Total PAH, (ppb, wet weight)	No. of stations	Depth (m)
North America			
Penobscot Bay	286–8,974[a]	49	9.2–126.3
Casco Bay	215–14,425	32	2–43
Gulf of Maine	543[a]	1	—
Murray Basin	540	1	282
Jordan Basin	500	1	265
Wilkinson Basin	540–870	1	215
Franklin Basin	200	1	225
North Atlantic			
Continental rise	160	1	4,150
Continental slope	120	1	1,830
Abyssal plain	18–97	2	5,250, 5,465
Abyssal plain	55[a]	1	—
Charles River, MA	87,000[a]	1	—
	120,000	1	—
Massachusetts Bay	160–3,400	3	90, 130, 155
Boston Harbor, MA	8,500	1	6
Buzzards Bay, MA	800	1	17
	4,000–5,000	3	—
	803[a]	1	17
Falmouth Marsh, MA	800	1	Intertidal
New Bedford Harbor, MA	63,000	1	—
Pettaquamscutt River, RI	10,000	1	—
New York Bight	5,830[a]	1	28
Pennsylvania Creek (FW)	100	4	0.3
Lake Erie (FW)	530–3,750	7	—
Adirondack Lakes (FW)	4,070–12,807	2	—
Alaska	5–113[a]	2	Intertidal
Mono Lake, California (FW)	157–399[a]	2	5–10
Europe			
Tamar Estuary (FW)	4,900	8	—
Southampton Estuary	91,000–1,791,000	19	—
Severn Estuary drainage system	1,600–25,700[a]	9	Intertidal
Mediterranean	198–372	2	6
Côte Bleue	1,232–232,000	3	3–10
Les Embiez	13,000–15,000	2	3–10
Monaco	5,200–12,100	2	3–10
Baltic Sea	258[a]	1	164
South Baltic Sea	50–2,550[a]	7	—
Gulf of Finland	437[a]	1	60
Western Norway	284–99,452[a]	6	—
Neckar, Rhine, and Danube Rivers (FW)	600–44,560[a]	73	—
Other			
Walvis Bay, Africa	68[a]	1	—
Cariaco Trench	1756[a]	1	—
Amazon River system (FW)	ND–544	4	—
South Georgia Island	100	1	18

[a] ppb, dry weight.

From Johnson, A. C., Larsen, P. F., Gadbois, D. F., and Humason, A. W., *Mar. Environ. Res.*, 15, 1, 1985. With permission.

Table 5.3—7
CONCENTRATIONS OF SELECTED PAH COMPOUNDS IN SEDIMENTS FROM CHESAPEAKE BAY[a]

Station and Year

Compound	LE5.5[b]					WE4.2					LE3.6				
	79S[c]	79F[d]	84	85[e]	86	79S	79F	84	85	86	79S	79F	84	85	86
Phenanthrene	11	47	22	100	25	5	8	26	32	28	10	24	28	29	27
Fluoranthene	29	52	51	410	54	26	16	54	58	60	16	59	63	56	51
Pyrene	34	46	40	380	52	21	18	49	67	57	12	58	64	55	48
Benzo(a)fluorene	13	25	16	130	23	7	4	13	13	21	3	13	24	15	18
Benzo(a)anthracene	18	30	21	140	19	12	9	28	17	23	5	30	29	16	20
Chrysene/triphenylene	37	47	35	170	31	18	16	39	34	37	7	39	44	29	34
Benzo(e)pyrene	2	1	23	93	16	2	11	25	17	24	5	2	27	17	23
Benzo(a)pyrene	22	18	23	130	19	18	12	26	19	33	4	35	33	19	31
Pyrene	26	8	42	36	9	21	22	44	34	38	11	39	46	21	42
Benzo(g,h,i)perylene	15	6	18	46	9	15	8	31	23	20	3	17	28	12	26

Compound	LE2.3				CB5.1				CB4.3C				CB3.3C[f]			
	79S	79F	84	86	79S	79F	84	86	79S	79F	84	86	79S	79F	84	86
Phenanthrene	19	42	54	64	17	49	47	50	44	68	26	11	280	220	300	240
Fluoranthene	34	70	89	88	35	81	85	74	60	82	42	14	370	220	370	300
Pyrene	29	57	72	87	33	73	79	66	43	70	38	4	360	220	370	290
Benzo(a)fluorene	10	16	24	34	10	27	24	23	11	38	13	2	120	98	150	77
Benzo(a)anthracene	10	23	28	25	10	32	26	20	8	25	16	>1	100	92	120	94
Chrysene/triphenylene	14	35	51	48	15	46	51	39	12	41	96	13	150	140	210	150
Benzo(e)pyrene	9	13	31	27	8	3	35	24	5	2	19	3	3	89	150	99
Benzo(a)pyrene	8	13	39	37	7	33	36	27	1	31	20	3	64	100	150	110
Pyrene	9	14	43	46	5	37	59	51	9	65	260	150	110	220	220	140
Benzo(g,h,i)perylene	7	5	27	21	5	18	35	19	4	21	21	10	38	56	96	79

[a] Concentrations in μg/kg.
[b] Southernmost sampling site.
[c] S = Summer.
[d] F = Fall.
[e] Northern Chesapeake Bay not sampled in 1985.
[f] Northernmost sampling site.

Reprinted from Huggett, R. J., de Fur, P. O., and Bieri, R. H., *Mar. Pollut. Bull.*, 19, 454, 1988. With kind permission from Pergamon Press Ltd., Headington Hill Hall, Oxford OX3 OBW, U.K.

Table 5.3—8
SELECTED PAH AND TOTAL PAH CONCENTRATIONS IN SEDIMENTS FROM THE HUDSON-RARITAN ESTUARY AND THE NEW YORK BIGHT AND IN SEWAGE SLUDGE[a]

Material and location	Naphthalene	Phenanthrene	Anthracene	Benzo(a)anthracene	Total PAH
Sediment					
Hudson-Raritan estuary					
15 km north of the Battery	60	120	60	330	2,000
Pierhead Channel	200	300	200	500	3,200
Gowanus Canal	100	1,000	500	3,000	16,400
Newtown Creek	120,000	14,600	9,600	5,600	182,000
Lower Bay	100	600	300	2,000	9,900
New York Bight Region					
Christiaensen Basin	800	500	300	1,000	6,000
Sewage sludge dumpsite	80	70	40	200	1,100
Outer Bight	0.6	3	N.D.[b]	3	22
Sewage sludge	2,200	4,400	1,100	1,000	20,400

[a] Concentrations in ng/g, dry wt.
[a] Not detected.

From O'Connor, J. M., Klotz, J. B., and Kneip, T. J., in *Ecological Stress and the New York Bight: Science and Management*, Mayer, G. F., Ed., Estuarine Research Federation, Columbia, SC, 1982, 631. Reprinted with permission of Estuarine Research Federation, copyright 1982.

Table 5.3—9
SELECTED PAH AND TOTAL PAH CONCENTRATIONS IN FINFISH AND SHELLFISH FROM THE HUDSON RIVER AND NEW YORK BIGHT[a]

Species (location)	Naphthalene	Phenanthrene	Anthracene	Biphenyl	Total PAH
Atlantic mackerel (*Scomber scombrus*) (New York Bight apex)	ND	10	ND	ND	10
Winter flounder (*Pseudopleuronectes americanus*) (Christiaensen Basin)	2	ND	ND	6	8
Winter flounder (*P. americanus*) (Raritan Bay)	2	1	ND	ND	5
Striped bass (*Morone saxatilis*) (Montauk Point)	7	ND	ND	ND	19
Striped bass (*M. saxatilis*) (Hudson River)	4	ND	ND	4	8
Lobster (*Homarus americanus*) (New York Bight)	7	ND	ND	ND	7
Lobster (*H. americanus*) (Raritan Bay)	5	5	ND	ND	25
Lobster (*H. americanus*) (Raritan Bay)	7	ND	ND	ND	77
Blue mussel (*Mytilus edulis*) (Sandy Hook)	6	6	ND	4	250
Blue mussel (*M. edulis*) (Shark River)	20	10	1	40	120

[a] Concentrations in ng/g, dry wt; ND, not detected.

From O'Connor, J. M., Klotz, J. B., and Kneip, T. J., in *Ecological Stress and the New York Bight: Science and Management*, Mayer, G. F., Ed., Estuarine Research Federation, Columbia, SC, 1982, 631. Reprinted with permission of Estuarine Research Federation, copyright 1982.

Table 5.3—10
PAH CONCENTRATIONS (μg/kg) IN SELECTED SHELLFISH AND FINFISH FROM ESTUARINE AND COASTAL MARINE WATERS

Taxonomic group, compound, and other variables	Concentration	Ref. (sources from Eisler, 1987)
Shellfish		
Rock crab, *Cancer irroratus*		
Edible portions, 1980		
New York Bight		
Total PAH	1600 FW	Humason and Gadbois, 1982
BaP	1 FW	
Long Island Sound		
Total PAH	1290 FW	
BaP	ND	
American oyster		
Crassostrea virginica, soft parts,		
South Carolina, 1983,		
residential resorts		
Total PAH		
Spring months		
Palmetto Bay	520 FW	Marcus and Stokes, 1985
Outdoor Resorts	247 FW	
Fripp Island	55 FW	
Summer months		
Palmetto Bay	269 FW	
Outdoor Resorts	134 FW	
Fripp Island	21 FW	
American lobster, *Homarus americanus*		
Edible portions, 1980		
New York Bight		
Total PAH	367 FW	Humason and Gadbois, 1982
BaP	15 FW	
Long Island Sound		
Total PAH	328 FW	
BaP	15 FW	
Softshell clam, *Mya arenaria*		
Coos Bay, Oregon, 1978–1979		
Soft parts		
Contaminated site		
Total PAH	555 FW	Mix, 1982
Phenanthrene = PHEN	155 FW	
FL	111 FW	
Pyrene = PYR	62 FW	
BaP	55 FW	
Benz(a)anthracene = BaA	42 FW	
Chrysene = CHRY	27 FW	
Benzo(b)fluoranthene = BbFL	12 FW	
Others	<10 FW	
Uncontaminated site		
Total PAH	76 FW	
PHEN	12 FW	
FL	10 FW	
Others	<10 FW	
Bay mussel, *Mytilus edulis*		
Oregon, 1979–1980		
Soft parts, total PAH		
Near industrialized area	106–986 FW	Mix and Schaffer, 1983
Remote site	27–274 FW	
Sea Scallop, *Placopecten magellanicus*		
Baltimore Canyon, East Coast U.S.		

Table 5.3—10 (continued)
PAH CONCENTRATIONS (μg/kg) IN SELECTED SHELLFISH AND FINFISH FROM ESTUARINE AND COASTAL MARINE WATERS

Taxonomic group, compound, and other variables	Concentration	Ref. (sources from Eisler, 1987)
Shellfish		
Muscle		
BaA	1 FW	Brown and Pancirov, 1979
BaP	<1 FW	
PYR	4 FW	
New York Bight, 1980		
Edible portions		
Total PAH	127 FW	Humason and Gadbois, 1982
BaP	3 FW	
Clam, *Tridacna maxima*		
Australia, 1980–1982, Great Barrier Reef		
Soft parts, total PAH		
Pristine areas	<0.07 FW	Smith et al., 1984
Powerboat areas	Up to 5 FW	
Mussel, *Mytilus* sp.		
Greenland		
Shell	60 FW	Harrison et al., 1975
Soft parts	18 FW	
Italy		
Shell	11 FW	
Soft parts	130–540 FW	
Bivalve mollusks, 5 spp.		
Edible portions	6(max.36)FW	Stegeman, 1981
Decapod crustaceans, 4 spp.		
Edible portions	2(max.8)FW	
Softshell clam, *Mya arenaria,* soft parts		
Coos Bay, Oregon		
1976–1978		
Near industrialized areas	6–20 FW	Mix and Schaffer, 1983
Remote areas	1–2 FW	
1978–1979		
Near industrialized areas	9 FW	
Remote areas	4 FW	
Finfish		
Fish, muscle		
Baltimore Canyon,		
East Coast, U.S., 5 spp.		
BaA	Max. 0.3 FW	Brown and Pancirov, 1979
BaP	Max. <5 FW	
PYR	Max. <5 FW	
Smoked		
FL	3 FW	
PYR	2 FW	EPA 1980
Nonsmoked		
FL	Max. 1.8 FW	
PYR	Max. 1.4 FW	
Winter flounder, *Pseudopleuronectes americanus*		
Edible portions, 1980		
New York Bight		
Total PAH	315 FW	Humason and Gadbois, 1982
BaP	21 FW	
Long Island Sound		
Total PAH	103 FW	
BaP	ND	

Table 5.3—10 (continued)
PAH CONCENTRATIONS (μg/kg) IN SELECTED SHELLFISH AND FINFISH FROM ESTUARINE AND COASTAL MARINE WATERS

Taxonomic group, compound, and other variables	Concentration	Ref. (sources from Eisler, 1987)
Windowpane, *Scopthalmus aquosus*		
Edible portions, 1980		
New York Bight		
Total PAH	536 FW	
BaP	4 FW	
Long Island Sound		
Total PAH	86 FW	
BaP	ND	
Red hake, *Urophycus chuss*		
Edible portions, 1980		
New York Bight		
Total PAH	412 FW	
BaP	22 FW	
Long Island Sound		
Total PAH	124 FW	
BaP	5 FW	
Fish		
Marine, edible portions	Max. 3 FW	
9 spp.	15 FW	Stegeman, 1981
Greenland	65 FW	Harrison et al., 1975
Italy	5–8 DW	
Steak, charcoal broiled	11 DW	Barnett, 1976
Ribs, barbecued		

Note: ND, Not detected; FW, Fresh weight; DW, Dry weight.

Modified from Eisler, R., Polycyclic Aromatic Hazards to Fish, Wildlife, and Invertebrates: A Synoptic Review, Biol. Rep. 85(1.11), U.S. Fish and Wildlife Service, Washington, D.C., 1987.

Table 5.3—11
TOXICITIES OF SELECTED PAH TO AQUATIC ORGANISMS

PAH compound, organism, and other variables	Concentration in medium[a]	Effect
Benzo(a)pyrene		
Sandworm, *Neanthes arenceodentata*	>1,000	LC_{50} (96 h)
Chrysene		
Sandworm	>1,000	LC_{50} (96 h)
Dibenz(a,h)anthracene		
Sandworm	>1,000	LC_{50} (96 h)
Fluoranthene		
Sandworm	500	LC_{50} (96 h)
Fluorene		
Grass shrimp, *Palaemonetes pugio*	320	LC_{50} (96 h)
Amphipod, *Gammarus pseudoliminaeus*	600	LC_{50} (96 h)
Sandworm	1,000	LC_{50} (96 h)
Sheepshead minnow, *Cyprinodon variegatus*	1,680	LC_{50} (96 h)
Naphthalene		
Copepod, *Eurytemora affinis*	50	LC_{30} (10 d)
Pink salmon, *Oncorhynchus gorbuscha,* fry	920	LC_{50} (24 h)
Dungeness crab, *Cancer magister*	2,000	LC_{50} (96 h)
Grass shrimp	2,400	LC_{50} (96 h)
Sheepshead minnow	2,400	LC_{50} (24 h)
Brown shrimp, *Penaeus aztecus*	2,500	LC_{50} (24 h)
Amphipod, *Elasmopus pectenicrus*	2,680	LC_{50} (96 h)
Coho salmon, *Oncorhyncus kisutch,* fry	3,200	LC_{50} (96 h)
Sandworm	3,800	LC_{50} (96 h)
Mosquitofish, *Gambusia affinis*	150,000	LC_{50} (96 h)
1-Methylnaphthalene		
Dungeness crab, *Cancer magister*	1,900	LC_{50} (96 h)
Sheepshead minnow	3,400	LC_{50} (24 h)
2-Methylnaphthalene		
Grass shrimp	1,100	LC_{50} (96 h)
Dungeness crab	1,300	LC_{50} (96 h)
Sheepshead minnow	2,000	LC_{50} (24 h)
Trimethylnaphthalenes		
Copepod, *Eurytemora affinis*	320	LC_{50} (24 h)
Sandworm	2,000	LC_{50} (96 h)
Phenanthrene		
Grass shrimp	370	LC_{50} (24 h)
Sandworm	600	LC_{50} (96 h)
1-Methylphenanthrene		
Sandworm	300	LC_{50} (96 h)

[a] Concentrations in μg/l.

Modified from Eisler, R., Polycyclic Aromatic Hazards to Fish, Wildlife, and Invertebrates: A Synoptic Review, Biol. Rep. 85(1.11), U.S. Fish and Wildlife Service, Washington, D.C., 1987.

5.3—12
FISH AND INVERTEBRATE SPECIES COMMONLY USED FOR ACUTE TOXICITY TESTS

- Freshwater
  - Vertebrates
    - Rainbow trout, *Salmo gairdneri*
    - Brook trout, *Salvelinus fontinalis*
    - Fathead minnow, *Pimephales promelas*
    - Channel catfish, *Ictalurus punctatus*
    - Bluegill, *Lepomis macrochirus*
  - Invertebrates
    - Daphnids, *Daphnia magna, D. pulex, D. pulicaria*
    - Amphipods, *Gammarus lacustris, G. fasciatus, G. pseudolimnaeus*
    - Crayfish, *Orconectes* sp., *Cambarus* sp., *Procambarus* sp., or *Pacifastacus ieniusculus*
    - Midges, *Chironomus* sp.
    - Snails, *Physa integra*
- Saltwater
  - Vertebrates
    - Sheepshead minnow, *Cyprinodon variegatus*
    - Mummichog, *Fundulus heteroclitus*
    - Longnose killifish, *Fundulus similis*
    - Silverside, *Menidia* sp.
    - Threespine stickleback, *Gasterosteus aculeatus*
    - Pinfish, *Lagodon rhomboides*
    - Spot, *Leiostomus xanthurus*
    - Sand dab, *Citharichthys stigmateus*
  - Invertebrates
    - Copepods, *Acartia tonsa, A. clausi*
    - Shrimp, *Penaeus setiferus, P. duorarum, P. aztecus*
    - Grass shrimp, *Palaemonetes pugio, P. vulgaris*
    - Sand shrimp, *Crangon septemspinosa*
    - Mysid shrimp, *Mysidopsis bahia*
    - Blue crab, *Callinectes sapidus*
    - Fidler crab, *Uca* sp.
    - Oyster, *Crassotrea virginica, C. gigas*
    - Polychaetes, *Capitella capitata, Neanthes* sp.

From Parrish, P. R., in *Fundamentals of Aquatic Toxicology: Methods and Applications,* Rand, G. M. and Petrocelli, S. M., Eds., Hemisphere Publishing, New York, 1985, 31. With permission.

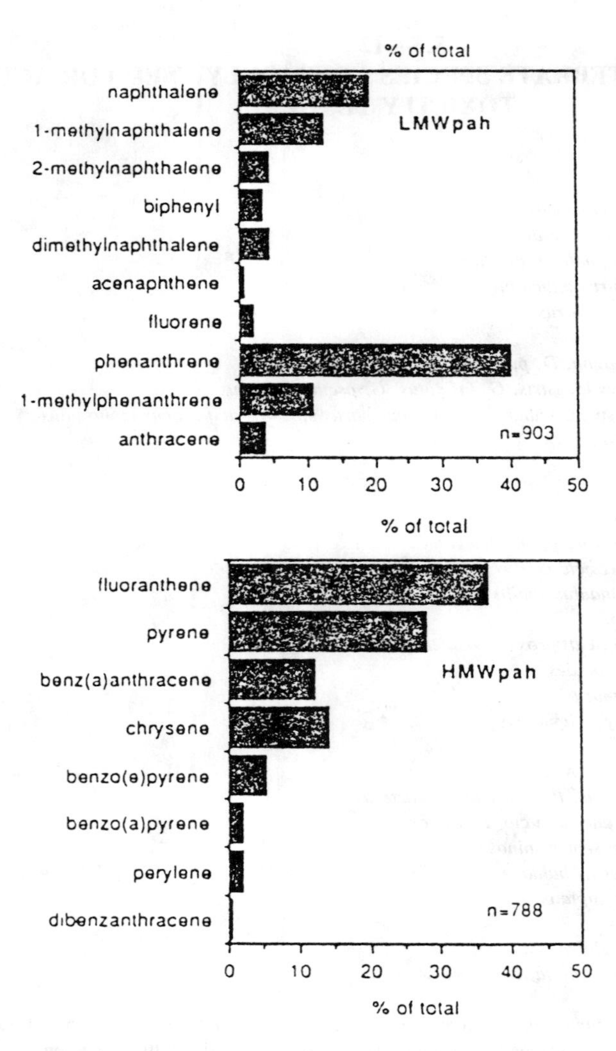

FIGURE 5.3—1. Mean percentage contributions of individual PAH compounds to low-molecular weight (LMWpah) and high-molecular weight (HMWpah) contaminant classes. (From National Oceanic and Atmospheric Administration, A Summary of Data on Tissue Contamination from the First Three Years (1986–1988) of the Mussel Watch Project, NOAA Tech. Mem. NOS OMA 49, National Oceanic and Atmospheric Administration, Rockville, MD, 1989.

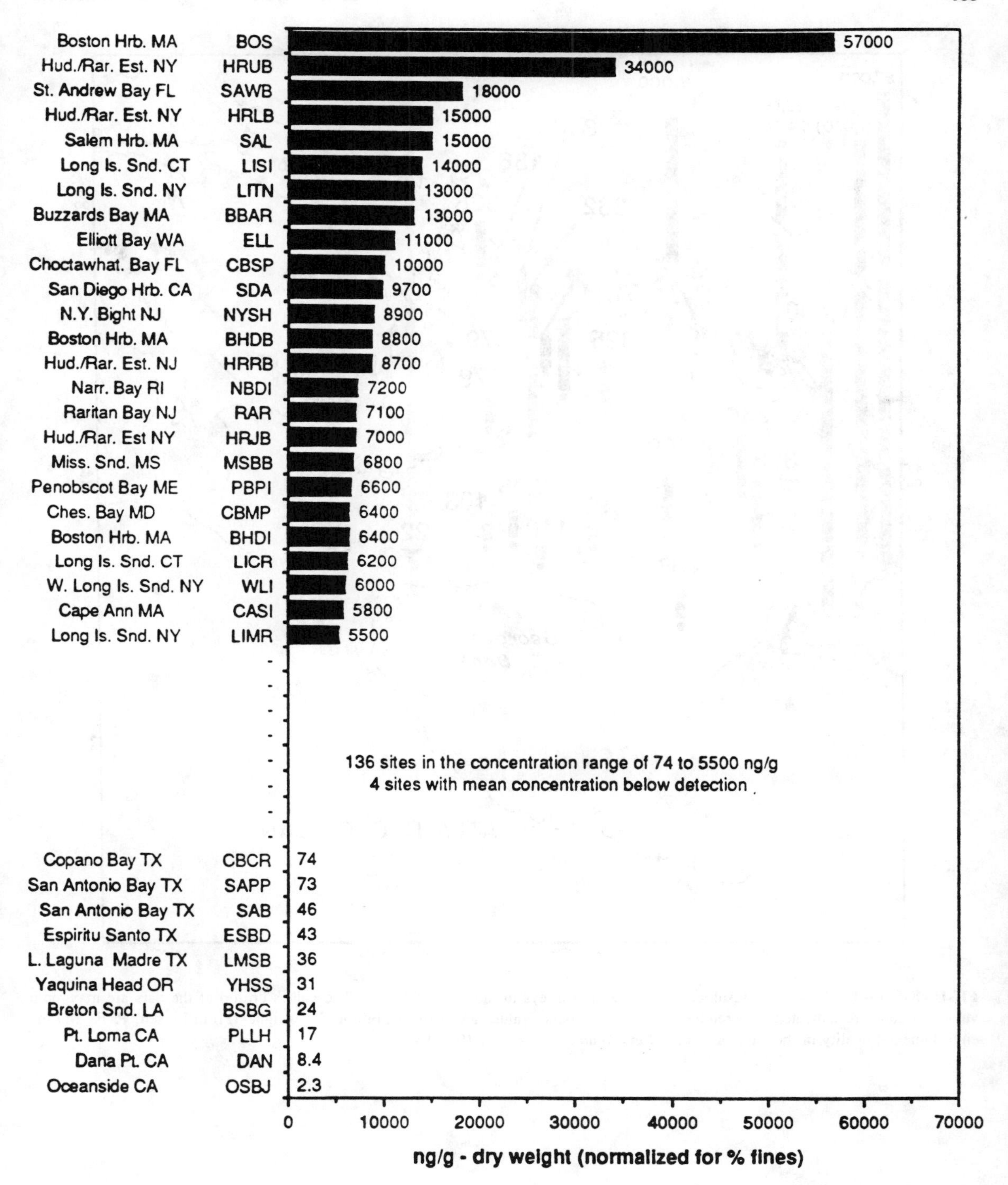

FIGURE 5.3—2. Total PAH concentrations in sediments of various estuarine and coastal systems of the U.S. (From NOAA, A Summary of Data on Individual Organic Contaminants in Sediments Collected During 1984, 1985, 1986, and 1987, Tech. Mem. NOS OMA 47, National Oceanic and Atmospheric Administration, Rockville, MD, 1989.)

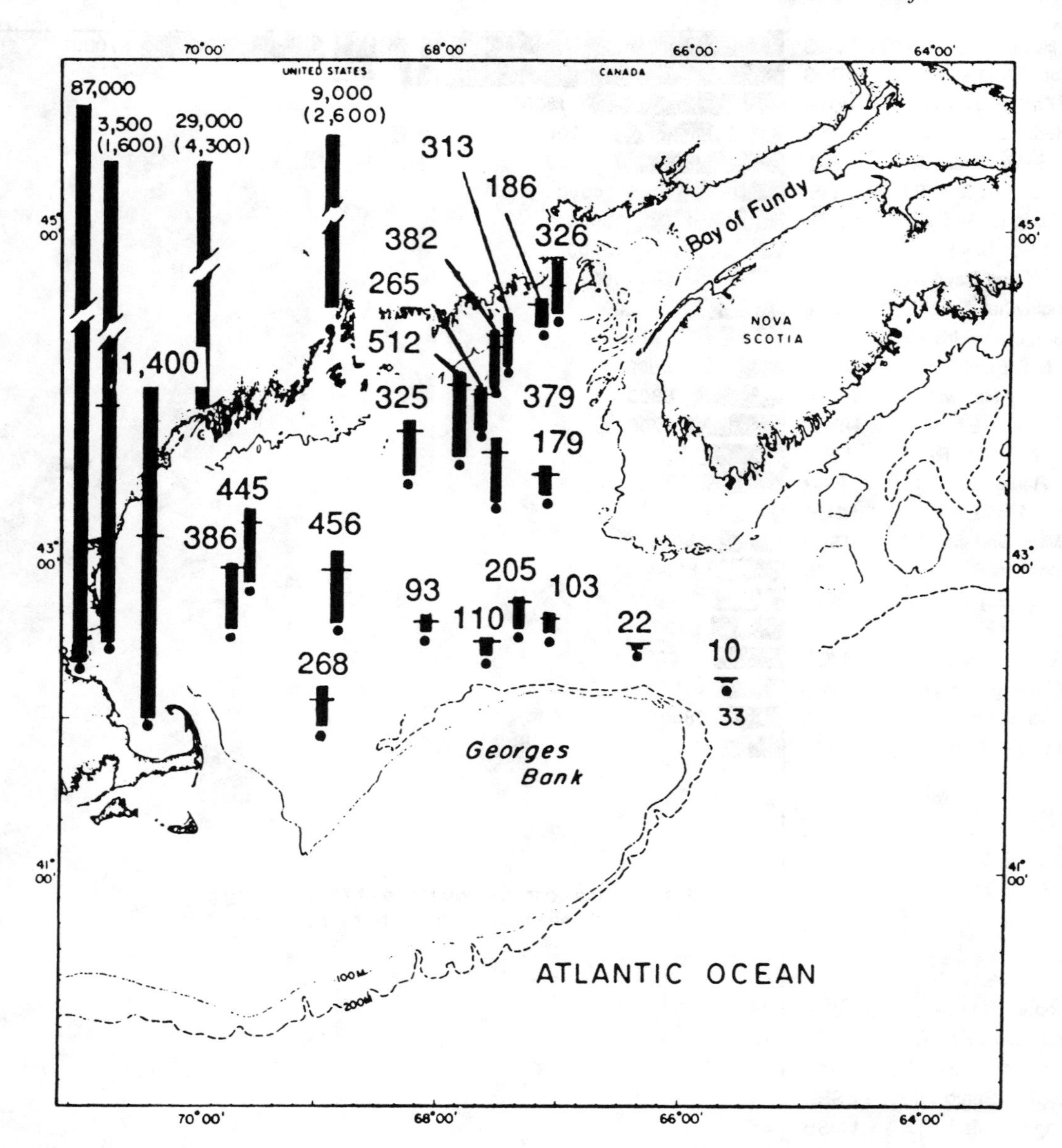

FIGURE 5.3—3. Composite results of several PAH surveys in the Gulf of Maine. The values on top of the bars are maximum values. Means are indicated in parentheses or as cross bars. Values are parts per billion dry weight. (From Larsen, P. F., Marine environmental quality in the Gulf of Maine, *Rev. Aquat. Sci.*, 6, 67, 1992.)

5.4 POLYCHLORINATED BIPHENYLS (PCB) AND ORGANOCHLORINE PESTICIDES

Table 5.4—1
LIST OF TARGET CHEMICAL RESIDUES SURVEYED BY MEARNS ET AL.[a] IN U.S. COASTAL FISH AND SHELLFISH

Residue	Use and occurrence
Polychlorinated biphenyls (PCB)	Dielectric fluid in capacitors; transformer fluid; lubricants; hydraulic fluids; plasticizers; cutting oil extenders; carbonless paper; banned in 1976; total is either sum of chlorination mixtures (Arochlors®) or number
Total PCB	
Arochlor® 1016	
Arochlor® 1242	
Arochlor® 1248	
Arochlor® 1254	
Arochlor® 1260	
PCB by chlorination number (2–10)	
DDT and structurally related chemicals	
DDE (o-p and p-p′)	Insecticides; DDT metabolites
DDD (TDE: o-p and p-p′)	Insecticides; DDT metabolites
DDT (o-p and p-p′)	Insecticides; parents of DDD/DDE
Total DDT	Sum of parent and metabolites
Kelthane (Dicofol)	Acaracide; parent of DDE
Methoxychlor	Insecticide
Cyclodiene pesticides	
Technical chlordane	Insecticide; mix of constituents
Chlordane (*trans*- and *cis*)	Insecticides; major constituents of technical chlordane
Nonachlor (*trans*- and *cis*-)	Insecticides; minor constituents of technical chlordane
Oxychlordane	Chlordane metabolite
Heptachlor	Insecticide; minor constituent of technical chlordane
Heptachlor epoxide	Metabolite of heptachlor
Endosulfan (I and II)	Insecticides; 7:3 mixture of stereoisomers
Endosulfan sulfonate	Metabolite of endosulfans
Aldrin	Insecticide
Dieldrin	Insecticide; Aldrin metabolite
Endrin	Insecticide
Hexachlorocyclohexane insecticides	
a-BHC	Constituent of BHC insecticide mix
Y-BHC (Lindane)	Insecticide; BHC constituent
Hexachlorocyclopentadiene pesticides	
Kepone	Acaricide, larvicide, fungicide, ant bait
Mirex	Insecticide (fire ant control)
Chlorinated camphenes	
Toxaphene	Insecticide (cotton)
Carboxylic acid derivatives	
2,4-D (2,4-DEP)	Weed herbicide (in cereals)
2,4,5-T	Wood plant herbicide
DCPA (dacthal)	Pre-emergence weed herbicide
Chlorinated benzenes and phenols	
HCB (hexachlorobenzene)	Fungicide
PCP (pentachlorophenol)	Wood preservative

[a] Modified from Schmitt, C. J., Zajiik, J. L., and Ribick, M. A., *Arch. Environ. Contam. Toxicol.*, 14, 225, 1985.

From Mearns, A. J., Matta, M. B., Simecek-Beatty, D., Buckman, M. F., Shigenaka, G., and Wert, W. A., PCB and Chlorinated Pesticide Contamination in U.S. Fish and Shellfish: a Historical Assessment Report, NOAA Tech. Memo. NOS OMA 39, Seattle, 1988.

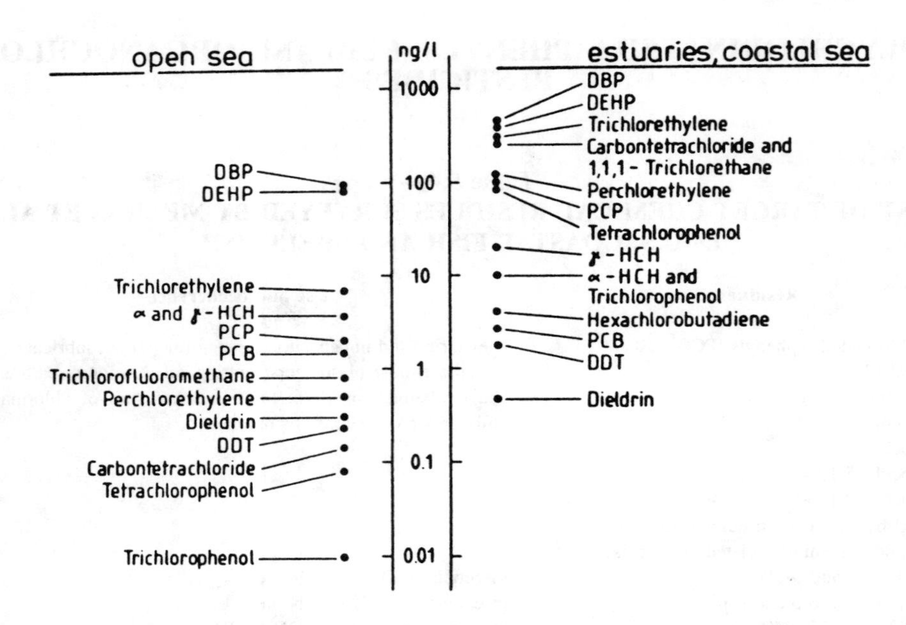

FIGURE 5.4—1. Concentration of various organochlorine pollutants in water of the open sea and of estuaries and coastal areas. (From Ernst, W., *Helgol. Wiss. Meeresunters,* 33, 302, 1980. With permission.)

Table 5.4—2
POSSIBLE PCB CONGENERS, MOLECULAR WEIGHTS, AND PERCENT CHLORINE OF THE VARIOUS PCB ISOMERS

Parent biphenyl	Number of isomers	Molecular weight	Chlorine (%)
Monochloro-	3	188.7	18.8
Dichloro-	12	223.1	31.8
Trichloro-	24	257.5	41.3
Tetrachloro-	42	292.0	48.6
Pentachloro-	46	326.4	54.3
Hexachloro-	42	360.9	58.9
Heptachloro-	24	395.3	62.8
Octachloro-	12	429.8	66.0
Nonachloro	3	464.2	68.7
Decachloro-	1	498.7	71.2
Total	**209**		

From Oliver, B. G. et al., in *Analysis of Trace Organics in the Aquatic Environment,* Afghan, B. K. and Chau, A. S. Y., Eds., CRC Press, Boca Raton, FL, 1989, 33.

Table 5.4—3
APPROXIMATE COMPOSITION (WEIGHT PERCENT) OF AROCLOR® PREPARATIONS

Homolog # Cl/ biphenyl	1221	1232	1016	1242	1248	1254	1260
0	10						
1	50	26	2	1			
2	35	29	19	13	1		
3	4	24	57	45	21	1	
4	1	15	22	31	49	15	
5				10	27	53	12
6					2	26	42
7						4	38
8							7
9							1
10							
Av. #Cl/molecule	1.15	2	3	3	4	5	6
Approx. wt. % Cl	21	32	42	42	48	54	60

From Oliver, B. G. et al., in *Analysis of Trace Organics in the Aquatic Environment,* Afghan, B. K. and Chau, A. S. Y., Eds., CRC Press, Boca Raton, FL, 1989, 34.

Table 5.4—4
RANGES IN PCB CONCENTRATIONS (ng/l) REPORTED FOR OPEN OCEANS, COASTAL WATERS, AND ESTUARIES OR RIVERS

Area	Location	Range in PCB (ng/l)	Reference (sources from Phillips, 1986)
Open oceans	North Atlantic	<1–150	Harvey et al. 1974
		0.4–41	Harvey et al., 1973
	Sargasso Sea	0.9–3.6	Bidleman and Olney, 1974
	North-South Atlantic	0.3–8.0	Harvey and Steinhauer, 1976
	Mediterranean Sea	0.2–8.6	Elder and Villeneuve, 1977
Coastal waters	Southern California	2.3–36	Scura and McClure, 1975
	Northwest Mediterranean	1.5–38	Elder et al., 1976
	Atlantic coast, U.S.	10–700	Sayler et al., 1978
	Baltic coasts	0.3–139	Brugmann and Luckas, 1978
		0.1–28	Ehrhardt, 1981
	Dutch coast	0.7–8	Duinker and Hillebrand, 1979
Estuaries/rivers	Wisconsin rivers, U.S.	<10–380	Dennis, 1976
	Rhine-Meuse system, Holland	10–200	Duinker and Hillebrand, 1979
	Tiber estuary, Italy	9–1000[a]	Puccetti and Leoni, 1980
	Brisbane River, Australia	ND–50	Shaw and Connell, 1980
	Hudson River, U.S.	$<100–2.8 \times 10^6$	Nisbet, 1976

Note: ND, not detectable (no limits quoted).

[a] As decachlorinated biphenyl equivalents.

From Phillips, D. J. H., *PCBs in the Environment,* Vol. 2, Waid, J. S., Ed., CRC Press, Boca Raton, FL, 1986, 130.

Table 5.4—5
ESTIMATED CONCENTRATION AND LOAD OF PCBs IN THE OPEN OCEAN ENVIRONMENT

Compartment mass PCB conc, load	North Pacific	South Pacific	North Atlantic	South Atlantic	Indian	Antarctic	Total load
Compartment mass[a]							
Air ($\times 10^{16}$ m^{1})[b]	70	76	38	35	58	46	
Water ($\times 10^{19}$ l)	30	33	15	14	23	19	
Sediment (dry, $\times 10^{15}$ g)[c]	70	76	38	35	58	46	
Plankton (wet, $\times 10^{15}$ g)[d]			48 (whole ocean)				
Fish (fresh, $\times 10^{14}$ g)[d]			26 (whole ocean)				
Mammals (fresh, $\times 10^{12}$ g)[d]			85 (whole ocean)				
PCB conc							
Air (ng/m^3)	0.3	0.2	0.5	0.2	0.2	0.1	
Water (ng/l)	0.2	0.1	0.6	0.1	0.1	0.05	
Sediment (dry, ng/g)	0.4	0.2	1.0	0.2	0.2	0.1	
Plankton (wet, ng/g)	2.0	1.0	5.0	1.0	1.0	0.5	
Fish (fresh whole, ng/g)	10	2.0	30	2.0	2.0	0.2	
Mammals (fresh whole, μg/g)[c]	2.0	0.5	5.0	0.5	0.5	0.05	
PCB load (t)							
Air	210	150	190	70	120	50	790
Water[f]	60,000	33,000	90,000	14,000	23,000	10,000	230,000
Sediment	28	15	38	7	12	5	105
Plankton[g]			20–240 (whole ocean)				130[h]
Fish[g]			1–80 (whole ocean)				40[h]
Mammals[g]			2–200 (whole ocean)				100[h]
Total							231,165

[a] Data on the surface area and mean depth in respective oceans employed for the calculation of air mass, water mass, and sediment mass were adapted from Sugimura, personal communications, 1983.

[b] Calculated in troposphere (10 km height).

[c] Calculated in upper 1-mm sediment layer. Sediment was assumed to contain about 50% water and have a mass density of 2.

[d] Compiled in consideration of the following moisture contents: plankton 95%, fish 75%, and mammals 65%.

[e] Values show the PCB concentrations regarding male specimens. PCB concentrations on whole-body basis were calculated following the relationship of PCB concentrations between blubber and whole body obtained from the striped dolphin.

[f] PCB loads in water were estimated on the assumption of their having vertically uniform concentrations in the water column as shown by their vertical profiles in Tanabe, S. and Tatsukawa, R., 1986.

[g] PCB loads in these organisms were estimated from the following probable concentrations: plankton 0.5–5 ng/g, fish 0.2–30 ng/g, and mammals (male) 0.05–5, μg/g. The PCB load in mammals took into account that female specimens generally have lower concentrations of PCBs in their bodies than males because of parturitional and lactational losses.

[h] Median values.

From Tanabe, S. and Tatsukawa, R., in *PCBs and the Environment,* Vol. 1, Waid, J. S., Ed., CRC Press, Boca Raton, FL, 1986, 156.

Table 5.4—6
PCB CONCENTRATIONS IN OPEN OCEAN SURFACE WATERS

Location	Year	N	PCB conc (ng/l) Range	Mean
North Atlantic				
Sargasso Sea	1973	8	<0.9—3.6	1.0
Sargasso Sea, New York	1973	9		0.8
9—55°N, 9—73°W	1973-1975	39	0.4—8.0	2.9
North Sea and Scottish coast	1974	5	<0.15—0.52	0.23
South Atlantic				
11—36°S, 2—33°W	1975	8	0.3—3.7	1.0
North Pacific				
Western Pacific (22—35°N, 141—154°E)	1975	13	0.25—0.56	0.41
Western Pacific (32—42°N, 133—143°E)	1976	8	0.29—1.1	0.54
Western Pacific (12—33°N, 129—138°E)	1978	6	0.23—0.59	0.35
Western Pacific (29—34°N, 137—146°E)	1979	5	0.27—0.38	0.33
Western Pacific (5—31°N, 107—152°E)	1980-1981	9	0.039—0.15	0.089
Bering Sea	1981	3	0.073—0.13	0.10
South Pacific				
Western Pacific (2—41°S, 152—156°E)	1981	5	0.081—0.21	0.12
Indian				
Eastern Indian (4—45°S, 104—123°E)	1980	6	0.057—0.25	0.14
Antarctic				
48—65°S, 124—163°E	1980-1981	7	0.042—0.072	0.058
Syowa Station (69°00′S, 39°35′E)	1981-1982	3	0.035—0.069	0.053

From Tanabe, S. and Tatsukawa, R., in *PCBs and the Environment,* Vol. 1, Waid, J. J., Ed., CRC Press, Boca Raton, FL, 1986, 147.

Table 5.4—7
FLUX OF PCB TO THE OCEAN SURFACE

	Flux (ng/m²/year)		
	Aroclor® 1242	**Aroclor® 1254**	**Total**
Particles			
Dry	16.0	41.0	57.0
Wet	250.0	650.0	900.0
Gas Phase			
Dry maximum	2,100.0	1,600.0	3,700.0
minimum	100.0	100.0	200.0[a])
Wet	9.0	9.5	19.0
Total	2375.0	2301.0	4676.0
Total flux to oceans (× 10^8 g/year)	8.6	8.3	16.9

[a] Based on seawater composition ≃ 0.5 Aroclors® 1242 and 0.5 1254.

From Atlas, E. et al., in *PCBs and the Environment,* Vol. 1, Waid, J. S., Ed., CRC Press, Boca Raton, FL, 1986, 94.

Table 5.4—8
ESTIMATED RESIDENCE TIME OF PCBs IN THE OPEN OCEAN MIXING LAYER (UPPER 100 m OF WATER COLUMN)

Ocean	Carbon productivity ($g/m^2/year$)	Residence time (day)
Oligotrophic		
Pacific Ocean (22°05′N, 145°02′E)	50–100	130–280
Eutrophic		
Southern Ocean (64°42′S, 124°15′E)	150–250	26–44

From Tanabe, S. and Tatsukawa, R., in *PCBs and the Environment,* Vol. 1, Waid, J. S., Ed., CRC Press, Boca Raton, FL, 1986, 157.

Table 5.4—9
PCB CONCENTRATIONS IN OPEN OCEAN PLANKTON

Location	Year	PCB conc (ng/g wet wt.)		
		N	Range	Mean
North Atlantic	1970		300–450	380
Northeast Atlantic	Before 1972	22	10–110	
North and South Atlantic	1970–1972	53		200
South Atlantic	1971	4	18–640	200
Western North Pacific	1981	1		1.1
Bering Sea	1982	3	1.0–1.6	1.3
Western South Pacific	1981	3	1.2–2.3	1.7
Antarctic (Ross Sea)	1972	1		<3
Antarctic (50–65°S, 124–126°E)	1981	3	0.2–1.0	0.5

From Tanabe, S. and Tatsukawa, R., in *PCBs and the Environment,* Vol. 1, Waid, J. S., Ed., CRC Press, Boca Raton, FL, 1986, 151.

Table 5.4—10
CONCENTRATIONS OF PCBs (RANGES, ng/l) IN THE DISSOLVED AND SUSPENDED FRACTIONS OF WATER SAMPLES FROM RIVERS DRAINING TO THE RHINE-MEUSE ESTUARY

Rivers	Date	PCB conc in water (ng/l)	
		Dissolved	Suspended
Lek	June 1974	14–80	40–300
Meuse	August 1976	13–26	25–40
Waal/Oude Maas	June 1974	40–100	50–230
	August 1976	120–200	130–160
Nieuwe Maas	June 1974	80–200	50–100
	December 1974	10–75	20–400
	August 1976	60–180	80
Estuary mixing zone	June 1974	50–120	80–200
	December 1974	20–100	40–100
	August 1976	40–65	25–60

Note: Filtration was by Whatman® GF/C filters.

From Phillips, D. J. H., in *PCBs and the Environment,* Vol. 2, Waid, J. S., Ed., CRC Press, Boca Raton, FL, 1986, 130.

Table 5.4—11
PARTITIONING VALUES OR CONCENTRATION FACTORS (PCBs in lipids/PCBs in water) FOR ZOOPLANKTON IN DIFFERENT REGIONS OF PUGET SOUND

Region	Conc factor × 10^6			Dominant fauna
	N = 4[a]	N = 5	N = 6	
Elliott Bay	1.06(±0.81)[b]	1.42(±1.06)	2.17(±1.64)	Euphausiids
Main Basin	1.07(±0.61)	1.90(±1.25)	3.18(±2.32)	Copepods
Whidbey Basin (Port Gardner)	0.80(±0.38)	1.09(±0.51)	1.47(±0.75)	Copepods/ euphausiids
Hood Canal	0.98(±0.56)	0.74(±0.52)	0.43(±0.32)	Ctenophores
Sinclair Inlet	1.12(±0.68)	3.61(±2.02)	6.90(±3.45)	Ctenophores
Admiralty Inlet and Straits of Juan de Fuca	0.34(±0.21)	0.28(±0.18)	0.29(±0.18)	Copepods

[a] N = number of chlorines/biphenyl.
[b] Values in parentheses are standard deviations.

From Phillips, D. J. H., in *PCBs and the Environment,* Vol. 2, Waid, J. S., Ed., CRC Press, Boca Raton, FL, 1986, 142.

Table 5.4—12
PCB CONCENTRATIONS IN OPEN OCEAN FISHES

Location/species	Year	N	Analyzed portion	PCB conc (ng/g fresh wt.) Range	Mean
North Atlantic					
Flying fish	1970—1971		Whole		50
Flying fish *(Cypselurus exsilens)*	1970—1971		Muscle		1.4
Flying fish *(Prognichthys rondeletii)*	1970—1971		Muscle		4
Trigger fish *(Canthidermis maculatus)*	1970—1971		Muscle		1.9
Western North Pacific (off Japan)					
Emmelichthys struhsakeri	1980	3	Whole	12-75	44
Tropidinis amoenus	1979	3	Whole	17-71	35
Priacanthus boops	1979	3	Whole	12-73	45
Bering Sea					
Herring *(Clupea pallasi)*	1973	1	Muscle		80
Walleye pollock *(Theragra chalcogramma)*	1973	2	Muscle	40-40	40
Flatfish *(Limanda aspera)*	1973—1974	7	Muscle	20-130	50
Chum salmon *(Oncorynchus keta)*	1980	1	Whole		5.0
Sockeye salmon *(Oncorynchus nerka)*	1980	1	Whole		15
Chum salmon *(Oncorynchus keta)*	1982	3	Whole	5.3-9.8	7.3
Walleye pollock *(Theragra chalcogramma)*	1982	4	Whole	9.8-13	11
Eastern South Pacific (off Chile)					
Cheiloductylus sp.	1978	5	Muscle	1.2-2.2	1.5
Merluccius australis	1977	5	Muscle	0.3-0.6	0.4
Brama sp.	1978	5	Muscle	0.6-1.6	0.9
Neophrynichthys marmoratus	1978	4	Muscle	0.2-0.3	0.2
Coelorhyncus fasciatus	1978	5	Muscle	0.1-0.2	0.2
Micromesistius australis	1977	3	Whole	0.11-0.33	0.19
Coelorhynchus fasciatus	1978	4	Whole	0.06-0.11	0.09
North Indian (Arabian Sea)					
Argyrops spinifer	1976	2	Whole	0.74-1.4	1.1
Thryssa vitrirostris	1976	3	Whole	0.93-2.0	1.6
South Indian (off Australia)					
Coryphaena hippurus	1980	5	Whole	0.02-0.05	0.03
Antarctic					
Two whole fish	1972				2.0
Pagothenia borchgrevinki	1981	21	Whole	0.18-0.77	0.31
Trematomus berenacchii	1981	5	Whole	0.12-0.24	0.17
Trematomus hansoni	1981	4	Whole	0.28-0.59	0.48
Trematomus newnesi	1981	2	Whole	0.08-0.33	0.21

From Tanabe, S. and Tatsukawa, R., in *PCBs and the Environment,* Vol. 1, Waid, J. S., Ed., CRC Press, Boca Raton, FL, 1986, 152.

Table 5.4—13
TOTAL PCB CONCENTRATIONS IN FISH LIVERS FROM NINE ESTUARINE AND COASTAL MARINE SYSTEMS[a]

Area	tPCB, ppm, ww	
	1976–77	1984
Western Long Island Sound	0.62	0.81
Lower Chesapeake Bay, Virginia	0.62	0.28
Duwamish River/Elliot Bay, Washington	26.7	4.23
Nisqually Reach, Washington	0.31	0.49
Columbia River, Oregon	0.24	0.20
Coos Bay, Oregon	<0.20	<0.15
Southern San Francisco Bay, California	0.22	1.23–2.30
Palos Verdes/San Pedro Canyon, California	18.63	2.27
Dana Point, California	0.07	0.38
Median	0.31	0.49
Range: minimum	<0.20	0.15
maximum	26.7	4.23

[a] Values in ppm wet weight.

From Mearns, A. J., Matta, M. B., Simecek-Beatty, D., Buchman, M. F., Shigenaka, G., and Wert, W. A., PCB and Chlorinated Pesticide Contamination in U.S. Fish and Shellfish: a Historical Assessment Report, NOAA Tech. Memo, NOS OMA 39, Seattle, 1988.

Table 5.4—14
PCB CONCENTRATIONS IN THE BLUBBER OF MALE PINNIPEDS AND CETACEANS

Location	Species	Year	PCB conc (μg/g fresh wt.) N	Range	Mean
Arctic					
Canada	Ringed seal	1972	4	0.05—1.5	0.58
	Ringed seal	1972	15	1—6	4.1
North Greenland	Atlantic walrus	1975—1977	8	0.16—1.1	0.36
North Atlantic					
Newfoundland	Harp seal	1970	1		26
Gulf of St. Lawrence	Harp seal	1971	7	6—22	13
Nova Scotia	Atlantic white-sided dolphin	1972	1		37
Rhode Island	Striped dolphin	1972	1		39
Caribbean	Long-snouted dolphin	1972	1		5.0
South Atlantic					
Uruguay	Franciscana dolphin	1974	5	3.2—18	6.8
North Pacific					
Bering Sea	Dall's porpoise	1980	4	3.5—6.8	5.2
Japan	Striped dolphin	1978	3	22—23	22
	Finless porpoise	1968—1975	2	64—96	80
	Pilot whale		1		2.0
California	Common dolphin	1974—1976	10	80—300	120
	Pilot whale		1		14
Hawaii	Rough-toothed dolphin	1976	3	7.0—14	9.4
Eastern tropical	Striped dolphin	1973—1976	3	2.6—7.6	5.7
South Pacific					
Eastern tropical	Fraser's dolphin	1973—1976	1		5.2
	Striped dolphin		1		5.0
New Zealand	Dusky dolphin	1980	1		1.4
Antarctic					
Syowa Station	Weddell seal	1981	1		0.038

From Tanabe, S. and Tatsukawa, R., in *PCBs and the Environment,* Vol. 1, Waid, J. S., Ed., CRC Press, Boca Raton, FL, 1986, 153.

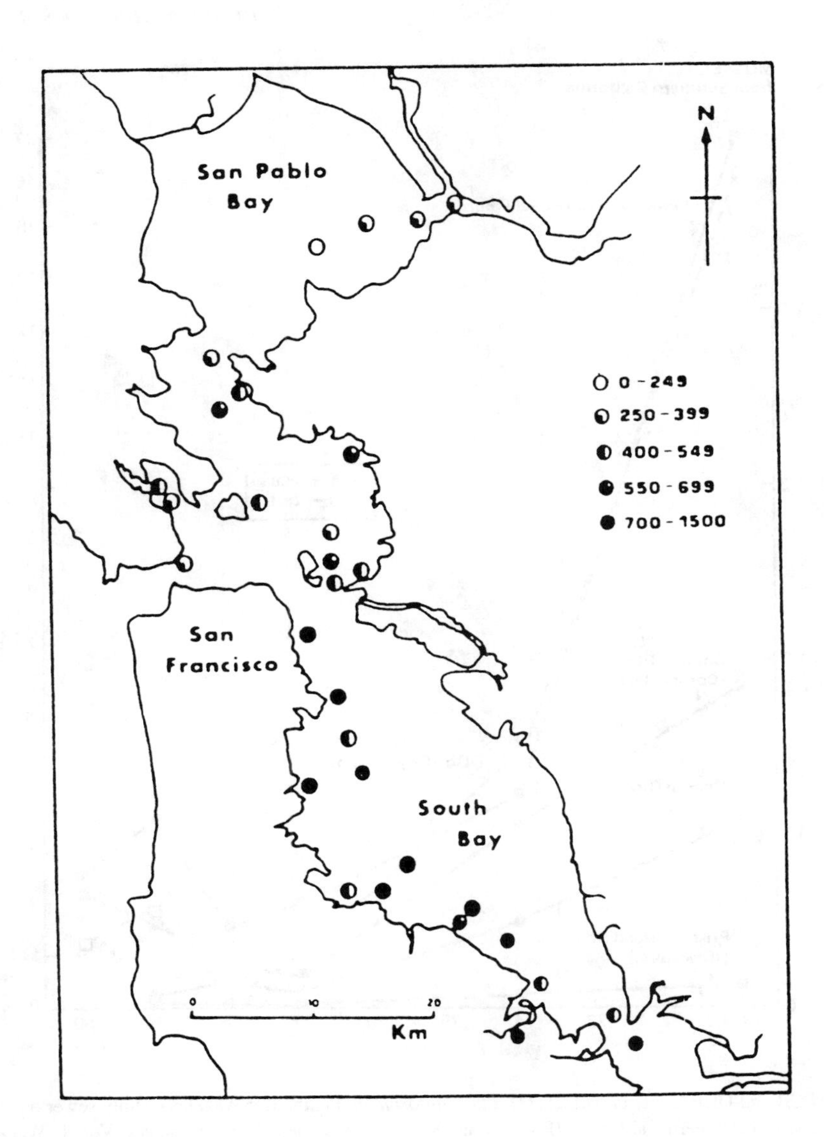

FIGURE 5.4—2. PCB concentrations (ng/g dry wt) in whole soft parts of native bay mussels (*Mytilis edulis*) in San Francisco Bay. (From Risebrough, R. W., Chapman, J. W., Okazaki, R. K., and Schmidt, T. T., Toxicants in San Francisco Bay and Estuary, Report of the Association of Bay Area Governments, Berkeley, CA, 1978.)

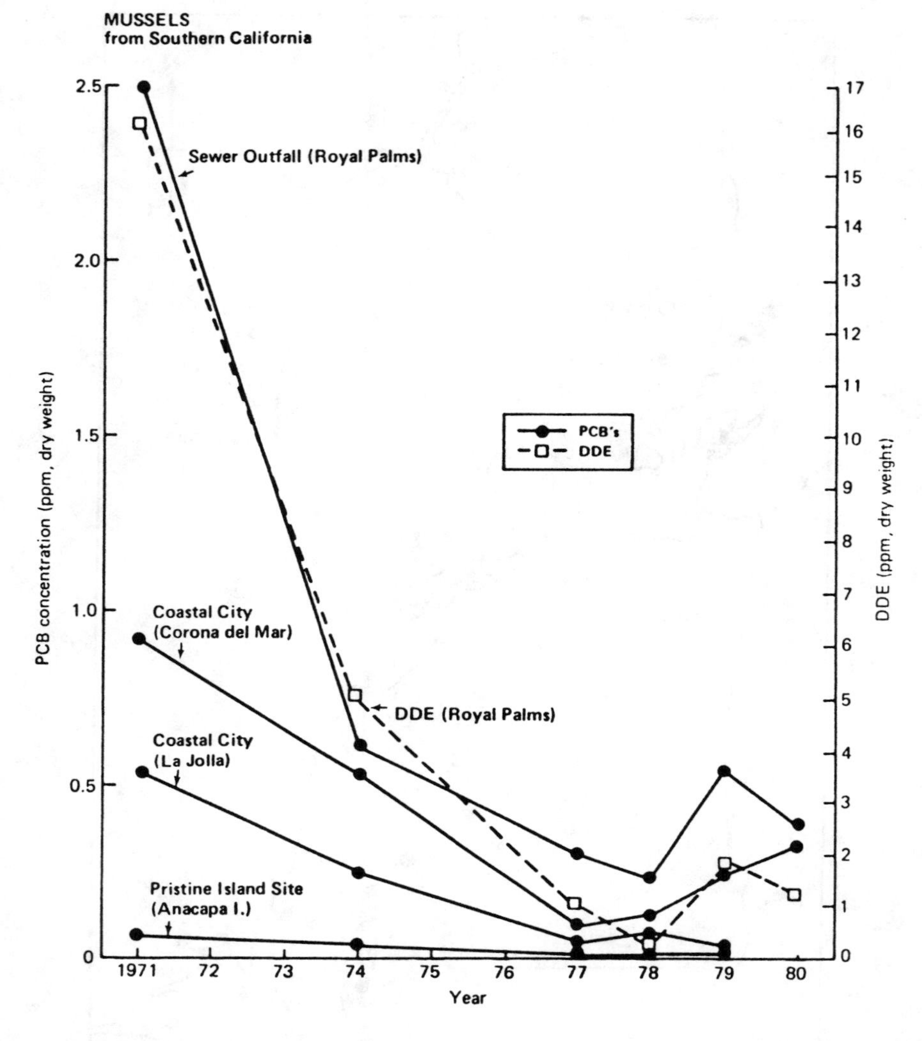

FIGURE 5.4—3. PCB and DDE levels in mussels (*Mytilus californianus*) at the Los Angeles County sewer outfall and PCB levels at three sites on the coast of Southern California. (From Stout, V. F., in *PCBs and the Environment,* Vol. 1, Waid, J. S., Ed., CRC Press, Boca Raton, FL, 1986, 188.

Table 5.4—15
PCB CONCENTRATIONS (μg/kg wet) IN MUSSELS[a] FROM THE MEDITERRANEAN SEA

Year	Region	PCB range	Reference (sources from Fowler, 1986)
1970	Ebro Delta	400–1800	Baluja et al., 1973
1980	Ebro Delta	34–200[b]	Risebrough et al., 1983
1972	French Coast	110–1920	DeLappe et al., 1972
1973–74	French Coast	40–2700[b]	Marchand et al., 1976
1974	French Coast	200–1100	Risebrough et al., 1976
~1978	French Coast	128[b]	Monod and Arnoux, 1979
1977–78	Ligurian Sea (Italy)	93–233[c]	Contardi et al., 1979
1976–78	Tuscan Iss. (Italy)	20–59	Bolognari et al., 1979
1976–79	North Adriatic (Italy)	41–100[c]	Fossato and Craboledda, 1981
1974–76	North Adriatic (Yugoslavia)	12–71	Nazansky et al., 1979
1977–78	Central Adriatic (Yugoslavia)	10–179	Dujmov et al., 1979
1977–78	South Adriatic (Yugoslavia)	2.2–148[b]	Vilicic et al., 1979
1976–77	Ionian Sea (Italy)	61–100	Amico et al., 1979
1975–79	North Aegean (Greece)	261–575	Kilikidis et al., 1981
1983	Algeria	7.3–630	Fowler and Villeneuve, unpublished

[a] Refers to *Mytilus galloprovincialis* which is often considered to be the same species as *M. edulis*. In the case of Algeria some samples were *Perna perna*.

[b] Values calculated using 0.2 dry/wet weight ratio.

[c] Range of mean values.

From Fowler, S. W., in *PCBs and the Environment*, Vol. 3, Waid, J. S., Ed., CRC Press, Boca Raton, FL, 1986, 222.

Table 5.4—16
AVERAGE LEVELS OF PCBs (μg/kg wet) IN MUSSELS MEASURED DURING THE MED-POL III POLLUTANT MONITORING PROGRAM

MED POL region	No. of samples	PCB $\bar{X}$	PCB σ	PCB Range
Northwestern basin	17	307	266	22–1200
Tyrrhenian Sea	13	95	114	5–420
Adriatic Sea	159	84	221	5–2662
Aegean Sea	12	62	12	40–80

From Fowler, S. W., in *PCBs and the Environment*, Vol. 3, Waid, J. S., Ed., CRC Press, Boca Raton, FL, 1986, 222.

Table 5.4—17
AVERAGE LEVELS OF PCBs (μg/kg wet) IN MUSCLE OF THE MULLET *MULLUS BARBATUS* MEASURED DURING THE MED-POL III POLLUTANT MONITORING PROGRAM

MED POL region	No. of samples	PCB $\bar{X}$	PCB σ	PCB Range
Northwestern basin	33	813	1496	30–8000
Tyrrhenian Sea	33	477	770	50–3950
Adriatic Sea	86	234	473	<1–3117
Aegean Sea	51	113	204	0–1110
North Levantine basin	6	9	19	0.4–52
South Levantine basin	42	69	75	0–284

From Fowler, S. W., in *PCBs and the Environment,* Vol. 3, Waid, J. S., Ed., CRC Press, Boca Raton, FL, 1986, 225.

Table 5.4—18
PCB CONCENTRATIONS (ng/kg wet) IN MUSCLE TISSUE OF FISH FROM THE MEDITERRANEAN

Species	Location	Date	PCB[a]
Mullus barbatus	Ligurian Sea (Italy)	1977	188–1,486
	North and Central Adriatic (Italy)	1976–79	69–211
	Central Adriatic (Yugoslavia)	1975–79	<1–497
	North Adriatic (Yugoslavia)	1973	3
	Sicily coast	1976–77	17–373
	Augusta Bay, Sicily	1980	300
	North Aegean	1975–79	703
	Saronikos Gulf	1975–82	8–138[b]
		1975–76	4–1,100
	Eastern Turkey	~1980	2
	Israel	1975–79	60
M. surmuletus	Israel	1975–79	69
Mugil auratus	Eastern Turkey	~1980	10
M. cephalus	North and Central Adriatic (Italy)	1972	870
Sardina pilchardus	North and Central Adriatic (Italy)	1970	37–1,060
		1972	620
	North Adriatic (Yugoslavia)	1973	2–19
	France	1975	51–309
	Spain	1970	540–6,900
	August Bay, Sicily	1980	2,300–6,100

Table 5.4—18 (continued)
PCB CONCENTRATIONS (ng/kg WET) IN MUSCLE TISSUE OF FISH FROM THE MEDITERRANEAN

Species	Location	Date	PCB[a]
Engraulis encrasicholus	Ligurian Sea (Italy)	1977–78	88–232
	Sicily coast	1976–77	9–176
	North and Central Adriatic (Italy)	1970	510–960
		1972	370
		1976–79	119–162
	North Adriatic (Yugoslavia)	1977	11–23
Thunnus thynnus	France	1975	95–407
		1977	6–89
	Sicily coast	1976–77	9–44
	North and Central Adriatic (Italy)	1976–79	344
	North Aegean	1975–79	2,613
Euthynnus alletteratus	Ligurian Sea (Italy)	1977–78	191–1,020
Sarda sarda	Ligurian Sea (Italy)	1977–78	1,133–14,020
Xiphias gladius	North Aegean	1975–79	364
Boops boops	Israel	1975–79	74
B. salpa	Libya	1982	2.5
Scorpaena scrofa	Libya	1982	3.9
Sprattus sprattus	North and Central Adriatic (Italy)	1970	620–920
Gobius paganellus		1972	100
Pleuronectes flesus		1972	250
Squalus acanthias		1972	720
Esox lucius		1972	350
Anguilla anguilla		1972	720
	Augusta Bay, Sicily	1980	2,500
Saurida undosquamis	Israel	1975–79	236
Merluccius merluccius		1975–79	16
Trachurus mediterraneus		1975–79	63
Upeneus moluccensis		1975–79	151
Pagellus acarne		1975–79	151
P. erythrinus		1975–79	188
Maena maena		1975–79	91
Dentex macrophthalmus		1975–79	195

[a] In some cases the ranges given are ranges of mean values. Single values represent either means of several determinations or a concentration based on a single measurement of a pooled sample containing tissues from several individuals.

[b] Computed from original data using 0.3 dry/wet weight ratio.

From Fowler, S. W., in *PCBs and the Environment,* Vol. 3, Waid, J. S., Ed., CRC Press, Boca Raton, FL, 1986, 224.

Table 5.4—19
PCB (μg/g WET) IN EGGS AND TISSUES OF MEDITERRANEAN SEA BIRDS

Species	Location	Tissue	No. (n)	PCBs Range	$\bar{X}$
Cory's shear water *(Calonectris diomedea)*	Crete	Eggs	2	3.2–4.1	3.65
Manx shearwater *(Puffinus puffinus)*	Gibraltar	Muscle	2	0.99–1.1	1.02
Shag *(Phalacrocorax aristotelis)*	Gibraltar	Muscle	3	4.1–7.1	5.62
Lesser black-backed gull *(Larus fuscus)*	Gibraltar	Muscle	1		0.52
Herring gull *(Larus argentatus)*	Madeira	Muscle	1		0.4
	Gibraltar	Muscle	3	0.82–3.8	2.12
	Balearic Iss.	Egg	1		11.0
	Rhone delta	Eggs	6	16–160	52
	Crete	Egg	1		5.32
	Cyprus	Eggs	2	1.2–4.2	2.7
	Tuscan Iss.	Eggs	25		~8.4
	Po delta	Eggs	4		4.0[a]
Audouin's gull *(Larus augouini)*	Morocco	Eggs	3	3.1–4.2	3.82
	Balearic Iss.	Eggs	4	12.8–20.6	16.8
Black-headed gull *(Larus ridibundus)*	Rhone delta	Eggs	4	1.6–7.8	5.45
	Po delta	Muscle	8	0.61–3.6	1.65
	N. Adriatic	Eggs	17		3.2[a]
Little gull *(Larus minutus)*	Malta	Liver	1		3.6
Common tern *(Sterna hirundo)*	Po delta	Muscle	2	0.43–8.0	4.21
	Po delta	Eggs	22		3.2[a]
	N. Adriatic	Eggs	13		4.1[a]
Little tern *(Sterna albifrons)*	Sardinia	Eggs	12		~4.0
	N. Adriatic	Eggs	16		2.6[a]
	C. sardinia	Eggs	6		2.8[a]
	S. Sardinia	Eggs	6		4.9[a]
Black tern *(Chlidonias nigra)*	Po delta	Muscle	1		<0.01
Coot *(Fulica atra)*	Po delta	Muscle	1		0.21
	Malta	Liver	1		0.1
	Po delta	Eggs	3		0.19[a]
Gull-billed tern *(Gelochelidon nilotica)*	S. Sardinia	Eggs	7		1.1[a]
	N. Adriatic	Eggs	15		1.7[a]
Slender-billed gull *(Larus genei)*	S. Sardinia	Eggs	33		3.5[a]
Black-winged stilt *(Himantopus himantopus)*	N. Adriatic	Eggs	5		0.14[a]
Avocet *(Recurvirostra avocetta)*	N. Adriatic	Eggs	5		0.12[a]
Little egret *(Egretta garzetta)*	Po delta	Eggs	9		2.3[a]
Night heron *(Nycticorax nycticorax)*	Po delta	Eggs	8		2.3[a]

[a] Computed from dry weight values using wet/dry ratio of 3 for terns, 4.3 for gulls, and 3.5 for others.

From Fowler, S. W., in *PCBs and the Environment,* Vol. 3, Waid, J. S., Ed., CRC Press, Boca Raton, FL, 1986, 228.

Table 5.4—20
ACCUMULATION OF PCBs IN BIRDS FROM FIELD EXPERIMENTS

Avian species	Residue level in prey	Residue level (ppm, wet weight)				
		Whole body	Muscle	Liver	Brain	Eggs
Great-crested grebe	0.17 (0.16–0.18)	—	—	11	—	—
(Podiceps cristatus)	0.12 (0.04–0.35)	32	—	—	—	—
	0.8–1.3	—	—	—	—	13
Brown pelican	0.19 (0.17–0.23)	—	—	—	—	2.2
(Pelecanus occidentalis)						
White pelican	0.05–0.11	2.3	3.1	4.5	—	1.7
Double-crested cormorant		3.6	2.3	2.0	—	5.7
(Phalacrocorax auritus)						
White-crested cormorant	1.6	6.2	—	—	4.4	2.9
(Phalacrocorax carbo)						
White-tailed sea eagle	0.03–0.18	—	150–240	130	29–70	—
(Haliaetus albicilla)						
Common guillemot	0.27 (0.01–1.0)	—	—	—	—	7.9–21
(Uria aalge)	0.01–2.0	3.4	—	0.4	—	—
Herring gull	2.2	—	—	—	—	124–157
(Larus argentatus)						

From Peakall, D. B., in *PCBs and the Environment,* Vol. 2, Waid, J. S., Ed., CRC Press, Boca Raton, FL, 1986, 34.

Table 5.4—21
PCB LEVELS IN HERRING GULLS AND THEIR EGGS

Location	Tissue	Year of collection	Residual level (ppm wet weight)
Norway (west coast)	Egg	1972	3.1–12.6
Baltic (Gdansk Bay)	Muscle	1975–1976	23–150
Finland	Muscle	1972–1974	0.68–38
East Scotland	Muscle	1971–1975	0.2–1.2
Camague, France	Egg	1972	16–160
New Brunswick, Canada	Egg	1969–1972	3.1–8.2
Lake Ontario	Egg	1974–1975	74–261
Maine	Egg	1977	(0–32.0)
			7.76 (30)
Virginia	Egg	1977	(0.13–16.70)
			9.06 (28)
East Frisian Is., Germany	Egg	1975	26.5
		1971	5.5
Denmark (North Sea)	Egg	1971	2.1 (1.3–2.6)
Baltic		1972	92 (21–199)

Adapted from Peakall, D. B., in *PCBs and the Environment,* Vol. 2, Waid, J. S., Ed., CRC Press, Boca Raton, FL, 1986, 43.

Table 5.4—22
PCB LEVELS IN DUCK WINGS (1969–79)[a]

Species	Flyway	Year	Pools	PCBs (mean (ppm))		
				Incidence	Detect.[b]/all	SE
Black	Atlantic	1969	42		(1.4)	0.16
		1972	44		(1.4)	0.15
		1976	32	100[a]	0.52	0.08
		1979	24	100[a]	0.63	0.09
Mallard	Atlantic	1969	19		(1.3)	0.46
		1972	21		(1.2)	0.23
		1976	20	100[a]	0.52	0.18
		1979	29	100[a]	0.45	0.07
	Mississippi	1969	51		(0.44)	0.061
		1972	61		(0.66)	0.30
		1976	69	61[a]	0.23/ 0.14	0.03
		1979	64	98[b]	0.11/ 0.11	0.02
	Central	1969	49		(0.20)	0.039
		1972	56		(0.10)[c]	0.013
		1976	56	13[a]	0.15/ 0.02	0.01
		1979	54	90[b]	0.06/ 0.05	0.01
	Pacific	1969	51		(0.20)	0.014
		1972	55		(0.11)[c]	0.009
		1976	50	14[a]	0.16/ 0.02	0.04
		1979	44	93[b]	0.07/ 0.07	0.02

Note: ppm: mg/g wet weight; each sample consisted of 25 wings.

[a] Data from Cain, B. W., *Pest Monit. J.*, 15, 128, 1981; White, D. H. and Heath, R. G., *Pest Monit. J.*, 9, 176, 1976.

[b] Arithmetic means: for 1976 and 1979 ''Detect.'' Values include only samples with detectable PCB levels; ''all'' values are means using 0.00 for undetected values. Parentheses indicate means for 1969 and 1972 not comparable to later years because of differences in analysis. a,b: Incidences differ ($p < 0.05$) within a flyway when letters (a,b) differ.

[c] 1972 mean differed ($p < 0.05$) from 1969 mean in same flyway; means of detectable levels in 1976 and 1979 not significantly different ($p > 0.05$).

From Stout, V. F., in *PCBs in the Environment,* Vol. 1, Waid, J. S., Ed., CRC Press, Boca Raton, FL, 1986, 194.

Table 5.4—23
RANGES IN CONCENTRATION OF PCBs (ng/g DRY WEIGHT) REPORTED FOR SEDIMENTS IN AREAS RANGING FROM RELATIVELY UNCONTAMINATED TO HIGHLY CONTAMINATED

Area	PCB conc.	Ref. (sources from Phillips, 1986)
Mediterranean Sea	0.8–9	Elder et al., 1976
Gulf of Mexico	0.2–35	Nisbet, 1976
Chesapeake Bay	4–400	Sayler et al., 1978
Lake Superior	5–390	Eisenreich et al., 1979
Tiber estuary	28–770	Puccetti and Leoni, 1980
Rhine-Meuse estuary	50–1,000	Duinker and Hillebrand, 1977
New York Bight	0.5–2,200	West and Hatcher, 1980
Palos Verdes Peninsula	30–7,900	Pavlou et al., 1977
Hudson River	tr–6,700	Nisbet, 1976
Escambia Bay	190–61,000	Nisbet, 1976

Note: tr, trace.

From Phillips, D. J. H., *PCBs and the Environment,* Vol. 2, Waid, J. S., Ed., CRC Press, Boca Raton, FL, 1986, 132.

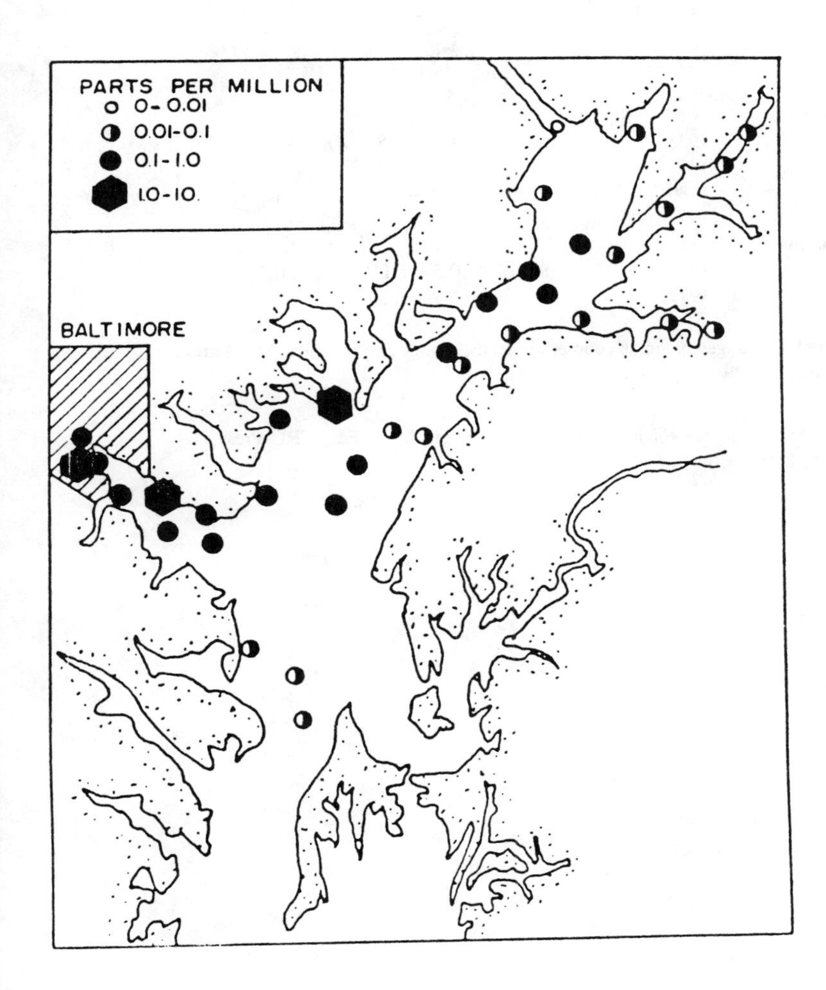

FIGURE 5.4—4. PCB concentrations in the surface sediments of the upper Chesapeake Bay. (From Munson, T. O., in Upper Bay Survey, Vol. 2, Final Report to the Maryland Department of Natural Resources, Westinghouse Electric Corporation, Ocean Sciences Division, Annapolis, MD, 1976. With permission.)

Table 5.4—24
PCB CONCENTRATIONS (ng/g DRY) IN COASTAL SURFACE SEDIMENTS FROM THE MEDITERRANEAN SEA

Region	Year	Sample no.	Range	$\bar{X}$[a]
Northwest Mediterranean				
French coast	1973	12	0.8—1165	—
Ile des Embiez	<1979	1	12	—
Golfe du Lion	1972	9	40—1,200	340
	1975	21	120—780	390
Marseille	1978—79	50	5—15,815	—
Monaco	1981	3	11—61	34
Spain	1977	17	0.3—100	19
Tyrrhenian Sea				
Tiber estuary	1976—77	2	46—770	—
Tiber offshore	1976—77	2	28—73	—
Naples offshore	1980	38	9—170	33
Naples Bay	1980	38	6—3,200	248
Sicily	1980—81	36	0.6—82	12
Ionian Sea				
Sicily	1980—81	36	0.8—49	12
Augusta Harbor	1980—81	9	130—457	253
Adriatic Sea				
Yugoslavia (north)	1977—78	7	<1.0—2.8	—
Yugoslavia (south)	1977—78	3	13—17	—
Aegean Sea				
Greece	<1973	15	1.3—775	—
Eastern basin				
Turkey	<1980	7	2—4	3
Egypt	1977	1	1.9	—
Central & western basins				
Tunisia	1977	8	0.5—1.1	0.7
Algeria	1977	5	7—323	—

[a] Mean not computed: (1) ranges given as geometric means, (2) too few samples, or (3) range includes single exceptionally high value.

From Fowler, S. W., *PCBs and the Environment,* Vol. 3, Waid, J. S., Ed., CRC Press, Boca Raton, FL, 1986, 219.

Table 5.4—25
PCB CONCENTRATIONS (ng/g DRY) IN SURFACE SEDIMENTS FROM THE OPEN MEDITERRANEAN SEA

Year	Region	Sample no.	Range	$\overline{X}$
1975	Algero-Provencal basin	5	0.8–9.0	4.0
	Gibraltar sill & Siculo-Tunisian sill	2	0.8	0.8
	Algerian margin	1	0.8	0.8
	Ionian Sea	3	0.8–5.1	2.8
1976[a]	Ligurian Sea	14	1.5–33	11
1977[b]	Tyrrhenian Sea	2	0.8–1.3	1.1
	Ionian Sea	3	1.2–1.6	1.4
	Aegean Sea	2	0.6	0.6
	Levantine & central basin	5	0.6–8.9	3.8

[a] Grab samples.
[b] Undisturbed sediment cores (top 1 cm).

From Fowler, S. W., in *PCBs in the Environment,* Vol. 3, Waid, J. S., Ed., CRC Press, Boca Raton, FL, 1986, 221.

Table 5.4—26
PCB CONCENTRATIONS IN THE OPEN OCEAN ATMOSPHERE

Location	Year	PCB conc (ng/m³)		
		N	Range	Mean
North Atlantic				
Bermuda	1973	4	0.15—0.50	0.30
Bermuda	1973	8	0.21—0.65	0.51
Bermuda, U.S.	1973	4	0.72—1.6	0.99
Grand Banks (45°N, 52°W)	1973	5	0.05—0.16	0.086
Newfoundland	1977	6	0.042—0.15	0.12
Gulf of Mexico	1977	10	0.17—0.79	0.35
Barbados	1977-1978	17	<0.005—0.37	0.057
North Pacific				
Enewetak Atoll (12°N, 162°E)	1979	14	0.35—1.0	0.54
Western Pacific (3-35°N, 105-151°E)	1980-1981	7	0.089—0.74	0.25
Western Pacific (43-53°N, 154-172°E)	1981	2	0.041—0.061	0.051
Western Pacific (41-46°N, 144-174°E)	1982	5	0.022—0.095	0.043
Bering Sea	1981	3	0.026—0.059	0.041
South Pacific				
Western Pacific (1-46°S, 151-157°E)	1981	5	0.083—0.50[a]	0.27
Indian				
Eastern Indian (1-44°S, 104-125°E)	1980	5	0.066—0.33[a]	0.15
Western Indian (20-54°S, 48-57°E)	1982	4	0.060—0.24	0.16
Antarctic				
53-65°S, 125-161°E	1980-1981	5	0.056—0.18	0.091
54-68°S, 38-58°E	1982	4	0.076—0.11	0.091
Syowa Station (69°00′S, 39°35′E)	1981-1982	11	0.017—0.17	0.061

[a] Excluding the coastal regions.

From Tanabe, S. and Tatsukawa, R., in *PCBs and the Environment,* Vol. 1, Waid, J. S., Ed., CRC Press, Boca Raton, FL, 1986, 145.

Table 5.4—27
PCBs IN MISCELLANEOUS TISSUES AND FLUIDS OF VARIOUS HUMAN POPULATIONS

	Av. conc (ng/m^3) as Aroclor®		
	Total	**1242**	**1254**
Urban			
Denver, CO (1980)	2.25	1.80	0.45
Columbia, SC (1977–80)	4.70	3.20	1.50
Houston, TX	—	—	3.00
Minneapolis, MN (1978–79)	7.10	5.60	1.56
Chicago, IL (1975–76)	8.00	6.70	1.28
Madison, WI (1978)	7.50	6.50	1.05
Milwaukee, WI (1977)	2.25	1.64	0.61
Ontario, several cities			
1979	5.90	—	—
1980	0.21	—	—
Stockholm, Sweden (1983–84)	0.34	0.23	0.11
Jacksonville, FL 1975)	4.70	—	—
Gainesville, FL (1977)	20.00	—	—
Rural/suburban/coastal			
Pigeon Key, FL (1978)	—	—	0.41
North Inlet Estuary, SC (1977–79)	0.44	0.25	0.19
Texas Gulf Coast (1976)	—	—	0.52
Texas Gulf Coast (1980)	—	—	0.067
College Station, TX (1979)	—	—	0.29
White Sands, NM (1981)	—	—	0.11
Ontario Province			
1979	5.40	—	—
1980	0.19	—	—
Rural Sweden (1984)	0.07	0.04	0.03
Rural Norway (1982)	—	—	0.02[a]
Great Lakes			
Lake Michigan (1977)	0.87	0.65	0.22
Lake Superior (1978–80)	1.20	0.86	0.34
Marine			
Gulf of Mexico (1976)	—	—	0.16
Newfoundland (1977)	—	—	0.096
Barbados (1977–78)	—	—	0.065
Enewetak Atoll (1979)	0.11	0.061	0.049
American Samoa (1981)	—	—	0.012
Peruvian Coast (1981)	—	—	0.012
Bermuda (1976–77)	—	—	0.075
E. Indian Ocean (1980–81)	0.15	—	—
W. Pacific	0.34	—	—
Antarctic/Southern Ocean	0.11	—	—
Arctic Ocean (1982–83)	—	—	0.02[a]

Note: — , Not measured or not reported.

[a] Pentachlorobiphenyls.

From Mes, J., in *PCBs and the Environment*, Vol. 3, Waid, J. S., Ed., CRC Press, Boca Raton, FL, 1986, 51.

Table 5.4—28
PCBs IN MISCELLANEOUS TISSUES AND FLUIDS OF VARIOUS HUMAN POPULATIONS

					PCBs (mg/kg)			
					Fat		Tissue	
Country	Region	Year of sample collection	No. of samples	Type of tissue or fluid	Average	Max.[a]	Average	Max.[a]
Brazil	Sao Paulo	1978	15	Gastric mucosa	5.72	11.32		
				Epiploon	1.43	3.22		
Denmark	Funen	1976–1978	22	Basal ganglia			0.080	
				M. psoas			0.097	
				Spleen			0.038	
				Liver			0.144	
				Kidney			0.058	
				Ovary[b]			0.027	
				Testis[c]			0.069	
Finland	Pori	1974–1975	41	Liver	3.44		0.17	
				Brain	1.19		0.12	
	Jyväskylä		32	Liver	1.75		0.13	
Germany		1982	5	Bone marrow	6.4	11.8		
Israel		1977	7	Uterine muscle	14.0968			
			9	Placenta	5.0267			
			4	Amniotic fluid	124.7529			
Japan		1972	15	Hair			0.74	1.71
	Kyoto	1973	33	Hair			0.32	0.71
		1975–1976	61	Hair	0.31	0.87		
		1977	2	Cerebrum			0.027	0.043
			3	Heart			0.141	0.20
			3	Kidney			0.037	0.076
			2	Liver			0.058	0.099
Norway		1977	10	Liver	1.87	2.29		
U.S.		1976	4[d]	Hair			0.61	0.76

[a] Maximum value observed.
[b] Five samples.
[c] Seventeen samples.
[d] Composition samples of five or more individuals.

From Mes, J., in *PCBs and the Environment,* Vol. 3, Waid, J. S., Ed., CRC Press, Boca Raton, FL, 1986, 51.

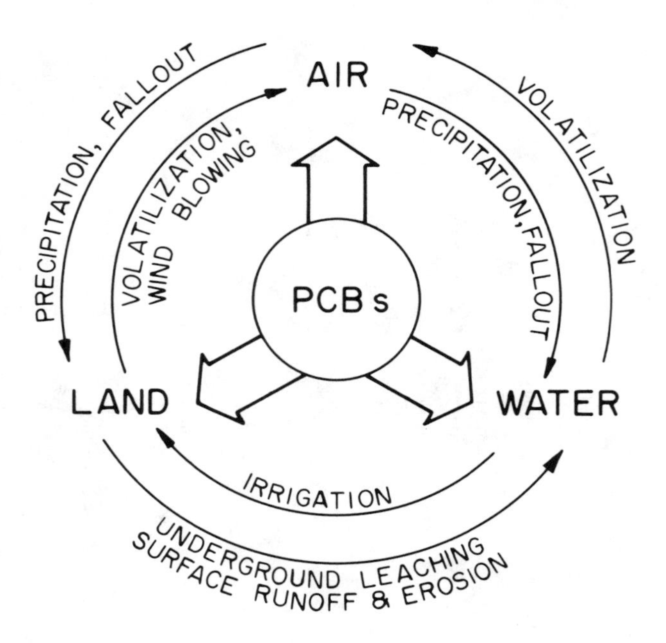

FIGURE 5.4—5. PCB environmental cycle. (From Lauber, J. D., in *PCBs and the Environment,* Vol. 3, Waid, J. S., Ed., CRC Press, Boca Raton, FL, 1986, 86.)

Table 5.4—29
PCBs AND CHLORINATED PESTICIDES LISTED IN TERMS OF STATUS OF KNOWLEDGE AND RECOMMENDATIONS FOR SURVEILLANCE IN FISH AND SHELLFISH ON A NATIONWIDE BASIS

1. Contaminants of continuing concern; maintain surveillance
   - PCBs
   - DDT (mainly DDE)
   - Dieldrin
   - alpha (*cis*-)-chlordane
   - *trans*-nonachlor
2. Contaminants of uncertain national concern
   - Endosulfan
   - Toxaphene
   - PCP (pentachlorophenol)
   - DCPA (dacthal)
   - Endrin
   - Kelthane
   - Kepone
3. Contaminants apparently not of national concern; reduce surveillance frequency
   - Aldrin
   - Methoxychlor
   - Heptachlor and Heptachlor epoxide
   - Mirex
   - HCB (hexachlorobenzene)
   - Lindane

From Mearns, A. J., Matta, M. B., Simecek-Beatty, D., Buchman, M. F., Shigenaka, G., and Wert, W. A., PCB and chlorinated pesticide contamination in U.S. fish and shellfish: a historical assessment report, NOAA Tech. Mem. NOS OMA 39, Seattle, WA, 1988.

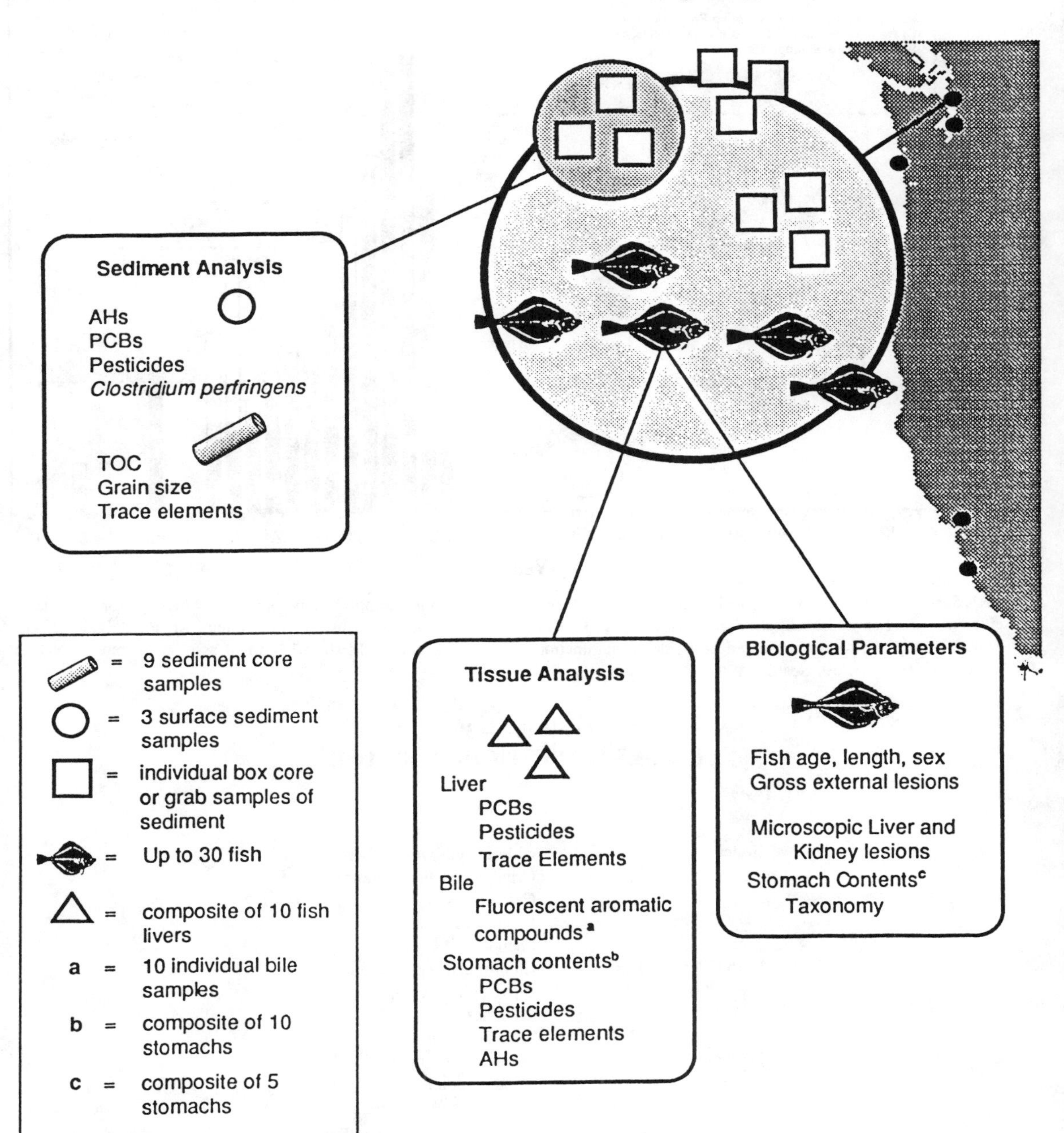

FIGURE 5.4—6. The sampling strategy of the National Benthic Surveillance Project. Each sampling site consists of three stations at which grab samples of sediment were taken. (From Varanasi, U., Chan, S-L., McCain, B. B., Schiewe, M. H., Clark, R. C., Brown, D. W., Myers, M. S., Landahl, J. T., Krahn, M. M., Gronlund, W. D., and MacLeod, W. D., Jr., National Benthic Surveillance Project: Pacific Coast. Part I. Summary and Overview of the Results for Cycles I to III 1984–86), NOAA Tech. Rept., Seattle, WA, 1988.)

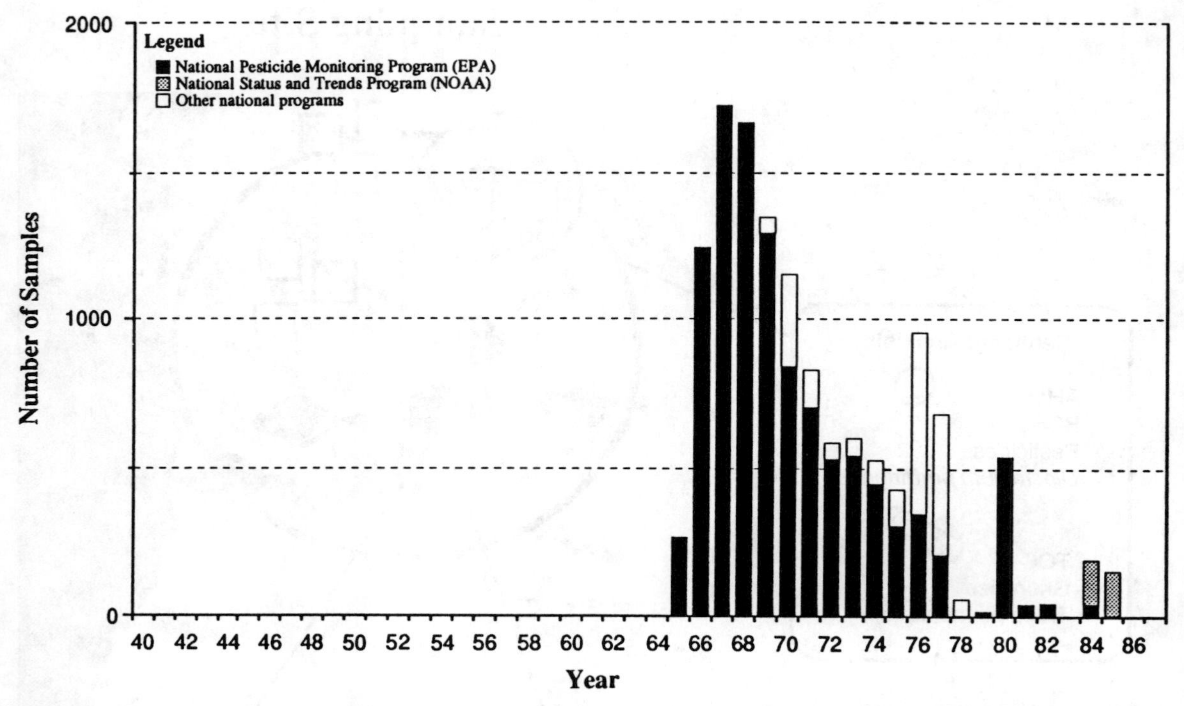

FIGURE 5.4—7. Numbers of samples by year, 1940–85, from national and other comprehensive surveys for PCB and/or chlorinated pesticides in U.S. fish and shellfish. (From Mearns, A. J., Matta, M. B., Simecek-Beatty, D., Buchman, M. F., Shigenaka, G., and Wert, W. A., PCB and chlorinated pesticide contamination in U.S. fish and shellfish: a historical assessment report, NOAA Tech. Mem. NOS OMA 39, Seattle, WA, 1988.)

Table 5.4—30
CLASSES OF PESTICIDES BY USE

Class	Example
Insecticides	Chlorinated hydrocarbons
	Organophosphorus compounds
	Carbamates
	Pyrethroids
	Phenols
Herbicides	Chlorophenoxy compounds
	Bipyridylium compounds
	Triazines
	Thiocarbamates
Fungicides	Inorganic compounds
	Organometallic compounds
	Antibiotics
	Chloroalkylthio compounds
	Quinones
	Dithiocarbamates
Rodenticides	Fluoroacetate derivatives
	Thioureas
	Antivitamin K compounds
Fumigants/nematocides	Hydrocyanic acid
	Carbon disulfide
Synergists	Piperonyl butoxide
	Sulfoxide

From Richardson, M. L., Ed., *Chemistry, Agriculture, and the Environment,* The Royal Society of Chemistry, Cambridge, 1991. With permission.

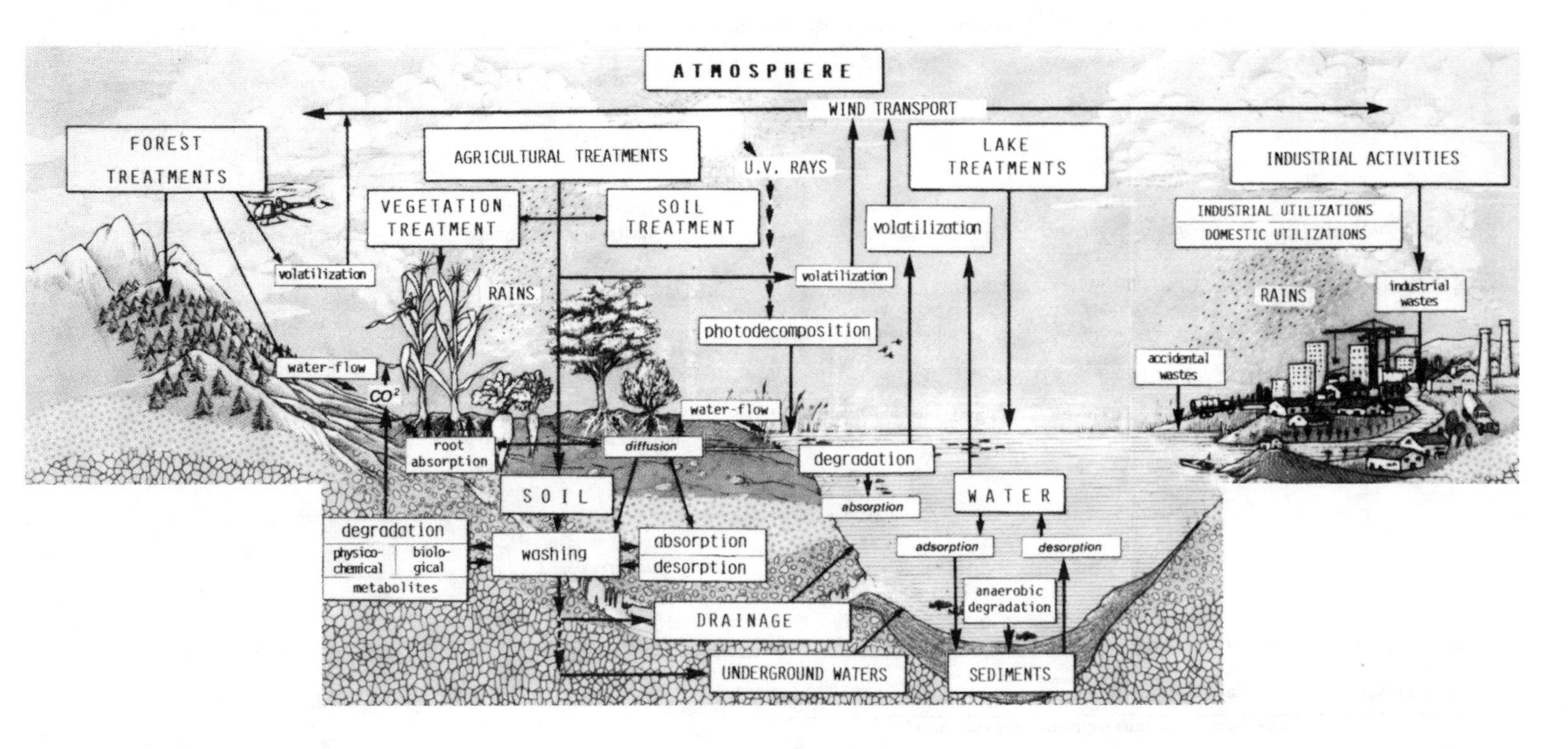

FIGURE 5.4—8. Pesticide dispersion in the environment. (From Boudou, A. and Ribeyre, F., Eds., *Aquatic Ecotoxicology: Fundamental Concepts and Methodologies*, CRC Press, Boca Raton, FL, 1989.)

Table 5.4—31
CHEMICAL PESTICIDES IN COMMON USE (U.S.) WITH SOME EXAMPLES OF THEIR EFFECTIVENESS

Type	Name	Evaluation for mosquito control — some examples			
		Date	Chemical	Mosquito species	Mortality (%)
1. Organophosphates	Parathion, Malathion, Fenthion (Baytex), Trichlorfan, Fenitrothion, Chloropyrifos (Dursban®), Naled (Dibrom), Diazinon, Temephos (Abate)	1973	Abate	*Aedes vigilax, Culex sitiens*	71–99
		1978	Dibrom	*Ae. taeniorhynchus*	65–85
Carbamates	Propoxur (Baygon), Carbaryl (Sevin®)	1978	Baygon	*Ae. taeniorhynchus*	56–72
Botanicals	Pyrethrum, synthetic pyrethroids (Resmethrin, Permethrin)				
2. Insect growth regulators (IGR)	Methoprene (Altosid®), Diflubenzuron (Dimilin)[a]	1976	Methoprene, Dimilin	*Anopheles quadrimaculatus, Cx. nigripalpus, Ae. taeniorhynchus*	>90
		1976	Dimilin, Altosid®	*Ae. taeniorhynchus*	96–100
		1979	Methroprene	*Ae. taeniorhynchus*	95
		1982	Stauffer®, MU678	*Ae. sollicitans, Cx. salinarius*	100
3. Surface films	Monomolecular organic surface films, e.g., 1SA-20E (kerosene, diesel oil, etc. are occasionally, but not commonly used in the U.S. or Australia)	1981	1SA-20E	*Ae. taeniorhynchus*	100 (eventually)

[a] By 1982, this had not been registered for use in the U.S.

From Dale, P. E. R. and Hulsman, K., A critical review of salt marsh management methods for mosquito control, *Rev. Aquat. Sci.*, 3, 281, 1990.

Table 5.4—32
SOME CHRONIC EFFECTS OF PESTICIDES WHICH LED TO THEIR RESTRICTION IN THE U.S.

Pesticide	Effects
Aldrin/dieldrin	Environmental persistence and oncogenicity
Chlordane/heptachlor	Environmental persistence and oncogenicity
Endrin	Oncogenicity, teratogenicity, and reduction of nontarget species
EPN (*O*-ethyl *O*-4-nitrophenyl phenylphosphonothioate)	Delayed neurotoxicity and acute toxicity
Lindane	Oncogenicity, fetotoxicity, and acute toxicity
Mirex	Oncogenicity, teratogenicity, and reduction of nontarget species
DBCP (dibromochloropropane)	Reproductuve effects and oncogenicity
Dimethoate	Oncogenicity, mutagenicity, fetotoxicity, and reproductive effects
EDB (ethylene dibromide)	Oncogenicity, mutagenicity, and reproductive effects
Aldicarb	Chronic toxicity, delayed neurotoxicity, reproductivity toxicity

From Richardson, M. L., Ed., *Chemistry, Agriculture, and the Environment,* The Royal Society of Chemistry, Cambridge, 1991. With permission.

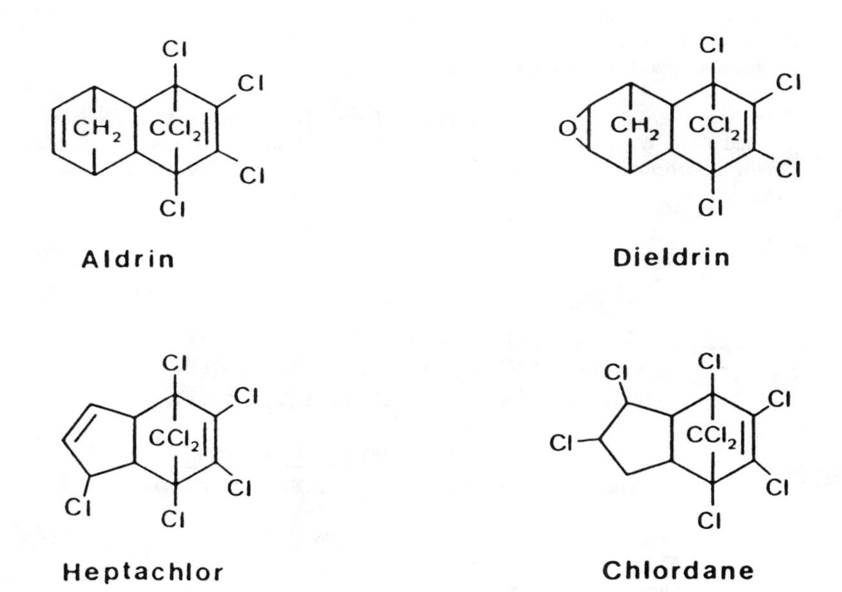

FIGURE 5.4—9. Structures of some cyclodiene insecticides. (From Richardson, M. L., Ed., *Chemistry, Agriculture, and the Environment,* The Royal Society of Chemistry, Cambridge, 1991. With permission.)

Table 5.4—33
MEDIAN OR GEOMETRIC MEAN DDT CONCENTRATIONS FOR SEVERAL NATIONAL SURVEY EVENTS

Organism: substrate	tDDT or DDE, ppm ww by sampling period			
	1965–72	1972–75	1976–77	1984–86
Bivalves	0.024[1]		0.01[2]/0.001[3]	0.003[10]
Fish, whole juvenile		0.014[4]	ND	ND
Fish, muscle		0.110[5]	0.012[6]	
Fish, liver			0.220[6]	0.054[7]
Fish, whole F.W.	0.7–1.1[8]	0.4–0.6[8]	0.370[9]	

Note: See Mearns et al. for original source material.

[1] From Butler (1973); median of 8,180-site means composited from 7,839 samples.
[2] From original data supporting Butler et al. (1978); median of 89-site means composited from 188 samples.
[3] From Farrington et al. (1981); median of 80-site values or site means.
[4] From original data supporting Butler and Schutzmann (1978); median of 144-site means composited from 1,524 composites.
[5] From original data supporting Stout and Beezhold (1981) and Stout (1980); median of area or site means from samples.
[6] From original data supporting Butler (1978) and Sherwood (1982); median of 19-site means from samples.
[7] From original data supporting Zdanowicz et al. (1986), Malins et al. (1986), and Hanson et al. (1986); median of 42-site medians from 126 composites.
[8] From Schmitt et al. (1983).
[9] From Schmitt et al. (1985).
[10] Preliminary calculations from NOAA (1987).

From Mearns, A. J., Matta, M. B., Simecek-Beatty, D., Buchman, M. F., Shigenaka, G., and Wert, W. A., PCB and Chlorinated Pesticide Contamination in U.S. Fish and Shellfish: a Historical Assessment Report, NOAA Tech. Mem. NOS OMA 39, Seattle, WA, 1988.

Table 5.4—34
RELATIVE PROPORTIONS OF DDT COMPONENTS IN FISH AND OTHER ORGANISMS FROM NORTH ATLANTIC

Group	Components of Σ DDT (%)		
	p,p'-DDE	*p,p'*-DDD	*p,p'*-DDT
Bivalves	51	36	13
Crustacea	84	12	4
Groundfish	45	14	41
Pelagic fish	45	17	38

From Murty, A. S., *Toxicity of Pesticides to Fish,* Vol. 1, CRC Press, Boca Raton, FL, 1986, 50.

Table 5.4—35
THE CONTAMINATION OF VARIOUS SPECIES OF PELAGIC SEABIRDS BY DDT, ITS METABOLITES, AND PCBs

Species	Place of capture (place of reproduction)	Tissues analyzed	DDT and metabolites (ppm)	PCBs (ppm)
Fulmarus glacialis	California (Alaska)	Whole bird	7.1	2.3
Puffinus creatopus	Mexico (Chile)	Whole bird	3.0	0.4
P. griseus	California (New	Fats	11.3	1.1
	Zealand)	Fats	40.9	52.6
P. gravis	New Brunswick (Southern Atlantic)	Fats	70.9	104.3
Pterodroma cahow	Bermudas (same)	Whole bird	6.4	—
Oceanodroma leuchorhoea (Leach's Petrel)	California (same)	Fats ex ovo	953	351
Oceanites oceanicus (Wilson petrel)	New Brunswick (Antarctica)	Fats	199	697

From Ramade, F., in *Aquatic Ecotoxicology: Fundamental Concepts and Methodologies,* Boudou, A. and Ribeyre, F., Eds., CRC Press, Boca Raton, FL, 1989, 156.

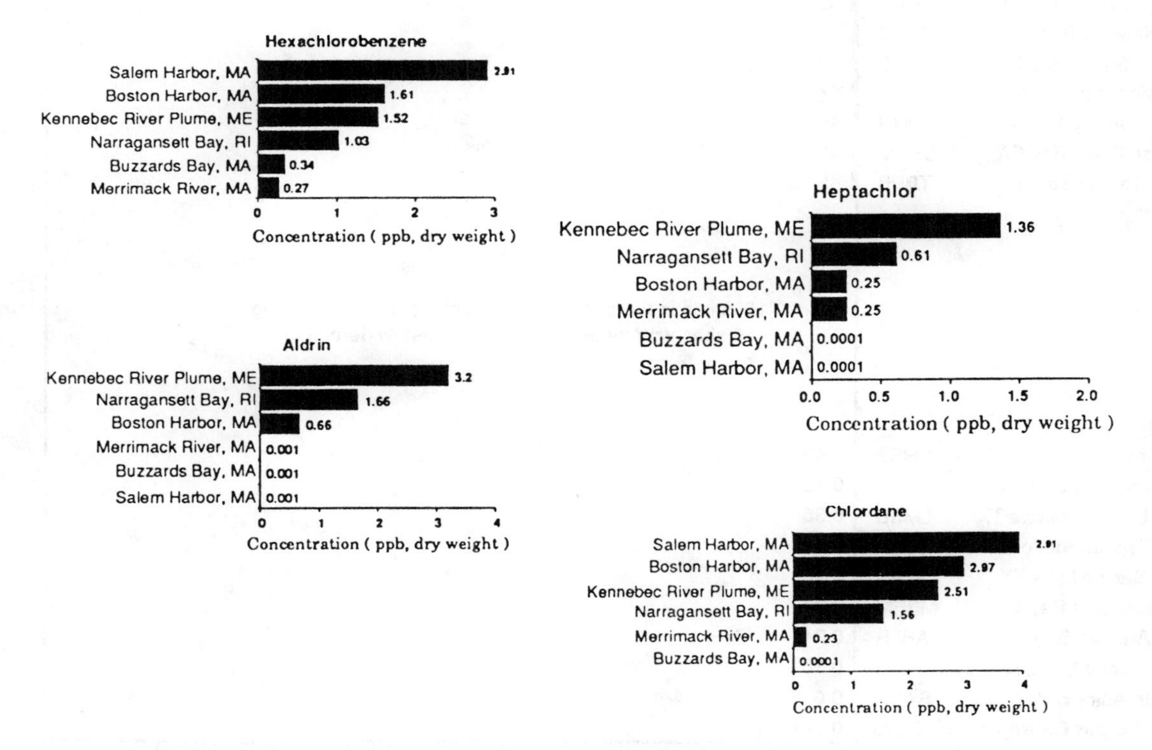

FIGURE 5.4—10. Concentrations of individual pesticides in New England marine sediments. (From Larsen, P. F., Marine environmental quality in the Gulf of Maine, *Rev. Aquat. Sci.*, 6, 67, 1992.)

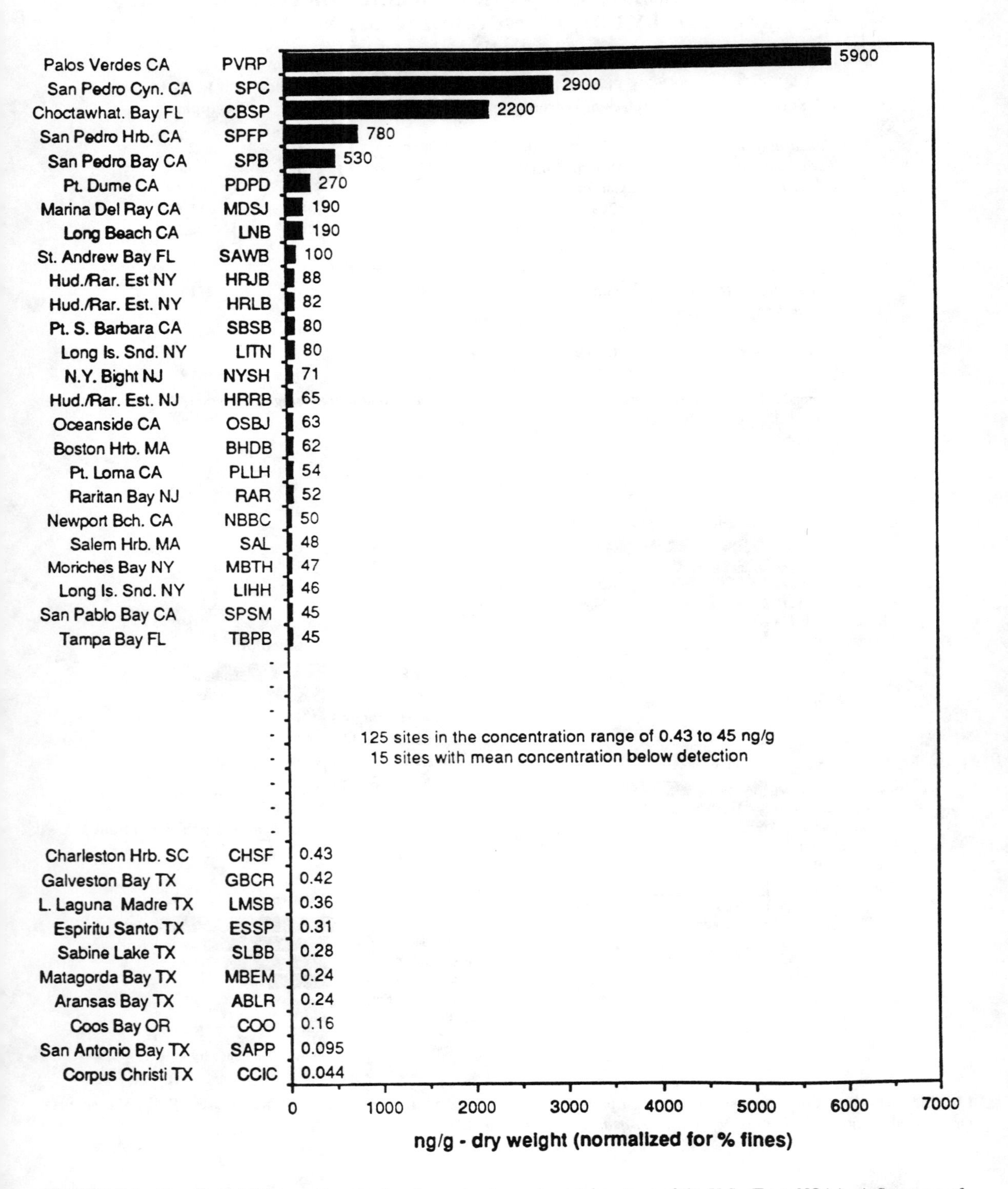

FIGURE 5.4—11. Total DDT in sediments of various estuarine and coastal systems of the U.S. (From NOAA, A Summary of Data on Individual Organic Contaminants in Sediments Collected During 1984, 1985, 1986, and 1987, NOAA Tech. Mem. NOS OMA 47, Rockville, MD, 1989.)

Table 5.4—36
REPRESENTATIVE BIOCONCENTRATION FACTORS (BCF) OF SOME PESTICIDES

Compound	Species	Conc. in water		Duration of exposure	BCF
DDT	Fathead minnow	0.5—2	μg/ℓ	56—112 days	100,000
	Golden shiner	0.3	μg/ℓ	15 days	100,000
	Green fish	0.11—0.33	μg/ℓ	15 days	17,500
	Pinfish and Atlantic croaker	0.1—1	μg/ℓ	15 days	10,000 to 38,000
	Daphnia magna	8—50	μg/ℓ	24 hr	16,000 to 23,000
	Salmo gairdneri	133—176	μg/ℓ	84 days	21,300 to 51,300
Dieldrin	Sculpin	0.017—8.6	μg/ℓ	32 days	300 to 11,000
	Guppy	0.8—4.3	μg/ℓ	18 days	49,307
	Bluegill (static test)	1	μg/ℓ	48 hr	2,441
	Bluegill (flow-through test)	1.5	μg/ℓ	48 hr	1,727
Heptachlor	Sheepshead minnow	6.5—21	μg/ℓ	96 hr	7,400 to 21,300
	Pinfish	0.32—32	μg/ℓ	96 hr	3,800 to 7,700
	Spot	1.2—3.7	μg/ℓ	96 hr	3,000 to 13,800
Mirex	Juvenile pinfish	25—46	μg/ℓ	3 days	3,800
Kepone®	Sheepshead minnow juveniles	41—780	μg/ℓ	21 days	2,600
	Adult male	41—780	μg/ℓ	21 days	7,600
	Adult female	41—780	μg/ℓ	21 days	5,700
HCH	Pinfish	18.4—31.3	μg/ℓ	4 days	218
	Sheepshead minnow	41.9—108.3	μg/ℓ	4 days	490
	Guppy	10—1400	μg/ℓ	24 hr	500
Technical HCH	Pinfish edible part	1.4—36	μg/ℓ	24 hr	500
	Pinfish offal	1.4—36	μg/ℓ	28 days	175
Chlordane	Goldfish	3.4	μg/ℓ	4 days	67 to 162
	Sheepshead minnow	15—51	μg/ℓ	4 days	12,600 to 18,700
	Pinfish	5.4—15.2	μg/ℓ	4 days	3,000 to 7,500
HCB	Killifish	160—380	μg/ℓ	6.68 hr	65 to 710
	Fathead minnow	—		—	16,200
	Green sunfish	—		—	21,900
	Rainbow trout	—		—	5,500
Permethrin	Juvenile Atlantic salmon	1.4—12	μg/ℓ	89 hr	73
Cypermethrin	Juvenile Atlantic salmon	1.4—12	μg/ℓ	96 hr	3 to 7
Fenvalerate	Juvenile Atlantic salmon	1	μg/ℓ	96 hr	200
PCP	Trout	2	mg/ℓ	24 hr	100
	Killifish	57—120	μg/ℓ	168 hr	10 to 64
	Killifish	100—610	μg/ℓ	240 hr	8 to 50
Trichlorophenol	*Poecilia* female	610	μg/ℓ	36 hr	12,180
	Poecilia male	350	μg/ℓ	36 hr	7,000
Leptophos	Bluegill sunfish	240	mg/ℓ	10 days	750
Fenitrothion	Killifish	800	μg/ℓ	10 days	53
	Coho salmon under-yearling	560	μg/ℓ	24 hr	16

Table 5.4—36 (continued)
REPRESENTATIVE BIOCONCENTRATION FACTORS (BCF) OF SOME PESTICIDES

Compound	Species	Conc. in water	Duration of exposure	BCF
Diazinon	Topmouth gudgeon	10 μg/ℓ	7 days	152
	Silver crucian carp	10 μg/ℓ	7 days	37
	Carp	10 μg/ℓ	7 days	65
	Guppy	10 μg/ℓ	7 days	18
Fluridone	Fathead minnow	140 mg/ℓ	10 weeks	64
2,4-D	Bluegill sunfish	3 mg/ℓ	8 days	1
2,4,5-T	Bluegill sunfish	3 mg/ℓ	8 days	1
MCPA	Trout	10—100 mg/ℓ	10—28 days	1
Fosamine ammonium	Channel catfish	1.1 mg/ℓ	4 weeks	1
Hexamethyl phosphoramide	Sheepshead minnow	0.5 mg/ℓ	28 days	1
Benthiocarb	Fathead minnow	28 μg/ℓ	2.5 days	446 to 471
	Channel catfish	29 μg/ℓ	3 days	120
	Bluegill	28 μg/ℓ	5 days	91
	Longear sunfish	99 μg/ℓ	1—5 days	280 to 300

From Murty, A. S., *Toxicity of Pesticides to Fish,* Vol. 1, CRC Press, Boca Raton, FL, 1986, 70.

Table 5.4—37
CONCENTRATIONS OF VARIOUS ORGANOCHLORINES (MEANS ± SE, μg/g DRY WEIGHTS) IN SIX MALE AND SIX FEMALE COMPOSITES OF MUSSELS (*MYTILUS CALIFORNIANUS*) FROM TWO SITES IN CALIFORNIAN WATERS

	Royal Palm (outfall region)		Point Sal (control region)	
Compound	Male	Female	Male	Female
Total DDT	4.100 ± 0.700	6.40 ± 0.320	0.110 ± 0.0040	0.110 ± 0.0070
p,p'-DDT	0.040 ± 0.005	0.06 ± 0.005	0.008 ± 0.0005	0.007 ± 0.0006
o,p'-DDT	0.030 ± 0.003	0.04 ± 0.004	0.004 ± 0.0001	0.004 ± 0.0005
p,p'-DDE	3.500 ± 0.640	5.40 ± 0.280	0.077 ± 0.0030	0.067 ± 0.0030
o,p'-DDE	0.410 ± 0.060	0.73 ± 0.050	0.015 ± 0.0010	0.026 ± 0.0070
p,p'-DDD	0.110 ± 0.010	0.18 ± 0.007	0.009 ± 0.0006	0.008 ± 0.0004
Total PCB	0.640 ± 0.060	0.94 ± 0.040	0.044 ± 0.0060	0.054 ± 0.0020
1242 PCB	0.120 ± 0.010	0.19 ± 0.010	0.016 ± 0.0030	0.019 ± 0.0010
1254 PCB	0.530 ± 0.060	0.76 ± 0.040	0.028 ± 0.0040	0.035 ± 0.0030

Note: Each composite was 5 individuals of shell length 40—60 mm; both samples were collected in late June 1974.

From Phillips, D. J. H., in *PCBs and the Environment,* Vol. 2, Waid, J. S., Ed., CRC Press, Boca Raton, FL, 1986, 167.

Table 5.4—38
BIOACCUMULATION AND TOXICITY OF TOXAPHENE IN ESTUARINE ORGANISMS

Species	96-h LC_{50} mg/l	96-h BCF[b] (tissue/water)
Pink shrimp *Penaeus duorarum*	1.4	3,100–20,600
Grass shrimp *Palaemonetes pugio*	4.4	3,100–20,600
Sheepshead minnow *Cyprinodon variegatus*	1.1	400–1,200
Pinfish *Lagodon rhomboides*	0.5	400–1,200
Longnose killifish *Fundulus similis*	0.9–1.4[a]	4,200–60,000 (whole body)

[a] 28 days.
[b] BCF, bioconcentration factor.

From Sergeant, D. B. and Onuska, F. I., in *Analysis of Trace Organics in the Aquatic Environment,* Afghan, B. K. and Chau, A. S. Y., Eds., CRC Press, Boca Raton, FL, 1989, 75.

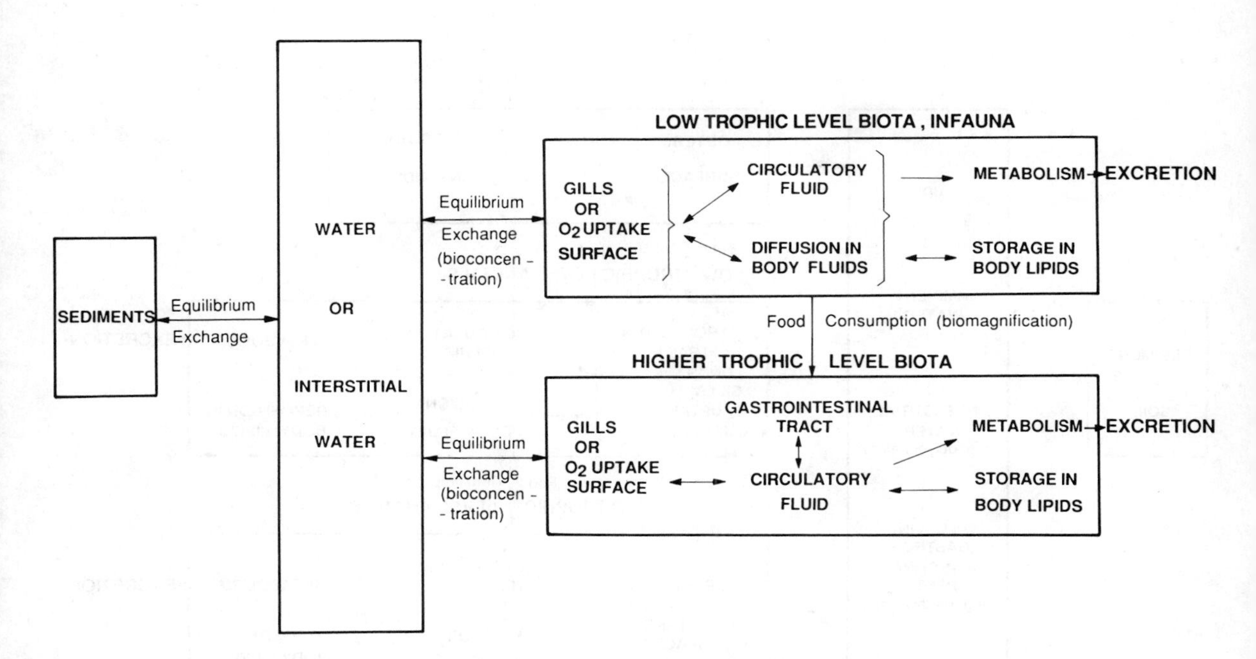

FIGURE 5.4—12. Diagrammatic representation of routes of uptake and clearance of lipophilic chemicals by aquatic biota. (From Connell, D. W., Ed., *Bioaccumulation of Xenobiotic Compounds,* CRC Press, Boca Raton, FL, 1989, 100.)

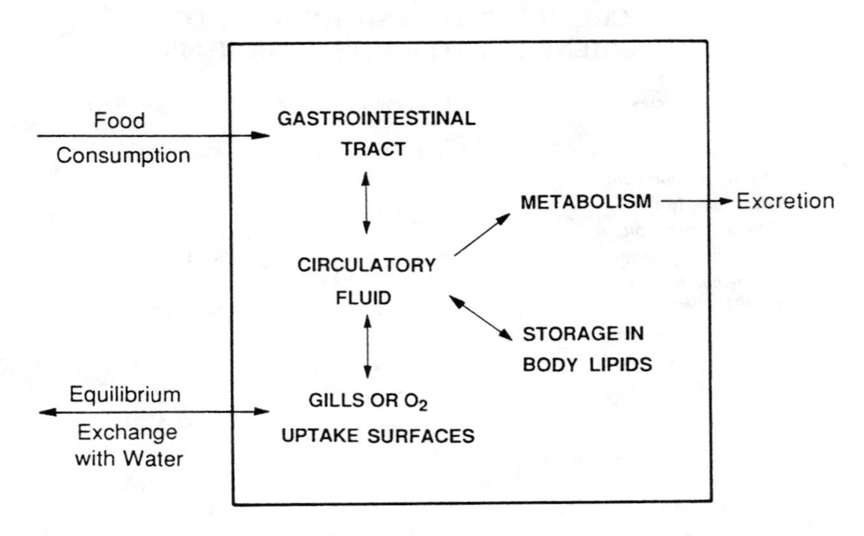

FIGURE 5.4—13. A diagrammatic illustration of the patterns of bioaccumulation of a xenobiotic chemical in an aquatic organism. (From Connell, D. W., Ed., *Bioaccumulation of Xenobiotic Compounds,* CRC Press, Boca Raton, FL, 1989, 162.)

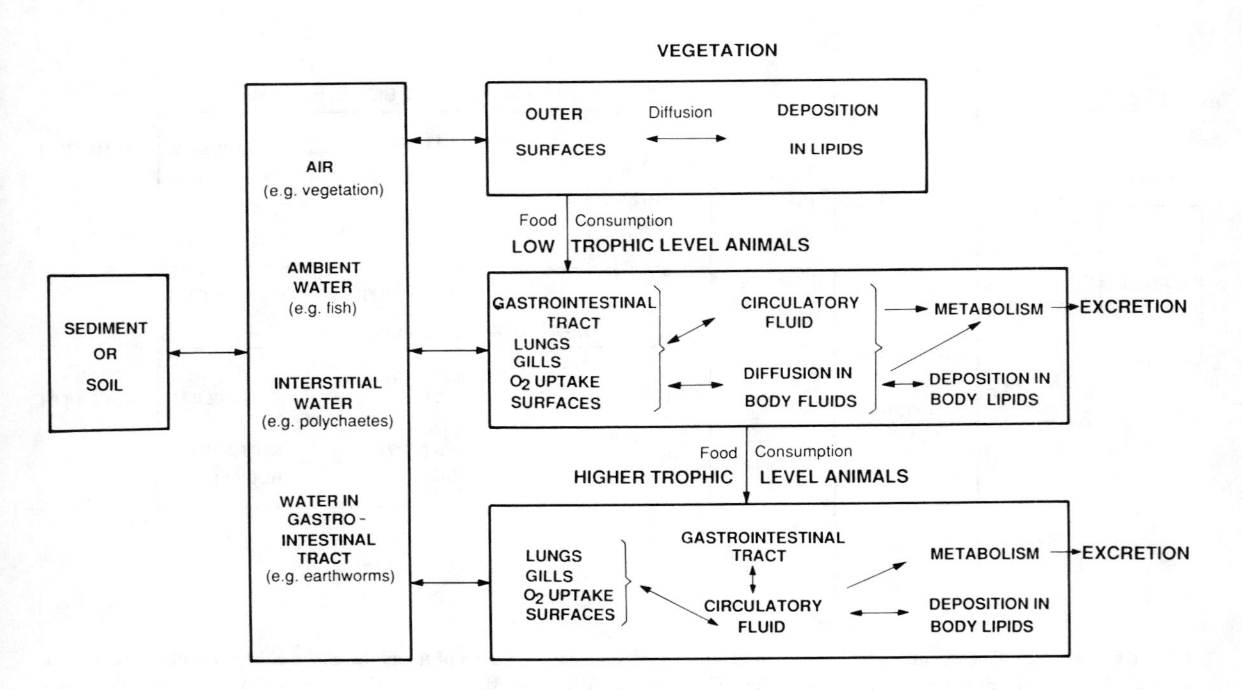

FIGURE 5.4—14. Diagrammatic illustration of the bioaccumulation pathways of xenobiotic chemicals in the environment. (From Connell, D. W., Ed., *Bioaccumulation of Xenobiotic Compounds,* CRC Press, Boca Raton, FL, 1989, 211.)

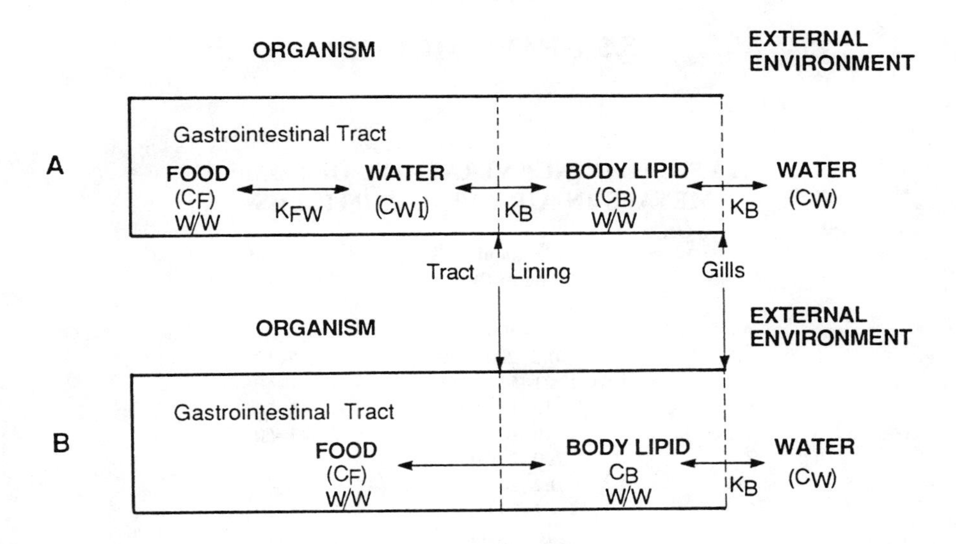

FIGURE 5.4—15. Two exchange processes involved in the biomagnification of persistent lipophilic compounds in food by aquatic biota. (From Connell, D. W., Ed., *Bioaccumulation of Xenobiotic Compounds,* CRC Press, Boca Raton, FL, 1989, 166.)

5.5 HEAVY METALS

Table 5.5—1
TYPICAL CONCENTRATIONS OF SOME METALS IN AIRBORNE PARTICLES

Metal	Urban air (ng m^{-3})	Rural air (ng m^{-3})
As	5–300	1–20
Cd	0.5–200	0.5–10
Ni	1–500	1–50
Pb	10–10,000	5–500
V	10–100	3–50
Zn	200–2,000	5–100
Co	0.2–20	0.1–5
Cr	2–200	1–20
Cu	10–1,000	2–100
Fe	100–10,000	100–10,000

From Harrison, R. M., Ed., *Pollution: Causes, Effects, and Control,* 2nd ed., Royal Society of Chemistry, Cambridge, 1990. With permission.

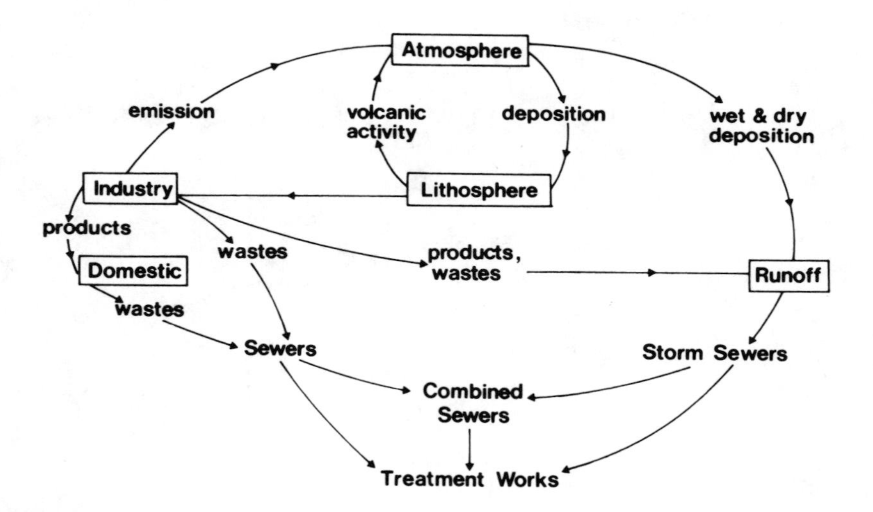

FIGURE 5.5—1. Sources and pathways of heavy metals entering wastewater-treatment processes. (From Stephenson, T., *Heavy Metals in Wastewater and Sludge Treatment Processes,* Vol. 1, Sources, Analysis, and Legislation, Lester, J. N., Ed., CRC Press, Boca Raton, FL, 1987, 33.)

Table 5.5—2
SOURCES OF HEAVY METALS IN DOMESTIC WASTEWATER BY PRODUCT TYPE

	Al	Sb	As	Be	Bi	Cd	Cr	Co	Cu	Fe	Pb	Mn	Hg	Mo	Ni	Se	Ag	Sn	Ti	V	Zn
Automotive products	X		X	X			X	X		X	X			X					X		X
Caulking compounds	X			X			X	X		X	X								X		X
Cleaners	X						X		X	X									X		X
Cosmetics	X			X	X	X		X	X	X	X	X	X		X	X	X	X	X		X
Disinfectants													X								
Driers	X																			X	
Fillers	X																				X
Fire extinguishers	X						X			X											X
Fuels				X					X		X										
Pesticides	X		X			X	X		X	X	X	X	X	X							X
Inks	X								X										X		X
Lubricants				X			X				X			X							X
Medicine	X	X	X		X			X	X	X											X
Oils				X					X		X										X
Ointments	X				X			X	X				X								X
Paints	X		X	X			X	X		X	X	X	X						X		X
Photography	X						X			X	X		X				X	X			
Pigments	X	X	X	X	X	X	X	X	X	X	X	X	X		X	X		X	X		X
Polish	X			X					X								X		X		X
Powders	X									X											X
Preservatives										X	X										X
Suppositories					X																X
Water treatment	X								X	X		X									X

From Stephenson, T., in *Heavy Metals in Wastewater and Sludge Treatment Processes, Vol. 1, Sources, Analysis, and Legislation,* Lester, J. N., Ed., CRC Press, Boca Raton, FL, 1987, 46.

Table 5.5—3
HEAVY METALS IN WASTEWATERS FROM DIFFERENT CATEGORIES OF INDUSTRIES

	Al	Sb	As	Bi	Cd	Cr	Co	Cu	Fe	Pb	Mn	Hg	Mo	Ni	Se	Ag	Te	Tl	Sn
Metal industries																			
Power plants (steam generation)						X													
Foundries — ferrous		X	X	X	X	X	X	X	X	X	X	X	X	X	X		X		X
Foundries — Nonferrous	X	X	X	X	X	X		X		X	X	X			X	X			
Plating	X				X	X		X				X		X		X		X	
Chemical industries																			
Cement and glass			X			X									X			X	
Organics and petrochemicals	X		X		X	X			X	X		X							X
Inorganic chemicals	X		X		X	X			X	X		X							X
Fertilizers	X		X		X	X		X	X	X				X					
Oil refining	X		X		X	X		X	X	X				X					
Others																			
Paper						X		X		X		X		X					
Leather						X													
Textiles		X				X													
Electronics		X						X				X			X		X		X

From Stephenson, T., in *Heavy Metals in Wastewater and Sludge Treatment Processes, Vol. 1, Sources, Analysis, and Legislation*, Lester, J. N., Ed., CRC Press, Boca Raton, FL, 1987, 51.

Table 5.5—4
TYPICAL BACKGROUND CONCENTRATIONS OF SOME METALS IN RIVERS AND OCEANS

Metal	River (ppb)	Ocean (ppb)
Ag	0.3	0.1
Al	400	5
As	1	2.3
Cd	0.03	0.05
Co	0.2	0.02
Cr	1	0.6
Cu	5	3
Fe	670	3
Hg	0.07	0.05
Mn	5	2
Mo	1	10
Ni	0.3	2
Pb	3	0.03
Sb	1	0.2
Se	0.2	0.45
Sn	0.04	0.01
V	1	1.5
Zn	10	5

From Bryan, G. W., in *Marine Pollution,* Johnston, R. B., Ed., 1976, chap. 3. Copyright Academic Press. With permission.

Table 5.5—5
TYPICAL METAL BIOCONCENTRATION FACTORS OF SELECTED AQUATIC ORGANISMS

	Marine organisms					Freshwater organisms		
Metal	Phytoplankton	Zooplankton	Macrophytes	Molluscs	Fish	Macrophytes	Molluscs	Fish
Ti	2,700	—	—	—	—	—	—	—
Cr	7,800	—	2,880	21,800	—	—	267	10
Mn	3,800	3,900	—	2,300	373	1,450	—	23
Fe	28,300	114,600	—	14,400	—	3,642	—	190
Ni	570	560	1,050	4,000	235	—	650	85
Co	—	—	—	—	50	1,367	300	90
Cu	2,800	1,800	2,890	3,800	127	158	1,500	60
Zn	5,500	8,800	7,000	27,300	533	318	2,258	228

From Williams, S. L., Aulenbach, D. B., and Clesceri, N. L., in *Aqueous Environmental Chemistry of Metals,* Rubin, A. J., Ed., 1976, Chap. 2. Copyright Ann Arbor Science Publishers. With permission.

Table 5.5—6
SOLUBILITIES OF HEAVY METALS IN RIVERS

River	Heavy-metal soluble fraction (%)										
	Sb	Cd	Cr	Co	Cu	Fe	Pb	Mn	Hg	Ni	Zn
Ruhr, FRG	—	100	96	100	74	56	62	69	—	—	86
Schuylkill, U.S.	—	21	15	—	36	—	5	—	—	42	33
Thames, U.K.	72	—	22	30	—	3	—	—	23	53	31
Yarra, Australia	—	80	—	—	75	83	58	—	—	—	90
Elbe, FRG	—	22	—	—	50	4	—	69	—	—	48

From Stephenson, T., in *Heavy Metals in Wastewater and Sludge Treatment Processes, Vol. 2, Treatment and Disposal,* Lester, J. N., Ed., CRC Press, Boca Raton, FL, 1987, 74.

Table 5.5—7
CLASSIFICATION OF SELECTED METALS AND ASSOCIATED LIGANDS

Classification		
Hard acceptor/acid	**Intermediate**	**Soft acceptor/acid**
K^+, Na^+, Be^{2+}, Ca^{2+}, Mg^{2+}, Mn^{2+}, Al^{3+}, As^{3+}, Co^{3+}, Cr^{3+}, Fe^{3+}	Co^{2+}, Cu^{2+}, Fe^{2+}, Ni^{2+}, Pb^{2+}, Zn^{2+}	Ag^+, Au^+, Cu^+, Tl^+, Cd^{2+}, Hg^{2+}
Hard donor/base	**Intermediate**	**Soft donor/base**
H_2O, OH^-, F^-, Cl^-, SO_4^{2-}, CO_3^{2-}, O^{2-}, PO_4^{3-}	Br^-, NO_2^-, SO_3^{2-}	SH^-, RS^-, CN^-, SCN^-, S^{2-}, CO, R_2, S, RSH
Class A metals	**Borderline**	**Class B metals**
K^+, Na^+, Ba^{2+}, Be^{2+}, Ca^{2+}, Mg^{2+}, Al^{3+}	Cd^{2+}, Co^{2+}, Cr^{2+}, Cu^{2+}, Fe^{2+}, Mn^{2+}, Ni^{2+}, Pb^{2+}, Sn^{2+}, V^{2+}, Zn^{2+}, Fe^{3+}, As(III), Sb(III), Sn(IV)	Ag^+, Cu^+, Tl^+, Hg^{2+}, Bi^{3+}, Tl^{3+} Pb(IV)
Class A ligands	**Borderline**	**Class B ligands**
H_2O,OH^-, NO_3^-, F^-, SO_4^{2-}, CO_3^{2-}, O^{2-}, HPO_4^{2-}, PO_4^-, ROH, $RCOO^-$, ROR	Br^-, Cl^-, N_3^-, NO_2^-, O_2^-, SO_3^{2-}, O_2^{2-}, O_2, NH_3, N_2, RNH_2, R_2NH, R_3N, CONR	H^-, I^-, R^-, CN^-, RS^-, S^{2-}, CO, R_2S, R_3As

Table 5.5—7 (continued)
CLASSIFICATION OF SELECTED METALS AND ASSOCIATED LIGANDS

Essential	Possibly beneficial	No apparent metabolic function
Animals: Co, Cr, Cu, Fe, Mn, Mo, Ni, Se, Sn, V, Zn Plants: B, Cu, Fe, Mn, Mo, Se, Zn	As, Ba	Bi, Cd, Hg, Pb, Tl
Noncritical	**Toxic/accessible**	**Toxic/insoluble 13**
Al, Ca, Fe, K, Mg, Na	Ag, As, Be, Bi, Cd, Co, Cu, Hg, Ni, Pb, Sb, Se, Sn, Te, Tl, Zn	Ba

Note: Some elements in the original classifications have been omitted for clarity.

From Rudd, T., in *Heavy Metals in Wastewater and Sludge Treatment Processes, Vol. 1, Sources, Analysis, and Legislation,* Lester, J. N., Ed., CRC Press, Boca Raton, FL, 1987, 3.

Table 5.5—8
TOXIC ELEMENTS OF IMPORTANCE IN MARINE POLLUTION BASED ON POTENTIAL SUPPLY AND TOXICITY, LISTED IN ORDER OF DECREASING TOXICITY[a]

	Rate of mobilization (10^9 g/year)				Relative critical index (km^3/year)	
Element	A (man) fossil fuels	B (natural) river flow	C total	Toxicity D (μg/l)	A/D	C/D
Mercury	1.6	2.5	4.1	0.1	16,000	41,000
Cadmium	0.35	?	3.0	0.2	1,750	15,000
Silver	0.07	11	11.1	1	70	11,100
Nickel	3.7	169	164	2	1,350	82,000
Selenium	0.45	7.2	7.7	5	90	1,540
Lead	3.6	110	113.6	10	360	11,360
Copper	2.1	250	252.1	10	210	25,210
Chromium	1.5	200	201.5	10	150	20,150
Arsenic	0.7	72	72.7	10	70	7,270
Zinc	7	720	727	20	330	36,350
Manganese	7.0	250	257	20	350	12,850

[a] Equals the volume of seawater that would be contaminated annually to the indicated level of toxicity by the specified rates of addition, both by natural processes and anthropogenic activities.

From Ketchum, B. H., Marine industrial pollution, in *Oceanography: The Past,* Sears, M. and Merriman, D., Eds., Springer-Verlag, New York, 1980, 397. With permission.

Table 5.5—9
METAL CONCENTRATIONS IN MARINE FAUNA (PPM)

Metal	Geometric mean and location	Phyto-plankton	Algae	Mussels	Oysters	Gastropods	Crustaceans	Fish	Seals, mammals
Arsenic	Geometric mean	—	20	15	10	20	30	10	
	Newfoundland	—	9.8–17 (b)	1.6–5	—	4.0–11.5	3.8–7.6	0.4–0.8	
	England	—	26.0–54 (b)	1.8–15	2.6–10	8.1–38	16	1.7–8.7	
	Greenland	—	36 (b)	14–17	—	—	63–80	14.7–307	
Cadmium	Geometric mean	2	0.5	2	10	6	1	0.2	
	Spain	—	0.8–4 (b)	0.5–8	2.9–3.5	1.1–9	0.7–32	<0.4–4.3	
	England	—	0.2–53 (b)	3.7–65	6–54	3.5–1.120	2.8–33	0.06–3.96+	2.2–11.6+
	Australia	—	—	4.2–83	9–174	2.8–30	--	0.05–0.4+	—
	Norway	—	1.0–13 (b)	1.9–140	—	0–51	1.9–7	<0.01–0.03+	
Copper	Geometric mean	7	15	10	100	60	70	3	
	Spain	—	5–26 (g)	6–14	120–435	5–50	110–435	<0.6–10	
	England	—	4–141 (b)	7–15	20–6,480	0–1,750	6–64+	0.5–14.6+	
	California	—	—	7–77	10–2,100	3–177	(4–150)	(16–29.3)	14.5–386 (m)
	Norway	—	9–170 (g)	3–120	—	17–190	2–90	—	
Lead	Geometric mean	4	4	5	3	5	1	3	
	Spain	—	4–20 (g)	2–15	4–11	10–27	<1.2–11	<1.2–2.2	
	England	—	16–66 (g)	7–19	5–17	0.2–0.8	8	14–28	0–4+
	California	—	—	0.3–42	—	0.6–21	—	<0.001–5.3	0.3–34.2 (s)
	Norway	—	3–1,200 (g)	2–3,100	—	0–39	8.3	—	—
Mercury	Geometric mean	0.17	0.15	0.4	0.4	0.2	0.4	0.4	0.6–103 (m)
	Hawaii	—	—	—	—	0–0.03	0.03–0.12	0.02–23	0.1–700 (s)
	California	—	—	—	—	<0.01–0.07+	0.02–0.04+	0.02–0.2	
	Atlantic	0.2–5.3	<0.01–0.07+	<0.01–0.13+	0.02–180+	—	<0.05–0.6+	0.1–9.0	—
	Mediterranean	—	<0.5–0.7+	0.25–0.4	—	0.1–3.5+	0.3–4.5+	0.1–29.8+	—
	Australia	—	—	0.05–0.23+	1.5–8.2	0.32–0.65	—	0.3–16.5	0.1–106 (m)
	Norway	0.5–25.2	—	0.24–0.84	—	0.61	0.31–0.39	0.14–7.3	0.4–225 (s)
	England	—	<0.01–25.5 (g)	0.64–1.86	0.56–1.2	0.02–1.84	0.98	0.02–1.8	
Nickel	Geometric mean	3	3	3	1	2	1	1	
	England	—	4–33 (g,b)	5–12	2–174	8.8–12.3	1.1–12.3	0.5–10.6	
	California	—	—	3.3–20	—	1.8–18.5	—	—	
Silver	Geometric mean	0.2	0.2	0.3	—	1	0.4	0.1	
	California	—	—	0.7–46	—	0.4–10.7	—	0.1–1.2 (m)	

Zinc	Geometric mean	38	90	100	1,700	200	80	80
	South Africa	0.6–710	5.6+ (g)	73–113	400–886	12+	17+	3.2–7.2+
	California	0.1–725	46–244	70–8,430	1.7–288	—	—	78–875
	Spain	—	63–345 (g)	190–370	310–920	60–120	79–330	21–220
	Australia	—	—	170–1,350	3,740–38,700	56–1,050	—	4–375
	England	—	28–1,240 (g)	12–779	1,830–99,200	9.7–1,500	36–82	2–342
	Norway	—	20–2,310 (g)	105–2,370	—	87–2,900	12–32	—

Note: +, ppm wet weight; all other values in ppm dry weight; b, brown algae; g, green algae m, mammals, s, seals.

From Förstner, U., in *Chemistry and Biogeochemistry of Estuaries,* Olausson, E. and Cato, I., Eds., John Wiley & Sons, Chichester, U.K., 1980, 307. With permission.

Table 5.5—10
RANKINGS, ON FOUR GEOGRAPHICAL SCALES, OF METAL CONCENTRATIONS IN FISH LIVERS FROM THE KENNEBEC RIVER PLUME

Region (no. sites)	Pb	Ag	Zn	Cu	Sn	Cd	Ni	Cr	Hg
Country (43)	1	3	4	10	15	18	18	18	22
East Coast (14)	1	1	1	1	4	3	5	5	7
New England (6)	1	1	1	1	1	3	1	3	3
Gulf of Maine (4)	1	1	1	1	1	1	1	3	2

From Larsen, P. F., Marine environmental quality in the Gulf of Maine, *Rev. Aquat. Sci.,* 6, 67, 1992.

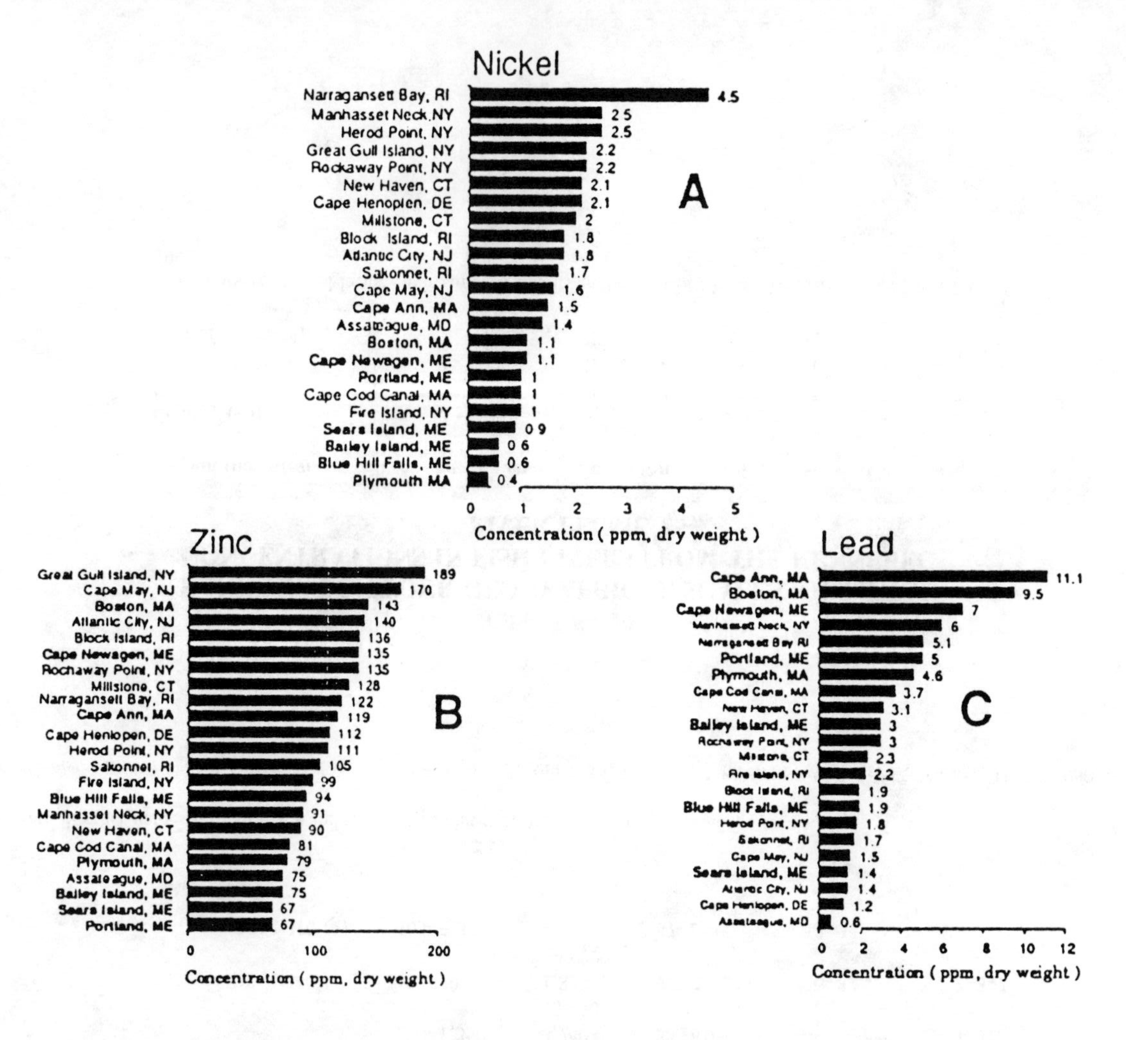

FIGURE 5.5—2. Concentrations of metals in mussels sampled in the EPA Mussel Watch Program. Gulf of Maine sites are in bold print. (From Larsen, P. F., Marine environmental quality in the Gulf of Maine, *Rev. Aquat. Sci.*, 6, 67, 1992.)

Table 5.5—11
SUBLETHAL EFFECTS OF METALS — GROWTH

Species	Metal concentration (μg/l)								Response
	Ag	As	Cd	Cu	Hg	Ni	Pb	Zn	
Phytoplankton									
Natural assemblage				0.3					Reduced ^{14}C fixation
Natural assemblage			1.0	1.0	1.0				Reduced ^{14}C fixation
Natural assemblage				10.0					Reduced ^{14}C fixation
Natural assemblage		23	112	6.4	<6.0	60[a]	20	20	Reduced growth
Monochrysis lutheri				21.6 (0.07 as Cu^{2+})					Reduced division
Natural assemblage					0.8				Reduced growth
Natural assemblage					1.0				Reduced productivity
Natural assemblage								15	Reduced photosynthesis
Natural assemblage		5							Reduced growth
Macroalgae									
Laminaria saccharina									
Sporeling				10	0.5 (0.5)[b]			100	Reduced growth
Sporophyte				50	50 (5.0)[b]			1000	Reduced growth
Hydroids									
Campanularia flexuosa			195	14.3	1.6			740	Reduced growth
Mollusks									
Mytilus edulis			10	3	0.3	>200[a]	>200[a]	10	Reduced shell growth
Mercenaria mercenaria	32			16	15	5700		195	Reduced growth[c]
Crassostrea virginica	25			33	12	1200			Reduced growth[c]
Fish									
Pleuronectes platessa			5	10					Reduced growth

[a] No effect observed at this concentration.

[b] Methylmercury.

[c] For larvae: concentrations also represent LC_{50} values (8 to 12 d).

From Langston, W. J., in *Heavy Metals in the Marine Environment*, Furness, R. W. and Rainbow, P. S., Eds., CRC Press, Boca Raton, FL, 1990, 107.

Table 5.5—12
SUBLETHAL EFFECTS OF METALS — MORPHOLOGY

Species	Metal concentration (μg/l)										Response
	Ag	As	Cd	Cr	Cu	Hg	Ni	Pb	Se	Zn	
Hydroids											
Eirene viridula		300	100		10	3		300	3,000	1,500	Altered hydranth morphology[a]
Mollusks											
Crassostrea gigas	22	326	611	4538	5	7	349	476	>10,000	199	50% abnormal larvae
Mytilus edulis	14	>3,000	1200	4469	6	6	891	758	>10,000	175	50% abnormal larvae
Crassostrea virginica	24					11				206	50% abnormal larvae
Fish											
Myoxocephalus quadricornis		32	0.5		0.8	0.1		1.2		5.3	Increase in vertebral deformities[b]

[a] Lowest reported threshold concentration.
[b] Metals applied as a mixture.

From Langston, W. J., in *Heavy Metals in the Marine Environment,* Furness, R. W. and Rainbow, P. S., Eds., CRC Press, Boca Raton, FL, 1990, 109.

Table 5.5—13
SUBLETHAL EFFECTS OF METALS — REPRODUCTION AND DEVELOPMENT

Species	Metal concentration (μg/l)									Response
	Ag	As	Cd	Cr	Cu	Hg	Ni	Pb	Zn	
Macroalgae										
Champia parvula		60								Inhibition of sexual reproduction
Hydroids										
Campanularia flexuosa					0.05	0.01			500	Increase in gonozooid frequency[a]
Polychaetes										
Neanthes arenaceodentata			1000	50				3100	320	Reductions in reproduction
Capitella capitata			560	100				200	560	Reductions in reproduction
Ctenodrilus serratus			2500	50	100	50	500	1000	500	Reductions in reproduction
Bivalves										
Mytilus edulis					50				200	Development of oocytes supressed
Spisula solidissima (germ cells)	9.5									Impaired embryogenesis
Crustaceans										
Pontoporeia affinis			5.5					4.9		Fecundity reduced
Rhithropanopeus harrisii			50						25	Hatch time increased
Tigriopus japonicus			44		6.4					Generation time doubled
Echinoderms										
Sea urchin eggs (various spp.)		1500	600	1000	10	10	600	1000	30	Fertilization and development arrested[a]
Fish										
Spring-spawning herring			5		10				10	Reduced fertilization
Fundulus heteroclitus						10				Reduced hatch
Leiostomus xanthurus (eggs)					0.064[b]					Reduced hatch
Menidia menidia (eggs)					0.025[b]					Reduced hatch

[a] Lowest reported threshold concentration.
[b] Calculated as free ion (Cu^{2+}).

From Langston, W. J., in *Heavy Metals in the Marine Environment*, Furness, R. W. and Rainbow, P. S., Eds., CRC Press, Boca Raton, FL, 1990, 111.

Table 5.5—14
EFFECTS OF TRIBUTYLTIN (TBT) ON MARINE AND ESTUARINE ORGANISMS

Species	Concentration of TBT in water (ng/1 as Sn)	Effect
Nucella lapillus, dogwhelk	1—10	Imposex, impaired reproduction
Crassostrea gigas, oyster[a]	8	Shell thickening
	20	Reduced growth, viability
Mytilus edulis, mussel[a]	40	Reduced viability
Venerupis decussata, clam[a]	40	Reduced growth, viability
Gammarus oceanicus, amphipod[a]	120	Reduced growth
Homarus americanus, lobster[a]	400	Reduced viability
Pavlova lutheri, *Skeletonema costatum*, *Dunaliella tertiolecta* — microalgae	40—400	Reduced growth

[a] Larvae.

From Langston, W. J., in *Heavy Metals in the Marine Environment,* Furness, R. W. and Rainbow, P. S., Eds., CRC Press, Boca Raton, FL, 1990, 117.

Table 5.5—15
CLASSIFICATION OF ESSENTIAL ELEMENTS

Element group	Subgroup	Number	Members	Tissue conc.
Major	Bulk elements	5	C,H,N,O,S	g/kg
	Major ions	6	Na,K,Ca,Mg,P,Cl	g/kg
Trace	Essential	9	Co,Cu,I,Fe,Mn,Mo,Se,Si, and Zn	µg—mg/kg
	Desirable	6	As,Cr,F,Ni,Sn,V	µg/kg

From Langston, W. J., in *Heavy Metals in the Marine Environment,* Furness, R. W. and Rainbow, P. S., Eds., CRC Press, Boca Raton, FL, 1990, 125.

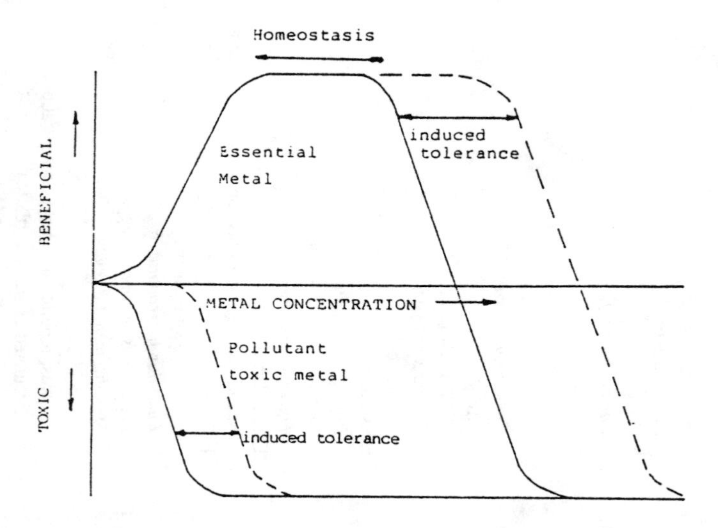

FIGURE 5.5—3. Effects of essential and pollutant metals on cellular metabolism. Intracellular concentrations of essential trace metals are controlled by homeostatic mechanisms. Once this capacity is exceeded, enzyme inhibition often occurs and the metal becomes toxic. Most nonessential (pollutant) metals are inhibitory. Inducible tolerance (detoxification systems) raises the inhibitory thresholds. (From George, S. G., in *Heavy Metals in the Marine Environment,* Furness, R. W. and Rainbow, P. S., Eds., CRC Press, Boca Raton, FL, 1990, 125.)

Table 5.5—16
METAL CONCENTRATIONS (μg/g) IN CONTAMINATED AND UNCONTAMINATED SEDIMENTS[a]

Sediment	Ag	As	Cd	Co	Cr	Cu	Fe	Hg	Mn	Ni	Pb	Sn	V	Zn
Firth of Clyde, Scotland														
Control area (mean)	<0.2	8	3.4	34	64	37	5.3×10^4	0.1	1,100	50	86	19	250	160
Sewage-sludge (dumpsite (max.)	5	24	7	40	310	210	6.1×10^4	2.2	1,000	70	320	100	400	830
California coast														
Control area (median)	0.2	—	0.33	—	22	8.3	—	0.043	—	9.7	6.1	—	—	43
Los Angeles wastewater outfall (max.)	27	—	66	—	1,500[b]	940	—	5.4	—	130	580	—	—	2,900
Eastern England														
Humber Estuary TiO_2 and smelting (max.)	—	—	—	30	200	160	9.2×10^4	—	1,100	63	220	—	2,000[b]	430
Southwestern England estuaries														
Control (Avon) (typical)	0.1	13	0.3	10	37	19	1.9×10^4	0.12	420	28	39	28	—	98
Restronguet Creek acid mine waste (max.)	4.1	2,500[b]	1.2	22	37	2,500	5.8×10^4	0.22	560	32	290	1,700[b]	—	3,500
Tasmania														
Derwent Estuary Zn refinery and chloralkali (Hg) (max.)	—	—	860[b]	140	200	>400	16×10^4	100[b]	8,900	42	1,000	—	—	>10,000
Norway														
Sorfjord smelting (max.)	190[b]	—	850	—	—	12,000[b]	—	—	—	—	31,000[b]	1,350	—	12,000[b]
Pacific Ocean														
Pelagic clay (>4000 m depth) (max.)	—	—	—	150[b]	—	1,200	7.5×10^4	—	38,000[b]	1,300[b]	—	—	—	390

[a] Reported values rounded to two significant digits.
[b] Highest value for each element.

From Bryan, G. W., in *Wastes in the Ocean,* Vol. 6, Nearshore Waste Disposal, Ketchum, B. H., Capuzzo, M. J., Burt, W. V., Duedall, I. W., Park, P. K., and Kester, D.R., Eds., John Wiley & Sons, New York, 1985, 41. With permission.

Table 5.5—17
EXAMPLES OF METAL CONCENTRATIONS (μg/g) IN CONTAMINATED SEDIMENTS[a]

Metal	Baltic Sea[1] (various sources)	Bristol Channel/ Severn estuary, U.K.[2] (industry, sewage)	Mersey estuary U.K.[3] (sewage, industry including chlor-alkali)	Los Angeles outfall, California[4] (sewage)	Derwent estuary Tasmania[5] (refinery, chlor-alkali)	Restronguet Creek, U.K.[2] (mining)	Port Pirie, Australia[6] (smelter)
As		8	71			2520 (13)	151 (1.0)
Cd	8.1 (<0.01)	1.1	3.9	66 (0.3)	862	1.2 (0.3)	267 (0.5)
Cu	283 (1.0)	54	144	940 (8.3)	>400	2,540 (19)	122 (3.0)
Hg	9 (0.01)	0.48	6.2	5.4 (0.04)	1,130	0.22 (0.12)	8
Ni	920 (1.0)	33	44	130 (9.7)	42	32 (28)	19.4 (12)
Pb	400 (2)	88	205	580 (6.1)	>1,000	400 (2)	5,270 (2)
Zn	2,090 (6)	255	255	2,900 (43)	>10,000	2,090 (6)	16,667 (11)

[a] Maximum concentrations shown together with local background values (in parentheses), where given.

From Langston, W. J., in *Heavy Metals in the Marine Environment,* Furness, R. W. and Rainbow, P. S., Eds., CRC Press, Boca Raton, FL, 1990, 115.

REFERENCES

1. **Brugman, L.,** *Mar. Pollut. Bull.,* 12, 214, 1981.
2. **Bryan, G. W. et al.,** *J. Mar. Biol. Assoc. U.K.,* Pub. 4, 1985.
3. **Langston, W. J.,** *Estuarine Coastal Shelf Sci.,* 23, 239, 1986.
4. **Mason, A. Z. and Simkiss, K.,** *Exp. Cell Res.,* 139, 383, 1982.
5. **Kojoma, Y. and Kagi, J. H. R.,** *Trends Biochem. Sci.,* 3, 403, 1978.
6. **Ward, T. J. et al.,** in Environmental Impacts of Smelters, Nriagu, J. O., Ed., John Wiley & Sons, New York, 1984, 1.

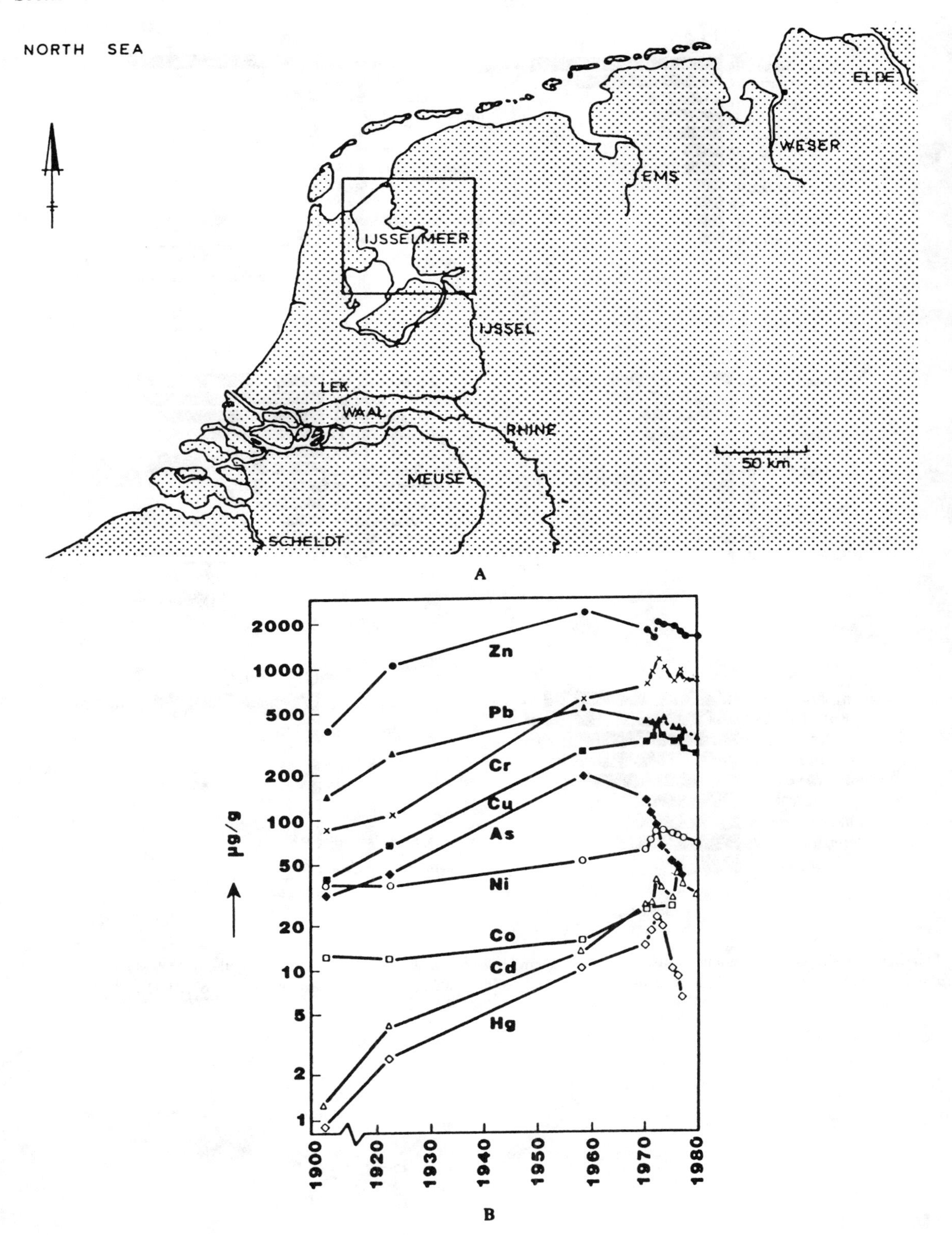

FIGURE 5.5—4. (A) Map showing the Ijsselmeer; (B) history of metal pollution of the Rhine River as reflected in its sediments in the Netherlands. (From Salomons, W., in *Aquatic Ecotoxicology: Fundamental Concepts and Methologies,* Vol. 1, Boudou, A. and Ribeyre, F., Eds., CRC Press, Boca Raton, FL, 1989, 187.)

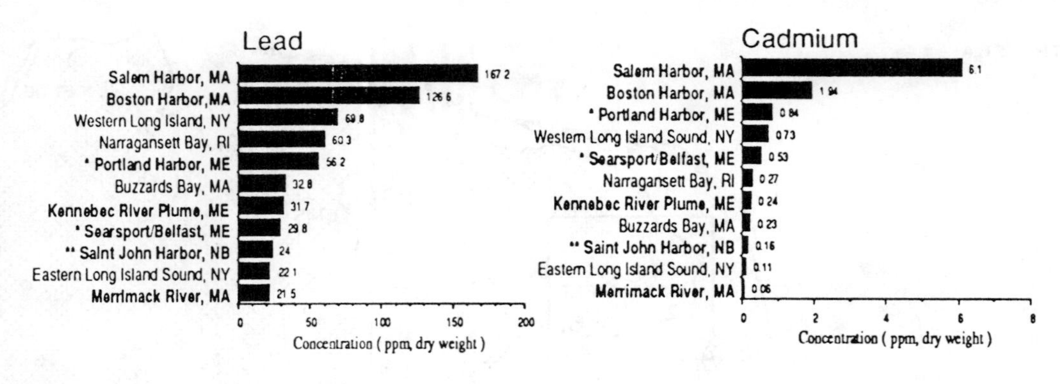

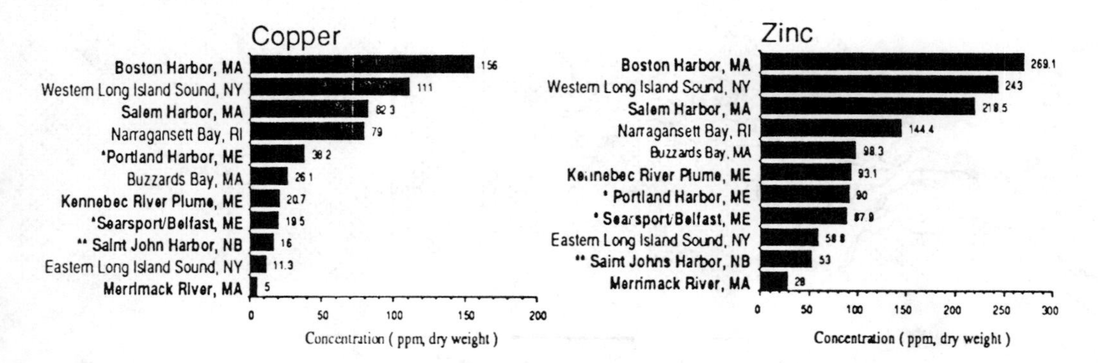

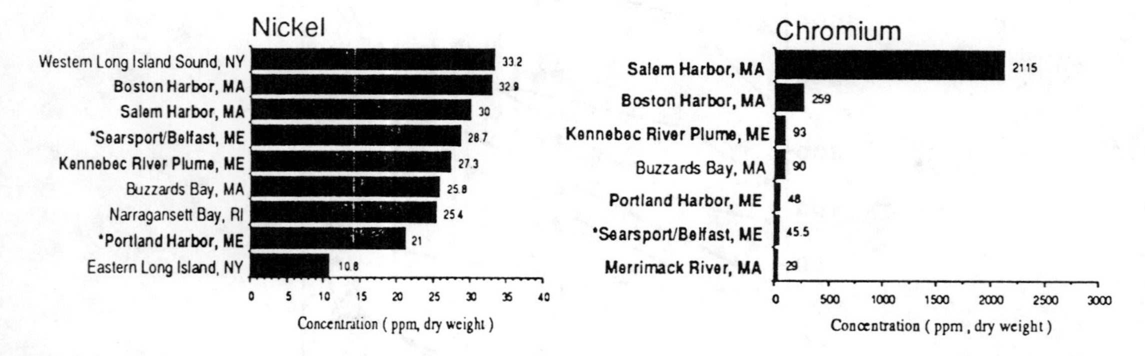

FIGURE 5.5—5. Concentrations of metals in northeastern U.S. coastal sediments as determined by the NOAA Status and Trends Program. (From Larsen, P. F., Marine environmental quality in the Gulf of Maine, *Rev. Aquat. Sci.*, 6, 67, 1992.)

Table 5.5—18
ENRICHMENT FACTORS (AVERAGE OR TYPICAL RANGE) FOR CHESAPEAKE BAY AND OTHER AREAS

Area	Cr	Mn	Ni	Cu	Zn	Pb
Chesapeake Bay (whole)	1	2	1	1	5	5
Upper (0–75 km)	1	3–6	—	2	6–8	4–7
Middle (75–200 km)	1	1	—	1	4–6	3–4
Lower (200–300 km)	1	<1	—	1	2–4	2–4
Baltimore Harbor	4	1	1	—	8	—
Hampton Roads	1	1	1	2–4	5–10	7–21
Susquehanna River	0.6	6	3	2	8	7
Offshore[a]	0.7	0.8	0.6	0.2	2	3
Delaware Bay	3	—	12	2	10	16
San Francisco Bay	—	<1	2	2	2	—

[a] Proposed disposal site on the inner Virginian continental shelf approximately 27 km off the Chesapeake Bay mouth.

From Sinex, S. A. and Wright, D. A., *Mar. Pollut. Bull.*, 19, 425, 1988. With permission.

5.6 RADIOACTIVITY

Table 5.6—1
CHARACTERISTICS OF COMMON NUCLEAR RADIATIONS

Radiation	Rest mass	Charge	Typical energy range	Path length (order of magnitude)		General comments
				Air	Solid	
α	4.00 amu	2+	4–10 MeV	5–10 cm	25–40 μm	Identical to ionized He nucleus
β (negatron)	5.48×10^4 amu 0.51 meV	−	0–4 MeV	0–1 m	0–1 cm	Identical to electron
Positron (β positive)	5.48×10^4 amu 0.51 meV	+	—	0–1 m	0–1 cm	Identical to electron except for charge
Neutron	1.0086 amu 939.55 MeV	0	0–15 MeV	0–100 m	0–100 cm	Free half-life: 16 min
γ (e.m. photon)	—	0	10 keV–3 MeV	0.1–10 m^4	1 mm–1 m	Photons from nuclear transitions

[a] Exponential attenuation in the case of electromagnetic radiation.

From Eichholz, G. G., *Environmental Aspects of Nuclear Power,* Ann Arbor Science Publishers, Ann Arbor, MI, 1976. With permission.

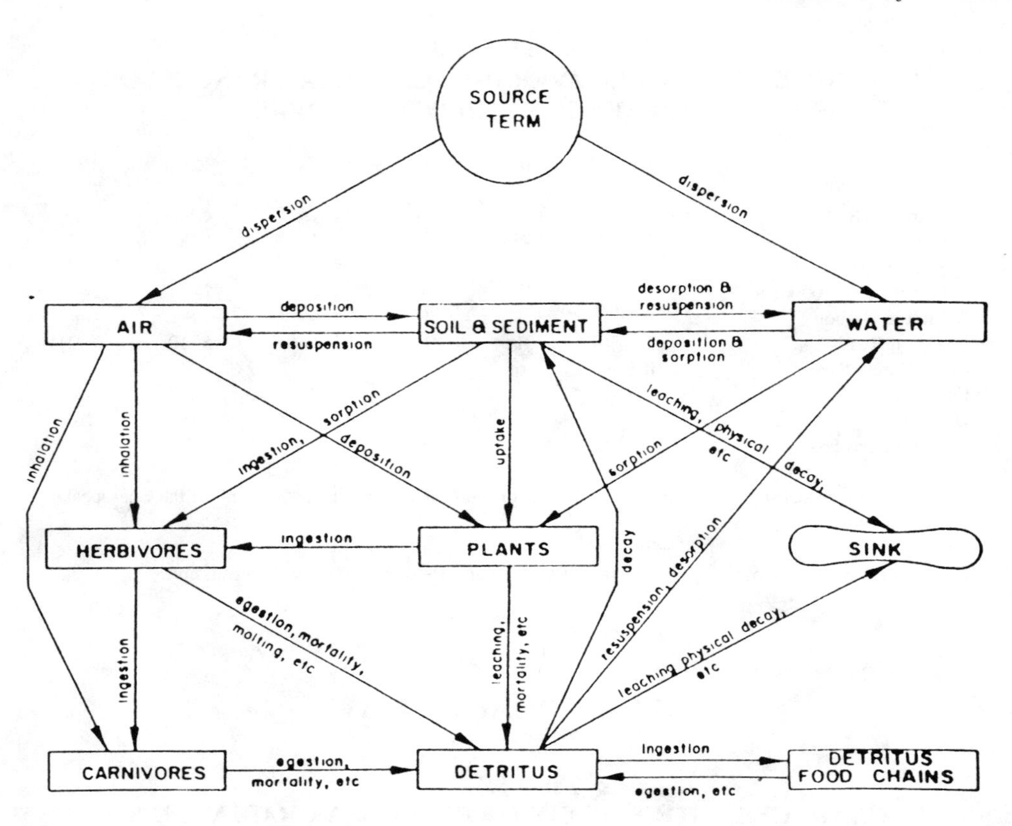

FIGURE 5.6—1. General transport processes operating on radionuclides in ecosystems. Boxes define ecosystem components, and arrows depict the flow of the materials through functional processes. (From Whicker, F. W. and Schultz, V., *Radioecology: Nuclear Energy and the Environment,* Vol. 1, CRC Press, Boca Raton, FL, 1982.)

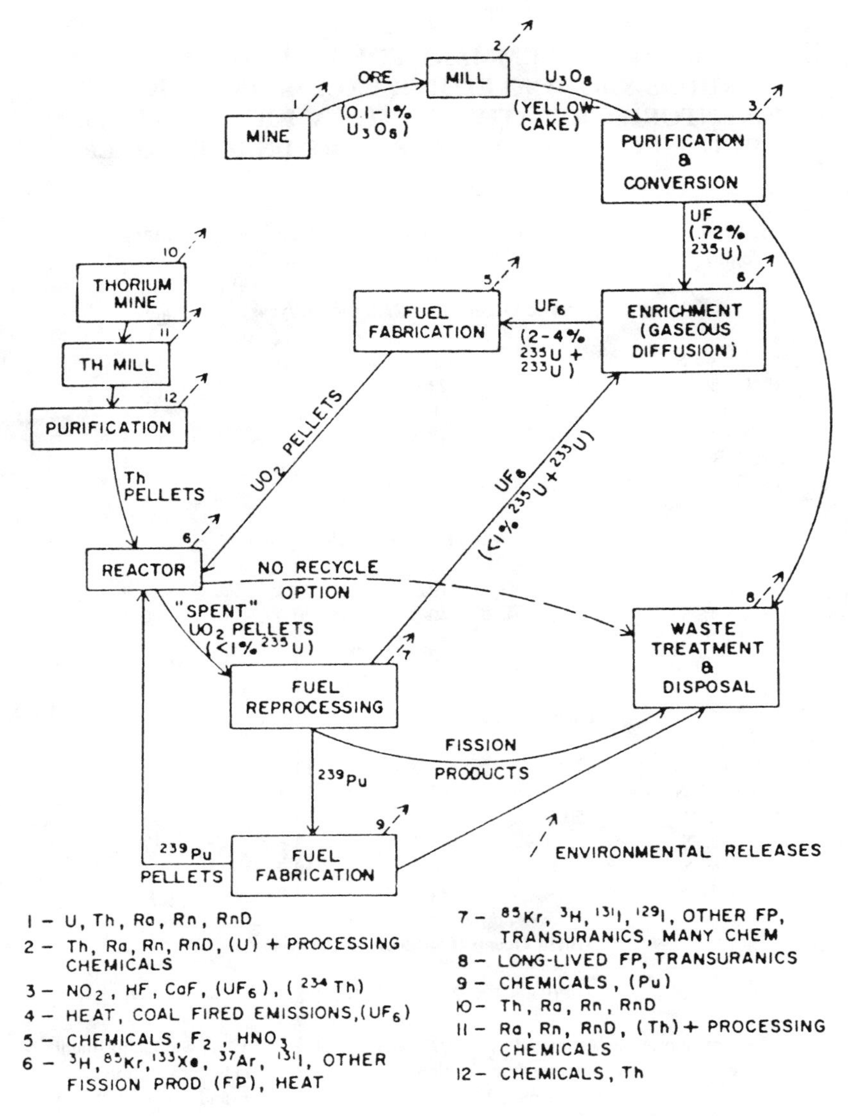

FIGURE 5.6—2. Basic elements and options within the nuclear fuel cycle, illustrating radioactive releases to the environment. (From Whicker, F. W. and Schultz, V., *Radioecology: Nuclear Energy and the Environment,* Vol. 1, CRC Press, Boca Raton, FL, 1982.)

5.6—2
BOMB-CARBON SUMMARY FOR INDIVIDUAL OCEAN REGIONS SHOWING EITHER EXCESS OR DEFICIENT INVENTORIES RELATIVE TO THE AMOUNT EXPECTED IF THE CO_2 INVASION RATE WERE UNIFORM OVER THE OCEAN[a]

Latitude band	Water area (10^{12} m^2)	Inventory (RCU)	Input (RCU)	Inventory-input[b] (RCU)
	Atlantic Ocean (I = 22.3 mol/m^2/year)			
80°N to 40°S	18.6	26.6	19.7	+6.9
40°N to 20°N	15.8	23.0	14.5	+8.5
20°N to 20°S	26.7	10.8	23.1	−12.3
20°S to 45°S	18.4	18.5	14.2	+4.3
45°S to 80°S	15.1	5.2	12.6	−7.4
80°N to 80°S	94.6	84.1	84.1	0.0
	Indian Ocean (I = 19.4 mol/m^2/year)			
25°N to 15°S	27.0	13.2	24.1	−10.9
15°S to 45°S	29.8	40.5	20.3	+20.2
45°S to 70°S	20.7	15.9	25.2	−9.3
25°N to 70°S	77.5	69.6	69.6	0.0
	Pacific Ocean (I = 19.2 mol/m^2/year)			
65°N to 40°N	15.1	7.3	13.4	−6.1
40°N to 15°N	35.0	35.9	27.8	+8.1
15°N to 10°S	50.0	23.7	39.1	−15.4
10°S to 55°S	63.0	61.3	43.6	+17.7
55°S to 80°S	13.8	6.6	10.9	−4.3
65°N to 80°S	176.9	134.8	134.8	0.0
	World Ocean (I = 20.1 mol/m^2/year)			
	349.0	288.5	288.5	0.0

[a] All inventories are estimated as of 1972, and are subject to 5% upward adjustment to compensate for implicit fractionation correction. RCU, radiocarbon units.

[b] The mean CO_2 invasion rate, I, is adjusted for each ocean and for the "World Ocean" so that ocean-average input matches inventory. This requires that like the inventories, the invasion rate estimates are also subject to 5% upward adjustment, which when applied become I = 23.5, 20.4, 20.2, and 21.1 mol/m^2/year for the Atlantic, Indian, Pacific, and World Oceans.

From Lassey, K. R., Manning, M. R., and O'Brien, B. J., An overview of oceanic radiocarbon: its inventory and dynamics, *Rev. Aquat. Sci.*, 3, 117, 1990.

Table 5.6—3
OCEANIC RADIOCARBON INVENTORIES[a]

Source	Inventory (RCU)
Cosmogenic radiocarbon	
Fairhall et al.	1.94×10^4
Killough and Emanuel	2.00×10^4
Recommended value	1.97×10^4
Bomb-carbon (as of ~1972)	
Stuiver et al. (extrapolated)	310
Broecker et al.	303
Recommended value	303

[a] Inventories are normalized (''recorrected'') to $\delta^{13}C = 0$.

From Lassey, K. R., Manning, M. R., and O'Brien, B. J., An overview of oceanic radiocarbon: its inventory and dynamics, *Rev. Aquat. Sci.*, 3, 117, 1990. With permission.

Table 5.6—4
BOMB-CARBON INVENTORIES ACCORDING TO BROECKER ET AL. FOR MID-LATITUDE ATLANTIC AND PACIFIC OCEANS[a]

	Atlantic Ocean		Pacific Ocean	
Latitude band	Col. inventory (10^{12} atoms/m^2)	Inventory (RCU)	Col. inventory (10^{12} atoms/m^2)	Inventory (RCU)
50°N–10°N	122	36.5	93	49.4
10°N–10°S	30	3.8	50	20.1
10°S–50°S	86	24.0	94	54.7
50°N–50°S	91	64.3	82	124.2

[a] All inventories estimated are as of 1972, and all are subject to 5% upward adjustment to compensate for an implicit fractionation correction. RCU, radiocarbon units.

From Lassey, K. R., Manning, M. R., and O'Brien, B. J., An overview of oceanic radiocarbon: its inventory and dynamics, *Rev. Aquat. Sci.* 3, 117, 1990.

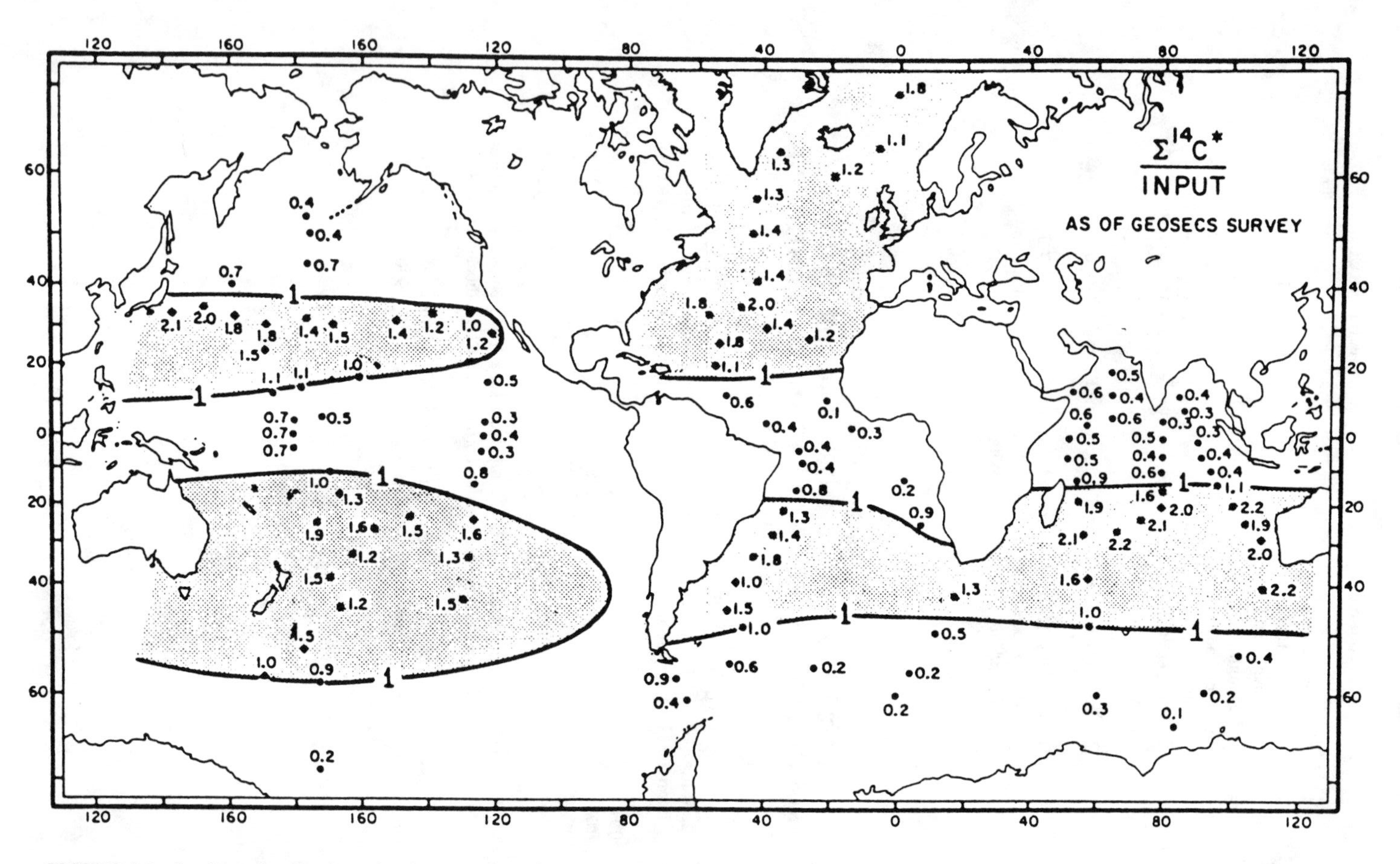

FIGURE 5.6—3. Map showing the ratio of water column inventory to net input of bomb radiocarbon for the stations occupied during the GEOSECS program. Regions with excess inventory over input are shaded. The ratio averaged over each of the three main oceans is unity. (From Broecker, W. S. et al., *J. Geophys. Res.*, 90, 6953, 1985. With permission.)

Table 5.6—5
NUMBER OF NUCLEAR TESTS CONDUCTED BY MAJOR NUCLEAR POWERS FOR THE 10-YEAR PERIOD FOLLOWING THE LIMITED NUCLEAR TEST BAN TREATY OF 1963, AND FOR THE PERIOD 1945–1973

Period	Environment	U.S.	U.S.S.R.	U.K.	France	China	Total
1963–1973	Atmosphere	0	0	0	29	14	43
	Subsurface	261	121	2	9	1	394
	Total	261	121	2	38	15	437
1945–1973	Atmosphere	193	161	21	33	14	422
	Subsurface	372	124	4	13	1	514
	Total	565	285	25	46	15	963

From Barnaby, F., *IAEA Bull.*, 15(4), 13, 1973. With permission of Stockholm Int. Peace Res. Inst. (SIPRI).

Table 5.6—6
SOME FISSION PRODUCT RADIONUCLIDES OF POTENTIAL BIOLOGICAL IMPORTANCE

Radionuclide	Fission yield (%)[a]	Radiation	Half-life	Important element analogs
^{3}H	0.01	β	12 years	H
^{85}Kr	0.29	β,γ	10 years	—
^{90}Sr	5.77	β	28 years	Ca
^{89}Sr	4.79	β	51 d	Ca
^{137}Cs	6.15	β,γ	27 years	K
^{131}I	3.1	β,γ	8.1 d	I
^{129}I	0.9	β,γ	1.7×10^{7} years	I
^{144}Ce[b]	6.0	β,γ	285 d	—
^{103}Ru[b]	3.0	β,γ	40 d	—
^{106}Ru[b]	0.38	β,γ	1.0 years	—
^{95}Zr[b]	6.2	β,γ	65 d	—
^{140}Ba[b]	6.32	β,γ	12.8 d	Ca
^{91}Y	5.4	β,γ	58 d	—
^{141}Ce[b]	5.7	β,γ	33 h	—
^{147}Nd[b]	2.7	β,γ	11 d	—

[a] Based upon thermal neutron fission of ^{235}U.
[b] Decay to radioactive daughters.

From Whicker, F. W. and Schultz, V., *Radioecology: Nuclear Energy and the Environment*, Vol. 1, CRC Press, Boca Raton, FL, 1982.

Table 5.6—7
AVERAGE AND RANGE OF RADIOACTIVITY IN THE NORTH ATLANTIC OCEAN IN THE EARLY 1970S DUE TO RADIOACTIVE FALLOUT FROM NUCLEAR WEAPONS TESTS

Radioisotope	Radioactivity (pCi/l)	
Tritium (^3H)	48	(31–74)
^{137}Cs	0.21	(0.03–0.80)
^{90}Sr	0.13	(0.02–0.50)
^{14}C	0.02	(0.01–0.04)
^{239}Pu		(0.0003–0.0012)

From Clark, R. B., *Marine Pollution,* 2nd ed., Clarendon Press, Oxford, 1989. With permission of Oxford University Press.

Table 5.6—8
NATURAL LEVELS OF RADIOACTIVITY IN SURFACE SEAWATER

Radionuclide	Concentration (pCi/l)
^{40}K	320
Tritium (^3H)	0.6–3.0
^{87}Rb	2.9
^{234}U	1.3
^{238}U	1.2
^{14}C	0.2
^{228}Ra	$(0.1–10.0) \times 10^{-2}$
^{210}Pb	$(1.0–6.8) \times 10^{-2}$
^{235}U	5×10^{-2}
^{226}Ra	$(4.0–4.5) \times 10^{-2}$
^{210}Po	$(0.6–4.2) \times 10^{-2}$
^{222}Rn	2×10^{-2}
^{228}Th	$(0.2–3.1) \times 10^{-3}$
^{230}Th	$(0.6–14.0) \times 10^{-4}$
^{232}Th	$(0.1–7.8) \times 10^{-4}$

From Clark, R. B., *Marine Pollution,* 2nd ed., Clarendon Press, Oxford, 1989. With permission of Oxford University Press.

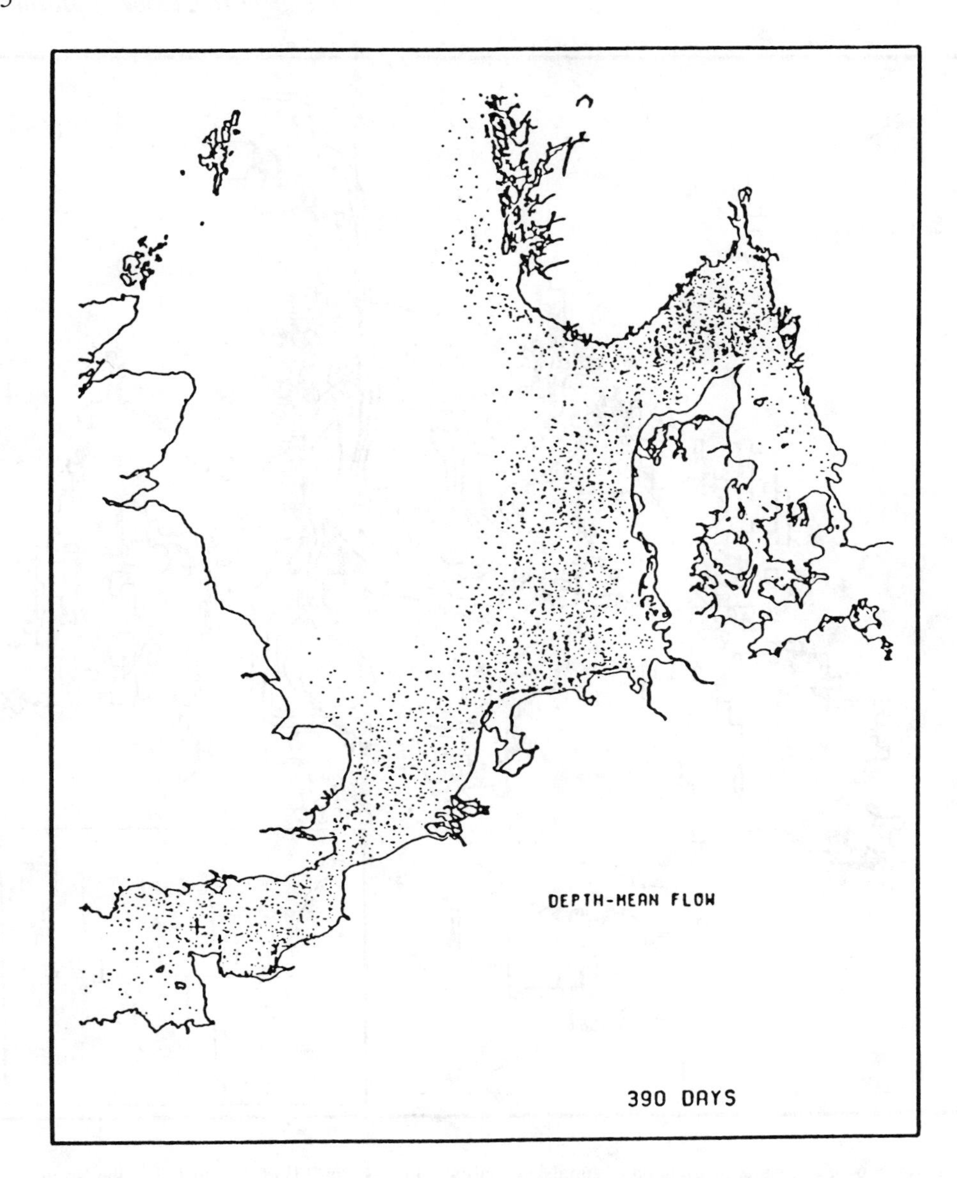

FIGURE 5.6—4. Particle pattern for continuous discharge (one particle per day) at the nuclear-reprocessing plant at Cap de la Hague for a particle "age" (time after release) of 390 d; period: 1971 to 1981. (From Backhaus, J. O., in *Modeling Marine Systems,* Vol. 1, Davies, A. M., Ed., CRC Press, Boca Raton, FL, 1990, 132.)

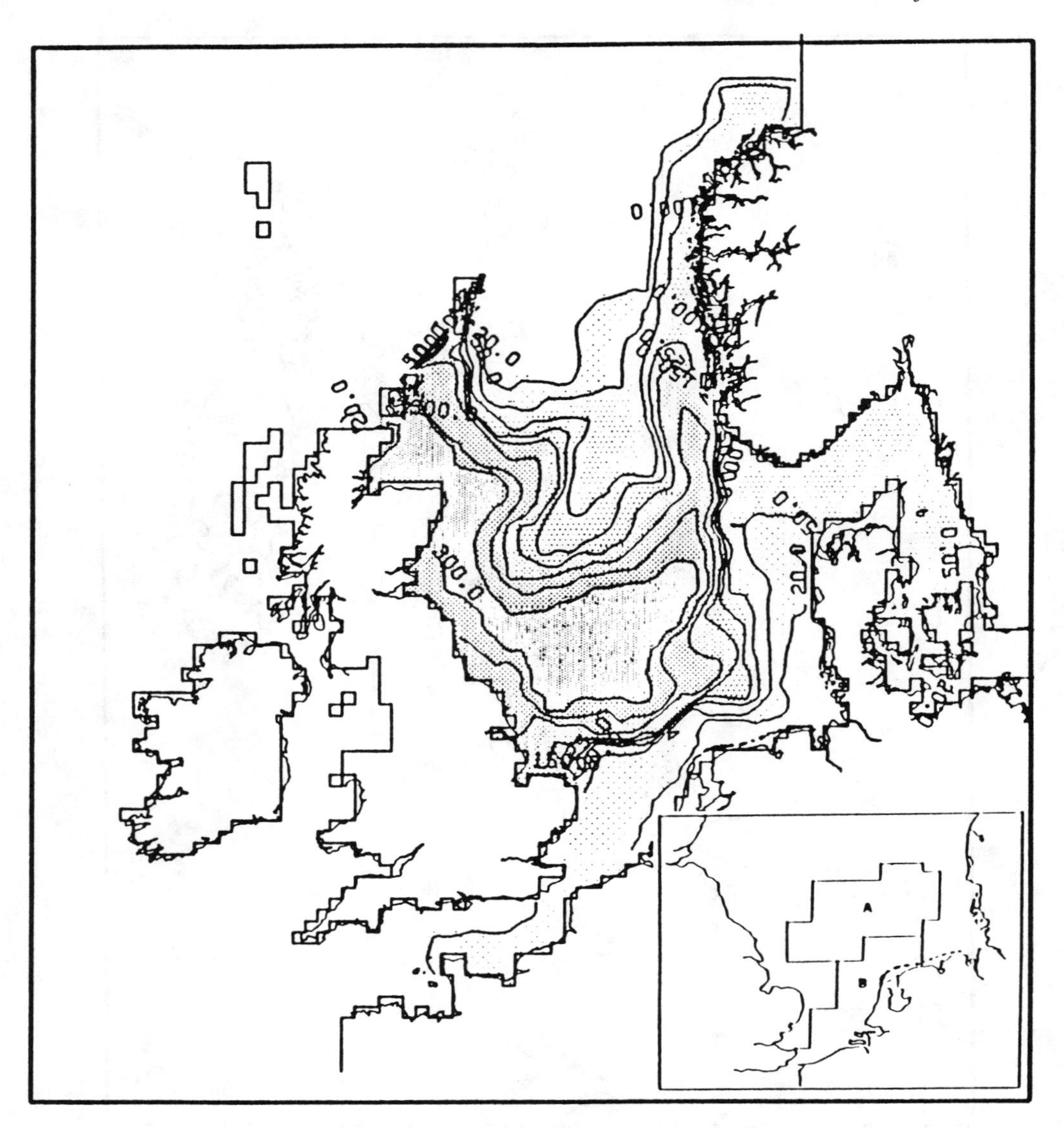

FIGURE 5.6—5. Spatial distribution of simulated concentrations (Bq/m**3) of cesium-137 in the North Sea for February 1978. Inset map: locations of subregions A and B. (From Backhaus, J. D., in *Modeling Marine Systems*, Vol. 1, Davies, A. M., Ed., CRC Press, Boca Raton, FL, 1990, 124.)

Table 5.6—9
ESTIMATES OF ANNUAL DOSES (mrad/year) RECEIVED BY MARINE ORGANISMS FROM NATURAL SOURCES OF RADIATION

Source	Taxonomic group	Marine 20 m depth)
Cosmic	Phytoplankton	4.4
	Zooplankton	4.4
	Mollusca	4.4
	Crustacea	4.4
	Fish	4.4
Water	Phytoplankton	3.5
	Zooplankton	1.8
	Mollusca	0.9
	Crustacea	0.9
	Fish	0.9
Sediment (β^+)	Phytoplankton	0
	Zooplankton	0
	Mollusca	27–324
	Crustacea	27–324
	Fish	0–324
Internal	Phytoplankton	17–64
	Zooplankton	23–138
	Mollusca	65–131
	Crustacea	69–188
	Fish	24–37
Sum of natural sources	Phytoplankton	25–72
	Zooplankton	29–168
	Mollusca	97–460
	Crustacea	101–517
	Fish	29–366

From Whicker, F. W. and Schultz, V., *Radioecology: Nuclear Energy and the Environment,* Vol. 1, CRC Press, Boca Raton, FL, 1982.

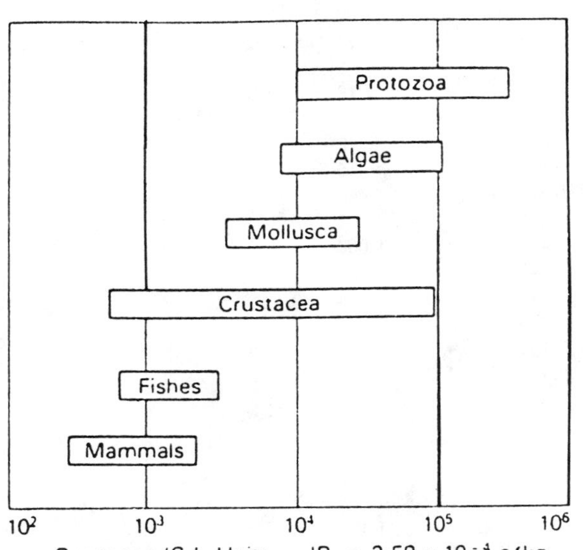

FIGURE 5.6—6. LD_{50} ranges for various groups of organisms irradiated by X- or γ-rays. (From Langford, T. E., *Electricity Generation and the Ecology of Natural Waters,* Liverpool University Press, Liverpool, U.K., 1983. With permission.)

Table 5.6—10
CONCENTRATION FACTORS (C.F.) FOR DIFFERENT CLASSES OF MARINE ORGANISMS

Element	Group	C.F. range	Mean C.F.
Cs	Plants	17–240	51
	Mollusca	3–28	15
	Crustacea	0.5–26	18
	Fish	5–244	48
Sr	Plants	0.2–82	21
	Mollusca	0.1–10	1.7
	Crustacea	0.1–1.1	0.6
	Fish	0.1–1.5	0.43
Mn	Plants	2,000–20,000	5,230
	Mollusca	170–150,000	22,080
	Crustacea	600–7,500	2,270
	Fish	35–1,800	363
Co	Plants	60–1,400	553
	Mollusca	1–210	166
	Crustacea	300–4,000	1,700
	Fish	20–5,000	650
Zn	Plants	80–2,500	900
	Mollusca	2,100–33,000	47,000
	Crustacea	1,700–15,000	5,300
	Fish	280–15,500	3,400
Fe	Plants	300–6,000	2,260
	Mollusca	1,000–13,000	7,600
	Crustacea	1,000–4,000	2,000
	Fish	600–3,000	1,800
I	Plants	30–6,800	1,065
	Mollusca	20–20,000	5,010
	Crustacea	20–48	31
	Fish	3–15	10
Ce	Plants	120–4,500	1,610
	Mollusca	100–350	240
	Crustacea	5–220	88
	Fish	0.3–538	99
K	Plants	4–31	13
	Mollusca	3.5–10	8
	Crustacea	8–19	12
	Fish	6.7–34	16
Ca	Plants	1.8–31	10
	Mollusca	0.2–112	16.5
	Crustacea	0.5–250	40
	Fish	0.5–7.6	1.9
Cu	Plants	—	1,000
	Mollusca	—	286
	Fish	0.1–5	2.55
Mo	Plants	12–42	23
	Mollusca	11–27	17
	Crustacea	8.9–17.3	13
	Fish	7.6–23.8	17
Mn	Plants	15–2,000	448
	Mollusca	1–3.6	2.2
	Crustacea	1–100	38
	Fish	0.4–26	6.6
Ar-Nb	Plants	170–2,900	1,119
	Mollusca	8–165	81
	Crustacea	1–100	51
	Fish	0.05–247	86

From Eisenbud, M., *Environmental Radioactivity,* 3rd ed., Academic Press, New York, 1987. With permission.

5.7 DREDGING AND DREDGED-SPOIL DISPOSAL

Table 5.7—1
TYPES OF DREDGING DEVICES AND THEIR RELATIONSHIP TO SEDIMENT TYPE AND DISPOSAL METHOD

Dredge type	Sediment type	Disposal conveyance
Mechanical devices		
Dipper dredge	Blasted rock	Vessel
Bucket dredge	Coarse grain size	Vessel
Ladder dredge	Fine grain size	Vessel
Agitation dredge	Mud, clay	Prevailing current
Hydraulic devices		
Agitation dredge	Mud, clay	Prevailing current
Hopper dredge	Fine grain size	Vessel
Suction dredge	Soft mud, clay	Pipeline
Cutterhead dredge	Consolidated, coarse grain size	Pipeline
Dustpan dredge	Sand	Pipeline
Sidecasting dredge	Fine grain size	Short pipe

From Kester, D. R., Ketchum, B. H., Duedall, I. W., and Park, P. K., in *Wastes in the Ocean,* Vol. 2, Dredged Material Disposed in the Ocean, Kester, D. R., Ketchum, B. H., Duedall, I. W., and Park, P. K., Eds., John Wiley & Sons, New York, 1983, 3. Reproduced by permission of copyright © owner, John Wiley & Sons Inc.

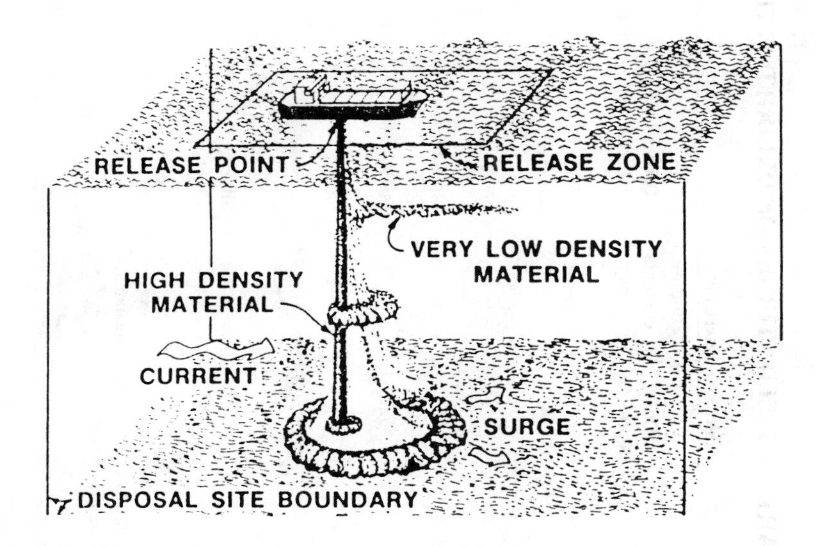

FIGURE 5.7—1. Sediment transport processes operating during open-water dredge disposal. (From Pequegnat, W. E., Pequegnat, L. H., James, B. M., Kennedy, E. A., Fay, R. R., and Fredericks, A. D., Procedural Guide for Designation Surveys of Ocean Dredged-Material Disposal Sites, Tech. Rep. EL-81-1, U.S. Army Corps of Engineers, Washington, D.C., 1981.)

Table 5.7—2
SUMMARY OF FIELD STUDIES OF DREDGED MATERIAL BEHAVIOR DURING OPEN-WATER DISPOSAL

	Site characteristics		Dredging/disposal characteristics					
Site	**Water depth (m)**	**Bottom currents (cm/s)**	**Dredged sediment**	**Dredge type**	**Disposal type**	**Typical volume (cu/m)**	**Monitoring technique/device**	**Sediment in upper water column (% of original)**
Long Island Sound	18–20	6–30	Silt-clay	Clamshell	Scow	900–2300	Transmissometer	1
Carquinez[a]	14	9–24	Silt-clay	Trailing Suction hopper	Hopper	1000	Transmissometer and gravimetric	1–5
Ashtabula (Lake Erie)	15–18	0–21	Sandy silt	Trailing Suction hopper	Hopper	690	Transmissometer and gravimetric	1[b]
New York Bight	26	6–24	Marine silt	Trailing Suction hopper	Hopper	6000	Transmissometer and gravimetric	1[b]
Saybrook (Long Island Sound)	52	21–70	Marine silt	Clamshell	Scow	1100	Transmissometer and gravimetric	1[b]
Elliot Bay	67	0–21	Sandy silt	Clamshell	Scow	380–535	Transmissometer and gravimetric	1[b]
Rochester (Lake Ontario)	17–45	0–21	Riverine silt	Trailing Suction hopper	Hopper	690	Transmissometer and gravimetric	1[b]
New York Bight	15–25	N/R	Silt-clay	Clamshell	Scow	1375–3000	Mass balance	3.7
Duwamish Waterway	20–21	6	Silt-clay	Clamshell	Scow	840	Gravimetric and mass balance	2–4

[a] Limited data at two additional sites included.
[b] Synthesis of all sites reported.

From Truitt, C. L., *J. Coastal Res.*, 4, 489, 1988. With permission.

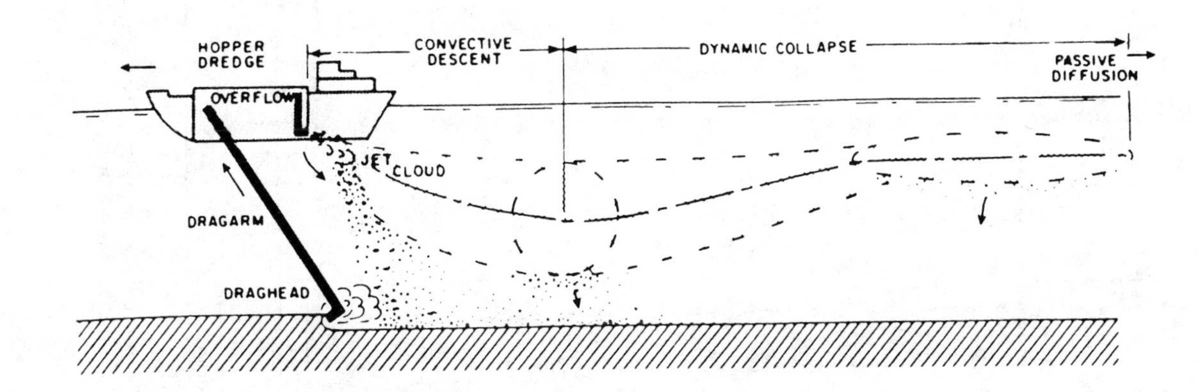

FIGURE 5.7—2. Hopper dredging operations in Chesapeake Bay showing an upper turbidity plume produced by overflow discharge and a near-bottom turbidity plume produced by draghead agitation and settling of particulates from the upper plume. The conceptual model depicts three transport phases for hopper overflow discharge: convective descent, dynamic collapse, and passive diffusion. (From Nichols, M., Diaz, R. J., and Schaffner, L. C., *Environ. Geol. Water Sci.*, 15, 31, 1990. With permission.)

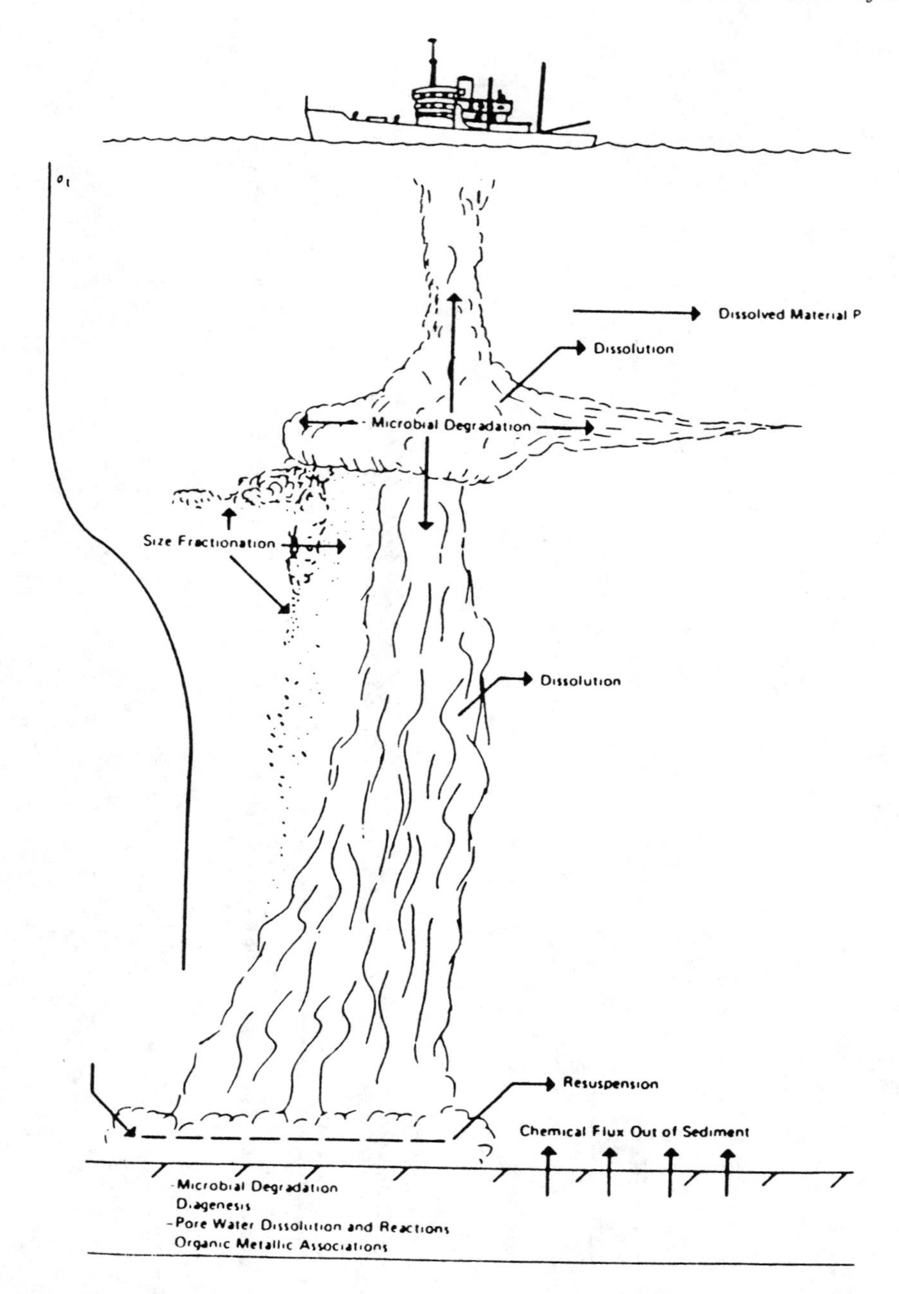

FIGURE 5.7—3. The processes influencing the distribution and fate of organic pollutants associated with dredged material. (From U.S. Army Corps of Engineers, Dredged-Material Research Program, Vicksburg, MS, 1973–1978.)

5.8 ELECTRIC GENERATING STATIONS

Table 5.8—1
ESTIMATED COOLING WATER REQUIREMENTS FOR A 1000-MWe STEAM-ELECTRIC PLANT OPERATING AT FULL LOAD

	Type of plant			
	Fossil		Nuclear	
	(1980)	(1990)	(1980)	(1990)
Plant heat rate,[a] Btu/k Wh	9500	9000	10,500	10,000
Condenser flows-cms, for various temperature rises across the condenser				
10°F (5.5°C)	58.9	53.5	82.6	76.7
15°F (8.3°C)	39.3	35.7	55.2	51.2
20°F (11.1°C)	29.4	26.9	41.2	38.5
Consumptive use, cms for various types of cooling				
Once through	0.34	0.28	0.48	0.42
Cooling ponds[b]	0.45	0.40	0.62	0.57
Cooling towers[c]	0.79	0.74	1.13	0.99

[a] For fossil-fueled plants in operation in 1970, a heat rate of 10,000 Btu/kWh and a temperature rise of 13°F (7.2°C) were assumed, except where reported heat rate data were available.
[b] Where appropriate, an additional allowance was made for natural evaporation from the pond surface.
[c] Evaporative towers; includes blowdown and drift.

From Eichholz, G. G., *Environmental Aspects of Nuclear Power,* Ann Arbor Science Publishers, Ann Arbor, MI, 1976. With permission.

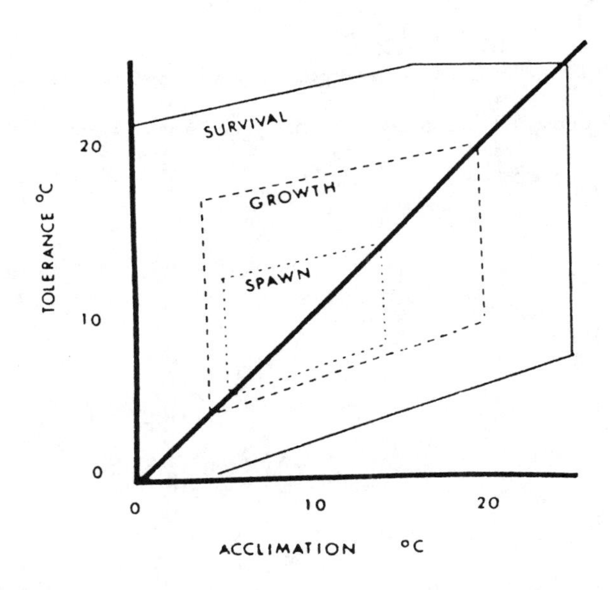

FIGURE 5.8—1. Temperature tolerance polygon for aquatic organisms. (From Brett, J. R., in *Water Pollution,* Taft, R. A., Ed., San. Eng. Center Tech. Rep., unpublished manuscript, 1960, 110. With permission.)

5.8—2
LETHAL TEMPERATURE LIMITS FOR ADULT MARINE, ESTUARINE, AND FRESHWATER FISHES

Species	Acclimation temp.[a]					
	°C	°F	°C	°F	°C	°F
Alewife	—	—	—	—	26.7–32.2	80.0–90.0
Bass, striped	—	—	6.0–7.5	42.8–45.5	25.0–27.0	77.0–80.0
California killifish	14.0–28.0	57.2–82.4	—	—	32.3–36.5	90.1–97.7
Common silverside	7.0–28.0	44.6–82.4	1.5–8.7	34.8–47.8	22.5–32.5	73.3–90.3
Flounder, winter	21.0–28.0	69.8–82.4	1.0–5.4	33.8–41.6	—	—
	7.0–28.0	44.6–82.4	—	—	22.0–29.0	71.6–84.2
Herring	—	—	−1.0	30.2	19.5–21.2	67.1–70.1
Northern swellfish	14.0–18.0	57.2–82.4	8.4–13.0	47.1–55.4	—	—
	10.0–28.0	50.0–82.4	—	—	28.2–33.0	82.9–90.4
Perch, white	4.4	40.0	—	—	27.8	82.0
Salmon (general)	—	—	0.0	32.0	26.7	80.0
Bass, largemouth	20.0	68.0	5.0	41.0	32.0	89.6
	30.0	86.0	11.0	51.8	34.0	93.2
Bluegill	15.0	59.0	3.0	37.4	31.0	87.8
	30.0	86.0	11.0	51.8	34.0	93.2
Catfish, channel	15.0	59.0	0.0	32.0	30.0	86.0
	25.0	77.0	6.0	42.8	34.0	93.2
Perch, yellow	5.0	41.0	—	—	21.0	69.8
(winter)	25.0	77.0	4.0	39.2	30.0	86.0
(summer)	25.0	77.0	9.0	48.2	32.0	89.6
Shad, gizzard	25.0	77.0	11.0	51.8	34.0	93.2
	35.0	95.0	20.0	68.0	37.0	98.6
Shiner, common	5.0	41.0	—	—	27.0	80.6
	25.0	77.0	4.0	39.2	31.0	87.8
	30.0	86.0	8.0	46.4	31.0	87.8
Trout, brook	3.0	37.4	—	—	23.0	73.4
	20.0	68.0	—	—	25.0	77.0

[a] Values are LD_{50} temperature tolerance limits, i.e., water temperatures survived by 50% of the test fish after 1- to 4-d acclimation.

From the Federal Water Pollution Control Administration, Industrial Waste Guide on Thermal Pollution, Federal Water Pollution Control Administration, Corvallis, OR, 1968.

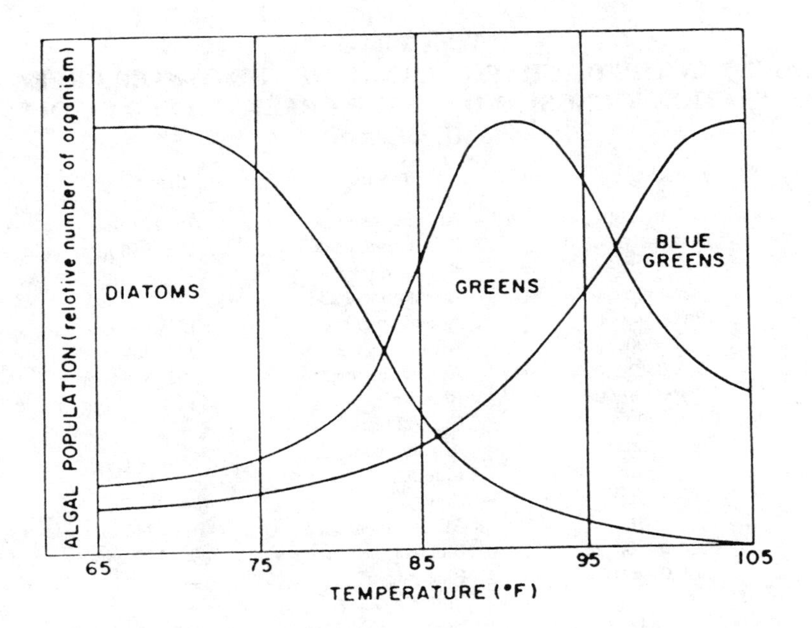

FIGURE 5.8—2. Population shifts of algae with changes in temperature. (From Eichholz, G. G., *Environmental Aspects of Nuclear Power,* Ann Arbor Science, Ann Arbor, MI, 1976. With permission.)

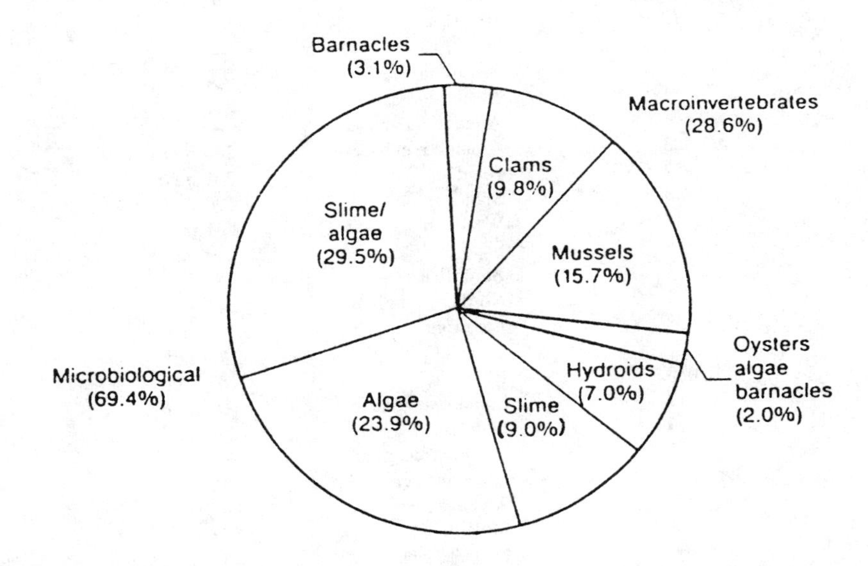

FIGURE 5.8—3. Biofouling organisms in cooling-water systems based on a survey of 365 units. (From Chow, W., in *Proceedings: Condenser Biofouling Control — State-of-the-Art Symposium,* Chow, W. and Massalli, Y. G., Eds., Copyright ©1985, Electric Power Research Institute, EPRI CS-4339, Palo Alto, CA, 1-1. Reprinted with permission.)

Table 5.8—3
FISH MORTALITY IN THE DISCHARGE CANAL OF THE OYSTER CREEK NUCLEAR GENERATING STATION (OCNGS) AND OYSTER CREEK CAUSED BY OPERATION OF THE OCNGS

Date	No.	Species	Size range	Probable cause
1/29/72	100,000–1,000,000	Atlantic menhaden	76–127 mm	Thermal shock
1/5/73	18,000–1,200,000	Atlantic menhaden	102–356 mm	Thermal shock
1/8/73	20	Bay anchovy	—	Thermal shock
2/16/73–2/21/73	Several thousand	Atlantic menhaden	—	Thermal shock
8/9/73	2,000–4,000	Atlantic menhaden	127–356 mm	Thermal shock
1/7/74	500	Atlantic menhaden	203–280 mm	Chlorine
1/11/74–1/15/74	9,900–180,000	Atlantic menhaden	102–356 mm	Thermal shock
	100–3,600	Bluefish	228–356 mm	Thermal shock
10/9/74	200	Crevalle jack	—	Thermal shock
2/4/75	100	Atlantic menhaden	—	Thermal shock
	50–100	Bluefish	—	Thermal shock
11/24/75	7–100	Crevalle jack	—	Thermal shock
12/29/75	15–100	Atlantic menhaden	100–250 mm	Thermal shock
	3–200	Bluefish	90–170 mm	Thermal shock
10/21/77	120–200	Blue runner	—	Thermal shock
		Crevalle jack	—	Thermal shock
1/15/79	682	Atlantic menhaden	165–225 mm	Thermal shock
8/2/79	50–100	Striped bass	34–44 mm	Thermal shock
		Northern puffer	Missing parts	
		Goosefish	Missing parts	
		Tautog		
12/17/79	Unknown	Unknown	Unknown	Unknown
12/20/79	12	Bluefish	—	Unknown
	1	Weakfish	—	
	1	Sea robin	—	
	1	Black sea bass	—	
	1	Atlantic menhaden	—	
1/5/80	5,483	Atlantic menhaden	240 mm[a]	Thermal shock
	952	Bluefish	295 mm[a]	
	544	Spot	120 mm[a]	
	43	Weakfish	501 mm[a]	
	5	Scup	200 mm[a]	
	1	Butterfish	—	
	1	Northern kingfish	240 mm[a]	
11/22/80	3,638	Blue runner	206 mm[a]	Thermal shock
		Crevalle jack	173 mm[a]	
	1,038	Bluefish	267 mm[a]	
	17	Smooth dogfish	601 mm[a]	
	3	Ladyfish	298 mm[a]	
	2	Northern kingfish	—	
	1	Gray snapper	118 mm[a]	
	1	American eel	—	
	1	Mojarra	221 mm[a]	
12/9/82–12/10/82	166	Crevalle jack	110–204 mm	Thermal shock
	80	Blue runner	171–218 mm	
	76	Bluefish	274–476 mm	
	28	Atlantic needlefish	250–661 mm	
	9	Scup	205–247 mm	
	2	American eel	—	
	1	Conger eel	—	
	1	Northern kingfish	185 mm	
	1	Ladyfish	410 mm	

[a] Mean size.

From Kennish, M. J., Roche, M. B., and Tatham, T. R., in *Ecology of Barnegat Bay, New Jersey,* Kennish, M. J. and Lutz, R. A., Eds., Springer-Verlag, New York, 1984, 318. With permission.

Table 5.8—4
MONTHLY ESTIMATES OF THE NUMBER OF WHITE PERCH IMPINGED AT ALL HUDSON RIVER POWER PLANTS COMBINED FOR 1974 AND 1975 YEAR CLASSES

		Year class			
		1974		1975	
		Number of years of vulnerability		Number of years of vulnerability	
Age (years)	Month	2	3	2	3
0	6	0		0	
	7	3,486		8,898	
	8	14,887		97,910	
	9	26,239		83,980	
	10	112,957		93,888	
	11	245,492		239,150	
	12	607,434		348,596	
	1	415,724		589,206	
	2	270,571		182,891	
	3	139,751		130,261	
	4	609,090		111,820	
	5	91,910		40,151	
1	6	37,242	18,621	27,014	13,507
	7	22,126	11,063	13,835	6,918
	8	14,122	7,061	6,770	3,385
	9	19,924	9,962	13,791	6,896
	10	19,534	9,767	25,676	12,838
	11	28,005	14,002	12,552	6,276
	12	7,803	3,902	48,102	24,051
	1	38,078	19,039	143,010	71,505
	2	9,293	4,646	43,558	21,779
	3	12,444	6,222	49,579	24,790
	4	14,103	7,052	38,692	19,346
	5	7,612	3,806	56,365	28,183
2	6		13,057		35,710
	7		6,918		8,805
	8		3,385		12,662
	9		6,896		8,736
	10		12,838		17,362
	11		6,276		19,145
	12		24,051		10,890
	1		71,505		
	2		21,779		
	3		24,790		
	4		19,346		
	5		28,182		

From Van Winkle, W., Barnthouse, L. W., Kirk, B. L., and Vaughan, D. S., Evaluation of Impingement Losses of White Perch at the Indian Point Nuclear Station and Other Hudson River Power Plants, Tech. Rep., ORNL/NUREG/TM-361, Oak Ridge National Laboratory, Oak Ridge, TN, 1980.

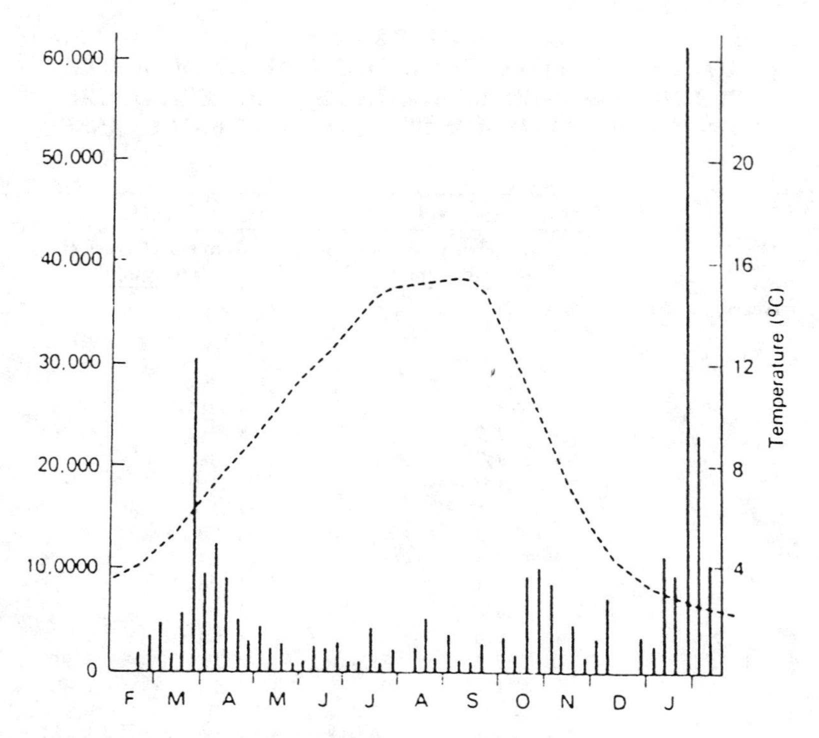

FIGURE 5.8—4. Weekly impingement of fish on intake screens of the Fawley Generating Station, England, in 1973. Vertical bars depict impingement. Dashed line gives water temperature. (From Langford, T. E., *Electricity Generation and the Ecology of Natural Waters,* Liverpool University Press, Liverpool, U.K., 1983. With permission.)

Table 5.8—5
AVERAGE YEARLY ENTRAINMENT AND IMPINGEMENT LOSSES AT THE BRUNSWICK STEAM ELECTRIC PLANT[a]

Species	Entrainment[b]	Impingement[c]	Impingement[d]
Spot	186,000	724	350
Croaker	123,000	356	235
Shrimp	171,000	675	760
Flounder	7,200	21	12
Mullet	5,200	34	18
Trout	38,500	169	205
Menhaden	32,500	9,744	4,000
Anchovy	913,000	2,748	1,600

[a] Number of organisms $\times 10^3$.

[b] Computed from weekly average plant flows from September 1976 through August 1978 and 5-year average entrainment densities for the same weekly period. These flows are close to the full plant flow.

[c] Two-year averages of measured impingement losses from September 1976 through August 1978.

[d] Five-year averages of measured impingement losses from January 1974 through August 1978.

From Lawler, J. P., Hogarth, W. T., Copeland, B. J., Weinstein, M. P., Hodson, H. G., and Chen, H. Y., in *Issues Associated With Impact Assessment, Proc. 5th Natl. Workshop on Entrainment and Impingement,* Jensen, L. D., Ed., EA Communications, Sparks, MD, 1981, 159.

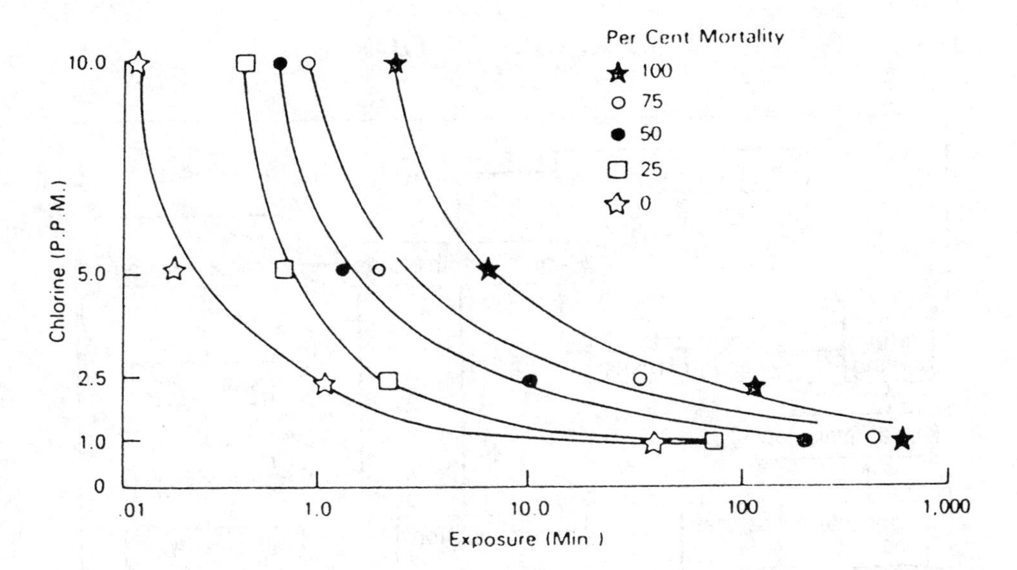

FIGURE 5.8—5. Mortality data (response isopleths) for the calanoid copepod, *Acartia tonsa,* exposed to chlorine. (From Morgan, R. P., II and Carpenter, E. J., in *Power Plant Entrainment: a Biological Assessment,* Schubel, J. R. and Marcy, B. C., Eds., Academic Press, New York, 1978, 95. With permission.)

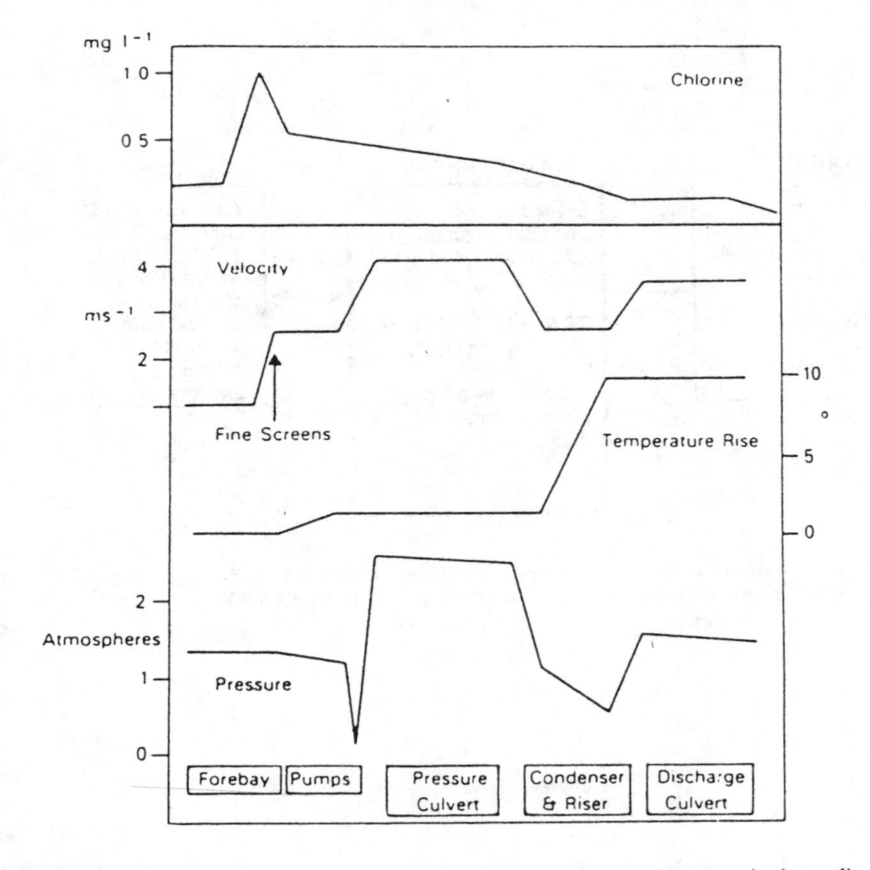

FIGURE 5.8—6. Typical changes in chlorine as well as velocity, temperature, and pressure in the cooling-water system of an electric generating station. (From Langford, T. E., *Electricity Generation and the Ecology of Natural Waters,* Liverpool University Press, Liverpool, U.K., 1983. With permission.)

5.9 AIR POLLUTION

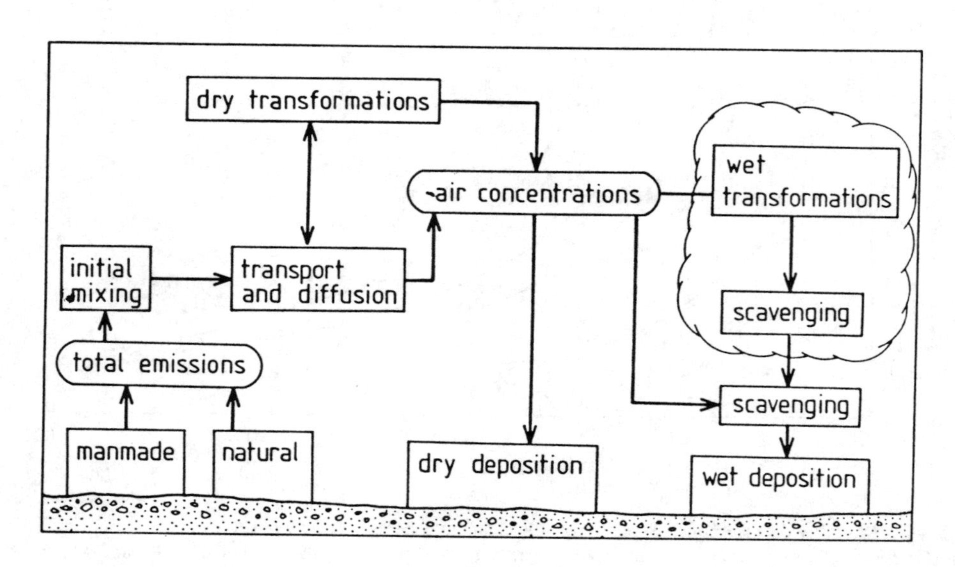

FIGURE 5.9—1. Typical atmospheric cycle of a pollutant. (From Harrison, R. M., Ed., *Pollution: Causes, Effects, and Control,* 2nd ed., Royal Society of Chemistry, Cambridge, 1990. With permission.)

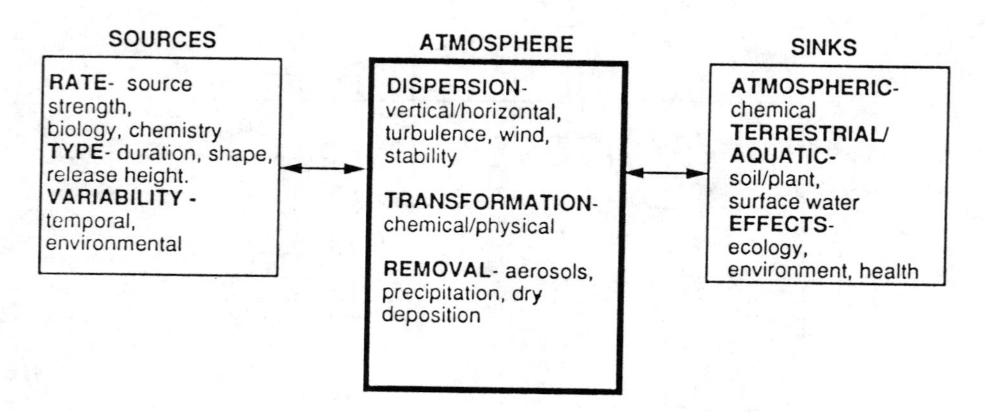

FIGURE 5.9—2. Factors influencing gaseous emissions and distributions. (From Harrison, R. M., Ed., *Pollution: Causes, Effects, and Control,* 2nd ed., Royal Society of Chemistry, Cambridge, 1990, 135. With permission.)

Table 5.9—1
SUMMARY OF COMMONLY EMPLOYED METHODS FOR MEASUREMENT OF AIR POLLUTANTS

Pollutant	Measurement technique	Sample collection period	Response time (continuous technique)[a]	Minimum detectable concentrations
Sulfur dioxide	Absorption in H_2O_2 and titration	24 h		2 ppb
	Absorption in tetrachloromercurate/ spectrophotometry	15 min		10 ppb
	Flame photometric analyzer		25 s	0.5 ppb
	Gas phase fluorescence		2 min	0.5 ppb
Oxides of nitrogen	Chemiluminescent reaction with ozone		1 s	0.5 ppb
Total hydrocarbons	Flame ionization analyser		0.5 s	10 ppb C
Specific hydrocarbons	Gas chromatography/flame ionization detector	b		<1 ppb
Carbon monoxide	Catalytic methanation FID	c		10 ppb
	Electrochemical cell		25 s	1 ppm
	Nondispersive infrared		5 s	0.5 ppm
Ozone	Chemiluminescent reaction with ethene		3 s	1 ppb
	UV absorption		30 s	1 ppb
Peroxyacetyl nitrate	Gas chromatography/electron capture detection	c		1 ppb
Particulates	High volume sampler	24 h		5 $\mu g\ m^{-3}$
	β-gauge	min/hours		5 $\mu g\ m^{-3}$

[a] Time taken for a 90% response to an instantaneous concentration change.
[b] Grab samples of air collected in an inert container and concentrated prior to analysis.
[c] Instantaneous concentrations measured on a cyclic basis by flushing the contents of a sample loop into the instrument.

From Harrison, R. M., Ed., *Pollution: Causes, Effects, and Control,* 2nd ed., Royal Society of Chemistry, Cambridge, 1990. With permission.

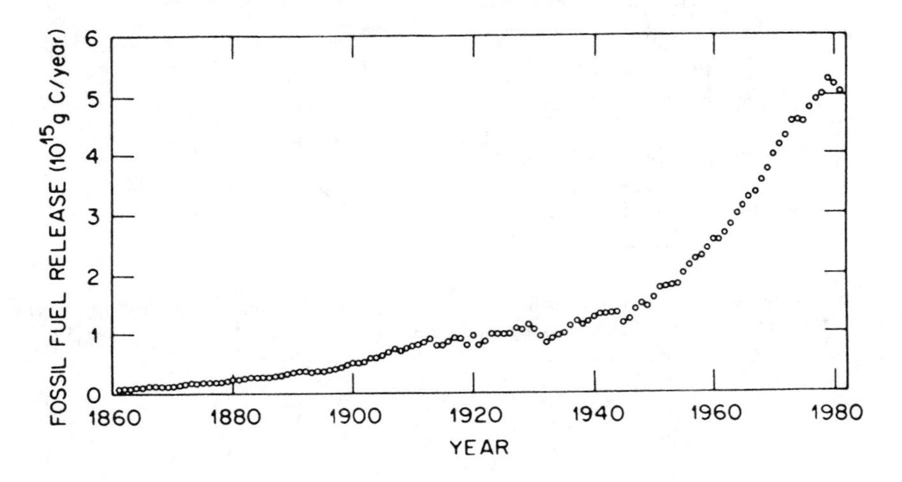

FIGURE 5.9—3. Fossil-fuel CO_2 emission ($\times 10^9$ tons/year) since the Industrial Revolution: 1860–1982. (From Walsh, J. J., *On the Nature of Continental Shelves,* Academic Press, New York, 1989. With permission.)

Table 5.9—2
NET CO_2 EXCHANGE ACROSS THE AIR-SEA INTERFACE[a]

Region	Area $\times 10^6$ km²	(%)	ΔpCO_2[b] (μatm)	Flux[c] ($\times 10^9$ tons C yr^{-1})
Temperate gyres	219	(61.0)	−14	−2.3
Antarctic (>50°S)	62	(17.0)	−7	−0.3
Equatorial (10°N–20°S)	40	(11.0)	+30	+0.9
Norwegian-Labrador-Greenland Seas	7.2	(2.0)	−100	−0.4
Mediterranean Sea	3.0	(0.8)	0	0
Red Sea	0.4	(0.1)	+25	+0.01
Asiatic seas	8.0	(2.0)	?	?
Arctic seas	14.0	(4.0)	?	?
Other seas	7.4	(2.1)	?	?
Total	361	(100.0)	−8	−2.1

[a] After T. Takahashi, personal communication (1982).
[b] $\Delta pCO_2 = pCO_2$ (seawater) $- pCO_2$ (air).
[c] Flux assumes a gas-exchange coefficient of 0.0615 mol CO_2 m^{-2} μatm^{-1} yt^{-1}.

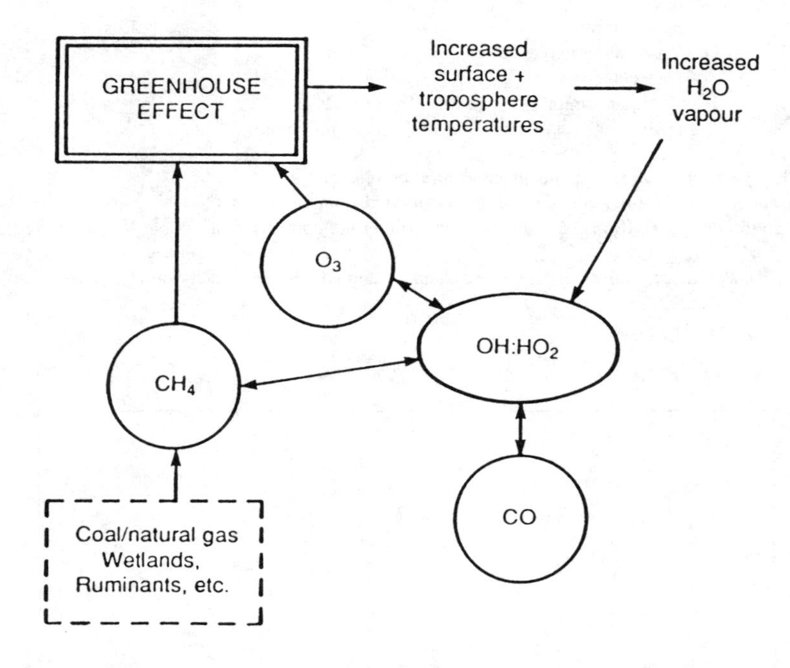

FIGURE 5.9—4. Simplified scheme indicating interactions of methane with other atmospheric constituents and global warming. (From Harrison, R. M., Ed., *Pollution: Causes, Effects, and Control,* 2nd ed., Royal Society of Chemistry, Cambridge, 1990, 137. With permission.)

5.10 SUBSTANCES CONTROLLED

Table 5.10—1
SUBSTANCES CONTROLLED BY THE LONDON DUMPING CONVENTION

Black List: Annex I

1. Organohalogen compounds.
2. Mercury and mercury compounds.
3. Cadmium and cadmium compounds.
4. Persistent plastics and other persistent synthetic materials; for example, netting and ropes, which may float or may remain in suspension in the sea in such a manner as to interfere materially with fishing, navigation, or other legitimate uses of the sea.
5. Crude oil, fuel oil, heavy diesel oil, lubricating oils, hydraulic fluids, and mixtures containing any of these, taken on board for the purpose of dumping.
6. High-level radioactive wastes or other high-level radioactive matter, defined on public health, biological, or other grounds, by the competent international body in this field, at present the International Atomic Energy Agency, as unsuitable for dumping at sea.
7. Materials in whatever form (such as solids, liquids, semiliquids, gases, or in a living state) produced for biological and chemical warfare.
8. The preceding paragraphs of this annex do not apply to substances which are rapidly rendered harmless by physical, chemical, or biological processes in the sea provided they do not: (1) make edible marine organisms unpalatable, or (2) endanger human health or that of domestic animals. The consultative procedure provided for under Article XIV should be followed by a party if there is doubt about the harmlessness of the substance.
9. This annex does not apply to wastes or other materials (such as sewage sludges and dredged spoils) containing the matters referred to in paragraphs 1 to 5 above as trace contaminants. Such wastes shall be subject to the provisions of Annexes II and III as appropriate.
10. Paragraphs 1 and 5 of the annex do not apply to the dispersal of wastes or other matter referred to in these paragraphs by means of incineration at sea. Incineration of such wastes or other matter at sea requires a prior special permit. In the issue of special permits for incineration the contracting parties shall apply the Regulations for the Control of Incineration of Wastes and Other Matter at Sea set forth in the addendum to this annex (which shall constitute an integral part of this annex) and take full account of the Technical Guidelines on the Control of Incineration of Wastes and Other Matter at Sea adopted by the contracting parties in consultation.

Grey List: Annex II

The following substances and materials require special permits, issued only according to the articles of the London Dumping Convention.

A. Wastes containing significant amounts of the matters listed below:
   - Arsenic, lead, copper, zinc, and their compounds
   - Organosilicon compounds
   - Cyanides
   - Fluorides
   - Pesticides and their by-products not covered in Annex I

B. In the issue of permits for the dumping of large quantities of acids and alkalis, consideration shall be given to the possible presence in such wastes of the substances listed in paragraph A, and to beryllium, chromium, nickel, vanadium, and their compounds.

C. Containers, scrap metal, and other bulky wastes liable to sink to the sea bottom which may present a serious obstacle to fishing or navigation.

D. Radioactive wastes or other radioactive matter not included in Annex I. In the issue of permits for the dumping of this matter, the contracting parties should take full account of the recommendations of the competent international body in this field, at present the International Atomic Energy Agency.

E. In the issue of special permits for the incineration of substances and materials listed in this annex, the contracting parties shall apply the Regulations for the Control of Incineration of Wastes and Other Matter at Sea set forth in the addendum to Annex I and take full account of the Technical Guidelines on the Control of Incineration of Wastes and Other Matter at Sea adopted by the contracting parties in consultation, to the extent specified in these regulations and guidelines.

From the Final Act of the London Dumping Convention, Office of the London Dumping Convention, International Maritime Organization, London.

INDEX

A

B

C

D

E

F

G

H

I

J

K

L

M

Q

R

S

T

U

V

W

X

Y

Z